Y0-DYE-667

ENZYMES

IN THE ENVIRONMENT

BOOKS IN SOILS, PLANTS, AND THE ENVIRONMENT

Soil Biochemistry, Volume 1, edited by A. D. McLaren and G. H. Peterson
Soil Biochemistry, Volume 2, edited by A. D. McLaren and J. Skujiņš
Soil Biochemistry, Volume 3, edited by E. A. Paul and A. D. McLaren
Soil Biochemistry, Volume 4, edited by E. A. Paul and A. D. McLaren
Soil Biochemistry, Volume 5, edited by E. A. Paul and J. N. Ladd
Soil Biochemistry, Volume 6, edited by Jean-Marc Bollag and G. Stotzky
Soil Biochemistry, Volume 7, edited by G. Stotzky and Jean-Marc Bollag
Soil Biochemistry, Volume 8, edited by Jean-Marc Bollag and G. Stotzky
Soil Biochemistry, Volume 9, edited by G. Stotzky and Jean-Marc Bollag

Organic Chemicals in the Soil Environment, Volumes 1 and 2, edited by C. A. I. Goring and J. W. Hamaker
Humic Substances in the Environment, M. Schnitzer and S. U. Khan
Microbial Life in the Soil: An Introduction, T. Hattori
Principles of Soil Chemistry, Kim H. Tan
Soil Analysis: Instrumental Techniques and Related Procedures, edited by Keith A. Smith
Soil Reclamation Processes: Microbiological Analyses and Applications, edited by Robert L. Tate III and Donald A. Klein
Symbiotic Nitrogen Fixation Technology, edited by Gerald H. Elkan
Soil–Water Interactions: Mechanisms and Applications, Shingo Iwata and Toshio Tabuchi with Benno P. Warkentin
Soil Analysis: Modern Instrumental Techniques, Second Edition, edited by Keith A. Smith
Soil Analysis: Physical Methods, edited by Keith A. Smith and Chris E. Mullins
Growth and Mineral Nutrition of Field Crops, N. K. Fageria, V. C. Baligar, and Charles Allan Jones
Semiarid Lands and Deserts: Soil Resource and Reclamation, edited by J. Skujiņš
Plant Roots: The Hidden Half, edited by Yoav Waisel, Amram Eshel, and Uzi Kafkafi
Plant Biochemical Regulators, edited by Harold W. Gausman

Maximizing Crop Yields, N. K. Fageria
Transgenic Plants: Fundamentals and Applications, edited by Andrew Hiatt
Soil Microbial Ecology: Applications in Agricultural and Environmental Management, edited by F. Blaine Metting, Jr.
Principles of Soil Chemistry: Second Edition, Kim H. Tan
Water Flow in Soils, edited by Tsuyoshi Miyazaki
Handbook of Plant and Crop Stress, edited by Mohammad Pessarakli
Genetic Improvement of Field Crops, edited by Gustavo A. Slafer
Agricultural Field Experiments: Design and Analysis, Roger G. Petersen
Environmental Soil Science, Kim H. Tan
Mechanisms of Plant Growth and Improved Productivity: Modern Approaches, edited by Amarjit S. Basra
Selenium in the Environment, edited by W. T. Frankenberger, Jr., and Sally Benson
Plant–Environment Interactions, edited by Robert E. Wilkinson
Handbook of Plant and Crop Physiology, edited by Mohammad Pessarakli
Handbook of Phytoalexin Metabolism and Action, edited by M. Daniel and R. P. Purkayastha
Soil–Water Interactions: Mechanisms and Applications, Second Edition, Revised and Expanded, Shingo Iwata, Toshio Tabuchi, and Benno P. Warkentin
Stored-Grain Ecosystems, edited by Digvir S. Jayas, Noel D. G. White, and William E. Muir
Agrochemicals from Natural Products, edited by C. R. A. Godfrey
Seed Development and Germination, edited by Jaime Kigel and Gad Galili
Nitrogen Fertilization in the Environment, edited by Peter Edward Bacon
Phytohormones in Soils: Microbial Production and Function, William T. Frankenberger, Jr., and Muhammad Arshad
Handbook of Weed Management Systems, edited by Albert E. Smith
Soil Sampling, Preparation, and Analysis, Kim H. Tan
Soil Erosion, Conservation, and Rehabilitation, edited by Menachem Agassi
Plant Roots: The Hidden Half, Second Edition, Revised and Expanded, edited by Yoav Waisel, Amram Eshel, and Uzi Kafkafi
Photoassimilate Distribution in Plants and Crops: Source–Sink Relationships, edited by Eli Zamski and Arthur A. Schaffer
Mass Spectrometry of Soils, edited by Thomas W. Boutton and Shinichi Yamasaki
Handbook of Photosynthesis, edited by Mohammad Pessarakli
Chemical and Isotopic Groundwater Hydrology: The Applied Approach, Second Edition, Revised and Expanded, Emanuel Mazor
Fauna in Soil Ecosystems: Recycling Processes, Nutrient Fluxes, and Agricultural Production, edited by Gero Benckiser
Soil and Plant Analysis in Sustainable Agriculture and Environment, edited by Teresa Hood and J. Benton Jones, Jr.
Seeds Handbook: Biology, Production, Processing, and Storage: B. B. Desai, P. M. Kotecha, and D. K. Salunkhe
Modern Soil Microbiology, edited by J. D. van Elsas, J. T. Trevors, and E. M. H. Wellington
Growth and Mineral Nutrition of Field Crops: Second Edition, N. K. Fageria, V. C. Baligar, and Charles Allan Jones
Fungal Pathogenesis in Plants and Crops: Molecular Biology and Host Defense Mechanisms, P. Vidhyasekaran
Plant Pathogen Detection and Disease Diagnosis, P. Narayanasamy
Agricultural Systems Modeling and Simulation, edited by Robert M. Peart and R. Bruce Curry
Agricultural Biotechnology, edited by Arie Altman
Plant–Microbe Interactions and Biological Control, edited by Greg J. Boland and L. David Kuykendall

Handbook of Soil Conditioners: Substances That Enhance the Physical Properties of Soil, edited by Arthur Wallace and Richard E. Terry

Environmental Chemistry of Selenium, edited by William T. Frankenberger, Jr., and Richard A. Engberg

Principles of Soil Chemistry: Third Edition, Revised and Expanded, Kim H. Tan

Sulfur in the Environment, edited by Douglas G. Maynard

Soil–Machine Interactions: A Finite Element Perspective, edited by Jie Shen and Radhey Lal Kushwaha

Mycotoxins in Agriculture and Food Safety, edited by Kaushal K. Sinha and Deepak Bhatnagar

Plant Amino Acids: Biochemistry and Biotechnology, edited by Bijay K. Singh

Handbook of Functional Plant Ecology, edited by Francisco I. Pugnaire and Fernando Valladares

Handbook of Plant and Crop Stress: Second Edition, Revised and Expanded, edited by Mohammad Pessarakli

Plant Responses to Environmental Stresses: From Phytohormones to Genome Reorganization, edited by H. R. Lerner

Handbook of Pest Management, edited by John R. Ruberson

Environmental Soil Science: Second Edition, Revised and Expanded, Kim H. Tan

Microbial Endophytes, edited by Charles W. Bacon and James F. White, Jr.

Plant–Environment Interactions: Second Edition, edited by Robert E. Wilkinson

Microbial Pest Control, Sushil K. Khetan

Soil and Environmental Analysis: Physical Methods, Second Edition, Revised and Expanded, edited by Keith A. Smith and Chris E. Mullins

The Rhizosphere: Biochemistry and Organic Substances at the Soil–Plant Interface, Roberto Pinton, Zeno Varanini, and Paolo Nannipieri

Woody Plants and Woody Plant Management: Ecology, Safety, and Environmental Impact, Rodney W. Bovey

Metals in the Environment, M. N. V. Prasad

Plant Pathogen Detection and Disease Diagnosis: Second Edition, Revised and Expanded, P. Narayanasamy

Handbook of Plant and Crop Physiology: Second Edition, Revised and Expanded, edited by Mohammad Pessarakli

Environmental Chemistry of Arsenic, edited by William T. Frankenberger, Jr.

Enzymes in the Environment: Activity, Ecology, and Applications, edited by Richard G. Burns and Richard P. Dick

Additional Volumes in Preparation

Plant Roots: The Hidden Half, Third Edition, Revised and Expanded, edited by Yoav Waisel, Amram Eshel, and Uzi Kafkafi

Handbook of Postharvest Technology, edited by A. Chakraverty, Arun S. Mujumdar, G. S. V. Raghavan, and H. S. Ramaswamy

Biological Control of Major Crop Plant Diseases, edited by Samuel S. Ganamanickam

Handbook of Plant Growth, edited by Zdenko Rengel

ENZYMES IN THE ENVIRONMENT

Activity, Ecology, and Applications

edited by

Richard G. Burns
University of Kent
Canterbury, Kent
England

Richard P. Dick
Oregon State University
Corvallis, Oregon

MARCEL DEKKER, INC. NEW YORK • BASEL

ISBN: 0-8247-0614-5

This book is printed on acid-free paper.

Headquarters
Marcel Dekker, Inc.
270 Madison Avenue, New York, NY 10016
tel: 212-696-9000; fax: 212-685-4540

Eastern Hemisphere Distribution
Marcel Dekker AG
Hutgasse 4, Postfach 812, CH-4001 Basel, Switzerland
tel: 41-61-261-8482; fax: 41-61-261-8896

World Wide Web
http://www.dekker.com

The publisher offers discounts on this book when ordered in bulk quantities. For more information, write to Special Sales/Professional Marketing at the headquarters address above.

Current printing (last digit):
10 9 8 7 6 5 4 3 2 1

PRINTED IN THE UNITED STATES OF AMERICA

Dedicated to the memory of A. Douglas McLaren

Preface

Enzymes that function within plants, animals, and microorganisms are fundamental to life, and their contributions to metabolic pathways and processes have been studied extensively. For over 100 years there has been interest in what today is called ecological or environmental enzymology. This aspect of enzymology originates from the work of Woods, who, in 1899, wrote about the survival and function in soil of plant peroxidases following their release from decaying plant roots. Environmental enzymologists recognize that the measured activity may be a composite of reactions taking place in different locations and at different rates. Thus, in addition to being intracellular, enzymes can be extracellular and attached to the external surfaces of cells, associated with microbial and plant debris, diffused or actively excreted into the solution phase, and complexed with minerals and organic compounds.

Although most extracellular enzymes released from cells are rapidly denatured or degraded, some will survive in solution, if only for short periods. This allows them to complex with adjacent and appropriate substrates and hydrolyze molecules that are too large or insoluble to pass through the cell wall or for which there are no uptake mechanisms. These soluble, low molecular mass products can then be utilized as carbon and/or energy sources by the cell. What this means is that the catalysis of a substrate does not necessarily represent a homogeneous enzymatic reaction but may be the result of isoenzymes derived from plants, microorganisms, or animals, and found in various locations within the soil or sediment matrix.

Much ecological enzymology research is driven by the need to understand the biological processes that are important for essential aquatic and terrestrial ecosystem functions. These include: organic matter decomposition in relation to both local and global biogeochemistry; mineralization and the release of inorganic nutrients for use by microbes, plants, and animals; complex combinations of reactions that determine and maintain soil fertility and soil productivity; and the response to and recovery of soil and aquatic systems from various natural and anthropogenic perturbations.

Until very recently there have been two large but rather separate camps in the study of ecological enzymology: those involved with aquatic environments and those who have concentrated on soil. In aquatic systems the early work included that by Fermi in 1906, who showed proteolysis activity in stagnant pools, and Harvey in 1925, who suggested that seawater had catalase and oxidase activity. Subsequent researchers, such as Kreps, Elster, and Einsele in the 1930s, showed that aquatic bacteria could excrete enzymes into solution and that these retained a portion of their catalytic activity. Pioneering soil science work by Rotini and Waksman, among others, was focused on catalase, although the 1940s saw a surge in influential papers on urease by Conrad and phosphatases by Rogers.

Until the 1950s ecological enzyme research made incremental progress. However, since then there has been an ever-increasing research output on ecological enzymology, and in the past 20 years well over 1000 papers have been published. On the aquatic side, this rapid growth of research was initiated by Overbeck and Reichardt, who demonstrated the role of extracellular phosphatases from bacteria in the mineralization of organic P compounds. They showed that the released phosphate was then used by algae that lacked the ability to directly utilize organic P, thereby showing an important microbial ecological mechanism for extracellular enzyme activity in aquatic systems. They also carried out pioneering research on the temporal and spatial distribution of enzyme activities in lake water. On the soil science front, pioneering work in the 1960s by, among others, McLaren, Kiss, Ross, Galstyan, Voets, and their coworkers gave an impetus that still drives much of today's research. Soil enzymology up to the late 1970s was summarized in the book *Soil Enzymes* (Academic Press, 1978).

Ecological enzymology can be divided into two broad and overlapping divisions that are both well represented in this book. The first can be categorized as microbial ecology and biochemistry: the study of enzymatic activities in order to better comprehend the processes or mechanisms that are operating in a given system (Chapters 1, 2, and 3). This research may have fundamental objectives targeted toward a greater understanding of highly complex environments such as the rhizosphere (Chapter 4), plant leaves and shoots (Chapter 6), soil surfaces (Chapter 11), or biofilms (Chapter 12). On the other hand, it may have explicit applied goals related to manipulating or preserving the environment in question. These applications include: microbe–plant symbioses (Chapter 5), controlling plant pathogens (Chapter 7), understanding organic matter decomposition and its impact on local and global carbon and nitrogen cycles (Chapters 8, 9, and 10), and environmental remediation of contaminated soils and sediments (Chapters 18, 19, and 20).

The second category of ecological enzymology research includes the use of enzymes (or microbial cells) as sensors to detect microbial activity and stresses due to pollution, management, or climatic changes in aquatic and terrestrial ecosystems (Chapters 15, 16, and 17). In this mode, enzymes can be used to assess nutrient turnover, soil health and the presence of plant pathogens, and the progress of remediation of polluted soils and waters. Conventional enzyme assays are attractive as sensors because their integrative nature, specificity of reactions, and relatively simple methodology make them feasible for adoption by commercial environmental laboratories. Alternatively, molecular methods using reporter systems linked to enzymatic processes are being developed for assessing microbial diversity and function (Chapter 14).

This book, in part, was the result of the historic conference "Enzymes in the Environment: Ecology, Activity and Applications," held in Granada, Spain, in July 1999. This meeting of over 200 scientists from 34 countries was unique because it brought together scientists from diverse backgrounds around the world who do not normally interact or attend the same professional meetings. Those enjoying the busy sessions included biochemists and microbial ecologists who study terrestrial or aquatic systems, and environmental and agronomic scientists. Some of the research presented at this meeting was published in a special issue of *Soil Biology & Biochemistry* (Vol. 32, Issue 13, 2000). There will be a follow-up conference in Prague in July 2003.

An interesting observation arising from the Granada conference was that research into such diverse microbial ecosystems as plant surfaces, soil aggregates, and biofilms of aquatic systems or populations at 1000 meters below the surface of the ocean presented strikingly similar methodological challenges and difficulties in the interpretation of the information derived (Chapter 21). How do you get a representative environmental sample? What are the appropriate assay conditions? What do the measured activities tell us about processes in the environment? What is the microbial and macroecological significance of extracellular enzymes? Are there commercial applications of extracellular enzymes in remediation and nutrient provision? And are there lots of microbes and enzymes out there waiting to be discovered and exploited (Chapter 13)? All these questions and more were heard frequently. The multidisciplinary group also discussed the ''big'' issues and responsibilities of current and future developments in environmental enzymology. Two of the most pressing of these are adequate and sustainable food production in terrestrial and aquatic ecosystems and counteracting global warming through carbon sequestration and other processes in soils and aquatic systems. This book presents 21 reviews by international experts who attempt to address all these questions and issues. Research progress in ecological enzymology in terrestrial and aquatic ecosystems is brought into the twenty-first century.

Richard Burns wishes to thank his wife, Wendy, for her support through this and other writing adventures and Hugo Z., who continues to give a sense of perspective to this confusing life. Richard Dick acknowledges Joan Sandeno for her editing assistance.

Richard G. Burns
Richard P. Dick

Contents

Contributors

Sergio Alvarez Department of Ecology, Universidad Autónoma de Madrid, Madrid, Spain

Carol Arnosti Department of Marine Sciences, University of North Carolina, Chapel Hill, North Carolina

Mark J. Bailey Molecular Microbial Ecology Group, Centre for Ecology and Hydrology, Oxford, England

Tamar Barkay Department of Biochemistry and Microbiology, Cook College, Rutgers University, New Brunswick, New Jersey

Marie-Hélène Baron Laboratoire de Dynamique, Interactions et Réactivité, Centre National de la Recherche Scientifique, Université Paris VI, Thiais, France

John A. Baross School of Oceanography, University of Washington, Seattle, Washington

Lee A. Beaudette Department of Environmental Biology, University of Guelph, Guelph, Ontario, Canada

Jean-Marc Bollag Laboratory of Soil Biochemistry, Center for Bioremediation and Detoxification, The Pennsylvania State University, University Park, Pennsylvania

Margaret M. Carreiro Department of Biology, University of Louisville, Louisville, Kentucky

Michael B. Cassidy Department of Environmental Biology, University of Guelph, Guelph, Ontario, Canada

Leonid Chernin Department of Plant Pathology and Microbiology, Faculty of Agriculture, The Hebrew University of Jerusalem, Rehovot, Israel

Ilan Chet Department of Plant Pathology and Microbiology, Faculty of Agriculture, The Hebrew University of Jerusalem, Rehovot, Israel

Ryszard Jan Chróst Department of Microbial Ecology, University of Warsaw, Warsaw, Poland

Julia R. de Lipthay Department of Geochemistry, Geological Survey of Denmark and Greenland, Copenhagen, Denmark

Jody W. Deming School of Oceanography, University of Washington, Seattle, Washington

Warren A. Dick School of Natural Resources, The Ohio State University, Wooster, Ohio

Robert S. Dungan George E. Brown, Jr., Salinity Laboratory, USDA-ARS, Riverside, California

William T. Frankenberger, Jr. Department of Environmental Sciences, University of California–Riverside, Riverside, California

José Manuel García-Garrido Department of Soil Microbiology, Estacíon Experimental del Zaidín, CSIC, Granada, Spain

Inmaculada García-Romera Department of Soil Microbiology, Estacíon Experimental del Zaidín, CSIC, Granada, Spain

Liliana Gianfreda Department of Chemical and Agricultural Sciences, University of Naples Federico II, Portici, Naples, Italy

Marc Habash Department of Environmental Biology, University of Guelph, Guelph, Ontario, Canada

Lars H. Hansen Department of General Microbiology, University of Copenhagen, Copenhagen, Denmark

Gerhard F. Herndl Department of Biological Oceanography, Netherlands Institute of Sea Research (NIOZ), Den Burg, The Netherlands

Hans-Georg Hoppe Section of Marine Ecology, Institute of Marine Science, Kiel, Germany

Jana Jass Department of Molecular Biology, Umeå University, Umeå, Sweden

Shung-Chang Jong Department of Microbiology, American Type Culture Collection, Manassas, Virginia

Ellen Kandeler Institute of Soil Science, University of Hohenheim, Stuttgart, Germany

Ken Killham Department of Plant and Soil Science, University of Aberdeen, Aberdeen, Scotland

Annelise H. Kjøller Department of General Microbiology, University of Copenhagen, Copenhagen, Denmark

Sabine Kuhbier Institute of Applied Microbiology, Justus Liebig University, Giessen, Germany

Hilary M. Lappin-Scott Department of Biological Sciences, Exeter University, Exeter, England

Hung Lee Department of Environmental Biology, University of Guelph, Guelph, Ontario, Canada

Hans-Joachim Lorch Institute of Applied Microbiology, Justus Liebig University, Giessen, Germany

James M. Lynch School of Biological Sciences, University of Surrey, Guildford, Surrey, England

Anne Kirstine Müller Department of General Microbiology, University of Copenhagen, Copenhagen, Denmark

Paolo Nannipieri Scienza del Suolo e Nutrizione Della Planta, Universitá degli Studi de Firenze, Firenze, Italy

David C. Naseby Department of Biosciences, University of Hertfordshire, Hatfield, Hertfordshire, England

Juan Antonio Ocampo Department of Soil Microbiology, Estacíon Experimental del Zaidín, CSIC, Granada, Spain

Johannes C. G. Ottow Institute of Applied Microbiology, Justus Liebig University, Giessen, Germany

Hervé Quiquampoix Laboratoire Sol et Environnement, Institut National de la Recherche Agronomique, Montpellier, France

Lasse Dam Rasmussen Department of General Microbiology, University of Copenhagen, Copenhagen, Denmark

Sara K. Roberts Department of Periodontics, College of Dentistry, University of Illinois–Chicago, Chicago, Illinois

Des J. Ross Landcare Research, Palmerston North, New Zealand

Pacifico Ruggiero Dipartimento di Biologie e Chemical Agro-Forestale et Anibieritale, Universitá degli Studi di Bari, Bari, Italy

Sylvie Servagent-Noinville Laboratoire de Dynamique Interactions et Réactivité, Centre National de la Recherche Scientifique, Université Paris VI, Thiais, France

Robert L. Sinsabaugh Department of Environmental Science, University of Toledo, Toledo, Ohio

Waldemar Siuda Department of Microbial Ecology, University of Warsaw, Warsaw, Poland

Søren J. Sørensen Department of General Microbiology, University of Copenhagen, Copenhagen, Denmark

Tom W. Speir Institute of Environmental Science and Research, Porirua, New Zealand

William J. Staddon Department of Biological Sciences, Eastern Kentucky University, Richmond, Kentucky

Sten Struwe Department of General Microbiology, University of Copenhagen, Copenhagen, Denmark

M. Ali Tabatabai Department of Agronomy, Iowa State University, Ames, Iowa

Robert L. Tate III Department of Environmental Science, Rutgers University, New Brunswick, New Jersey

Ian P. Thompson Molecular Microbial Ecology Group, Centre for Ecology and Hydrology, Oxford, England

Jack T. Trevors Department of Environmental Biology, University of Guelph, Guelph, Ontario, Canada

ENZYMES

IN THE ENVIRONMENT

1

Enzyme Activities and Microbiological and Biochemical Processes in Soil

Paolo Nannipieri
Universitá degli Studi di Firenze, Firenze, Italy

Ellen Kandeler
University of Hohenheim, Stuttgart, Germany

Pacifico Ruggiero
Universitá degli Studi di Bari, Bari, Italy

I. INTRODUCTION

It is well known that soil organisms, particularly microbiota, play an essential role in the cycling of elements and stabilization of soil structure (1,2). The mineralization of organic matter is carried out by a large community of microorganisms and involves a wide range of metabolic processes. For this reason, it is important to relate ecosystem structure and function to species and functional diversity (3). However, the relationships between genetic diversity and taxonomic diversity are not well understood and even less is known about the manner in which these two properties affect microbial functional diversity (3–5). Microbial species diversity is related to richness (i.e., the number of different species), evenness (i.e., the relative contribution that individuals of all species make to the total number of organisms present), and composition (i.e., the type and relative contribution of particular species present) (3). According to the well-known and much used BIOLOG approach, microbial functional diversity is related both to the rates of substrate utilization and to the presence or absence of utilization of specific substrates. A decrease in microbial diversity may reduce microbial functionality of soil if ''keystone species,'' such as nitrifiers and nitrogen-fixing microorganisms, are negatively affected (6). However, this is not generally the rule because rarely are there only a few species that perform a singular function. Several processes, such as organic carbon mineralization, are carried out by a large number of microbial species and a reduction in any group of species has little effect on overall soil processes since other organisms can fulfill these functions (7–9). This spare capacity or resilience is a feature of most soil ecosystems.

The molecular techniques used today have identified myriad microbial genes. However, the ecological importance of these genes is largely unknown because it is difficult to quantify their biochemical and microbial expression in situ (10). One significant advance using comparatively recent developments in molecular techniques (based on deoxyribonucleic acid [DNA] extraction, purification, amplification, and analysis) has allowed recording and monitoring the so-called nonculturable microorganisms (11). However, the determination of microbiological activities requires detecting metabolic gene transcripts (messenger ribonucleic acids [mRNAs]) in conjunction with modern sensitive assays of metabolites and mRNA translation products, such as enzymes. These approaches are rapidly improving our understanding of the distribution and extent of microbe-mediated processes in the field. It has been proposed that the monitoring of enzyme activities can be used to determine the effect of genetically modified microorganisms on soil metabolism (12,13).

Enzymes are proteins many of whose activities can be measured in soil. The assays of soil enzymes are generally simple, accurate, sensitive, and relatively rapid. A range of enzyme activities, and a large number of samples, can be analyzed over a period of a few days using small quantities of soil. It is well known that changes in enzyme activities depend not only on variations of gene expression but also on changes of environmental factors affecting the considered activity (14,15). The expression of genes may occur in natural samples, but numerous factors might effectively prevent the actual enzyme process from taking place (Fig. 1). It has been hypothesized that the microbial composition of a soil determines its potential for substrate catalysis since most of the processes occurring in soil are microbe-mediated and are carried out by enzymes (14).

The objective of this chapter is to discuss the potential of soil enzyme assays to determine soil microbial functional or process diversity and, when possible, identify future research needs. The carbon substrate utilization approach is compared with enzyme measurements for monitoring soil microbial functional diversity. Since soil microbial functional diversity encompasses several metabolic activities, this approach requires assays of many hydrolytic and oxidative enzymes. Therefore, the discussion includes the use of composite indices or multivariate statistical analysis for integrating enzyme data sets for comparing soil samples. Methods that distinguish between the contributions of extracellular and intracellular enzyme reactions are discussed.

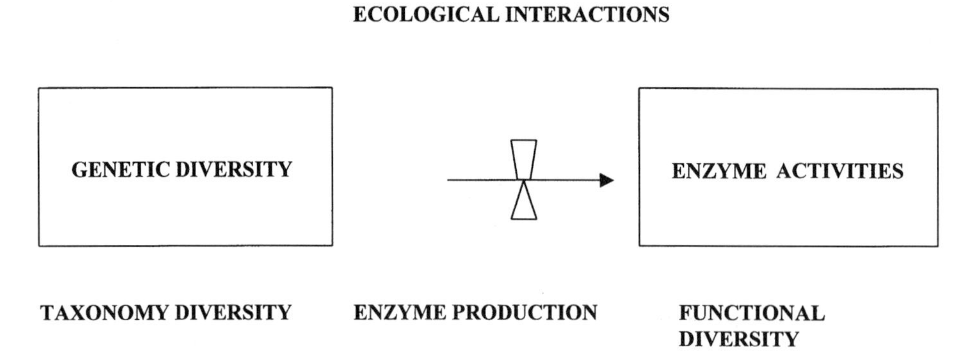

Figure 1 Scheme showing the possible relations among taxonomic diversity, genetic diversity, functional diversity, and enzyme activity. (From Ref. 14.)

II. CARBON SUBSTRATE UTILIZATION PATTERNS AND MICROBIAL FUNCTIONAL DIVERSITY

The BIOLOG system, originally developed to assist with the taxonomic description of bacteria in axenic culture, is based on the ability of bacterial cells to oxidize up to 95 different carbon substrates in microtiter plates (16). The plates are incubated for a suitable period of time (generally 72 h), and the oxidation of the substrate is monitored by measuring the reduction of a tetrazolium dye. The rate and density of the color change depend upon the number and activity of microbial cells in the well of the microtiter plate. The BIOLOG system also has been applied to assess the functional metabolic diversity of microbial communities from different habitats (17) and soil types (18–20), including the rhizosphere (21,22) and grassland soils (23). Microbial communities produced habitat-specific and reproducible patterns of carbon source oxidation, and thus this technique was suggested to be sensitive for detecting temporal and spatial differences among soil microbial communities (24). The carbon substrate utilization profiles have been shown to be sensitive to heavy metal pollution (25), organic pollutants (26), soil types (26,27), crop type, and crop management (28). The last mentioned study showed that, of the expectants tested, 0.01 *M* NaCl yielded the highest well color development (28).

The BIOLOG approach presents several advantages. First, there is no doubt that the utilization of carbon is a key factor in governing microbial growth in soil, and, for this reason, this technique has been considered ecologically important (Table 1). In addition, the technique is very rapid and simple. Thus it has become a very popular method of assessing soil metabolic diversity and therefore soil functionality. In 1997 an international meeting was devoted almost completely to the subject (29).

However, there are several drawbacks that limit the accuracy of the method in assessing the metabolic diversity of soil microbiota (Table 1). First, the inoculum is a mixture of organisms extracted from soil rather than a single species from axenic culture. Thus, only the culturable minority of microorganisms is capable of oxidizing the organic substrates (30). It is well established that only 1–10% of soil microflora are culturable by a range of conventional techniques and media (31). Furthermore, reproducible results can be obtained only if replicates present a similar inoculation density (24). Because BIOLOG plates were developed for a different type of microbiological analysis, not all of the 95 organic substrates offered are ecologically relevant, and it could be important to choose organic compounds that are appropriate for the microorganisms in their specific habitats. For example, organic carbon compounds commonly present in root exudates were tested to discriminate between the functional activities of microbial communities in rhizosphere

Table 1 Advantages and Disadvantages of the BIOLOG Technique

Advantages	Disadvantages
Simple and rapid	Fungi are not involved in the carbon substrate utilization profiles
Use of organic carbon, a key factor in governing microbial growth in soil	Only culturable microorganisms provide information
Patterns of carbon source oxidation are reproducible and habitat-specific	Changes in the microbial community structure may occur during incubation

soil (32). Insam (33) reduced the number of substrates to 31 of those reported in soil to allow three replicates of each substrate (plus a control) in a 96-well plate. Another problem is that the multivariate statistics used (principal component analyses, discriminant analyses, and detrended correspondence) to analyze BIOLOG data may mask the analytical problems described. Changes in the microbial composition of the inoculum may occur during the incubation (Table 1). Indeed, not all constituents of the inoculum contribute to the color development, and significant changes in the community structure occur over a 72-h period (34). By comparing DNA melting profiles (denaturing gradient gel electrophoresis and temperature gradient gel electrophoresis) at the beginning and at the end of the incubation, it was shown that the structure of the microbial community changed in potato rhizosphere soil. In particular, it was observed that fast-growing bacteria become dominant during the incubation period. However, the changes were not observed in an activated sludge reactor amended with glucose or peptone (34). It was postulated that in this case the dominant bacteria had been selected for rapid growth on readily utilizable carbon sources because the activated sludge had been continuously fed with glucose and peptone. In contrast, the dominant bacterial population of the potato rhizosphere was selected in an environment characterized by a much lower content of organic substrates than in activated sludge. Therefore, the dominant bacterial population (equivalent to K strategists) of the rhizosphere inoculum was displaced by a more competitive microbial population (equivalent to r strategists). Another significant limitation is that the community profiles obtained with the use of the BIOLOG procedure do not include fungi because of their comparatively slow growth rate (19).

Degens and Harris (35) have attempted to overcome the many problems of the BIOLOG approach by measuring the patterns of in situ catabolic potential of microbial communities. Differences in the individual short-term respiration responses (or substrate induced respiration [SIR]) of soils to the addition to 36 simple substrates were used to assess microbial communities. Despite the fact that this approach seems more accurate than the BIOLOG technique, it has not been used widely since it was first published in 1997.

III. ENZYME ACTIVITIES AND SOIL MICROBIAL FUNCTIONAL DIVERSITY

A. Limitations of a Single Enzyme Activity as an Index of Microbial Activity

Enzyme activities can be measured and used as an index of microbiological functional diversity if they reflect changes in microbial activities. Since microbial functional diversity includes many different metabolic processes, theoretically a large number of different enzymes should be measured. Data should be integrated in an attempt to calculate an index of microbial functional diversity and to compare the microbial functional diversity of different soil samples. Because this task is not possible, a representative set of enzyme activities is needed. One approach might be to measure only the enzyme activities that control the key metabolic pathways. Generally, these are rate-limiting exergonic steps and are the targets of metabolic regulation. In the case of glycolysis (one of the main metabolic pathways present in almost every microbial cell and transforming 1 mole of glucose to 2 moles of pyruvate), phosphofructokinase 1 is one of the 10 enzymes involved and catalyzes the conversion of fructose–6-phosphate to fructose–1,6-bisphosphate. This regulates the rate of glycolysis (36). Therefore, determination of the phosphofructokinase-1 activity should be an indication of the potential rate of glycolysis in soil.

The amount of soil N available to plants depends on many processes, but N mineralization-immobilization turnover is considered to play the main role (37). The immobilization process (the conversion of N-NH_4^+ to organic N) includes several reactions catalyzed by different enzymes. The first reaction converts N-NH_4^+ to an amino acid and involves the formation of glutamine from glutamate catalyzed by glutamine synthetase (GS). The deamination of 2-oxoglutarate results in the formation of glutamate, which is catalyzed by glutamate dehydrogenase (GDH) followed by the amination of aspartate with the formation of asparagine by asparagine synthetase (38). The K_m of GS is lower than that of the other two enzymes, and at an ammonia ion concentration < 0.1 m*M*, ammonia is incorporated into glutamine only, according to the reaction catalyzed by GS (39). Therefore, the determination of GS activity can give useful information on the potential rate of the N immobilization in soil.

It has been suggested (42) that β-glucosidase activity is a sensitive indicator for assessing the effect of long-term burning and fertilization on the biological activity of tallgrass prairie soil. According to Staddon and coworkers (43), acid phosphatase can be used as an index for assessing the impact of fire on soils. Usually such hypotheses derive from the fact that the measured enzyme activity was significantly correlated with some general microbial parameter such as respiration or microbial biomass (15,44–46). In the 1950s and 1960s, invertase, protease, asparaginase, urease, phosphatase, and catalase activities were assumed to be valid indicators of soil fertility (40). According to Skujins (40), such suggestions produced conflicting and confusing data, not only because the adopted methodologies were sometimes questionable but also because it is conceptually wrong to use a single enzyme activity to determine plant productivity, microbiological activity, or soil fertility. However, the use of a single enzyme activity, whether or not in a crucial metabolic process, as an index for the microbiological activity or fertility of soil has been criticized (40,41). Nannipieri (15) explains why reliance on a single assay is wrong.

1. Enzyme activities catalyze specific reactions and generally are substrate-specific. Thus, they cannot be related to the overall microbiological activity of soil, which includes a broad range of enzyme reactions. The synthesis of a particular enzyme can be repressed by a specific compound while the overall microbiological activity of soil or crop productivity is not affected. Chunderova and Zubers (47) showed that after 4 years of cropping, high P concentrations depressed phosphatase activity. Microbial phosphatase is a repressible enzyme, as shown by its decrease when microorganisms are transferred from a deficient to a normal phosphate medium (48).

2. Measurements of microbial biomass have been used as indexes of microbial activity in soil. Of course, biomass size and activity are different properties, but they often are confused in soil studies (41). Nonetheless, microbial biomass can be correlated with microbial processes, such as respiration rate, or with specific enzyme activities under particular environmental conditions, but this does not mean that this relationship is valid for every soil under every environmental condition.

3. As discussed later, the overall activity of any single enzyme in soil may depend on enzymes in different locations including extracellular enzymes immobilized by soil colloids. The activity of these immobilized enzymes may not be as sensitive to environmental factors as are these directly associated with microbiological activity. For example, enzymes in humic complexes are more resistant to thermal denaturation and proteolysis than are the respective free enzymes (41).

There is the potential to use an integrated measurement of a number of enzyme activities (as discussed later), in conjunction with other physical, chemical, and microbiological measurements, in assessing soil quality. Some hydrolase activities do not show wide seasonal variation, probably because of the large amount of activity associated with enzymes stabilized by soil colloids. This provides a great advantage over microbiological measurements such as respiration, which vary so widely within a season or on a year-to-year basis and make it difficult to find trends or identify the impacts of different management systems.

Dehydrogenase activity is an intracellular process that occurs in every viable microbial cell and is measured to determine overall microbiological activity of soil (40,41). The problem with this is that the electron acceptors (2,3,5-triphenyltetrazolium chloride [TTC] (see also p. 19) or 2-p-Iodophenyl-3-p-nitrophenyl-5-phenyltetrazolium chloride [INT] (see also p. 13)) used in the assays are not very effective, and thus the measurements may underestimate the true dehydrogenase activity (41).

Another potentially confusing aspect of these studies arises as a consequence of soil collection and pretreatment. According to Nannipieri and coworkers (41), enzyme activity measurements are carried out after removal of visible animals and plant debris and on sieved soil samples under laboratory conditions. Thus the measured activities of these samples may depend on the metabolic processes or enzyme activities associated only with the microbial cells. Conversely, when rates of metabolic processes, such as respiration, are measured in the field, the contributions of living roots and animals as well as macroscopic organic debris are recorded. In other words, total biological rather than microbiological activities are measured (41).

B. Enzyme Activities: Methodology and Interpretation

According to the review by Skujins (40), Woods in 1899 suggested that extracellular enzymes could be present and active in soil. The first measurements of enzyme activities in soil were done on catalase and peroxidase activities from 1905 to 1910 (40). Since then, the activity of dozens of enzymes has been detected in soil. Obviously the number of enzymes is considerably greater than those measured because of the multitude of microbial, faunal, and plant species inhabiting soil (46). In addition, the activity measured by many assays cannot be ascribed to the action of a single enzyme. Thus dehydrogenase activity is determined by the multiple enzyme reduction of a synthetic substrate (TTC or INT) due to an oxidation of generally unknown endogenous substrates whose concentration is also unknown (46). Casein-hydrolyzing activities are measured without specific identification of the bond hydrolyzed or of all products formed. It is important to emphasize that even when all the components of the reaction are known, for example, in the case of urease assays, different enzymes from different sources (microbial, plant, or animal cells) catalyze the same reaction.

Tables 2 and 3 (99–105) report a range of activities of enzymes commonly investigated in soil. The ranges are generally very broad, possibly as a result of differences in methods and soil types. In the late 1990s, several authors suggested standardized procedures (106–108) based on conventional enzymology protocols. Tscherko and Kandeler (109) proposed a classification for microbiological variables based on the activity classes of the 30th and 70th percentiles. Using 30th and 70th percentiles gave similar widths of the three categories, which were attributed to low, normal, and high activities. Different sites could be classified according to the land use.

Table 2 Some Enzyme Activities Involved in Organic C and N Transformations with Their Range of Activities in Soil

Enzyme	Range	Reference
Enzymes involved in C transformations		
Xylanase	1.33–3125.00 μmol glucose g^{-1} 24 h^{-1}	(14, 49, 50, 51, 52, 53, 54, 55)
Cellulase	0.4–80.0 μmol glucose g^{-1} 24 h^{-1}	(14, 56, 57)
Invertase	0.61–130 μmol glucose g^{-1} h^{-1}	(52, 53, 58, 59)
β-Glucosidase	0.09–405.00 μmol *p*-nitrophenol g^{-1} h^{-1}	(42, 56, 57, 60, 61, 62, 63, 64, 65, 66)
β-Galactosidase	0.06–50.36 μmol *p*-nitrophenol g^{-1} h^{-1}	(56, 57, 61)
Enzymes involved in N transformations		
Protease (casein-hydrolyzing proteases)	0.5–2.7 μmol. *p*-tyrosine g^{-1} h^{-1}	(14, 50, 51, 67, 68, 69, 70)
Dipeptidase	0.08–1.73 μmol leucine g^{-1} h^{-1}	(60)
Arginine deaminase	0.07–0.86 μmol N-NH_3 g^{-1} h^{-1}	(14, 42, 69, 71, 72, 73, 74)
L-Asparaginase	0.31–4.07 μmol N-NH_3 g^{-1} h^{-1}	(60, 65, 75, 76)
Amidase	0.24–12.28 μmol N-NH_3 g^{-1} h^{-1}	(60, 65, 75)
L-Glutaminase	1.36–2.64 μmol N-NH_3 g^{-1} h^{-1}	(75, 77)
Urease	0.14–14.29 μmol N-NH_3 g^{-1} h^{-1}	(42, 69, 78, 79, 80, 81, 82, 83, 84)
Nitrate reductase	1.86–3.36 μg N g^{-1} h^{-1}	(14)

Many enzyme activities have been detected in soil, but a reliable assay either has not been developed or has been developed, but long after the initial report. For example, hydrolysis of laminarin and inulin occurs in soil (110–112), but there is no specific assay protocol. L-glutaminase, which catalyzes the hydrolysis of L-glutamine, yielding L-glutamic acid and NH_3, was first detected in soil by Galstyan and Saakyan (113), but a simple and rapid method was developed much later by Frankenberger and Tabatabai (77). L-Asparaginase activity, which catalyzes the hydrolysis of L-asparagine producing L-aspartic acid and NH_3, was detected in soil by Drobni'k (114), but the simple and sensitive method was developed much later (76). In fact, Tabatabai and coworkers have been responsible for the development of many assays for enzyme activities in soil (91,115,116).

The enzyme assay has to be simple and rapid, but above all sensitive and accurate (117,118). This requires an efficient extraction and then an accurate determination of the substrate or the reaction products from soil. Since most of the procedures for determining either product formation or substrate disappearance are based on colorimetric reactions, it is preferable to use buffers, which, in general, do not extract organic matter from soil. Appropriate substrate concentration, pH, and temperature and optimal pH have to be found for the assay of any soil enzyme (119). At a substrate concentration exceeding the value limiting the reaction rate, the period of time selected should assure a linear substrate disappearance or product formation and should be the shortest one (only a few hours at

Table 3 Range of Some Hydrolase and Oxidase Activities in Soil

Enzyme	Range	Sources
Enzymes involved in organic S transformations		
Arylsulfatase	0.01–42.50 μmol *p*-nitrophenol g^{-1} h^{-1}	(14, 42, 43, 63, 69, 85, 86, 87, 88, 89, 90, 91)
Enzymes involved in organic P transformations		
Alkaline phosphatase	6.76–27.34 μmol *p*-nitrophenol g^{-1} h^{-1}	(14, 42, 43, 51, 86, 92, 93, 94, 95)
Phosphatase at pH 6.5	6.76–27.34 μmol *p*-nitrophenol g^{-1} h^{-1}	(96, 97)
Acid phosphatase	0.05–86.33 μmol *p*-nitrophenol g^{-1} h^{-1}	(42, 43, 69, 86, 94, 95, 98, 99, 100, 101)
Phospholipase C	5.02–8.15 μg *p*-nitrophenol g^{-1} h^{-1}	(14)
Other enzyme activities		
Dehydrogenase	0.002–1.073 μmol TPF g^{-1} 24 h^{-1}[a]	(14, 42, 50, 61, 69, 92, 93, 102, 103, 104, 105)
	0.003–0.051 μmol INF g^{-1} 24 h^{-1}[a]	(102, 105)
Fluorescein diacetate hydrolysis	0.12–0.52 μmol fluorescein g^{-1} h^{-1}	(60)
Catalase	61.2–73.9 μmoles O_2 g^{-1} 24 h^{-1}	(63)

[a] TPF, triphenyl formazan; INF, iodonitrotetrazolium formazan.

most) to produce a measurable value of activity. Complications due to microbial growth, intracellular catalysis, and new enzyme synthesis can occur in assays with long incubation times. The buffer solution must maintain the pH to that required for optimal activity throughout the incubation period (120).

It is important interpreting of enzyme activities to understand that potential rather than in situ activity is often being determined. This is because the incubation conditions are chosen to ensure a rapid rate of substrate catalysis (41,120). In addition, enzyme assays employ soil slurries to reduce diffusional limitations. These assay conditions are very different from those occurring in soil, where moisture and temperature fluctuate widely, the substrate concentration is generally not in excess, and pH is rarely optimal. When possible, it is worthwhile to compare potential enzyme activities measured by the enzyme assays involving soil slurries with hydrolysis of natural substrates added to soil samples and incubated under the realistic in situ conditions. In the case of urea hydrolysis, it is easy to compare urease activity determined by the conventional enzyme assay with the rate of hydrolysis of solid urea added to soil and incubated in a range of relevant temperatures and in the absence of buffer. This might be useful for predicting urea fertilizer hydrolysis under field conditions for specific soil types.

It has been discussed in several reviews that the foremost problem in interpreting measurements of enzyme activities is to distinguish among many components contributing to the overall activity (15,41,121,122). The activity of any particular enzyme in soil depends on enzymes that can have different locations: (1) active and present intracellularly in living cells, (2) in resting or dead cells, (3) in cell debris, (4) extracellularly free in the

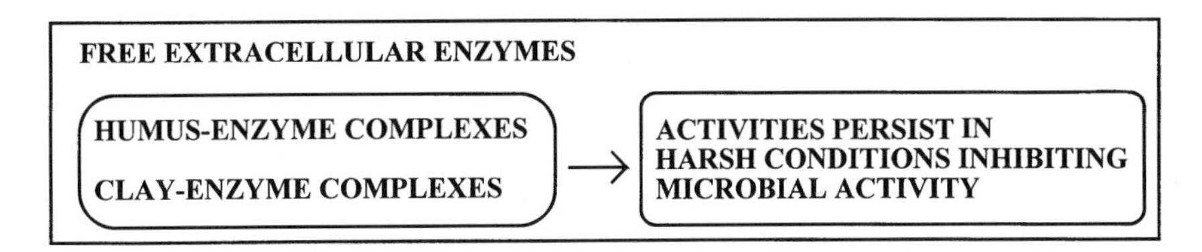

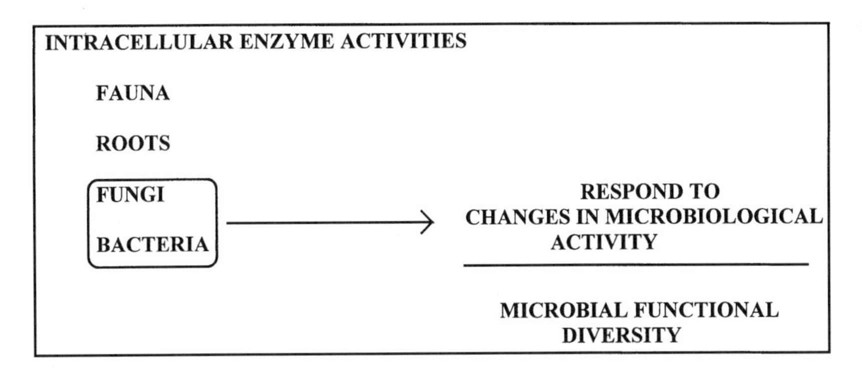

Figure 2 Various activities contributing to the overall enzyme activity measured in soil with those affecting the microbial functional diversity in soil.

soil solution, (5) adsorbed by inorganic colloids, or (6) associated in various ways with humic molecules. In addition, abiotic transformations, the so-called enzyme-like reactions, can contribute to the overall activity (Fig. 2). Intracellular enzymes are present in plant, animal, and microbial cells; however, since visible animals and vegetable remains are largely removed prior to an assay and those that have been released from lysed cells are rapidly degraded by microorganisms, it is reasonable to suppose that the most important intracellular enzymes of soil are those in living microbial cells. Thus, the determination of intracellular enzyme activity can give important information about the processes mediated by the current microbial inhabitants. By determining the intracellular enzyme activities of soil samples, it is possible to have information on the microbial functional diversity of soil. On the other hand, of the three extracellular locations (free enzymes, enzymes adsorbed by inorganic complexes, or those associated with organic colloids) the first are supposed to be short-lived (15,46,121), whereas the other two are characterized by a marked resistance to thermal and proteolytic degradation (123). With the present methods it is difficult if not impossible to determine the different locations of the enzyme activities.

C. Soil Minerals as Catalysts (Pseudoenzymes) in Biochemical Reactions

Soil minerals can affect the fate of biochemical compounds in soil in at least three main ways: (1) incorporation of N-, P-, and S-bearing organics into the structural network of mineral colloids and adsorption of these organics to their surface: consequently the dynamics and bioavailability of these nutrients may be modified (124); (2) adsorption of enzymes on clay and/or clay-organic complexes: these immobilized enzymes are active and stable, but they exhibit activities quite different from those of free enzymes (123,125); and (3) abiotic transformation of natural organic components: this means that the mineral component should be considered not only as a support for adsorption and binding of organics (enzymes and substrates), but also as a reactive surface for many transformation processes.

Clays demonstrate the ability to catalyze electron transfer reactions. In the same way, metal oxides/hydroxides, such as Fe^{3+} and $Mn^{3+,4+}$ oxides, quite common in soils, are able to catalyze oxidation reaction of organic compounds. Birnessite (δ-MnO_2), in particular, is considered to be an "electron pump" for a wide range of redox reactions (126). In summary, clay minerals and metal oxides and hydroxides are reactive in promoting abiotic degradation of natural organics.

A large part of the research into the catalytic role of clay minerals has been devoted to studying the oxidative transformation of phenols and polyphenols and other natural organic components and the subsequent formation of humic substances (127,128). It is noteworthy that phenolic acids are degradation products of lignin and constitute an important fraction of the exudates released by plants under stress conditions (129).

Shindo and Huang (130) found that montmorillonite, vermiculite, illite, and kaolinite accelerated the synthesis of humic substances from hydroquinone, as the precursor, whereas non-tronite- (Fe^{3+}–bearing smectite), in the presence of O_2, showed a synergistic effect that greatly enhanced the polymerization of hydroquinone (131,132). Similar results were obtained when pyrogallol was used as the precursor (133). In systems containing phenolic compounds and amino acids, Ca-illite catalyzed the formation of N-containing humic acids (134). The yields and nitrogen contents depended on the kind of amino acids used. In 1997 Bosetto and colleagues (135) studied the formation of humiclike polymers from L-tyrosine and homoionic clays and showed that the amount produced depended on the type of interlayer cation.

Besides the clay minerals, $Mn^{2+/4+}$ and Fe^{3+} oxides oxidize phenolic compounds rapidly. Their relative effectiveness in the synthesis of humic substances has been studied extensively (127,128,136). Hydroquinone, resorcinol, and catechol were used as substrates. The catalytic effect of birnessite was higher than that of Fe oxide; however, the relatively high content of the Fe oxides in soils suggests that their role in the abiotic formation of humic substances should not be overlooked. The synthesis of humic substances was obtained also by using aluminas as catalysts (137). McBride and associates (138) proposed an oxidation mechanism by which soluble Al tended to stabilize *o*-semiquinone radicals at low pH, directing subsequent radical polymerization. Pyrogallol-derived polymer formation was strongly promoted by birnessite (139) and the cross-polarization, magic angle spinning-nuclear magnetic resonance ^{13}C-nuclear magnetic resonance (CPMAS-^{13}C NMR) spectrum of humic acids formed resembled those of humic acids extracted from natural soils. Birnessite also was able to cleave the ring structure of pyrogallol, releasing CO_2. The abiotic ring cleavage of polyphenols might, in part, form aliphatic fragments contributing to the aliphatic nature of humic substances in the environment. The amount of CO_2 released from the ring cleavage of pyrogallol and the quantity of humic polymers formed were directly related to the catalytic activity of clay minerals such as nontronite, bentonite, and kaolinite saturated which Ca^{2+} (139).

Further research has compared the activity of biotic and abiotic catalysts in the transformation of phenolic substrates. Mn oxide influenced the darkening of hydroquinone and resorcinol to a larger extent than did the enzyme tyrosinase, whereas the reverse was true for catechol (140). The yields of humic acids were influenced significantly by the kind of catalyst and the type of diphenols. The comparison between biotic and abiotic catalysts has also been conducted by examining the reaction products that resulted from the transformation reactions. Birnessite, bentonite, tyrosinase, horseradish peroxidase, and the laccases of *Trametes versicolor* and *Rhizoctonia praticola* were used separately in the oxidation of guaiacol (141). All reaction products were analyzed for the presence of guaiacol and guaiacol-derived oligomers. The same seven products (five dimers and two trimers) were produced by using the various oxidizing agents, whereas the amounts of the respective oligomers varied. The differences in the amounts obtained were likely due to different reaction conditions.

The rate of 2,6-dimetoxyphenol transformation was considerably higher in reactions catalyzed by tyrosinase or laccase than in those by birnessite (141). At the same time, significant differences in oxygen consumption were observed. When continuous additions of catechol as substrate were employed, it was shown that tyrosinase and laccase transformed catechol after each addition, whereas birnessite was active only in the first incubation with catechol.

The reaction products of catechol oxidation by tyrosinase had a higher degree of aromatic ring condensation relative to those of the birnessite-catechol system (142). In addition, the products of birnessite catalysis contained various fragments, including aliphatic components, with lower molecular weights than those produced by tyrosinase. In contrast, Birkel and Niemeyer (143) obtained similar reaction products when comparing enzymatic reaction products of humic precursors with products obtained at clay mineral surfaces (montmorillonite) through abiotic reactions.

Phenolic acids with higher methoxy substituents were oxidized more rapidly in the presence of Mn and Fe oxides (144,145). By enzymatic polymerization of syringic and vanillic acids, soluble oligomers (from dimers to hexamers) were found as oxidation products (146,147), whereas tests with ferulic acid, incubated with MnO_2, showed that the soluble products obtained did not contain any oligomers because they were rapidly sorbed to MnO_2 surfaces (144). In 1998 Hames and coworkers (148) reported the first efficient modification/degradation of in situ lignin by MnO_2 and oxalic acid, either produced by fungi or resulting from oxidative degradation of cell wall components. The MnO_2/oxalate system appeared to oxidize the lignin macromolecule selectively and to play an important role in the transformation of the lignin polymer into humus and/or its precursors.

The results of these studies indicate that the catalytic effects of soil minerals and enzymes on the oxidation of phenols need more investigation. The similarities and differences when comparing both activities might be due to the reaction conditions (pH, substrate/catalyst ratio, rate of oxidation) and/or to the reaction mechanisms. It is not clear, for instance, whether the metal oxides meet the definition of a catalyst. A true catalyst cannot be altered by the reaction it catalyzes. For both abiotic and biotic agents, ions such as Cu^{2+}, Mn^{4+}, and Fe^{3+} act as electron acceptors. However, whereas in the protein complex (enzyme) an electron is transferred from the reduced metal to oxygen in a cyclic manner, in metal oxides, in particular birnessite (MnO_2), the metal remains in the reduced form (Mn^{2+}) even for long periods of incubation and Mn^{2+} recovers to its initial state fairly slowly in the presence of O_2. The result is that the enzymes behave as a catalyst and birnessite as an oxidizing agent.

Phenols, together with amino acids, have been used to synthesize nitrogenous polymers resembling humic acids in the presence of both polyphenol oxidases and birnessite and nontronite (149–151). It was assumed that ring cleavage of phenol and release of

CO_2 and NH_3 occurred and were followed by polycondensation of semiquinone radicals, aliphatic fragments, and amino acids to form humic polymers. Similar mechanisms have been proposed in systems involving phenoloxidase enzymes.

The deamination of amino acids, such as serine, phenylalanine, proline, methionine, and cysteine by birnessite, and the role of pyrogallol in influencing their mineralization have been investigated (152,153). Nitrogen mineralization was inhibited by pyrogallol, whereas S mineralization of S-containing amino acids was not, except in the iron oxide–methionine system. Deamination and decarboxylation of amino acids as catalyzed by soil minerals may constitute a pathway of C turnover and N transformation in nature.

Some natural organic compounds, other than polyphenolics, also have been shown to be oxidized by metal oxides. The reactions of malic acid, an important constituent of root exudates, with the hydrous oxides of Mn and Fe were studied (154). The reactions followed two pathways, depending on the pH-controlled adsorption of oxalacetic acid (the first product of the malate oxidation) on the oxide surfaces. The production of NH_4^+ in the glutamic acid–treated birnessite suspensions was attributed to direct chemical oxidative deamination of glutamic acid by the manganese oxide (155).

In the presence of Na-montmorillonite, the isocitric acid was oxidized and transformed into α-ketoglutaric acid (156). Isocitrate oxidative decarboxylation comprised several steps but always started with a protonation reaction. A certain analogy exists between enzymatic and clay mineral catalysis, provided that in both cases transformation began with a protonation of a chemical function. For enzymatic catalysis, it was principally the coenzyme (NADH + H^+) of the isocitrate dehydrogenase that supplied the proton. Nevertheless, the reaction rate in the presence of clay was very much lower than in the presence of enzyme.

Glutamic acid was selectively deaminated by a combination of pyridoxal phosphate (PLP), a cofactor in enzyme systems important in amino acid metabolism, and Cu^{2+}-smectite with production of ammonia and α-ketoglutaric acid (157). The system exhibited specificity for glutamic acid in comparison to some other amino acids. One possible explanation was that glutamic acid reacted with PLP to form a Schiff base, which then complexed with Cu^{2+} on the mineral surface. This was followed by hydrolysis to ammonia, α-ketoglutaric acid, and regenerated PLP. The catalyst system was not stereoselective because it was equally effective for both L- and D- glutamic acid. The PLP-Cu^{2+}-smectite has acted as a ''pseudoenzyme'' wherein the PLP was active and independent of the protein matrix of the enzyme and the silicate structure substituted for the apoenzyme (157).

The deamination of glutamic acid also occurred in the presence of montmorillonite saturated with various cations and in the absence of any cofactor. The reaction products were α-hydroxyglutaric acid and traces of butyric acid, whereas in the presence of the enzyme glutamate dehydrogenase and oxidized nicotinamide-adenine dinucleotide (NAD^+), the glutamic acid yielded α-ketoglutaric acid (158). The catalytic activity of the montmorillonite surface depended on pH and remained lower than that of the enzyme. Nevertheless, the montmorillonite had an advantage over the enzyme as it displayed a larger activity pH range and a nonspecific activity. The abiotic catalytic role of montmorillonite also was demonstrated in deamination of aspartic acid by aspartase-Ca-montmorillonite systems (159). Deamination of L- and D-glutamic and aspartic amino acids and of their DL racemic mixtures in the presence of Na-montmorillonite showed a stereoselectivity of the clay mineral for the L-isomer and implicitly a structural chirality character of the clay mineral (160).

In conclusion, the decomposition of plant and animal residues and the stabilization of organic C in soils depend not only on microbial activity and enzyme diversity but also on the nature and properties of soil minerals. Abiotic reactions of natural organics include polymerization, polycondensation, ring cleavage, decarboxylation, and deamination. Other reactions that may be pathways of turnover of organic C, N, P, and S need to be investigated (161).

Under natural conditions, the importance of an abiotic reaction involving an organic substrate depends upon relative rates of other competitive biodegradation processes. It is likely that under microenvironment conditions hostile toward microbial life, abiotic transformation may dominate. However, no accurate methods are available to distinguish enzyme from enzymelike reactions. Sterilization of soil has been used to denature enzymes in soil and to measure the contribution of enzymelike reactions (46,117). However, even after sterilization, accumulated enzymes may express a residual activity and sterilization also can affect the activity of soil colloids responsible for enzymelike reactions (46,125).

D. Determination of the Extracellular Enzyme Activity in Soil

Bacteriostatic agents have been used in enzyme assays to prevent microbial growth and thus synthesis of intracellular enzymes during incubation. This approach has been unsatisfactory because of changes in cell wall permeability and induction of plasmolysis with release of intracellular enzymes. In addition, the effect of bacteriostatic agents (usually toluene), and irradiation with gamma rays or electron beams have shown variable effects, depending on soil and enzyme assay conditions (46,120,162,163).

The kinetic analysis of the decline in the activity of L-histidine NH_3-lyase during a 96-h incubation period with a biostatic agent (toluene or Na azide) was described as a two-component model in which each component declined according to the first-order kinetics (164,165). The decline of the labile and the stable components suggested that both depended upon continuous enzyme synthesis and thus were under microbial control.

Other approaches have tried to distinguish the extracellular from the intracellular enzyme activity by monitoring both microbial numbers and the enzyme activity during microbial growth after substrate amendment and the subsequent microbial decline when substrate is exhausted. Paulson and Kurtz (166) determined microbial number, ureolytic microorganisms, and urease activity in a surface silty clay loam soil treated with a C (dextrose) and an N (ammonium sulfate or urea) source. A stepwise multiple regression equation was used to calculate the main pools of urease activities contributing to the overall activity (Y) determined by the assay

$$Y = \alpha + \beta_1 X_1 + \beta_2 X_2 + \beta_3 X_3 + \beta_4 X_4$$

where α represents the activity of extracellular ureases adsorbed or encapsulated by soil colloids, $\beta_1 X_1$ represents the urease activity present inside ureolytic microorganisms, $\beta_2 X_2$ represents the urease activity released by dead ureolytic microorganisms, $\beta_3 X_3$ is the stabilized extracellular urease activity inactivated during the incubation, and $\beta_4 X_4$ represents the urease activity released from living microorganisms. It was shown that the amount of urease activity ($\beta_4 X_4$) released by living microorganisms was insignificant. The greatest contribution was due to the stabilized extracellular urease activity, which accounted for 79–89% of the overall enzyme activity under normal conditions (166). When microbial growth was induced, this percentage temporarily decreased until a new state was reached.

Interesting and indirect evidence of the presence of extracellular urease stabilized by humic molecules was shown by Norstadt and associates (167) in turf infected by the fungus *Marasmus oreades* (Bolt), which produced concentric rings of affected soil. Urease activities of infected rings were 10–40% of the values determined in unaffected soil and a decrease in humic content also was observed in infected rings.

McLaren and Pukite (168) correlated the urease activity and the number of ureolytic microorganisms reported by Paulson and Kurtz (166). Both variables were significantly correlated; by plotting urease activity against the number of ureolytic microorganisms, the extrapolation to zero population produced a positive intercept of urease activity, which was assumed to be the extracellular soil urease activity. The same approach was followed by Nannipieri and coworkers (169), who demonstrated that 100 m*M* sodium pyrophosphate at pH 7.1 extracted 30–40% of the extracellular urease of the soil. This approach to determining the extracellular component of the overall enzyme activity may be defined as the *physiological response* (PR) method, because it uses the response of microbial enzyme activity to substrate amendments to provide an estimate of the extracellular activity. Similarly, the substrate induced respiration (SIR) method utilizes the respiration response of soil organisms to substrate amendments to provide an estimate of soil microbial biomass (170–172). The difference in the SIR and PR methods is the time scale of measurements. In the SIR method, the enhanced respiration is monitored for 6–8 h at most before the increase in respiration due to cell division and growth. In the PR method measurements are taken during the microbial growth (Fig. 3).

Unfortunately, the approach followed by Paulson and Kurtz (166) and by Nannipieri et al. (169) was based on the determination of ureolytic microorganisms by conventional plate counts, which, as mentioned before, give only a partial estimate of the total microbial population of soil. By considering that the adenosine triphosphate (ATP) content of soil has been shown to give an estimate of soil microbial biomass (173), Nannipieri and colleagues (96) monitored the ATP content and phosphatase activity at pH 6.5 in fresh and dry soils that were treated annually either with (50 or 100 t ha^{-1}) or without industrial sewage sludges. Both parameters were measured after the treatment of soil with glucose (3 mg C g^{-1} dry wt soil) and potassium nitrate (250 μg N g^{-1} dry wt soil), with samplings

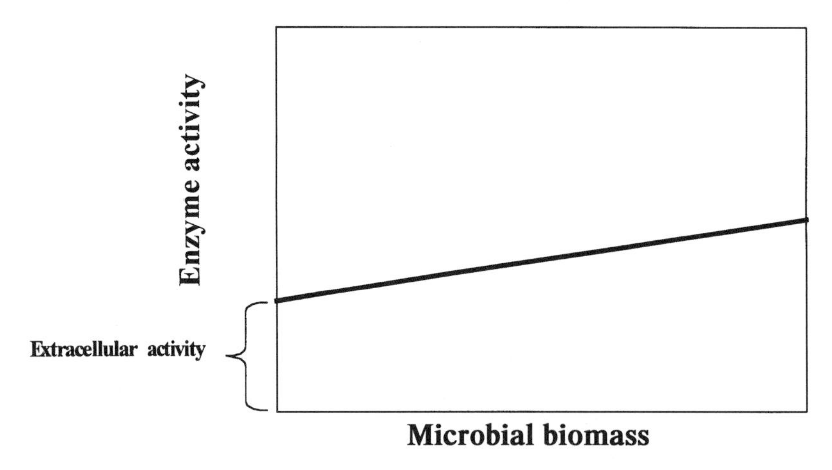

Figure 3 Extracellular activity determined by the intercept of the regression plot when the enzyme activity is correlated with microbial biomass.

at 36, 48, 60, 72, 84, and 96 h. Phosphatase activity and the ATP content were significantly correlated ($P < 0.05$) in moist soils. At zero ATP, the intercept on the ordinate gave positive values of 2.086, 0.741, and 0.597 mg of *p*-nitrophenol g^{-1} dry wt soil for the 0-, 50-, and 100-t sewage sludge ha^{-1} soil treatments, respectively. In the case of dry soils, significant correlations between the phosphatase activity and the ATP content were observed at 0- ($p < 0.05$) and 100- ($p < 0.01$) t sewage sludge ha^{-1} treatments (96). Negative intercepts were observed at 0- and 50-t sewage sludge ha^{-1} treatments, and they were postulated to be due to the absence of phosphatase activity in some of the microorganisms inhabiting the soil or to the underestimation of the intracellular enzyme activity. Air drying of soil and the 2-year storage before laboratory measurements may have facilitated the formation of nonproliferating cells (fungal spores, protozoan cysts, etc.) and microorganisms surrounded by polysaccharide-based gums able to survive under dry conditions and in the absence of fresh substrates (174). In both cases, the detection of the intracellular enzyme activity may have been problematic because the external barrier of these cells may have impeded the diffusion of the substrate to the intracellular phosphatases during the assay.

Several investigations have monitored the ATP content and enzyme activities during microbial growth promoted by the addition of glucose and an N source to soil (70,96,175–180). Asmar and coworkers (175) and Watanabe and Hayano (181) found that extractable enzyme activity increased more than did total activity, but probably only as a small part of the overall enzyme activity of soil. The ATP content and protease activity were correlated after substrate addition (175).

Dilly and Nannipieri (182) correlated the ATP content with the activities of phosphatase or urease on two Canadian grasslands Dark Brown Chernozemic soils that had a neutral pH and 2% organic C content but different textures. These parameters were measured during microbial growth induced by the addition of glucose and nitrate to the two soils (177). In spite of the different textures, the extracellular phosphatase activity recorded by the intercepts gave similar values in the two soils. However, these calculations are unrealistic because they are based on only limited data.

In addition to its inaccuracy in dry soils, the PR method has the following drawbacks:

1. Only a proportion of the soil microbiota respond to the addition of glucose and nitrate to soil. This proportion of glucose-utilizing microorganisms can vary with soil type, management, and pollution (182).
2. The ATP content determined in soils immediately after sampling may not be an accurate index of microbial biomass since it represents the microbiological activity of soil (41). A linear relationship has been found between microbial biomass and ATP content when both measurements were carried out in soils preincubated for 7–10 days at constant temperature and at 50–60% of the water-holding capacity after the sampling (173). Under these conditions, the intercept on the ordinate (equivalent to zero population) gave a zero value for ATP, which indicated that extracellular ATP was insignificant. It has been suggested that during preincubation, aerobic metabolism prevails, ATP of plant debris disappears rapidly, and the preincubated soil contains only microbial ATP (183). Determinations during the daytime indicated that the changes in ATP content reflected microbiological activity rather than the microbial biomass (41,183). Such changes in the surface soil layer were very high within a few hours (four-

fold increase from 7 A.M. to 5 P.M.) and depended on changes in temperature. Considering the long average generation time of microbes in soil, similar changes in microbial biomass were unlikely (41). Evidence supporting the use of ATP as an index of microbiological activity was obtained by Nannipieri and associates (177), who analyzed the ATP immediately after soil sampling from a laboratory experiment wherein both fungal growth and bacterial growth were stimulated by the addition of glucose and other nutrients.

3. The response of enzyme activities to glucose and nitrate additions may not coincide with microbial growth. In the latter case, there may not be a significant correlation between the microbial growth and the increase in enzyme activity. In glucose-treated soils, the increase in casein-hydrolyzing activities was markedly delayed with respect to microbial growth, regardless of the C source added to soil (70,179). Microbial growth and the increase in enzyme activity can coincide when constitutive enzymes predominate among the intracellular enzymes catalyzing the same reaction in soil. However, enzymes are not only constitutive, but also inducible and repressible. For example, the presence of constitutive and repressible microbial ureases has been reported (184). McCarty and coworkers (185) showed that in glucose-amended soils, urease activity decreased by increasing the NH_4^+ and NO_3^- concentration. However, the repression of urease synthesis depended on the N product derived by microbial uptake of the inorganic form. Indeed, L- but not D-isomers of alanine, arginine, asparagine, aspartate, and glutamine repressed urease production in glucose-amended soil. As reported, phosphatase activities are repressed by the presence of inorganic P (48). Carbon-nitrogen lyases (such as L-histidine NH_3-lyase) eliminate NH_3 and form a double bond, thus eliminating the necessity for a redox carrier. Histidine or urocanate induced the synthesis of L-histidine NH_3-lyase activity in soil (165). The stimulatory effect of the enzyme synthesis by urocanate was eliminated completely by the addition of glucose-C at 4000 μg g^{-1}. This agrees with the report that the induction of the synthesis of many catabolic enzymes is repressed by the addition of a superior catabolite such as glucose (165).

A similar approach to the PR method was carried out by Grierson and Adams (186), who correlated the ergosterol (an index of fungal biomass) or microbial P with acid phosphatase in soil sampled through the year from two eucalypt forests: Sampling was stratified in relation to where Jarrah (*Eucalyptus marginata* Donn ex Sm.), characterized by an extensive surface system of fine lateral roots with ectomycorrhizal associations, occurred alone (Jarrah) or in association with the understory species *Branksia grandis* Willd. (Jarrah + Branksia). In contrast to the ectomycorrhizal roots of Jarrah, *B. grandis* produced a mat of nonmycorrhizal cluster (proteoid) roots (0–20 cm) in the late winter and early spring. Significant correlations were found in both Jarrah + Branksia and Jarrah soils regardless of the season, and the relative regression plots produced positive intercepts. Intercepts of the relationship between the enzyme activity and ergosterol ranged from 2.1 to 7.9 μmol *p*-nitrophenol g^{-1} h^{-1} in the Jarrah and from 2.9 to 8.5 μmol *p*-nitrophenol g^{-1} h^{-1} in the Jarrah + Branksia soils. Intercepts calculated from the regression of acid phosphatase and microbial P ranged from 1.4 to 8.6 and from 3.8 to 9.1 μmol *p*-nitrophenol g^{-1} h^{-1}, respectively. The method used to determine microbial biomass by Grierson and Adams (186) is more accurate than that used by Nannipieri and associates (96); it is based on the lysis of microbial cells caused by liquid $CHCl_3$ (187) and gives a better estimation of microbial biomass than the ATP content measured immediately after sampling. However, the positive intercepts obtained with the relationships with ergosterol can be due to both extracellular and intracellular acid phosphatase present in bacterial or plant cells. Several measurements were carried out over a period of 2 years (186). The positive intercepts obtained when acid phosphatase was correlated with microbial P probably represent the average extracellular acid phosphatase of the forest soil and the activities of root-derived enzymes due to the high density of cluster and mycorrhizal roots in the soil that was sampled. Positive intercepts varied seasonally but were not significantly different ($P < 0.05$) in terms of species composition except at the break of the dry season. In this case, the intercept for Jarrah + Branksia (with the greatest root density and dominated by cluster roots) was significantly greater than for Jarrah (186). This approach has the advantage of eliminating the problem of repressible or inducible enzymes because it is not based on the induction of microbial growth, unlike the PR method. It may be worthwhile to test the approach of Grierson and Adams (186) in the laboratory under controlled conditions to assess changes in enzyme activity and microbial biomass during prolonged incubations under controlled conditions.

Both the PR method and the approach followed by Grierson and Adams (186) assume that the enzyme activity depends on the contribution of both intracellular and extracellular enzymes. The opposite hypothesis is at the basis of the approach used by Klose and Tabatabai (80) to distinguish the extracellular from the intracellular enzyme activity of soil. According to these approaches, the short-term enzyme assays mainly give the extracellular activity. The total contribution of intracellular and extracellular enzyme activities can be determined after cell lysis by $CHCl_3$ fumigation (80) or ultrasonic treatment (Badalucco, personal communication). The difference in enzyme activity in the values determined before and after cells lysis is supposed to give the intracellular enzyme activity of soil. The $CHCl_3$ fumigation method (CF) has been applied to determine the microbial intracellular and extracellular urease and arylsulfatase activity (80,89). The fumigation period lasted 24 h; commercial purified enzymes also were fumigated to take into account the denaturing effect of fumigation on released enzymes by cell lysis. Surface soils were sampled from corn, soybeans, oats, and alfalfa and analyzed for arylsulfatase activity (89). The highest total extracellular and intracellular enzyme activities were obtained in soils under cereal meadow rotations (sampled under oats or alfalfa) and the lowest under continuous cropping systems. About 45% (range 36.4–56.4%) of the overall enzyme activity was extracellular; the remaining 55% (range 43.6–63.6%) was located intracellularly in microbial cells (89). The intracellular arylsulfatase to microbial biomass C ratio ranged from 279 to 602 μg *p*-nitrophenol mg^{-1} microbial biomass C h^{-1} (89). Intracellular urease activity determined by the CF method ranged from 37.1% to 73.1% (average value 54%) of the overall enzyme activity, whereas the remaining extracellular urease activity ranged from 26.9% to 65.9% (average value 46.6%) (80). The analyzed soils were field-moist surface samples characterized by a broad range of properties. The intracellular urease activity to microbial biomass C ratio ranged from 80 to 419 μg NH_4^+-N mg^{-1} microbial biomass C 2 h^{-1} (80).

The main problems of the CF technique are the following:

1. The chloroform fumigation of soil for 24 h is not effective in killing all microbial cells (188,189). Transmission electron microscope examinations of fumigated soils showed that microbes embedded (2–3 μm) in the mucigel or deep in pores were not lysed by $CHCl_3$ (190,191).

2. The efficiency of $CHCl_3$ in lysing cells is markedly affected by soil structural properties (188); thus, the amount of cells lysed by $CHCl_3$ fumigation can vary according to the soil texture.
3. It has been hypothesized that the enzyme activity determined by short-term (a few hours) assays measures only the contribution of the extracellular activity. The response of the enzyme activity to substrate addition to soil, which stimulates microbial growth, seems to indicate that the short-term enzyme assays also determine the contribution of the intracellular enzymes.

The application of ultrasonic treatments also presents disadvantages. Ultrasonic treatments can be more efficient than $CHCl_3$ in lysing microbial cells in soil, but they also may increase the activity of extracellular enzymes stabilized by soil colloids. For example, enzymes encapsulated in a humic matrix with pores that are not large enough to allow the diffusion of the substrates to the active sites can be released. In this new state, the active sites of enzymes may be more accessible to their substrates.

Selective inhibition of bacterial or fungal growth in remoistened, overdried, inoculated soil has been used to study the microbial sources of enzymes in soil. Thus, it was ascertained that fungi such as *Mortierella* and *Actimucor* species were the main sources of β-glucosidase in an Andosol of a greenhouse field of tomato (192) and bacteria such as *Bacillus* species were the main sources of benzyloxycarbonyl-L-phenylalanyl L-leucine–hydrolyzing activity and casein-hydrolyzing activity in Andosols sampled under sweet potatoes (193). The drawbacks in using microbial inhibitors, such as streptomycin and cycloheximide, in soil are due to their nontarget effects (194).

E. Enzyme Indexes and Multivariate Analysis

As mentioned, microbial functional diversity depends on many metabolic reactions and interactions of microbiota. Therefore, it is unrealistic to assume that a simple relationship exists between measurements of a single enzyme activity and microbial functional diversity of soil. The simultaneous measurement of the activities of a range of enzymes is needed. The next step is proper analysis to allow interpretation of enzyme activities for assessing microbial functional diversity across different soil types that have been affected by pollution or soil disturbances. There are two different approaches to solve the problem:

1. To integrate the simultaneous measurement of the activities of a range of enzymes in a single index
2. To carry out the proper statistical analysis of the enzyme measurements so as to calculate the microbial functional diversity of soil

A few attempts have been conducted to integrate different enzyme activities in a single index. The Biological Index of Fertility (BIF) proposed by Stefanic and associates (195) is calculated by the following expression:

$$BIF = \frac{DA + KCA}{2}$$

where DA and CA represent dehydrogenase (expressed as milligrams of 2,3,5-triphenyl formazan [TPF]) and catalase (expressed as percentage of O_2 released from H_2O_2), whereas K is a proportional coefficient. The Enzyme Activity Number (EAN) proposed by Beck (196) takes into consideration dehydrogenase, catalase, alkaline phosphatase, protease, and amylase activities, according to the following expression:

$$EAN = 0.2\left(\text{mg } TPF + \frac{\text{catalase}}{10} + \frac{\mu\text{g phenol}}{40} + \frac{\mu\text{g amino-N}}{2} + \frac{\text{amylase}}{20}\right)$$

Dehydrogenase, alkaline phosphatase, and protease activities were expressed as milligrams of TPF, micrograms of phenol, and micrograms of amino-N released from soil during the enzyme assay, respectively. Catalase and amylase activities were expressed as percentages of O_2 released from H_2O_2 and percentages of starch hydrolyzed, respectively. Higher EAN values were found in forest than in arable soils (196).

Both indexes were calculated by Perucci (197), using a loamy soil treated with municipal refuse. The EAN index was correlated with amylase, arylsulfatase, deaminase, dehydrogenase, alkaline phosphomonoesterase, and protease activity but not with catalase and phosphodiesterase activity. The BIF was only significantly correlated with catalase activity. Thus, it appears that the EAN can give a more realistic indication of the microbiological activities of soil than the BIF. The EAN takes into consideration enzyme activities related to C, N, and P dynamics as well as oxidoreductase activities of soil. Nannipieri (15) argued that the methods used by Stefanic et al. (195) and by Beck (196) were not as sensitive as those currently used to determine enzyme activities in soil. For example, according to Benefield and associates (102) INT is a better competitor for liberated electrons than TTC (see also p. 6).

An empirical relationship relating total soil nitrogen to microbial biomass C, mineralized nitrogen and phosphatase, β-glucosidase, and urease activities was proposed in 1998 by Trasar-Cepeda and coworkers (198). The following relationship relates total N to the measured parameters:

$$\begin{aligned}\text{Total N} = {} & (0.38 \times 10^{-3})\text{ microbial biomass C} + (1.4 \times 10^{-3})\text{ mineralized N} \\ & + (13.6 \times 10^{-3})\text{ phosphatase} + (8.9 \times 10^{-3})\ \beta\text{-glucosidase} \\ & + (1.6 \times 10^{-3})\text{ urease}\end{aligned}$$

In these examples of indexes, the reasons for choosing the biochemical and microbiological parameters are not always clear. Obviously, the determination of urease activity can give information on the potential activity of the soil to hydrolyze urea. However, the determination of a number of proteolytic and deaminase activities may give much more valid information on the potential N mineralization rate of soil. In the case of the EAN index, the choice of amylase rather than cellulase activity can be criticized on the grounds that cellulose is quantitatively more important than starch in plant residues (199). The measurement of microbial biomass can be useful if long-term changes in soil quality or microbial functional diversity need to be monitored. On a short-term basis, microbial activity can show rapid and high changes that do not correlate to changes in microbial biomass, whereas on a long-term basis, durable changes in microbiological activity can reflect similar changes in microbial biomass (41).

A more accurate and focused selection of enzyme activities was carried out by Sinsabaugh and colleagues (200), who measured six enzyme (β-1,4-glucosidase, β-1,4-endoglucanase or endocellulase, β-1,4-exoglucanase or exocellulase, β-xylosidase, phenol oxidase, and peroxidase) activities involved in the degradation of lignocellulose material, which is quantitatively the most important component of plant debris. Temporally integrated lignocellulase activity, a single factor calculated from the six measured enzyme activities using principal component analysis, was significantly correlated with the mass loss of white birch incubated in situ.

An alternative approach to the use of single indexes integrating measurements of more enzyme activities is to analyze the measured enzyme activities by multivariate statistics and to calculate microbial functional diversity. In a study by Kandeler and colleagues (14), microbial biomass, respiration N mineralization, potential nitrification, actual nitrification and 11 different enzymes (dehydrogenase, xylanase, cellulase, β-glucosidase, protease, arginine deaminase, urease, nitrate reductase, alkaline phosphatase, phospholipase C, and arylsulfatase) were measured in three different soils (Calcaric Phaeozem, Eutric Cambisol, and Dystric Lithosol) at four levels of metal contamination: uncontaminated controls, light pollution (300 ppm Zn, 100 ppm Cu, 50 ppm Ni, 50 ppm V, and 3 ppm Cd), medium pollution (twofold greater concentration), and heavy pollution (threefold greater concentration). The dendogram of cluster analysis showed that the controls were clearly distinguished from the polluted soils (Fig. 4). This behavior was confirmed by discriminant analysis; the first discriminant axis, axis 1, showed a significant difference between controls and polluted samples. The same type of analysis showed that arylsulfatase and dehydrogenase activities were the most important variables, explaining more than 94% of the total variance of the data set. The same approach was used to explain the data set variability due to different soil types (109), farmyard manure amendment (201), or tillage management (51). The latter study investigated the effect of conventional, minimum, and reduced

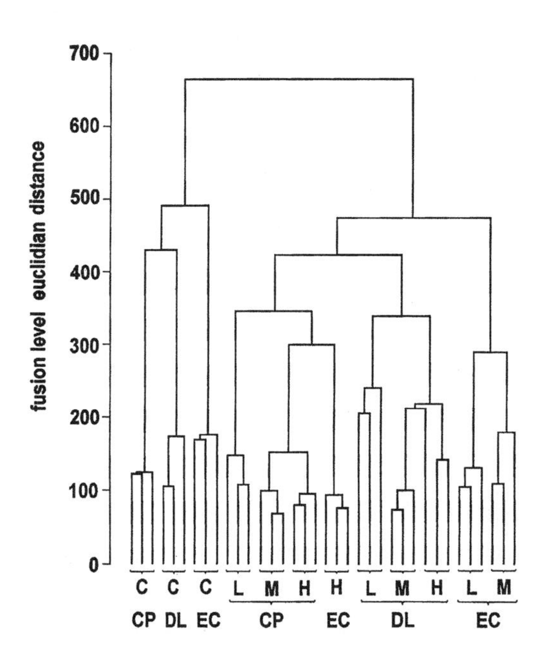

Figure 4 Compositional relationships among the Calcaric Phaeocem, Eutric Cambisol, and Dystric Lithosol, differing in extent of heavy metal pollution. Dendogram classified the 36 soil containers according to their enzymatic activities (three soil types, four heavy metal pollution treatments, three replicates). C, control; L, light pollution; M, medium pollution; H, high pollution; CF, Calcaric Phaeocem; EC, Eutric Cambisol; DL, Dystric Lithosol. (From Ref. 14.)

tillage on microbial biomass (determined by SIR), N mineralization, potential nitrification, xylanase, protease (casein-hydrolyzing activity), and alkaline phosphatase in a fine-sandy loamy Haplic Chernozem. In the soil sampled in 1989, reduced and minimum tillage could be distinguished from conventional tillage management on the basis of the parameters along axis 1 (Fig. 5). The discriminant function 1 explained 89% of the total variance of the data set and was dominated by xylanase, potential nitrification, and microbial biomass. The same type of analysis of the soil sampled 8 years later confirmed that the discrimina-

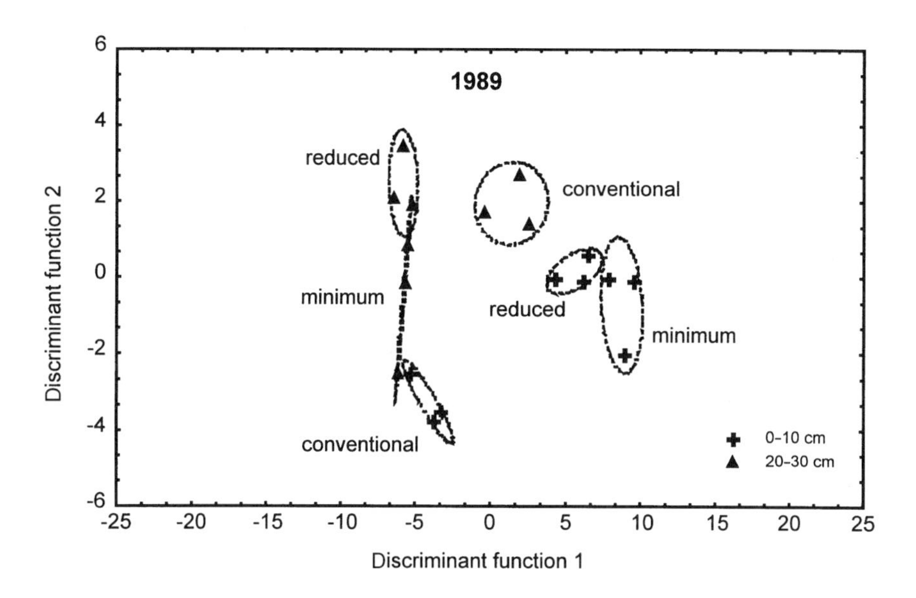

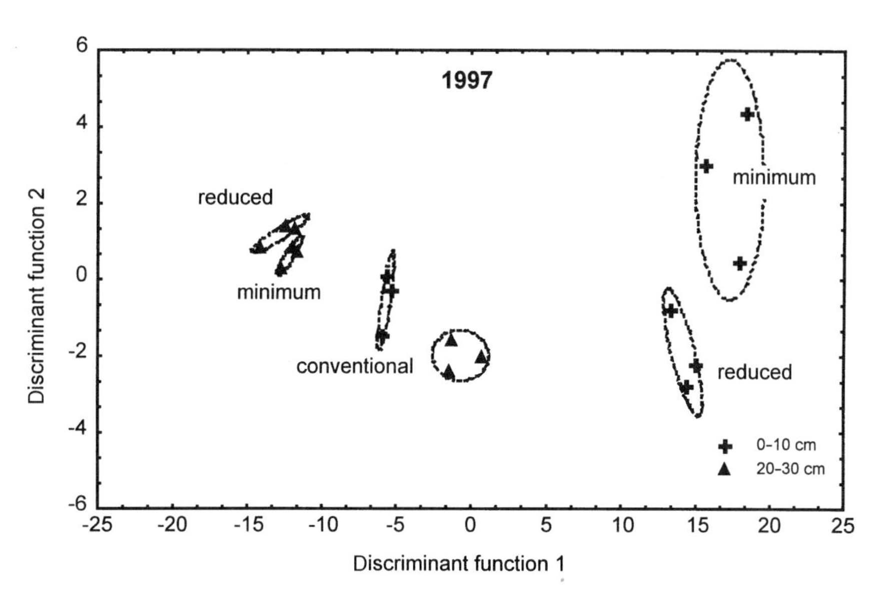

Figure 5 Two-dimensional plot of the discriminant analysis of soil microbiological and biochemical parameters from soil under different types of tillage. (From Ref. 51.)

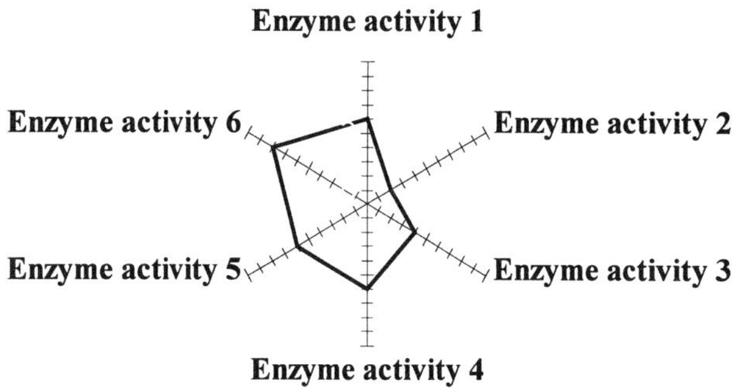

Enzyme activity 1
Enzyme activity 6
Enzyme activity 2
Enzyme activity 5
Enzyme activity 3
Enzyme activity 4

Figure 6 Generalized sun ray plot of different enzyme activities.

tion along axis 1 still was dominated by xylanase and microbial biomass; however, the potential nitrification was replaced by protease activity among the dominant parameters. Xylanase activity was higher in soil under minimum or reduced tillage than in those from conventional tillage, because both reduced tillage and minimum tillage increased the 200- to 2000-μm particles of soil; these particles contain particulate organic matter and are higher in xylanase activity than smaller soil particles (53,202). It is well established that organic matter is more uniformly distributed along the soil profile in conventional tillage, whereas organic matter is concentrated in the surface soil layer of reduced or minimum tillage plots and largely present as crop residues (1,2,203–205).

In theory, an index integrating the enzyme activities that catalyze the reactions limiting the rate of the main metabolic processes could be used to measure microbial functional diversity. The choice of the enzyme activities can be restricted to those involved in a particular process, such as the degradation of organic matter, N mineralization, or nitrification.

Combining enzyme activities in a single index is useful for a quantification of the overall soil microbiological functionality but one loses information on specific enzymes. A sun ray plot that displays the measured biochemical parameters of each measured parameter shows this difference (206) (Fig. 6). Area and shape of the star may provide integrated fingerprinting for assessing microbial functional diversity. Also, the star shape allows a visual comparison of the measured parameters with a rapid overview of the sample features.

III. CONCLUSIONS

Microbially mediated processes are central to the functions that soils perform. They are at the basis of C, N, P, and S cycling in soil; contribute to the decontamination of soil by degrading organic pollutants or immobilizing heavy metals; participate in the formation of soil structure, and have negative (i.e., plant pathogens) or positive (i.e., plant growth promoting rhizobacteria) effects on plant growth. Enzyme catalysis, largely under the direct or indirect control of microorganisms, mediates this wide range of soil functions. For this reason, it is reasonable to suggest that the determination of enzyme activities in soil may be used as a research tool to assess microbial functional diversity, to study biochemical processes, to investigate microbial ecology, and to provide indicators of soil quality. But to be meaningful in these contexts enzymes must be chosen carefully and the results correctly interpreted. Limitations that must be considered are as follows:

1. The enzyme assays currently used cannot distinguish unequivocally between the contribution of activities associated with enzymes adsorbed by inorganic or organic colloids, activities due to abiotic transformations, and activities of intracellular enzymes in viable cells. The third group of activities are those directly responsible for the microbial functional diversity of soil.
2. To determine microbial functional diversity, it is necessary to measure a range of enzymes that reflect the soil functions of interest. A reasonable approach may be to measure the activity of the key enzymes catalyzing the reaction or reactions that limit the rate of the overall metabolic pathway or process.
3. Multienzyme indexes may be used for quantifying microbial functionality and for comparing different soils or impacts on a single soil. Multivariate statistical analysis can be useful in determining the dominant parameters that account for the variability among soils or for soil management systems. Combining enzyme activities in a single index is useful for quantifying overall microbiological functionality of soil, but it can mask information on specific enzyme reactions or processes. Sun ray plots displaying the measured biochemical parameters of each tested sample can show these differences. Area and shape of the star may provide an integral signal for the assessment of the microbial functional diversity.

Spatial and temporal variability also must be taken into account when using enzyme assays (15). A statistical approach that can overcome this variability and assist interpretation of enzyme assays needs to be developed. One method that holds potential for the use of enzyme assays and soil quality indicators is fuzzy logic modeling (207). The first results, based on 2520 data entities from different arable, grassland, and forest soils of Central Europe, indicated that fuzzy classification is more realistic than traditional classification, which has sharply defined boundaries, and that this approach may allow decision-making processes to be modeled more effectively.

REFERENCES

1. JN Ladd, RC Forster, P Nannipieri, JM Oades. Soil structure and biological activity. In: G Stotzky, J-M Bollag, eds. Soil Biochemistry. Vol. 9. New York: Marcel Dekker, 1996, pp 23–79.
2. JM Oades. The role of biology in the formation, stabilization and degradation of soil structure. Geoderma 56:377–400, 1993.
3. BS Griffiths, K Ritz, RE Wheatly. Relationship between functional diversity and genetic diversity in complex microbial communities. In: H Insam, A Rangger, eds. Microbial Communities: Functional Versus Structural Approaches. Berlin: Springer Verlag, 1997, pp 1–18.
4. R Ohtonen, S Aikio, H Väre. Ecological theories in soil biology. Soil Biol Biochem 29: 1613–1619, 1997.
5. DA Wardle, KE Giller. The quest for a contemporary ecological dimension to soil biology. Soil Biol Biochem 12:1549–1554, 1997.
6. DC Coleman. Compositional analysis of microbial communities: Is there room in the middle? In: K Ritz, J Dighton, KE Giller, eds., Beyond the Biomass: Compositional and Functional

Analysis of Soil Microbial Communities. Chichester: John Wiley and Sons, 1993, pp 201–220.
7. PC Brookes. The use of microbial parameters in monitoring soil pollution by heavy metals. Biol Fertil Soils 16:269–279, 1995.
8. KE Giller, E Witter, SP McGrath. Toxicity of heavy metals to microorganisms and microbial processes in agricultural soils: A review. Soil Biol Biochem 30:1389–1414, 1998.
9. P Nannipieri, L Badalucco, L Landi, G Pietramellara. Measurement in assessing the risk of chemicals to the soil ecosystem. In: JT Zelikoff, ed. Ecotoxicology: Responses, Biomarkers and Risk Assessment. Fair Haven, NJ: SOS Publications, 1997, pp 507–534.
10. JAN Morgan. Molecular biology: New tools for studying microbial ecology. Sci Prog Edinburgh 75:265–278, 1991.
11. JT Trevors, JD van Elsas. Introduction to nucleic acids in the environment: Methods and applications. In: JT Trevors, JD van Elsas, eds. Nucleic Acids in the Environment. Berlin: Springer Verlag, 1995, pp 1–7.
12. JL Mawdsley, RG Burns. Inoculation of plants with *Flavobacterium* P25 results in altered rhizosphere enzyme activities. Soil Biol Biochem 26:871–882, 1994.
13. DC Naseby, JM Lynch. Rhizosphere soil enzymes as indicators of perturbations caused by enzyme substrate addition and inoculation of a genetically modified strain of *Pseudomonas fluorescens* as wheat seeds. Soil Biol Biochem 29:1353–1362, 1997.
14. E Kandeler, C Kampichler, O Horak. Influence of heavy metals on the functional diversity of soil microbial communities. Biol Fertil Soils 23:299–306, 1996.
15. P Nannipieri. The potential use of soil enzymes as indicators of productivity, sustainability and pollution. In: CE Pankhurst, BM Doube, VVSR Gupta, PR Grace, eds. Soil Biota: Management in Sustainable Farming Systems. Adelaide: CSIRO, 1994, pp 238–244.
16. B Bochner. Breathprints at the microbial level. ASM News 55:536–539, 1989.
17. JL Garland, AL Mills. Classification and characterization of heterotrophic microbial communities on the basis of patterns of community-level-sole-carbon-source utilization. Appl Environ Microbiol 57:2351–2359, 1991.
18. DH Bossio, KM Scow. Impact of carbon and flooding on the metabolic diversity in soils. Appl Environ Microbiol 61:4043–4050, 1995.
19. SK Haack, H Garchow, MJ King, LJ Forney. Analysis of factors affecting the accuracy, reproducibility, and interpretation of microbial community carbon source utilization patterns. Appl Environ Microbiol 60:1458–1468, 1995.
20. A Winding. Fingerprinting bacterial soil communities with BIOLOG microtiter plates. In: K Ritz, J Dighton, KE Giller, eds. Beyond the Biomass: Compositional and Functional Analysis of Soil Microbial Communities. Chichester: John Wiley and Sons, 1994, pp 85–94.
21. JL Garland. Pattern of potential C source utilisation by rhizosphere communities. Soil Biol Biochem 28:223–230, 1996.
22. SJ Grayston, S Wang, CD Campbell, AC Edwards. Selective influence of plant species on microbial diversity in the rhizosphere. Soil Biol Biochem 30:369–378, 1998.
23. JC Zak, MR Willig, DL Moorhead, HG Wildman. Functional diversity of microbial communities: A quantitative approach. Soil Biol Biochem 26:1101–1108, 1994.
24. H Insam. Substrate utilization tests in microbial ecology. A preface to the special issue of the J Microbiol Meth 30:1–2, 1997.
25. CD Campbell, J Van Gelder, MS Davidson, CM Cameron. Use of sole carbon source utilisation pattern to detect changes in soil microbial communities affected by Cu, Ni and Zn. In: RD Wilken, U Forstner, A Knochel, eds., International Conference on Heavy Metals in the Environment. Edinburgh: CEP Consultants, 1995, pp 447–450.
26. MV Gorlenko, TN Majorova, PA Kozhevin. Disturbances and their influence on substrate utilization patterns in soil microbial communities. In: H Insam, A Rangger, eds. Microbial Communities. Functional Versus Structural Approaches. Berlin: Springer Verlag, 1997, pp 84–93.

27. A Filessbach, P Mader. Carbon source utilization by microbial communities in soils under organic and conventional farming practices. In: H Insam, A Rangger, eds. Microbial Communities: Functional Versus Structural Approaches. Berlin: Springer Verlag, 1997, pp 109–120.
28. S Kreitz, TH Anderson. Substrate utilization patterns of extractable and non-extractable bacterial fractions in neutral and acidic beech forest soil. In: H Insam, A Rangger, eds. Microbial Communities: Functional Versus Structural Approaches. Berlin: Springer Verlag, 1997, pp 149–160.
29. H Insam, A Rangger. Microbial Communities: Functional Versus Structural Approaches. Berlin: Springer Verlag, 1997.
30. A Winding, NB Hendriksen. Biolog substrate utilisation assay for metabolic fingerprints of soil bacteria: Incubation effects. In: H Insam, A Rangger, eds. Microbial Communities: Functional Versus Structural Approaches. Berlin: Springer Verlag, 1997, pp 195–250.
31. V Torsvik, J Goksoyr, FL Daae, R Sorheim, J Michalsen, K Salte. Use of DNA analysis to determine the diversity of soil communities. In: K Ritz, J Dighton, KE Giller, eds. Beyond the Biomass: Compositional and Functional Analysis of Soil Microbial Communities. Chichester: John Wiley and Sons, 1994, pp 39–48.
32. CD Campbell, SJ Grayston, DJ Hirst. Use of rhizosphere carbon source in sole carbon source tests to discriminate soil microbial communities. J Microbiol Methods 30:33–41, 1997.
33. H Insam. A new set of substrates proposed for community characterization in environmental samples. In: H Insam, A Rangger, eds. Microbial Communities: Functional Versus Structural Approaches. Berlin: Springer Verlag, 1997, pp 259–260.
34. K Smalla, U Wachtendorf, H Heuer, W-T Liu, L Forney. Analysis of BIOLOG GN substrate utilization patterns by microbial communities. Appl Environ Microbiol 64:1220–1225, 1998.
35. BP Degens, JA Harris. Development of a physiological approach to measuring the catabolic diversity of soil microbial communities. Soil Biol Biochem 29:1309–1320, 1997.
36. AL Lehninger, DL Nelson, MM Cox. Principles of Biochemistry. New York: Worth Publisher, 1993.
37. DS Powlson, D Barraclough. Mineralisation and assimilation in soil-plant systems. In: R Knowles, TH Blackburn, eds. Nitrogen Isotopes Technique, 1993. San Diego: Academic Press, 1993, pp 209–242.
38. LJ Reitzer, B Magasanik. Ammonia assimilation and the biosynthesis of glutamine, glutamate, aspartate, asparagine, L-alanine and D-alanine. In: FC Neidhardt, JH Ingham, KB Low, B Magasanik, M Schaecher, HE Umbarger, eds. *Escherichia coli* and *Salmonella typhimurium*: Cellular and Molecular Biology. Washington DC: American Society of Microbiology, 1987, pp 302–320.
39. P Nannipieri, L Badalucco, L Landi. Holistic approaches to the study of populations, nutrient pools and fluxes: Limits and future research needs. In: K Ritz, J Dighton, KE Giller, eds. Beyond the Biomass: Compositional and Functional Analysis of Soil Microbial Communities. Chichester: John Wiley and Sons, pp 231–238.
40. J Skujins. History of abiomtic soil enzyme research. In: RG Burns, eds. Soil Enzymes. New York: Academic Press, 1978, pp 1–49.
41. P Nannipieri, B Ceccanti, S Grego. Ecological significance of the biological activity in soil. In: J-M Bollag, G Stotzky, eds. Soil Biochemistry. Vol 6. New York: Marcel Dekker, 1990, pp 293–355.
42. HA Ajwa, CJ Deil, CW Rice. Changes in enzyme activities and microbial biomass of tallgrass prairie soil as related to burning and nitrogen fertilization. Soil Biol Biochem 31:769–777, 1999.
43. WJ Staddon, LC Duchesne, JT Trevors. Acid phosphatase, alkaline phosphatase and arylsulfatase activities in soils from a jack pine (*Pinus banksiana Lamb*) ecosystem after clear-cutting, prescribed burning, and scarification. Biol Fertil Soils 27:1–4, 1998.
44. J Skujins. Dehydrogenase: As an indicator of biological activity in arid soils. Bull Ecol Res Comm 17:235–241, 1973.

45. JT Trevors. Electron transport system: Activity in soil, sediment and pure culture. Crit Rev Microbiol 11: 83–99, 1986.
46. JN Ladd. Soil enzymes. In: D Vaughan, RE Malcom, eds. Soil Organic Matter and Biological Activity. Dordrecht, Netherlands: Matinus Nijhoff, 1985, pp 175–221.
47. AT Chunderova, T Zubers. Phosphatase activity in dernopodzolic soils. Pochovedenie 11: 47–53, 1969.
48. VP Hollander. Acid phosphatase. In: PD Boyer, ed. The Enzymes. Vol. II. 3rd ed. New York: Academic Press, 1971, pp 449–498.
49. E Kandeler, G Eder. Effect of cattle slurry in grassland on microbial biomass and on activities of various enzymes. Biol Fertil Soils 16:249–254, 1993.
50. E Kandeler, E Murer. Aggregate stability and soil microbial processes in a soil under different cultivations. In: L Brussard, MI Koistra, eds. International Workshop on methods of research on soil structure/soil biota interrelationships. Geoderma 56:503–513, 1993.
51. E Kandeler, D Tscherko, H Spiegel. Long-term monitoring of microbial biomass, N mineralisation and enzyme activities of a Chernozem under different tillage management. Biol Fertil Soils 28:343–351, 1999.
52. F Schinner, W von Mersi. Xylanase, CM-cellulase and invertase activity in soil: An improved method. Soil Biol Biochem 22:511–515, 1990.
53. M Stemmer, MH Gerzabek, E Kandeler. Organic matter and enzyme activity in particle-size fractions of soils obtained after low-energy sonication. Soil Biol Biochem 30:9–17, 1998.
54. W von Mersi, R Kuhnert-Finkernagel, F Schinner. The influence of rock powers on microbial activity of three forest soils. Z Pflanzenernaehr Bodenkd 155:29–33, 1992.
55. S Zechmeister-Boltenstern, K Spandinger, H Kinzel. Bodenenzymatische Untersuchungen in verschieden stark belasteten Buchenwaldstandorten. In: R Albert, K Burian, H Kinzel, eds. Zustandserhebung Wienerwald. Pflanzenphysiologische und bodenökologische Untersuchungen zur Bioindikation. Wien: Verlag der Österreichischen Akademie der Wissenschaften, 1991.
56. C Serra-Wittling, S Houot, E Burriuso. Soil enzymatic response to addition of municipal solid-waste compost. Biol Fertil Soils 20:226–236, 1995.
57. SP Deng, MA Tabatabai. Effect of tillage and residue management on enzyme activities in soils. II. Glycosidases. Biol Fertil Soils 22:208–213, 1996.
58. DJ Ross. Studies on a climosequence of soils in tussock grasslands. New Z J Soil Sci 18: 527–534, 1975.
59. DJ Ross. Distribution of invertase and amylase activities in pasture topsoil fractions isolated by ultrasonic dispersion in nemagon and a surfactant. Soil Biol Biochem 8:485–490, 1976.
60. JZ Burket, RP Dick. Microbial and soil parameters in relation to N mineralization in soils of diverse genesis under differing management systems. Biol Fertil Soils 27:430–438, 1998.
61. M Curci, MDR Pizzigallo, C Crechio, R Mininni, P Ruggiero. Effects of conventional tillage on biochemical properties of soils. Biol Fertil Soils 25:1–6, 1997.
62. F Eizavi, MA Tabatabai. Factors affecting glucosidase and galactosidase activities in soils. Soil Biol Biochem 22:891–897, 1990.
63. C Garcia, T Hernandez. Biological and biochemical indicators in derelict soils subject to erosion. Soil Biol Biochem 29:171–177, 1993.
64. RG Kuperman, MM Carriero. Soil heavy metal concentrations, microbial biomass and enzyme activities in a contaminated grassland ecosystem. Soil Biol Biochem 29:179–190, 1997.
65. M Miller, RP Dick. Thermal stability and activities of soil enzymes as influenced by crop rotations. Soil Biol Biochem 27:1161–1166, 1995.
66. SU Sarathchandra, KW Perrott. Assay for β-glucosidase activity in soils. Soil Sci 138:15–19, 1984.
67. L Badalucco, PJ Kuikman, P Nannipieri. Protease and deaminase activities in wheat rhizosphere and their relation to bacterial and protozoan populations. Biol Fertil Soils 23:99–104, 1996.

68. JK Friedel, JC Munch, WR Fischer. Soil microbial properties and assessment of available soil organic matter in a haplic luvisol after several years of different cultivation and crop rotation. Soil Biol Biochem 28:479–488, 1996.
69. RJ Haynes, PH Williams. Influence of stock camping behaviour on the soil microbiological and biochemical properties of grazed pastoral soils. Biol Fertil Soils 28:253–258, 1999.
70. JN Ladd, EA Paul. Changes in enzyme activities and distribution of acid-soluble, amino acid nitrogen in soil during nitrogen immobilization and mineralization. Soil Biol Biochem 5: 825–840, 1973.
71. K Alef, D Kleiner. Applicability of arginine ammonification as indicator of microbial activity in different soils. Biol Fertil Soils 5:148–151, 1987.
72. O Dilly, JC Munch. Microbial biomass and activities in partly hydromorphic agricultural and forest soils in the Bornhöved Lake region of Northern Germany. Biol Fertil Soils 19: 343–347, 1995.
73. AJ Franzluebbers, DA Zuberer, FM Hons. Comparison of microbiological methods for evaluating quality and fertility of soil. Biol Fertil Soils 19:135–140, 1995.
74. RJ Haynes, R Tregurtha. Effects of increasing periods under intensive arable vegetable production on biological, chemical and physical indices of soil quality. Biol Fertil Soils 28:259–266, 1999.
75. SP Deng, MA Tabatabai. Effect of tillage and residue management on enzyme activities in soils. I. Amidohydrolases. Biol Fertil Soils 22:202–207, 1996.
76. WT Frankenberger Jr, MA Tabatabai. L-Asparaginase activity of soils. Biol Fertil Soils 11: 6–12, 1991.
77. WT Frankenberger Jr, MA Tabatabai. L-Glutaminase activity of soils. Soil Biol Biochem 23:869–874, 1991.
78. JM Bremner, RL Mulvaney. Urease activity in soils. In: RG Burns, ed. Soil Enzymes. London: Academic Press, 1978, pp 149–195.
79. MC Dash, PC Mishra, RK Mohanty, N Bhatt. Effects of specific conductance and temperature on urease activity in some Indian soils. Soil Biol Biochem 13:73–74, 1981.
80. S Klose, MA Tabatabai. Urease activity of microbial biomass in soils. Soil Biol Biochem 31:205–211, 1999.
81. JW McGarity, MG Myers. A survey of urease activity in soils of northern New South Wales. Plant Soil 27:217–238, 1967.
82. KI Sahrawat. Relationships between soil urease activity and other properties of some tropical wetland rice soils. Fertil Res 4:145–150, 1983.
83. ZN Senwo, MA Tabatabai. Aspartase activity in soils: Effects of trace elements and relationships to other amidohydrolases. Soil Biol Biochem 31:213–219, 1999.
84. MA Tabatabai, JM Bremner. Assay of soil urease activity in soil. Soil Biol Biochem 4:479–487, 1972.
85. MR Appiah, Y Ahenkorah. Arylsulphatase activity of different latosol soils of Ghana cropped to cocoa (*Theobroma cacao*) and coffee (*Coffea canephora var. robusta*). Biol Fertil Soils 5:186–190, 1989.
86. SP Deng, MA Tabatabai. Effect of tillage and residue management on enzyme activities in soils. III. Phosphatases and arylsulfatase. Biol Fertil Soils 24:141–146, 1997.
87. RE Farrell, VVSR Gupta, JJ Germida. Effects of cultivation on the activity of arylsulfatase in Saskatchewan soils. Soil Biol Biochem 8:1033–1040, 1994.
88. L Haastra, P Doelman. An ecological dose-response model approach to short- and long-term effects of heavy metals on arylsulphatase activity in soil. Biol Fertil Soils 11:18–23, 1991.
89. S Klose, JM Moore, MA Tabatabai. Arylsulfatase activity of microbial biomass in soils as affected by cropping systems. Biol Fertil Soils 29:46–54, 1999.
90. MC Press, J Henderson, JA Lee. Arylsulphatase activity in peat in relation to acidic deposition. Soil Biol Biochem 17:99–103, 1985.

91. MA Tabatabai, JM Bremner. Arylsulfatase activity of soils. Soil Sci Soc Am Proc 34:225–229, 1970.
92. L Beyer, K Sieling, K Pingpank. The impact of a low humus level in arable soils on microbial properties, soil organic matter quality and crop yield. Biol Fertil Soils 28:156–161, 1999.
93. K Chander, S Goyal, MC Mundra, KK Kapoor. Organic matter, microbial biomass and enzyme activity of soils under different crop rotations in the tropics. Biol Fertil Soils 24:306–310, 1997.
94. F Eizavi, MA Tabatabai. Phosphatases in soils. Soil Biol Biochem 9:167–172, 1997.
95. KY Kim, D Kordan, GA McDonald. *Enterobacter agglomerans*, phosphate solubilizing bacteria, and microbial activity in soil: Effect of carbon sources. Soil Biol Biochem 30:995–1003, 1998.
96. P Nannipieri, I Sastre, L Landi, MC Lobo, G Pietramellara. Determination of extracellular neutral phosphomonoesterase activity in soil. Soil Biol Biochem 28:107–112, 1996.
97. PCK Pang, H Kolenko. Phosphomonoesterase activity in forest soils. Soil Biol Biochem 18: 35–40, 1986.
98. VC Baligar, RJ Wright, MD Smedley. Acid phosphatase activity in soils of the Appalachian region. Soil Sci Soc Am J 52:1612–1616, 1998.
99. NG Juma, MA Tabatabai. Distribution of phosphomonoesterases in soils. Soil Sci 129:101–108, 1978.
100. A Saa, C Trasar-Cepeda, F Gil-Sotres, T Carballas. Changes in soil phosphorus and acid phosphatase activity immediately following forest fires. Soil Biol Biochem 23:1223–1230, 1993.
101. R Margesin, F Schinner. Phosphomonoesterase, phosphodiesterase, phosphotriesterase, and inorganic pyrophosphatase activities in forest soils in an alpine area: Effect of pH on enzyme activity and extractability. Biol Fertil Soils 18:320–326, 1994.
102. CB Benefield, PJA Howard, DM Howard. The estimation of dehydrogenase activity in soil. Soil Biol Biochem 9:67–70, 1977.
103. C Marzadoni, C Ciavatta, D Montecchio, C Gessa. Effects of lead pollution on different soil enzyme activities. Biol Fertil Soils 22:53–58, 1996.
104. P Perucci, U Bonciarelli, R Santilocchi. Effect of rotation, nitrogen fertilization and management of crop residues on some chemical, microbiological and biochemical properties of soil. Biol Fertil Soils 24:311–316, 1997.
105. JT Trevors. Dehydrogenase activity in soil: A comparison between the INT and TTC assay. Soil Biol Biochem 16:673–674, 1984.
106. K Alef, P Nannipieri. Methods in Applied Soil Microbiology and Biochemistry. London: Academic Press, 1995.
107. F Schinner, R Öhlinger, E Kandeler, R Margesin. Methods in Soil Biology. Berlin: Springer Verlag, 1996.
108. RW Weaver, S Angle, P Bottomley, D Bezdicek, S Smith, A Tabatabai, A Wollum. Methods in Soil Analysis. Part 2. Microbiological and Biochemical Properties. Madison, WI: Soil Science Society of America, 1996.
109. D Tscherko, E Kandeler. Biomonitoring of soils-microbial biomass and enzymatic processes as indicators for environment change. Bodenkultur 50:215–226, 1999.
110. G Hoffmann. Investigations on the effect of enzymes in soil. Z Pflanzenernaehr Dung Bodenkde 85:193–201, 1959.
111. D Jones, DM Webley. A new enrichment technique for studying lysis of fungal cell walls in soil. Plant Soil 28:147–157, 1968.
112. S Kiss, S Peterfi. Presence of carbohydrases in the peat of a Salicea community. Stud Cercet Biol Cluj 335–340, 1961.
113. AS Galstyan, EG Saakyan. Determination of soil glutaminase activity. Doklady Adademii Nauk SSJR 209:1201–1202, 1973.

114. J Drobni'k. Degradation of asparagine by soil enzyme complex. Czeskoslov Mikrobiol 1: 47, 1956.
115. MA Tabatabai, JM Bremner. Use of p-nitrophenyl phosphate for assay of soil phosphatase activity. Soil Biol Biochem 1:301–307, 1969.
116. MA Tabatabai, BB Singh. Rhodanase activity of soils. Soil Sci Soc Am J 40:381–385, 1976.
117. P Nannipieri, L Landi. Soil enzymes. In: ME Summer, ed. Handbook of Soil Science. Boca Raton, FL: CRC Press, 1999, pp C129–C137.
118. MA Tabatabai. Soil enzymes. In: RW Weaver, S Angle, D Bezdicek, S Smith, MA Tabatabai, A Wollum, eds. Methods in Soil Analysis. Part 2. Microbiological and Biochemical Properties. Madison, WI: Soil Science Society of America, 1994, pp 775–833.
119. RE Malcolm. Assessment of phosphatase activity in soils. Soil Biol Biochem 15:403–408, 1983.
120. RG Burns. Enzyme activity in soil: Some theoretical and practical considerations. In: RG Burns, ed. Soil Enzymes. London: Academic Press, 1978, pp 295–340.
121. RG Burns. Enzyme activity in soil: Location and a possible role in microbial ecology. Soil Biol Biochem 14:423–427, 1982.
122. L Gianfreda, J-M Bollag. Influence of natural and anthropogenic factors on enzyme activity in soil. In: J-M Bollag, G Stotzky, eds. Soil Biochemistry. Vol. 9. New York: Marcel Dekker, 1996, pp 123–193.
123. P Nannipieri, P Sequi, P Fusi. Humus and enzyme activity. In: A Piccolo. ed., Humic Substances in Terrestrial. Amsterdam: Elsevier, 1996, pp 293–328.
124. G Stotzky. Influence of soil mineral colloids on metabolic processes, growth, adhesion, and ecology of microbes and viruses. In: PM Huang, M Schnitzer, eds. Interactions of Soil Minerals with Natural Organics and Microbes, SSSA Special Publication No. 17. Madison, WI: Soil Science Society of America, 1986, pp 305–428.
125. P Ruggiero, J Dec, J-M Bollag. Soil as a catalytic system. In: J-M Bollag, G Stotzky, eds. Soil Biochemistry. Vol. 9. New York: Marcel Dekker, 1996, pp 79–122.
126. MB McBride. Surface chemistry of soil minerals. In: JB Dixon, SB Weed, eds. Minerals in Soil Environments. 2nd ed. SSSA, Book Series No 1. Madison, WI: Soil Science Society of America, 1989, pp 35–88.
127. TSC Wang, PM Huang, CH Chou, JH Chen. The role of soil minerals in the abiotic polymerization of phenolic compounds and formation of humic substances. In: PM Huang, M Schnitzer, eds. Interactions of Soil Minerals with Natural Organics and Microbes. SSSA Special Publication No. 17. Madison, WI: Soil Science Society of America, 1986, pp 251–281.
128. PM Huang. Role of soil minerals in transformations of natural organics and xenobiotics in soil. In: J-M Bollag, G Stotzky, eds. Soil Biochemistry. Vol. 6. New York: Marcel Dekker, 1990, pp 29–115.
129. JC Brown, JE Ambler. ''Reductants'' released by roots of Fe-deficient soybeans. Agron J 65:311–314, 1973.
130. H Shindo, PM Huang. The catalytic power of inorganic components in the abiotic synthesis of hydroquinone-derived humic polymers. Appl Clay Sci 1:71–81, 1985.
131. MC Wang, PM Huang. Catalytic polymerization of hydroquinone by nontronite. Can J Soil Sci 67:867–875, 1987.
132. MC Wang, PM Huang. Catalytic power of nontronite, kaolinite, and quartz and their reaction sites in the formation of hydroquinone-derived polymers. Appl Clay Sci 4:43–57, 1988.
133. MC Wang, PM Huang. Pyrogallol transformations as catalyzed by nontronite, bentonite, and kaolinite. Clays Clay Miner 37:525–531, 1989.
134. TSC Wang, JH Chen, WH Hsiang. Catalytic synthesis of humic acids containing various amino acids and dipeptides. Soil Sci 140:3–10, 1985.
135. M Bosetto, P Arfaioli, OL Pantani, GG Ristori. Study of the humic-like compounds formed from L-tyrosine on homoionic clays. Clay Miner 32:341–349, 1997.

136. H Shindo, PM Huang. Role of Mn(IV) oxide in abiotic formation of humic substances in the environment. Nature 298:363–365, 1982.
137. TSC Wang, MC Wang, PM Huang. Catalytic synthesis of humic substances by using aluminas as catalysts. Soil Sci 136:226–230, 1983.
138. MB McBride, FJ Sikora, LG Wesselink. Complexation and catalyzed oxidative polymerization of catechol by aluminum in acidic solution. Soil Sci Soc Am J 52:985–993, 1988.
139. MC Wang, PM Huang. Significance of Mn(IV) oxide in the abiotic ring cleavage of pyrogallol in natural environments. Sci Tot Environ 113:147–152, 1992.
140. H Shindo, PM Huang. Comparison of the influence of Mn(IV) oxide and tyrosinase on the formation of humic substances in the environment. Sci Total Environ 117/118:103–109, 1992.
141. J-M Bollag, C Meyers, S Pal, PM Huang. The role of abiotic and biotic catalysts in the transformation of phenolic compounds. In: PM Huang, J Berthelin, J-M Bollag, WB McGill, AL Page, eds. Environmental Impact of Soil Components Interactions. Vol. I. Boca Raton, FL: CRC Press/Lewis Publishers, 1995, pp 299–310.
142. A Naidja, PM Huang, J-M Bollag. Comparison of reaction products from the transformation of catechol catalyzed by birnessite or tyrosinase. Soil Sci Soc Am J 62:188–195, 1998.
143. U Birkel, J Niemeyer. Montmorillonite-catalyzed formation of bound-residue-precursors of catechol and p-chloroaniline. Chemie Erde Geochem 59:47–55, 1999.
144. RG Lehmann, HH Cheng. Reactivity of phenolic acids in soil and formation of oxidation products. Soil Sci Soc Am J 52:1304–1309, 1988.
145. RG Lehmann, HH Cheng, JB Harsh. Oxidation of phenolic acids by soil iron and manganese oxides. Soil Sci Soc Am J 51:352–356, 1987.
146. SY Liu, RD Minard, J-M Bollag. Oligomerization of syringic acid, a lignin derivative, by a phenoloxidase. Soil Sci Soc Am J 45:1100–1105, 1981.
147. J-M Bollag, SY Liu, RD Minard. Enzymatic oligomerization of vanillic acid. Soil Biol Biochem 14:157–163, 1982.
148. BR Hames, B Kurek, B Pollet, C Lapierre, B Monties. Interaction between MnO_2 and oxalate: Formation of a natural and abiotic lignin oxidizing system. J Agric Food Chem 46:5362–5367, 1998.
149. H Shindo, PM Huang. Significance of Mn(IV) oxide in abiotic formation of nitrogen complexes in natural environments. Nature 308:57–58, 1984.
150. MC Wang, PM Huang. Polycondensation of pyrogallol and glycine and the associated reactions as catalyzed by birnessite. Sci Tot Environ 62:435–442, 1987.
151. MC Wang, PM Huang. Nontronite catalysis in polycondensation of pyrogallol and glycine and the associated reactions. Soil Sci Soc Am J 55:1156–1161, 1991.
152. MC Wang, CH Lin. Enhanced mineralization of amino acids by birnessite as influenced by pyrogallol. Soil Sci Soc Am J 57:88–93, 1993.
153. MC Wang. Influence of pyrogallol on the catalytic action of iron and manganese oxides in amino acid transformation. In: PM Huang, J Berthelin, J-M. Bollag, WB McGill, AL Page, eds. Environmental Impact of Soil Components Interactions. Vol. I. Boca Raton, FL: CRC Press/Lewis Publishers, 1995, pp 169–175.
154. MA Jaureguei, HM Reisenauer. Dissolution of oxides of manganese and iron by root exudate components. Soil Sci Soc Am J 46:314–317, 1982.
155. SJ Traina, HE Doner. Copper-manganese (II) exchange on a chemically reduced birnessite. Soil Sci Soc Am J 49:307–313, 1985.
156. A Naidja, B Siffert. Oxidative decarboxylation of isocitric acid in the presence of montmorillonite. Clay Miner 25:27–37, 1990.
157. MM Mortland. Deamination of glutamic acid by pyridoxal phosphate-Cu-smectite catalyst. J Mol Catal 27:143–155, 1984.
158. A Naidja, B Siffert. Glutamic acid deamination in the presence of montmorillonite. Clay Miner 24:649–661, 1989.

159. A Naidja, PM Huang. Deamination of aspartic acid by aspartase-Ca-montmorillonite complex. J Mol Catal 106:255–265, 1996.
160. B Siffert, A Naidja. Stereoselectivity of montmorillonite in the adsorption and deamination of some amino acids. Clay Miner 27:109–118, 1992.
161. SA Boyd, MM Mortland. Enzyme interactions with clays and clay-organic matter complexes. In: J-M Bollag, G Stotzky, eds. Soil Biochemistry. Vol. 6. New York: Marcel Dekker, 1990, pp 1–28.
162. WT Frankenberger Jr, JB Johanson. Use of plasmolytic agents and antiseptics in soil enzyme assays. Soil Biol Biochem 18:209–213, 1986.
163. S Kiss, M Dracan-Bularda, D Radulesu. Biological significance of enzymes accumulated in soil. Adv Agron 27:25–87, 1975.
164. DL Burton, WB McGill. Role of enzyme stability in controlling histidine deaminating activity in soil. Soil Biol Biochem 21:903–910, 1989.
165. DL Burton, WB McGill. Inductive and repressive effects of carbon and nitrogen on L-histidine ammonia lyase activity in a black Chernozemic soil. Soil Biol Biochem 23:939–946, 1991.
166. KN Paulson, LT Kurtz. Locus of urease activity. Proc Am Soil Sci Soc 33:897–901, 1969.
167. FA Norstadt, CR Frey, H Sigg. Soil urease activity: Paucity in the presence of the fairy ring fungus *Marasmus oreades* (Bolt.). Soil Sci Soc Am Proc 37:880–885, 1973.
168. AD McLaren, A Pukite. Ubiquity of some soil enzymes and isolation of soil organic matter with urease activity. In: D Povoledo, ML Goltermaan, eds. Humic Substances and Function in the Biosphere. Wageningen: Pudoc, 1973, pp 187–193.
169. P Nannipieri, B Ceccanti, S Cervelli, P Sequi. Use of 0.1 M pyrophosphate to extract urease from a podzol. Soil Biol Biochem 6:359–362, 1974.
170. JPE Anderson, KH Domsch. Use of selective inhibitors in the study of respiratory activities and shifts in bacterial and fungal populations in soil. Ann Microbiol 24:189–194, 1974.
171. JPE Anderson, KH Domsch. A physiological method for the quantitative measurement of microbial biomass in soil. Can J Microbiol 21:314–322, 1975.
172. GP Sparling. The substrate-induced respiration. In: K Alef, P Nannipieri, eds., Methods in Applied Soil Microbiology and Biochemistry. London: Academic Press, 1995, pp 397–404.
173. DS Jenkinson. Determination of microbial carbon and nitrogen in soil. In: JE Wilson, ed. Advances in Nitrogen Cycling and Agricultural Ecosystems. Wallingford, UK: CAB International, 1988, pp 368–386.
174. JF Parr, WR Gardner, LF Elliot. Water Potential Relationships in Soil Microbiology. Madison, WI: Soil Science Society of America, 1981.
175. F Asmar, F Einland, NE Nielsen. Interrelationship between extracellular enzyme activity, ATP content, total counts of bacteria and CO_2 evolution. Biol Fertil Soils 17:32–38, 1992.
176. AMK Falik, M Wainwright. Microbial and enzyme activity in soils amended with a natural source of easily available carbon. Biol Fertil Soils 21:177–183, 1996.
177. P Nannipieri, RL Johnson, EA Paul. Criteria for measurement of microbial growth and activity in soil. Soil Biol Biochem 10:223–229, 1978.
178. P Nannipieri, F Pedrazzini, PG Arcara, C Piovanelli. Changes in amino acids, enzyme activities and biomass during soil microbial growth. Soil Sci 127:26–34, 1979.
179. P Nannipieri, L Muccini, C Ciardi. Microbial biomass and enzyme activity: Production and persistence. Soil Biol Biochem 15:679–685, 1983.
180. GP Sparling, BG Ord, D Vaughan. Microbial biomass and activity in soils amended with glucose. Soil Biol Biochem 13:99–104, 1981.
181. K Watanabe, K Hayano. Mechanisms of production of soil protease by proteolytic *Bacillus substilis* in wetland rice soil. Biol Fertil Soils 21:109–113, 1996.
182. O Dilly, P Nannipieri. Intracellular and extracellular enzyme activity in soil with reference to elemental cycling. Z Pflanzenernahr Bodenk 161:242–248, 1998.
183. F Einland. Determination of adenosine triphosphate (ATP) and adenylate energy charge

(AEC) in soil and use of adenine nucleotides as measures of soil microbial biomass and activity. Dan J Plant Soil Sci S: 1777:1–193, 1985.
184. HLT Mobley, RP Hausinger. Microbial urease: Significance, regulation and molecular characterization. Microbiol Rev 53:85–108, 1989.
185. GW McCarty, DR Shogren, JM Bremner. Regulation of urease production in soil by microbial assimilation of nitrogen. Soil Biol Biochem 24:261–264, 1992.
186. PF Grierson, MA Adams. Plant species affect acid phosphatase, ergosterol and microbial P in a jarrah (*Eucalyptus marginata* Donn ex Sm.) forest in south-western Australia. Soil Biol Biochem 32:1817–1828, 2000.
187. K Kouno, Y Tichiya, T Ando. Measurement of soil microbial biomass phosphorus by anion exchange membrane method. Soil Biol Biochem 27:1353–1357, 1995.
188. K Arnebrant, J Schnurer. Changes in ATP content during and after chloroform fumigation. Soil Biol Biochem 22:875–877, 1990.
189. L Zelles, A Palojarvi, E Kandeler, M von Lutzow, K Winter, QY Bay. Changes in soil microbial properties and phospholipid fatty acid fractions after chloroform fumigation. Soil Biol Biochem 22:875–877, 1990.
190. JK Martin, RC Foster. A model system for studying the biochemistry and biology of the soil-root interface. Soil Biol Biochem 17:261–269, 1985.
191. RC Foster. Microenvironments of soil microorganisms. Biol Fertil Soils 6:189–203, 1988.
192. K Hayano, K Tubaki. Origin and properties of β-glucosidase activity of tomato-field soil. Biol Fertil Soils 17:553–557, 1995.
193. K Watanabe, K Hayano. Estimate of the source of soil protease in upland fields. Biol Fertil Soils 17:553–557, 1994.
194. L Landi, L Badalucco, F Pomarè, P Nannipieri. Effectiveness of antibiotics to distinguish the contributions of fungi and bacteria to net nitrogen mineralisation, nitrification and respiration. Soil Biol Biochem 25:1771–1778, 1993.
195. C Stefanic, G Ellade, J Chirnageanu. Researches concerning a biological index of soil fertility. In MP Nemes, S Kiss, P Papacostea, C Stefanic, M Rusan, eds. Fifth Symposium on Soil Biology. Roman National Society of Soil Science, Bucharest, 1984, pp 35–45.
196. T Beck. Methods and application of soil microbial analyses at the Landensanstalt fur Bodenkultur und Pflanzenbau (LBB) in Munich for the determination of some aspects of soil fertility. In MP Nemes, S Kiss, P Papacostea, C Stefanic, M Rusan, eds. Fifth Symposium on Soil Biology. Roman National Society of Soil Science, Bucharest, 1984, pp 13–20.
197. P Perucci. Enzyme activity and microbial biomass in a field soil amended with a municipal refuse. Biol Fertil Soils 14:54–60, 1992.
198. C Trasar-Cepeda, C Leiros, F Gil-Sotres, S Seona. Towards a biochemical quality index for soils: An expression relating several biological and biochemical properties. Biol Fertil Soils 26:100–106, 1998.
199. FJ Stevenson. Cycles of soils, carbon, nitrogen, phosphorus, sulfur, micronutrients. New York: John Wiley & Sons, 1986.
200. RL Sinsabaugh, RK Antibus, AE Linkins, CA McClaugherty, L Rayburn, D Repert, T Weiland. Wood decomposition over a first-order watershed: mass loss as a function of lignocellulase activity. Soil Biol Biochem 24:743–749, 1992.
201. E Kandeler, M Stemmer, EM Klimanek. Response of soil microbial biomass, urease and xylanase within particle size fractions to long-term soil management. Soil Biol Biochem 31: 261–273, 1999.
202. M Stemmer, MH Gerzabek, E Kandeler. Invertase and xylanase activity of bulk soil and particle-size fractions during maize straw decomposition. Soil Biol Biochem 31:9–18, 1999.
203. D Angers, N Bissonnette, A Legère, N Samson. Microbial and biochemical changes induced by rotation and tillage in a soil under barley production. Can J Soil Sci 73:39–50, 1993.
204. MA Arshad, M Scnitzer, DA Anger, JA Rippmeester. Effects of till vs no-till on the quality of soil organic matter. Soil Biol Biochem 22:595–599, 1990.

205. CA Cambardella, ET Elliott. Carbon and nitrogen distribution in aggregates from cultivated and native grassland soil. Soil Sci Soc Am J 57:1071–1076, 1993.
206. O Dilly, HP Blume. Indicators to assess sustainable land use with references to soil microbiology. Adv GeoEcol 31:29–39, 1998.
207. D Tscherko. The response of soil microorganisms to environmental change. Ph.D. thesis, University of Agriculture, Vienna, Austria, p 87, 1999.

2

Ecology of Microbial Enzymes in Lake Ecosystems

Ryszard Jan Chróst and Waldemar Siuda
University of Warsaw, Warsaw, Poland

I. INTRODUCTION

During the past decade, an increasing number of ecological studies have considered the complexity of freshwater ecosystems. One major outcome of these studies has been an accelerated interest in the role of heterotrophic microorganisms (particularly bacteria) in the functioning of aquatic environments and the processes by which organic matter is made available to them (1–4). These heterotrophic microorganisms are the key trophic level at which the metabolism of the whole ecosystem is affected, i.e., organic matter decomposition, nutrient cycling, and structure of aquatic food webs. The demonstration of the importance of heterotrophic bacteria as a particulate carbon source for higher trophic levels and a major respiratory sink has created a renewed interest in the production and utilization of organic substrates by these microorganisms.

Most organic compounds produced in natural waters have a polymeric structure (5,6) and they are too large to be readily assimilated. The transport of organic molecules across microbial cell membranes is an active process mediated by specific enzymes called *permeases*. Only the low-molecular-weight organic molecules (monomers or oligomers) can therefore be taken up (7). In order to be available for microbial metabolism, polymeric compounds must be transformed into smaller molecules through enzymatic depolymerization.

Besides the physicochemical conditions of aquatic environments, the composition and availability of organic matter are the major factors that influence the development and activity of heterotrophic bacteria (8,9). The heterotrophic bacteria are the only biological populations capable of significantly altering both dissolved (DOM) and particulate (POM) organic matter. Microbial enzymes associated with these processes are the principal catalysts for a large number of biochemical transformations of organic constituents in aquatic environments. Many of these transformations can be mediated only by heterotrophic bacteria because the enzyme systems required for these reactions are not found in other organisms.

Microorganisms have adopted essentially two strategies that enable them to utilize macromolecular compounds. The macromolecule can be engulfed by the cytoplasmic membrane to form a vacuole within the cytoplasm. Enzymes are secreted into this vacuole and the polymeric compounds are hydrolyzed and subsequently taken up. Uptake of substrate solutions by this method is referred to as *pinocytosis*; uptake of particulate substrates is termed *phagocytosis*. However, many microbial cells are unable to carry out these processes, and therefore pinocytosis and phagocytosis are restricted to only those eukaryotic microorganisms that lack a cell wall, e.g., many protozoa. Those eukaryotic and prokaryotic microorganisms that possess a cell wall have developed an alternative strategy for the assimilation of polymeric substrates. Hydrolytic enzymes are secreted outside the cytoplasmic membrane, where they hydrolyze macromolecules in close vicinity to the cell. The resulting low-molecular-weight products are then transported across the cell membrane and utilized inside the cytoplasm.

The hydrolysis of polymers is an acknowledged rate-limiting step in the utilization of organic matter by microorganisms in aquatic and soil environments. Prior to incorporation into microbial cells, polymeric materials undergo stepwise degradation by a variety of cell surface–associated enzymes and/or enzymes secreted by intact living cells or liberated into the environment through the lysis of microorganisms. The importance of microbial enzymatic activities to the mobilization, transformation, and turnover of organic and inorganic compounds in freshwater and marine environments has been shown in many studies (10–18). Results of these studies have shown that studying enzymatic processes provides powerful information that helps in understanding basic processes of decomposition and microbial activity in both freshwater and marine ecosystems.

II. ORIGIN AND ASSOCIATION OF ENZYMES WITH AQUATIC MICROORGANISMS

Three common terms are used for the enzymes involved in the transformation and degradation of polymeric substrates outside the cell membrane: *ectoenzymes* (19), *extracellular enzymes* (20), and *exoenzymes* (21). In this chapter, the term *ectoenzyme* is used to refer to any enzyme that is secreted and actively crosses the cytoplasmic membrane and remains associated with its producer. Ectoenzymes are cell-surface-bound or periplasmic enzymes that react outside the cytoplasmic membrane with polymeric substrates that do not penetrate the cytoplasm. Extracellular enzymes occur in free form dissolved in the water and/or are adsorbed to surfaces (e.g., detrital particles, organic colloids, humic complexes, minerals in suspension). Extracellular enzymes in water may be secreted actively by intact viable cells, they can be released into the environment after cell damage or viral lysis, and/or they may result from zooplankton grazing on algal cells and from protozoan grazing on bacteria.

Ectoenzymes and extracellular enzymes (in contrast to intracellular enzymes) react outside the cell, and most of them are hydrolases. The ectoenzymes that cleave polymers by splitting the key linkages on the interior of the substrate molecule and form intermediate sized fragments are called *endoectoenzymes* (e.g., aminoendopeptidases act on the centrally located peptide bonds and liberate peptides) (22). Those ectoenzymes that hydrolyze the substrate by consecutive splitting of monomeric products from the end of the molecule are termed *exoectoenzymes* (e.g., aminoexopeptidases hydrolyze peptide bonds adjacent to terminal α-amino or α-carboxyl groups and liberate free amino acids) (23).

There are three pools of microbial enzymes in water samples: intracellular enzymes are located and react with substrates inside the cytoplasmic region and are mostly responsi-

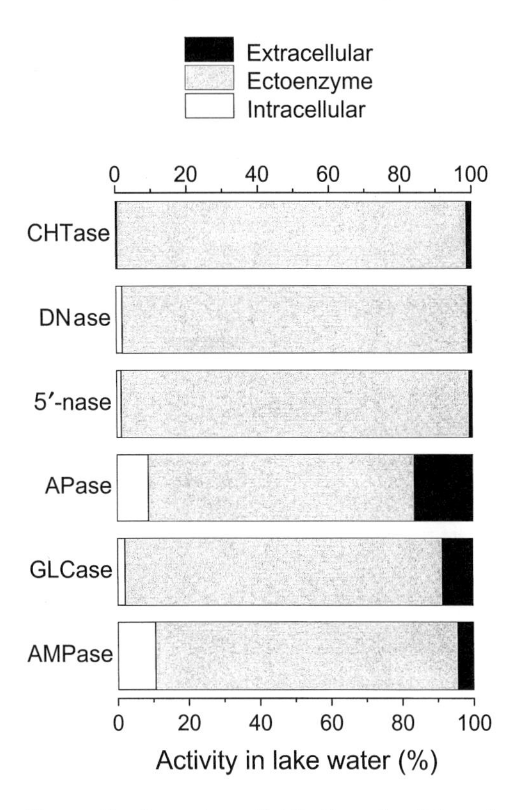

Figure 1 Percentage distribution of cell-bound and extracellular activity of microbial chitinase (CHTase), deoxyribonuclease (DNase), 5′-nucleotidase (5′-nase), alkaline phosphatase (APase), β-glucosidase (GLCase), and aminopeptidase (AMPase) in water samples from eutrophic Lake Mikołajskie. (Chróst, unpublished.)

ble for internal cell metabolism; extracellular enzymes are in the surrounding environment and catalyze reactions without control from their producers; and ectoenzymes are cell-surface-bound enzymes, mostly hydrolases, that degrade polymeric substrates, yielding readily utilizable monomers. All pools are composed of both endo- and exoenzymes.

Distribution between ecto- and extracellular activity for selected enzymes (aminopeptidase, β-glucosidase, alkaline phosphatase [APase], 5′-nucleotidase [5′-nase], deoxyribonuclease [DNase], and chitinase [CHTase]) has shown that ectoenzymes contributed on average from 75% (APase) to 98% (chitinase) of the total activity in lake water (Fig. 1). Activities of the intracellular and extracellular pool enzymes are low. Intracellular enzymes contributed from 0.5% (chitinase) to 10.7% (aminopeptidase) to the total activity of water samples. Activity of the extracellular enzymes, dissolved in the water, constituted from 1% (5′-nase) to 16.5% (APase) of the total activity. An interesting observation is that extracellular enzyme activity as a percentage of total activity is higher in lake sediments than in the water column (Table 1). This was particularly evident in the case of chitinase and lipase activities.

Enzyme activities bound to the 0.2- to 1.0-μm-size fraction of microplankton (mainly composed of bacteria) make up a greater fraction of activity by microorganisms in lake water. High ectoenzyme activity found in this size fraction has correlated with

Table 1 Percentage Contribution of Extracellular Enzyme Activities to the Total Activity of Lake Water and Lake Sediment Samples

Enzyme	Lake/trophic status	Percentage	
		Water	Sediment
Leucine-aminopeptidase	Plußsee/eutrophic	9.5 ± 3.7	12.6 ± 6.4
	Schöhsee/mesotrophic	8.4 ± 2.1	11.2 ± 5.1
α-Glucosidase	Plußsee/eutrophic	10.2 ± 4.9	15.1 ± 7.3
	Mikołajskie/eutrophic	8.7 ± 2.6	13.3 ± 3.8
	Szymon/hypereutrophic	10.8 ± 3.3	14.6 ± 4.9
Lipase	Jagodne/hypereutrophic	28.8 ± 9.6	41.0 ± 7.4
	Bełdany/eutrophic	21.4 ± 6.8	33.1 ± 8.3
	Kisajno/mesotrophic	22.4 ± 6.2	35.8 ± 9.1
Alkaline phosphatase	Plußsee/eutrophic	24.5 ± 6.3	32.6 ± 7.3
Chitinase	Plußsee/eutrophic	2.3 ± 0.7	22.8 ± 6.9
	Schöhsee/mesotrphic	1.7 ± 0.5	23.2 ± 8.4

± Standard deviation of an average value.
Source: Data from Chróst, unpublished.

bacterial abundance and/or bacterial production of lake water. A variety of microorganisms produce ectoenzymes in waters and sediments in freshwater ecosystems. However, many studies have reported that bacteria are the major producers of ectoenzymes among aquatic microorganisms (24–33).

III. CONTROL OF ECTOENZYME SYNTHESIS AND ACTIVITY

The conditions in the aquatic environment, as in the soil aqueous phase, are unfavorable for enzymes. First, the substrate concentration is usually very low and highly variable. Many substrates may be insoluble, exist in intimate association with other compounds, and/or be bound to humic substances, colloidal organic matter, and detritus. Therefore, these conditions are suboptimal for the coupling of an enzyme to its substrate. Second, an enzyme may be lost from the parent cell and may be bound to suspended particles and humic materials, or it may be exposed to a variety of inhibitors present in the water. Finally, an enzyme may be denaturated by physical and chemical factors in the aquatic environment or hydrolyzed by proteases.

Obviously, for an enzyme to be of benefit to its producer microorganism, it must avoid degradation long enough to associate with its substrate. Moreover, even if an enzyme overcomes these obstacles and binds with its substrate, the physical and chemical conditions of the reaction medium may be unsuitable for catalysis (e.g., nonoptimal pH or temperature, presence of inhibitors, absence of activators, suboptimal ionic strength). Nevertheless, there is strong evidence that various aquatic microorganisms produce ectoenzymes in freshwaters that encounter a number of polymeric substrates (31) and that microbial growth is dependent on the products of ectoenzymatic reactions (13,14,16,24, 34–36).

A microbial cell living in an aquatic ecosystem is influenced by a variety of environmental factors. The signal for appropriate gene expression and consequent ectoenzyme production within a cell is in response to the surrounding environment. Depending on the regulatory control of gene expression, two types of microbial enzymes are synthesized in waters and sediments: constitutive enzymes, whose synthesis is constant regardless of the presence or absence of the substrate in the environment, and inducible enzymes, whose rates of synthesis are strongly dependent on the presence of their substrates (or substrate derivatives). Many inducible enzymes are synthesized at a low basal rate (i.e., are constitutive) in the absence of a substrate. When the substrate is available in the environment, there is a dramatic increase in the production rate of the particular enzyme. Synthesis continues at this amplified rate until the inducer is removed and/or the product of enzymatic catalysis accumulates (24) and it then returns to the basal rate.

Most of the ectoenzymes synthesized by aquatic microorganisms are catabolic enzymes involved in degradation of polymeric substrates that are not continuously available in the water or sediments. Therefore, the constant synthesis of ectoenzymes in the absence of substrates is unnecessary, because it requires the expenditure of energy that otherwise may be channeled into other useful activities. Since microorganisms have been competing with each other for millions of years, the evolutionary advantages of induction are readily apparent. Most of the ectoenzymes found in freshwaters are inducible, and only a few have a constitutive nature (e.g., some amylases or proteases in bacteria).

A. Induction and Repression/Derepression of Ectoenzyme Synthesis

The efficient induction of ectoenzymes is more complicated than that of intracellular enzymes. First, many of the ectoenzyme substrates present in fresh water are polymeric compounds, and they are too large to enter the cell and serve as inducers of synthesis. Second, for an ectoenzyme to be secreted at appropriate rates, the microorganism must be able to monitor the activity of the ectoenzyme outside the cell. We suggest that these problems are overcome by a low constitutive rate of ectoenzyme secretion. If the substrate is present, then low-molecular-weight products accumulate to a certain level, enter the cell, and serve as the inducer (20). When environmental conditions inhibit an ectoenzyme activity (e.g., unsuitable pH, absence of activating cations Mg^{2+}, Zn^{2+}), the induction of its synthesis does not occur because the product of catalysis is not generated. However, since the microorganisms in freshwater ecosystems are in a complex relationship with a variety of readily utilizable compounds of autochthonous and allochthonous origin, the induction of a particular ectoenzyme by an end product resulting only from degradation of a single polymeric substrate seems to be questionable. Until now, it has appeared that one ectoenzyme may have several inducing compounds (24,37).

It is well documented that synthesis of many ectoenzymes produced by aquatic microorganisms is repressed by the end product that accumulates in the cell or in surrounding environment. The repression of alkaline phosphatase synthesis by inorganic phosphate (the end product of phosphomonoester hydrolysis) in microalgae and bacteria is probably one of the best-known examples (11,13,38,39). In Lake Plußsee, the specific activity of APase significantly decreased when the ambient orthophosphate concentrations were higher than 0.5 μM (13). APase activity was inversely related to the amount of intracellular phosphorus stored (P_{st}) in algal cells. When P_{st} constituted less than 10% of the total cellular phosphorus, the algae produced alkaline phosphatase with a high specific activity, and when P_{st} was higher than 15% and the ambient orthophosphate concentrations exceeded 0.6 μM, this activity rapidly decreased.

The synthesis of virtually all ectoenzymes in most aquatic microorganisms is re-

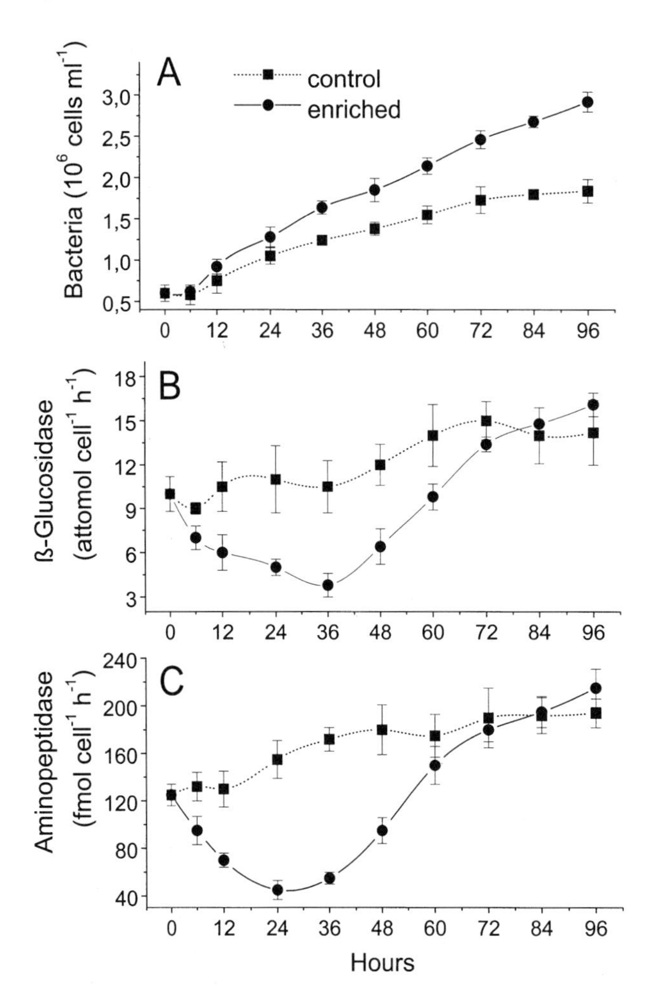

Figure 2 Effect of water supplementation with dissolved organic matter extracted from phytoplankton on growth of bacteria (A) and specific activity of bacterial β-glucosidase (B) and aminopeptidase (C) in Lake Mikołajskiet. (Chróst, unpublished.)

pressed when they are grown on sources of readily utilizable dissolved organic matter (UDOM). This mode of regulation is called *catabolic repression*. When water samples from eutrophic Lake Mikołajskie were supplemented with dissolved organic matter extracted from phytoplankton (both UDOM and polymeric compounds) the bacterial cell numbers increased markedly during 96 hours of incubation (Fig. 2A). Contrary to that in control samples, supplementation of lake water with phytoplankton organic matter resulted in a significant decrease in the rates of specific activities of bacterial β-glucosidase and aminopeptidase (calculated per bacterial cell) during the first period of bacterial growth (6–48 hours). However, in both of these enzymes, specific ectoactivity began to increase after 48 hours of bacterial growth. In control samples, where bacteria grew solely on naturally present DOM in lake water, the specific activity of these ectoenzymes increased within the incubation period (Figs. 2B, 2C).

The repression of ectoenzymes is tightly coupled to the availability of UDOM in lake water. Figures 2B and 2C show that ectoenzyme synthesis in DOM-enriched samples was no longer repressed when the concentration of the readily utilizable low molecular-weight molecules fell below a critical level, and polymeric substrates had to be used to support the growth and metabolism of bacteria. Similar in situ observations during phytoplankton bloom development and breakdown were reported for β-glucosidase activity in eutrophic Lake Plußsee (24), for β-glucosidase and aminopeptidase activities in mesotrophic Lake Schöhsee (25), and for lipase activity in eutrophic Lake Mikołajskie (40).

Despite the widespread occurrence of catabolic repression, with the exception of those for enteric bacteria, the molecular details of the repression are poorly understood. Some studies have indicated that cyclic adenosine monophosphate (cAMP), together with its receptor protein, may play a central role in control of catabolic repression (41,42). Using the repression strategy for ectoenzyme synthesis, microorganisms can avoid the wasteful production of inducible enzymes, which are not useful when their growth is not limited by UDOM (3,19,24,35).

B. Inhibition of Activity

It is important to consider that the repression/derepression of an ectoenzyme not be equated to the reversible inhibition of activity. Even if an ectoenzyme is synthesized, its activity may be inhibited by the accumulation of the end product or by high concentrations of the substrate (19). Two general types of reversible inhibition are known: competitive and noncompetitive inhibition.

Competitive inhibition occurs when an inhibiting compound is structurally similar to the natural substrate and, by mimicry, binds to the enzyme. In doing so, it competes with an enzyme's natural substrate for the active substrate-binding site. The hallmark of competitive inhibition of many ectoenzymes (e.g., alkaline phosphatase, β-glucosidase, aminopeptidase) is that it decreases the affinity of an ectoenzyme (an increase of the apparent Michaelis constant is observed) for the substrate and, therefore, inhibits the initial velocity of the reaction (Fig. 3) (13,26,37). Competitive inhibition is reversible and can be overcome by increased substrate concentration, and therefore the maximum velocity (V_{max}) of the reaction is unchanged (Fig. 3A).

Noncompetitive inhibition generally is characterized as an inhibition of enzymatic activity by compounds that bear no structural relationship to the substrate. Therefore, the inhibition cannot be reversed by increasing the concentration of the substrate. It may be reversed only by removal of the inhibitor. Unlike competitive inhibitors, reversible noncompetitive inhibitors cannot interact at the active site but bind to some other portion of an enzyme-substrate complex. This type of inhibition encompasses a variety of inhibitory mechanisms and is therefore not amenable to a simple description. Noncompetitive inhibition of the activity of exoproteases by Cu^{2+} ions (43) and inhibition of α-glucosidase, β-glucosidase, *N*-acetyl-glucosaminidase, and alkaline phosphatase by H_2S in natural waters have been described (12).

C. Environmental Control of the Synthesis and Activity of Ectoenzymes

The complex environmental regulation of ectoenzyme synthesis and activity has been demonstrated in studies of bacterial β-glucosidase and aminopeptidase in surface waters of lakes (24,25,37). The results of these studies demonstrated that the ectoenzyme synthe-

(A)

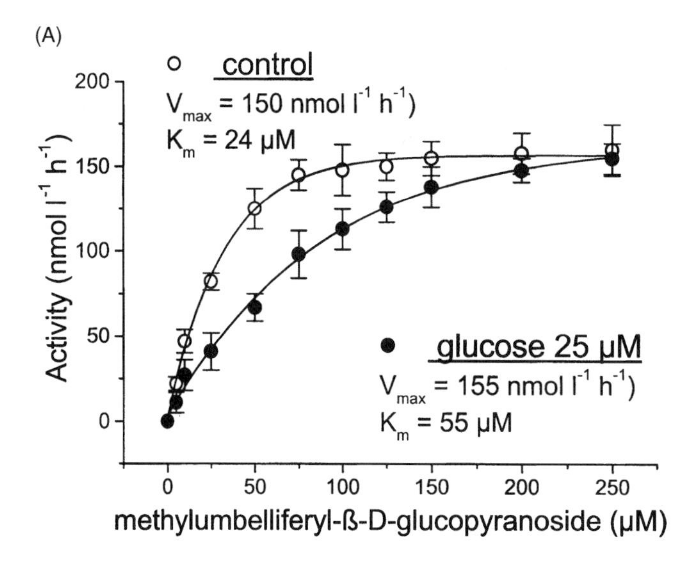

(B)

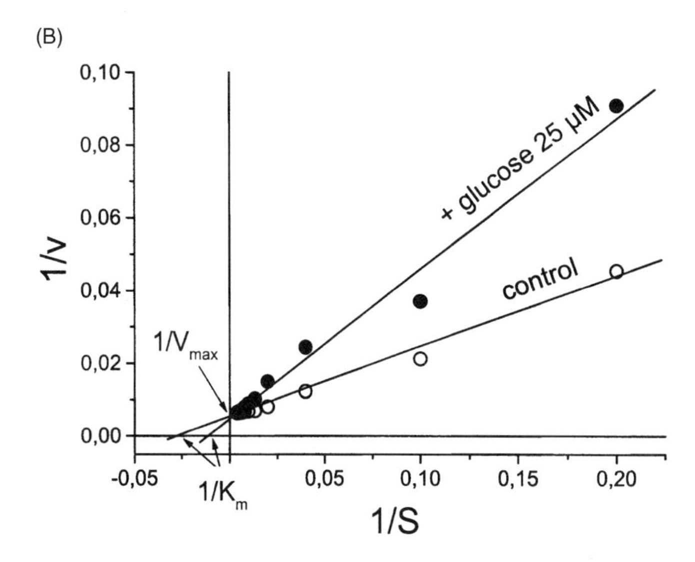

Figure 3 Competitive inhibition of β-glucosidase activity by glucose (end product of enzyme reaction) in water samples from eutrophic Lake Mikołajskie. (A) Hyperbolic relationship between enzyme activity and increasing substrate concentrations, (B) Lineweaver-Burk's linear transformation of the relationship between enzyme activities ($1/v$) and increasing concentrations of substrate (1/S). (Chróst, unpublished.)

sis and activity were under different control mechanisms, which were dependent on the physical-chemical conditions of the habitat. There is ample evidence for general catabolic repression of ectoenzyme synthesis in bacteria due to readily utilizable carbon sources (24,37,44,45) as well as more specific repression by end products of enzyme catalysis (34,46). However, control of aminopeptidases appears to be distinct and more complex than that of other ectoenzymes. In some bacteria, amino acids, peptides, and/or proteins seem to induce aminopeptidase synthesis (45,47). It is not known specifically how aminopeptidase induction operates, especially since amino acids are reported to act as inducers in some bacteria, rather than acting in their more predictable role as end-product inhibitors.

The ability of bacteria living in the euphotic zone of the lakes to produce ectoenzymes seems to be strongly affected by the availability of the low-molecular-weight, readily utilizable substrates exudated by algae (eg., excreted organic carbon-EOC), which are known to be excellent substrates for bacteria (48–50). Chróst and Rai (25) found that the rates of leucine-amino-peptidase and α-glucosidase production by aquatic bacteria strongly depend on bacterial organic carbon demand. When the amount of EOC fulfilled the bacterial organic carbon requirement, microorganisms did not synthesize enzymes needed for hydrolysis of the polymeric substrates because their utilization was unnecessary. Moreover, the specific activity of aminopeptidase correlated negatively to the rates of algal EOC.

During the active growth of phytoplankton, algal populations excrete into the water a variety of photosynthetic products, including easily assimilable low-molecular-weight substrates (51), which support bacterial growth and metabolism. These substrates inhibit the activity and repress the synthesis of ectoenzymes in bacteria. On the other hand, when low levels of readily available substrates limit bacterial growth and metabolism, bacteria produce ectoenzymes with high specific activity to degrade polymers and other nonlabile substrates. Such a situation occurs in lake water during the breakdown of phytoplankton bloom. Senescent algae liberate, through autolysis of cells, a high amount of polymeric organic compounds (polysaccharides, proteins, organophosphoric esters, nucleic acids, lipids, etc.), which induces synthesis of ectoenzymes. Another mechanism that causes repression cessation of enzyme synthesis is low level of directly utilizable organic compounds in the water during bloom breakdown (52).

Bacteria living in the profundal zone of the lakes are often substrate-limited (2,53) because the amount of substrate in deep waters depends primarily on the sedimentation rates of the organic matter that is produced in the euphotic zone. There is no direct supply of labile organic compounds exudated by algae. In the profundal zone, sedimentation provides labile monomeric organic compounds that are mostly polymers that are utilized by bacteria. Under such environmental circumstances, bacterial metabolism is strongly dependent on the presence and amount of polymeric substrates and the activity of synthesized ectoenzymes that catalyze the release of readily utilizable monomers.

Microbial ectoenzymatic activity in natural waters is also strongly dependent on environmental factors, such as temperature, pH, inorganic and organic nutrients, ultraviolet B (UV-B) radiation, and presence of activators and/or inhibitors (3,13,21,54–59). Several studies have shown that ectoenzymes display the highest activities in alkaline waters of pH 7.5 to 8.5 (24,40) or acid waters of pH 4.0 to 5.5 (55). In contrast to the pH response, many ectoenzymes exhibit no obvious adaptation to ambient temperature, because the optimal temperature is often considerably higher than in situ temperature of waters (13,33, 40). The optimal temperatures for alkaline phosphatase and β-glucosidase are unchanged when they are produced by planktonic microorganisms in lake water under different in situ temperatures (13,24).

In light of these aforementioned studies, the environmental regulation of ectoenzyme synthesis and activity is complex and usually no single factor is involved in this process. It is important to realize that environmental regulation of ectoenzymes, induction, synthesis repression, and inhibition are related to concentration, period of exposure, and such factors as temperature, pH, oxygen level, and chemical characteristics of regulatory molecules. The same molecule that is an inducer under one set of circumstances may be a repressor under other environmental conditions, or at different concentrations.

IV. ASSAYS OF ECTOENZYME ACTIVITY

A. Methods

There are significant difficulties in measuring ectoenzyme activities in heterogeneous environments such as natural waters and soil, which include questions about methodology and data interpretation. For example, should assays be performed according to the well-established principles of enzymology (e.g., excess substrate, optimal pH and temperature, shaking of reaction mixtures) or in situ conditions encountered in an aquatic environment (e.g., limiting and unevenly distributed substrate, suboptimal and fluctuating physical conditions, stationary incubation)? How are the optimal assays related to those assays done under more ''realistic'' conditions?

A variety of methods are available for monitoring the enzyme activities when working with microbial cultures or isolated enzymes in biochemical laboratories. However, most classical enzymatic methods cannot be applied directly in aquatic environments. The enzyme amount and activity in natural waters are usually much lower than those measured in cultures or in enzyme extracts, and therefore the classical biochemical methods often are inadequate for measuring low ectoenzyme reaction velocity. Furthermore, the environmental conditions of ectoenzyme assays in water samples often are suboptimal (e.g., unsuitable temperature, pH, presence of interfering compounds) and the choice of substrate used to study ectoenzymes of natural microbial assemblages in aquatic environments often is problematic.

Depending on the chemical nature of the ectoenzyme substrate, there are three categories of methods for measurement of ectoenzyme activity in aquatic environments: spectrophotometric, fluorometric, and radioactive. The most commonly used in the past were spectrophotometric methods (60–63). The major disadvantage of spectrophotometric methods is long incubation time necessary for enzyme reactions, which is due to their relatively low sensitivity (micromolar [μ*M*] to millimolar [m*M*] concentrations of the final product of enzyme reaction are required). However, spectrophotometric assays can be used when measuring high enzyme activity in samples, or when working with purified and/or concentrated enzymes.

During the last two decades, fluorometric methods have been widely used for enzyme activity determinations in aquatic environments (3,21,24,33,52,64,65). Fluorometric assays are very sensitive, and they measure the final products of enzymatic reactions in nanomolar (n*M*) to micromolar (μ*M*) concentrations. When using a modern spectrofluorometer to measure enzyme activity in water samples, the incubation time for monitoring substrate-enzyme reaction can be shortened to a few minutes. Several authors have applied radiometric methods for enzyme activity determination in aquatic environments

(66–69). Although these methods are extremely sensitive, they are seldom used because of greater costs and precautions needed when handling radioactive materials.

B. Substrates

When studying the significance of ectoenzyme activities in relation to in situ substrate turnover in aquatic (and in soil) environments, one should be able to determine the real rates of the process. In this case, the substrate should have an affinity for the ectoenzyme similar to that of the natural substrates in situ. Moreover, the enzyme should be assayed by using low substrate concentrations comparable to the concentrations of the natural substrate. Application of low substrate concentrations results in low levels and difficult detection of end products. Generally, in enzyme reactions, the length of incubation required for end product formation is related to product detection sensitivity.

A large variety of commercially produced ectoenzyme substrates are now available. Depending on their chemical structure and the enzyme assay, two types of organic compounds can serve as substrates: natural and artificial substrates. Natural substrates are native compounds (nonlabeled) or their chemical structure is only slightly modified by labeling with chromophores, fluorophores, or radiolabeling with ^{14}C, ^{3}H, ^{32}P, ^{35}S, or ^{125}I. Most natural substrates have an affinity for the enzyme that is complementary to that of the natural substrates in aquatic samples. Monitoring of enzyme activity by means of natural nonlabeled substrates requires a sensitive analytical method to measure the end product or substrate remaining after incubation time (70). Modern analytical methods offer precise and rapid determination of several natural compounds that can be used as enzyme substrates or products (e.g., amino acids, proteins, deoxyribonucleic acid [DNA], carbohydrates). The application of the labeled natural substrates requires very sensitive and accurate methods for quantitative determination of the label bound to the substrate molecule (e.g., spectrophotometry, fluorometry, radiometry). Most suitable natural substrates are radiolabeled compounds because their end products can be measured after a short incubation period (minutes) (66–69). Until now, this approach has been limited by the reduced availability of radiolabeled substrates and the high handling costs of radioactive materials. Except for some cases of analytical difficulties, natural substrates are promising for studying ectoenzymes in aquatic environments.

Artificial substrates are synthesized in laboratories and their chemical structure (e.g., chemical bonds) only mimics that of natural compounds. Ectoenzymes react with artificial substrates by splitting specific chemical bonds between an organic moiety and its chromophore or fluorophore, yielding colored or fluorescent products, respectively. Because these are not natural substrates, enzyme activities obtained are not necessarily identical to those measured by using natural substrates. However, their application allows for low costs and simpler and more rapid measurements of ectoenzymatic activity.

In the past, chromogenic artificial substrates were used intensively in the studies of ectoenzyme activity in fresh waters (10,11,60–63,71). It is advantageous to use chromogenic substrates because they can be measured easily by spectrophotometry. However, low sensitivity is a major disadvantage of this technique, and long incubation times of 72 to 96 hours often are required (62,71). This may result in microbial proliferation and ectoenzyme synthesis during the assay, changes, which must be prevented. They usually are avoided by adding plasmolytic or antiseptic agents to assays, such as toluene or chloroform (10,38,71). However, these agents change the membranes, thereby leading to release of ecto- and intracellular enzymes. In cases in which some enzymes are located intra- and extracellularly (e.g., phosphatase, arylsulfatase), ectoenzyme activity may be significantly overestimated (72).

Recently, fluorophore-labeled artificial substrates have been commonly used for sensitive assays of ectoenzyme activity in aquatic environments (13–16,21,25,33,36,37, 54,55,64,65) and are advantageous when it is necessary to perform a large number of assays. Fluorogenic substrates yield highly fluorescent, water-soluble products with optical properties significantly different from those of the substrate. Many substrates are degraded to products that have longer wavelength excitation or emission spectra. Therefore, these fluorescent products typically can be quantified in the presence of an unreacted substrate by using a fluorometer. Three types of substrates derived from water-soluble fluorophores are commercially available: blue, green, and red (73).

Hydroxy- and amino-substituted coumarins are the most widely used fluorogenic substrates. Coumarin-based substrates produce highly soluble, intensely blue fluorescent products. Phenolic dyes such as 7-hydroxycoumarin (umbelliferone) and the more common 7-hydroxy-4-methylcoumarin (methylumbelliferone) are not fully deprotonated and therefore not fully fluorescent unless the reaction mixture has $pH > 10$ (64). Substrates derived from these fluorophores are not often used for continuous measurement of enzymatic activity. Products of substrates containing aromatic amines, including the commonly used 7-amino-4-methylcoumarin, 7-amino-4-trifluoromethylcoumarin, and 6-aminoquinoline, are partially protonated at $pH < 5$ but fully deprotonated at neutral pH. Thus, their fluorescence is not subject to variability due to pH-dependent protonation/deprotonation when assayed near or above physiological pH.

Substrates derived from water-soluble green fluorophores, fluorescein, and rhodamine, provide significantly greater sensitivity in fluorescence-based enzyme assays. In addition, most of these longer-wavelength fluorophores have excitation coefficients that are 5 to 25 times that of coumarins, nitrophenols, or nitroanilines, making them potentially useful as sensitive chromogenic substrates (73). Substrates based on the derivatives of fluorescein and rhodamine usually incorporate two moieties, each of which serves as a substrate for the enzyme. Consequently, they are cleaved first to the monosubstituted analog and then to the free fluorophore. Because the monosubstituted analog often absorbs and emits light at the same wavelength as that of the ultimate hydrolysis product, this initial hydrolysis complicates the interpretation of hydrolysis kinetics (often biphasic kinetics are observed). However, when highly purified, the disubstituted fluorescein- and rhodamine-based substrates have virtually no visible-wavelength absorbance or background fluorescence, making them extremely sensitive substrates.

Substrates derived from water-soluble red fluorophores (long-wavelength fluorophores) often are preferred because background absorbance and autofluorescence generally are lower when longer excitation wavelengths are used. Substrates derived from the red fluorescent resorufin and dimethylacridinone contain only a single hydrolysis-sensitive moiety, thereby avoiding the biphasic kinetics.

The majority of fluorophore-labeled substrates produce very low background fluorescence and can be used without any loss in sensitivity at the high concentrations (millimolar) that are sometimes needed for enzyme saturation (65). It also is possible to work with substrate concentrations in the nanomolar range, close to the presumed range of natural substrate concentrations in aquatic environments. It has been shown that substrates linked to fluorophores provide a very sensitive system for detecting and quantifying many specific and nonspecific hydrolases in aquatic environments (21). The potential ectoenzymatic activity of water or sediment samples can be measured over a short incubation time without problems of microbial proliferation, low activity, and nonsaturation of the ectoenzyme. In spite of this advantage in using the fluorescent substrates in ectoenzyme assays, their use is controversial (as are chromogenic substrates), because of their unknown affinity for the ectoenzymes in comparison to that of natural substrates.

C. Potential Enzyme Activity—Kinetic Approach

If information about the potential activity of the ectoenzyme in the aquatic habitat is required, there are reasons for using high concentrations of the substrate in assays. The enzyme should be substrate-saturated. Possible competition with co-occurring natural substrates should be prevented, as should competitive inhibition of the substrate with inhibitors in samples (Fig. 3).

Many hydrolytic ectoenzymes follow Michaelis-Menten kinetics:

$$v = (V_{max} \times [S])/(K_m + [S])$$

where a plot of the initial velocity of reaction (v) against increasing concentrations of substrate ($[S]$) gives a rectangular hyperbola. For such assays, the kinetic approach is recommended; it allows calculation from the experimental data of the kinetic parameters characterizing an enzyme-substrate reaction. They are V_{max}, the maximal velocity of enzyme catalysis that theoretically is attained when the enzyme has been saturated by an infinite concentration of substrate, and K_m, the Michaelis constant, which is numerically equal to the concentration of substrate for the half-maximal velocity (V_{max}), which indicates the enzyme affinity to the substrate (74). The kinetic approach requires substrate concentrations ranging from low to high for first-order (the reaction velocity increases linearly with the increase in substrate concentrations) and zero-order (reaction velocity remains constant, not affected by the concentration of substrate) enzyme reactions.

A typical ectoenzyme kinetic experiment may be described as follows: Data are collected as a function of at least five triplicate reactant concentrations of substrate, and the experimental dependence on this function is determined and plotted graphically. The results depend essentially on the shape of the hyperbolic curve described by the data, thus making determination of V_{max} and K_m difficult (Fig. 3A). To obtain these kinetic parameters, the Michaelis-Menten equation often is rearranged to the linear form and V_{max} and K_m are obtained from the slope and intercept (Fig. 3B) (74).

Such graphical methods produce correct values for the parameters only in the absence of error. Unfortunately, all the measurements are subject to some degree of imprecision, and therefore use of linearized equations such as that of Lineweaver-Burk, Eadie-Hofstee, and Woolf may give inaccurate or biased experimental data (75,76). The best solution to this problem is to perform a nonlinear regression analysis on the original experimental data. The kinetic parameters then can be calculated from the direct plot of reaction velocity (v) versus substrate (S) concentration by using a computer program to determine the best fit of the rectangular hyperbola (77).

D. In Situ Enzyme Activity—Direct Approach

True ecological information requires the detection of environmental processes under in situ conditions, which cannot be fully controlled and, therefore, cannot be simulated in the laboratory. The composition of naturally occurring substrates in water samples usually is unknown, and concentrations may vary widely over short sampling times. This condition complicates the choice of the substrate concentration being monitored in ectoenzyme assays because of the potential interference or competition with natural substrates and/or inhibitors.

Ideally, to prevent these problems and to measure the real in situ rates of ectoenzyme activity, one should follow the decrease in naturally occurring substrate concentration or the increase in ectoenzyme product formation under in situ conditions. Because of the analytical difficulties, this approach is very seldom used in aquatic studies. Moreover, the increase in concentration of ectoenzyme product in samples simply cannot be measured because liberated product is simultaneously utilized by microorganisms. To overcome this problem and to be able to measure the amount of product released from its substrate, it is necessary to inhibit product assimilation by intact living microorganisms. Several inhibitory agents that do not inhibit enzyme activity can be used to prevent the microbial assimilation of low-molecular-weight products of ectoenzymatic hydrolysis of polymeric substrates (e.g., antibiotics or chemotherapeutics blocking active transport systems, some fixing agents such as sodium azide).

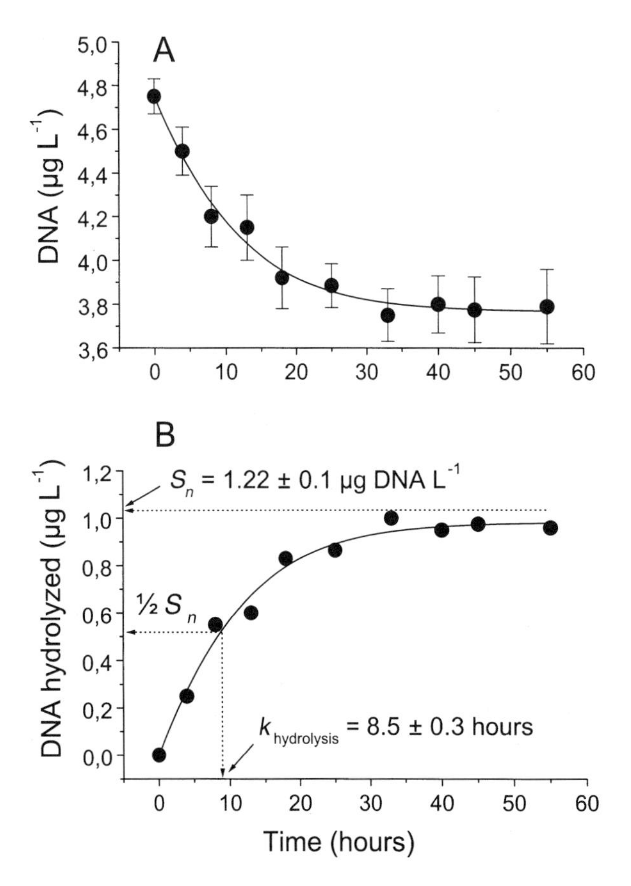

Figure 4 Direct estimation of enzymatic hydrolysis of natural DNA by means of decrease in substrate concentration in water samples from eutrophic Lake Mikołajskie. (A) Concentration of remaining DNA in samples after incubation times, (B) amount of DNA hydrolyzed (see text for description of kinetic parameters). (Chróst, unpublished.)

Figure 4 presents an example of direct estimation of natural DNA hydrolysis in eutrophic Lake Mikołajskie. To prevent microbial growth and utilization of products of DNA degradation in the course of its hydrolysis, water samples were fixed with 0.3% sodium azide. The concentration of DNA decreased from 4.75 ± 0.08 μg L^{-1} (at time zero) to a plateau of 3.79 ± 0.17 μg L^{-1} (after 55 hours) (Fig. 4A). A plot of the amount of DNA-hydrolyzed [s] versus time of hydrolysis [t] gave a rectangular hyperbola (Fig. 4B):

$$s = [S_n] \times [t])/(k_{\text{hydrolysis}} + [t])$$

By applying a nonlinear regression analysis to the experimental data it was possible to estimate hydrolysis parameters: [S_n], concentration of DNA naturally present in water samples, which theoretically is attained after infinite time of hydrolysis, and $k_{\text{hydrolysis}}$, the hydrolysis constant, i.e., the hydrolysis time of the half-concentration of natural DNA (S_n). The preceding data provide an example for determining hydrolysis parameters of a naturally occurring enzyme substrate by analysis of substrate concentration evolution in water samples during the course of its enzymatic degradation.

Using a direct approach, it also is possible to estimate the hydrolysis parameters characterizing in situ enzymatic degradation of natural substrates when the concentrations of the final product are determined during the course of hydrolysis. In situ hydrolysis of proteins by proteolytic enzymes yielded increasing concentrations of free, dissolved amino acids in lake water samples when microbial uptake was inhibited by 0.3% sodium azide (Fig. 5).

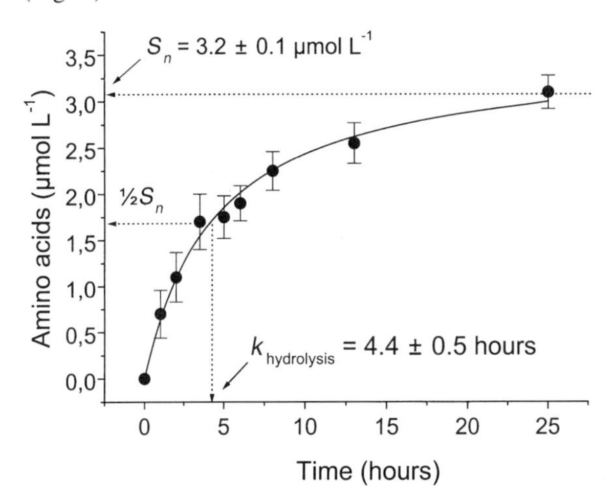

Figure 5 Direct estimation of enzymatic hydrolysis of natural protein by means of release of reaction products (amino acids) in water samples from eutrophic Lake Mikołajskie. (Siuda and Kiersztyn, unpublished.)

V. TEMPORAL AND SPATIAL DISTRIBUTION OF ECTOENZYME ACTIVITY

Both spatial and seasonal ectoenzymatic activities fluctuate markedly in lake waters (13,24,28,29,36,38). The production of ectoenzymes by microorganisms is strongly corre-

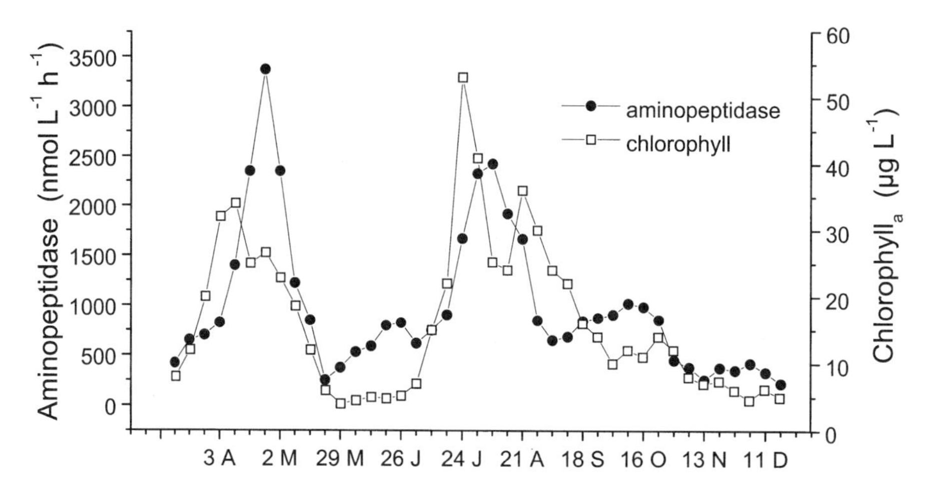

Figure 6 Seasonal aminopeptidase activity and chlorophyll$_a$ concentration in the surface water samples (0- to 1-m depth) from Lake Plußsee. (Chróst, unpublished.)

lated to the influx of polymeric organic matter and/or the depletion of readily utilizable UDOM in the environment (3,24,52).

Ectoenzyme production and activity show marked seasonal variation in both surface and deep waters of lakes. In surface waters, the maximal ectoenzymatic activities occur during the late stage of phytoplankton bloom and after its breakdown, and minimal activities occur during the clearwater phase in lakes (Fig. 6) (13,24,38,78).

Ectoenzymatic activity is especially lower during summer thermal stratification in the hypolimnion than in the epilimnion of a lake, and the activity is strongly dependent upon the sedimentation rates of detritus produced in the euphotic zone (Fig. 7). Usually, a lag period is observed between the maximal ectoenzymatic activity in surface and that in deep waters of a lake (Fig. 7B,D) (78).

There also are known high diurnal fluctuations in enzyme activity in lake water (78,79) because dynamic environmental factors affect enzyme production and/or activity, as well as microbial growth, metabolism, and biomass of microbial enzyme producers (3,25,36). In the summer epilimnion of eutrophic Lake Mikołajskie, APase activity varied by a factor 1.8 during 24 hours (Fig. 8). Minimal rates of APase activity were measured during the night period, and the enzyme activity continuously increased from morning to evening. The rates of enzyme activity were positively correlated with fluctuations in chlorophyll$_a$ concentrations, indicating that phytoplankton was a major producer of alkaline phosphatase in an epilimnion of the lake. In contrast to alkaline phosphatase activity, the highest rates of aminopeptidase activity in the surface water layer of eutrophic Lake Głębokie were recorded during the night when the maximal concentrations of the total and dissolved proteins were determined (78).

VI. ROLE OF PHOSPHOHYDROLASES IN PHOSPHORUS CYCLING

Current research clearly shows that orthophosphate ions (Pi) are major factors for controlling microbial primary and secondary production in many freshwater environments (80–

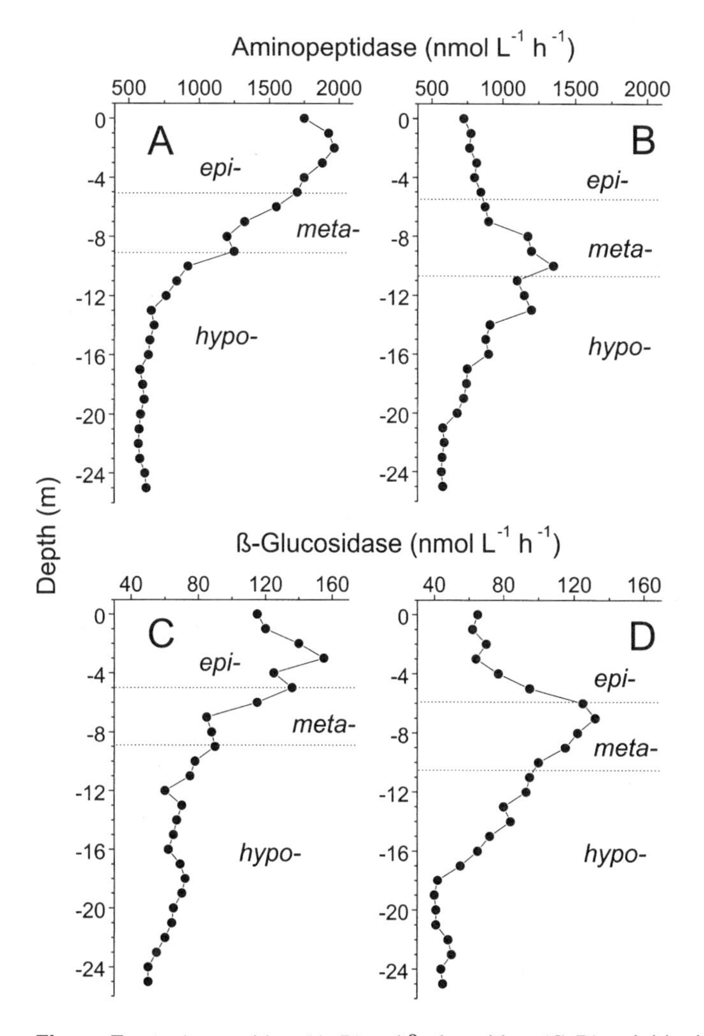

Figure 7 Aminopeptidase (A, B) and β-glucosidase (C, D) activities in the thermal stratified water column of Lake Mikołajskie during summer phytoplankton bloom (A, C) and after bloom breakdown (B, D) in the epilimnion (epi-), metalimnion (meta-), and hypolimnion (hypo-). (Chróst, unpublished.)

85). As confirmed by several independent approaches, the ambient Pi concentration is far too low to meet plankton phosphorus (P) requirements in the euphotic zone of lakes, and therefore most (80–90%) of the P used for production of microbial biomass originates from dephosphorylation of P organic compounds during their degradation.

A variety of aquatic organisms (bacteria, algae, cyanobacteria, protozoa, macrozooplankton, benthic animals, and aquatic angiosperms) release Pi from organic compounds. Although contributions of the last two groups of aquatic organisms to P cycling can be important in some fresh waters (86), these are not discussed here.

In this review, enzymatic microbial cycling of P is defined as the process of dephosphorylation of organic P compounds by hydrolytic enzymes produced by microorganisms that leads to the release of Pi into the environment surrounding microbial cells. This defi-

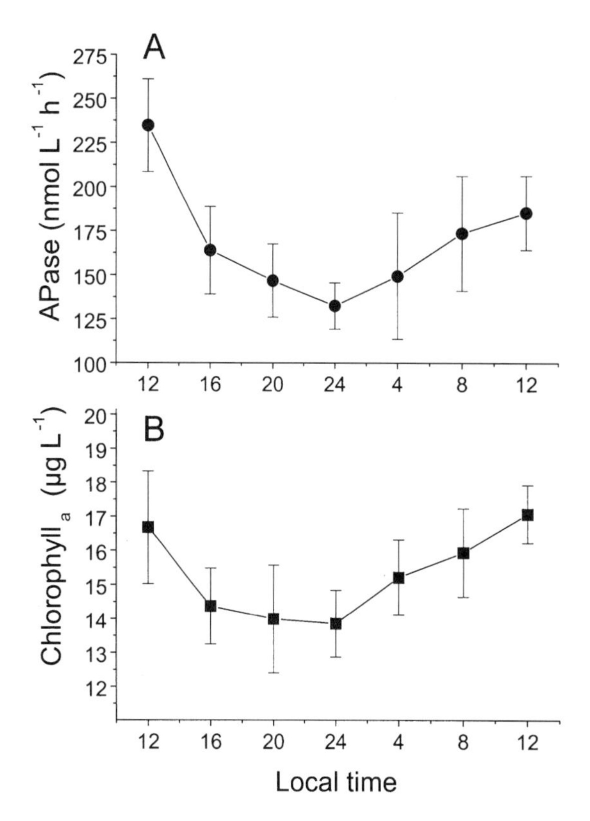

Figure 8 Summer diurnal fluctuations of alkaline phosphatase activity (A) and concentration of chlorophyll$_a$ in the surface water samples (0–0.5 m) from eutrophic Lake Głębokie. (Chróst, unpublished.)

nition excludes the release of Pi and dissolved organic phosphorus (DOP) compounds by zooplankton (87,88). However, since these planktonic animals can affect significantly the whole microbial community as well as dynamics of P compounds in aquatic environments, it is necessary to discuss selected aspects of their influence on enzymatic Pi release.

Only 30%, or less, of the total organic P pool in freshwaters is composed of easily hydrolyzable dissolved or colloidal P constituents (86). The remainder constitutes particulate organic P that is not directly available for microbial metabolism but can be utilized after ingestion of food particles by herbivorous zooplankton. Transformation of particulate P into DOP compounds (incomplete digestion of food particles, zooplankton grazing) effectively accelerates enzymatic Pi release by increasing the substrate pool for phosphohydrolases in an environment (89). One interesting and extensively studied aspect of P recycling by zooplankton entails the production of specific phosphohydrolases by these planktonic animals and the liberation of the intracellular phosphohydrolytic enzymes from grazed phytoplankton (67).

Almost all (except phosphoamides) natural DOP compounds in aquatic environments are chemically stable phosphate esters. Phosphorus in these compounds is not readily available for microorganisms because the majority of phosphate ester molecules cannot be transported directly through microbial cell membranes. Limited quantities of β-glycerophosphate and several phosphorylated monosaccharides can be taken up by some aquatic bacteria (90,91); orthophosphate ions, however, are the dominant forms of phosphorus for microbial assimilation. Therefore microbial utilization of Pi from almost all its organic compounds must be preceded by their enzymatic dephosphorylation (3,13,68).

Release of Pi into aquatic ecosystems is affected by a great variety of abiotic and biotic environmental factors and processes. The most important are activity of phosphohydrolytic enzymes (3,11), zooplankton grazing (92,93), viral and spontaneous lysis of microplankton cells (94,95), and UV light (96). However, it is now well founded that enzyme-mediated hydrolysis of naturally occurring phosphate esters is the most significant mechanism for P release in aquatic environments.

There are three main groups of hydrolytic enzymes responsible for Pi release: nonspecific and/or only partially specific phosphoesterases (mono- and diesterases), nucleotidases (mainly 5′-nucleotidase), and nucleases (exo- and endonucleases). Most of them are typical ectoenzymes (3,67). However, some of phosphohydrolytic enzymes are actively secreted by planktonic microorganisms into surrounding water (e.g., extracellular phosphoesterases and some nucleases) (Fig. 9).

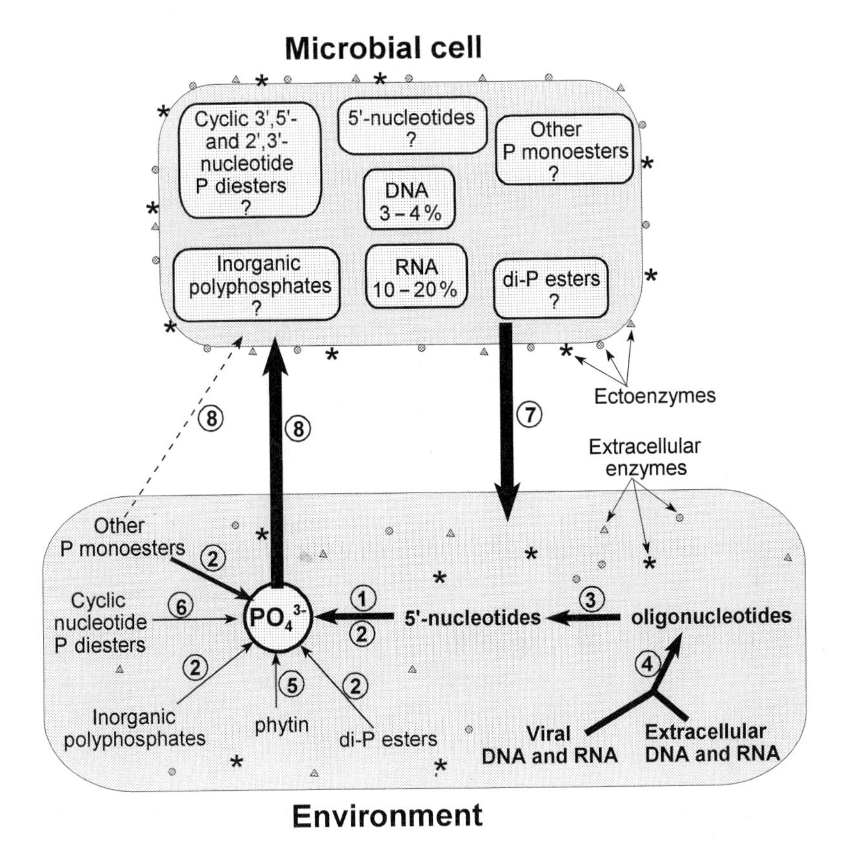

Figure 9 Conceptual model of enzymatic decomposition of various organic phosphorus compounds in lake water. Pathways that are crucial for Pi regeneration in scale of the whole ecosystem are illustrated by bold arrows. ‡@, 5′-nucleotidase; ‡A, alkaline and acid phosphatases; ‡B, exonucleases; ‡C, endonucleases; ‡D, phytase; ‡E, cyclic 3′,5-nucleotide phosphodiesterases and 2′,3-nucleotide phosphodiesterases; ‡F, liberation and release of DOP compounds from disrupted and living cells; ‡G, direct uptake of organic P source. (Siuda, unpublished.)

A. Phosphomonoesterases (Phosphatases)

Phosphatases (nonspecific phosphomonoesterases) are the most intensively studied group of phosphohydrolases participating in Pi release. The presence of active phosphohydrolytic enzymes that were excreted by zooplankton in water samples and their capacity for Pi release from organic P compounds were first mentioned by Steiner in 1938 (97). But the research of Overbeck and Reichardt (60,61,98,99) are the foundations for the current knowledge of the ecological role of phosphatases in aquatic environments. The results of hundreds of studies in the last 40 years have contributed greatly to the present knowledge of phosphatases, which are probably now the best-known phosphohydrolytic enzymes in aquatic ecosystems (3,10–13,26,37,39,55,60–63,68,71,98,99). A variety of alkaline and acid phosphatases are produced by almost all members of the plankton community, including bacteria, algae, cyanobacteria, fungi, protozoa, and zooplankton (11).

Alkaline phosphatases (APases) are the group of adaptative isoenzymes that react optimally in pH range 7.6–9.6. They liberate Pi from monophosphate esters of primary and secondary alcohols, sugar alcohols, cyclic alcohols, phenols, and amines but not from phosphodiesters (100). The rates of APase synthesis are regulated by repression/derepression mechanisms, and Pi acts as a repressor (3,11,13,101). APase activity also is regulated by competitive inhibition Pi (3,11,13).

Since the pH of lakes often is alkaline (pH 7.2–9.5), phosphatases that exhibit their maximal activity in acid water (acid phosphatases [AcPases]) probably have only a minor importance in alkaline lakes. Several studies reported high AcPase activities during hydrolysis of organic P compounds in acidified lakes (55,102). Contrary to APase, acid phosphatases comprise isoenzymes in which synthesis often is not repressed by Pi present in the aquatic environment (103).

Despite considerable knowledge of spatial and temporal changes in potential APase activities in lakes, the quantitative aspects of Pi release mediated in situ by phosphatases are poorly understood. Although APase activities in surface waters of mesotrophic and eutrophic lakes, measured by sensitive fluorometric methods, are commonly high (5–100 nmol PO_4^{3-} L^{-1} and 20–500 nmol PO_4^{3-} L^{-1}, respectively), they do not demonstrate the importance of Pi release in these environments. To estimate the real in situ rates of Pi release by APase, the ambient concentrations of Pi and natural APase substrates (phosphomonoesters [PMEs]) in water samples should be considered.

During the summer stratification period, Pi concentrations in water samples from eutrophic lakes extend over a wide range. They vary from 0 μM in surface samples to 15 μM, or even more, in the hypolimnion (Fig. 10). Concentration and chemical composition of PME in lake water usually are not known. However, some studies reported the concentrations of enzymatically hydrolyzable P (EHP) in aquatic ecosystems (104–107). Assuming that the majority of PME fraction is composed of EHP compounds, one can suppose that the PME pool does not exceed 0.2–0.4 μmol L^{-1}. High PME concentrations usually were found in eutrophic lakes during periods of breakdown of phytoplankton blooms.

Orthophosphate is a well-known competitive inhibitor of APase in aquatic environments (Fig. 11). Therefore, in situ activity of these enzymes is dependent on inhibitor/substrate ([I]/[S]) ratio. More than 80% of APase activity is inhibited when the value of the [I]/[S] ratio in lake water increases above 2.5 (107). Determination of ([I]/[S]) ratios in surface waters of various mesotrophic and eutrophic lakes showed that in the majority

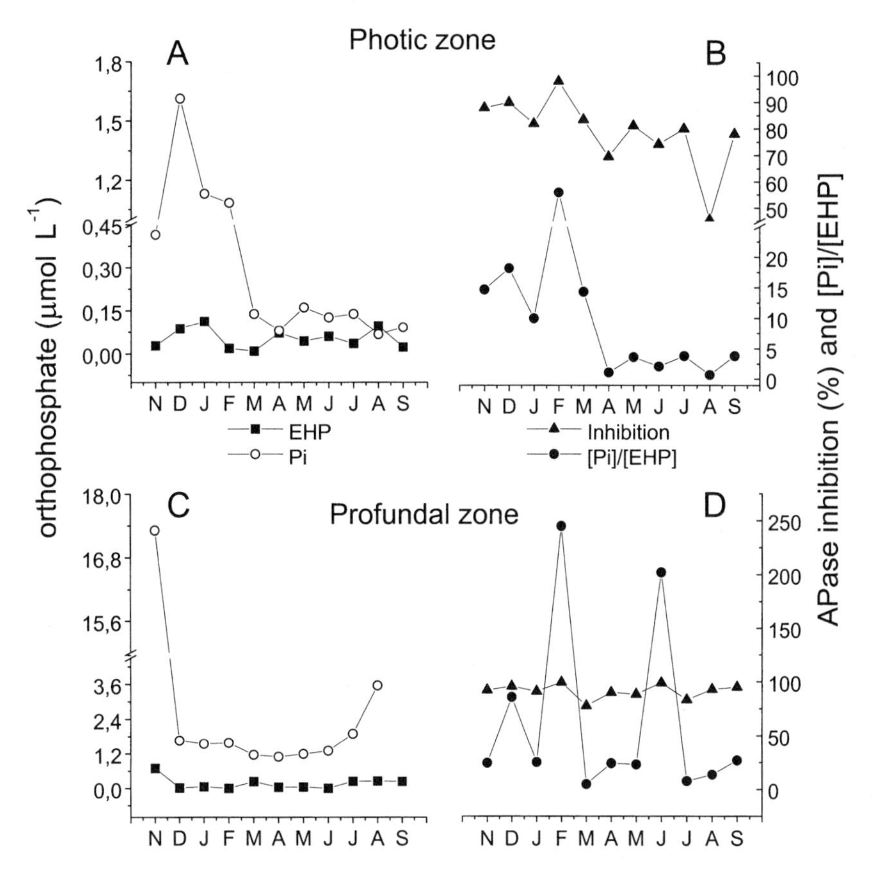

Figure 10 (A, C) Concentrations of orthophosphate ion (Pi) and enzymatically hydrolyzable phosphate (EHP), and (B, D) relationship between [Pi]/[EHP] ratio and percentage of competitive inhibition of alkaline phosphatase activity (APase) in the photic and profundal zone of eutrophic Lake Głębokie. (Data modified from Ref. 38.)

of the studied lakes, the [I]/[S] ratio varied from 3.3 to 29.1 and only occasionally dropped below 1 during the periods of maximal Pi depletion. Similar calculations were made for depth profiles of the lake during a summer stratification period. They showed that surface waters had [I]/[S] ratios that fluctuated around 1 and increased rapidly in the hypolimnion to ~252 (Table 2). These observations strongly suggested that efficient Pi regeneration by APase in deep stratified lakes probably is restricted exclusively to the thin layer of the surface waters and periods of maximal Pi depletion (67). In the water column of the moderately deep lakes (10- to 30-m depth), APase may have only minor importance for the decomposition of organic P compounds.

Phosphatases of lake microplankton are represented by a group of enzymes characterized by different biochemical properties (half-saturation constants, temperature and pH optima, substrate specificity, susceptibility to the presence of activators and/or inhibitors). It should be emphasized that the role of APase in Pi release processes in freshwater ecosystems probably is more complicated than may be expected from simple models of the synthesis and activity regulation based on mechanisms of the repression/derepression and

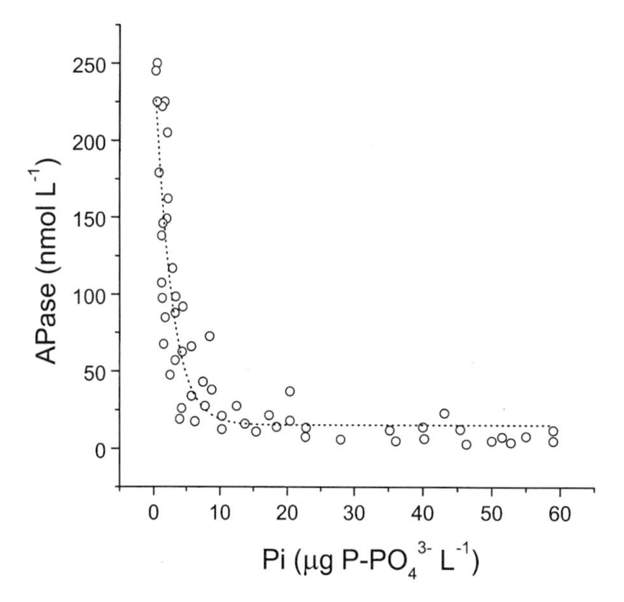

Figure 11 Relationship between alkaline phosphatase activity (APase) and orthophosphate (Pi) concentration in the surface water of eutrophic Lake Głębokie. (Data modified from Ref. 38.)

Table 2 Competitive Inhibition of Alkaline Phosphatase (APase) Activity by Orthophosphate (Pi) in the Water Column of Mesotrophic Lake Constance and Eutrophic Lake Schleinsee

Lake	Depth (m)	Pi (μg P-PO_4^{3-} L^{-1})	EHP (μg P-PO_4^{3-} L^{-1})	Pi/EHP	APase inhibition[a] (%)
Constance					
	2	1.39	1.64	0.85	77
	10	1.80	1.67	1.08	80
	12	1.79	1.05	1.70	82
	20	6.70	2.08	3.22	91
	50	28.49	0.80	35.61	97
	190	59.30	4.53	13.09	94
Schleinsee					
	1	1.30	2.51	0.52	20
	3	1.80	2.20	0.82	40
	5	2.98	0.99	3.01	85
	7	161.51	0.93	173.67	98
	9	297.60	2.08	143.07	96
	12	327.63	1.30	252.03	99

APase, alkaline phosphatase; Pi, orthophosphate ion; EHP, enzymatically hydrolyzable P.
[a]The decrease of APase activity was calculated from enzyme kinetics and inhibitor/substrate ratio. APase was measured by means of methylumbelliferyl-phosphate (MUFP) as a substrate under saturation condition; the affinity of MUFP and EHP to APase was assumed to be the same.
Source: Adapted from Ref. 26.

competitive inhibition by Pi. Information on alternative mechanisms of the regulation of APase activity and synthesis in freshwaters is limited. A few studies described high APase activity in deep oceanic waters in the presence of high (~3.5-μ*M*) Pi concentrations (108,109). Similar observations were found in the profundal zone of deep eutrophic lakes (Siuda, unpublished).

It is commonly believed that APase activity in deep regions of the lakes originates from the surface water and is exported down to the hypolimnion by rapidly sinking particles. There is also some evidence that APase activity in Pi-rich layers of an aquatic ecosystem originates from bacteria producing Pi-resistant APase (109,110). Quantitative participation of bacterial, Pi-resistant APase in Pi release in freshwaters is unknown and needs further intensive investigation. Several studies suggested that bacterial APase, contrary to algal APase that is produced under Pi limitation, has multifunctional properties that can alter both organic P decomposition and C and N mineralization from dissolved organic compounds (13,50,107,109).

B. 5′-Nucleotidase and Nucleases

Free nucleic acids dissolved in water (DNA and ribonucleic acid [RNA]) represent probably the most significant reservoir of P potentially available for planktonic microorganisms in aquatic ecosystems. The distribution of extracellular, dissolved DNA (dDNA) in both freshwater and marine environments is relatively well known. Minear (111) found from 4 to 30 μg dDNA L^{-1} in oligotrophic and eutrophic lowland ponds. Similar dDNA concentrations in various oligotrophic and mesotrophic environments (0.2–44.0 μg L^{-1} and 0.5–25.6 μg L^{-1}, respectively) were documented by later studies (32,112–115). An extremely high content of dDNA (88 μg L^{-1}) was found by Karl and Bailiff (116) in an eutrophic Hawaiian pond. RNA, one of the basic cell components, constitutes about 10–20% of cell biomass; free dissolved RNA, however, has not been found in lake water. It suggests that turnover time of dissolved RNA is extremely short because of its rapid enzymatic hydrolysis by ribonuclease (RNase).

The absolute quantities of Pi release resulting from extracellular nucleic acids are subject to considerable uncertainty as a result of methodological limitations. Nevertheless, they may be important Pi sources in many aquatic environments. Cautious calculations show that enzymatically hydrolyzable extracellular DNA contributes from 10% to 60% of P to the total dissolved organic P pool (32,117). Therefore, enzymatic liberation of Pi from dDNA by DNase and subsequently by 5′-nucleotidase action can be one of the most effective pathways of Pi release.

Three main groups of microplankton enzymes mediate the processes of Pi liberation from nucleic acids: nucleases (exo- and endonucleases), 5′-nucleotidase (5′-nase), and phosphatases. Literature information on nucleases and their activity and distribution among various microplankton components in freshwaters is scarce. Siuda and Güde (117) found that in the epilimnion of Lake Constance, DNase activity was mainly extracellular and/or similar to 5′-nase activity in being coupled to the plankton size fraction (0.1–1.0 μm). In a eutrophic part of Tokyo Bay, Maeda and Taga (118) concluded that DNA-hydrolyzing bacteria were distributed widely in seawater and that DNase activity was bound mainly to suspended particles or microbial cells.

DNA-degrading capacity of microbial communities is usually estimated by measuring the loss of DNA integrity added to water samples in quantities similar to those naturally present in the environment. Kinetic data give only rough approximation of the velocity of nucleic acid degradation in situ. Various independent approaches suggest that half-life of extracellular DNA is relatively short and usually varies between 4.2 and 15 hours in oligotrophic and eutrophic ecosystems (113,117). Considering the relatively high extracellular DNA concentration in aquatic environments and the fact that DNase activity remains relatively unchanged during the whole summer period, constant supplementation of lake waters with nucleotides can be expected (117).

In earlier studies, only APase was regarded as the main enzyme responsible for enzymatic Pi release in natural waters. However, Azam and Hodson (119) showed the potential role of 5′-nase activity in Pi release in marine environments. This bacterial, membrane-bound enzyme is largely specific for various 5′-nucleotides and does not dephosphorylate other phosphate esters. In contrast to activity of APase, the activity of 5′-nase is not dependent on Pi concentrations (67,120,121). Although the function of 5′-nase in Pi release processes in marine environments is relatively well known (67,119,122,123), few papers have discussed the role and importance of this enzyme in fresh waters (3, 117,124,125). Orthophosphate ions released by the action of 5′-nase may be either immediately taken up by bacteria producing this enzyme or mixed with the bulk water. The fate of the released Pi strongly depends on ambient Pi concentration in the environment and on Pi demand of microplankton. In oligotrophic and other Pi-poor waters, release of Pi from 5′-nucleotides by 5′-nase is tightly coupled with its uptake and more than 50% of released Pi can be taken up by the bacteria (3,19). Under high Pi concentrations, in highly polluted waters or in deep regions of the lakes, however, only 10–15% of the Pi resulting from 5′-nase activity is assimilated by microorganisms; an excess of enzymatically liberated Pi mixes with the existing Pi in the bulk phase (67,107,124,125).

Estimation of the quantitative contribution of APase and 5′-nase activities to Pi release into aquatic environments is difficult. Studies of various DOP compounds suggest that P nucleotide generally is assimilated by aquatic bacteria more efficiently than P bound to other phosphate esters. Moreover, comparative studies on kinetic parameters of both enzymes showed that liberation of Pi by 5′-nase appears more efficient than Pi release by Apase, especially in Pi-rich environments (Table 3) (107,117). However, it should be stated that in situ 5′-nase activity may be substantially affected by the rate of 5′-nucleotide

Table 3 Comparison of the Turnover Time of 5′-Nucleotidase Substrates (Adenosine Monophosphate, Adenosine Triphosphate, and Alkaline Phosphatase Substrate glucose-6-phosphate) in Surface Water Samples from Mesotrophic Lake Constance

	5′-Nucleotidase		Alkaline phosphatase
Method	AMP	ATP	G6P
^{32}P	1.11	0.45	12.22
Colorimetric	0.30	2.40	19.20

AMP, adenosine monophosphate; ATP, adenosine triphosphate; G6P, glucose-6-phosphate.
Source: Recalculated data from Ref. 107.

supply (117) and that hydrolysis of extracellular nucleic acids must be regarded as the limiting step for the enzymatic regeneration of Pi from the nucleotide pool.

C. Other Phosphohydrolases

A substantial part of the high-molecular-weight DOP fraction (>10,000 D) is thought to be composed of inositol phosphates bound to proteins, lipids, or fulvic acid. Moreover, some of low-molecular-weight DOP compounds (<1,000 D) in aquatic ecosystems frequently have been identified as inositol phosphates (126–130). These findings suggest that phytase may be another (probably adaptative) phosphohydrolytic enzyme that potentially may be important in Pi release processes. Herbes and associates (127) found that up to 50% of DOP in lake water was hydrolyzed by phytase under its optimal pH conditions. On the other hand, studies of DOP decomposition in the natural pH of lake waters (pH 7.2–8.6) showed that phytase activity decomposed only 15% of the total DOP within 30 days (129). Since phytase has an optimal pH around 5.0 (127), its activity may presumably support Pi release mainly in naturally or artificially acidified environments.

As with phytase, the ecological significance of cyclic nucleotide phosphodiesterase is poorly elucidated. According to Barfield and Francko (69), cyclic nucleotide phosphodiesterase activity in lake water is a result of activities of a group of seasonally different isoenzymes with an optimal pH between 7.0 and 8.0. Although participation of this enzyme in Pi release in lakes seems to be almost insignificant from a qualitative perspective, the cyclic nucleotide phosphodiesterase probably is one of the most important enzymes controlling cAMP concentration in aquatic environments, thus potentially affecting a variety of physiological processes of microplankton mediated by cAMP (42).

D. Enzymatic Release of Pi in Lake Water—Conclusions

Since the amount of readily assimilable Pi in the majority of nonpolluted lakes does not fulfill P requirements for microplankton, it must be released by microorganisms from organic P compounds. For this purpose, aquatic microorganisms developed two main enzymatic Pi release systems (Fig. 9).

The first, adaptative mechanism, activated relatively rapidly (by induction/derepression) during Pi limitation periods, is based on activity of nonspecific phosphomonoesterases and (APases) produced by almost all members of the plankton community. In eutrophic environments, this mechanism is mediated mainly by algal phosphatases. In oligotrophic and mesotrophic ecosystems, however, APase of bacterial origin seems to play a more important role. As APase activity is strongly dependent on dynamically changing Pi concentrations in the environment, its participation in Pi release processes in lake water is restricted to the trophogenic zone of the lake and short intervals of Pi depletion during the summer stratification period.

The second mechanism, exclusively bacterial, involves interaction of various types of nucleases and 5′-nucleotidase that liberate Pi from nucleic acids and from nucleotides. Release of Pi from nucleic acids must be preceded by endo- and exonuclease reactions that liberate 5′-nase substrates (5′-nucleotides). Although 5′-nucleotides can be hydrolyzed by both APase and 5′-nase, it seems that the role of APase in their decomposition is of minor importance. Orthophosphate release is mediated by nucleases and 5′-nase probably is constitutive in the majority of aquatic environments. And those processes are more effective than Pi release by APase because they are not inhibited by Pi in lake water. From this point of view, nucleic acids and their degradation products might be the basic source of released Pi in aquatic ecosystems.

VII. SIGNIFICANCE OF ENZYMES FOR METABOLISM OF LAKE MICROPLANKTON

Currently, there is a large amount of scientific information indicating that besides biotic relations, the diversity of microbial communities, their biomass, and their metabolic activity depend strongly on the composition and biological availability of inorganic nutrients and/or organic substrates in aquatic ecosystems (131–134). It is widely accepted that microorganisms, especially heterotrophic bacteria, play a crucial role in nutrient cycling and organic matter decomposition. The majority of these processes rely on the ectoenzymatic capacity of microorganisms.

The review of literature showed that microbial phosphohydrolases are very important in P mineralization and provision of Pi for aquatic microorganisms. Also, a great variety of organic labile compounds that serve as bacterial substrates are produced in the course of stepwise enzymatic degradation of polymeric organic matter. Heterotrophic bacteria and their hydrolytic enzymatic activities are of the utmost importance in this process. Polymers, in this sense, are biologically the most important nutritional sources. Compounds such as proteins, polysaccharides, lipids, and nucleic acids constitute a major fraction of dissolved organic matter that is hydrolyzed and consequently utilized by aquatic microheterotrophs.

One of the important sources of dissolved organic matter in lake waters arises from enzymatic solubilization of detrital particulate organic matter (POM) (78,135–137). Different microorganisms (bacteria, fungi, protozoa) are known to be closely associated with organic particles, and through stepwise enzymatic hydrolysis of insoluble particulate material, they liberate a substantial portion of solutes (17,138).

There is some evidence that attached bacteria are highly effective in ectoenzyme production. The specific activity of some ectoenzymes, calculated per bacterium, may be 2 to 20 times higher in attached cells than in free-living bacteria (3,19,138). Our studies indicate that the specific activity of β-glucosidase and aminopeptidase produced by bacteria attached to particles (>2 μm) was on average 3.4 and 4.9 times higher, respectively, than the enzyme activity associated with free-living cells of 0.2- to 2.0-μm size fraction. Moreover, the specific uptake rates of glucose (end product of β-glucosidase hydrolysis) and leucine (end product of aminopeptidase catalysis) of attached bacteria were on average 3.6 and 4.2 times lower, respectively, than that of free-living bacteria (3). Similar observations were reported by Jacobsen and Azam (139). Moreover, Azam and Cho (140) suggested that attached bacteria become 'baby machines' and release their progeny in the particle's vicinity, which is enriched by DOM liberated from POM. It appears that colonized POM, through the ectoenzymatic action of attached bacteria, can become a source of both DOM and free-living bacteria.

An interesting but still controversial issue is whether microorganisms hydrolyze organic matter in a biochemically controlled process and assimilate most of the hydrolysis products, ''or whether much of the hydrolysate diffuses into the environment as DOM'' (140). Several studies have demonstrated that ectoenzymatic hydrolysis of dissolved organic polymers and uptake of the low-molecular-weight products of hydrolysis are tightly coupled processes (14,16–17,19,24,58,66,67,108,141). Microorganisms producing ectoenzymes and utilizing the final products of polymer hydrolysis must have very effective "hydrolysis-uptake coupling systems."

The microorganisms that produce ectoenzymes are presumably superior competitors for organic and inorganic nutrients in aquatic environments. Through the production of ectoenzymes, they are capable of utilizing a variety of polymeric compounds, which otherwise cannot be utilized. Ectoenzymes transform nonutilizable compounds of the organic matter pool to readily transportable and utilizable substrates (UDOM) for microorganisms. Because these very efficient degradative enzyme systems are located at the cell surface (142); they make UDOM directly available for substrate-transporting systems.

This gives an increased chance of survival to microorganisms in aquatic environments when readily available energy and nutrient sources become limiting and also enables them to increase their growth and biomass production. The production of ectoenzymes close to the cell, which degrade substrates resistant to attack by other microbial species, may give the producer a nutritional advantage and allow it to dominate an ecosystem. Moreover, bacteria producing the ectoenzymes form a vital part of aquatic ecosystems in which they not only make C, P, and N available to themselves by organic matter degradation, but also interact with algae supplying them with nutrients (13,50).

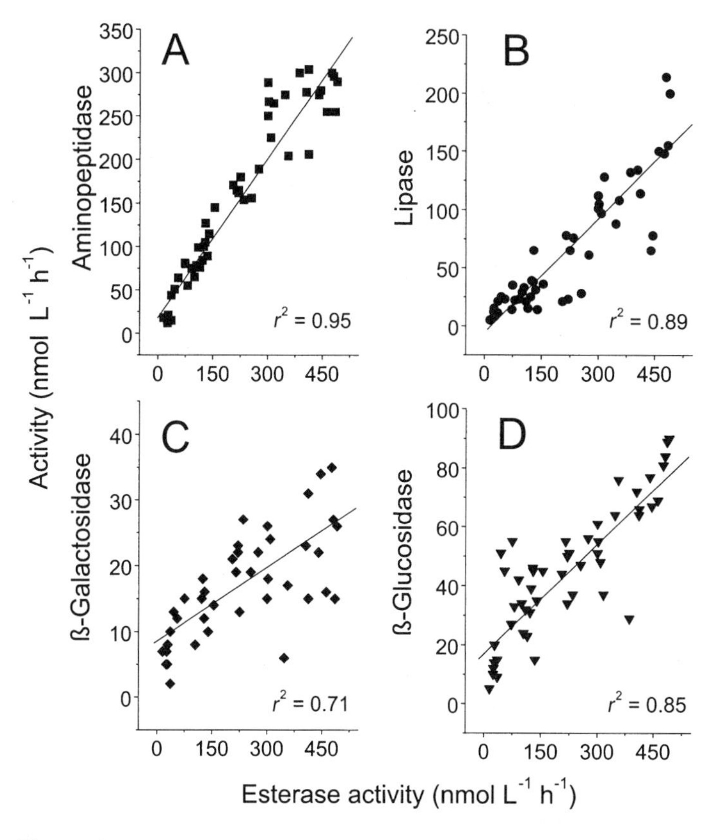

Figure 12 Relationship between activity of aminopeptidase (A), lipase (B), β-galactosidase (C), β-glucosidase (D), and overall esterase activity in lake water samples (Mazurian Lake District, Poland). (Chróst, unpublished.)

Several reports concluded that the measurement of enzymatic activities of microorganisms could be used as a powerful methodological key in aquatic ecology (14,19,21, 36,39,59,136,138). Depending on the enzyme studied, it is now possible to estimate a potential role of an enzyme that is involved in a number of biochemical transformations of organic and inorganic compounds mediated by microorganisms in aquatic ecosystems. Many of the intensively studied ectoenzymes (e.g., 5′-nase, APase, lipase, DNase, several proteolytic enzymes) have displayed their capacity to hydrolyze ester bonds in a variety of natural constituents of organic matter (33). Our studies have shown that activity of microbial esterases in lake water was significantly positively correlated to activities of aminopeptidase, β-glucosidase, β-galactosidase, and lipase (Fig. 12). Microbial ectoenzymes exhibiting esterase activities are involved in many different processes in fresh waters. Therefore, we suggest that esterase activity might be a very useful parameter for the determination of overall potential activity of microbial enzymes in biochemical transformations of organic matter (33).

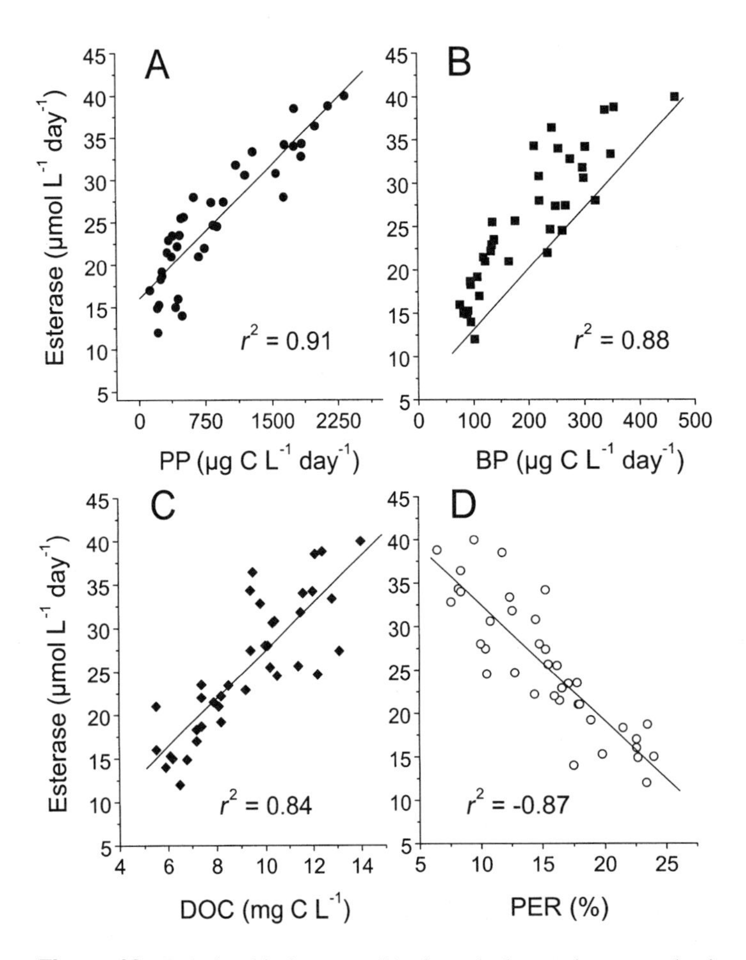

Figure 13 Relationship between (A) phytoplankton primary production (PP), (B) bacterial secondary production (BP), (C) dissolved organic carbon concentration (DOC), (D) percentage of release (PER) of photosynthetic organic carbon by phytoplankton, and overall esterase activity in lake water samples (Mazurian Lake District, Poland). (Chróst, unpublished.)

Currently, it is evident that microorganisms form complex microbial food webs in all aquatic ecosystems, and that their activities and metabolisms often are tightly coupled and/or mutually affected (132,143,144). Therefore, it is not surprising that enzymatic properties and activities of different components creating the microbial food webs in lake ecosystems have demonstrated close relationships. Several reports have documented the strong dependency of bacterial secondary production on ectoenzyme activities of aquatic microorganisms (2–4,16,17,19,25,28,29,33,36,59). There often is a significant correlation between phytoplankton primary production and activities of different ectoenzymes in freshwater ecosystems (25,28,29,33,52).

Our studies in lakes of differing degrees of eutrophication have shown microbial esterase activity to be positively correlated to phytoplankton primary production, bacterial secondary production, and concentration of dissolved organic carbon (DOC) (Fig. 13). We have found a significant negative relationship between enzyme activity and the percentage of phytoplankton extracellular release (PER) of photosynthetic organic carbon in the studied lakes. This negative correlation between PER and esterase activity indicated that enzyme synthesis was partially inhibited in bacteria by low-molecular-weight photosynthetic products of phytoplankton that were readily utilized by these microheterotrophs: i.e., catabolic repression of esterase synthesis was found in lakes characterized by high PER of phytoplankton (29,33).

VIII. ECTOENZYME ACTIVITY AND LAKE WATER EUTROPHICATION

The importance of organic matter as a variable for evaluating the trophic status of lakes has been recognized since the beginning of the 20th century (145,146). Increasing concentrations of organic constituents in water are the distinct indicators of accelerated eutrophication processes in many lakes (147–149). Our studies clearly demonstrated that enzyme activities were significantly positively proportional to DOC content of lakes (Fig. 13C). As described earlier in this chapter, several microbial ectoenzymes are responsible for rapid transformation and degradation of both dissolved organic matter and POM in freshwater ecosystems. Therefore, we hypothesize that an ''enzymatic approach'' can be very useful in the studies of lake eutrophication.

Several reports pointed out that microbial enzymatic activities were closely related to the indices of water eutrophication and/or the trophic status of aquatic ecosystems (25,27,29,31,33,38,52,58,62,78). Our studies along the trophic gradient of lakes (from oligo/mesotrophic to hypereutrophic lakes [Fig. 14A] support our hypothesis (and the assumptions of others) that selected enzymatic microbial activities are very practical for a rapid recognition of the current trophic status of lakes. Activities of alkaline phosphatase, esterase, and aminopeptidase increased exponentially along a trophic gradient and correlated significantly with the trophic state index of the studied lakes (Fig. 14B,C,D). We also found a strong relationship between activities of ectoenzymes and phytoplankton primary production in these lakes. Rapid increases in ectoenzyme activities were observed especially in a range of gradually eutrophic lakes when the value of Carlson's trophic state index (TSI) was above 55 (150) (Fig. 14).

Moreover, the ratio between ectoenzymatic activities and rates of bacterial production gradually increased with the trophic gradient of the lakes studied (29), suggesting that more organic matter had to be decomposed through enzymatic hydrolysis to fulfill bacterial metabolism and to reach a certain level of bacterial production in eutrophic lakes

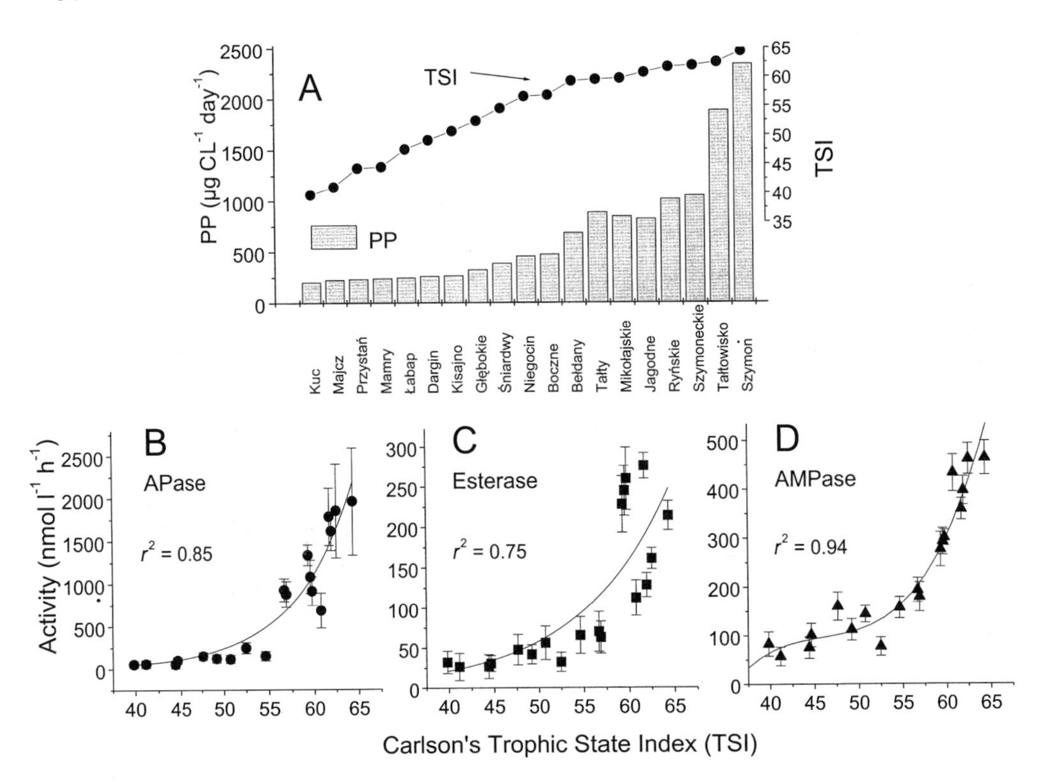

Figure 14 (A) Phytoplankton primary production (PP) and Carlson's trophic state index (TSI) and relationship between (B) activity of alkaline phosphatase (APase), (C) esterase, and (D) aminopeptidase (AMPase) and the trophic state index of lakes of differing degrees of eutrophication (Mazurian Lake District, Poland). (Chróst, unpublished.)

than in mesotrophic systems. Since the percentage of extracellular release of phytoplankton photosynthetic products falls with the degree of eutrophication (49,151), it seems that nutritional requirements of bacteria might have been fulfilled by labile photosynthetic products to a greater extent in oligo/mesotrophic lakes than in eutrophic ecosystems. As a consequence, ectoenzymes in eutrophic lakes play a much more important role in bacterial metabolism than in oligo/mesotrophic environments.

IX. CONCLUSIONS

The concept of the microbial loop (152,153) has generated new information about and stimulated studies of composition and mechanisms of utilization of organic matter in aquatic ecosystems (132,140). From 10% to 50% of the total organic carbon formed by the primary production of phytoplankton is transformed into bacterial biomass (2,25), channeled to the microbial loop, and subsequently transferred to higher trophic levels through the grazing food chains. The metabolic regulations within the trophic levels and their interspecific relationships are the most important mechanisms steering the flow of organic matter in aquatic food webs.

Heterotrophic bacteria operating on every level in aquatic food webs are responsible for the predominant production of a variety of enzymes involved in the release of nutrients and the transfer of C and energy within the whole lake ecosystem. Above all, bacteria are the only single component of aquatic biota that converts the DOM to the particulate phase (bacterial biomass). Moreover, heterotrophic bacteria colonize the detrital POM and, through stepwise ectoenzymatic hydrolysis of insoluble polymeric material, liberate a substantial portion of solutes. Bacteria associated with organic detrital particles channel energy into hyperproduction of hydrolytic ectoenzymes that rapidly solubilize POM and support DOM pool in a microenvironment. Moreover, bacteria, by colonizing detrital POM, enrich the particles with their cell components (proteins, polysaccharides, nucleic acids, lipids, etc.) and significantly increase the nutritional value of the POM when it is consumed by particle-feeding animals (zooplankton, benthic fauna, fishes, etc.). On the other hand, attached bacteria operating on different levels of the aquatic food webs compete with these animals by rapidly solubilizing POM. The particles become a source of DOM that is not accessible to animals.

These unique functions of heterotrophic bacteria, obviously mediated by their high enzymatic activities, are of the utmost significance for the flow and partitioning of organic matter that transmits energy to the higher trophic levels of the lake ecosystems. DOM-utilizing bacteria now are regarded as an important trophic link between the detrital and grazing food chains for energy and nutrient transfer. Heterotrophic bacteria have a very high respiratory potential for oxidation of organic matter, thereby releasing CO_2, PO_4^{3-}, NH_4^+, and other nutrients that are required by phytoplankton. Heterotrophic bacteria, through their enzymatic activity, are important contributors to the steady supply of algal nutrients and promote a steady state of algal biomass and primary production in the absence of other sources of nutrients. When the importance of microbial enzymes in organic matter transformation, degradation, and utilization is considered, it can be concluded that these enzymes operate at the molecular level in aquatic environments, but they affect the function of the entire aquatic ecosystem.

ACKNOWLEDGMENTS

This work was supported by grants 6 P04F 044 11, 6 P04F 049 11, and 6 P04F 030 16 from the Committee for Scientific Research (Poland).

REFERENCES

1. U Münster, RJ Chróst. Origin, composition and microbial utilization of dissolved organic matter. In: J Overbeck, RJ Chróst, eds. Aquatic Microbial Ecology: Biochemical and Molecular Approaches. New York: Springer-Verlag, 1990, pp 8–46.
2. RJ Chróst, H Rai. Bacterial secondary production. In: J Overbeck, RJ Chróst, eds. Microbial Ecology of Lake Plußsee. New York: Springer-Verlag, 1994, pp 92–117.
3. RJ Chróst. Microbial enzymatic degradation and utilization of organic matter. In: J Overbeck, RJ Chróst, eds. Microbial Ecology of Lake Plußsee. New York: Springer-Verlag, 1994, pp 118–174.
4. M Søndergaard. Bacteria and dissolved organic carbon in lakes. In: K Sand-Jensen, O Pedersen, eds. Freshwater Biology. Priorities and Development in Danish Research. Copenhagen: G.E.C. Gad Publishers, 1997, pp 138–161.

5. U Münster. Investigations about structure, distribution and dynamics of different organic substrates in the DOM of Lake Plußsee. Arch Hydrobiol Suppl 70:429–480, 1985.
6. U Münster, D Albrecht. Dissolved organic matter: Analysis of composition and function by a molecular-biochemical approach. In: J Overbeck, RJ Chróst, eds. Microbial Ecology of Lake Plußsee. New York: Springer-Verlag, 1994, pp 24–62.
7. HJ Rogers. The dissimilation of high molecular weight organic substrates. In: IC Gunsalus, RY Stanier, eds. The Bacteria. New York: Academic Press, 1961, pp 261–318.
8. G Bratbak. Production, amount and fate of bacterial biomass in marine waters. PhD dissertation, Bergen University, Bergen, Norway, 1988.
9. LJ Tranvik. Bacterioplankton in humic lakes. PhD dissertation, Lund University, Lund, Sweeden, 1989.
10. GZ Hałemejko, RJ Chróst. The role of phosphatases in phosphorus mineralization during decomposition of lake phytoplankton blooms. Arch Hydrobiol 101:489–502, 1984.
11. W Siuda. Phosphatases and their role in organic phosphorus transformation in natural waters: A review. Pol Arch Hydrobiol 31:207–233, 1984.
12. HG Hoppe. Degradation in sea water. In: HJ Rehm, G Reed, eds. Biotechnology. Weinheim: VCH Verlagsgesellschaft, 1986, pp 453–474.
13. RJ Chróst, J Overbeck. Kinetics of alkaline phosphatase activity and phosphorus availability for phytoplankton and bacterioplankton in Lake Plußsee (north German eutrophic lake). Microb Ecol 13:229–248, 1987.
14. HG Hoppe, SJ Kim, K Gocke. Microbial decomposition in aquatic environments: Combined processes of extracellular enzyme activity and substrate uptake. Appl Environ Microbiol 54: 784–790, 1988.
15. LM Mayer. Extracellular proteolytic enzyme activity in sediments of an intertidal mudflat. Limnol Oceanogr 34:973–981, 1989.
16. RJ Chróst. Significance of bacterial ectoenzymes in aquatic environments. Hydrobiologia 243/244:61–70, 1992.
17. HG Hoppe, H Ducklow, B Karrasch. Evidence for dependency of bacterial growth on enzymatic hydrolysis of particulate organic matter in the mesopelagic ocean. Mar Ecol Prog Ser 93:277–283, 1993.
18. DL Kirchman, J White. Hydrolysis and mineralization of chitin in the Delaware Estuary. Aquat Microb Ecol 18:187–196, 1999.
19. RJ Chróst. Microbial ectoenzymes in aquatic environments. In: J Overbeck, RJ Chróst, eds. Aquatic Microbial Ecology: Biochemical and Molecular Approaches. New York: Springer-Verlag, 1990, pp 47–78.
20. FG Priest. Extracellular Enzymes. Wokingham: Van Nostrand Reinhold, 1984, pp 1–79.
21. HG Hoppe. Significance of exoenzymatic activities in the ecology of brackish water: Measurements by means of methylumbelliferyl-substrates. Mar Ecol Prog Ser 11:299–308, 1983.
22. M Murgier, C Pelissier, A Lazdunski, C Lazdunski. Existence, location and regulation of the biosynthesis of amino-endopeptidase in Gram-negative bacteria. Eur J Biochem 65:517–520.
23. BA Law. Transport and utilization of proteins by bacteria. In: JW Payne, ed. Microorganisms and Nitrogen Sources. New York: John Wiley & Sons, pp 381–409.
24. RJ Chróst. Characterization and significance of β-glucosidase activity in lake water. Limnol Oceanogr 34:660–672, 1989.
25. RJ Chróst, H Rai. Ectoenzyme activity and bacterial secondary production in nutrient-impoverished and nutrient-enriched freshwater mesocosms. Microb Ecol 25:131–150, 1993.
26. W Siuda, H Güde. A comparative study on 5′-nucleotidase (5′-nase) and alkaline phosphatase (APA) activities in two lakes. Arch Hydrobiol 131:211–229, 1994.
27. AJ Gajewski, RJ Chróst. Microbial enzymatic decomposition of β-linked glycosidic biopolymers in four lakes of the Mazurian Lake District, Poland. Pol Arch Hydrobiol 41:405–414, 1994.

28. AJ Gajewski, RJ Chróst. Production and enzymatic decomposition of organic matter by microplankton in a eutrophic lake. J Plankton Res 17:709–728, 1995.
29. AJ Gajewski, RJ Chróst. Microbial enzyme activities and phytoplankton and bacterial production in the pelagial of the Great Mazurian Lakes (north-eastern Poland) during summer stratification. Ekol Pol 43:245–265, 1995.
30. AJ Gajewski, AKT Kirschner, B Velimirov. Bacterial lipolytic activity in a hypertrophic dead arm of the river Danube in Vienna. Hydrobiologia 344:1–10, 1997.
31. R Wcisło, RJ Chróst. Selected enzymatic properties of bacterioplankton in lakes of various degrees of eutrophication. Acta Microbiol Polon 47:203–218, 1997.
32. W Siuda, RJ Chróst, H Güde. Distribution and origin of dissolved DNA in lakes of different trophic states. Aquat Microb Ecol 15:89–96, 1998.
33. RJ Chróst, A Gajewski, W Siuda. Fluorescein-diacetate (FDA) assay for determining microbial esterase activity in lake water. Arch Hydrobiol Spec Issues Advanc Limnol 54:167–183, 1999.
34. RJ Chróst, J Overbeck. Substrate-ectoenzyme interaction: Significance of β-glucosidase activity for glucose metabolism by aquatic bacteria. Arch Hydrobiol Beih Ergeb Limnol 34: 93–98, 1990.
35. RJ Chróst. Ectoenzymes in aquatic environments: Microbial strategy for substrate supply. Int Ver Theor Angew Limnol Verh 24:2597–2600, 1991.
36. M Middelboe, M Søndergaard. Bacterioplankton growth yield: Seasonal variations and coupling to substrate lability and β-glucosidase activity. Appl Environ Microbiol 59:3916–3921, 1993.
37. RJ Chróst. Environmental control of the synthesis and activity of aquatic microbial ectoenzymes. In: RJ Chróst, ed. Microbial Enzymes in Aquatic Environments. New York: Springer-Verlag, 1991, pp 29–59.
38. RJ Chróst, W Siuda, GZ Hałemejko. Long-term studies on alkaline phosphatase activity (APA) in a lake with fish-aquaculture in relation to lake eutrophication and phosphorus cycle. Arch Hydrobiol Suppl 70:1–32, 1984.
39. W Siuda, RJ Chróst. The relationship between alkaline phosphatase (APA) activity and phosphate availability for phytoplankton and bacteria in eutrophic lakes. Acta Microbiol Polon 36:247–257, 1987.
40. AJ Gajewski, RJ Chróst, W Siuda. Bacterial lipolytic activity in an eutrophic lake. Arch Hydrobiol 128:107–126, 1993.
41. JL Botsford. Cyclic nucleotides in prokaryotes. Microbiol Rev 45:620–645, 1981.
42. D Francko. Phytoplankton metabolism and cyclic nucleotides. II. Nucleotide-induced perturbations of alkaline phosphatase activity. Arch Hydrobiol 100:409–421, 1984.
43. JE Little, RE Sjogren, GR Carson. Measurement of proteolysis in natural waters. Appl Environ Microbiol 37:900–908, 1979.
44. RS Boethling. Regulation of extracellular protease secretion in *Pseudomonas maltophilia*. J Bacteriol 123:954–961, 1975.
45. CD Litchfield, JM Prescott. Regulation of proteolytic enzyme production by *Aeromonas proteolytica*. II. Extracellular aminopeptidase. Can J Microbiol 16:23–27.
46. AR Glenn. Production of extracellular proteins by bacteria. Annu Rev Microbiol 30:41–62.
47. MCC Daatselaar, W Harder. Some aspects of the regulation of the production of extracellular proteolytic enzymes by a marine bacterium. Arch Hydrobiol 101:21–34, 1874.
48. JJ Cole, WH McDowell, GE Likens. Sources and molecular weight of dissolved organic carbon in an oligotrophic lake. Oikos 42:1–9, 1984.
49. RJ Chróst. Plankton photosynthesis, extracellular release and bacterial utilization of released dissolved organic carbon (RDOC) in lakes of different trophy. Acta Microbiol Polon 32: 275–287, 1983.
50. RJ Chróst. Algal-bacterial metabolic coupling in the carbon and phosphorus cycle in lakes.

In: F Megusar, M Gantar, eds. Perspectives in Microbial Ecology. Ljubljana: Slovene Soc Microbiol, 1986, pp 360–366.
51. RJ Chróst, MA Faust. Organic carbon release by phytoplankton: Its composition and utilization by bacterioplankton. J Plankton Res 5:477–493, 1983.
52. RJ Chróst, U Münster, H Rai, D Albrecht, PK Witzel, J Overbeck. Photosynthetic production and exoenzymatic degradation of organic matter in euphotic zone of an eutrophic lake. J Plankton Res 11:223–242, 1989.
53. RJ Chróst, J Overbeck, R Wcisło. Evaluation of the [3H]thymidine method for estimating bacterial growth rates and production in lake water: Re-examination and methodological comments. Acta Microbiol Polon 37:95–112, 1988.
54. JR Christian, DM Karl. Ectoaminopeptidase specificity and regulation in Antarctic marine pelagic microbial communities. Aquat Microb Ecol 15:303–310, 1998.
55. M Jansson. Induction of high phosphatase activity by aluminum in acid lakes. Arch Hydrobiol 93:32–44, 1981.
56. JG Rueter. Alkaline phosphatase inhibition by copper: Implications to phosphorus nutrition and use as a biochemical marker of toxicity. Limnol Oceanogr 28:743–748, 1983.
57. GJ Herndl, G Müller-Niklas, J Frick. Major role of ultraviolet-B in controlling bacterioplankton growth in the surface layer of the ocean. Nature 361:717–719, 1993.
58. HG Hoppe, HC Giesenhagen, K Gocke. Changing patterns of bacterial substrate decomposition in a eutrophication gradient. Aquat Microb Ecol 15:1–13, 1998.
59. M Middelboe, M Søndergaard, Y Letarte, NH Borch. Attached and free-living bacteria: Production and polymer hydrolysis during a diatom bloom. Microb Ecol 29:231–248, 1995.
60. J Overbeck. Die Phosphatasen von *Scenedesmus quadricauda* und ihre ökologische Bedeutung. Int Ver Theor Angew Limnol Verh 14:226–231, 1961.
61. J Overbeck, HD Babenzien. Nachweis von freien Phosphatasen, Amylase und Saccharase im Wasser eines Teiches. Naturwissenschaften 50:571–572, 1963.
62. JG Jones. Studies on freshwater microorganisms: Phosphatase activity in lakes of differing degrees of eutrophication. J Ecol 60:777–791.
63. D Wynne, M Gophen. Phosphatase activity in freshwater zooplankton. Oikos 37:369–376, 1981.
64. RJ Chróst, HJ Krambeck. Fluorescence correction for measurements of enzyme activity in natural waters using methylumbelliferyl-substrates. Arch Hydrobiol 106:79–90, 1986.
65. U Münster, P Einio, J Nurminen. Evaluation of the measurements of extracellular enzyme activities in a polyhumic lake by means of studies with 4-methylumbelliferyl-substrates. Arch Hydrobiol 115:321–337, 1989.
66. JT Hollibaugh, F Azam. Microbial degradation of dissolved proteins in seawater. Limnol Oceanogr 28:1104–1116, 1983.
67. JW Ammerman, F Azam. Bacterial 5′-nucleotidase in aquatic ecosystems: A novel mechanism of phosphorus regeneration. Science 227:1338–1340, 1985.
68. RJ Chróst. Phosphorus and microplankton development in an eutrophic lake. Acta Microbiol Polon 37:205–225, 1988.
69. J Barfield, DA Francko. Cyclic nucleotide phosphodiesterase activity in epilimnetic lake water. In: RJ Chróst, ed. Microbial Enzymes in Aquatic Environments. New York: Springer-Verlag, 1991, pp 239–248.
70. SE Jones, MA Lock. Peptidase activity in river biofilms by product analysis. In: RJ Chróst, ed. Microbial Enzymes in Aquatic Environments. New York: Springer-Verlag, 1991, pp 144–154.
71. T Berman. Alkaline phosphatases and phosphorus availability in Lake Kinneret. Limnol Oceanogr 15:663–674, 1970.
72. WT Frankenberger, AJB Johanson. Use of plasmolytic agents and antiseptics in soil enzyme assays. Soil Biol Biochem 18:209–214, 1986.

73. RP Haugland. Handbook of Fluorescent Probes and Research Chemicals. 6th ed. Leiden: Molecular Probes Europe BV, 1996, pp 201–250.
74. FB Armstrong. Biochemistry. 2nd ed. New York: Oxford University Press, 1983, pp 129–162.
75. JE Dowd, DS Riggs. A comparison of estimates of Michaelis-Menten kinetic constants from various linear transformations. J Biol Chem 240:863–869, 1965.
76. A Lundin, P Arner, J Hellmer. A new linear plot for standard curves in kinetic substrate assays extended above the Michaelis-Menten constant: Application to a luminometric assay of glycerol. Anal Biochem 177:125–131, 1989.
77. RJ Leatherbarrow. Enzfitter. A Non-Linear Regression Data Analysis Program for the IBM PC. Cambridge: Elsevier Biosoft, pp 1–91.
78. GZ Halemejko, RJ Chróst. Enzymatic hydrolysis of proteinaceous particulate and dissolved material in an eutrophic lake. Arch Hydrobiol 107:1–21, 1986.
79. W Siuda, RJ Chróst, R Wcisło, M Krupka. Factors affecting alkaline phosphatase activity in a lake (short-term experiments). Acta Hydrobiol 24:3–20, 1982.
80. RA Vollenveider. Input-output models with special reference to the phosphorus loading concepts in limnology. Schweiz Z Hydrol 37:58–84, 1975.
81. DW Schindler. Evolution of phosphorus limitation in lakes. Science 195:260–262, 1977.
82. DW Schindler. Predictive eutrophication models. Limnol Oceanogr 23:1080–1081, 1978.
83. PJ Dillon, FH Rigler. The phosphorus chlorophyll relationship in lakes. Limnol Oceanogr 19:767–773, 1974.
84. PJ Dillon, FH Rigler. A test of simple nutrient budget model predicting the phosphorus concentration in lake water. J Fish Res Board Can 31:1771–1778, 1974.
85. HW Pearl, DRS Lean. Visual observations of phosphorus movement between algae, bacteria, and abiotic particles in lake waters. J Fish Res Board Can 33:2805–2813, 1976.
86. RG Wetzel. Limnology. 2nd ed. Philadelphia: Saunders, 1975, pp 255–297.
87. OK Andersen, JC Goldman, DA Caron, MR Dennett. Nutrient cycling in a microflagellate food chain. III. Phosphorus Dynamics. Mar Ecol Prog Ser 31:47–55, 1986.
88. H Güde. Influence of phagotrophic processes on the regeneration of nutrients in two-stage continuous culture system. Microb Ecol 11:193–204, 1985.
89. R Peters, DRS Lean. The characterization of soluble phosphorus released by limnetic zooplankton. Limnol Oceanogr 18:270–279, 1973.
90. M Argast, W Boos. Co-regulation in *Escherichia coli* of a novel transport system for sn-glycerol-3-phosphate and outer membrane protein Ic (e, E) with alkaline phosphatase and phosphate binding protein. J Bacteriol 143:142–150, 1980.
91. RT Heath, AC Edinger. Uptake of 32P-phosphoryl from glucose-6-phosphate by plankton in an acid bog lake. Int Ver Theor Angew Limnol Verh 24:210–213, 1990.
92. JT Lehman. Release and cycling of nutrients between planktonic algae and herbivores. Limnol Oceanogr 25:620–632, 1980.
93. JT Lehman. Nutrient recycling as an interface between algae and grazers in freshwater communities. In: WC Kerfood, ed. Evolution and Ecology of Zooplankton Communities. Hannover: University Press, 1980, pp 251–263.
94. G Bratbak, M Heldal, S Norland, TF Thingstad. Viruses as partners in spring bloom microbial trophodynamics. Appl Environ Microbiol 56:1400–1405, 1990.
95. W Reisser, S Grein, C Krambeck. Extracellular DNA in aquatic ecosystems may in part be due to phycovirus activity. Hydrobiologia 252:199–201, 1993.
96. MN Scully, DRS Lean. The attenuation of ultraviolet radiation in temperate lakes. Arch Hydrobiol Beih Ergebn Limnol 43:135–144, 1994.
97. M Steiner. Zur Kenntnis des Phosphatkreislaufes in Seen. Naturwissenschaften 26:723–724, 1938.
98. W Reichardt, J Overbeck, L Steubing. Free dissolved enzymes in lake water. Nature 216: 1345–1347, 1967.

99. J Overbeck. Early studies on ecto- and extracellular enzymes in aquatic environments. In: RJ Chróst, ed. Microbial Enzymes in Aquatic Environments. New York: Springer-Verlag, 1991, pp 1–5.
100. J Feder. The phosphatases. In: EJ Griffith, A Beeton, JM Spencer, DT Mitchell, eds. Environmental Phosphorus Handbook. New York: John Wiley & Sons, 1973, pp 475–508.
101. A Torriani. Influence of inorganic phosphate in the formation of phosphatases by *Escherichia coli.* Biochim Biophys Acta 38:460–469, 1960.
102. H Olsson. Phosphatases in lakes—characterization, activity and ecological implications. PhD dissertation, Uppsala University, Uppsala, Sweeden, 1988.
103. MH Kuo, HJ Blumenthal. Absence of phosphatase repression by inorganic phosphate in some microorganisms. Nature 190:29–31, 1961.
104. RT Heath, GD Cooke. The significance of alkaline phosphatase in a eutrophic lake. Int Ver Theor Angew Limnol Verh 19:959–965, 1975.
105. L Solorzano. Soluble fractions of phosphorus compounds and alkaline phosphatase activity in Loch Creran and Loch Etive, Scotland. J Exp Mar Biol Ecol 34:227–232, 1978.
106. RJ Chróst, W Siuda, D Albrecht, J Overbeck. A method for determining enzymatically hydrolysable phosphate (EHP) in natural waters. Limnol Oceanogr 31:662–667, 1986.
107. W Siuda, H Güde. The role of phosphorus and organic carbon compounds in regulation of alkaline phosphatase activity and P regeneration processes in eutrophic lakes. Pol Arch Hydrobiol 41:171–187, 1994.
108. HG Hoppe. Relations between bacterial extracellular enzyme activity and heterotrophic substrate uptake in a brackish water environment. GERBAM Deuxième Colloque de Bacteriology Marine CNRS, IFREMER, Actes Colloq 3:119–128, 1986.
109. HG Hoppe, S Ulrich. Profiles of ectoenzymes in the Indian Ocean: Phenomena of phosphatase activity in the mesopelagic zone. Aquat Microb Ecol 19:139–148, 1999.
110. N Taga, H Kobori. Phosphatase activity in eutrophic Tokyo Bay. Mar Biol 49:223–229, 1978.
111. RA Minear. Characterization of naturally occurring dissolved organophosphorus compounds. Environ Sci Technol 6:431–437, 1972.
112. MF DeFlaun, JH Paul, D Davis. Simplified method for dissolved DNA determination in aquatic environments. Appl Environ Microbiol 52:654–659, 1986.
113. JH Paul, WH Jeffrey, AW David, MF DeFlaun, LH Cezares. Turnover of extracellular DNA in eutrophic and oligotrophic freshwater environments of southwest Florida. Appl Environ Microbiol 55:1823–1828, 1989.
114. A Maruyama, M Oda, T Higashihara. Abundance of virus-sized non-DNase-digestible DNA (coated DNA) in eutrophic seawater. Appl Environ Microbiol 59:712–716, 1993.
115. MG Weinbauer, D Fuks, S Puskatic, P Peduzzi. Diel, seasonal, and depth-related variability of viruses and dissolved DNA in the northern Adriatic Sea. Microb Ecol 30:25–41, 1995.
116. DM Karl, MD Bailiff. The measurement and distribution of dissolved nucleic acids in aquatic environments. Limnol Oceanogr 34:543–558, 1989.
117. W Siuda, H Güde. Evaluation of dissolved DNA and nucleotides as a potential source of phosphorus for plankton organisms in Lake Constance. Arch Hydrobiol Spec Issues Adv Limnol 48:155–162, 1996.
118. M Maeda, N Taga. Occurrence and distribution of deoxyribonucleic acid-hydrolyzing bacteria in sea water. J Exp Mar Biol Ecol 14:157–169, 1974.
119. F Azam, RE Hodson. Dissolved ATP in the sea and its utilization by marine bacteria. Nature 267:696–698, 1977.
120. C Bengis-Garber, DJ Kushner. Purification and properties of 5′-nucleotidase from the membrane of *Vibrio costicola*, a moderately halophilic bacterium. J Bacteriol 146:24–32, 1981.
121. C Bengis-Garber, DJ Kushner. Role of membrane-bound 5′-nucleotidase in nucleotide uptake by the moderate halophile *Vibrio costicola*. J Bacteriol 149:808–815, 1982.

122. T Taminen. Dissolved organic phosphorus regeneration by bacterioplankton: 5′-nucleotidase activity and subsequent phosphate uptake in mesocosm enrichment experiment. Mar Ecol Prog Ser 58:89–100, 1989.
123. TF Thingstad, EV Skjodal, RA Bohne. Phosphorus cycling and algal bacterial competition for phosphorus in Sadsfjord, Western Norway. Mar Ecol Prog Ser 99:239–259, 1993.
124. JB Cotner, RG Wetzel. 5′-Nucleotidase activity in a eutrophic lake. Appl Environ Microbiol 57:1306–1312, 1991.
125. JB Cotner, RG Wetzel. Uptake of dissolved inorganic and organic phosphorus compounds by phytoplankton and bacterioplankton. Limnol Oceanogr 37:232–243, 1992.
126. SJ Eisenreich, DE Armstrong. Chromatographic investigation of inositol phosphate esters in lake waters. Environ Sci Technol 11:497–501, 1977.
127. SE Herbes, HE Allen, RH Mancy. Enzymatic characterization of soluble organic phosphorus in lake water. Science 187:432–434, 1975.
128. DA Francko, RT Heath. Functionally distinct classes of complex phosphorus compounds in lake water. Limnol Oceanogr 24:463–473, 1979.
129. JE Cooper, J Early, AJ Holding. Mineralization of dissolved organic phosphorus from a shallow eutrophic lake. Hydrobiologia 209:89–94, 1991.
130. RJ Stevens. Different forms of phosphorus in freshwater. Anal Proc 17:375–376, 1980.
131. T Berman, C Béchemin, SY Maestrini. Release of ammonium and urea from dissolved organic nitrogen in aquatic ecosystems. Aquat Microb Ecol 16:295–302, 1999.
132. J Pinhassi, F Azam, J Hemphälä, RA Long, J Martines, UL Zweifel, Ao Hagström. Coupling between bacterioplankton species composition, population dynamics, and organic matter degradation. Aquat Microb Ecol 17:13–26, 1999.
133. E Granéli, P Carlsson, JT Turner, PA Tester, C Béchemin, R Dawson, E Funari. Effects of N : P : Si ratios and zooplankton grazing on phytoplankton communities in the northern Adriatic Sea. I. Nutrients, phytoplankton biomass, and polysaccharide production. Aquat Microb Ecol 18:37–54, 1999.
134. F Touratier, L Legendre, A Vézina. Model of bacterial growth influenced by substrate C : N ratio and concentration. Aquat Microb Ecol 19:105–118, 1999.
135. RL Sinsabaugh, AE Linkins. Enzymic and chemical analysis of particulate organic matter from a boreal river. Freshwater Biol 23:301–309, 1990.
136. RL Sinsabaugh, MP Osgood, S Findlay. Enzymatic models for estimating decomposition rates of particulate detritus. J North Am Benthol Soc 13:160–169, 1994.
137. CR Jackson, CM Foreman, RL Sinsabaugh. Microbial enzyme activities as indicators of organic matter processing rates in a Lake Erie coastal wetland. Freshwater Biol 34:329–342, 1995.
138. HG Hoppe. Microbial extracellular enzyme activity: A new key parameter in aquatic ecology. In: RJ Chróst, ed. Microbial Enzymes in Aquatic Environments. New York: Springer-Verlag, 1991, pp 60–83.
139. TR Jacobsen, F Azam. Role of bacteria in copepod fecal pellet decomposition: Colonization, growth rates and mineralization. Bull Am Sci 35:495–502, 1985.
140. F Azam, BC Cho. Bacterial utilization of organic matter in the sea. In: M Fletcher, TRG Gray, JG Jones, eds. Ecology of Microbial Communities. Cambridge: Cambridge University Press, 1987, pp 261–281.
141. M Somville, G Billen. A method for determining exoproteolytic activity in natural waters. Limnol Oceanogr 28:190–193, 1983.
142. J Martinez, F Azam. Periplasmic aminopeptidase and alkaline phosphatase activities in a marine bacterium: Implications for substrate processing in the sea. Mar Ecol Prog Ser 92: 89–97, 1993.
143. B Riemann, K Christoffersen. Microbial trophodynamics in temperate lakes. Marine Microbial Food Webs 7:69–100, 1993.
144. JT Turner, JC Roff. Trophic levels and trophospecies in marine plankton: Lessons from the microbial food web. Marine Microbial Food Webs 7:225–248, 1993.

145. EA Brige, C Juday. Organic content of lake water. Bull Bur Fish 42:185–204, 1926.
146. EA Birge, C Juday. Particulate and dissolved organic matter in inland lakes. Ecol Monogr 4:440–474, 1934.
147. R Bertoni, C Callieri. Organic carbon trend during the oligotrophication of Lago Maggiore. Mem Istit Ital Idrobiol 52:191–205, 1993.
148. GE Likens. Eutrophication and aquatic ecosystems. Limnol Oceanogr Spec Symp 1:3–13, 1972.
149. B Riemann, M Søndergaard. Carbon Dynamics in Eutrophic, Temperate Lakes. Amsterdam: Elsevier, 1986.
150. RE Carlson. A trophic state index for lakes. Limnol Oceanogr 22:261–269, 1977.
151. U Larsson, Å Hagström. Fractionated phytoplankton primary production, exudate release and bacterial production in a Baltic eutrophication gradient. Mar Biol 67:57–70, 1982.
152. F Azam, T Fenchel, JG Field, JS Gray, LA Meyer-Reil, F Thingstad. The ecological role of water-column microbes in the sea. Mar Ecol Prog Ser 10:257–263, 1983.
153. Ao Hagström, F Azam, A Anderson, J Wikner, F Rassoulzadegan. Microbial loop in an oligotrophic pelagic marine ecosystem: Possible roles of cyanobacteria and nannoflagellates in the organic fluxes. Mar Ecol Prog Ser 49:171–178.

3

Ecological Significance of Bacterial Enzymes in the Marine Environment

Hans-Georg Hoppe
Institute of Marine Science, Kiel, Germany

Carol Arnosti
University of North Carolina, Chapel Hill, North Carolina

Gerhard F. Herndl
Netherlands Institute of Sea Research (NIOZ), Den Burg, The Netherlands

I. INTRODUCTION

The general biochemical role of extracellular enzymes in the sea is similar to that in other aquatic environments. However, in order to understand the ecological significance of enzymes in the sea specific marine environmental factors have to be considered, *e.g.*, a) seawater is a highly diluted medium interspersed with hot spots of organic matter concentration, aggregation, and decomposition (1). b) as a result of hydrographic conditions, the oceans are characterized by distinct horizontal and vertical zonations. The deep sea, in particular, with its enormous volume, depends entirely on substrate supply from sinking material produced in the surface layer (2). c) in the sea, low-molecular-weight organic matter that persists as dissolved organic carbon (DOC) is less bioreactive and more strongly diagenetically altered than the bulk of high-molecular-weight matter (3,4). d) the deep seabed receives very little of the total surface-derived primary productivity, and much of this organic matter is strongly altered. Finally, at chemical and hydrographic discontinuities in the sea (e.g., fronts, boundary layers, and oxygen, nutrient, and salinity gradients), drastic changes occur in microbial species diversity and enzymatic properties.

Organic material, prone to bacterial degradation on nongeological time scales, is actively involved in biogeochemical cycles. Such material originates from phytoplankton primary production and from 'sloppy' feeding of zooplankton as well as from the excretions of all kinds of organisms. Currently, the spectrum of extracellular enzymes investigated in the sea is relatively limited, comprising principally hydrolytic enzymes such as proteases, glucosidases, chitinase, lipase, and phosphatase. A larger variety has been investigated in limnetic systems (5).

The principal focus of this chapter is on the ecological significance of extracellular enzymes in marine waters and sediments ranging from microscales to oceanwide scales.

Investigations of extracellular enzymes from marine animals and enzymes isolated from prokaryotes are considered only if a clear connection to marine ecology is established. The term *extracellular enzymes* is used throughout this chapter, whereas Chróst (5) distinguishes between ectoenzymes and extracellular enzymes. *Ectoenzymes* are defined by Chróst (5) and in Chapter 2 as enzymes located in the periplasmic space or attached to the outer membrane of the bacterial cell. Extracellular enzymes are enzymes freely dissolved in the water or attached to particles other than the enzyme-synthesizing cell. In this chapter, however, the term *extracellular enzymes* refers to both ectoenzymes and extracellular enzymes, unless otherwise stated.

Early studies on the fate of organic aggregates and dissolved polymers in the sea were presented by Riley (6), Walsh (7), and Khailov and Finenko (8). Overbeck (9) reviewed the early studies on extracellular enzyme activity in the aquatic environment.

II. ECOLOGICAL PRINCIPLES OF ENZYMATIC PATTERNS IN THE SEA

A. The Concept of the Microbial Loop and the Role of Extracellular Enzymes

The microbial loop (10) encompasses the combined activities of autotrophic and heterotrophic—eukaryotic as well as prokaryotic—organisms smaller than 20 μm. These organisms, represented by bacteria, nanoflagellates, ciliates, and phototrophic prochlorophytes, as well as cyanobacteria, form a food web of their own, loosely connected to the food web of the larger grazers. In general, the nutritional basis of the microbial food web is provided by the pool of dissolved organic matter (DOM) and particulate organic matter (POM). The DOM pool is a priori reserved for bacterial utilization, whereas competition with metazoans occurs for POM. This competition is determined by the bacterial potential for enzymatic dissolution of POM on the one hand and the feeding activity of the metazoans on the other hand. The bulk of both the dissolved and particulate resources, however, requires enzymatic hydrolysis prior to uptake by bacteria (Fig. 1). Thus the enzymatic activities of bacteria initiate organic carbon (C) remineralization and define the type and quantity of substrate available to the total microbial food web and, to certain extent, also to the top predators in the system.

B. Free and Attached Enzyme Activity

Generally, extracellular enzymes may be bound to the cell (defined as ectoenzymes by Chróst [5]) or in the free and adsorbed state (11,12). Most of the total enzyme activity in seawater has been found to be associated with the particle size class dominated by bacteria (>0.2 μm–3μm) (13,14) (Table 1). Dissolved enzymes (15) and large particles >8 μm generally contribute only minor parts to the total enzyme activity. In estuaries, however, which are characterized by strong gradients and fluctuations of turbidity and salinity, total enzyme activity can be dominated by particle size classes >3 μm. Enzyme activity measured in such particles originated mainly from attached bacteria, leading to the conclusion that particle-attached bacteria accounted for most of POM degradation in these estuaries (16). Another example of the dominance of particle-associated enzyme activity is marine snow. Although bacterial production was not enhanced on snow particles, enzyme activities (α- and β-glucosidases, leucine amino peptidase) of marine snow–

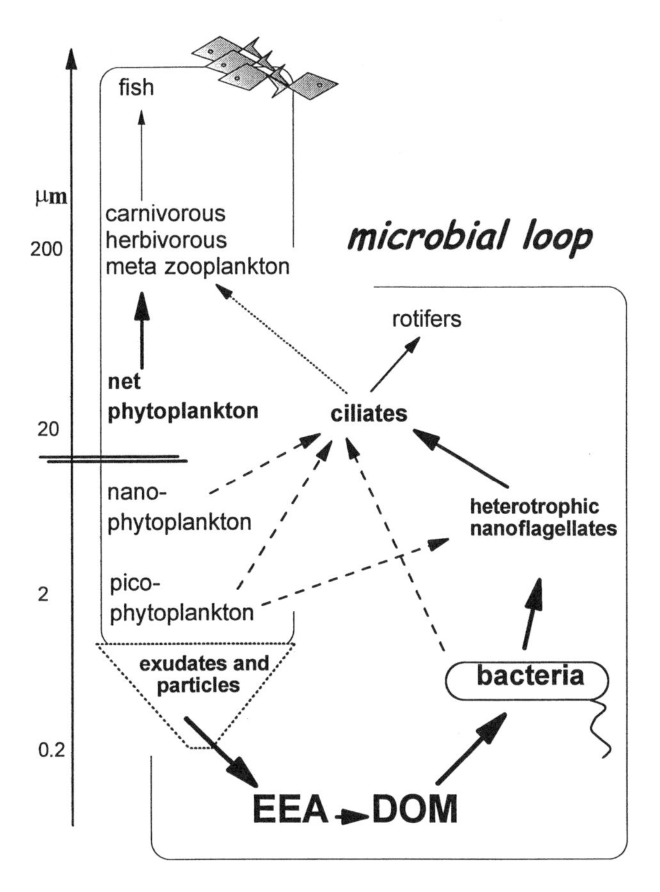

Figure 1 Extracellular enzyme activity (EEA), the initial step of the microbial loop. The enzymatic conversion of particulate organic matter and macromolecular exudates to dissolved organic matter (DOM) triggers the microbial loop. Arrows indicate the pathways of degradation, grazing, and predation.

attached bacteria were significantly higher than those of free-living bacteria, in terms of both absolute and per-cell rates (17). Similar observations, with respect to the relationship between enzyme activities and amino acid incorporation of particle-associated bacteria in the San Francisco Bay, were reported by Murrell, et al. (18). The enzyme activities of bacteria associated with the recently explored transparent exopolymer particles (TEPs)-which can harbor 2% to 25% of total bacteria in the sea (19,20)-have not yet been examined.

Significant differences in the extracellular enzyme activity per cell between particle-attached and free-living bacteria frequently have been reported although these differences are not always observed and most likely depend on the quality and composition of the particles as well as on the nature of the colonizing bacteria (17,21–26) (Table 2). Metabolically active particle-attached bacteria commonly have a larger polysaccharidic capsule than do free-living bacteria (27). Because the majority of extracellular enzymes are embedded in the capsular envelope of metabolically active bacteria, a larger capsule potentially could harbor a greater quantity of enzymes. It also has been observed that the capsular

Table 1 Particle-Attached and Free (<0.2-µm) Extracellular Enzyme Activity of Different Enzymes in Different Habitats

Environment	Conditions	Enzyme, substrate	Size class (µm)	Percentage of total activity	Size class (µm)	Percentage of total activity	Size class (µm)	Percentage of total activity	Reference
Estuary	Salinity-turbidity gradient	leu-AMP			<3	~8–23	>3	~77–92	(16)
		β-glucosidase			<3	~5 to ~50	>3	~50 to ~95	
Northern Adriatic Sea	Zooplankton enzyme release	leu-AMP	<0.2	60–86					(29)
		α-glucosidase		4–11					
		β-glucosidase		0.6–10					
California Bight	Surface sea water	leu-AMP	<0.2	0.2	<1	40–80	>1	~20–60	(37)
North Sea		leu-AMP	<0.2	0–30					(15)
Santa Monica Basin	Above 100 m	leu-AMP	<0.2	<30	0.2–0.8	70–75			(14)
San Francisco Bay	Spring and summer	leu-AMP			>1	47–76 av. 65			(18)
		β-glucosidase			>1	15–87 av. 56			
Kiel Fjord,	Mesotrophic	phosphatase	<0.2	33	0.2–3	14	3–150	53	(219)
		chitobiase	<0.2	4	0.2–3	96	3–150	0	
Northern Red Sea	Oligotrophic	phosphatase	0.2–2	50–71	2–20	12–27	>20	3–37	(77)

Table 2 Specific Extracellular Enzyme Activity per Bacterial Cell (amol cell^{-1} h^{-1}) of Enzymes in Different Habitats

Environment	Conditions	leu-AMP	Lipase	P-ase	α-Glucosidase	β-Glucosidase	Chitobiase	Reference
Trophic gradient	Eutrophic	31.6	10.3		1.69	0.18		(113)
	Mesotrophic						8.1–9.3	
	Oligotrophic	75.6	35.1		0.4	0.06		
Santa Monica Basin	Oligotrophic	~78–618						(14)
Adriatic Sea	Marine snow	432–4996			7–40	6–140		(17)
Selected aggregates	Experimental	av. 242 ± 493						(40)
Seawater		av. 52.5 ± 15						
California Bight	44 Isolates from marine sources	4–3810	0.2–584	0.7–410	0–8	0–35	0–559	(148)
Baltic Sea	Tank incubations			2–14				(220)
Baltic Sea	Summer	0.3–5			0.1–3.3	0.2–3.7	0.7–3.3	(115)
	Autumn	20–237			0.1–1	0.2–2	0.5–5.8	
Arabian Sea	Euphotic zone	6.6–23.2		0.4–3.6		0.16–0.22		(61)
	Deep water	33–118		5.6–23.4		0.27–1.18		
Oman coast, upwelling	Euphotic zone	12.6–46.9		1.2–8.3		0.02–1.2		(61)
	Deep water	455–1817		10.8–86.2		7.7–52.5		
Coastal lagoon	Hypertrophic	LT 218–478				6.9–25.0		(221)
		HT 188–625				5.4–12.9		
San Francisco Bay	Cells <1 μm	7.2–12				0.16–0.57		(18)
	All cells	16–31				0.47–1.60		
Uranouchi Inlet, Japan	Surface water	23.2–1017						(114)
	Bottom water	21.1–270						

LT, low tide; HT, high tide.

layer is continuously renewed by the bacteria (28). Metabolically inactive bacteria, in contrast, are usually devoid of a capsule (27), and a larger fraction of active bacteria has been found in marine snow than in free-living bacteria (27). Consequently, a higher proportion of active, particle-associated bacteria might result in an overall higher extracellular enzyme activity per cell.

The pool of free dissolved enzymes (i.e., enzymes that pass through 0.2-μm-pore-size filters) is fueled by various sources. In addition to enzymes released by bacteria, they can be derived from protozoa such as flagellates and ciliates and from mesozooplankton (e.g. chitinase). Obviously, during periods of high zooplankton grazing activity, selected enzymes can contribute the majority of the bulk activity (29) (Table 1), but this is not a common feature. Special patterns of distribution were recorded for phosphatases (13,30,31), which are generated not only by bacteria but also by phytoplankton, cyanobacteria (32,33), and macroalgae (34).

C. The Particle Decomposition Paradox and the Biological C Pump

Organic particles represent the nutritional basis for bacteria, and life in general, in the aphotic zone of the marine environment. However, microscopic analysis has revealed that particles are frequently less heavily colonized by bacteria than expected. Nevertheless, below the euphotic zone, particle decomposition has to supply the entire microbial community including the free-living bacteria. A key to understanding this paradox lies in the enhanced individual enzyme activity of the attached bacteria (18,21,35,36) and probably also in the extracellular release of endo-enzymes by these bacteria (37,38). By hydrolyzing macromolecular linkages in an endo- fashion (i.e., hydrolyzing the nonterminal linkages in a polymer), these enzymes are able to break up the complex polymers inside the particles. Both processes potentially create a surplus of dissolved monomeric or oligomeric hydrolysis products from the particles that are not entirely taken up by the attached bacteria (loose hydrolysis-uptake coupling). Escaping into the surrounding water, these substrates support the nutrition of free-living bacteria (39,40). A loose hydrolysis-uptake coupling frequently has been reported for particle-attached bacteria, whereas tight coupling has been reported between hydrolysis of DOM and uptake of the resulting monomers (17,22,40).

Other studies, however, have not revealed a difference in the hydrolysis-uptake coupling between attached and free-living bacteria (24,36). In two recently published investigations on the extracellular enzyme activity of marine snow–associated bacteria, no evidence was found that glucosidase and aminopeptidase activity in marine snow–associated bacteria were less tightly coupled to the uptake of the respective monomers than in free-living bacteria (23,26). Furthermore, in a number of studies using thymidine and leucine incorporation into bacterial deoxyribonucleic acid (DNA) and protein (41,42), respectively, as an estimate of bacterial C production, growth and the hydrolytic activities of attached and free-living bacteria were compared (17,24,36,40). On the basis of such bacterial production estimates and concurrently measured hydrolytic activity, it was concluded that the C demand of attached bacteria was lower than the amount of C cleaved by enzymatic activity, hence indicating a loose hydrolysis-uptake coupling (43). However, it is well known that leucine and especially thymidine are efficiently adsorbed to polysaccharides. This adsorbed (radiolabeled) thymidine and leucine is taken up at significantly lower rates than their nonadsorbed, truly dissolved counterparts (44). Since determinations of the saturating substrate concentrations are rarely made in such investigations (for logistic reasons), the amount of radiolabeled thymidine and leucine actually available for bacteria in the free form, and consequently available for rapid uptake, remains unknown. This adsorption and the concurrent lower availability for bacterial uptake might cause an underestimation of the actual bacterial production on and in polysaccharide-rich material such as marine snow (44), relative to bacterial enzyme activity.

The coupling between hydrolysis and uptake of DOM in particle-associated and free bacteria is still not fully understood. The reasons why the attached bacteria benefit so little from their strong hydrolytic activities, if there are no limiting factors interfering with the uptake of enzymatic hydrolysis products, are unknown. This fundamental discrepancy should be more thoroughly investigated in order to improve understanding of the biogeochemical flux of organic matter and the role of bacteria in the cycling of DOM in the ocean. In any case, it is well accepted that particle decomposition (45) contributes significantly to the loss of organic material from settling particles during sinking and thus determines the efficiency of the biological C pump (organic matter transport from the sea surface to the seabed).

D. Environmental Factors Influencing Enzymatic Activity

The magnitude of the main extracellular enzyme activities in marine water is frequently in the order aminopeptidase > phosphatase > β-glucosidase > chitobiase > esterase > α-glucosidase. However, exceptions may occur, as observed by Christian and Karl (46) in the equatorial Pacific, where β-glucosidase was about four times higher than aminopeptidase. This suggests that there may be factors regulating activities on a large scale. However, knowledge of global regulating factors is scarce. Christian and Karl (47) found that histidine and phenylalanine inhibited aminopeptidase expression in Antarctic waters. Likewise, Kim and Lipscomb (48) suggested that metals may be regulating factors for proteases (leucine amino peptidase seems to be principally a Zn^{2+}-dependent enzyme). This was especially due to Zn^{2+} (which is rare in marine waters), but Mn^{2+}, Co^{2+}, Fe^{2+}, and Mg^{2+} might also play a role (47–50). In the surface layer of the ocean, ultraviolet-B radiation can be important, mainly through photochemical degradation of the extracellular enzymes (51,52). With respect to phosphatase activity, the abundance of inorganic P is regarded as a regulating factor, particularly for the P-limited regions in the oceans (53–55). However, dissolved organic phosphorus (DOP) and particulate organic P also should be considered (56). Furthermore, mechanisms of phosphatase regulation are different for bacteria and phytoplankton. While the phosphatases of phytoplankton seem to be regulated strictly by inorganic P concentrations (49,57–59), this mechanism is not so clear for bacterial phosphatases. The latter may target C and N rather than P supply, as pointed out for the limnetic environment by Siuda and Güde (60) and for the deep and C-limited, but phosphate-replete, ocean by Hoppe and Ullrich (61). In any case, regardless of environmental factors, variation of species composition within the bacterial community can significantly influence the distribution of enzyme activities in the sea (62,63).

The effects of environmental factors on enzyme regulation are reflected by the diversity of extracellular enzymes, as expressed in the possible ranges of K_m and the patterns of individual cell-specific enzyme potentials (Table 2, Table 3). Information on the K_m values of marine bacteria, however, is scarce. Proteinase affinities seem to be higher in oligotrophic than in eutrophic regions. K_m values observed in Antarctic regions at in situ temperature were similar to those in warmer regions; the relationship does not seem to hold for Arctic environments (Table 3). Cell-specific enzyme activities vary over a wide range. They are low in eutrophic waters, but relatively high in oligotrophic waters and

Table 3 K_m Values (Apparent) of Extracellular Enzymes in Different Marine Habitats

Environment	Conditions	Enzyme, substrate	K_m (μmol L^{-1})	Reference
Trophic gradient	Eutrophic	leu-AMP	47.6	(113)
	Oligotrophic		0.71	
Antarctica	−1.7°C to +2°C	leu-AMP	67–132	(47)
	−1.7°C to +20°C		48–218	
Mesocoms	Before storm event	leu-AMP	32.5–51.7	(117)
	After storm induced	β-D-glucosidase	21.9–57.6	
Laptev Sea	Arctic	leu-AMP	3.3	(222)
Lena River plume	Arctic, eutrophic	leu-AMP	28.6–83.3	(222)
		β-D-glucosidase	14.3–40	
		Phosphatase	8–28.6	
North Sea	Coastal zone water	Proteinase	6.67	(223)

particularly high on organic particles and in deep water (Table 2). In general, the characteristics of these variables indicate (in some cases clearly) a dependency on the prevailing environmental conditions.

III. FUNCTIONALITY OF EXTRACELLULAR ENZYMES SUBSTRATES IN MARINE ENVIRONMENTS

A. The Size Continuum of Organic Matter from DOM to POM

Dissolved organic matter (DOM) in the ocean is recognized as one of the three main reservoirs of organic matter on the planet, equal to the organic matter stored in terrestrial plants or soil humus (64). DOM of natural waters is chemically complex: less than 40% of the oceanic DOM pool is chemically characterized. The concentration of DOM is, therefore, frequently measured as dissolved organic carbon (DOC). DOC in the ocean typically decreases from the euphotic zone with concentrations ranging from 100 to 150 μM C to around 40 μM C in the ocean's interior (65). Despite the lower concentrations in the deep ocean, the major fraction of the DOM is found in the aphotic zone of the ocean, comprising ~90% of the total oceanic DOC (66).

Chemical characterization of oceanic DOM is hampered by both the low concentrations of DOM and the high salt content of ocean waters, which interferes with chemical analysis. Typically, 20–30% of oceanic DOC is recovered via 1000-Da ultrafiltration (67). In estuarine environments, recoveries of DOM are usually higher (up to 70%), as a result of the higher average molecular weight of the DOM in fresh water (68). On the basis of size fractionation studies performed over the past two decades on oceanic DOM, it appears that most of the DOM retained by ultrafiltration through 1000-Da filter cartridges is composed of compounds in the size range of 1000 to 30,000 Da (67–71). This high-molecular-weight DOM has been shown to be of contemporary origin (67,69) and derived from release processes taking place during photosynthesis of phytoplankton, grazing, and lysis of organisms (72–75). Phytoplankton extracellular materials have a similar carbohydrate signature to that of the oceanic DOM in surface layers (76–78). Phytoplankton activity and mortality (grazing and viral lysis) have been suggested to be the major source of oceanic DOM (70,74,75,78–82).

The molecular weight fraction of the DOM larger than 1000 Da but smaller than 0.2 μm is also frequently termed *colloidal organic matter* (COC), in contrast to the truly dissolved DOM of <1000 Da. Freshly produced high-molecular-weight DOM consists mostly of carbohydrates, as indicated also by overall C:N ratios ranging from 15 to 25 (67,69,70). In addition to polysaccharides, proteins and lipids are present as chemically characterizable DOM components. Polysaccharides, however, are by far the most abundant macromolecular class of oceanic DOM.

DOM is likely present as a size continuum in seawater; molecular weight and hydrodynamic volume of DOM may vary with specific environmental conditions. For example, the fibrillar structure of polysaccharides allows them to form bundles of molecules bound together via cationic bridges mediated by Mg and Ca (83). Thus, coagulation processes of DOM, and particularly of the polysaccharides, are likely to be more important in oceanic seawater than in freshwater systems, as a result of the higher ionic concentrations in seawater. These coagulation processes lead to the formation of colloidal and ultimately microparticulate organic material; thus DOM may be transformed to POM. In this coagulation process, polysaccharides play a major role as a result of their relatively high concentration and the physicochemical characteristics of the fibrillar structure (84–86). In 1998 Chin, Orellana, and Verdugo showed that even low-molecular-weight DOM has the potential to coagulate spontaneously to form polymeric gels (87). These gels represent condensed organic matter at a higher concentration relative to that of the surrounding water and might therefore be of considerable importance for bacterioplankton (1) and enzymatic hydrolysis. Furthermore, these microgels might interact with other colloidal matter, forming distinct submicrometer particles that are ubiquitously present in seawater at concentrations of up to $\sim 10^9$ ml^{-1} (88–93). Whereas these submicrometer particles are not colonized by bacteria, the larger transparent exopolymer particles (TEPs) are frequently densely colonized by bacteria (94,95). TEPs have been shown to originate mainly from phytoplankton blooms and their decay (96). In addition to these polysaccharidic TEPs, protein particles have been reported to be abundant in the surface layers of the ocean (97).

At the upper end of the size continuum of condensed colloidal organic matter, marine snow is commonly present, although at highly varying concentrations, in the surface as well as in the deep waters. The structural frame of this marine snow is also provided by polysaccharides: they are highly hydrated structures larger than 0.5 mm and range up to meters in diameter as observed in the subpycnocline layers of the Adriatic and Mediterranean Seas and in the deep ocean (24,98–100).

Whether there are close links between different size categories of condensed organic matter and whether smaller aggregates are really the precursors for the next larger group of particles remain unclear at the moment. There are indications, however, based on the common chemical signatures of the polysaccharide pool (which dominates the macromolecular fraction of all these particles), that there is a link between them and that they are derived mainly from auto- and heterotrophic microorganisms.

Irrespective of the exact relationships between submicrometer particles, TEP, and marine snow, all of this polysaccharide-based condensed matter interacts with the surrounding chemical environment by adsorbing inorganic and organic nutrients. This results in a higher nutrient concentration on these particles than in the ambient water (99,24).

The nutrient-enriched zones might be in the micrometer range, similar to the microzones proposed by Azam (1) and elegantly visualized by Blackburn et al. (101), or marine snow of meters in diameter. In any case, they are attractive to indigenous bacteria because of nutrient concentrations up to three orders of magnitude higher than in the ambient water (102,103). Similarly, bacteria also have been reported to be enriched by up to three orders of magnitude on these particles (24).

B. Enzyme Activity and DOM/POM Reactivity

Contemporary DOM of high molecular weight has been shown to be efficiently utilized by bacterioplankton (3,4), while the majority of oceanic low-molecular-weight DOM, which persists long enough to be measured, can be considered as refractory. This finding led to the formulation of the size-reactivity model (3,4) proposing that the majority of the low-molecular-weight DOM pool is the consequence of chemical and biological degradation (4). Since bacterioplankton can take up molecules only smaller than 600 Da without prior cleavage by extracellular enzymes, the efficient utilization of this high-molecular-weight DOM indicates the importance of bacterial extracellular enzymes. This enzyme pool includes both endo- and exohydrolases (104). Endohydrolases cleave polymers into oligomeric compounds, and, subsequently, exohydrolases generate monomers, which are taken up by bacteria. In order to cleave complex molecules, several endohydrolases act in concert, as has been demonstrated for the cellulase complex (105). With commonly used fluorogenic substrate analogs, such as methylumbelliferyl derivatives (13,21), only the final step of the cleavage of the monomer (i.e., the exohydrolase activity) can be measured. The majority of this cleavage activity by hydrolases is bound to the cell wall or occurs in the periplasmic space of Gram-negative bacteria (106), and only a small percentage of freely dissolved enzymatic activity can be detected (see Sec. III. A.). Fluorescently labeled high-molecular-weight substrates (see Section IV. F.) can be used to measure endohydrolase activities.

C. Significance of Enzyme Activity for Substrate Supply

Bacteria have different possibilities to respond to nutrient limitation (107) because they can use inorganic as well as combined and monomeric organic molecules to supply their cellular demands for energy, growth, and maintenance. The relative contribution of different sources to bacterial nutrition depends essentially on the availability of inorganic nutrients (108) and on the C:N:P ratios (e.g., 106:16:1; the *Redfield ratio*) of organic matter (109), which determine its nutritional value after hydrolysis. Examples of the utilization of the N pools are presented in Table 4. Using $^{15}NH_4^+$ techniques, Tupas and Koike (110) demonstrated that natural bacterial assemblages in nutrient-enriched seawater cultures fueled 50–88% of their N demands for growth by NH_4^+ even in the presence of large amounts of DON. This DON consisted mostly of combined amino acids and contributed, together with NH_4^+ uptake, 70–260% of bacterial N production. However, an average of 80% of the DON used was subsequently remineralized to NH_4^+ by the bacteria (110). In general, information is too scarce to derive principles about the preferences of bacteria for specific N sources in the sea, however, hydrolysis of dissolved combined amino acids is always a prominent feature.

Table 4 Contribution of Different N Sources to Bacterial N Demand

Environment	Conditions	DFAA (% of N demand)	DCAA (% of N demand)	NH_4^+ (% of N demand)	Experiment	Reference
Delaware Bay estuary	Exponential growth	<34 (C% < 14)	<24 (C% < 10)	n.d.	Batch cultures enriched with C and N	(224)
	C-limited	37–62	4–10	27–59		
	N-limited	78	14	8		
Sargasso Sea	Surface water, depth profiles	<20	20–65	n.d.	Radiotracer incubations	(118)
Aburatsubo Bay and Ohtsuchi Bay	Enriched seawater cultures	n.d.	70–260	50–88	$^{15}NH_4^+$ techniques	(110)
Santa Rosa Sound Gulf of Mexico	Seawater incubations	DFAA dominant N source	DCAA secondary N source	NH_4^+ tertiary N source	^{14}C-AA, enzyme essay	(111)
		Uptake(nM h^{-1})	Uptake(nM h^{-1})	Uptake(nM h^{-1})		
Delaware Estuary	Salinity gradient	114	117	n.d.	^{14}C-AA, algal protein	(171)
		LMW-DOC % uptake d^{-1}	HMW-DOC % uptake d^{-1}			
Gulf of Mexico, tropical lagoons	Seawater incubations	3–6.6	4.5–22.5		Tangential flow DOM fractionation	(4)

Generally, the combined substrates DCAA, dissolved combined amino acids, require enzymatic hydrolysis, and their utilization therefore reflects indirectly the contribution of enzyme activity to bacterial growth. DFAA, dissolved free amino acids; LMW-DOC, low-molecular-weight dissolved organic carbon; HMW-DOC, high-molecular-weight dissolved organic carbon.

IV. DISTRIBUTION OF EXTRACELLULAR ENZYME ACTIVITIES IN SEAWATER

A. Enzymes in Coastal Regions, Lagoons, and Estuaries

Coastal regions represent the transition zone between land and the open sea. Thus they are frequently characterized by local morphological and hydrographical patterns and by gradients of salinity, eutrophication, pollution, and sediment resuspension. These conditions are clearly reflected by specific patterns of extracellular organic matter degradation. In general, the enzymatic potential within the coastal regions affects the export of organic matter to the adjacent open sea, which would otherwise only be supplied by autochthonous primary production.

Studies along trophic gradients have shown that extracellular enzyme activities react in a specific manner, a response that differs from that of other microbial variables. In a comparative study of a moderately eutrophic estuary and an open-water ecosystem, Jørgensen et al. (111) observed a 2.5 times higher cell-specific leucine-aminopeptidase activity but a 2.4 to 18 times higher cell-specific free amino acid assimilation in the eutrophic system. Likewise, in a trophic gradient in the Adriatic Sea, Karner et al. (112) found positive trends for leucine-aminopeptidase whereas α-glucosidase did not exhibit such a clear trend. Trophic conditions also were clearly reflected by the patterns of a variety of enzyme activities in a gradient at the Atlantic Barrier Reef off Belize. Particularly, K_m values of leucine-aminopeptidase showed a much higher substrate affinity in oligotrophic water than in the eutrophic region of the gradient (113). This corresponded to much higher per-cell activity in the oligotrophic environment. In the salinity gradient between the Sacramento River and the central San Francisco Bay, increasing salinity was positively correlated with aminopeptidase activity and negatively correlated to β-D-glucosidase activity (18).

In coastal regions, environmental factors, such as seasonal and diurnal variability in temperature and nutrient supply as well as stratification, are highly important for enzymatic activity patterns. In waters of the Uranouchi Inlet (Japan), phosphatase activity was similar throughout the entire water column during the mixing period, whereas it was 20 times higher near the surface than in bottom waters during thermal stratification (114). This finding suggests that phosphatase was limited by temperature and dissolved oxygen concentration in the deep. In the Pomeranian Bight (Baltic Sea), phosphatase activity was 184–270 nmol L^{-1} h^{-1} in summer at about 18°C and 5.7 nmol L^{-1} h^{-1} in winter (0°C). Similar temperature effects were observed for α- and β-glucosidase and chitobiase. In contrast, peptidase activity was 9 to 72 times higher in autumn than in summer (115). In a coastal station of the Ligurian Sea (Mediterranean Sea), Karner and Rassoulzadegan (116) measured the short-term variability of different extracellular enzymes, α- and β-glucosidase exhibited particularly strong diurnal variation, but such a variation was not observed for aminopeptidase. Sediment resuspension by storms in the coastal region also can have a strong impact, particularly on enzyme activity in bottom water. This was shown in mesocosm experiments in which aminopeptidase and β-glucosidase were 24% and 43% higher, respectively, after a simulated storm event compared to the calm period (117).

B. Enzymes in the Open Sea

Although the oceans generally have low concentrations of organic matter per volume of water, they play, because of their huge dimensions, a dominating role in organic matter production and decomposition of the biosphere. The production and transformations of organic matter in the surface ocean and the burial of organic materials in the deep sediments contribute significantly to the efficiency of the biogeochemical cycles and climate change on Earth. Nevertheless, enzyme investigations in truly offshore regions and on transoceanic cruises rarely have been conducted.

In an extremely oligotrophic domain of Sargasso Sea, dissolved combined amino acids dominated the N pools and also contributed the largest part (20–65%) of the bacterial N demand in surface water. A modified form of protein (glucosylated protein) accounted for the highest portion of bacterial N demand in deeper waters (118) (Table 4). Comparing the enzymatic properties of three oceanographic provinces of the Pacific (northern subtropical, equatorial, and the Southern Ocean down to Antarctica), Christian and Karl (46) found significant variations in aminopeptidase and β-glucosidase activities and their temperature characteristics. The relative relationship between aminopeptidase and β-glucosidase shifted from 0.3 at the equator to 593 in Antarctic waters. This change was due mainly to an increase of β-glucosidase activity from 0.44 in Antarctica to 1519 nmol L^{-1} d^{-1} at the equator. The authors hypothesized that there was a longitudinal trend in bacterial utilization of polysaccharides relative to amino acids and proteins. Investigating the northern Pacific from 45°N 165°E down to the south edge of the equatorial zone (8°S 160°E), Koike and Nagata (119) also found an increase of β-glucosidase activity in the surface layer from 0.11 to 3.1 nmol L^{-1} h^{-1}; this change, however, was far less dramatic as observed by Christian and Karl (46). In the northern Indian Ocean (Arabian Sea), enzyme (aminopeptidase, β-glucosidase, phosphatase) activities together with other microbial activity measurements reflected clearly the specific hydrographic conditions created by the SW Monsoon (61). β-Glucosidase activity was not enhanced at the equator (~0.5 nmol L^{-1} h^{-1}) but increased strongly in the upwelling regions off the coast of Oman (2 nmol L^{-1} h^{-1}). In the same direction, the relationships between aminopeptidase and β-glucosidase activities increased from ~10 to ~50.

Aminopeptidase activity measured along a N-S Atlantic transect (54° N–62° S) (120) suggests a strong dependency of this variable on the global current system and climate zones. Lowest values were measured in the northern and southern subtropical gyres (<10 nmol L^{-1} d^{-1}). Increased enzyme activity was detected near the southern edge of the North Equatorial Current (7° N) and the northern edge of the South Equatorial Current (10° S), which are fueled by the Canary Current and the Benguela upwelling regions, respectively. At the continental shelf edge (Patagonain Shelf) at 40° S (where the subtropical Brazil Current and the subantarctic Falkland Current meet and establish the Subtropical Convergence), aminopeptidase activity increased instantaneously by factors of 6 to 11. Surprisingly, aminopeptidase activities in the Antarctic Weddell Sea were nearly as low as in the warm but nutrient-depleted subtropical regions (120).

C. Enzymes in Extreme Marine Environments

Extreme environments in the marine biosphere are represented by the high-pressure waters of the deep sea and the permanently cold polar regions, together with some zones influenced by hot vent and anoxic (or even sulfidic) conditions.

Extracellular enzyme activities (aminopeptidase, α- and β-glucosidase, phosphatase) in sea ice were similar to those reported from eutrophic, temperate marine environments. Though psychrophilic isolates showed temperature optima of their enzymes that were similar to those of mesophilic strains (~30°C), their activity at low temperature was relatively much higher. Activities of some enzymes in polar sediments, however, have temperature optima much lower (ca. 16 °C) than those reported for temperate sites (121).

Enzyme activity in the melted ice at 1°C was generally much higher than in the water underneath the ice, 3.2–1702 times for aminopeptidase, and 2.4–42.2 times for phosphatase. In poorly colonized ice cores, enzyme activity can be similar to or lower than in the water (122,123). K_m values for aminopeptidase of bacterial communities from Antarctic regions, measured at in situ temperature, were comparable to those observed for a variety of aquatic environments (47). Studies on enzyme activities in waters of the deep sea (mesopelagic and bathypelagic strata) are rare. Koike and Nagata (119) measured a very strong decrease of (particle-associated) α- and β-glucosidase activity down to 4000 m depth in the central Pacific Ocean, where activities were generally less than 1% of the surface values. In contrast, phosphatase activity at depth was up to 50% of that near the surface. This observation was confirmed by Hoppe and Ullrich (61), who found very low glucosidase activity (in unfractionated samples) in the lower mesopelagic zone of the Indian Ocean. At greater depths, aminopeptidase activity was equal to or much lower than the activity measured near the surface, but the phosphatase activities were up to seven times higher than near the surface. These phenomena were interpreted to be i) a result of enzyme export by sinking particles and ii) a consequence of severe C limitation of deep sea bacteria meeting (partly) their C demands by the hydrolysis of organic P compounds (61). In deep coastal waters (Santa Monica Basin), proteolytic activities correlated significantly with bacterial abundance and bacterial growth down to the sea floor at 900-m depth (14). Effects of sulfidic conditions (H_2S) on enzyme activities of bacterial communities from the Baltic Sea were investigated by Hoppe et al. (124). In comparison to oxic control treatments, reduction of activity was particularly strong for peptidase and the glucosidases and to a lesser degree for chitinase and phosphatase (124).

D. Enzymatic Properties of Marine Bacterial Species and Other Organisms

The ability of marine bacterial isolates to support growth via freely released enzymes on particulate organic substances (amylopectin, chitin, animal hide) was measured by Vetter and Deming (125). Under these conditions, the bacteria were able to grow, albeit at rates generally lower than rates reported for growth on dissolved organic substances. The interactions between corals and bacteria living in the mucus of corals were investigated by Santavy et al. (126) in the coral *Colpophyllia natans*. If the corals were under stress by infection (black band disease), they produced more mucus and the activity of the associated bacteria was generally enhanced. Phosphatase activity was especially enhanced, and the measurement of phosphatase activity was recommended as a tool to quantify stress metabolism of corals. Chitobiase (*N*-acetyl-β-D-glucosaminidase) activity, expressed during the premolt phase of crustaceans, was used as a measurement of copepod (*Temora longicornis*) secondary production in the sea (127). Other studies suggest that phagotrophic nanoflagellates can contribute significantly to the pool of free α-glucosidases and aminopeptidases in marine environments (128), and the occurrence of phosphatase activity in red-tide dinoflagellates has been shown by Vargo and Shanley (129). From the applied aspect, Sawyer et al. (130) investigated enzymatic properties (including that of phosphatase; it was not clear, however, whether extracellular or intracellular enzymes were studied) of potentially pathogenic amoebae (*Acanthamoeba* sp.) from the sediments of marine sewage dumping sites. Morphologically similar species could be distinguished from each other by the diversity of their enzymatic capabilities. Dramatic effects of a genetically engineered bacterium with enhanced phosphatase activity on the growth of natural marine phytoplankton were recorded by Sobecky et al. (131).

V. ENZYMES IN MARINE SEDIMENTS

On a global scale, marine sediments fulfill a critical function as a sink for reduced C: a small fraction of the CO_2 fixed by phytoplankton in the surface ocean sinks through the water column as fast-settling particles. An even smaller fraction of these particles escapes the efficient remineralization processes in sediments and is ultimately buried. Burial of this reduced organic C represents long-term removal of CO_2 from the ocean (64). Marine sediments are also diverse and variable environments and serve as habitats for a wide range of benthic organisms. The nutritional basis for many of these organisms is the rain of particles from the surface ocean, so successful existence in the benthos is directly related to the ability to gain C and energy from POM flux.

Marine sediments are also dynamic systems, subject to physical as well as chemical changes on a wide range of time scales. The flux of particles from the surface ocean is variable in time and space; surface sediments may be resuspended as a result of slumping or turbidity currents. Activities of infaunal and epifaunal organisms lead to formation of tubes and burrows and extensive bioturbation. Frequently, the net result is a ''patchy'' environment characterized by physical and chemical discontinuities, redox gradients, and zonation of microbial activities, where biological and chemical parameters can vary greatly on small spatial and temporal scales. Effectively assessing these variations and discerning the patterns that may underlie this variability are major challenges in studying sedimentary environments.

A. Seasonal and Spatial Patterns in Coastal and Temperate Sediments

Coastal temperate regions are subject to significant seasonal changes in a range of physical and chemical parameters, including light, temperature, and nutrient availability. Cycles of productivity respond to these seasonal changes; variations in extracellular enzyme activities in shallow and temperate sediments have in turn been linked to seasonal variations in input of organic matter. Reichardt (132) observed that an approximately twofold increase in extracellular enzyme activity in surface sediments followed the sedimentation of a phytoplankton bloom. Protease activity, measured by using hide powder azure, was particularly high. At the same site, Meyer-Reil (133) observed annual maxima in leucine amino peptidase activity in surface sediments, which occurred simultaneously with the main sedimentation events in Kiel Bight. Pantoja and Lee (38) likewise observed an annual maximum in peptide hydrolysis rate constants in Flax Pond sediments, which also correlated with sedimentary input of fresh organic matter. Changes in enzyme activities have also been correlated with annual temperature cycles in intertidal sediments in Maine (134). Pantoja and Lee (38), however, found that there was no direct relationship between temperature and peptide hydrolysis rate constant over an annual cycle at another coastal site, although rate constants in the upper 10 cm of the sediments were higher in the spring/summer than in the winter.

Because C influx to the sediments occurs via settling particles, surface metabolism is frequently stimulated above the levels observed in deeper layers of the sediments (135). Although organic matter in these deeper layers is often considered to be of lower nutritional quality, since it has already been reworked by the surface community, fresh organic matter can be rapidly mixed to considerable depths by infaunal organisms (136), providing the subsurface community with fresh organic C. The depth trends observed in enzyme activities generally correspond to these patterns. Activities in surface layers of the sediment are greater than those in lower depths (133,134,137). Shallow subsurface maxima, however, also have been observed; such maxima may correspond to ''hot spots'' of microbial activity and/or be related to infaunal organisms (38,134,138). Potential hydrolysis rates of carbohydrates and proteins measured at a wide range of coastal and temperate sediments vary over several orders of magnitude (Tables 5, 6, and 7). Intertidal sediments in Maine include the highest carbohydrate-hydrolysis rates reported to date (137). The wide range of rates King (137) observed for a suite of monosaccharide substrates illustrates the fact that extracellular enzymes in natural systems are sensitive to structural variations among closely related substrate analogs.

B. Deep Ocean Environments

In deep ocean sediments, enzyme activity likewise frequently is maximal in surface intervals and decreases with depth in the sediments (139), although shallow subsurface maxima also have been reported (140), perhaps also linked to biogenic structures and macrofaunal tubes (141,142). Surveys of the deep ocean consistently have shown patterns of decreasing overall enzyme activities with increasing water column depth (139,143). No evidence has

Table 5 Carbohydrate-Hydrolyzing Enzyme Activity in Coastal and Shallow Marine Sediments

Rate (nmol cm^{-3} h^{-1})	Substrate	Depth	Site	Reference
90	MUF-β-fucose	Intertidal sediments	Maine	(137)
114	MUF-β-arabinose	Intertidal sediments	Maine	(137)
216	MUF-β-xylose	Intertidal sediments	Maine	(137)
240	MUF-β-mannose	Intertidal sediments	Maine	(137)
660	MUF-β-galactose	Intertidal sediments	Maine	(137)
1392	MUF-β-glucose	Intertidal sediments	Maine	(137)
1.6–8.3[a]	MUF-β-glucose	Coastal	Kiel Bight	(225)
0.19–1.5[a]	Pullulan	17–35 m	Kiel Bight	(138)[b]
2.0–3.5[a]	Laminarin	17–35 m	Kiel Bight	(160)[b]
0.41–1.2[a]	Pullulan	10 m	Cape Lookout Bight	(226)[b]
0.58–0.94[a]	Laminarin	10 m	Cape Lookout Bight	(226)[b]
0.41–0.86[a]	Xylan	10 m	Cape Lookout Bight	(226)[b]
4.7	Pullulan	190 m	Skagerrak	(162)[b]
2.6	Laminarin	190 m	Skagerrak	(162)[b]
0.90	Chondroitin sulfate	190 m	Skagerrak	(162)[b]
0.41	Arabinogalactan	190 m	Skagerrak	(162)[b]
0.64	Xylan	190 m	Skagerrak	(162)[b]
0.22	Fucoidan	190 m	Skagerrak	(162)[b]

Measurements made in sediment slurries, except as noted.
[a] Measured in intact cores, range of values.
[b] Recalculated from reference (Arnosti, in prep.)

Table 6 Carbohydrate-Hydrolyzing Enzyme Activity in Polar and Deep Sea Sediments

Rate (nmol cm^{-3} h^{-1})	Substrate	Depth (m)	Site	Reference
0.67–2.4[a]	Pullulan	115–175	Arctic Ocean	(161)[b]
1.2–3.8[a]	Laminarin	115–175	Arctic Ocean	(161)[b]
0.04–0.33[a]	Xylan	115–175	Arctic Ocean	(161)[b]
0.02–0.35	MUF-α-glucose	37–3427	Arctic continental slope	(143)
0.02–3.02	MUF-β-glucose	37–3427	Arctic continental slope	(143)
0.11	MUF-α-glucose	1000	Arctic continental slope	(146)
0.30	MUF-β-glucose	1000	Arctic continental slope	(147)
81–266[c]	MUF-β-glucose	439–567	Ross Sea	(202)
0.2–0.17	MUF-α-glucose	135–1680	Atlantic continental shelf	(139)
0.1–0.7	MUF-β-glucose	135–1680	Atlantic continental shelf	(139)
0–0.79	MUF-α-glucose	193–4617	Mediterranean	(140)
0.08–2.13	MUF-β-glucose	193–4617	Mediterranean	(140)
0–0.6	MUF-α-glucose	1920–4420	Arabian Sea	(228)
0.3–0.9	MUF-β-glucose	1920–4420	Arabian Sea	(228)
0.0049–0.0082	MUF-α-glucose	2879–4919	Northeast Atlantic	(145)
0.002–0.016	MUF-β-glucose	2879–4919	Northeast Atlantic	(145)
ca. 0.005–0.006	MUF-α-glucose	4500	Northeast Atlantic	(142)
0.010–0.090	MUF-β-glucose	4500	Northeast Atlantic	(142)
ca. 0.50	MUF-α-glucose	4500	Northeast Atlantic	(214)
ca. 0.30	MUF-β-glucose	4500	Northeast Atlantic	(214)

Measurements made with sediment slurries, except as noted.
[a] Measured in intact cores; range of values.
[b] Recalculated from reference (Arnosti, in prep.).
[c] nmol g^{-1} dry weight sediment h^{-1}.

Table 7 Protein Hydrolysis in Coastal, Polar, and Deep Marine Sediments

Rate (nmol cm^{-3} h^{-1})	Water depth	Site	Reference
0.011–0.37	Coastal	Kiel Bight	(133)
0.002–0.03[a]	Coastal	Kiel Bight	(225)
8–116	37–3427 m	Arctic continental slope	(143)
10–60	135–1680	Northeast Atlantic	(139)
14–284	193–4517 m	Mediterranean	(140)
144	1000 m	Arctic continental slope	(147)
6–145	1920–4420	Arabian Sea	(228)
1.7–3.8	2879–4919 m	Northeast Atlantic	(145)
5–16	4500 m	Northeast Atlantic	(142)
ca. 22	4500 m	Northeast Atlantic	(214)
1312–2728[b]	439–567 m	Ross Sea	(202)
10–300[b]	Continental shelf	Arctic	(227)

Measured with L-leucine-4-methylcoumarinyl-7-amide (leu-MCA). Measurements made in sediment slurries, except as noted.
[a] Measurements made in intact cores, range of values.
[b] nmol g^{-1} dry weight sediment h^{-1}.

been found for pressure effects on enzyme activities when slurries of deep sea sediments have been incubated in parallel under atmospheric and in situ pressure (144,145), although these experiments have been conducted with sediments that were first taken to the surface (depressurized) and subsequently repressurized. The general decrease in surface sediment activity with increasing water column depth has been linked with decreasing organic matter input, i.e., with substrate limitation for the microbial communities (144,146). This model of substrate limitation is consistent with the results of Boetius et al. (140), who found that a deep Mediterranean trench showed much higher activity than the surrounding abyssal plain. Analyses of sedimentary organic matter also suggested that "fresher" organic C was concentrated in the trench. Peptidase activities have occasionally been observed to increase with water column depth, in contrast to the patterns observed for activities of other enzymes (143,147).

Comparisons of relative levels of enzyme activities, using a suite of fluorescent substrate analogs, have frequently shown that peptidase activity exceeds glucosidase activity. Boetius and Damm (143) found that peptidase activity was generally one to two orders of magnitude higher than the hydrolysis rates of α- and β-glucosidase activity and fluorescein diacetate activity in deep ocean sediments. Poremba (145) likewise found that the top horizon of deep sea sediments showed activities decreasing in the order aminopeptidase > esterase > chitobiase > β-glucosidase > α-glucosidase with relative ratios of 687:174:11:3:1. The extent to which potential hydrolysis rates measured with substrate analogs represent hydrolysis of actual macromolecules is open to question, however, as discussed later. Martinez and colleagues (148) also have noted that potential hydrolysis rates of α- and β-glucopyranoside are typically low in comparison to activities measured with other substrate analogs and suggested that these substrates may in fact underestimate actual enzyme activities.

VI. METHODS USED IN WATER AND SEDIMENT

The determination of the activity of extracellular enzymes in seawater requires the application of highly sensitive methods, particularly with respect to conditions in the open and deep sea. In rich marine environments such as estuaries and tropical lagoons, less sensitive applications may be appropriate. Several approaches that fulfill these requirements are currently in use.

A. Application of Fluorogenic Substrate Analogs and Fluorescently Labeled Substrates

The technique of enzyme activity measurement by fluorogenic substrate analogs was derived from biochemical approaches and adopted to the marine field. Suitable compounds have been reviewed by Manafi et al. (149) and Haugland and Johnson (150).

1. Community Enzyme Activity

Several highly sensitive fluorogenic substrate analogs are in use. These are mainly combined methylumbelliferyl (MUF) molecules (13,151,152) or β-naphthylamide molecules (14,153). In these substrates, the fluorescent molecule is bound, via a specific linkage, with one or more molecules of sugar, amino sugar, amino acid, fatty acid, or with inorganic components such as sulfate or phosphate. These substrates are used as analogs to measure the hydrolysis of the most abundant combined molecules in the sea: carbohydrates, proteins, and lipids. 4-Methylumbelliferyl-β-D-glucuronide (MUG) has been used successfully to determine coliform bacteria in seawater (154). MUF substrates screened against to a variety of commercially available enzymes showed relatively few nonspecific reactions (155). The sensitivity of the fluorogenic substrates for the measurement of enzyme activity is generally in the nanomolar range. Martinez and Azam (106) have demonstrated, with these substrates, that activity originates from periplasmic and other extracellular enzymes and not from cytoplasmic enzymes. An interesting and promising method for the separation of such enzymes by capillary electrophoresis was presented by Arrieta and Herndl (156) (see also section VII).

The fluorescent substrate analogs also are suited for automation of enzyme activity measurements. Continuous measurements, for instance during cruises, can provide very detailed information of the spatial distribution of enzyme activity. A device for such measurements was constructed and applied by Ammerman and Glover (157). A technique for the determination of proteolytic activity by flow injection analysis of fluorogenic substrate analogs was presented by Delmas et al. (158).

New methods using fluorescently labeled high-molecular-weight soluble substrates have been applied to investigations of enzymatic hydrolysis rates. Fluorophores are covalently linked to polysaccharides of differing structure (137) and peptides of varying length and composition (159). This hydrolysis is measured during the course of an experiment as the change with time in molecular weight or size of the initial substrate. These new substrates have the advantage that they are actually oligomers and polymers compounds, rather than analogs for these structures, therefore directly dealing with questions about the extent to which a given substrate actually represents an oligo- or polymer compound in solution (38,121). These oligopeptide and polysaccharide substrates also can be used to differentiate among a spectrum of specific enzyme activities. This allows determination of hydrolysis rates for peptides differing in size by a few amino acids (38) and among polysaccharides differing in molecular weight, linkage position, monomer composition, and/or anomeric linkages (160–162). A further technical advantage is that the fluorophores used for labeling polysaccharides (fluorescein-amine, isomer II) and peptides (4-amino-3,6-disulfo-1,8-naphthalic anhydride) have excitation/emission maxima of 490/530 nm and 424/550 nm, respectively, far removed from the natural fluorescence of DOC (ca. 350/450 nm) (163). Measurements in high DOC porewaters, problematic with MUF-labeled substrates, can easily be made with these newer fluorophores. These substrates can be synthesized by the individual investigator, following published procedures (159,160).

2. Bacterial Colony and Single Cell Activity

From the ecological viewpoint, the determination of the individual enzyme activity is very important because it reflects most clearly the individual response to the prevailing environmental conditions. However, the requirements with respect to sensitivity and other substrate qualities are very high. Enzymatic properties of bacterial colonies obtained from marine sources can be detected easily by the application of filter pads soaked with fluorogenic substrate analogs, such as MUF substrates, and placed on the colonized agar surface (164). Enzymatic properties of single cells (bacteria and algae) can be screened by a new type of fluorogenic compound, ELF-97 (165,166), which is combined with sugar, amino acid, fatty acid, or inorganic compounds such as sulfate or phosphate. The fluorescent moiety of these substrates precipitates after hydrolysis of the combined molecule around the cells, which can be detected microscopically with a special combination of excitation and emission filters.

B. Application of Radiolabeled Substrates

Radioactive chitin, prepared by acetylation of chitosan with tritiated acetic anhydride, was used by Molano et al. (167) as a substrate for measuring chitinase activity. Kirchman and White (168) isolated ^{14}C-chitin from a specially grown culture of the fungus *Paeosphaeria spatinicola* and demonstrated that significant amounts of the chitin were released as DOM. In a similar fashion, ^{3}H-chitin has also been isolated from marine fungi and used in studies of organic matter degradation (169,170).

Algal ^{14}C-labeled proteins were used by Coffin (171) for the determination of bacterial uptake of combined amino acids. Keil and Kirchman (172) demonstrated that the chemical characteristics of radiolabeled proteins significantly affected their utilization by marine bacteria. Proteins prepared by methylation of amine groups were hydrolyzed to the same extent as proteins prepared by iodination of tyrosine residues, but very little of the methylated protein was ultimately assimilated or respired. Taylor (173) measured the degradation of sorbed and dissolved proteins in seawater by the hydrolysis of [methyl-^{3}H]–ribulose-1,5-biphosphate carboxylase-oxygenase (Rubisco). In another study, [^{32}P]–adenoshine triphosphate (ATP) (174) and the two substrates ([^{14}C]Hb) methemoglobin and [^{124}I]BSA (37) were used for the detection of 5′-nucleotidase and proteinase, respectively.

C. Application of Chromophoric Substrates

Chromatogenic substrates such as para-nitrophenyl (*p*-NP) compounds and azur dyes are frequently not sensitive enough for measurements of enzyme activity in seawater. However, they have been successfully applied in tropical lagoons (175,176) and even with plankton communities in the Red Sea (measured in 10-cm cuvettes) (77). High-molecular-weight solid substrates such as hide powder azure, cellulose, chitin, and agar stained with remazol brilliant blue R have also been used to investigate enzyme activities in extracts from sediments (132).

VII. DIVERSITY OF EXTRACELLULAR ENZYMES

The problem of the diversity of extracellular enzymes is closely linked to species diversity. Bacterioplankton diversity is an issue that has only been recently addressed with the introduction of molecular tools in aquatic microbial ecology. Species diversity is defined by at least two terms, the richness of the species and the evenness of the distribution of the number of individuals per species (178). Almost all the molecular tools available thus far for community characterization are methods based on polymerase chain reaction (PCR) and therefore allow only the determination of the richness (179,180), but not of the evenness, because of the potential selective amplification of the DNA in the PCR. An exception are hybridization techniques, such as fluorescent in situ hybridization, although these techniques have limitations under open ocean conditions (181). It is likely that bacterioplankton exhibit species succession similar to that of phytoplankton, although there are relatively few reports addressing this ecologically important question (182,183). These successions of bacterioplankton species might be, among other factors, also the response to qualitative and quantitative changes in the DOM. Hence, one would expect to detect shifts in the enzymatic activity and in its kinetic parameters during such successions. As discussed previously in this chapter (also see Table 3), increases in the K_m values for different extracellular enzymes from eutrophic to oligotrophic conditions have been reported (113).

In 1999 Pinhassi et al. (184) investigated the response of seawater mesocosms to protein enrichment. They detected a shift in bacterial community composition that paralleled changes in enzyme activities, as measured by several fluorescent substrate analogs. Their data supported the hypothesis that changes in levels of enzyme activities were related to changes in microbial population composition, not simply due to enzyme induction in a static population. This notion also is supported by the work of Riemann et al. (185), who found significant changes in microbial community composition and parallel changes in potential enzyme activities during the colonization and decay of a phytoplankton bloom.

Biphasic kinetics of extracellular enzyme activity have been reported for bacterioplankton (25,186). Using bacterial isolates, however, such a bi- or multiphasic mode of the enzymes has not been detected. However, different bacterial strains isolated from the water column all exhibited different enzyme characteristics in terms of K_m and V_{max} (Arrieta, unpublished observations 1999). Thus, one might assume that along with shifts in the species composition of the bacterioplankton community over time, a shift in the characteristics of the particular extracellular enzymes takes place, by analogy with the shifts detected from eutrophic to oligotrophic conditions.

Investigations of the diversity of extracellular enzymes in natural aquatic environments are rather limited. In an attempt to characterize the diversity of bacterial β-glucosidase in the water column of the Adriatic Sea, Rath and Herndl (187) detected only two different β-glucosidases. However, using capillary electrophoresis, up to 8 different bacterial β-glucosidases were detected in a single sample and a total of 11 β-glucosidases during the wax and wane of the spring phytoplankton bloom in the coastal North Sea (188). Major changes in the diversity of the β-glucosidases were accompanied by shifts in the species composition of the bacterioplankton (188). This novel technique allows not only the separation of the different bacterial extracellular enzymes but also the simultaneous determination of the kinetic parameters of the individual enzymes. A wider application of this promising new tool probably will lead to a better understanding of the diversity of bacterial extracellular enzymes in the natural aquatic environment. It might also provide new insights into the dynamics of the transformation of the different compound classes of the DOM.

The need to improve our understanding of substrate specificity among extracellular enzymes, as well as the diversity of organisms that produce them, is also highlighted by a comparative study in 2000 of enzyme activity in bottom water and sediments from Skagerrak (North Sea/Baltic Sea transition) (162). Potential hydrolysis rates of six polysaccharides (pullulan, laminarin, xylan, fucoidan, arabinogalactan, chondroitin sulfate) were compared in bottom water and underlying surface sediments. Three of the six polysaccharides (pullulan, arabinogalactan, fucoidan) were not hydrolyzed in bottom water, although they were (and in the case of pullulan, rapidly) hydrolyzed in sediments (162). Molecular biological investigations have revealed fundamental differences between free-living bacteria and bacteria concentrated on particles and in sediments (63,189); the extracellular enzymes expressed by these communities may differ as well.

VIII. MOLECULAR BIOLOGICAL AND BIOTECHNOLOGICAL ASPECTS OF MARINE-DERIVED ENZYMES

Until recently, investigations of marine-derived enzymes were hampered by the necessity of isolating a microbe in pure culture and then obtaining sufficient quantities of an enzyme for biochemical and structural investigations. These requirements are now a less formidable obstacle, thanks to developments in molecular biology. In particular, the ability to obtain and amplify DNA from cultured or uncultured microbes, and to express foreign genes in host organisms, has greatly increased the range of enzymes that can be sought and characterized.

Cottrell and colleagues (190) investigated the chitinases of uncultured marine microbes by such a strategy. They extracted DNA from seawater, then used a lambda phage cloning vector to produce libraries of genomic DNA. These libraries were screened for chitin-hydrolyzing activity by using MUF-β-*d*-*N*, *N*′-diacetylchitobioside, and chitobiase activity was then assayed in protein extracts prepared from the positive clones. The chitinases of marine bacteria were also studied by constructing PCR primers based on chitinase genes of four members of the γ-proteobacteria (191). These PCR primers were used to amplify DNA collected from Pacific Ocean surface waters and the Delaware Bay estuary. Although some of the chitinase genes obtained from uncultured marine organisms were similar or identical to genes from cultured members of the *Roseobacter* sp. group (α-proteobacteria), none was identical to genes of the γ-proteobacteria. Approaches similar to those of Cottrell and colleagues should prove to be a promising means of investigating and identifying enzymes available to the vast majority of marine organisms that have not been isolated in pure cultures.

Investigations of marine-derived enzymes over the past several decades have focused particularly on enzymes from thermophilic and hyperthermophilic organism, and their potential use in biotechnological and industrial processes. Probably the best-known examples of marine-derived enzymes in commercial applications are the DNA polymerases used in PCR, such as *Pfu* (derived from *Pyrococcus furiosus*; Stratagene) and Vent (New England Biolabs), derived from *Thermococcus litoralis*. Hyperthermophilic enzymes are also potentially important for industrial conversion of starch, as well as for pulp and paper processing. For example, heat-stable pullulanase has been cloned from *Pyrococcus woesei* (192) and amylopullulanase from *P. furiosus* (193). Both enzymes were expressed in *Escherichia coli*, and the recombinant enzymes were purified and characterized.

A further major focus has been the biochemical basis of enzyme thermostability (194–197). Comparing the structures of thermophilic and hyperthermophilic enzymes to those of their mesophilic counterparts has shown that relatively subtle changes in a small set of amino acids can confer significant structural stability to an enzyme. Cold-adapted enzymes have received less attention than their heat-adapted counterparts, but these enzymes also are potentially useful for biotechnological applications, as well as for fundamental investigations of protein function and structure. The molecular basis of cold adaptation of α-amylase and triosphosphate isomerase enzymes from Antarctic bacteria has been investigated in detail (198). Very few crystal structures are available for cold-adapted enzymes, but use of modeling programs allows investigations of enzyme conformation. In general, high homology with mesophilic enzymes has been observed for amino acid sequences corresponding to substrate binding and catalytic centers (199). Bioactivity screening of cold-adapted enzymes from a range of Antarctic organisms was reviewed by Nichols and colleagues in 1999 (200).

IX. CONCLUSIONS

A number of new questions and insights have begun to emerge from a diverse range of recent studies. The hydrolysis of high-molecular-weight substrates to sequentially smaller sizes has been monitored directly by a combination of liquid chromatography and nuclear magnetic resonance spectroscopy (201), as well as liquid chromatography and fluorescence detection (38,137,159). Direct detection of bond cleavage and systematic changes in substrate molecular weight have shown that enzymatic hydrolysis rates may, in certain cases, actually outpace terminal remineralization processes. In enrichments from anaerobic marine sediments, production of oligosaccharides from polysaccharides clearly exceeded uptake of the newly formed oligosaccharides (201). Field studies also support these results: a comparison of amino acid turnover and peptide hydrolysis rates showed that production of amino acids potentially exceeded uptake by a factor of ca. 8 in coastal sediments (38). Likewise, a comparison of potential enzyme activities and sedimentary carbohydrate and amino acid inventories (202) suggested that potential hydrolysis rates on time scales of hours were fundamentally mismatched with sedimentary inventories of carbohydrates and amino acids. In deep sea sediments, potential enzyme activities could theoretically exceed total C input by a factor of 200 (145). Kirchman and White (168) found that hydrolysis rates of ^{14}C-chitin in waters of the Delaware Estuary also generally exceeded chitin mineralization rates.

In some cases, therefore, extracellular enzymes clearly have the potential to hydrolyze suitable substrates very rapidly. In the water column, the persistence of carbohydrate-containing DOC (see Sec. IV. C). demonstrates that not all organic C that can be measured chemically as carbohydrates is equally available to microbial enzymes. The persistence of significant quantities of organic C in sediments, including both carbohydrate- and amino acid–containing components, over time scales of hundreds to many thousands of years (203,204) additionally underscores this point. These geochemical results also support studies that suggest that deep sea and polar microbial benthic communities are limited by substrate availability, not by permanently low temperatures and subsequently slow enzyme activities (121,144,161,202,205).

If bacteria in marine environments have a suite of enzymes that potentially can function rapidly, what impedes turnover of organic matter, and why might a microbial community be substrate-limited? Perhaps the bacteria lack the ''right tool to fit the job,'' possibly as a result of geochemical transformations of organic matter that may occur on very short time scales (206,207). In addition, the substrates and substrate analogs currently used to measure enzyme activities in marine systems most likely inadequately represent the diversity of structures actually present in marine environments. Bacteria capable of hydrolyzing fluorogenic substrate analogs are not necessarily able to use the high-molecular-weight polymers these analogs are supposed to represent; likewise, some polymer degraders do not hydrolyze the substrate analogs (122). Additionally, such substrate analogs may measure activities of periplasmic as well as other extracellular enzymes (106), whereas true macromolecules are too large to enter the periplasm prior to hydrolysis (208). Kinetic measurements derived from model substrates and actual polymers also differ significantly (209). Since small substrate analogs likely preferentially measure the activities of exo-acting enzymes, a lack of correspondence with endo-acting enzymes, which cleave polymers at midchain, is not surprising. Bacteria typically can express a range of both exo- and endo-acting extracelllar enzymes. These enzymes typically have distinct domains for binding and cleavage (210), which may not be met by the substrate analogs. Data suggesting that water column and sedimentary microbial communities differ fundamentally in their abilities to hydrolyze polysaccharides (162) (see Sect. IV. G) highlight some of the gaps in our knowledge of marine microbial enzymes.

Limitations in our current experimental approaches also are reflected by the fact that total bacterial number or bacterial production frequently exhibits variable or weak correlations with enzyme activities (134,140,211). An obstacle to establishing these correlations is the probability that only specific members of the total microbial community are producing the enzymes in question. Establishing correlations between enzymatic hydrolysis and uptake of low-molecular-weight radiolabeled substrates is also complicated by the possibility that a limited subset of bacteria may produce a given extracellular enzyme, whereas a wider range of organisms take up the simple substrates used to measure incorporation. This strategy of substrate utilization has been observed in anaerobic cultures of rumen bacteria (212). In addition, some marine organisms cannot grow on monosaccharides, although they are capable of growth on disaccharides and polysaccharides (213). Under these circumstances, measurements of monosaccharide turnover would underestimate carbohydrate uptake among marine bacteria.

The difficulties in determining the major factors controlling enzyme activity is illustrated by the range of responses obtained from a variety of enrichment experiments. Meyer-Reil and Köster (144) found that fluorescein diacetate hydrolysis (used as a measure of intracellular esterase activity) increased in a sediment slurry in response to addition of boiled plankton debris. Addition of POM and DOM derived from net plankton to sediment slurries, however, yielded a mixed response, including increases, decreases, and no change in activities, as measured by a range of substrate analogs for different enzyme activities (214). In a further set of experiments, addition of a suite of specific macromolecules (protein, starch, cellulose, chitin) to sediment slurries yielded enhanced enzyme activities for some but not all enzyme proxies. In general, β-glucosidase activity was induced by substrate addition, whereas leucine aminopeptidase activity declined at high substrate concentrations (147,215). Clearly, control of enzyme production at the cellular level is complex, and not well understood in natural systems.

Establishing correlations between microbial communities and enzyme activities is further complicated by the possibility that free enzymes may contribute significantly to substrate hydrolysis. Bacteria in sediments may use enzymes to "forage" for substrates: i.e., release of free enzymes could yield a net positive return for particle-attached microbes (125,216). Such enzymes also might remain active in sediments after burial to greater depth, further contributing to a mismatch between measures of total microbial populations and activities of specific classes of enzymes. More specific detection of enzyme activities, and correlations with the organisms responsible for production of those specific enzymes, will be required in order to gain a thorough understanding of the connections between extracellular enzyme production and substrate utilization.

New approaches also are needed to investigate the particulate-dissolved transition of organic matter: sedimentary metabolism is fueled by particle input, and methods to investigate this critical phase transition are still sparse. One notable study in this regard was carried out by Vetter and Deming (125), who determined that bacteria physically isolated from a solid substrate could release enough enzymes to provide sufficient C for growth. The solid-dissolved transition may represent one bottleneck in C remineralization. The persistence of carbohydrate-containing components of DOC in seawater (66), which has an average age of ca. 6000 years (217), however, demonstrates that phase transition is not the only factor slowing remineralization of organic C. Our inability to characterize marine DOM structurally in detail is paralleled by a dearth of information about the actual macromolecules available to bacteria (218) and the specific substrates that they metabolize in marine waters and sediments.

REFERENCES

1. F Azam. Microbial control of oceanic carbon flux: The plot thickens. Science 280:694–696, 1998.
2. H-G Hoppe, H Ducklow, B Karrasch. Evidence for dependency of bacterial growth on enzymatic hydrolysis of particulate organic matter in the mesopelagic ocean. Mar Ecol Prog Ser 93:277–283, 1993.
3. RMW Amon, R Benner. Rapid cycling of high-molecular-weight dissolved organic matter in the ocean. Nature 369:549–552, 1994.
4. RMW Amon, R Benner. Bacterial utilization of different size classes of dissolved organic matter. Limnol Oceanogr 41:41–51, 1996.
5. RJ Chróst. Microbial ectoenzymes in aquatic environments. In: J Overbeck, RJ Chróst, eds. Aquatic Microbial Ecology: Biochemical and Molecular Approaches. New York: Brock/Springer, 1990, pp 47–78.
6. G Riley. Organic aggregates in seawater and the dynamics of their formation and utilization. Limnol Oceanogr 8:372–381, 1963.
7. GE Walsh. Studies on dissolved carbohydrate in Cape Cod waters. I. General survey. Limnol Oceanogr 10:570–576, 1965.
8. KM Khailov, ZZ Finenko. Organic macromolecular compounds dissolved in sea-water and their inclusion into the food chains. In: JH Steele, ed. Marine Food Chains. Edinburgh: Oliver & Boyd, 1970, pp 6–18.
9. J Overbeck. Early studies on ecto- and extracellular enzymes in aquatic environments. In: RJ Chróst, ed. Microbial Enzymes in Aquatic Environments. Berlin: Springer Verlag, 1991, pp 1–5.
10. F Azam, T Fenchel, JG Field, JS Gray, L-A Meyer-Reil, F Thingstad. The ecological role of water-column microbes in the sea. Mar Ecol Prog Ser 10:257–263, 1983.
11. RJ Chróst. Environmental control of the synthesis and activity of aquatic microbial ectoenzymes. In: RJ Chróst, ed. Microbial, Enzymes in Aquatic Environments. Berlin: Springer Verlag, 1991, pp 29–59.
12. RG Wetzel. Extracellular enzymatic interactions: Storage, redistribution, and interspecific communication. In: RJ Chróst, ed. Microbial Enzymes in Aquatic Environments. Berlin: Springer Verlag, 1991, pp 6–28.
13. H-G Hoppe. Significance of exoenzymatic activities in the ecology of brackish water: Measurements by means of methylumbelliferyl-substrates. Mar Ecol Prog Ser 11:299–308, 1983.
14. AL Rosso, F Azam. Proteolytic activity in coastal oceanic waters: depth distribution and relationship to bacterial populations. Mar Ecol Prog Ser 41:231–240, 1987.
15. JV Rego, G Billen, A Fontigny, M Somville. Free and attached proteolytic activity in water environments. Mar Ecol Prog Ser 21:245–249, 1985.
16. BC Crump, JA Baross, CA Simenstad. Dominance of particle-attached bacteria in the Columbia River estuary, USA. Aquat Microb Ecol 14:7–18, 1998.

17. M Karner, GJ Herndl. Extracellular enzymatic activity and secondary production in free-living and marine snow associated bacteria. Mar Biol 113:341–347, 1992.
18. MC Murrell, JT Hollibaugh, MW Silver, PS Wong. Bacterioplankton dynamics in northern San Francisco Bay: Role of particle association and seasonal freshwater flow. Limnol Oceanogr 44:295–308, 1999.
19. U Passow, AL Alldredge. Distribution, size and bacterial colonization of transparent exopolymer particles (TEP) in the ocean. Mar Ecol Prog Ser 113:185–198, 1994.
20. MD Kumar, VVS Sarma, N Ramaiah, M Gauns, SN de Sousa. Biogeochemical significance of transparent exopolymer particles in the Indian Ocean. Geoph Res Lett 25(1):81–84, 1998.
21. H-G Hoppe. Microbial extracellular enzyme activity: A new key parameter in aquatic ecology. In: RJ Chróst, ed. Microbial Enzymes in Aquatic Environments. New York: Springer Verlag, 1991, pp 60–80.
22. H-G Hoppe, S-J Kim, K Gocke. Microbial decomposition in aquatic environments: Combined process of extracellular enzyme activity and substrate uptake. Appl Environ Microbiol 54:784–790, 1988.
23. M Agis, M Unanue, J Iriberri, GJ Herndl. Bacterial colonization and ectoenzymatic activity in artificial marine snow. Part II. Cleavage and uptake of carbohydrates. Microb Ecol 36: 66–74, 1998.
24. G Müller-Niklas, S Schuster, E Kaltenböck, GJ Herndl. Organic content and bacterial metabolism in amorphous aggregations of the northern Adriatic Sea. Limnol Oceanogr 39:58–68, 1994.
25. M Unanue, B Ayo, M Agis, D Slezak, GJ Herndl, J Iriberri. Ectoenzymatic activity and uptake of monomers in marine bacterioplankton described by a biphasic kinetic model. Microb Ecol 37:6–48, 1999.
26. M Unanue, I Azúa, JM Arrieta, A Labirua-Iturburu, L Egea, J Iriberri. Bacterial colonization and ectoenzymatic activity in phytoplankton-derived model particles: Cleavage of peptides and uptake of amino acids. Microb Ecol 35:136–146, 1998.
27. A Heissenberger, GG Leppard, GJ Herndl. Relationship between the intracellular integrity and the morphology of the capsular envelope in attached and free-living marine bacteria. Appl Environ Microbiol 62:4521–4528, 1996.
28. K Stoderegger, GJ Herndl. Production and release of bacterial capsular material and its subsequent utilization by marine bacterioplankton. Limnol Oceanogr 43:877–884, 1998.
29. AB Bochdansky, S Puskaric, GJ Herndl. Influence of zooplankton grazing on free dissolved enzymes in the sea. Mar Ecol Prog Ser 121:53–63, 1995.
30. T Berman, D Wynne, B Kaplan. Phosphatases revised: Analysis of particle-associated enzyme activities in aquatic systems. Hydrobiologia 207:287–294, 1990.
31. JW Ammerman. Role of ecto-phosphohydrolases in phosphorus regeneration in estuarine and coastal ecosystems. In: RJ Chróst, ed. Microbial Enzymes in Aquatic Environments. Berlin: Springer Verlag, 1991, pp 165–186.
32. CM Yentsch, SC Yentsch, J Perras. Alkaline phosphatase activity in the tropical marine blue-green algae *Oscillatoria erythrea* (''*Trichodesmium*''). Limnol Oceanogr 17:772–773, 1972.
33. AL Huber, KS Hamel. Phosphatase activities in relation to phosphorus nutrition in *Nodularia spumigena* (Cyanobacteriaceae). 2. Laboratory studies. Hydrobiologia 123:81–88, 1985.
34. I Hernández, M Christmas, JM Yelloly, BA Whitton. Factors affecting surface alkaline phosphatase activity in the brown alga *Fucus spiralis* at a North Sea intertidal site (Tyne Sands, Scotland). J Phycol 33:569–575, 1997.
35. S-J Kim. Untersuchungen zur heterotrophen Stoffaufnahme und extrazellulären Enzymaktivität von freilebenden und angehefteten Bakterien in verschiedenen Gewässerbiotopen. Dissertation, University of Kiel, 1985, p 203.
36. DC Smith, GF Steward, RA Long, F Azam. Bacterial mediation of carbon fluxes during a diatom bloom in a mesocosm. Deep Sea Res II 42:75–97, 1995.

37. JT Hollibaugh, F Azam. Microbial degradation of dissolved proteins in seawater. Limnol Oceanogr 28:1104–1116, 1983.
38. S Pantoja, C Lee. Peptide decomposition by extracellular hydrolysis in coastal seawater and salt marsh sediment. Mar Chem 63:273–291, 1999.
39. F Azam, DC Smith. Bacterial influence on the variability in the ocean's biogeochemical state: A mechanistic view. In: S Demers, ed. Particle Analysis in Oceanography. Berlin, Heidelberg: Springer, 1991, pp 213–236.
40. DC Smith, M Simon, AL Alldredge, F Azam. Intense hydrolytic enzyme activity on marine aggregates and implications for rapid particle dissolution. Nature 359:139–142, 1992.
41. JA Fuhrman, F Azam. Thymidine incorporation as a measure of heterotrophic bacterioplankton production in marine surface waters: evaluation and field results. Mar Biol 66:109–120, 1982.
42. D Kirchman, E K'Ness, R Hodson. Leucine incorporation and its potential as a measure of protein synthesis by bacteria in natural aquatic systems. Appl Environ Microbiol 49:599–607, 1985.
43. BC Cho, F Azam. Major role of bacteria in biogeochemical fluxes in the ocean's interior. Nature 332:441–443, 1988.
44. S Schuster, JM Arrieta, GJ Herndl. Adsorption of dissolved free amino acids on colloidal DOM enhances colloidal DOM utilization but reduces amino acid uptake by orders of magnitude in marine bacterioplankton. Mar Ecol Prog Ser 166:99–108, 1998.
45. DM Karl, GA Knauer, JH Martin. Downward flux of particulate organic matter in the Ocean: A particle decomposition paradox. Nature 332:438–441, 1988.
46. JR Christian, DM Karl. Bacterial ectoenzymes in marine waters: Activity ratios and temperature responses in three oceanographic provinces. Limnol Oceanogr 40:1042–1049, 1995.
47. JR Christian, DM Karl. Ectoaminopeptidase specificity and regulation in Antarctic pelagic microbial communities. Aquat Microb Ecol 15:303–310, 1998.
48. H Kim, WN Lipscomb. Structure and mechanism of bovine lens leucine aminopeptidase. Adv Enzymol 68:153–213, 1994.
49. RB Rivkin, E Swift. Characterization of alkaline phosphatase and organic phosphorus utilization in the oceanic dinoflagellate *Pyrocystis noctiluca*. Mar Biol 61:1–8, 1980.
50. VI Krupyanko, AI Kudryavtseva, AY Valiakhmetov, AI Severin, GV Abramochkin, AM Zyakun, YN Lysogorskaya, IY Philippova, IS Kulaev, VM Stepanov. Substrate specificity of neutral metalloproteinase of the enzyme preparation of lysomidase from the culture fluid of Pseudomonadaceae. Biochemistry (Moscow) 54:1140–1149, 1998.
51. GJ Herndl, G Müller-Niklas, J Frick. Major role of ultraviolet-B in controlling bacterioplankton growth in the surface of the ocean. Nature 361:717–719, 1993.
52. G Müller-Niklas, A Heissenberger, S Pukaric, GJ Herndl. Ultraviolet-B radiation and bacterial metabolism in coastal waters. Aquat Microb Ecol 9:111–116, 1995.
53. E Paasche, SR Erga. Phosphorus and nitrogen limitation in the Oslofjord (Norway). Sarsia 73:229–243, 1988.
54. MD Krom, N Kress, S Brenner. Phosphorus limitation of primary production in the eastern Mediterranean Sea. Limnol Oceanogr 36:424–432, 1991.
55. JB Cotner, JW Ammerman, ER Peele, E Bentzen. Phosphorus-limited bacterioplankton growth in the Sargasso Sea. Aquat Microb Ecol 13:141–149, 1997.
56. T Zohary, RD Robarts. Experimental study of microbial P limitation in the eastern Mediterranean. Limnol Oceanogr 43:387–395, 1998.
57. MJ Perry. Alkaline phosphatase activity in subtropical Central North Pacific waters using a sensitive fluorometric method. Mar Biol 15:113–119, 1972.
58. S Myklestad, E Sakshaug. Alkaline phosphatase activity of *Skeletonema costatum* populations in the Trondheimsfjord. J Plankton Res 5:557–564, 1983.
59. ST Dyhrman, BP Palenik. The identification and purification of a cell-surface alkaline phos-

phatase from the dinoflagellate *Prorocentrum minimum* (Dinophyceae). J Phycol 33:602–612, 1997.
60. W Siuda, H Güde. The role of phosphorus and organic carbon compounds in regulation of alkaline phosphatase activity and P regeneration processes in eutrophic lakes. Pol Arch Hydrobiol 41:171–187, 1994.
61. H-G Hoppe, S Ullrich. Profiles of ectoenzymes in the Indian Ocean: Phenomena of phosphatase activity in the mesopelagic zone. Aquat Microb Ecol 19:129–138, 1999.
62. F Azam. Microbial response to mucilage in the Gulf of Trieste. Eos (suppl), Transactions, American Geographical Union, Vol. 79, No. 1, poster abstract No. OS22A-5, 1998.
63. EF DeLong, DG Franks, AL Alldredge. Phylogenetic diversity of aggregate-attached vs. free-living marine bacterial assemblages. Limnol Oceanogr 38:924–934, 1993.
64. JI Hedges. Global biogeochemical cycles: progress and problems. Mar Chem 39:67–93, 1992.
65. DA Hansell, CA Carlson. Deep-ocean gradients of dissolved organic carbon. Nature 395: 443–453, 1998.
66. R Benner, JD Pakulski, M McCarthy, JI Hedges, PG Hatcher. Bulk chemical characteristics of dissolved organic matter in the ocean. Science 255:1561–1564, 1992.
67. R Benner, B Biddanda, B Black, M McCarthy. Abundance, size distribution, and stable carbon and nitrogen isotopic compositions of marine organic matter isolated by tangential-flow filtration. Mar Chem 57:243–263, 1997.
68. L Guo, PH Santschi, KW Warnken. Dynamics of dissolved organic carbon (DOC) in oceanic environments. Limnol Oceanogr 40:1392–1403, 1995.
69. L Guo, PH Santschi, LA Cifuentes, SE Trumbore, J Southon. Cycling of high-molecular-weight dissolved organic matter in the Middle Atlantic Bight as revealed by carbon isotopic (^{13}C and ^{14}C) signatures. Limnol Oceanogr 41:1242–1252, 1996.
70. B Biddanda, R Benner. Carbon, nitrogen, and carbohydrate fluxes during the production of particulate and dissolved organic matter by marine phytoplankton. Limnol Oceanogr 42:506–518, 1997.
71. M McCarthy, J Hedges, R Benner. Major biochemical composition of dissolved high molecular weight organic matter in seawater. Mar Chem 55:281–297, 1996.
72. JA Fuhrman. Marine viruses and their biogeochemical and ecological effects. Nature 399: 541–548, 1999.
73. I Obernosterer, GJ Herndl. Phytoplankton extracellular release and bacterial growth: Dependence on the inorganic N:P ratio. Mar Ecol Prog Ser 116:247–257, 1995.
74. SL Strom, R Benner, S Ziegler, MJ Dagg. Planktonic grazers are a potentially important source of marine dissolved organic carbon. Linmol Oceanogr 42:1364–1374, 1997.
75. MG Weinbauer, P Peduzzi. Effect of virus-rich high molecular weight concentrates of seawater on the dynamics of dissolved amino acids and carbohydrates. Mar Ecol Prog Ser 127: 245–253, 1995.
76. LI Aluwihare, DJ Repeta. A comparison of the chemical characteristics of oceanic DOM and extracellular DOM produced by marine algae. Mar Ecol Prog Ser 186:105–117, 1999.
77. LI Aluwihare, DJ Repeta, RF Chen. A major biopolymeric component to dissolved organic carbon in surface water. Nature 387:166–169, 1997.
78. A Biersmith, R Benner. Carbohydrates in phytoplankton and freshly produced dissolved organic matter. Mar Chem 63:131–144, 1998.
79. CJ Gobler, DA Hutchins, NS Fisher, EM Cospe, SA Sañudo-Wilhelmy. Release and bioavailability of C, N, P, Se, and Fe following viral lysis of a marine chrysophyte. Limnol Oceanogr 42:1492–1504, 1997.
80. AG Murray. Phytoplankton exudation: exploitation of the microbial loop as a defense against algal viruses. J Plankton Res 17:1079–1094, 1995.

81. TF Thingstad, JR Dolan, JA Fuhrman. Loss rate of an oligotrophic bacterial assemblage as measured by ^{3}H-thymidine and $^{32}PO_4$: good agreement and near-balance with production. Aquat Microb Ecol 10:29–36, 1996.
82. P le B Williams. The importance of losses during microbial growth: Commentary on the physiology, measurement and ecology of the release of dissolved organic material. Mar Microb Food Webs 4:175–206, 1990.
83. AW Decho. Microbial exopolymers secretions in ocean environments:Their role(s) in food webs and marine processes. Oceanogr Mar Biol Ann Rev 28:73–153, 1990.
84. GG Leppard. Evaluation of electron microscope techniques for the description of aquatic colloids. In: J Buffle, HP von Leeuwen, ed. Environmental Particles. Boca Raton, FL: Lewis Publishers, 1992, pp 231–289.
85. GG Leppard, BK Burnison, J Buffle. Transmission electron microscopy of the natural organic matter of surface waters. Anal Chim Acta 232:107–121, 1990.
86. GG Leppard, A Heissenberger, GJ Herndl. Ultrastructure of marine snow. I. Transmission electron microscopy methodology. Mar Ecol Prog Ser 135:289–298, 1996.
87. WC Chin, MV Orellana, P Verdugo. Spontaneous assembly of marine dissolved organic matter into polymer gels. Nature 395:568–572, 1998.
88. I Koike, S Hara, K Terauchi, K Kogure. Role of sub-micrometre particles in the ocean. Nature 345:242–244, 1990.
89. ML Wells, ED Goldberg. Occurrence of small colloids in sea water. Nature 353:342–344, 1991.
90. ML Wells, ED Goldberg. Marine submicron particles. Mar Chem 40:5–18, 1992.
91. ML Wells, ED Goldberg. Colloid aggregation in seawater. Mar Chem 41:353–358, 1993.
92. ML Wells, ED Goldberg. The distribution of colloids in the North Atlantic and Southern Oceans. Limnol Oceanogr 39:286–302, 1994.
93. ML Wells. A neglected dimension. Nature 391:530–531, 1998.
94. AL Alldredge, U Passow, BE Logan. The existence, abundance, and significance of large transparent exopolymer particles in the ocean. Deep Sea Res I 40:1131–1140, 1993.
95. S Schuster, GJ Herndl. Formation and significance of transparent exopolymeric particles in the northern Adriatic Sea. Mar Ecol Prog Ser 124:227–236, 1995.
96. U Passow, A Alldredge. Aggregation of a diatom bloom in a mesocosm: The role of transparent exopolymer particles (TEP). Deep Sea Res II 42:99–109, 1995.
97. RA Long, F Azam. Abundant protein-containing particles in the sea. Aquat Microb Ecol 10: 213–221, 1996.
98. GJ Herndl, P Peduzzi. Ecology of amorphous aggregations (marine snow) in the Northern Adriatic Sea. I. General considerations. PSZNI Mar Ecol 9:79–90, 1988.
99. E Kaltenböck, GJ Herndl. Ecology of amorphous aggregations (marine snow) in the Northern Adriatic Sea. IV. Dissolved nutrients and the autotrophic community associated with marine snow. Mar Ecol Prog Ser 87:147–159, 1992.
100. MW Silver, AL Alldredge. Bathypelagic marine snow: deep sea algal and detrital community. J Mar Res 39:501–530, 1981.
101. N Blackburn, T Fenchel, J Mitchell. Microscale nutrient patches in planktonic habitats shown by chemotactic bacteria. Science 282:2254–2256, 1998.
102. AL Alldredge, MW Silver. Characteristics, dynamics and significance of marine snow. Prog Oceanogr 20:41–82, 1988.
103. GJ Herndl. Marine snow in the northern Adriatic Sea: Possible causes and consequences for a shallow ecosystem. Mar Microb Food Webs 6:149–172, 1992.
104. FG Priest. Extracellular Enzymes: Aspects of Microbiology 9. Wokingham: Van Nostrand Reinhold, 1984, p 79.
105. LM Robson, GH Chambliss. Cellulases of bacterial origin. Enzyme Microb Technol 11:626–644, 1989.

106. J Martinez, F Azam. Periplasmic aminopeptidase and alkaline phosphatase activities in a marine bacterium: Implications for substrate processing in the sea. Mar Ecol Prog Ser 92: 89–97, 1993.
107. W Harder, L Dijkhuizen. Physiological responses to nutrient limitation. Annu Rev Microbiol 37:1–23, 1983.
108. RB Rivkin, MR Anderson. Inorganic nutrient limitation of oceanic bacterioplankton. Limnol Oceanogr 42:730–740, 1997.
109. RE Hecky, P Campbell, LL Hendzel. The stoichiometry of carbon, nitrogen, and phosphorus in particulate matter of lakes and oceans. Limnol Oceanogr 38:709–724, 1993.
110. L Tupas, I Koike. Amino acid and ammonia utilization by heterotrophic marine bacteria grown in enriched seawater. Limnol Oceanogr 35:1145–1155, 1990.
111. NOG Jørgensen, N Kroer, RB Coffin, MP Hoch. Relations between bacterial nitrogen metabolism and growth efficiency in an estuarine and an open-water ecosystem. Aquat Microb Ecol 18:247–261, 1999.
112. M Karner, D Fuks, GJ Herndl. Bacterial activity along a trophic gradient. Microb Ecol 24: 243–257, 1992.
113. J Rath, C Schiller, GJ Herndl. Ectoenzymatic activity and bacterial dynamics along a trophic gradient in the Caribbean Sea. Mar Ecol Prog Ser 102:89–96, 1993.
114. AB Patel, K Fukami, T Nishijima. Regulation of seasonal variability of aminopeptidase activities in surface and bottom waters of Uranouchi Inlet, Japan. Aquat Microb Ecol 21:139–149, 2000.
115. M Nausch, F Pollehne, E Kerstan. Extracellular enzyme activities in relation to hydrodynamics in the Pomeranian Bight (Southern Baltic Sea). Microb Ecol 36:251–258, 1998.
116. M Karner, F Rassoulzadegan. Extracellular enzyme activity: Indications for high short-term variability in a coastal marine ecosystem. Microb Ecol 30:143–156, 1995.
117. RJ Chróst, B Riemann. Storm-stimulated enzymatic decomposition of organic matter and bacterial secondary production in benthic/pelagic coastal mesocosms. Mar Ecol Pog Ser 108: 185–192, 1994.
118. RG Keil, DL Kirchman. Utilization of dissolved protein and amino acids in the northern Sargasso Sea. Aquat Microb Ecal 18:293–300, 1999.
119. I Koike, T Nagata. High potential activity of extracellular alkaline phosphatase in deep waters of the central Pacific. Deep Sea Res II 44:2283–2294, 1997.
120. H-G Hoppe, K Gocke. The influence of global climate and hydrography on microbial activity in the ocean: Results of a N-S Atlantic transect. Proceedings of International Symposium on Environmental Microbiology. Seoul National University, 1993, 93–110.
121. C Arnosti, BB Jørgensen, J Sagemann, B Thamdrup. Temperature dependence of microbial degradation of organic matter in marine sediments: polysaccharide hydrolysis, oxygen consumption, and sulfate reduction. Mar Ecol Prog Ser 159:59–70, 1998.
122. E Helmke, H Weyland. Effect of temperature on extracellular enzymes occurring in permanently cold marine environments. Kieler Meeresforsch Sonderh 8:198–204, 1991.
123. E Helmke, H Weyland. Bacteria in sea ice and underlying water of the eastern Wedell Sea in midwinter. Mar Ecol Prog Ser 117:269–287, 1995.
124. H-G Hoppe, K Gocke, J Kuparinen. Effect of H_2S on heterotrophic substrate uptake, extracellular enzyme activity and growth of brackish water bacteria. Mar Ecol Prog Ser 64:157–167, 1990.
125. YA Vetter, JW Deming. Growth rates of marine bacteria isolates on particulate organic substances solubilized by freely released extracellular enzymes. Microb Ecol 37:86–94, 1999.
126. DL Santavy, WL Jeffrey, RA Anyder, J Campbell, P Malouin, L Cole. Microbial community dynamics in the mucus of healthy and stressed corals hosts. Bull Mar Sci 54:1077–1087, 1994.
127. SS Oosterhuis, MA Baars, WCM Klein-Breteler. Release of the enzyme chitobiase by the

copepod *Temora longicornis*: Characteristics and potential tool for estimating crustacean biomass production in the sea. Mar Biol Prog Ser 196:195–206, 2000.
128. M Karner, C Ferrier-Pagès, F Rassoulzadegan. Phagotrophic nanoflagellates contribute to occurrence of α-glucosidase and aminopeptidase in marine environments. Mar Ecol Prog Ser 114:237–244, 1994.
129. GA Vargo, E Shanley. Alkaline phosphatase activity in the red-tide dinoflagellate, *Ptychodiscus brevis*. PSZNI Mar Ecol 6:251–264, 1985.
130. TK Sawyer, TS Nerad, PM Daggett, SM Bodammer. Potentially pathogenic protozoa in sediments from oceanic sewage-disposal sites. In: JM Capuzzo, DR Kester, eds. Oceanic Processes in Marine Pollution. Vol. 1. Biological Processes and Wastes in the Ocean. Malaba: Krieger, 1987, pp 183–194, 1987.
131. PA Sobecky, MA Schell, MA Moran, RE Hodson. Impact of a genetically engineered bacterium with enhanced alkaline phosphatase activity on marine phytoplankton communities. Appl Environ Microbiol 62:6–12, 1996.
132. W Reichardt. Enzymatic potential for decomposition of detrital biopolymers in sediments from Kiel Bay. Ophelia 26:369–384, 1986.
133. L-A Meyer-Reil. Seasonal and spatial distribution of extracellular enzymatic activities and microbial incorporation of dissolved organic substrates in marine sediments. Appl Environ Microbiol 53:1748–1755, 1987.
134. LM Mayer. Extracellular proteolytic enzyme activity in sediments of an intertidal mudflat. Limnol Oceanog 34:973–981, 1989.
135. SM Henrichs. Early diagenesis of organic matter in marine sediments: Progress and perplexity. Mar Chem 39:119–149, 1992.
136. NE Blair, LA Levin, DJ DeMaster, G Plaia. The short-term fate of fresh algal carbon in continental slope sediment. Limnol Oceanogr 41:1208–1219, 1996.
137. GM King. Characterization of β-glucosidase activity in intertidal marine sediments. Appl Environ Microbiol 51:373–380, 1986.
138. C Arnosti. Measurement of depth- and site-related differences in polysaccharide hydrolysis rates in marine sediments. Geochim Cosmochim Acta 59:4247–4257, 1995.
139. K Poremba, H-G Hoppe. Spatial variation of benthic microbial production and hydrolytic enzymatic activity down the continental slope of the Celtic Sea. Mar Ecol Prog Ser 118: 237–245, 1995.
140. A Boetius, S Scheibe, A Tselepides, H Thiel. Microbial biomass and activities in deepsea sediments of the Eastern Mediterranean: Trenches are benthic hotspots. Deep Sea Res I 43: 1439–1460, 1996.
141. M Köster, P Jensen, L-A Meyer-Reil. Hydrolytic activities of organisms and biogenic structures in deep-sea sediments. In: RJ Chróst, ed. Microbial Enzymes in Aquatic Ecosystems. Berlin: Springer-Verlag, 1991, pp 298–310.
142. A Boetius. Microbial hydrolytic enzyme activities in deep-sea sediments. Helgoländer Meeresuntersuchungen 49:177–187, 1995.
143. A Boetius, E Damm. Benthic oxygen uptake, hydrolytic potentials and microbial biomass at the Arctic continental slope. Deep Sea Res I 45:239–275, 1998.
144. L-A Meyer-Reil, M Köster. Microbial life in pelagic sediments: The impact of environmental parameters on enzymatic degradation of organic material. Mar Ecol Prog Ser 81:65–72, 1992.
145. K Poremba. Hydrolytic enzymatic activity in deep-sea sediments. FEMS Microb Ecol 16: 213–222, 1995.
146. JW Deming, PL Yager. Natural bacterial assemblages in deep-sea sediments: Towards a global view. In: GT Rowe, V Pariente, ed. Deep-Sea Food Chains and the Global Carbon Cycle. Dordrecht: Kluwer, 1992, pp 11–27.
147. A Boetius, K Lochte. Effect of organic enrichments on hydrolytic potentials and growth of bacteria in deep-sea sediments. Mar Ecol Prog Ser 140:239–250, 1996.
148. J Martinez, DC Smith, GF Steward, F Azam. Variability in ectohydrolytic enzyme activities

of pelagic marine bacteria and its significance for substrate processing in the sea. Aquat Microb Ecol 10:223–230, 1996.

149. M Manafi, W Kneifel, S Bascomb. Fluorogenic and chromatogenic substrates used in bacterial diagnostics. Microb Rev 55:335–348, 1991.
150. RP Haugland, ID Johnson. Detecting enzymes in living cells using fluorogenic substrates. J Fluoresc 3:119–127, 1993.
151. M Somville. Measurement and study of substrate specificity of exoglucosidase activity in eutrophic water. Appl Environ Microbiol 48:1181–1185, 1984.
152. H-G Hoppe. Use of fluorogenic model substrates for extracellular enzyme activity (EEA) measurement of bacteria. In: PF Kemp, BF Sherr, EB Sherr, JJ Cole, eds. Current methods in aquatic microbial ecology. Boca Raton, FL: CRC Press, 1993, pp 423–431.
153. M Somville, G Billen. A method for determining exoproteolytic activity in natural waters. Limnol Oceanogr 28:190–193, 1983.
154. G Müller-Niklas, GJ Herndl. Activity of fecal coliform bacteria measured by 4-methylumbelliferyl-β-D-glucuronide substrate in the northern Adriatic Sea with special reference to marine snow. Mar Ecol Prog Ser 89:305–309, 1992.
155. M Köster, S Dahlke, L-A Meyer-Reil. Microbiological studies along a gradient of eutrophication in a shallow coastal inlet in the Southern Baltic Sea (Nordrügensche Bodden). Mar Ecol Prog Ser 152:27–39, 1997.
156. JM Arrieta, GJ Herndl. Variability of β-glucosidases in seawater. Abstract book, Symposium on Ectoenzymes in the Environment, Granada, Spain, 1999.
157. JW Ammerman, WB Glover. Continuous underway measurement of microbial ectoenzyme activities in aquatic ecosystems. Mar Ecol Prog Ser 201:1–12, 2000.
158. D Delmas, C Legrand, C Bechemin, C Collinot. Exoproteolytic activity determined by flow injection analysis: Its potential importance for bacterial growth in coastal marine ponds. Aquat Living Res 7:17–24, 1994.
159. S Pantoja, C Lee, JF Marecek. Hydrolysis of peptides in seawater and sediments. Mar Chem 57:25–40, 1997.
160. C Arnosti. A new method for measuring polysaccharide hydrolysis rates in marine environments. Organic Geochem 25:105–115, 1996
161. C Arnosti. Rapid potential rates of extracellular enzymatic hydrolysis in Arctic sediments. Limnol Oceanogr 43:315–324, 1998.
162. C Arnosti. Substrate specificity in polysaccharide hydrolysis: Contrasts between bottom water and sediments. Limnol Oceanogr 45:1112–1119, 2000.
163. SA Green, NV Blough. Optical absorption and fluorescence properties of chromophoric dissolved organic matter in natural waters. Limnol Oceanogr 39:1903–1916, 1994.
164. S-J Kim, H-G Hoppe. Microbial extracellular enzyme detection on agar-plates by means of fluorogenic methylumbelliferyl-substrates. GERBAM-Deuxième Colloque de Bactériology marine-CNRS, IFREMER, Actes de Colloques 3:175–183, 1986.
165. S Gonzáles-Gil, BA Keafer, RVM Jovine, A Aguilera, S Lu, DM Anderson. Detection and quantification of alkaline phosphatase in single cells of phosphorus-starved marine phytoplankton. Mar Ecol Prog Ser 164:21–35, 1998.
166. RA Long, LB Fandino, GF Steward, P Del Negro, P Ramani, B Cataletto, C Welker, A Puddu, S Fonda, F Azam. Microbial response to mucilage in the Gulf of Trieste. Eos (suppl), Transactions, American Geographical Union, Vol. 79, No. 1, poster abstract No. OS22A-5, 1998.
167. J Molano, A Duran, E Cabib. A rapid and sensitive assay for chitinase using tritiated chitin. Anal Biochem 83,648–656, 1977.
168. DL Kirchman, J White. Hydrolysis and mineralization of chitin in the Delaware Estuary. Aquat Microb Ecol 18:187–196, 1999.
169. AL Svitil, SMN Chadhain, JA Moore, DL Kirchman. Chitin degradation proteins produced

by the marine bacterium *Vibrio harveyi* growing on different forms of chitin. Appl Environ Microbiol 63:408–413, 1997.
170. MT Cottrell, DN Wood, L Yu, DL Kirchman. Selected chitinase genes in cultured and uncultured marine bacteria in the α- and γ-subclasses of the Proteobacteria. Appl Environ Microbiol 66:1195–1201, 2000.
171. RB Coffin. Bacterial uptake of dissolved free and combined amino acids in estuarine waters. Limnol Oceanogr 34:531–542, 1989.
172. RG Keil, DL Kirchman. Bacterial hydrolysis of protein and methylated protein and its implications for studies of protein degradation in aquatic systems. Appl Environ Microbiol 58: 1374–1375, 1992.
173. GT Taylor. Microbial degradation of sorbed and dissolved protein in seawater. Limnol Oceanogr 40:875–885, 1995.
174. JW Ammerman, F Azam. Bacterial 5′-nucleotidase activity in estuarine and coastal marine waters: Characterization of enzyme activity. Limnol Oceanogr 36:1427–1436, 1991
175. H-G Hoppe, K Gocke, D Zamorano, R Zimmermann. Degradation of macromolecular organic compounds in a tropical lagoon (Cienaga Grande, Colombia) and its ecological significance. Int Rev Ges Hydrobiol 68:811–824, 1983.
176. R Panosso, FA Esteves. Phosphatase activity and plankton dynamics in two tropical lagoons. Arch Hydrobiol 146:341–354, 1999.
177. H Li, MJW Veldhuis, AF Post. Alkaline phosphatase activities among planktonic communities in the northern Red Sea. Mar Ecol Prog Ser 173:107–115, 1998.
178. C Pedros-Alio. Diversity of bacterioplankton. TREE 8:86–90, 1993.
179. SJ Giovannoni, TB Britschgi, CL Moyer, KG Field. Genetic diversity in Sargasso Sea bacterioplankton. Nature 345:60–63, 1990.
180. MM Moeseneder, JM Arrieta, G Muyzer, C Winter, GJ Herndl. Optimization of terminal-restriction fragment length polymorphism analysis for complex marine bacterioplankton communities and comparison with denaturing gradient gel electrophoresis. Appl Environ Microbiol 65:3518–3525, 1999.
181. AE Murray, KY Wu, CL Moyer, DM Karl, EF DeLong. Evidence for circumpolar distribution of planktonic Archaea in the Southern Ocean. Aquat Microb Ecol 18:263–273, 1999.
182. J Pinhassi, UL Zweifel, Å Hagström. Dominant marine bacterioplankton species found among colony-forming bacteria. Appl Environ Microbiol 63:3359–3366, 1997.
183. A-S Rehnstam, S Bäckman, DC Smith, F Azam, Å Hagström. Blooms of sequence-specific culturable bacteria in the sea. FEMS Microbiol Ecol 102:161–166, 1993.
184. J Pinhassi, F Azam, J Hemphala, RA Long, J Martinez, UL Zweifel, Å Hagström. Coupling between bacterioplankton species composition, population dynamics, and organic matter decomposition. Aquat Microb Ecol 17:13–26, 1999.
185. L Riemann, GF Steward, F Azam. Dynamics of bacterial community composition and activity during a mesocosm bloom. Appl Environ Microbiol 66:578–587, 2000.
186. V Talbot, M Bianchi. Bacterial proteolytic activity in sediments of the Subantarctic Indian Ocean sector. Deep Sea Res II 44:1069–1084, 1997.
187. J Rath, GJ Herndl. Characteristics and diversity of β-D-glucosidase (EC 3.2.1.21) activity in marine snow. Appl Environ Microbiol 60:807–813, 1994.
188. JM Arrieta, GJ Herndl in press. A new approach to assess the diversity of marine bacterial β-glucosidase using capillary electrophoresis zymography. Appl Environ Microbiol.
189. E Llobet-Brossa, R Rossello-Mora, R Amann. Microbial community composition of Wadden Sea sediments as revealed by fluorescence *in situ* hybridization. Appl Environ Microbiol 64: 2691–2696, 1998.
190. MT Cottrell, JA Moore, DL Kirchman. Chitinases from uncultured marine microorganisms. Appl Environ Microbiol 65:2553–2557, 1999.

191. MT Cottrell, DL Kirchman. Natural assemblages of marine proteobacteria and members of the Cytophaga-Flavobacter cluster consuming low- and high-molecular-weight dissolved organic matter. Appl Environ Microbiol 66:1692–1697, 2000.
192. A Rudiger, PL Jorgensen, G Antranikian. Isolation and characterization of a heat-stable pullulanase from the hyperthermophilic archaeon *Pyrococcus woesei* after cloning and expression of its gene in *Escherichia coli*. Appl Environ Microbiol 61:567–575, 1995.
193. G Dong, C Vielle, A Savchenko, JG Zeikus. Cloning, sequencing, and expression of the gene encoding extracellular α-amylase from *Pyrococcus furiosus* and biochemical characterization of the recombinant enzyme. Appl Environ Microbiol 63:3569–3576, 1997.
194. MWW Adams. Enzymes and proteins from organisms that grow near and above 100°C. Ann Rev Microbiol 47:627–658, 1993.
195. MWW Adams, RM Kelly. Enzymes from microorganisms in extreme environments. Chem and Eng News, Dec. 18, 1995, pp. 32–42.
196. MJ Danson, DW Hough. Structure, function, and stability of exoenzymes from the Archaea Trends Microbiol 6:307–314, 1998.
197. JG Zeikus, C Vielle, A Savchenko. Thermozymes: Biotechnology and structure-function relationships. Extremophiles 2:179–183, 1998.
198. G Feller, E Narinx, JL Arpigny, M Aittaleb, E Baise, S Genicot, C Gerday. Enzymes from psychrophilic organisms. FEMS Microbiol Rev 18:189–202, 1996.
199. G Feller, C Gerday. Psychrophilic enzymes: molecular basis of cold adaptation. Cell Mol Life Sci 53:830–841, 1997.
200. D Nichols, J Bowman, K Sanderson, C Nichols, T Lewis, T McMeekin, PD Nichols. Developments with Antarctic microorganisms: culture collections, bioactivity screening, taxonomy, PUFA production and cold-adapted enzymes. Curr Opin Biotechnol 10:240–246, 1999.
201. C Arnosti, DJ Repeta, NV Blough. Rapid bacterial degradation of polysaccharides in anoxic marine systems. Geochim Cosmochim Acta 58:2639–2652, 1994.
202. M Fabiano, R Danovaro. Enzymatic activity, bacterial distribution, and organic matter composition in sediments of the Ross Sea (Antarctica). Appl Environ Microbiol 64:3838–3845, 1998.
203. GL Cowie, JI Hedges, FG Prahl, GJ De Lange. Elemental and major biochemical changes across an oxidation front in a relict turbidite: An oxygen effect. Geochim Cosmochim Acta 59:33–46, 1995.
204. SG Wakeham, C Lee, JI Hedges, PJ Hernes, ML Peterson. Molecular indicators of diagenetic status in marine organic matter. Geochim Cosmochim Acta 61:5363–5369, 1997.
205. PA Wheeler, M Gosselin, E Sherr, D Thibault, DL Kirchman, R Benner, TE Whitledge. Active cycling of organic carbon in the central Arctic Ocean. Nature 380:697–699, 1996.
206. RG Keil, DL Kirchman. Abiotic transformation of labile protein to refractory protein in seawater. Mar Chem 45:187–196, 1994.
207. NH Borch, DL Kirchman. Protection of protein from bacterial degradation by submicron particles. Aquat Microb Ecol 16:265–272, 1999.
208. R Benz. Structure and function of porins from gram-negative bacteria. Annu Rev Microbiol 42:359–393, 1988.
209. G Feller, F Payan, F Theys, M Qian, R Haser, C Gerday. Stability and structural analysis of α-amylase from the Antarctic psychrophile *Alteromonas haloplanctis* A23. Eur J Biochem 222:441–447, 1994.
210. RAJ Warren. Microbial hydrolysis of polysaccharides. Annu Rev Microbiol 50:183–212, 1996.
211. A Boetius, T Ferdelman, K Lochte. Bacterial activity in sediments of the deep Arabian Sea in relation to vertical flux. Deep Sea Res II 47:2835–2875, 2000.
212. MA Cotta. Interaction of ruminal bacteria in the production and utilization of maltooligosaccharides from starch. Appl Environ Microbiol 58:48–54, 1992.

213. WM de Vos, SWM Kengen, WGB Voorhorst, J van der Oost. Sugar utilization and its control in hyperthermophiles. Extremophiles 2:201–205, 1998.
214. A Boetius, K Lochte. Regulation of microbial enzymatic degradation of organic matter in deep-sea sediments. Mar Ecol Prog Ser 104:299–307, 1994.
215. A Boetius, K Lochte. High proteolytic activities of deep-sea bacteria from oligotrophic polar sediments. Arch Hydrobiol 48:269–276, 1996.
216. YA Vetter, JW Deming, PA Jumars, BB Krieger-Brockett. A predictive model of bacterial foraging by means of freely released extracellular enzymes. Microb Ecol 36:75–92, 1998.
217. PM Williams, ERM Druffel. Radiocarbon in dissolved organic matter in the central North Pacific Ocean. Nature 330:246–248, 1987.
218. JI Hedges, G Eglinton, PG Hatcher, DL Kirchman, C Arnosti, S Derenne, RP Evershed, I Kögel-Knabner, JW de Leeuw, R Littke, W Michaelis, and J Rullkötter. The molecularly-uncharacterized component (MUC) in nonliving organic matter in natural environments. Organ Geochem 31:954–958, 2000.
219. H-G Hoppe. Relations between bacterial extracellular enzyme activity and heterotrophic substrate uptake in a brackish water environment. GERBAM-Deuxième Colloque de Bactériology marine-CNRS, IFREMER, Actes de Colloques 3:119–128, 1986.
220. M Nausch. Alkaline phosphatase activities and the relationship to inorganic phosphate in the Pomeranian Bight (Southern Baltic Sea). Aquat Microb Ecol 16:87–94, 1998.
221. MA Cunha, MA Almeida, F Alcântara. Patterns of ectoenzymatic and heterotrophic bacterial activities along a salinity gradient in a shallow tidal estuary. Mar Ecol Prog Ser 204:1–12, 2000.
222. A Saliot, G Cauwet, G Cahet, D Mazaudier, R Daumas. Microbial activities in the Lena River Delta and Laptev Sea. Mar Chem 53:247–254, 1996.
223. A Fontigny, G Billen, J Vives-Rego. Some kinetic characteristics of exoproteolytic activity in coastal seawater. Estuar Coast Shelf Sci 25:127–133, 1987.
224. M Middelboe, NH Borch, DL Kirchman. Bacterial utilization of dissolved free amino acids, dissolved combined amino acids and ammonium in the Delaware Bay estuary: Effects of carbon and nitrogen limitation. Mar Ecol Prog Ser 128:109–120, 1995.
225. LA Meyer-Reil. Measurement of hydrolytic activity and incorporation of dissolved organic substrates by microorganisms in marine sediments. Mar Ecol Prog Ser 31:143–149, 1986.
226. C Arnosti, M Holmer. Carbohydrate dynamics and contributions to the carbon budget of an organic-rich coastal sediment. Geochim Cosmochim Acta 63:393–403, 1999.
227. YA Vetter, JW Deming. Extracellular enzyme activity in the Arctic Northeast Water Polynya. Mar Ecol Prog Ser 114:23–34, 1994.
228. A Boetius, T Ferdelmann, and K Lochte. Bacterial activity in sediments of the deep Arabian Sea in relation to vertical flux. Deep Sea Research II 47:2835–2875, 2000.

4

Enzymes and Microorganisms in the Rhizosphere

David C. Naseby
University of Hertfordshire, Hatfield, Hertfordshire, England

James M. Lynch
University of Surrey, Guildford, Surrey, England

I. INTRODUCTION

Although the term rhizosphere was coined by Hiltner (1) to describe specifically the interaction between the roots of legumes and bacteria, the usage of the term has broadened. Today we consider the rhizosphere to be the zone of influence of all plant root systems, and it includes the various cell layers of the root itself (the endorhizosphere), which microorganisms can colonize; the root surface (the rhizoplane); and the region surrounding the root (the ectorhizosphere). There has been steady progress in our understanding of the rhizosphere during the last 100 years and an increasing realization that it can have many influences on crop productivity (Table 1). However, there has been a sudden increase in interest and investment in rhizosphere research in the past few years. This is largely due to the growth of biotechnology in relation to agriculture and, in particular, our increasing ability to manipulate plants by using recombinant deoxyribonucleic acid (DNA). Even with the genetic information of the plant modified, however, there will still be a highly influential microbial population associated with the roots. Therefore, the possibility exists that the balance between beneficial and harmful rhizosphere microorganisms might be improved not only by modification of the plant genotype but also by inoculation with useful and compatible microorganisms. The source of these useful bacteria and fungi could be either other soils and rhizospheres, hence amplifying the natural population, or introduction of "foreign" organisms that have been genetically modified to elevate their useful properties and/or increase their competence to colonize the rhizosphere. Given current public opinion, the latter is unlikely in the near future.

The general properties of rhizospheres have been described in several books (2–5). Not surprisingly two highly integrated microbe-plant root associations have attracted the most intense study. The first of these involves the activities of symbiotic dinitrogen-fixing bacteria. These bacteria express the nitrogenase enzyme complex and have received much attention in recent years as the various *nif* genes have been fully characterized (6–8). The other major group of microbe-plant symbionts in soil, the mycorrhiza, which promote the

Table 1 Influences of Rhizosphere Microorganisms on Crop Productivity

Beneficial effects	Harmful effects
Nutrient and water uptake enhancement	Nutrient competition
Dinitrogen fixation	Denitrification
Phytohormone production	Phytotoxin production
Symbiosis	Pathogenesis
Disease and pest control	Oxygen depletion
Soil stabilization	Degradation of soil structure

solubilization and uptake of phosphorus and other plant nutrients by a variety of biotic and abiotic mechanisms (9,10), have also been studied extensively (see Chapter 5).

Rhizosphere enzymology has been particularly exploited in biomonitoring and impact analysis. Most attempts to monitor the effects of microorganisms in the rhizosphere have centered on the enumeration of specific populations. However, for a significant perturbation to be measured, changes of between 100% and 300% are necessary to produce a significant impact. Standard population measurements, assessing the survival and dissemination and effect on total indigenous populations, do not give an indication of the functioning of the ecosystem. There is very little literature regarding the functional impact of microbes on ecosystems, for instance, the effect upon nutrient cycling. In this review functional methodology is assessed for its application as an indicator of the impact on the soil ecosystem. In 1997 a range of soil enzyme assays was used as an alternative to, or in combination with, population measurements (34). These appear quite sensitive, as impacts of less than 20% on ecosystem disturbance could be significantly detected. Enzyme functions of this kind in the rhizosphere are potentially exploitable in processes such as biocontrol and bioremediation, and these are the major focus of this review.

II. SOIL ENZYMES AS FUNCTIONAL INDICATORS

Soil enzyme measurements are extremely important in assessing the status or the condition of the soil environment. This is because enzymes are essential to the cycling of nutrients in soil and are thus critical to the availability of nutrients to both microbiota and plants. Frequently enzymes are regulated and externalized as a response to exogenous soil conditions, such as phosphatase secretion due to phosphate deficiency (37,38) or plant chitinase as a result of fungal or insect attack (11), and are thus good indicators of change or perturbation. The second example shows that soil enzymes are responding to the functional status of the biota, in addition to the chemical environment as measured by nutrient cycle measurements (Fig. 1). Many extracellular enzymes are adsorbed to, complexed with, or entrapped within soil clays and humics (12). As a consequence, they may have a long-term stability and thus can provide an indication of the history of the sample and not just a snapshot from the time of sampling. However, as Figure 1 indicates, the ultimate assessment of the effect of the gene products is the plant bioassay (especially if the human perspective is considered), by which productivity and nutritional values are measured.

Many enzymes are secreted actively into the soil environment by microbes and by

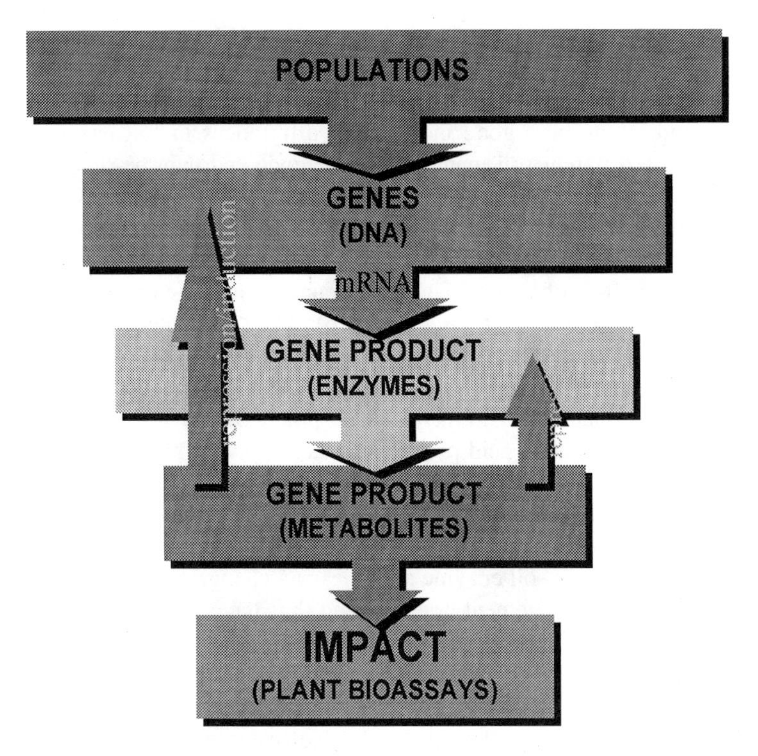

Figure 1 A sequence illustrating the impact of a rhizosphere microorganism. The first functional product of the deoxyribonucleic acid (DNA) that can be monitored is messenger ribonucleic acid (RNA) which produces the gene products (enzymes and metabolites).

plant roots and function by solubilizing nutrients; others are released into the soil by the lysis of microbial and plant cells. As mentioned, these extracellular enzymes often are immobilized by soil components and thus protected from degradation. There are several mechanisms by which enzymes can become immobilized in the soil, including adsorption on and within clay particles and organic matter, chemical complexing with organic matter, and entrapment within microbial polymers or soil aggregates (13). Once immobilized, the enzymes still may be active in the soil, but at variably reduced rates caused by factors such as steric hindrance, diffusional resistance, blocking of active sites, electrostatic effects, or reversible denaturation (14).

If soil enzymes can be assessed and shown to vary with different soil treatments, then such measurements could be useful indicators of perturbation and of changes in functional diversity. Land management practices have been shown to have a significant influence on soil enzymes; for example, it has been demonstrated that cultivation can have a large effect on the soil arylsulfatase activity (15). Thus, long-term grassland soil cultivation (69 years) caused a 66% reduction in arylsulfatase activity, whereas cultivation of a forest soil for 47 years resulted in an 88% decrease. Large differences in sulfatase activity were found between a cultivated forest soil (63% decrease) and a similar forest soil that had been left fallow for 5 years (30% decrease).

In the same study (15) changes in the kinetic properties of the soil sulfatases were investigated. After cultivation, a reduction was found in the V_{max} (74% in the grassland soil and 90% in the forest soil) and the Michaelis constant (K_m), which decreased with the duration (years) of cultivation. Variations in K_m values obtained from the forest and grassland soils suggest that the origins of the enzymes were different in the two soils. Therefore, such kinetic studies can be a good indicator of differences in soil enzymes with soil type and under very different soil management methods and may be sensitive enough to be applied to assessing the changes caused by less drastic soil treatments. Overall, sulfatase activity was extremely sensitive to the soil management system and sulfatase may be one enzyme that is a good indicator of small-scale perturbations.

A second example of the effect of land management practices on soil enzyme activities is provided by the work of Jordan and associates (16). They evaluated microbial measurements (including acid and alkaline phosphatase activities) as indicators of soil quality in long-term cropping practices in historical fields. Differences in phosphatase activities among soils could be related to land management practices and soil properties (especially organic matter). For example, acid phosphatase activity was 150% higher in soil under continuous corn with no-till and receiving full fertility treatment than under conventional tillage with no fertility treatment. Alkaline phosphatase activity was 50% higher in soil under continuous corn with full fertility treatment than without the fertility treatment. This study again shows that soil enzyme activities can change greatly after the large perturbations caused by soil management practices and thus have great potential as a general indicator of smaller-scale perturbation.

In another study concentrating on the impact of crop and land management systems on soil enzyme activities from a coconut-based multistoried cropping system (17), decreases in urease and an unspecified phosphatase (but not dehydrogenase) with increasing depth in rhizosphere soil were found. Differences were found also in urease and phosphatase with the different crops grown. Cocoa and pineapple produced contrasting results to the coconut rhizosphere (higher phosphatase activity in the coconut rhizosphere and higher urease activity with multistoried systems over a coconut monocrop). Depth and crop system appear to have a large influence on rhizosphere enzyme activities.

The effect of pesticides on the activity of various soil enzymes has been investigated by a number of researchers (18,19). A prime example of this is the work of Satpathy and Behera (20), who examined the effect of malathion on cellulase, protease, urease, and phosphatase activities in a tropical grassland soil. They found that the significantly depressed cellulase and protease activity recovered to almost the same level as that of the control 21 days after malathion application, whereas urease and, to a larger extent, phosphatase activities did not show such a rapid recovery.

A similar situation to the addition of pesticides to soil is the input of inorganic pollutants, which can have major effects not just on enzymes but on the whole soil ecosystem. Decreases in cellulase activity were found with increasing sulfur, nitrogen, and heavy metal pollution (21). The reduction in cellulase activity was correlated with a decrease in respiration along the same pollution gradient. However, cellulase activity was a more precise indicator of the level of pollution than were respiration measurements.

III. MICROBIAL INOCULATION AND SOIL ENZYME ACTIVITIES

A proposed target for genetic manipulation is the insertion or enhancement of genes encoding specific enzymes; a prime candidate for this is chitinase because of its postulated role in the biocontrol of fungal crop pathogens. Ridout and colleagues (22) investigated the protein production induced in a *Trichoderma* species when cell wall fragments of the crop plant pathogen *Rhizoctonia solani* (to which the *Trichoderma* species was an antagonist) were used as the sole carbon source. Both β-glucanase and chitinase were found to be important components of the inducible extracellular proteins analyzed by electrophoretic profiles. The sequence of chitin degradation (and pathogen biocontrol) was extremely complex with several other inducible enzymes probably involved.

Much of the interest in the enzymology of *Trichoderma* sp. pathogen biocontrol has focused on chitinase and 1,3-β-D-glucanase, although proteases could be involved (22,23). Furthermore, many authors have suggested that chitinase may be the most important enzyme in biocontrol by *Trichoderma* sp., and because the enzyme can be purified and characterized readily (24), it has been a good target for transformation (25). As yet, however, it has not been possible to demonstrate that introduction of the gene into a nonproducing strain increases biocontrol potential. However, much has been learned about gene expression and its regulation during mycoparasitism of fungal pathogens by *Trichoderma* sp. by using gene probes and markers (26–28).

Other rhizosphere microorganisms produce chitinase and exert biocontrol effects; notable among these are the actinomycetes (29). To detect these chitinase genes, a *chiA* DNA probe is available (30). It should be remembered, however, that detection of a gene does not necessarily mean the gene will be expressed, and therefore, in the example here, a positive signal from the probe does not necessarily mean that chitinase is expressed—merely that the genotype or potential exists.

The situation in which genes encoding enzymes (especially extracellular) are the target of manipulation gives an added emphasis to the use of soil enzyme activities as a measure of perturbations caused by the introduction of such microorganisms into the soil and rhizosphere. In an attempt to understand the effect of manipulated enzyme production on a soil microorganism, one study involved rhizosphere growth of *Pseudomonas solanacearum*, which was genetically altered in extracellular enzyme production (endopolygalactouronase A and endoglucanase) (31). The strains that had been enhanced in terms of enzyme production had greatly reduced fitness in the rhizosphere. However, the authors did not use the opportunity to study the effect on soil extracellular enzyme activity with the inoculation of strains with such functional modifications.

There are few data regarding the use of soil enzymes as a measure of perturbations caused by the introduction of extraneous microorganisms into the soil or the rhizosphere, and much of the available information is contradictory. Doyle and Stotzky (32) evaluated a number of enzymes, including arylsulfatase, dehydrogenase, and acid and alkaline phosphatase, for the detection of changes in the microbial ecological characteristics of the soil caused by the introduction of *Escherichia coli* but found no consistent significant differences in any of the enzyme activities. It must be noted, however, that this experiment (as with a great many others) did not reflect ''ecologically relevant'' conditions, as the strains used were not natural soil organisms and the experiment was conducted in bulk soil in the absence of plants.

Commercial inoculants often are designed to live and perform functions in the rhizosphere and respond to root exudates (a major source of carbon and nutrients). Therefore, it is not surprising that the work of Doyle and Stotzky (32) found few effects upon enzyme activities. The survival and metabolic activity of the *E. coli* strain were likely to be low since there was no additional substrate and indigenous carbon would have been unsuitable or utilized by the preexisting microflora. Furthermore, *E. coli* is not a soil organism and is unlikely to establish a viable population after introduction into soil. In contrast,

Mawdsley and Burns (33) introduced a soil *Flavobacterium* species into the rhizosphere of wheat. They found that the microbial inoculant caused increased α-galactosidase, β-galactosidase, α-glucosidase, and β-glucosidase activities in the more ecologically and agriculturally relevant conditions of the wheat rhizosphere.

A number of simple enzyme assays for the detection of perturbations resulting from different soil treatments, including the introduction of a modified *Pseudomonas fluorescens* SBW25 strain and substrate amendments, have been designed (34). The aim of the experiment was to deduce whether these assays were sensitive enough to measure perturbations caused by microbial inoculation. Specific attention was paid to the validation of soil biochemical techniques as a method of monitoring the effects of inoculation. Differences in wheat rhizosphere soil biomass (measured by adenosine triphosphate [ATP] content) and several key soil enzyme activities with microbial inoculation and/or modified conditions (addition of the enzyme substrates, chitin, urea, and glycerophosphate, to soil) were measured.

Microbial biomass, as well as enzyme activities (with the exception of that of acid phosphate), decreased with depth. The addition of a substrate mixture of urea, colloidal chitin, and glycerophosphate to soil significantly increased *N*-acetyl glucosaminidase, chitobiosidase, arylsulfatase, and urease activities but did not cause a change in acid and alkaline phosphatase and phosphodiesterase activities. Inoculation of wheat seeds with *P. fluorescens* resulted in significant increases in rhizosphere chitobiosidase and urease activities at 5- to 20-cm depth and a significant decrease in alkaline phosphatase activity. The 0- to 5-cm depth activities were unchanged. Inoculation with *P. fluorescens* in combination with the substrate mixture had effects opposite to those of treatments without substrate mix addition: chitobiosidase, arylsulfatase, and urease activities were significantly lower and alkaline phosphatase was significantly higher at the 5- to 20-cm depth interval with inoculation of bacteria. Biomass values for the combined bacteria and substrate mix treatment were significantly higher than for the substrate mix alone treatment.

IV. SOIL ENZYME ACTIVITIES AND GENETICALLY MODIFIED ORGANISMS

Naseby and Lynch (35) used enzymatic analysis, combined with microbial population measurements and other soil indicators, to investigate the effect of wild-type and genetically modified microbes in the rhizosphere. They studied whether the impact of a *P. fluorescens* strain, a genetically modified microbe (GMM) altered for kanamycin resistance and lactose utilization, could be enhanced by adding lactose and kanamycin to the soil (prior to/subsequent to amendment). Lactose addition decreased the shoot-to-root ratio and both soil amendments increased the populations of total culturable bacteria and the inoculated GMM. Only kanamycin perturbed the bacterial community dynamics, causing a shift toward slower-growing organisms. The community structure with the GMM inoculum, in the presence of kanamycin, showed the only impact of the GMM compared to that the wild-type inoculum. The shift toward K strategists (i.e., slower-growing organisms) found in the other kanamycin-amended treatments was reversed with the GMM inoculation. Lactose amendment increased acid and alkaline phosphatase, phosphodiesterase, and carbon cycle enzyme activities, whereas the kanamycin addition affected only the alkaline phosphatase and phosphodiesterase activities. None of the soil enzyme activities was affected by the GMM under any of the soil amendments.

Although this study involved the use of a genetically modified microbe, the modifications were not intended to have a functional impact; they were inserted as genetic markers. A second study comparing the effect of the same genetically marked strain to that of a functionally modified strain showed effects that are more interesting (36). The aim of this work was to determine the impact in the rhizosphere of wild type along with functionally and nonfunctionally modified *Pseudomonas fluorescens* strains. The wild-type F113 strain carried a gene encoding the production of the antibiotic 2,4-diacetylphloroglucinol (DAPG), useful in plant disease control, and was marked with a *lacZY* gene cassette. The first modified strain was a functional modification of strain F113 with repressed production of DAPG, creating the DAPG negative strain F113 G22. The second paired comparison was a nonfunctional modification of wild-type (unmarked) strain SBW25, constructed to carry marker genes only, creating strain SBW25 EeZY-6KX.

Significant perturbations were recorded in the indigenous bacterial population structure; the F113 (DAPG+) strain caused a shift toward slower-growing colonies (K strategists) compared with the non-antibiotic-producing derivative (F113 G22) and SBW25 strains. The DAPG+ strain also significantly reduced, in comparison with those of the other inocula, the total *Pseudomonas* sp. populations, but did not affect the total microbial populations. The survival of F113 and F113 G22 was an order of magnitude lower than that of the SBW 25 strains. The DAPG+ strain caused a significant decrease in the shoot-to-root ratio in comparison to that of the control and other inoculants, indicating plant stress. F113 increased soil alkaline phosphatase, phosphodiesterase, and arylsulfatase activities (Table 2) compared to those of the controls. The other inocula reduced the same enzyme activities when compared to the control. In contrast, the β-glucosidase, β-galactosidase, and *N*-acetyl glucosaminidase activities decreased with the inoculation of the DAPG+ strain (Table 1). These results indicate that soil enzymes are sensitive to the impact of GMM inoculation.

Increased available (soluble) inorganic phosphate is known to decrease soil phosphatase activity (37,38). Therefore, the F113 (DAPG+) strain must have caused a decrease in the available phosphate, thus causing an overall increase in activity. The decrease in available P may have taken the form of an increase in the available carbon in the rhizosphere (by stimulation of root exudation or leakage, as there was a decrease in the shoot/root ratio). Other studies have highlighted changes in root exudation caused by biocontrol *P. fluorescens* strains (39). Therefore, increasing the ratio of C to P available would increase the microbial P demand.

Inverse trends were found with the C and N cycle enzymes in comparison to the general trend found in the P and S cycle enzymes. The F113 (DAPG+) strain was associated with the lowest acid β-galactosidase activity, which was significantly lower than that of the SBW25 WT treatment. The F113 (DAPG+) strain produced the lowest β-glucosidase activity, which was significantly smaller than that of the control, the SBW25 WT, and the F113 G22 treatments, which all had similar activities. As with the β-galactosidase and β-glucosidase activities the F113 (DAPG+) strain produced the lowest *N*-acetyl glucosaminidase activity. This was a significantly lower activity than that of the control, the SBW25 WT, and the F113 G22 treatments, which all had comparable activities. All three carbon cycle enzyme activities, therefore, also indicate an increase in carbon availability. Subsequent work with a genetically modified DAPG overproducer found effects similar to those of the wild-type DAPG producer (40).

Further, the carbon fractions in the rhizosphere of pea plants inoculated with strains F113 and F113 G22 were examined (41). Both strains significantly increased the water-

Table 2 Soil Enzyme Activities in the Rhizosphere of Pea Plants Inoculated with Wild-Type and Genetically Modified *Pseudomonas fluorescens* Strains

Treatment[a]	Acid phosphatase[b]	Alkaline phosphatase[b]	Phosphodiesterase[b]	Arylsulfatase[b]	Acid β-galactosidase[b]	β-glucosidase[b]	NAGase[b]
Control	7.91ab ± 0.40	1.23bc ± 0.15	0.11a ± 0.018	0.09a ± 0.01	0.52ab ± 0.04	1.02b ± 0.10	0.32b ± 0.03
F113 G22	8.24ab ± 0.69	0.70a ± 0.16	0.09a ± 0.03	0.08a ± 0.02	0.1ab ± 0.05	1.02b ± 0.14	0.29b ± 0.03
F113	7.06a ± 0.40	1.52c ± 0.13	0.17b ± 0.01	0.16b ± 0.02	0.45a ± 0.04	0.62a ± 0.07	0.18a ± 0.02

[a] Treatments: control, no inocula; F 1113 G22, inoculated with *lacZY* marked DAPG− (Tn5 mutated) *P. fluorescens* F113 G22; F113, inoculated with *lac*ZY marked DAPG+ *P. fluorescens* F113.

[b] mg pNP Released g^{-1} dry soil. Standard errors for means ($n = 7$) indicated. Values followed by the same letter within a column are not significantly different at $p = 0.05$ level.

soluble carbohydrates and the total water-soluble carbon in the rhizosphere soil (Table 3). Strain F113 significantly increased the soil protein content relative to that of the control but not in relation to the F113 G22 treatment. The F113 treatment had a significantly greater organic acid content than the control and F113 G22 treatments; the F113 G22 treatment also had significantly greater content than the control. Both inocula resulted in significantly lower phosphate content than the control. Both inocula increased carbon availability; however, antibiotic production by strain F113 reduced the utilization of organic acids released from the plant, resulting in differing effects of the two strains on nutrient availability, plant growth, soil enzyme activities, and *Pseudomonas* sp. populations.

These studies highlight the importance of the use of soil enzyme activities in impact studies. They also illustrate how the combination of a number of enzymatic measurements from different nutrient cycles can be used as a diagnostic tool of the processes involved in such perturbations. The enzymatic methods worked well in small microcosms and large intact soil cores where environmental variation is minimized; the same methods then were tested in field scale trials of the F113 strain.

Soil enzyme activities were used to investigate the impact of strain F113 in the rhizosphere of field-grown sugar beet (42). There were distinct trends in rhizosphere enzyme activities in relation to soil chemical characteristics (measured by electroultrafiltration). For example, the activities of enzymes from the phosphorus cycle (acid phosphatase, alkaline phosphatase, and phosphodiesterase) and of arylsulfatase were negatively correlated with the amount of readily available P (Table 4), whereas urease activity was positively correlated with available P. Significant correlations between electroultrafiltration nutrient levels and enzyme activity in the rhizosphere were obtained, highlighting the usefulness of enzyme assays to document variations in soil nutrient cycling. Contrary to the results of previous microcosm studies, which did not investigate plants grown to maturity, the biocontrol inoculant had no obvious effect on enzyme activity or on soil chemical characteristics in the rhizosphere. The results show the importance of homogenous soil microcosm systems, used in previous work, in risk assessment studies in which inherent soil variability is minimized and an effect of the pseudomonad on soil enzymological features could be detected. This study also highlights the fact that the effect of spatial variability is far greater than the effect of any microbial inocula.

Following this study of the effects of a biocontrol agent on soil enzyme activities in a field trial, the soil enzyme methodology was used to assess the impact of genetically modified plants on soil biochemical features (D. C. Naseby, A. Greenland, and J. M.

Table 3 Carbon Fractions of Pea Rhizosphere Soil Inoculated with DAPG-Producing and -Nonproducing *Pseudomonas fluorescens* Strain F113

C fraction[a]	Control[b]	F113 G22[b]	F113[b]
Carbohydrates (H_2O)	139.44a ± 8.57	200.04b ± 8.36	209.41b ± 8.29
Water-soluble C	1855.32a ± 139.7	2731.78b ± 77.73	2669.73b ± 54.92
Protein	56.16a ± 4.78	69.56ab ± 4.66	78.38b ± 5.88
Organic acids	6.09a ± 0.36	7.18b ± 0.38	9.92c ± 0.66

[a] Expressed as ppm g^{-1} dry soil. Standard errors for means ($n = 8$) indicated. Values followed by the same letter within a column are not significantly different at $p = 0.05$ level.

[b] Treatments: control, no inocula; F113 G22, inoculated with *lacZY* marked DAPG− (Tn5 mutated) *P. fluorescens* F113 G22; F113, inoculated with *lacZY* marked DAPG+ *P. fluorescens* F113.

Table 4 Coefficients of Correlation Between Alkaline Phosphatase Activity and EUF Levels (Soil Chemicals) in a Sugar Beet Field Trial Inoculated with *Pseudomonas fluorescens* Strain F113

Treatment	Activity (mg pNP g^{-1} soil)	Org N	Ca	P	K	Mg (μg g^{-1} soil)	Na	Mn	pH	NO_3
Non Rhiz	1.658	17	519	38	118	19	32	0.5	7.0	4.0
Cont	0.671	17	549	53	388	27	59	1.2	7.2	7.2
Cont	2.261	12	480	30	208	19	27	1.2	6.9	4.7
Cont	1.624	17	396	41	210	19	24	0.6	6.9	5.4
F113	1.241	15	665	44	211	27	15	1.6	6.9	6.1
F113	2.526	14	593	31	197	25	23	1.5	7.0	5.8
F113	1.878	21	548	31	321	25	45	1.4	7.0	9.0
r value		−0.348	−0.14	−0.95**	−0.51	−0.40	−0.49	−0.11	−0.53	−0.24

** Significant at $P = 0.01$; $n = 7$.

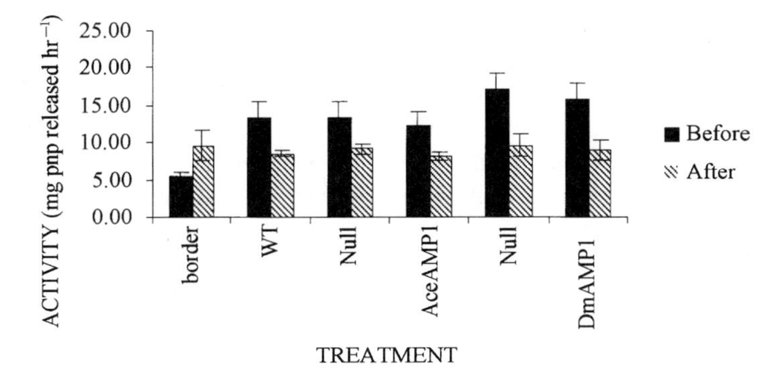

Figure 2 Alkaline phosphatase activity in the rhizosphere of genetically modified (GM) and non-GM oilseed rape. WT, wild-type control (var. Westar); border: different variety of oilseed rape; DM AMP1, transgenic, expressing antifungal protein derived from dahlia, preceded by its divergent null line (not expressing antifungal proteins); Ace AMP1, transgenic, expressing antifungal protein derived from allium, preceded by its divergent null line (not expressing antifungal proteins).

Lynch, 1999 unpublished data). The two modified oil seed rape lines used (*Brassica napus* var. Westar) produced small cysteine-rich proteins with antifungal activity, specifically expressing either the DmAMP1 gene from *Dahlia merckii* or the AceAMP1 gene from *Allium cepa*. The field trial consisted of these transgenic lines compared to a number of controls, their divergent null lines, a wild-type control (Westar), and a different variety of oil seed rape. Sampling of this trial for enzymatic analysis consisted of taking the rhizosphere soil of 10 replicated plants for each treatment. Sampling occurred over 2 days: i.e., five replicates of each treatment were taken on the first day and five on the second day of sampling. A range of soil enzyme activities was measured, and the available soil nutrients were analyzed.

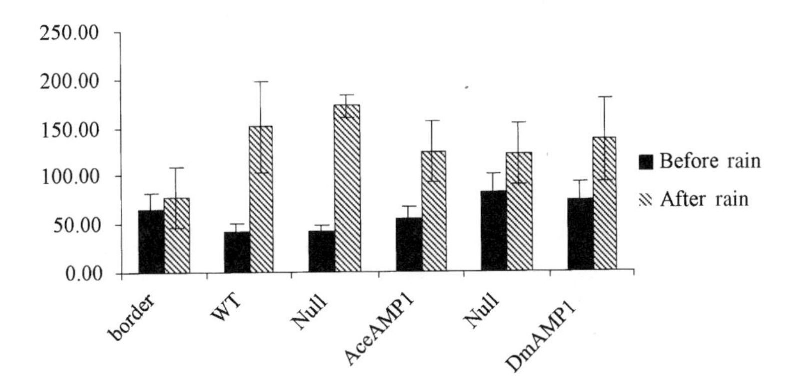

Figure 3 Available nitrate (ppm g^{-1} soil) in the rhizosphere of genetically modified (GM) and non-GM oilseed rape. WT, wild-type control (var. Westar); border, samples taken from border rows of a different variety of oilseed rape; DM AMP1, transgenic, expressing antifungal protein derived from dahlia, preceded by its divergent null line (not expressing antifungal proteins); Ace AMP1, transgenic, expressing antifungal protein derived from allium, preceded by its divergent null line (not expressing antifungal proteins).

The results showed large differences between the 2 days of sampling in soil enzyme activities (e.g., alkaline phosphatase, Fig. 2) and available soil nutrients (e.g., nitrate, Fig. 3). Differences were found also between the various oil seed rape varieties with most soil enzymes measured and with the available soil nutrients. However, there was little difference between the enzyme activities in the rhizosphere of the GM and non-GM plants. The major factor influencing the enzyme activities and soil nutrients between the two sampling days was the soil moisture content, which was increased by overnight rain. Therefore, in this field trial, the differences between soil enzyme activities were not attributable to plant genetic modification, but to environmental variation and to differences in plant variety.

V. CONCLUSIONS

Clearly enzyme activities are useful in determining perturbations in the soil environment brought about by changes in agricultural practices, the use of agrochemicals, pollution events, or the exploitation of genetically modified organisms. Biocontrol of pests and diseases is a means by which enzyme function has been exploited (43), but there is even greater opportunity to monitor and manipulate enzymes as generations of plant nutrients, plant-growth-promoting agents, soil structure stimulants, and bioremediation catalysts.

Although bioremediation has had less attention than biocontrol, the potential for exploitation is enormous (44). Most research has been focused on microbial inoculants (bioaugmentation), but it is equally relevant to consider how to optimize the function of the indigenous organisms (biostimulation). Phytoremediation, by plant roots themselves or associated microbiota (rhizoremediation), is becoming an increasingly interesting cleanup solution for soils. Most attention has been paid to heavy metal decontamination, and whereas there is inevitably some enzyme involvement, little has been characterized. However, rhizosphere microorganisms produce enzymes that have the capacity to catabolize a wide range of organic pollutants. Microbial dehalogenation is described in detail in Chapters 18 and 19, but of special interest are hydrogen cyanide and other nitriles. Not only is the xenobiotic source of cyanides from industrial pollution important, but some microorganisms themselves produce cyanides in the rhizosphere (45), and these are potentially phytotoxic. A variety of cyanide catabolism enzymes have been characterized. For example, *Fusarium* and *Trichoderma* spp. that can colonize the rhizosphere have strain-dependent capacities to catabolize cyanide with the enzymes formamide hydrolase and rhodanase (J. M. Lynch, M. Ezzi, 1999 unpublished). In comparisons, rhizosphere bacteria such as *Enterobacter* spp. produce β-cyanoalanine synthase (46). Thus, in such a rhizosphere process, the root supplies the carbon and energy for the microorganisms to produce the bioremediation, although plants themselves may also have cyanide-catabolizing potential.

The harnessing and exploitation of rhizosphere enzymes are at a very early stage of development; however, such use of rhizosphere enzymes seems set to advance rapidly. Much of the emphasis of the most recent papers on rhizosphere enzymes has concerned the biocontrol properties of rhizosphere enzymology, for example, the production of serine protease by *Stenotrophomonas maltophilia* (47). However, no relationship was found between enzyme production from fluorescent pseudomonads in the antifungal *Cylindrocladium* sp. response reported in the rhizosphere of bananas (48).

Azospirillum lipoferum is a good rhizosphere colonist and can stimulate plant growth; it produces a laccase-type polyphenol oxidase in the rhizosphere of rice, a factor that may be involved in the colonization of roots (49). However, endochitinase continues to be implicated in the biocontrol of *Pythium ultimum* by *Trichoderma* spp. (50,51).

Immunological detection of diazotrophic enterobacterial strains with growth-promoting properties, using the double-antibody sandwich enzyme-linked immunosorbent assay, has been reported (52). Such detection systems for rhizosphere diagnostics have probably not had the attention that they warrant.

Enzyme production in field soil is attracting interest. In a forested landscape (53), measurements of β-glucosidase, chitinase, phenol oxidase, and acid phosphatase with explanatory path analysis indicated that the four enzymes could help resolve spatial dependencies at a range of scales and could also be used to develop a scale-independent metric to be used for the regional analyses in a geographical information system (GIS) environment, such systems are of great interest at present to environmental engineers and provide a very exciting opportunity for collaboration with soil biologists and biochemists (see Chapter 15).

REFERENCES

1. L Hiltner. Uber neuere Erfahrungen und Problem auf dem Gebeit der Bodenbakteriologie and unter besonderer Berucksichtigung der Grundunging und Brache. Arb Dtsch Landwist Ges 98:59–78, 1904.
2. EA Curl, B Truelove. The Rhizosphere. Berlin: Springer-Verlag, 1986.
3. JM Lynch, ed. The Rhizosphere. Chichester: John Wiley, 1990.
4. R Pinton, Z Varanini, P Nannipieri, eds. The Rhizosphere: Biochemistry of Organic Substances at the Soil-Plant Interface. New York: Marcel Dekker, 2000.
5. PB Tinker, PH Nye. Solute Movement in the Rhizosphere. New York: Oxford University Press, 2000.
6. J Frazzon, IS Schrank. Sequencing and complementation analysis of the nifUSV genes from Azospirillum brasilense. FEMS Microbiol Lett 159:151–158, 1998.
7. S Lei, L Pulakat, N Gavini. Genetic analysis of nif regulatory genes by utilizing the yeast two-hybrid system detected formation of a NifL-NifA complex that is implicated in regulated expression of nif genes. J Bacteriol 181:6535–6539, 1999.
8. Mazurier, G Laguerre. Unusual localization of nod and nif genes in *Rhizobium leguminosarum* bv *viciae*. Can J Microbiol 43:399–402, 1997.
9. X Lu, T Koide. Avena-fatua 1 seed and seedling nutrient dynamics as influenced by mycorrhizal infection of the maternal generation. Plant Cell Environ 14:931–939, 1991.
10. JC Tarafdar, H Marschner. Efficiency of VAM hyphae in utilization of organic phosphorus by wheat plants. Soil Sci Plant Nutr 40:593–600, 1994.
11. XM Yu, SX Guo. Progress on plant chitinase induced by fungi. Prog Biochem Biophys 27: 40–44, 2000.
12. RG Burns. Enzyme-activity in soil—location and a possible role in microbial ecology. Soil Biol Biochem 14:423–427, 1982.
13. HH Weetall. Immobilised enzymes and their application in the food and beverage industry. Process Biochem 10:3–24, 1975.
14. G Haska. Activity of bacteriolytic enzymes adsorbed to clays. Microb Ecol 7:331–341, 1981.
15. RE Farrell, VVSR Gupta, JJ Germida. Effects of cultivation on the activity and kinetics of arylsulfatase in Saskatchewan soil. Soil Biol Biochem 26:1033–1040, 1994.
16. D Jordan, RJ Kremer, L Stanley, WA Bergfield, KY Kim, VN Cacnio. An evaluation of microbial methods as indicators of soil quality in long-term cropping practices in historical fields. Biol Fertil Soils 19:297–308, 1995.

17. BM Bopaiah, HS Shetty. Soil microflora and biological activities in the rhizospheres and root regions of coconut-based multi-storeyed cropping and coconut monocropping systems. Soil Biol Biochem 23:89–94, 1991.
18. T Endo, K Taiki, T Nobatsura. S Michihiko. Effect of insecticide cartap hydrochloride on soil enzyme activities, respiration and on nitrification. J Pesticide Sci 7:101–110, 1982.
19. PC Mishra, SC Pradhan. Seasonal variation in amylase, invertase, cellulase activity and carbon dioxide evolution in a tropical protected grassland of Orissa, India, sprayed with carbyl insecticide. Environ Pollution 43:291–300, 1987.
20. G Satpathy, N Behera. Effect of malathion on cellulase, protease, urease and phosphatase activities from a tropical grassland soil of Orissa, India. J Environ Biol 14:301–310, 1993.
21. R Ohtonen, P Lahdesmaki, AM Markkola. Cellulase activity in forest humus along an industrial pollution gradient in Oulu, Northern Finland. Soil Biol Biochem 26:97–101, 1994.
22. CJ Ridout, JR Coley-Smith, JM Lynch. Enzyme activity and electrophoretic profile of extracellular protein induced in *Trichoderma* spp. by cell walls of *Rhizoctonia solani*. J Gen Microbiol 132:2345–2352, 1986.
23. CJ Ridout, JR Coley-Smith, JM Lynch. Fractionation of extracellular enzymes from a mycoparasitic strain of *Trichoderma harzianum*. Enzyme Microb Technol 10:180–187, 1988.
24. EE Deane, JM Whipps, JM Lynch, JF Peberdy. Purification and characterization of a *Trichoderma harzianum* exochitinase. Biochem Biophys Acta 1383:105–110, 1998.
25. EE Deane, JM Whipps, JM Lynch, JF Peberdy. Transformation of *Trichoderma reesei* with a constitutively expressed heterologous fungal chitinase gene. Enzyme MicrobTechnol 24: 419–424, 1999.
26. M Lorito, RL Mach, P Sposato, J Strauss, CK Peterbauer, CP Kubicek. Mycoparasitic infection relieves binding of Cre 1 carbon catabolite repressor protein to promote sequence of *ech-42* (endochitinase—encoding) gene of *Trichoderma harzianum*. Proc Natl Acad Sci USA 93: 14868–14872, 1996.
27. RL Mach, CK Peterbauer, K Payer, S Jaksits, SL Woo, S Zelinger, CM Kullnig, M Lorito, CP Kubicek. Expression of two major chitinase genes of *Trichoderma atrovivide (T. harzianum* Pl) is triggered by different regulatory signals. Appl Environ Microbiol 65:1858–1863, 1999.
28. S Zeilinger, C Galhaup, K Payer, SL Woo, RL Mach, C Fekete, M Lorito, CP Kubicek. Chitinase gene expression during mycoparasitic interation of *Trichoderma harziamum* with its host. Fungal Genet Biol 26:131–140, 1999.
29. DL Crawford, JM Lynch, JM Whipps, MA Ousley. Isolation and characterization of actinomycete antagonists of a fungal root pathogen. Appl Environ Microbiol 59:3899–3905, 1993.
30. R Kimura, JM Lynch, K Katoh, K Miyashita. Enumeration of specifically functional soil bacteria by the DNA probe method. In: Trends in Microbial Ecology. R Guerrero, C Pedros-Alio, eds. Barcelona: Spanish Society for Microbiology, 1992, pp 655–658.
31. JW Williamson, PG Hartel. Rhizosphere growth of *Pseudomonas solanacearum* genetically altered in extracellular enzyme production. Soil Biol Biochem 23:453–458, 1991.
32. JD Doyle, G Stotzky. Methods for the detection of changes in the microbial ecology of soil caused by the introduction of micro-organisms. Microb Release 2:63–72, 1993.
33. JL Mawdsley, RG Burns. Inoculation of plants with Flavobacterium P25 results in altered rhizosphere enzyme activities. Soil Biol Biochem 26:871–882, 1994.
34. DC Naseby, JM Lynch. Rhizosphere soil enzymes as indicators of perturbation caused by a genetically modified strain of *Pseudomonas fluorescens* on wheat seed. Soil Biol Biochem 29:1353–1362, 1997.
35. DC Naseby, JM Lynch. Establishment and impact of *Pseudomonas fluorescens* genetically modified for lactose utilisation and kanamycin resistance in the rhizosphere of pea. J Appl Microbiol 84:169–175, 1998.
36. DC Naseby, JM Lynch. Soil enzymes and microbial population structure to determine the impact of wild type and genetically modified *Pseudomonas fluorescens* in the rhizosphere of pea. Mol Ecol 7:617–625, 1998.

37. MA Tabatabai. Soil enzymes. In: AL Page, RH Miller, DR Keeney, eds. Methods in Soil Analysis. Part 2. Chemical and Microbiological Properties. 2nd ed. American Society of Agronomy, 1982, pp 903–948.
38. T Tadano, K Ozowa, M Satai, M Osaki, H Matsui. Secretion of acid phosphatase by the roots of crop plants under phosphorus-deficient conditions and some properties of the enzyme secreted by lupin roots. Plant Soil 156:95–98, 1993.
39. A Mozafar, F Duss, JJ Oertli. Effect of *Pseudomonas fluorescens* on the root exudates of two tomato mutants differently sensitive to Fe chlorosis. Plant Soil 144:167–176, 1992.
40. DC Naseby, JM Lynch. Effect of 2,4 diacetylphloroglucinol producing, overproducing and non-producing *Pseudomonas fluorescens* F113 in the rhizosphere of pea. Microbial Ecology 2001.
41. DC Naseby, J Pascual, JM Lynch. Carbon fractions in the rhizosphere of pea inoculated with 2,4 diacetylphloroglucinol producing and non-producing *Pseudomonas fluorescens* F113. J Appl Microbiol 87:173–181, 1999.
42. DC Naseby, Y Moënne-Loccoz, J Powell, F O'Gara, JM Lynch. Soil enzyme activities in the rhizosphere of field-grown sugar beet inoculated with the biocontrol agent *Pseudomonas fluorescens* F113. Biol Fertil Soils 27:39–43, 1998.
43. JM Lynch, A Wiseman, eds. Environmental Biomonitoring: The Biotechnology Ecotoxicology Interface. London: Cambridge University Press, 1998.
44. RL Crawford, DL Crawford, eds. Bioremediation: Principles and Applications. New York: Cambridge University Press, 1994.
45. AM Dartnell, RG Burns. A sensitive method for measuring cyanide and cyanogenic glucosidases in sand culture and soil. Biol Fertil Soils 5:141–147, 1987.
46. CJ Knowles, AW Bunch. Microbial cyanide metabolism. Adv Mircob Physiol 27:73–111, 1986.
47. C Dunne, Y Moenne-Loccoz, FJ de Bruijn, F O'Gara. Overproduction of an inducible extracellular serine protease improves biological control of *Pythium ultimum* by *Stenotrophomonas maltophilia* stains W81. Microbiology, 146:2069–2078, 2000.
48. L Sutra, JM Risede, L Gardan. Isolation of fluorescent pseudomonads from the rhizosphere of banana plants antagonistic towards root necrosing fungi. Letts Appl Microbiol 31:289–293, 2000.
49. G Diamantidis, A Effosse, P Potier, R Bally. Purification and characterization of the first bacterial laccase in the rhizospheric bacterium *Azospirillum lipoferum*. Soil Biol Biochem 32: 919–927, 2000.
50. C Thrane, DF Jensen, A Tromso. Substrate colonization, strain competition, enzyme production *in vitro*, and biocontrol of *Pythium ultimum* by *Trichoderma* spp isolates P1 and T3. J Plant Pathol 106:215–225, 2000.
51. DC Naseby, JA Pascual, JM Lynch. Effect of five Trichoderma strains on plant growth, Pythium ultimum populations, soil microbial communities and soil enzyme activities. J Appl Microbiol 88:161–169, 2000.
52. R Remas, S Ruppel, HJ Jacob, C Hecht-Buchholz, W Merbach. Colonization behaviour of two enterobacterial strains on cereals. Biol Fertil Soils 30:550–557, 2000.
53. KLM Decker, REJ Boerner, SJ Morris. Scale-dependent patterns of soil enzyme activity in a forested landscape. Can J For Res 29:232–241, 1999.

5

Enzymes in the Arbuscular Mycorrhizal Symbiosis

José Manuel García-Garrido, Juan Antonio Ocampo, and Inmaculada García-Romera
Estación Experimental del Zaidín, CSIC, Granada, Spain

I. INTRODUCTION

Terrestrial fungi can adopt different life strategies to exploit nutrient sources. They grow as saprotrophs on simple or complex organic substrates, or they can establish a nutritional relationship with higher plants, either as biotrophs or as necrotrophs. Mycorrhizal associations are the most important mutualistic biotrophic interactions (1). Over 80% of vascular flowering plants are capable of entering into symbiotic associations with arbuscular mycorrhizal (AM) fungi (2).

The fungi that form these associations are members of the zygomycetes, and the current classification places them all into one order, Glomales (3). They are strictly dependent on their host plant to complete their life cycle, whereas other mycorrhizal fungi, such as ericoid fungi, can be grown in pure culture (4–6). The AM association is a relatively nonspecific, highly compatible, long-lasting mutuality from which both partners derive benefit. The plant supplies the fungus with carbon, on which it is entirely dependent. The fungal contribution is more complex. Although it is clear that the fungi assist the plant with the acquisition of phosphate and other mineral nutrients from the soil, AM fungi also may influence the plant's resistance to invading pathogens (7). In addition to its ecological significance, the association also may have applications in agriculture. This is particularly important for developing more sustainable systems (8) because mycorrhizae create an intimate link between the soil and the plant and may be manipulated to improve plant nutrition efficiency and soil conservation.

The interaction begins when fungal hyphae arising from spores in the soil and on adjacent colonized roots or hyphae contact the root surface. Here they differentiate to form appressoria and penetrate the root. The formation of appressoria is one of the first morphological signs that recognition between the plant and the fungus has occurred. Once inside the root, the fungus may grow both inter- and intracellularly throughout the cortex, but AM fungi do not invade the vascular or the meristematic regions. The types of internal structures that develop depend on the plant/fungal combination and may include intracellular differentiated hyphae called *arbuscules* and/or *intracellular coils* (9). Wall-like material containing proteins and polysaccharides is deposited by the continuous host plasmalemma against the wall of the fungus, forming an interfacial matrix or apoplast (10). Although the fungal hypha penetrates the cortical cell wall to form arbuscules within the cell, it does not penetrate the plant plasma membrane, which extends to surround the arbuscule (11). Arbuscules die after a few days encased in host cell wall material. The senescence of arbuscules does not affect the development of the residual mycelia, which continue to grow and form arbuscules in other parenchymal cells. The complex interaction at the cellular and molecular level that has resulted in a functional AM symbiosis must be based on highly evolved physiological and genetic coordination between fungus and host.

The variety of factors that act immediately before and after contact of an AM fungus with a root surface and might influence the success of root colonization is quite broad. However, fungal development within the host may be modulated by the ability of fungus and host to produce enzymes. The purpose of this chapter is to discuss the role of enzymes in the penetration and development of the fungus inside the plant root.

II. ENZYMATIC MECHANISMS OF PENETRATION AND FORMATION OF THE SYMBIOSIS

The mechanisms by which endomycorrhizal fungi enter and spread through host tissues are unknown. Different steps in the infection process (e.g., formation of entry points, inter- and intracellular colonization) necessitate the growth of hyphae along the middle lamella or through cell walls of the host root. Only localized changes in wall texture have been observed as endomycorrhizal fungi penetrate epidermal cells or develop through the middle lamella of parenchymal tissue, suggesting that wall-degrading enzyme activities within host tissues are very limited (12).

Biotrophic fungi usually are thought to penetrate host tissues mechanically. It has been calculated that high pressure can be generated by appressoria of *Magnaporte grisea* (a nonmycorrhizal fungus) at the penetration point (13). This mechanical pressure allows the fungus to perforate the host wall through formation of a penetration peg. Some wall components, such as melanin, are considered to play an important role in the increase of hydrostatic pressure since they act to trap solutes within the appressoria, causing water to be absorbed because of the increasing osmotic gradient (13).

Most phytopathogenic fungi and bacteria are known to produce enzymes that degrade pectin, cellulolytic, and hemicellulolytic substances (14). These hydrolytic enzymes play a fundamental role in pathogenesis (15,16). Polygalacturonase plays multiple roles during infection; this enzyme allows the fungus to colonize the host tissues and to obtain nutrients from the degradation of complex pectic substrates. Concomitantly, polygalacturonases can produce oligogalacturonides, which elicit plant defense response (17).

Many of the enzymes that degrade pectic, cellulolytic, and hemicellulolytic substances are produced by the plants themselves, including the fruits, epicotyls, cotyledons, and other growing tissues (18–20). Research is scarce on these enzymes in plant roots, and on their mode of action in the process of penetration and development of symbiotic microorganisms (21). Infection of roots by other mutualistic microorganisms, such as *Rhizobium* and *Azospirillum* species, appears to be mediated by cell wall–hydrolyzing enzymes (22–24).

The observation that arbuscular mycorrhizal (AM) fungi penetrate the plant cell wall at the site of contact during the establishment of intracellular symbiosis (25) indicates that hydrolytic enzymes may be involved in the AM colonization process. However, since AM fungi have not yet been cultured axenically in the absence of plant roots, it is difficult to confirm the production of hydrolytic enzymes by AM fungi or their possible participation in the colonization of roots. This is because of the very low levels of enzyme produced, as occurs with the other mutualistic microorganisms (24). Investigations have demonstrated the production of pectinase, cellulase, and xyloglucanase (5,6,26–32) from the external hyphae and the mycorrhizal roots. It seems that mycorrhizal fungi colonize the root tissues of their host plant by a combination of mechanical and enzymatic mechanisms (33,34). A very weak and localized production of hydrolytic enzymes by AM fungi might ensure that viability of the host is maintained and defense responses are not triggered, allowing compatibility between plant and fungi (17).

The primary (growing) cell walls of plants are rigid yet dynamic structures composed of roughly equal quantities (around 30% for each) of cellulose, hemicellulosic, and pectic polysaccharides, plus about 10% glycoproteins (hydroxyproline-rich glycoproteins and enzymes) and a small proportion of phenolic compounds (35,36). The cell wall comprises a crystalline microfibrillar array of cellulose embedded in an amorphous mass of pectic and hemicellulose materials. The AM fungi hydrolyze these cellular complexes in a very organized manner to make their entry into the root cortical cells (37). The mode of action of some of the important enzymes and the role of these enzymes in the penetration of the fungus inside the plant root are discussed later.

A. Cellulases

Cellulose is the best known of all plant cell wall polysaccharides. It is particularly abundant in secondary cell walls and accounts for 20%–30% of the dry mass of most primary cell walls (38). Chemically, cellulose is a linear β-4-linked D-glucan that provides the mechanical strength of plant cell walls. Cellulose self-associates by intermolecular hydrogen bonding to form microfibrils of at least 36 glucan chains and becomes strongly associated with hemicellulose in the cell wall. Indeed, it has been suggested that the diameter of the cellulose microfibril may be determined, at least in part, by the binding of hemicellulose during cellulose synthesis, which prevents combining of small microfibrils into larger bundles (39).

Cellulases comprise a number of extracellular β-1,4-glucanases. Endohydrolases randomly disrupt internal linkages throughout β-1,4-glucan chains, producing glucose, cellobiose, and high-molecular-weight fractions. Exohydrolases or β-1,4-cellobiohydrolases act only on the exposed ends of β-1,4-glucan chains releasing the disaccharide cellobiose (17). β-Glucosidase and cellobiohydrolase also are part of the cellulase complex of some microorganisms. Because of its crystalline nature, native cellulose is degraded slowly. Plant pathologists generally have thought that cellulases are not particularly important in pathogenesis since extensive cellulose degradation typically occurs only late in infection, if at all. However, when the major endoglucanase genes of the phytopathogenic bacteria *Pseudomonas solanacearum* and *Xanthomonas campestris* pv. *compestris* were disrupted, virulence decreased (14).

Extracts of arbuscular mycorrhizal fungus (AMF) spores and external mycelium of *G. mosseae* have been shown to have endo- and exoglucanase activities (27). The enzyme activities in spores and external mycelium indicated which types of enzymes are found in mycorrhizae during root colonization. Endo- and exoglucanase activities increased in plants colonized by AMF when *G. mosseae* was in its logarithmic stage of growth (40). No relationship was found between number of vesicles and endo- and exoglucanase activities, although the maximum hydrolytic activities coincided with the beginning of entry point formation and arbuscule development (40).

Endoglucanases are present in noncolonized roots during growth and development (41). Several electrophoretic bands of endoglucanase activity observed in colonized plants had the same mobility as in noncolonized plants; however, some of these bands were present at earlier stages of plant growth in mycorrhizal plants than in nonmycorrhizal plants (27). The presence of bands different from those observed in nonmycorrhizal roots or external mycelia suggests that some of this activity may be induced by the fungus in the plant (Fig. 1). These findings indicate that endoglucanases produced by either the plant or the AM fungus may be involved in the process of host wall degradation. Some of the endoglucanase activity can be attributed to the extramatrical phase of the AM fungi since at least one of the endoglucanase activities found in the external mycelium and in the mycorrhizal root extracts showed the same electrophoretic mobility (42,43) (Fig. 1). Endoglucanase (EC 3.2.1.4) was purified from roots of onion (*Allium cepa*) colonized by *G. mosseae*. The endoglucanase has a relative molecular weight of about 27 kD and behaves as a monomer in its native form (44).

B. Pectinases

Pectins and related polysaccharides provide a protective material between plant cells. The term *pectin* encompasses a complex group of polysaccharides, some of which may be structural domains of larger, more complex molecules. The classic pectin fraction from

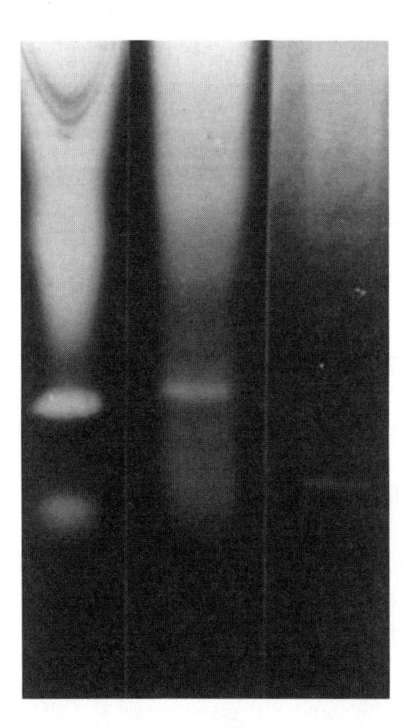

Figure 1 Nondenaturing polyacrylamide gradient gel electrophoresis of cellulase on 4–12% acrylamide. Lane 1, extract from non-AM onion roots; lane 2, extracts from AM onion roots; lane 3, extracts from external mycelium of *Glomus mosseae*.

oat seedlings contains 23% galacturonic acid and earlier it was thought that pectin consisted solely of α-D-1,4-linked galacturonic acid residues. Today, all evidence suggests that other sugars are covalently attached to the polygalacturonide backbone and that other sugars may even form an integral part of the main chain (45,46).

Pectinolysis is carried out by a complex of enzymes (pectinases), which include endo- and exopectate lyase (PL), endo- and exopolygalacturonase (PG), and pectin methylesterase (PME).

Degradation of pectin was reported for a sterile ericoid mycelium isolated from *Calluna vulgaris* by Perotto and associates (4) and Cairney and Burke (47). A wide range of ericoid fungi from different geographic regions was capable of growing on pectin as a sole carbon source. Ericoid fungi seem to use polygalacturonase during their saprotrophic life.

Attempts to demonstrate pectinase in extracts from AM tissues have not been successful (48). However, catabolic repression experiments by García-Romera and colleagues (49) showed that pectolytic enzymes may be involved in the process of root colonization by AM fungi. The spores and external mycelium of *G. mosseae* possess a complex of pectinolytic (pectin esterase [PE], pectin lyase [PL], pectatolyase [PNL], polygalacturonase [PG], and polymethylgalacturonase [PMG]) activities (26). The production of hydrolytic enzymes was studied during the process of penetration and development by *G. mosseae* in plant roots (29). The PE activity was consistently higher throughout the process of root colonization in plants inoculated with AM fungi than in controls. PE is thought to facilitate the action of the other pectinase enzymes (50). PMG and PNL (pectinolytic) activities were higher during the logarithmic stage of AM development in plants inoculated with the fungus than in nonmycorrhizal plants. The increase in fungal structures that penetrate the cell wall during the logarithmic stage of root colonization may explain the increase in PMG and PNL activities at this time. However, PG and PL (pectolytic) activities in AM plants were similar to those in controls throughout the experiment (29). The lack of differences in these degradative enzymes is not, however, conclusive evidence that they do not participate in the colonization process. It may indicate that PG and PL are involved during other stages of development (i.e., appressoria formation and penetration) in view of the presence of these enzymes in the extracts of spores and external mycelium of AM fungi (26).

The simultaneous presence of polygalacturonase produced by the fungus and of pectins secreted by the plant in the interfacial matrix suggests that the fungus might use pectins as a food source (31), as suggested by Dexheimer and coworkers (51).

Wall-degrading pectic enzymes uniquely associated with the interface of fine arbuscule branches may contribute to the interference of wall formation of the host plant. Active H^+ adenosine triphosphatase (ATPase) on fungal and plant membranes bordering the interface suggest that protons accumulate in the interfacial matrix and the resulting change in pH also could contribute to wall loosening (52).

C. Hemicellulases

Hemicelluloses are an integral part of all plant cell walls and form about 25% of the total dry weight of annuals and up to 40% in woody species. Hemicelluloses consist of chains of sugars in nonfibrillar organization that are linked to cellulose microfibrils by weak hydrogen bonds. In dicot primary walls, the major hemicelluloses are neutral xyloglucans and acidic arabinoxylans; in monocots they are acidic arabinoxylans and neutral β-(1-3,1-

4)-glucans (53). Xyloglucans are β-1,4-glucans with side chains that can hydrogen bond to cellulose microfibrils, cross-linking them and restraining cell expansion. In addition to a structural role, xyloglucans can be hydrolyzed by hydrolytic enzymes, and the oligosaccharides produced may act as signal molecules (15,54).

The plant cell wall contains glucanases and glycosidases that hydrolyze xyloglucan into monosaccharides. Endo-β-1,4-glucanase activity is responsible for the first step of degradation whereby the xyloglucan is endohydrolyzed into large fragments and exo-1,4-glucanase activity liberates low-molecular-weight fractions from the ends of long polysaccharide chains (41). The production of hemicellulolytic enzymes has been observed not only in parasites but also in mutualistic microorganisms such as *Rhizobium* species (24) and arbuscular mycorrhiza (28).

Endoxyloglucanase activity increases during growth and development of roots (55). This activity was consistently higher at the beginning of colonization and the logarithmic stage of development of mycorrhizal fungus (55). The increase in fungal structures that penetrate the cell wall during the logarithmic stage of root colonization may explain the increase in the different activities at this time (56). The evolution of endoxyloglucanase activities in plants paralleled the changes in the external mycelium. There were, however, bands of xyloglucanase activity in nonmycorrhizal roots that were absent in mycorrhizal roots; that may suggest qualitative inhibition by the fungus of some plant activity. Inhibition of plant protein synthesis by AM fungi has been observed in several plant–AM fungi associations (57,58).

III. ENZYMES IN THE PHYSIOLOGY OF THE ASSOCIATION

A. Phosphorus Uptake

It now is established that mycorrhizal colonization can enhance the uptake from soil of soluble inorganic P by plant roots (59). Although particularly important in low-P soils, an increased rate of P uptake can occur over a range of soil P levels even when mycorrhizal growth responses no longer occur. The enhanced P uptake by mycorrhizal plants is most likely the result of the external fungal hyphae's acting as an extension of the root system, thereby providing a more efficient (more extensive and better distributed) absorbing surface for uptake of nutrients from the soil and for translocation to the host root (60). External hyphae of AM fungi must absorb orthophosphate (Pi) by active transport (59,61). They have an active H^+-ATPase in the plasma membrane that would be capable of generating the required proton-motive force to drive H^+-phosphate cotransport, and P certainly is accumulated to high concentration (62).

Polyphosphate (poly-P) is a major P reserve in many fungi and it accumulates in vacuoles of AM fungi (63). Transfer of mycorrhizal roots from low- to high-P media results in a rapid accumulation of poly-P (64). Enzymes of poly-P synthesis have been found in mycorrhizal tissue (63,65). Polyphosphate kinase, which catalyzes the transfer of the terminal phosphate from ATP to poly-P, was detected in both external hyphae and mycorrhizal roots but not in uninfected roots, indicating that poly-P can be synthesized only by the fungal component of the mycorrhiza.

Although it now seems likely that P is translocated by protoplasmic streaming into the intraradical hyphae as poly-P (66), little is yet known of the biochemical mechanisms involved. The transport through the hyphae and unloading steps within the arbuscule may be linked to poly-P metabolism (Fig. 2). High proportion of long-chain poly-P to total

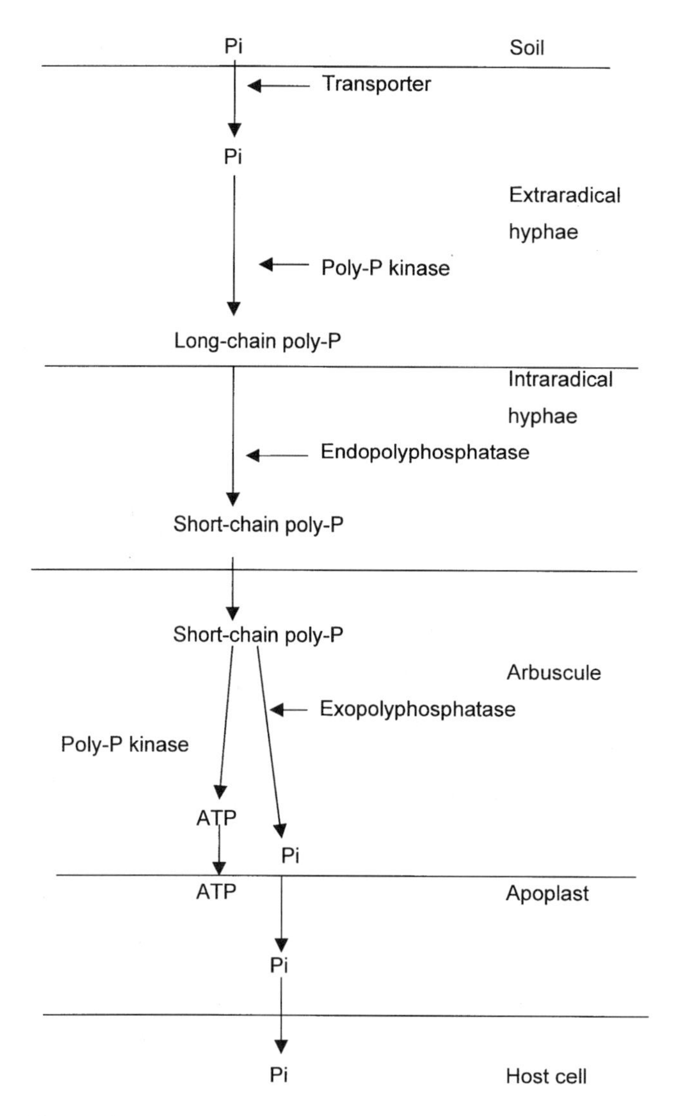

Figure 2 Enzymes involved in P transport in AM roots.

poly-P was observed in the external hyphae, and short-chain poly-P was higher in the internal hyphae (67). Long-chain poly-P seems to be more efficient in transporting Pi from the extraradical to the intraradical part of the fungi. Activity of enzymes of polyphosphate breakdown (exopolyphosphatase and endopolyphosphatase) is greater in mycorrhizal roots than in uninfected roots (65). Both enzymes have been detected in extracts of internal hyphae, but not those of external hyphae. The long-chain poly-P may be partly hydrolyzed into short-chain poly-P with endopolyphosphatase. Depolymerized short-chain poly-P may be hydrolyzed further with exopolyphosphatase to liberate Pi (67). Alternatively, the reaction catalyzed by polyphosphate kinase is readily reversible, so there is also the possibility that poly-P could be hydrolyzed, liberating ATP (63). The Pi (or ATP) so released in the arbuscule then would be transferred into the host (66).

The presence of intense ATPase activity indicates there is a carrier that mediates active transport mechanisms for Pi uptake at the host plasmalemma. ATPase activity has been observed in both plant and fungal plasma membranes and in the interfacial matrix associated with young arbuscules that decreased with senescence of arbuscules (68). A lack of H^+-ATPase activity in the host periarbuscular membrane surrounding nonfunctional arbuscules has been reported (69).

Other enzymes also have been implicated in P metabolism. Mycorrhizal-specific alkaline phosphatase is located in the vacuole of extraradical and intraradical hyphae (70–72). Maximum activity occurs while infections are young (100% arbuscular), coinciding with the start of the mycorrhizal growth response, but disappears with degeneration and collapse of the arbuscule. This enzyme appears to be of fungal origin. However, the role of alkaline phosphatase in Pi metabolism is still unknown (59,71).

The amount of P in soil available to plants is small, about 1% to 5% of the total P content. This finding has led to the suggestion that AM fungi are capable of utilizing insoluble P sources. Organic phosphates in soil may be utilized by plants through the action of phosphatases. Phosphatase activity in soil may originate from the plant roots or from microorganisms (73,74). High levels of acid and alkaline phosphatases have been found in the roots (70) and rhizosphere (75,76) of plants colonized by AM fungi. This increase in phosphatase activity would result in Pi's being liberated from organic phosphates immediately adjacent to the cell surface to be captured by the uptake mechanisms of mycorrhizal fungi. Some results have shown exudation of phosphatase by the external hypha and efficient hydrolysis of phytate-P by the phosphatase of mycorrhizal hyphae (76). Acid phosphatase activity release was visually shown as a red-colored ''hyphal print'' on filter paper treated with napthyl phosphate and Fast Red TR (diazotized 2-amino-5 chlorotoluene 1,5-naphthalene disulphonate) (77). However, other results indicated that the role of fungal phosphatases in P uptake from organic P is not clear: (1) extracellular phosphatase activity of mycorrhizal roots was stimulated in the presence of easily hydrolyzed substrates (76) but repressed by nonhydrolyzable forms of organic P (Po) (78), (2) no effect of mycorrhiza on specific activity of phosphatase was detected for clover grown in soil amended with ^{32}P-labeled organic matter (79), (3) the production of phosphatase varied with the choice of host plant and fungal endophyte (75,78), and no relationship between the level of AM colonization and phosphatase activity in different wheat cultivars has been found (80,81), (4) the addition of P fertilizer and $CaCO_3$ to soils decreased AM colonization but increased phosphatase activity in the plant rhizosphere (82); and (5) soil microorganims can mineralize organic P and AM hyphae may use Pi derived from their activity (83). Thus the results obtained are conflicting despite much effort (84).

One of the most important factors involved in controlling AM colonization of roots is soil and plant P. High P concentrations inhibit mycorrhizal colonization (60,85). The activity of mycorrhiza-specific alkaline phosphatases of *Glomus* species declines at high P levels (86,87). These observations suggest these enzymes would be involved in the regulation of mycorrhizal colonization of roots by P content of plants (85). However, high soil P concentration decreased AM colonization of roots by *Gigaspora* species but did not affect alkaline phosphatase activity (72). Thus the mechanism whereby the internal P content of the host regulates mycorrhizal infection is not clear.

B. Nitrogen Metabolism

Mycorrhizal plants sometimes improved nodulation and N fixation (88), an effect that may be due to enhanced P uptake (89). However, AM contribute to the N nutrition of the host by assimilation of soil nitrogen (N). The plant growth response to AM colonization may be greater in the presence of NH_4 than NO_3 (59).

Ammonium N and nitrate N have different pathways for metabolism, cation-carboxylate storage, and pH regulation, and hence they have rather different biochemical and physiological implications for the host (89). Nitrate is reduced, first to nitrite by nitrate reductase, then to ammonium by nitrite reductase. Ammonium N, once inside the cell, becomes directly incorporated into the various pathways for amino acid synthesis. Assimilation may be by glutamate dehydrogenase (GDH), or via glutamine synthetase (GS) and glutamate synthase (GOGAT) with the formation of glutamate. GS activity is increased in mycorrhizal root systems, partly as a result of a contribution from the fungi themselves; activity has been detected in fungal tissue separated from mycorrhizal roots (90). Improved P nutrition in the plants resulted in only a small increase in activity, confirming that the fungi have an important contribution and that GS is not related to P nutrition. In contrast, GDH activity showed no direct relationship with colonization (90). This limited evidence suggests that the fungi may have the capacity to assimilate NH_4^+ and, in consequence, N is likely to be transferred from fungi to plants in organic form.

Nitrate reductase activity has been detected in isolated spores of AM fungi (91,92). There are suggestions that either the AM fungi increase the nitrate reductase activity in the host plant (regardless of the P content) or the AM fungi have enzymatic activity per se (93). The fungal nitrate reductase messenger ribonucleic acid (mRNA) was detected in arbuscules but not in vesicles by in situ hybridization (94). The observation that AM fungi possess the gene coding for assimilatory nitrate reductase does not rule out the possibility that plant root cells mainly reduce nitrate in the AM symbiosis (95). The plant colonized by different AM fungi showed different nitrate reductase activity (93). Nitrate reductase and glutamine synthetase decreased with the age of mycorrhizal plants (96). Nitrite formation catalyzed by nitrate reductase was mainly reduced nicotinamide-adenine dinucleotide phosphate–(NADPH)-dependent in roots of AM colonized plants but not in those nonmycorrhizal plants, a finding consistent with the fact that the nitrate reductases of fungi preferentially utilized NADPH as the reductant (94). These investigations suggest that the fungus in AM-colonized root performs nitrate uptake and nitrate reduction to some degree. Because of its toxicity, the nitrite formed probably is not exported from the fungal to the plant cells. Other enzymes of nitrate assimilation have been described to occur in AM fungi (90). Thus nitrite reductase, glutamate synthetase, and glutamate synthase may transform nitrite, and N compounds (e.g., ammonium, glutamine, glutamate) probably are transferred from arbuscules to host cells.

C. Carbohydrate Assimilation

It is commonly accepted that the AM fungi are obligate symbionts and that carbohydrates are transferred from autotroph to heterotroph. It is likely that the fungus obtains the bulk of its carbon from host sugars; short-term $^{14}CO_2$ labeling experiments have shown transport of photosynthate from the host to the fungus (97). Most (70% to 90%) of the ^{14}C label present in both the roots and mycelium was in the form of soluble carbohydrates. The carbohydrates, predominantly sucrose, are delivered to the apoplast by the host cell. Then sucrose is hydrolyzed in the apoplastic interface by an acid invertase of plant origin, and the resulting hexoses are absorbed by the fungus (59), and used for trehalose synthesis. Trehalose has been shown to accumulate in both spores and external hyphae of AM fungi (66,98,99). Trehalose was detected in roots of colonized plants but not of control plants (56). Polyphosphates may serve in phosphorylation for the active transport of carbon skeletons into the arbuscule from the host either through the ATP produced by degradative polyphosphate kinase action or through a direct phosphorylation of sugars by enzymes of the polyphosphate glucokinase type (64). On the other hand, trehalase has been found in plants (100), and this enzyme increased upon mycorrhizal colonization (101). The biological function of plant trehalases is unknown, but they might be involved in the degradation of trehalose released from senescent AM fungus.

The possible metabolic pathways of carbon utilization in AM are largely uninvestigated. Dehydrogenases indicative of glycolysis are found in hyphae, vesicles, arbuscules, and spore germ tubes (102). From this, MacDonald and Lewis (102) have inferred that AM fungi employ the Embden–Meyerhof–Parnas glycolytic scheme, the hexose monophosphate shunt (or pentose phosphate cycle), and the tricarboxylic acid cycle.

A cyanide-insensitive respiratory pathway has been noted in AM roots (103). Such a pathway of electron transport to oxygen has been established in the sheath tissue of ectomycorrhizal roots (104). This pathway is not coupled to oxidative phosphorylation and may operate when oxidative phosphorylation is reduced by adenosine diphosphate (ADP) limitations. It is likely that the operation of such a pathway would increase the overall utilization of carbohydrates in mycorrhizal tissues (66).

IV. ENZYMES IMPLICATED IN THE HOST DEFENSE RESPONSE TO ARBUSCULAR MYCORRHIZAL FUNGAL COLONIZATION

Arbuscular mycorrhizal fungal penetration and establishment in the host roots involve a complex sequence of events and intracellular modifications that influence root colonization (25). Genotype and environmental factors influence the infection intensity or even the host compatibility and/or resistance (33,105).

The key to understanding the phenomenon of compatibility is to study recognition mechanisms and molecules involved in early stages of the AM interaction. In this sense, the formation of appressoria is one of the first morphological signs that recognition between the plant and the fungus has occurred. Some authors suggest that plant defense reactions may occur only after appresorium formation when the fungus has changed its state from saprophytic to infective (106).

Although AM fungi are considered as biotrophic microorganisms and biotrophs generally exhibit a high degree of host specificity, most AM fungi that have been studied show little or no specificity and are not thought to induce typical defense responses in host plants. Nevertheless, some plant resistance markers have been investigated in compatible symbiotic AM fungus–root interactions, and the early activation of certain plant defense genes has been shown (105). Since the plant host can elicit a weak defense response to the invading fungus, this may be a natural mechanism to control the number and/or location of infections. Furthermore, some phenomena of suppression of defense responses have been demonstrated in mycorrhizal roots (107,108). Whether this suppression is systemic or restricted to the infected area or whether products of symbiosis-related plant genes suppress the defense genes directly or through activation of fungal-derived suppressors remains to be elucidated. So far, it is not known how the induction/suppression of mechanisms associated with plant resistance could participate in the phenomenon of compatibility between plant roots and AM fungi. The investigation of early events and molecules involved in fungal–plant interactions is crucial for a better understanding of symbiosis.

In the following sections, some results obtained from studying the enzymatic activities produced by the host are reviewed and their contribution in the induction/suppression of mechanisms associated with plant resistance and in the control of intraradical fungal growth and maintenance of the symbiotic status is discussed. For ease of discussion the defense-related activities are divided into three classes based on their role in defense response. The first class involves hydrolases such as chitinases and β-1,3-glucanases that act directly as potent inhibitors of fungal growth. The second class involves enzymes related to oxidative stress such as catalases and peroxidases, and the third class consists of key enzymes that catalyze core reactions in phenylpropanoid metabolism.

A. Chitinases and β-1,3-Glucanases

The initiation of chitinase and other hydrolase activities is predominantly one of the coordinated and widespread mechanisms of plant defense against pathogen attack. There is good evidence that the action of the endohydrolases leads to detrimental effects, such as the inhibition of hyphal growth by invading fungi, as well as the probable release of signaling molecules (β-glucans and chitin/chitosan oligomers) that activate defense genes in the plants (109,110).

Most research into chitinase enzyme activity in plant roots colonized by the AM fungi has focused on the measurements of enzymatic activity during the different phases of mycorrhizal development. Several authors have shown that roots of infected plants contain enhanced levels of endochitinase activity at early stages of AM development. These results have been obtained with different combinations of plant and fungus (111–114). The peak of chitinase activity is followed by a period in which the enzyme activity is generally repressed to levels that are below those in nonmycorrhizal roots and that coincide with the extensive fungal development within roots (111–114). A similar result has been observed for β-1,3-glucanases. In bean and tomato mycorrhizal roots, β-1,3-glucanase activity was suppressed during the phase of rapid colonization (113,114).

Greater suppression of chitinase activity was observed in soybean roots under low P concentration and coincided with the period of maximum intraradical growth rate of the fungus (115). A correlation between the rate of suppression of chitinase activity in the inoculated roots and the infectivity of fungal isolates was observed (116). The maximum level of suppression was found in roots of soybean plants infected with the more infective isolate of *Glomus intraradices* (116).

In some particular plant–fungal combinations, the initial increase of chitinase activity was not followed by suppression, and higher levels of activity persisted (117). In some cases, no changes in chitinase and β-1,3-glucanase activities were detected between inoculated and noninoculated plants at all stages of mycorrhiza development (118–119).

Corroborating the biochemical data, differential gene expression of acidic and basic forms of chitinase and β-1,3-glucanase has been observed during mycorrhiza formation in different plant–fungal combinations (108,113,120,121). Studies of in situ localization of transcripts of bean endochitinases and β-1,3-glucanases in mycorrhizal roots showed that mRNAs accumulated predominantly in the vascular cylinder (121). Nevertheless, the accumulation of chitinase and β-1,3-glucanase transcripts has been observed around a number of cortical cells containing arbuscules (120,121), suggesting that the encoded enzyme might be involved in the control of intraradical fungal growth. The accumulation of β-1,3-glucanase mRNA in cells containing arbuscules was modulated by P concentration. The higher level of mRNA accumulation was obtained at a low level of P concentration (121).

The patterns of enzyme activity and mRNA accumulation suggest that chitinases and β-1,3-glucanases might be part of the early defense response by the plant to the invading fungus, which is then suppressed as symbiotic interactions develop. In this context, plant hydrolases may be involved in the regulation of AM development. Nevertheless, some experimental data revealed that it is not likely that plant chitinases and glucanases are essential to the control of the growth of AM fungi. Transgenic plants constitutively expressing high levels of different acidic forms of tobacco PRs (including chitinases and β-1,3-glucanases) became normally colonized by the AM fungi (122,123). The fact that chitinases and β-1,3-glucanases induced by the AM symbiotic fungi or by constitutive gene expression, do not prevent root colonization suggests that they are ineffective in controlling fungal development. The low enzymatic affinity for AM fungal components or inaccessibility of these enzymes to fungal cell wall components may cause this ineffectiveness (112).

Conversely, specific acidic forms of chitinase and β-1,3-glucanase are activated in several plants colonized by AM fungi. These symbiotic, specific isoenzymes have been reported in pea (124), tobacco (118), and tomato (125–127) roots and are different from pathogen-induced isoforms or constitutive enzymes. In addition, new chitosanase isoforms have been shown in pea (128) and tomato (126). Chitosanases are hydrolytic enzymes acting on chitosan, a derivative partially or fully deacetylated of chitin (129). Interestingly, the mycorrhizal-related chitinase isoform described in tomato-colonized roots appeared to display chitosanase activity. This bifunctional character was not found for the constitutive enzymes, or in *Phytophthora* sp.–induced chitinases (126). Mycorrhizal-specific plant chitinases are not active in pathogen-infected roots (118,124–125) or in *Rhizobium* sp. legume symbiosis (130), indicating a differential induction and function.

Although the precise function of plant hydrolase activities in the establishment of AM symbiotic interaction is still unclear, their stimulation seems to be a key point in the mechanism of recognition and signaling between plant roots and AM fungi. A regulatory role of these enzymes during establishment of AM and other root symbiosis has been proposed. Stimulation of specific plant chitinases has been reported in soybean/*Rhizobium* sp. (131) and ectomycorrhiza (132). It has been postulated that chitinases may be involved in the recognition of the rhizobial nodulation signals and, thus, in the regulation of the nodulation process (133). The data suggest a specific role for these enzymes, one that could be related in the AM symbiosis to the detection, modification, and/or release of chitin or chitosan oligomers from the fungal cell wall that can act as signaling compounds during the development of AM (Fig. 3). In this process of signal exchange, the modulation of chitinase activity by substrate specificity could be important (126). Furthermore, a putative role in bioprotection against fungal pathogens has been proposed for the new mycorrhiza-specific hydrolases. In this sense, the additional acidic and basic β-1,3-glucanase isoforms revealed during the interaction of *Phytophtora parasitica* and tomato plants pre-inoculated with *G. mosseae* could be implicated in the protective effect caused by AM symbiosis (127).

B. Catalases, Peroxidases, and Other Enzymes Related to Oxidative Stress

One of the major and rapid processes in the response of plant cells to environmental stresses is the generation of an oxidative burst, characterized by the release of active oxygen species (AOS) (134). This rapid response has been characterized in the hypersensi-

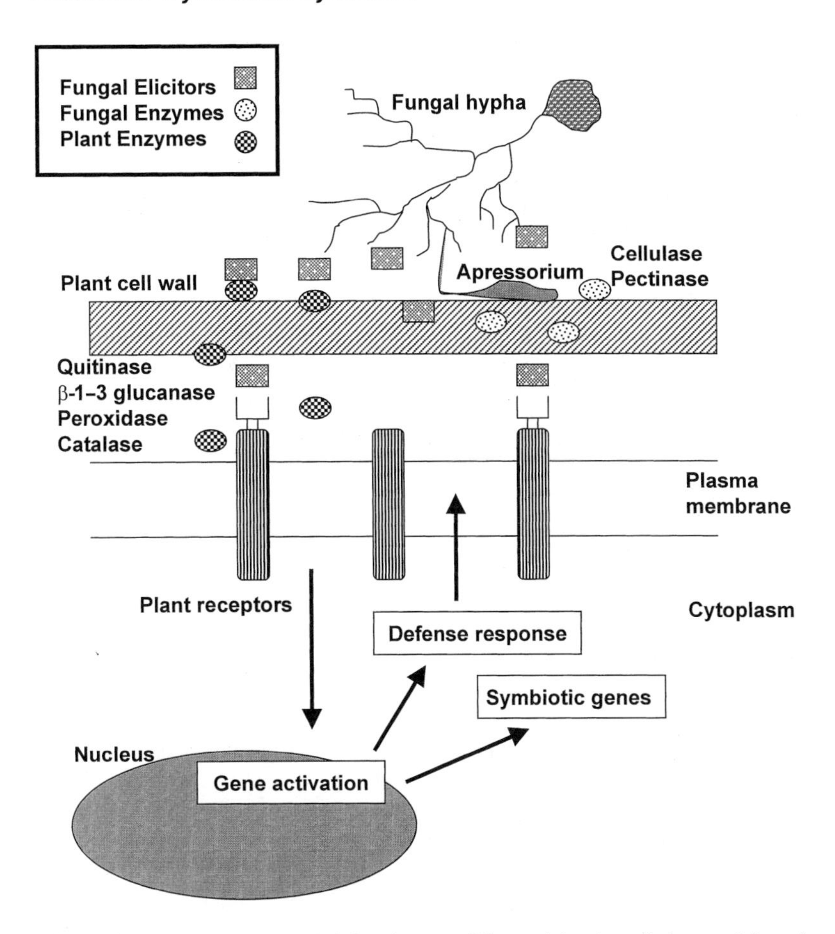

Figure 3 A speculative model showing possible participation of plant and fungal enzymes at early stages of plant–AM fungal interactions.

Fungal elicitors released during or after appressorium formation may be coupled to active plant receptors. In the process of release, perception of modification of these fungal signaling molecules could involve/stimulate plant enzymes. Alternatively, these plant enzymes may play a role facilitating changes in plasma membrane and cell wall architecture associated with early symbiotic interactions. Additionally, fungal enzymes involved in plant cell wall modification could facilitate the process of signal generation by cell wall architecture modification.

These initial reactions lead to the activation of a subset of elicitor-responsive genes, including defense and symbiotic genes. Possibly, gene activation is a consequence of a complex mechanism of signal transduction. The correct balance of the function of these induced genes is one of the keys of AM fungal–plant compatibility.

Unfortunately, most of the components of the signal recognition and transduction are unknown, and the assignment of components and sequence of events requires additional data.

tive response (HR) of plants to pathogens or elicitors (134,135). The predominant species detected in plant–pathogen interactions are superoxide anion (O_2^-), hydrogen peroxide (H_2O_2), and the hydroxyl radical (OH^-). The experimental data suggest different roles of the AOS in plant defense response (134,136). They can contribute to cell wall protein cross-linking and programmed host cell death during the hypersensitive response and may directly contribute to reduction of pathogen viability and growth. In addition, they have been proposed as mediators in pathways leading to defense-related gene expression (136).

The release of AOS in some plant–pathogen interactions can result in damage to the host tissues. Therefore, mechanisms that limit the duration of the oxidative burst and its toxic effects are necessary to minimize damage to the plant itself. One of these mechanisms is the action of endogenous antioxidant enzymes, such as superoxide dismutases, catalases, peroxidases, and glutathione peroxidases, which are capable of neutralizing the AOS.

During the establishment of a compatible plant–fungus AM symbiosis, the host plant showed little reaction at the cytological level to appressorium formation or infection hyphae. Occasionally some thickening was observed in epidermal cell walls at the point of contact with appressoria (105), and only a response similar to HR has been detected in Ri T-DNA–transformed roots of alfalfa colonized by *Gigaspora margarita* (137). Nevertheless, recent studies, using the diaminobenzidine (DAB) staining technique, revealed that a brownish stain, indicative of H_2O_2 accumulation, was present within cortical root cells in the space occupied by clumped arbuscules and around hyphal tips attempting to penetrate roots of *Medicago truncatula* colonized by *G. intraradices* (138). These results suggest that a locally restricted oxidative burst could be involved in the response of the plant to AM formation and development.

Relatively few data exist concerning the possible participation of antioxidant enzymes in the plant response to AM formation. A peak of cell wall–bound peroxidase was observed during the initial stages of fungal penetration in leek (*Allium porrum*) cells. Once infection was established, the activity decreased to the levels shown in nonmycorrhizal plants (139). In potato roots, the activity of extracellular peroxidase recovered in root leachates was not stimulated by AM infection; peroxidase activity per gram of fresh weight was significantly enhanced in AM roots (140). When potato plants were grown with higher P supply, extracellular peroxidase activity increased linearly with increasing P supply, suggesting a role of peroxidase in limiting AM infection in well-P-nourished plants (140).

The analysis of catalase and ascorbate peroxidase activities during the early stage of tobacco–*Glomus mosseae* interaction revealed transient enhancements of both enzymatic activities in the inoculated plants (141). These increases coincided with the stage of appressoria formation on root surfaces and the appearance of a peak of accumulation of free salicylic acid in inoculated roots (141). These data indicate the activation of catalase and peroxidase activities in root cells where the fungus forming appressoria might be part of the plant response to the invading fungus. The role of these enzymes in this response could be related to activation of a defensive mechanism or to a process of cell wall repair at the site of infection (Fig. 3). Alternatively, the early activation of catalase and peroxidase may play a role that facilitates changes in cell wall architecture associated with early symbiotic interactions. This hypothesis has been proposed for the *Rhizobium meliloti–Medicago truncatula* association, when a *Rhizobium* sp.–induced peroxidase gene is induced rapidly and transiently by compatible *R. meliloti*, and Nod factor. The transcript is localized to differentiating epidermal cells in the root zone that is subsequently infected by *Rhizobium* sp. (142).

Other important enzymes that participate in the primary defense against the AOS are the superoxide dismutases (SODs) that catalyze the disproportionation of the superoxide free radical (O_2^-) to H_2O_2 and O_2. Data suggest that SOD acts as an antioxidant system in the N_2 fixation process of nodules (143). Some changes in the isoenzymatic pattern and SOD activity in several plant–AM fungal symbioses have been reported (144–148).

The appearance of additional SOD isozymes in plant roots inoculated with *G. mosseae* has been reported in red clover (144,146) and onion plants (144). This new isozyme was not detected in noninoculated plants nor in plants inoculated with *G. intraradices*; thus SOD enzymes appearance was associated with specific fungal–plant interaction (144). So far, it is not known what role SOD enzymatic activity plays during the period of AM development; nevertheless some data suggest that the change in enzymatic activity was related to differences between AM inoculated and noninoculated plants in stress situations, including drought exposure (148) and plant senescence (144).

C. Enzymes That Catalyze Core Reactions in the Phenylpropanoid Metabolism

Phenylpropanoid compounds include a variety of chemical formulas with a wide range of biological roles in plant life cycles. The biochemical reactions and genetic regulation of phenylpropanoid metabolism are complex, because many compounds are constitutive or induced, depending on the plant species or tissues, and a broad variety of biotic and abiotic stresses can induce their accumulation (149).

Among phenylpropanoid compounds, flavonoid and isoflavonoid compounds are involved in diverse aspects of plant growth, development, and interactions with microorganisms, mainly in defense responses with the liberation of phytoalexins (150) and in *Rhizobium* sp.–legume interaction, when specific flavonoids act to initiate the symbiosis (151).

Flavonoids can stimulate spore germination, hyphal growth, and enhancements of AM colonization by AM fungi (152). Some of these compounds have been isolated and characterized. Their effect on different AM fungal species has been assayed, and the results show that the AM fungal growth response to root flavonoids is not uniform (153).

In *Medicago* species, transient increases in different flavonoid/isoflavonoid compounds were found, depending on the specific interaction of *Medicago* and fungal species (107,111,154,155). Some of these compounds stimulated hyphal growth (154). Formononetin was found to accumulate in *Medicago sativa* roots in the presence of the fungus *Glomus intraradix*, before fungal penetration and colonization (111). The analysis of enzymatic activities and accumulation of mRNA transcripts encoding enzymes of flavonoid/isoflavonoid metabolism revealed that the changes in compound accumulation correlate with increases in mRNA and enzyme activity. Increases in enzymatic activity and messenger ribonucleic acid (mRNA) accumulation of phenylalanine ammonia lyase (PAL), chalcone synthase (CHS), and chalcone isomerase (CHI) were observed in *Medicago* species colonized by AM fungi (107,155). Increases in PAL transcript accumulation also have been detected in rice roots inoculated with *G. mosseae* (156). This increase was concomitant with the accumulation of salicylic acid, a phenolic acid derived from phenylpropanoid metabolism that has been implicated in plant defense responses (157–159). Nevertheless, no changes in transcript level were observed in other mycorrhizal interactions, such as those involving bean (113) or parsley (160).

The increase, followed by decline, in transcript accumulation and enzyme activity of PAL and CHI in roots of alfalfa during infection by *G. intraradices* (107) has been interpreted as a mechanism of activation/suppression of the defense reaction elicited by *G. intraradices* in alfalfa (161). Although phenylpropanoid metabolism can be enhanced in roots during symbiotic interactions, this is not a general phenomenon, and the extent, timing, and enzymatic activities and compounds released appear to depend on the plant and fungal genotypes involved. In situ localization of transcripts encoding PAL and CHS in mycorrhizal roots showed that the transcripts were discretely localized in cells containing arbuscules (155). The expression of other gene encoding enzymes in the flavonoid/isoflavonoids pathway, such as CHI or isoflavone reductase (IFR), was not significantly affected in mycorrhizal roots (155).

Altogether, the available data suggest an activation of phenylpropanoid metabolism in mycorrhizal roots, characterized by the weak, localized, and uncoordinated induction of genes and accumulation of phytoalexin products, some of them at high levels (162). Nevertheless, no evidence exists to support a specific role for flavonoid/isoflavonoid in the AM symbiosis. Even though several of them can stimulate hyphal growth, some results suggest that they are not necessarily signaling compounds involved in the AM symbiosis (163).

V. ENZYMES AS A METABOLIC ACTIVITY INDEX

Mycorrhizal colonization of plant roots has been evaluated by using nonvital staining techniques. However, often there is no relationship between percentage mycorrhizal colonization and the effectiveness of a particular fungus in plant growth (164). Several authors have developed vital staining techniques to measure metabolic active fungal colonization of roots.

The alkaline phosphatase enzyme in intraradical hyphae was found to be related to the stimulation of the growth of the plants when colonized by AM fungi (165). Alkaline phosphatase can be histochemically visualized in external and intracellular mycelium (87,166). Fungal alkaline phosphatase activity diminished in plants growing under several adverse conditions in spite of the constant level of mycorrhizal colonization assessed with nonvital stains. Alkaline phosphatase has been proposed as a vital marker for root colonization (71). However, the application of a fungicide, which inhibits P uptake, and transfer to plants via hyphae do not affect alkaline phosphatase activity (167). More work needs to be conducted on the fungal phosphatases in order to identify whether these enzymes play a key role in the efficiency of the symbiosis.

MacDonald and Lewis (102) developed a histochemical technique to stain for succinate dehydrogenase (SDH) activity in AM fungi. Several authors showed that this enzymatic activity of the fungus was depressed when herbicides (168) or fungicides (169,170) were applied to mycorrhizal colonized plants, in spite of the small effect on the fungal structures visualized by nonvital staining techniques. A decrease in SDH activity also was observed (along with the formation of septa in the intraradical hypha) when mycorrhizal plants were subjected to the presence of the antagonistic fungus *Trichoderma koningii* (171). Some correlation between the frequency of SDH-active arbuscules and shoot mass of plants has been found (86,172). However, the proportion of the AM mycelium with SDH activity is not related to the effect of the fungus on plant growth (169,173,174). Thus SDH activity appears to be a sensitive parameter for measures effects of environmental stress on the fungi but is not a sensitive parameter for measuring growth.

Other enzymes, such as malate dehydrogenase (MDH) (175) and colonization-specific phosphatase (IPS) (176), have been proposed as markers of the fungal metabolic activity. A very high correlation coefficient has been found between the intensity of the fungal MDH electrophoretic bands and the AM colonization of roots measured by the concentration of glucosamine (175,177). The activity of this MDH was inhibited also in the presence of fungicides. Conversely, some IPS (probably a neutral phosphatase) was detected in mycorrhizal plants. The activity increased as the colonization rate increased and decreased at the stationary phase of the host growth when the colonization rate was still high (176).

VI. ENZYMES IN AM FUNGI IDENTIFICATION

The assessments of the biodiversity of AM fungi in ecosystems have relied on the isolation of the resting spores from soils. This approach does not necessarily supply useful information about functional ecological characteristics since the fungus that colonizes roots might not produce spores under certain conditions. One plant can be colonized by several AM fungi at the same time, but the intraradical mycelium of the different species of AM fungi shows little morphological variation (164). Molecular techniques have been used to identify AM fungi (178). However, with isozyme techniques it is possible to measure the metabolically active mycorrhizae; this is not yet possible with DNA-based techniques. AM fungi can be identified within roots by differences in the mobility of specific fungal enzymes on polyacrylamide gel electrophoresis (179–181). This method has been used to identify and quantify endophytes within a root in a competition experiment with different AM isolates (182). The staining intensity of esterase, glutamate oxaloacetate transaminase, and peptidase (measured as peak height on a densitometer trace) was correlated with the biomass of the fungus in the root sample, and so this offers a method of quantifying the contribution of a single fungus to a mixed colonization (183).

The use of MDH and esterase enzymes allows characterization of different AM fungi (184). The grouping of *Gigaspora* sp. isolates provided by the SSU sequence analysis was similar to the grouping of the isozyme profile of MDH (185).

VII. CONCLUSIONS

Fungi that form AM associations have not been cultured in the absence of host roots. Most of the enzymatic studies on AM symbiosis have been performed with extracts of mycorrhizal roots containing both plant and fungal enzymes (43). The small differences in hydrolytic enzyme activities between mycorrhizal and nonmycorrhizal plants, together with the low rate of production of cell wall hydrolytic enzymes, suggest that AM fungi penetrate the root surface mostly by mechanical force. Appressoria with well-melanized walls produce hyphae that tend to progress by growing between root epidermal cells rather than by crossing their outer walls. Once inside the roots, many AM fungi produce intercellular hyphae that run within huge air channels (186), and then cross the wall of cortical cells to become intracellular, producing penetration pegs and causing only limited and subtle changes in the structure of the host wall (187). These slight modifications suggest that they may produce at this stage very limited or localized amounts of hydrolytic enzymes. AM fungi seem to colonize the root tissues of their host plant by means of a combination of mechanical and enzymatic mechanisms. Very weak and localized production of enzymes might ensure that viability of the host is maintained, defense responses are not triggered, and a high degree of compatibility is reached.

Some plant defense responses have been shown to be activated in compatible AM fungus–root interactions. The role of these plant defense compounds could be to control intraradical fungal growth and maintenance of the symbiotic status. Plant defense chemicals and enzymes may be involved in the perception, modification, and/or release of fungal cell wall fragments that can act as signaling compounds during the process of recognition and formation of AM symbiosis.

The enhanced growth of plants colonized by AM fungi results from improved uptake of soil P (59). Enzymes are involved in P transport from the fungus to the host plant, but it is not clear whether AM enzymes are involved in P mobilization and fungal/plant uptake from soil. AM fungi use P from the soluble fraction of soil (59). However, there are indications of P mineralization from organic fractions of soil by fungal phosphatases that may represent another source of P uptake in P-deficient soils (76,77). The increase of N uptake by AM symbiosis has been attributed to better P nutrition of plants (164). Nevertheless, enzymes implicated in assimilation of ammonium N and nitrate N that have been found in arbuscular fungi that contribute to N nutrition of the host plant by assimilation of soil N regardless of P effect (90).

In spite of the importance of AM symbiosis to nutrient movement and soil conservation, very few studies on the impact of the fungal enzymes on soil have been carried out (188). There is a clear need to extend studies on the ecological role of AM enzymes in soil.

REFERENCES

1. JL Harley. Introduction: The state of art. In: JR Norris, DJ Read, AK Varma, eds. Methods in Microbiology. Vol 21. London: Academic Press, 1991, pp 1–24.
2. R Law, DH Lewis. Biotic environments and the maintenance of sex-some evidence from mutualistic symbioses. Biol J Linn Soc 20:249–276, 1983.
3. JB Morton, GL Benny. Revised classification of arbuscular mycorrhizal fungi (zygomycetes): A new order, Glomales, two new suborders, Glomineae and Gigasporaceae, with an amendation of Glomaceae. Mycotaxonomy 37:471–491, 1990.
4. S Perotto, V Bettini, P Bonfante. The biology of mycorrhiza in the Ericaceae. I. The isolation of the endophyte and synthesis of mycorrhizas in aseptic cultures. New Phytol 72:371–379, 1993.
5. S Perotto, R Peretto, A Schubert, A Varma, P Bonfante. Ericoid mycorrhizal fungi: Cellular and molecular basis of their interactions with host plant. Can J Bot 73:557–568, 1995.
6. A Varma, P Bonfante. Utilization of cell wall-related carbohydrates by ericoid mycorrhizal endophytes. Symbiosis 16:301–313, 1994.
7. KK Newsham, AH Fitter, AR Watterson. Arbuscular mycorrhiza protect an annual grass from root pathogenic fungi in the field. J Ecol 83:991–1000, 1995.
8. RP Schreiner, GJ Bethlenfalvay. Mycorrhizal interactions in sustainable agriculture. Crit Rev Biotechnol 15:271–287, 1995.
9. FA Smith, SE Smith. Structural diversity in (vesicular)-arbuscular mycorrhizal symbioses. New Phytol 137:373–388, 1997.
10. P Bonfante-Fasolo. Ultrastructural analysis reveals the complex interactions between root cells and arbuscular mycorrhizal fungi. In: S Gianinazzi, H Schüepp, eds. Impact of Arbuscular Mycorrhizas on Sustainable Agriculture. Basel, Switzerland: Birkahäuser Verlag, 1984, pp. 73–87.
11. MJ Harrison. Development of the arbuscular mycorrhizal symbiosis. Curr Opin Plant Biol 1:360–365, 1998.
12. V Gianinazzi-Pearson, S Gianinazzi. Protein and protein activities in endomycorrhizal symbioses. In: A Varma, B Hock, eds. Mycorrhiza: Structure, Function, Molecular Biology and Biotechnology. Berlin, Heidelberg, New York: Springer-Verlag, 1995, pp 251–267.

13. RJ Howard, MA Ferrari. Role of melanin in appressorium function. Exp Mycol 13:403–418, 1989.
14. JD Walton. Deconstructing the cell wall. Plant Physiol 104:1113–1118, 1994.
15. T Hayashi. Xyloglucans in the primary cell wall. Annu Rev Plant Physiol Plant Mol Biol 40:139–168, 1989.
16. MG Hahn, P Bucheli, F Cervone, SH Doares, RA O'Neill, A Darvill, P Arbesheim. The role of cell wall constituents in plant-pathogen interactions. In: E Nester, T Kosuge, eds. Plant-Microbe Interactions. New York: McGraw Hill, 1989, pp 131–181.
17. A Varma. Hydrolytic enzymes from arbuscular mycorrhizae: the current status. In: A Varma, B Hock, eds. Mycorrhiza: Structure, Function, Molecular Biology and Biotechnology. Berlin, Heidelberg, New York: Springer-Verlag, 1999, pp 372–389.
18. SC Fry, RC Smith, KF Renwick, DJ Martin, SK Hodge, KG Matthews. Xyloglucan endotransglycosylase, a new wall-loosening enzyme activity from plants. Biochem J 282:821–828, 1992.
19. G Maclachlan, C Brady. Multiple forms of 1,4-β-glucanase in ripening tomato fruits include a xyloglucanase activatable by xyloglucan oligosaccharides. Aust J Plant Physiol 19:137–146, 1992.
20. NC Carpita. Structure and biogenesis of the cell walls of grasses. Annu Rev Plant Physiol Plant Mol Biol 47:445–476, 1996.
21. A Collmer, DW Bauer, SW He, M Lindeberg, S Kelemu, P Rodríguez-Palenzuela, TJ Burr, AK Chatterjee. Pectic enzyme production and bacterial plant pathogenicity. In: H Hennecke, DPS Verma, eds. Advances in Molecular Genetics of Plant-Microbe Interactions. Vol. 1. Dordrecht, The Netherlands: Kluwer Academic Publishers, 1991, pp 65–72.
22. M Umali-Garcia, DH Hubbell, MH Gaskins, FB Dazzo. Association of *Azospirilum* with grass roots. Appl Environ Microbiol 39:219–226, 1980.
23. PF Mateos, JI Jimenez-Zurdo, J Chen, AS Squartini, SK Haack, E Martinez-Molina, DH Hubbell, FB Dazzo. Cell-associated pectinolytic and cellulolytic enzymes in *Rhizobium leguminosarum* biovar. *trifoli*. Appl Environ Microbiol 58:1816–1822, 1992.
24. JI Jimenéz-Zurdo, PF Mateos, FB Dazzo, E Martínez-Molina. Cell-bound cellulase and polygalacturonase production by *Rhizobium* and *Bradyrhizobium* species. Soil Biol Biochem 28:917–921, 1996.
25. P Bonfante-Fasolo, S Perotto. Plants and endomycorrhizal fungi: The cellular and molecular basis of their interaction. In: DPS Verma, ed. Molecular Signals in Plant-Microbe Interactions. Boca Raton, Florida: CRC, 1992, pp 445–470.
26. I García-Romera, JM García-Garrido, JA Ocampo. Pectolytic enzymes in the vesicular-arbuscular mycorrhizal fungus *Glomus mosseae*. FEMS Microbiol Lett 78:343–346, 1991.
27. JM García-Garrido, I García-Romera, JA Ocampo. Cellulase production by the vesicular-arbuscular fungus *Glomus mosseae* (Nicol. and Gerd.) Gerd. and Trappe. New Phytol 121: 221–226, 1992.
28. I García-Romera, JM García-Garrido, E Martinez-Molina, JA Ocampo. Possible influence of hydrolytic enzymes on vesicular arbuscular mycorrhizal infection of alfalfa. Soil Biol Biochem 22:149–152, 1990.
29. I García-Romera, JM García-Garrido, JA Ocampo. Pectinase activity in vesicular-arbuscular mycorrhiza during colonization of lettuce. Symbiosis 12:189–198, 1991.
30. JM García-Garrido, I García-Romera, JA Ocampo. Cellulase activity in lettuce and onion plants colonized by the vesicular-arbuscular mycorrhizal fungus *Glomus mosseae*. Soil Biol Biochem 25:503–504, 1992.
31. R Peretto, V Bettini, F Favaron, P Alghisi, P Bonfante. Polygalacturonase activity and localization in arbuscular mycorrhizal roots of *Allium porrum*. Mycorrhiza 5:157–164, 1995.
32. S Perotto, JD Coisson, I Perugini, V Cometti, P Bonfante. Production of pectin-degrading enzymes by ericoid mycorrhizal fungi. New Phytol 135:151–162, 1997.

33. P Bonfante, S Perotto. Strategies of arbuscular mycorrhizal fungi when infecting host plants. New Phytol 130:3–21, 1995.
34. V Gianinazzi-Pearson, E Dumas-Gaudot, G Armelle, A Tahiri, S Gianinazzi. Cellular and molecular defence-related root responses to invasion by arbuscular mycorrhizal fungi. New Phytol 133:45–58, 1996.
35. P Bonfante. The role of the cell wall in mycorrhizal association. In: S Scannerini, D Smith, P Bonfante-Fasolo, V Gianinazzi-Pearson, eds. NATO ASI series, series H, vol. 17. New York: Berlin, Heidelberg, Springer, 1988, pp 219–236.
36. DA Brummell, CC Coralie, AB Bennett. Plant endo-1,4-β-D-glucanases: Structure, properties and physiological function. Am Chem Soc Symp Ser 566:100–129, 1994.
37. M Saito. Enzyme activities of the internal hyphae of *Gigaspora margarita* isolated from onion root compared with those of the germinated spores. In: C Azcón-Aguilar, JM Barea, eds. Mycorrhizas in Integrated Systems from Genes to Plant Development. Brussels. Luxembourg: Official Publications of the European Communities, 1996, pp 260–262.
38. NC Carpita, DM Gibeaut. Structural models of primary cell walls in flowering plants: Consistency of molecular structure with the physical properties of the walls during growth. Plant J 3:1–30, 1993.
39. RL Fischer, AB Bennett. Role of cell wall hydrolases in fruit ripening. Annu Rev Plant Physiol Plant Mol Biol 42:675–703, 1991.
40. JM García-Garrido, MN Cabello, I García-Romera, JA Ocampo. Endoglucanase activity in lettuce plants colonized with the vesicular arbuscular mycorrhizal fungus *Glomus fasciculatum*. Soil Biol Biochem 24:955–959, 1992.
41. SC Fry. Polysaccharide-modifying enzymes in the plant cell wall. Annu Rev Plant Physiol Plant Mol Biol 46:497–520, 1995.
42. I García-Romera, JM García-Garrido, E Martinez-Molina, JA Ocampo. Production of pectolytic enzymes in lettuce root colonized with *Glomus mosseae*. Soil Biol Biochem 23:597–601, 1991.
43. I García-Romera, JM García-Garrido, A Rejón-Palomares, JA Ocampo. Enzimatic mechanisms of penetration and development of arbuscular mycorrhizal fungi in plants. In: SG Pandalai, ed. Recent Research Developments in Soil Biology and Biochemistry. Thuandrum, India: Research Signpost, 1997, pp 121–136.
44. JM García-Garrido, I García-Romera, MD Parra-García, JA Ocampo. Purification of an arbuscular mycorrhizal endoglucanase form onion root colonized by *Glomus mosseae*. Soil Biol Biochem 28:1443–1449, 1996.
45. PM Dey, JB Harborne. Plant biochemistry. London: Academic Press, 1995.
46. JSG Reid. Carbohydrate metabolism: Structural carbohydrate. In: PM Dey, JB Harborne, eds. Plant Biochemistry. London: Academic Press, 1995, pp 205–235.
47. JWG Cairney, RM Burke. Plant cell wall-degrading enzymes in ericoid and ectomycorrhizal fungi. In: C. Azcón-Aguilar, JM Barea, eds. Mycorrhizas in Integrated Systems from Genes to Plant Development. Brussels. Luxembourg: Official Publications of the European Communities, 1996, pp 218–221.
48. AJ Anderson. Mycorrhizae-host specificity and recognition. Am Phytopathol Soc 78:375–378, 1988.
49. I García-Romera, JM García-Garrido, JA Ocampo. Hydrolytic enzymes in arbuscular mycorrhiza. In: C Azcón-Aguilar, JM Barea, eds. Mycorrhizas in Integrated Systems from Genes to Plant Development. Brussels. Luxembourg: Official Publications of the European Communities, 1996, pp 234–237.
50. F Stutzenberger. Pectinase production. In: J Lederberg, ed. Encyclopedia of Microbiology. Vol III, San Diego: Academic Press, 1992, pp 327–337.
51. J Dexheimer, S Gianinazzi, V Gianinazzi-Pearson. Ultrastructural cytochemistry of the host-fungus interfaces in the endomycorrhizal association *Glomus mosseae/Allium cepa*. Z Pfanz 92:191–206, 1979.

52. A Gollotte, M-C Lemoine, V Gianinazzi-Pearson. Morphofuntional integration and cellular compatibility between endomycorrhizal symbionts. In: KG Mukerji, ed. Concepts in Mycorrhizal Research. Dordrecht, The Netherlands: Kluwer Academic Publisher, 1996, pp 91–111.
53. SC Fry. Cellulases, hemicellulases and auxin-stimulated growth: A possible relationship. Physiol Plant 75:532–536, 1989.
54. SC Fry. The structure and functions of xyloglucan. J Exp Bot 40:1–11, 1989.
55. A Rejón-Palomares, JM García-Garrido, JA Ocampo, I García-Romera. Presence of xyloglucan-hydrolyzing glucanases (xyloglucanases) in arbuscular mycorrhizal symbiosis. Symbiosis 21:249–261, 1996.
56. A Schubert, P Wyss. Trehalase activity in mycorrhizal and nonmycorrhizal root of leek and soybean. Mycorrhiza 6:401–404, 1995.
57. JM García-Garrido, N Toro, JA Ocampo. Presence of specific polypeptides in onion roots colonized by *Glomus mosseae*. Mycorrhiza 2:175–177, 1993.
58. E Dumas-Gaudot, P Guillaume, A Tahiri-Alaoui, V Gianinazzi-Pearson, S Gianinazzi. Changes in polypeptide patterns in tobacco roots colonized by two *Glomus* species. Mycorrhiza 4:215–221, 1994.
59. SE Smith, DJ Read. Mycorrhizal symbiosis. 2nd ed. New York: Academic Press. 1997.
60. PB Tinker. Effects of vesicular-arbuscular mycorrhizas on higher plants. Symp Soc Exp Biol 29:325–328, 1975.
61. MJ Harrison, ML Van Buuren. A phosphate transporter from the mycorrhizal fungus *Glomus versiforme*. Nature 378:626–629, 1995.
62. J Lei, G Becard, JG Catford, Y Piche. Root factors stimulate ^{32}P uptake and plasmalemma ATPase activity in a vesicular-arbuscular fungus, *Gigaspora margarita*. New Phytol 118: 289–294, 1991.
63. RE Beever, DJW Burn. Phosphorus uptake storage and utilization by fungi. Adv Bot Res 8:128–219, 1980.
64. JA Callow, LCM Capaccio, G Parish, PB Tinker. Detection and estimation of polyphosphate in vesicular-arbuscular mycorrhizas. New Phytol 80:125–134, 1978.
65. LCM Capaccio, JA Callow. The enzymes of polyphosphate metabolism in vesicular-arbuscular mycorrhizas. New Phytol 91:81–91, 1982.
66. KM Cooper. Physiology of VA mycorrhizal association. In: CL Powell, DJ Bagyaraj, eds. VA Mycorrhizae. Boca Raton, Florida: CRC, 1984, pp 155–203.
67. MZ Solaiman, T Ezawa, T Kojima, M Saito. Polyphosphates in intraradical and extraradical hyphae of an arbuscular mycorrhizal fungus *Gigaspora margarita*. Appl Environ Microbiol 65:5604, 1999.
68. V Gianinazzi-Pearson, SE Smith, S Gianinazzi, FA Smith. Enzymatic studies on the metabolism of vesicular-arbuscular mycorrhizas. V. Is H^+-ATPase a component of ATP-hydrolysing enzyme activities in plant-fungus interfaces? New Phytol 117:61–64, 1991.
69. V Gianinazzi-Pearson, A Gollote, J Leherminier, B Tisserant, P Franken, E Dumas-Gaudot, MC Lemoine, D Van Tuinen, S Gianinazzi. Cellular and molecular approaches in the characterization of symbiotic events in functional arbuscular mycorrhizal associations. Can J Bot 73:526–532, 1995.
70. S Gianinazzi, V Gianinazzi-Pearson, J Dexheimer. Enzymatic studies on the metabolism of vesicular-arbuscular mycorrhizas. III. Ultrastructural localisation of acid and alkaline phosphatases in onion roots infected by *Glomus mosseae* (Nicol. and Gerd.). New Phytol 82: 127–132, 1979.
71. V Gianinazzi-Pearson, E Dumas-Gaudot, S Gianinazzi. Proteins and proteins activities in endomycorrhizal symbioses. In: A Varma, B Hock, eds. Mycorrhiza. 2nd ed. Berlin: Springer-Verlag, 1999, pp 255–272.
72. CL Boddington, JC Dodd. Evidence that differences in phosphate metabolism in mycorrhizas formed by species of *Glomus* and *Gigaspora* might be related to their life-cycle strategies. New Phytol 142:531–538, 1999.

73. B Dinkelaker, H Marschner. *In vivo* demonstration of acid phosphatase activity in the rhizosphere of soil-grown plants. Plant Soil 144:199–205, 1992.
74. JC Tarafdar, N Classen. Organic phosphorus compounds as a phosphorus source for higher plants through the activity of phosphatases produced by plant roots and microorganisms. Biol Fertil Soils 5:308–312, 1988.
75. JC Dodd, CC Burton, RG Burns, P Jeffries. Phosphatase activity associated with the roots and the rhizosphere of plants infected with vesicular-arbuscular mycorrhizal fungi. New Phytol 107:163–172, 1987.
76. JC Tarafdar, H Marschner. Phosphatase activity in the rhizosphere of VA-mycorrhizal wheat supplied with inorganic and organic phosphorus. Soil Biol Biochem 26:387–395, 1994.
77. JC Tarafdar. Visual demonstration of *in vivo* acid phosphatase activity of VA mycorrhizal fungi. Curr Sci 69:541–543, 1995.
78. R Azcón, F Borie, JM Barea. Exocellular phosphatase activity of lavender and wheat roots as affected by phytate and mycorrhizal inoculation. In: S Gianinazzi, V Gianinazzi-Pearson, A Trouvelot, eds. Les Mycorrhizes: Biologie et Utilization. Dijon: INRA, 1982, pp 83–85.
79. E Joner, I Jakobsen. Uptake of ^{32}P from labelled organic matter by mycorrhizal and non-mycorrhizal subterranean clover (*T. subterraneum* L.). Plant Soil 172:221–227, 1995.
80. R Rubio, E Moraga, F Borie. Acid phosphatase activity and vesicular-arbuscular mycorrhizal infection associated with roots of four wheat cultivars. J Plant Nutr 13:585–598, 1990.
81. F Borie, R Rubio, I Martinez, C Castillo. Mycorrhizal and phosphatase activities of two wheat cultivars at different developmental stages. Second International Conference on Mycorrhiza, Uppsala, 1998, p 31.
82. R Rubio, F Borie, E Moraga, E Albornoz. Efecto del encalado sobre algunos parámetros biológicos y rendimiento de trebol rosado (*Trifolium pratense* L) en un suelo con alto contenido de aluminio. Agric Tec 52:394–397, 1992.
83. E Joner, J Magid, TS Gahoonia, I Jakobsen. Phosphorus depletion and activity of phosphatases in the rhizosphere of mycorrhizal and non-mycorrhizal cucumber (*Cucumis sativus* L.). Soil Biol Biochem 27:1145–1151, 1995.
84. I Jakobsen. Transport of phosphorus and carbon in arbuscular mycorrhizas. In: A Varma, B Hock, eds. Mycorrhiza. 2nd ed. Berlin: Springer-Verlag, 1999, pp 305–332.
85. SE Smith, V Gianinazzi-Pearson, R Koide, JWG Cairney. Nutrient transport in mycorrhizas: Structure, physiology and consequences for efficiency of the symbiosis. In AD Robson, N Malajczuk, LK Abbot, eds. Management of Mycorrhizas in Agriculture, Horticulture and Forestry, Dordrecht, The Netherlands: Kluwer Academic, 1994, pp 103–113.
86. JP Guillemin, MO Orozco, V Gianinazzi-Pearson, S Gianinazzi. Influence of phosphate fertilization on fungal alkaline phosphatase and succinate dehydrogenase activies in arbuscular mycorrhiza of soybean and pineapple. Agric Ecosyst Environ 53:63–69, 1995.
87. B Tisserant, S Gianinazzi, V Gianinazzi-Pearson. Relationships between lateral root order, arbuscular mycorrhiza development, and the physiological state of the symbiotic fungus in *Platanus acerifolia*. Can J Bot 74:1947–1955, 1996.
88. JM Barea, C Azcon-Aguilar. Mycorrhizas and their significance in nondulating nitrogen-fixing plants. In: NC Brady, ed. Advances in Agronomy 36, London: Academic Press, 1983, pp 1–54.
89. SE Smith. Mycorrhizas of autotrophic plants. Biol Rev 55:475–510, 1980.
90. SE Smith, SBT John, FA Smith, DJ Nicholas. Activity of glutamine synthetase and glutamate dehydrogenase in *Trifolium subterraneum* L and *Allium cepa* L.: Effects of mycorrhizal infection and phosphate nutrition. New Phytol 99:211–227, 1985.
91. I Ho, JM Trappe. Nitrate reducing capacity of two vesicular arbuscular mycorrhizal fungi. Mycologia 67:886–888, 1975.
92. P Sundaresan, NV Rata, P Gunasera, M Lakshman. Studies on nitrate reduction by VAM fungal spores. Curr Sci 57:211–227, 1985.

93. JM Ruiz-Lozano, R Azcón. Mycorrhizal colonization and drought stress as factors affecting nitrate reductase activity in lettuce plants. Agric Ecosyst Environ 60:175–181, 1996.
94. M Kaldorf, E Schmelzer, H Bothe. Expression of maize and fungal nitrate reductase genes in arbuscular mycorrhiza. Mol Plant Microbe Interact 11:439–448, 1998.
95. M Kaldorf, W Zimmer, H Bothe. Genetic evidence for the occurrence of assimilatory nitrate reductase in arbuscular mycorrhizal and other fungi. Mycorrhiza 5:23–28, 1994.
96. R Azcon, RM Tobar. Activity of nitrate reductase and glutamine synthase in shoot and root of mycorrhizal *Allium cepa*: effect of drought stress. Plant Sci 133:1–8, 1998.
97. DI Bevege, GD Bowen, MF Skinner. Comparative carbohydrate physiology of ecto- and endomycorrhizas. In: FE Sanders, B Mosse, P Tinker eds. Endomycorrhizas. London: Academic Press, 1975, pp 149–174.
98. F Amijee, DP Stribley. Soluble carbohydrates of vesicular-arbuscular mycorrhizal fungi. Micologist 21:20–21, 1987.
99. G Becard, LW Doner, DB Rolin, DD Douds, PE Pfeffer. Identification and quantification of trehalose in vesicular-arbuscular mycorrhizal fungi in vivo ^{13}C NMR and HPLC analyses. New Phytol 118:547–552, 1991.
100. J Muller, T Boller, A Wiemken. Trehalose and trehalase in plants: Recent developments. Plant Sci 112:1–9, 1995.
101. L Schellenbaum, J Muller, T Boller, A Wiemken, H Schuepp. Effects of drought on non-mycorrhizal and mycorrhizal maize: Changes in the pools of non-structural carbohydrates, in the activities of invertase and trehalase, and in the pools of amino acids and imino acids. New Phytol 138:59–66, 1998.
102. RM MacDonald, M Lewis. The occurrence of some acid phosphatases and dehydrogenases in the vesicular-arbuscular mycorrhizal fungus *Glomus mosseae*. New Phytol 80:135–141, 1978.
103. RK Antibus, JM Trappe, AE Linkins. Cyanide resistant respiration in *Salix nigra* endomycorrhizae. Can J Bot 58:14–20, 1980.
104. JL Harley, CC McCready, RT Wedding. Control of respiration of beech mycorrhizas during ageing. New Phytol 78:147, 1977.
105. V Gianinazzi-Pearson, E Dumas-Gaudot, A Gollotte, A Tahiri-Alaoui, S Gianinazzi. Cellular and molecular defense-related root responses to invasion by arbuscular mycorrhizal fungi. New Phytol 133:45–57, 1996.
106. M Giovannetti, C Sbrana, C Logi. Early processes involved in host recognition by arbuscular mycorrhizal fungi. New Phytol 127:703–709, 1994.
107. H Volpin, DA Phillips, Y Okon, Y Kapulnik. Suppression of an isoflavonoid phytoalexin defense response in mycorrhizal alfalfa roots. Plant Physiol 108:1449–1454, 1995.
108. R David, H Itzhaki, I Ginzberg, Y Gafni, G Galili, Y Kapulnik. Suppression of tobacco basic chitinase gene expression in response to colonization by the arbuscular mycorrhizal fungus *Glomus intraradices*. Mol Plant Microbe Interact 11:489–497, 1998.
109. DB Collinge, KM Kragh, JD Mikkelsen, KK Nielsen, U Rasmussen, K Vad. Plant chitinase. Plant J 3:31–40, 1993.
110. LS Graham, NB Steklen. Plant chitinases. Can J Bot 72:1057–1083, 1994.
111. H Volpin, Y Elkind, Y Okon, Y Kapulnik. A vesicular arbuscular mycorrhizal fungus (*Glomus intraradix*) induces a defense response in alfalfa roots. Plant Physiol 104:683–689, 1994.
112. P Spanu, T Boller, A Ludwig, A Wiemken, A Faccio, P Bonfante-Fasolo. Chitinase in roots of mycorrhizal *Allium porrum*: Regulation and localization. Planta 177:447–455, 1989.
113. MR Lambais, MC Mehdy. Suppression of endochitinase, β-1,3-endoglucanase, and chalcone isomerase expression in bean vesicular-arbuscular mycorrhizal roots under different soil phosphate conditions. Mol Plant Microbe Interac 6:75–83, 1993.
114. H Vierheilig, M Alt, U Mohr, T Boller, A Wiemken. Ethylene biosynthesis and activities of chitinase and β-1,3-glucanase in the roots of host and non-host plants of vesicular-arbuscular

mycorrhizal fungi after inoculation with *Glomus mosseae*. J Plant Physiol 143:337–343, 1994.
115. MR Lambais, MC Mehdy. Differential expression of defense-related genes in arbuscular mycorrhiza. Can J Bot 73:533–540, 1995.
116. MR Lambais. Expression of plant defense genes during the development of vesicular-arbuscular mycorrhiza. PhD Austin: Division of Biological Sciences, The University of Texas at Austin, 1993.
117. HW Dehne, F Schönbeck. The influence of endotrophic mycorrhiza on plant diseases. 3. Chitinase-activity and ornithine-cycle. J Plant Dis Proteins 85:666–678, 1978.
118. E Dumas-Gaudot, V Furlan, J Grenier, A Asselin. New acidic chitinase isoform induced in tobacco roots by vesicular-arbuscular mycorrhizal fungi. Mycorrhiza 1:133–136, 1992.
119. U Mohr, J Lange, T Boller, A Wiemken, R Vögeli-Lange. Plant defence genes are induced in the pathogenic interaction between bean roots and *Fusarium solani*, but not in the symbiotic interaction with the arbuscular mycorrhizal fungus *Glomus mosseae*. New Phytol 138:589–598, 1998.
120. KA Blee, AJ Anderson. Defense-related transcript accumulation in *Phaseolus vulgaris* L. colonized by the arbuscular mycorrhizal fungus *Glomus intraradices* Schenk & Smith. Plant Physiol 110:675–688, 1996.
121. MR Lambais, MC Mehdy. Spatial distribution of chitinases and β-1,3-glucanase transcripts in bean arbuscular mycorrhizal roots under low and high soil phosphate conditions. New Phytol 140:33–42, 1998.
122. H Vierheilig, M Alt, JM Neuhaus, T Boller, A Wiemken. Colonization of transgenic *Nicotiana silvestris* plants, expressing different forms of Nicotiana tabacum chitinase, by the root pathogen *Rhizoctonia solani* and by the mycorrhizal symbiont *Glomus mosseae*. Mol Plant Microbe Interact 6:261–264, 1993.
123. H Vierheilig, M Alt, J Lange, M Gut-Rella, A Wiemken, T Boller. Colonization of transgenics tobacco constitutively expressing pathogenesis-related proteins by vesicular-arbuscular mycorrhizal fungus *Glomus mosseae*. Appl Environ Microbiol 61:3031–3034, 1995.
124. B Dassi, E Dumas-Gaudot, A Asselin, C Richard, S Gianinazzi. Chitinase and β-1,3-glucanase isoforms expressed in pea roots inoculated with arbuscular mycorrhizal or pathogenic fungi. Eur J Plant Pathol 102:105–108, 1996.
125. MJ Pozo, E Dumas-Gaudot, S Slezack, C Cordier, A Asselin, S Gianinazzi, V Gianinazzi-Pearson, C Azcón-Aguilar, JM Barea. Induction of new chitinase isoforms in tomato roots during interactions with *Glomus mosseae* and/or *Phytophthora nicotianae* var, *parasitica*. Agronomie 16:689–697, 1996.
126. MJ Pozo, C Azcón-Aguilar, E Dumas-Gaudot, JM Barea. Chitosanase and chitinase activities in tomato roots during interactions with arbuscular mycorrhizal fungi or *Phytophthora parasitica*. J Exp Bot 49:1729–1739, 1998.
127. MJ Pozo, C Azcón-Aguilar, E Dumas-Gaudot, JM Barea. β-1,3-glucanase activities in tomato roots inoculated with arbuscular mycorrhizal fungi and/or *Phytophthora parasitica* and their possible involvement in bioprotection. Plant Sci 141:149–157, 1999.
128. E Dumas-Gaudot, J Grenier, V Furlan, A Asselin. Chitinase, chitosanase and β-1,3-glucanase activities in *Allium* and *Pisum* roots colonized by *Glomus* species. Plant Sci 84:17–24, 1992.
129. RL Monaaghan, DE Eveleigh, RP Tewari, ET Reese. Chitosanase, a novel enzyme. Nature New Biol 245:78–80, 1973.
130. S Slezack, B Dassi, E Dumas-Gaudot. Arbuscular mycorrhiza-induced chitinase isoforms. In: RAA Muzzarelli, ed. Chitin Enzymologie. Vol. 2. Ancona, Italy: Atec Edizioni, pp. 339–347, 1996.
131. C Staehelin, J Müller, RB Mellor, A Wiemken T Boller. Chitinase and peroxidase in effective (fix+) and ineffective (fix−) soybean nodules. Planta 187:295–300, 1992.
132. C Albrecht, A Asselin, Y Piché, F Lapeyrie. Comparison of *Eucalyptus* root chitinase patterns

following inoculation by ectomycorrhizal or pathogenic fungi *in vitro*. In: B Fritig, M Legrand, eds. Mechanisms of plant defence responses, developments in plant pathology. Vol 2 Dordrecht, The Netherlands: Kluber Academic, 1993, p 380.
133. RB Mellor, DB Collinge. A simple model based on known plant defence reactions is sufficient to explain most aspects of nodulation. J Exp Bot 46:1–18, 1995.
134. MC Mehdy. Active oxygen species in plant defense against pathogens. Plant Physiol 105: 467–472, 1994.
135. C Lamb, RA Dixon. The oxidative burst in plant disease resistance. Annu Rev Plant Physiol Plant Mol Biol 48:251–275, 1997.
136. A Levine, R Tehnhaken, RA Dixon. H_2O_2 from the oxidative burst orchestrates the plant hypersensitive disease resistance response. Cell 79:583–593, 1994.
137. DD Douds, L Galvez, G Bécard, Y Kapulnik. Regulation of arbuscular mycorrhizal development by plant host and fungus species in alfalfa. New Phytol 138:27–35, 1998.
138. P Salzer, H Corbière, T Boller. Hydrogen peroxide accumulation *in Medicago truncatula* roots colonized by the arbuscular mycorrhiza-forming fungus *Glomus mosseae*. Planta 208: 319–325, 1999.
139. P Spanu, P Bonfante-Fasolo. Cell wall-bound peroxidase activity in roots of mycorrhizal *Allium porrum*. New Phytol 109:119–124, 1988.
140. DAJ McArthur, NR Knowles. Resistance responses of potato to vesicular-arbuscular mycorrhizal fungi under varying abiotic phosphorus level. Plant Physiol 100:341–351, 1992.
141. I Blilou, P Bueno, JA Ocampo, JM García-Garrido. Induction of catalase and ascorbate peroxidase activities in tobacco roots inoculated with the arbuscular mycorrhizal fungus *Glomus mosseae*. Mycol Res 104:722–725, 2000.
142. D Cook, D Dreyer, D Bonnet, M Howell, E Nony, K VandenBosch. Transient induction of a peroxidase gene in *Medicago truncatula* precedes infection by *Rhizobium meliloti*. Plant Cell 7:43–55, 1995.
143. M Becana, FJ Paris, LM Sandalio, LA del Rio. Isoenzymes of superoxide dismutase in nodules of *Phaseolus vulgaris* L., *Pisum sativum* L., and *Vigna unguiculata* (L.) Walp. Plant Physiol 90:1286–1292, 1989.
144. J Martín, I García-Romera, JA Ocampo, JM Palma. Superoxide dismutase and arbuscular mycorrhizal fungi: Relationship between the isoenzyme pattern and the colonizing fungus. Symbiosis 24:247–258, 1998.
145. J Arines, A Vilariño, JM Palma. Involvement of the superoxide dismutase enzyme in the mycorrhization process. Agric Sci Fin 3:303–307, 1994.
146. JM Palma, MA Longa, LA del Rio, J Arines. Superoxide dismutase in vesicular arbuscular-mycorrhizal red clover plants. Physiol Plant 87:77–83, 1993.
147. J Arines, M Quintela, A Vilariño, JM Palma. Protein patterns and superoxide dismutase activity in non-mycorrhizal and arbuscular mycorrhizal *Pisum sativum* L. plants. Plant Soil 166:37–45, 1994.
148. JM Ruiz-Lozano, R Azcón, JM Palma. Superoxide dismutase activity in arbuscular mycorrhizal *Lactuca sativa* plants subjected to drought stress. New Phytol 134:327–333, 1996.
149. RA Dixon, NL Paiva. Stress-induced phenylpropanoid metabolism. Plant Cell 7:1085–1097, 1995.
150. RA Dixon, AD Choudhary, K Dalkin, R Edwards, T Fahrendorf, G Gowri, MJ Harrison, CJ Lamb, GJ Loake, CA Maxwell, J Orr, NL Paiva. Molecular biology of stress-induced phenylpropanoid and isoflavonoid biosynthesis in alfalfa. In: HA Stafford, RK Ibrahim, eds. Phenolic Metabolism in Plants. New York: Plenum Press, 1992, pp 91–138.
151. F Sánchez, JE Padilla, H Peréz, M Lara. Control of nodulin genes in root-nodule development and metabolism. Annu Rev Plant Physiol Plant Mol Biol 42:507–528, 1991.
152. V Gianinazzi-Pearson, B Branzanti, S Gianinazzi. *In vitro* enhancement of spore germination and early hyphal growth of a vesicular-arbuscular mycorrhizal fungus by host root exudates and plant flavonoids. Symbiosis 7:243–255, 1989.

153. H Vierheilig, B Bago, C Albrecht, MJ Poulin, Y Pichè. Flavonoids and Arbuscular-Mycorrhizal fungi. In: Manthey and Buslig, ed. Flavonoids in the living system. New York: Plenum Press, 1998, pp 9–33.
154. M Harrison, R Dixon. Isoflavonoid accumulation and expression of defense gene transcripts during the establishment of vesicular arbuscular mycorrhizal associations in roots of *Medicago truncatula*. Mol Plant Microbe Interact 6:643–659, 1993.
155. M Harrison, R Dixon. Spatial patterns of expression of flavonoid/isoflavonoid pathway genes during interactions between roots of *Medicago truncatula* and the mycorrhizal fungus *Glomus versiforme*. Plant J 6:9–20, 1994.
156. I Blilou. Aspectos moleculares de la respuesta de las plantas a la formación de la simbiosis micorriza-arbuscular. Granada, Spain: PhD, Division of Biological Sciences, The University of Granada, 1998.
157. DF Klessig, J Malamy. The salicylic acid signal in plants. Plant Mol Biol 26:1439–1458, 1994.
158. J Ryals, U Neuenschwander, MG Willits, A Molina, H-Y Steiner, MD Hunt. Systemic acquired resistance. Plant Cell 8:1809–1819, 1996.
159. MG Willits, J Ryals. Determining the relationship between salicylic acid levels and systemic acquired resistance induction in tobacco. Mol Plant Microbe Interact 11:795–800, 1998.
160. P Franken, F Gnädinger. Analysis of parsley arbuscular endomycorrhiza: infection development and mRNA levels of defense-related genes. Mol Plant Microbe Interact 7:612–620, 1994.
161. Y Kapulnik, H Volpin, H Itzhaki, D Ganon, S Galili, R David, O Shaul, Y Elad, I Chet, Y Okon. Suppression of defense response in mycorrhizal alfalfa and tobacco roots. New Phytol 133:59–64, 1996.
162. D Morandi. Effect of xenobiotics on endomycorrhizal infection and isoflavonoid accumulation in soybean roots. Plant Physiol Biochem 27:697–701, 1989.
163. G Bécard, LP Taylor, DD Douds, PE Pfeffer, LW Doner. Flavonoids are not necessary plant signal compound in arbuscular mycorrhizal symbiosis. Mol Plant Microbe Interact 8:252–258, 1995.
164. DS Hayman. The physiology of vesicular-arbuscular endomycorrhizal symbiosis. Can J Bot 61:944–963, 1983.
165. B Tisserant, V Gianinazzi-Pearson, S Gianinazzi, A Gollotte. Plant histochemical staining of fungal alkaline phosphatase activity for analysis of efficient arbuscular mycorrhizal infections. Mycol Res 97:245–250, 1993.
166. JKL Kough, V Gianinazzi-Pearson. Physiological aspects of VA mycorrhizal hyphae in root tissue and soil. In: V Gianinazzi-Pearson, S Gianinazzi, eds. Physiological and Genetical Aspects of Mycorrhizae. Paris: INRA, 1986, pp 223–226.
167. J Larsen, I Thingstrup, I Jakobsen, S Rosendahl. Benomyl inhibits phosphorus transport but not fungal alkaline phosphatase activity in a *Glomus*-cucumber symbiosis. New Phytol 132: 127–134, 1996.
168. JA Ocampo, JM Barea. Effect of carbamate herbicides on VA mycorrhizal infection and plant growth. Plant Soil 85:375–383, 1985.
169. JKL Kough, V Gianinazzi-Pearson, S Gianinazzi. Depressed metabolic activity of vesicular-arbuscular mycorrhizal fungi after fungicide applications. New Phytol 106:707–715, 1987.
170. M Kling, I Jakobsen. Direct application of carbendazim and propiconazole at field rates to the external mycelium of three arbuscular fungi species: effect on ^{32}P transport and succinate dehydrogenase activity. Mycorrhiza 7:33–37, 1997.
171. CB McAllister, I García-Romera, A Godeas, JA Ocampo. Interaction between *Trichoderma koningii, Fusarium solani* and *Glomus mosseae*: Effect on plant growth, arbuscular mycorrhizas and the saprophytic inoculants. Soil Biol Biochem 26:1363–1367, 1994.
172. SE Smith, S Dickson. Quantification of active vesicular-arbuscular mycorrhizal infection using image analysis and other techniques. Aust J Plant Physiol 18:637–648, 1991.

173. H Vierheilig, JA Ocampo. Relationship between SDH-activity and VA mycorrhizal infection. Agric Ecosyst Environ 29:439–442, 1989.
174. H Vierheilig, JA Ocampo. Receptivity of various wheat cultivars to infection by VA-mycorrhizal fungi as influenced by inoculum potential and the relation of VAM-effectiveness to succinic dehydrogenase activity of the mycelium in the roots. Plant Soil 133:291–296, 1991.
175. S Rosendahl. Influence of three vesicular-arbuscular mycorrhizal fungi (Glomaceae) on the activity of specific enzymes in the root system of *Cucumis sativus* L. Plant Soil 144:219–226, 1992.
176. T Ezawa, T Yoshida. Characterization of phosphatase in marigold roots infected with vesicular-arbuscular mycorrhizal fungi. Soil Sci Plant Nutr 40:225–264, 1994.
177. I Thingstrup, S Rosendahl. Quantification of fungal activity in arbuscular mycorrhizal symbiosis by polyacrylamide gel electrophoresis and densitometry of malate dehydrogenase. Soil Biol Biochem 26:1483–1489, 1994.
178. IR Sanders, M Alt, K Groppe, T Boller, A Viemken. Identification of ribosomal DNA polymorphisms among and within spores of the Glomales: Applications to studies on the genetic diversity of arbuscular mycorrhizal communities. New Phytol 130:419–427, 1995.
179. CM Hepper, R Sen, CS Maskell. Identification of vesicular-arbuscular mycorrhizal fungi in roots of leek (*Allium porrum* L.) and maize (*Zea mays* L.) on the basis of enzyme mobility during polyacrilamide gel electrophoresis. New Phytol 102:529–539, 1986.
180. R Sen, CM Hepper. Characterization of vesicular-arbuscular mycorrhizal fungi (*Glomus* spp) by selective enzyme staining following polyacrylamide gel electrophoresis. Soil Biol Biochem 18:29–34, 1986.
181. CM Hepper, R Sen, C Azcón-Aguilar, C Grace. Variation in certain isozymes amongst different geografical isolates of the vesicular-arbuscular mycorrhizal fungi *Glomus clarum, Glomus monosporum* and *Glomus mosseae*. Soil Biol Biochem 20:51–59, 1988.
182. CM Hepper, C Azcón-Aguilar, S Rosendahl, R Sen. Competition between three species of *Glomus* used as introduced or indigenous mycorrhizal inocula for leek (*Allium porrum*). New Phytol 110:207–215, 1988.
183. S Rosendahl, R Sen, CM Hepper, C Azcón-Aguilar. Quantification of three vesicular-arbuscular mycorrhizal fungi (*Glomus* spp) in roots of leek (*Allium porrum*) on the basis of activity of diagnostic enzymes after polyacrylamide gel electrophoresis. Soil Biol Biochem 21:519–522, 1989.
184. JC Dodd, S Rosendahl, M Giovannetti, A Broome, L Lanfranco, C Walker. Inter- and intraspecific variation within the morphologically similar arbuscular mycorrhizal fungus *Glomus mosseae* and *Glomus coronatum*. New Phytol 133:113–122, 1996.
185. B Bago, SP Bentivenga, V Brenac, JC Dodd, Y Piche, L Simon. Molecular analysis of *Gigaspora* (Glomales, Gigasporaceae). New Phytol 139:581–588, 1998.
186. MC Brundrett, B Kendrick. The roots and mycorrhizas of herbaceous woodland plants. II. Structural aspects of morphology. New Phytol 114:469–479, 1990.
187. P Bonfante, B Vian. Cell wall texture in symbiotic roots of *Allium porrum* L. Ann Sci Nat Bot Biol Veg 10:97–109, 1989.
188. A Camprubi, C Calvet, V Estaun. Growth enhancement of *Citrus reshni* after inoculation with *Glomus intraradices* and *Trichoderma aureoviride* and associated effects on microbial populations and enzyme activity in potting mixes. Plant Soil 173:233–238, 1995.

6

Microbes and Enzymes Associated with Plant Surfaces

Ian P. Thompson and Mark J. Bailey
Centre for Ecology and Hydrology, Oxford, England

I. INTRODUCTION

Plant–microbe interactions are vital to agricultural and plant productivity, which in turn have a direct and significant influence on the diversity, nature, and activity of colonizing microbial communities. Enzymes play a key role in these interactions and consequently have been the subject of intense interest. The main focus has been at the soil–plant interface (the rhizosphere), but plant litter decomposition is also of major importance. The ecological characteristics, ecosystem function, and mechanistic approaches to the study of phytopathogenicity all have aspects that involve enzymes. These subject areas are dealt with elsewhere in this volume (see Chapter 7). In this chapter, we concentrate on the interplay of microorganisms, how ''cross-talking'' and the regulation of gene expression influence the enzymes they produce, and how such processes influence community succession and functional diversity. We intend to draw on the limited data available to describe microorganisms that colonize the surface of leaves and, to a lesser extent, the inner tissues and other aerial parts of the plant such as the stems, fruits, flowers, and woody tissue (the phyllosphere). The chapter has been organized to describe the physical and chemical nature of the habitat, and the type and survival strategies of colonizing microbial populations. We discuss the nature and role of enzymes that have evolved to provide the genotypic and phenotypic plasticity necessary for microorganisms to survive under the selective pressures of this extreme habitat. The chapter is completed by discussion of developments in methodology that recently have become available to investigate the role and potential activity of enzymes in what can be extreme conditions, and the potential for their biotechnological exploitation.

In terms of surface area, leaves and other aerial parts of plants represent some of the largest terrestrial habitats available for microbial exploitation. Although the global extent of plant aerial surfaces is enormous, their exposed and fluctuating nature can make them the most inhospitable terrestrial habitats for microbial colonization. Leaves, for example, have evolved to be exposed to maximum intensities and durations of solar radiation. This exposure can raise temperatures on the leaf surface well above those in the surrounding air, rapidly evaporating any moisture that may be present. In order to survive these conditions, any microbial colonizers must be able to tolerate not only intense heat but also ultraviolet radiation and desiccation. Furthermore, physical conditions can alter rapidly, from the consequences of intense solar radiation to the abrasive conditions resulting from high winds and rain provided during storms. Few other terrestrial habitats are as exposed and susceptible to such violent environmental extremes. Despite this, leaves are typically colonized by large numbers and an extensive diversity of eubacteria, the archaea, and other eukaryotic microorganisms, including the fungi, yeasts, plants, and animals.

The range of microorganisms and the life-styles of these leaf colonizers are similarly diverse. Evolutionary solutions for survival range from epiphytic, opportunistic colonizers to pathogenic forms adapted to existence in the tissues of a specific plant species and even subspecies. Indeed, considerable plant–microbe coevolution and cospeciation are assumed (1). A key determinant in the success and persistence represented by these contrasting microbial life-styles is the leaf cuticle. The cuticle is a significant physical barrier that separates leaf tissues from the environment. For microbial saprophytes and epiphytes, this provides a surface for permanent attachment so that they can scavenge the scarce resources of nutrients and moisture. In contrast, pathogens often are less well adapted to the rigors of the leaf surface and respond by penetrating the cuticle in order to gain access to the nutrients and tissues within the leaf cortex.

The different challenges these contrasting life-styles represent clearly determine the survival strategies of leaf colonizers, and these in turn are dependent on the repertoire of enzymes required in order to ensure persistence and proliferation in extreme conditions. Additionally, the ability of these enzymes to function in extreme environmental conditions is of interest because they may prove valuable in biotechnological exploitation, where conditions can be equally unfavorable for microbial survival. However, before describing the range, role, and potential of enzymes produced by microorganisms that colonize aerial parts of plants, it is necessary to describe the physical and chemical nature of the habitat and its immediate vicinity. Since the volume of information on leaves is considerably greater than that on other aerial plant parts, this description concentrates on the leaf surface (the phylloplane).

II. LEAF STRUCTURE

The variation in physical characteristics and architecture of leaves and the aerial portion of plants makes it difficult to generalize on microbial adaptation in the phylloplane or phyllosphere. Leaf surface structures and chemical composition vary with plant species, cultivars of the same species, and leaves of the same plant and between upper and lower leaf surfaces. However, common habitat features are found on plant surfaces, and these are addressed in reviews elsewhere (2,3). For the purposes of this chapter the key structures are a cuticular wax outer layer overlying three deeper regions composed of cutin and wax, cutin and carbohydrate, and carbohydrate, respectively (Fig. 1). The principal exposed structure for microbial colonizers is the outer wax cuticle. It is made up of lipid membranes consisting of cuticular waxes and lipophilic cutin polymers that always contain polar polysaccharides (i.e., pectins, celluloses, polypeptides). Its thickness varies considerably (0.1 to 20 μm) and is dependent upon factors such as leaf age, plant health, and growth conditions. The cuticle protects the leaf from biotic and abiotic damage and regulates passage of water (4), inorganic ions (5), and organic solutes between the plant and the environment. This protection results from cuticular resistance, but mostly from physical resistance provided by epicuticular waxes (quantity, composition, structure, and arrangement).

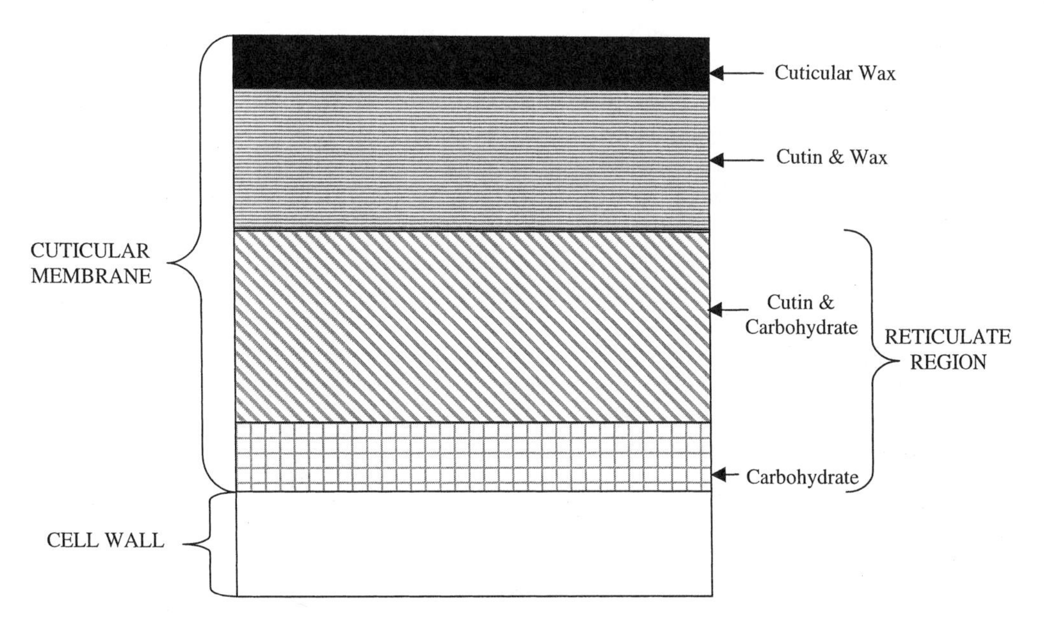

Figure 1 Schematic of leaf surface. Microbial saprophytes colonize the outer surface of the cuticular wax. Pathogens initially colonize the cuticular wax then penetrate through to the cell wall.

The degree of protection afforded by the wax is dependent on its thickness, which in turn is influenced by a range of external factors such as exposure to sunlight. For example, Steinmuller and Tevini (6) demonstrated that enhanced levels of ultraviolet-B (UV-B) radiation promote the synthesis of up to 25% more wax when compared with that of untreated control plants. However, although the cuticle is hard and dense (100 to 250 μg/cm), it can be damaged easily by physical abrasion from wind, attack by phytophagous insects, or microbial degradation that follows the release and activity of cutinase enzymes. Cuticular waxes constitute the main barrier to the transport of nutrients and water in leaves, even though they amount to only a small percentage of the mass of cuticles. This is readily shown by comparing cuticles before and after extraction of waxes, which increases water permeability up to 1500 times (7) and permeability to organic compounds by as much as four orders of magnitude (8).

III. LEAF AS A MICROBIAL HABITAT

The leaf provides an exposed and unstable habitat the harshness of which is dependent on its the architecture. However, it still possesses favourable features that make it attractive for microbial colonization. For instance, the leaf surface can be a rich source of nutrients. Significant quantities of leachates can diffuse through the stomata and the hydrophobic cuticular layer, particularly in young and emerging leaves of plants that form rosettes and in old senescing leaves (9,10). Apple leaves, for instance, may lose 25–30 kg of potassium and 800 kg of sugars per hectare per year. Similarly, amino acids, organic acids, hormones (e.g., indoleacetic acid), and phenolic compounds are lost (11). The exudation can be continuous but is particularly copious after long periods without rain (12). Loss of nutrients also may occur by diffusion and evaporation through the stomata and as a consequence of vascular and cellular damage brought about by wounding. These mechanisms accelerate the loss of nutrients and water but provide the presumed resources exploited by the heterotrophic and oligotrophic microbial colonists. However, there is little research regarding the chemical nature of leaf exudates, certainly not since the comprehensive studies of Tukey in the 1970s, although more recently it has been demonstrated that the complex mix of alcohols and esters is responsible for their distinctive odor (13). In addition, more complex substances, such as monoterpenes and terpenes, are produced, but as with most other leaf exudates, little is known of their potential influence on microbial colonizers as potential substrates or inhibitors.

The quality of habitats for microbial colonization varies among aerial plant parts. In many respects, the leaf is the most adverse habitat since it is exposed to a range of environmental extremes such as desiccation, intense radiation, and agitation from high winds and heavy rainfall. Furthermore, the outer cuticular wax layer is hard and hydrophobic and there is considerable evidence that the waxes contain microbial inhibitors (14,15). In contrast, the leaf presents a planar surface that potentially provides firm anchorage for epiphytic forms and, importantly, is a rich source of nutrients for microbial growth. This is reflected in the high densities of bacteria, yeasts, and fungi (1×10^4 to 1×10^6 cells/cm^2) that can occur on leaves. However, the unfavorable nature of the leaf is reflected in the uneven distribution of the colonizers that typically are concentrated in aggregates that occur in areas of the leaf, such as the ridge along the main leaf vein, which represents less than 1% of the surface. Competition for space and resources in these ''hot spots'' is likely to be extreme, although the survival rate of microbial cells within the aggregates has been demonstrated to be greater than that of single isolated individuals (16). It has also been suggested that epiphytic microorganisms may take an even more active approach to survival by producing exudates, including biosurfactants and enzymes, that increase the flux of nutrients from the plant to the colonizers (17,18). Similarly, the plant may not be passive but defend itself, particularly from potential invasion by microbial pathogens, by increasing the concentration within its cells of a range of enzymes including chitinases, glucanases, and phenoloxidases. These act as antimicrobial agents that, in the case of chitinase, can degrade the cell wall of fungal pathogens (19). Other broad-spectrum inhibitory compounds, such as a range of oxygen species, that oxidize the proteins of bacteria and viruses have also been identified (20,21).

IV. LEAF MICROBIAL COLONIZERS

The interaction between microorganisms and their host on the leaf surface is complex, and in most cases we know little of the processes and organisms involved. The leaves of most species of plants have been shown to be colonized by high densities of microorganisms. It is not the purpose of this chapter to assemble an exhaustive list of microbial species isolated from the leaf surface. However, in terms of the enzymes produced by microbial leaf colonizers it is important to acknowledge the two main microbial life-styles that exist. The phytopathogens (which can be highly specialized and colonize a narrow range, or indeed a single plant species) tend to be adapted to penetrate the leaf cuticle and exploit the potential wealth of nutrients in the deeper inner tissues. Because of their potential to destroy crop plants and impact on economic returns, considerably more is known of this group than of the more generalized and diverse epiphytic saprophytic forms.

This difference is also due partly to the limited number of phytopathological forms discovered to date. Among the 1600 different bacterial species known less than 1% are confirmed as significant pests (22). Similarly, we also know considerably more about enzymes involved in pathogenicity than saprophytic existence.

Despite the ubiquitous nature of epiphytic microbial forms on the leaf surface, we know little of their ecological characteristics and even less of the enzymes produced by this diverse group. This is certainly the case when compared with the more numerous studies of the diversity of rhizosphere microbial communities. However, there have been a few studies that have assessed the abundance and diversity of saprophytic microbial communities colonizing a range of plant species (23–26). These studies have revealed some commonality in the microbial diversity of different plant species. For instance, the group has been shown to be diverse, probably because of the large exposed surface that leaves represent and that form a sink for deposition of dust, debris, and rain; this characteristic is likely to favor opportunistic microbial colonizers. In general, plants in temperate zones are colonized by pink and white yeasts (mainly *Sporobolomyces* and *Cryptococcus* species) yeastlike fungi (e.g. *Aureobasidium* sp.), filamentous fungi (e.g., *Cladosporium* sp.), and bacteria (27).

Our studies of heterotrophic bacteria found on sugar beet (*Beta vulgaris* L.) leaves showed them to be a diverse and dynamic habitat. We have examined the bacterial community colonizing the leaves of sugar beet grown in the same field over 10 consecutive years. During the course of these studies, we have characterized, both phenotypically and genotypically, over 10,000 individual bacteria isolates and observed some striking features that probably are characteristic of saprophytic microbial communities in the phyllosphere. As other workers (26,27) have observed the community was found to be very diverse. In one growing season 1236 randomly selected bacterial isolates were phenotypically characterized and found to be composed of 78 known genera and 34 unknown groups. Most species were transient and detected on only one leaf type (immature, mature, or senescent) or at a single sampling occasion (9). Only one bacterial species was detected throughout the growing season and on all seven sampling occasions (April–October) and the three leaf types samples, and we believe this is attributable to the harsh and very selective nature of the leaf.

As previous workers (23,28) have demonstrated for other crops, the dominant population was composed of several species of pseudomonads, and indeed Gram-negative chromogenic forms tend to be the most abundant. The persistence of pseudomonads is of considerable interest since this group harbors both pathogenic and saprophytic forms that may enhance plant productivity. Intensive studies of pseudomonads suggest that their persistence is due to heterogeneity; the group is composed of an array of different genotypes, each one suited to specific environmental conditions (29). As conditions alter, dominant genotypes decline and are replaced rapidly by those types better able to proliferate in the prevailing conditions. Observations indicate that the duration of numerical dominance for the different genotypic populations on the leaf was in the region of 28 days but may be even shorter (30).

V. ENZYMES AS DETERMINANTS OF LIFE-STYLES

The diversity of taxa and the survival strategies used to persist and proliferate range from opportunistic (probably due to accidental colonization) to highly specialised coevolved pathogenic forms adapted to exist in the inner plant tissues. On the basis of the phenotypic traits required for organisms to proliferate in these contrasting life-styles and leaf habitats, it is likely each produces a repertoire of enzymes that are highly distinctive and even diagnostic. The importance of specific enzymes in determining the interaction of potential microbial colonizers on the plants' aerial surfaces is demonstrated elegantly in a study in which pathogenic potential of a strain was altered by introducing additional genes for enzyme production (31). In this study, the authors were able to transfer the genes from *Fusarium solani* F. sp. *pisi*, a pea pathogen, into *Mycosphaerella* sp., a parasitic fungus that affects papaya, but only if the fruit is mechanically breached before inoculation. Transfer of the genes resulted in transformants that could infect intact fruits, without prior mechanical damage. The enzyme involved was cutinase, an extracellular enzyme, the acquisition of which clearly altered the potential for invading the plant. This study demonstrates the effect that a single enzyme can have in altering specific microbe–plant interactions.

Similarly, other studies that take quite different but complementary approaches to that of recombinant technology also demonstrate the tight association between the life-style of microorganisms and their enzyme-production characteristics. Studies on enzyme versatility, relative to substrates, suggest that the more generalist saprophytes produce hydrolases, which have a broader spectrum of activity than pathogenic forms (32). The diversity of enzymes within the saprophytes enables them to have an opportunistic life-style, particularly the ability to exploit resources that become temporarily available. For example, the *Aspergillus* genus contain members that not only are saprophytic colonizers of plant surfaces but are also pathogenic to plants and insects. As would be expected, the enzyme diversity of the saprophytic forms is considerable.

These observations have led to speculation concerning the evolutionary origins of the pathogenic forms. Knogge (33) suggests that pathogens may be derived from ancestral saprophytes that have refined some traits, such as enzyme specificity, and developed additional characters, which aid host cell exploitation. In many cases it is difficult to determine which additional traits are developed, since candidates such as polysaccharides that may be required to gain entry into plant hosts are also produced by saprophytes. The issue is further complicated since some pathogens have a saprophytic stage and produce two forms (isozymes) of some enzymes, which potentially can play different roles in plant cell tissue exploitation. Cutinase, one such enzyme, has been extensively studied in *Alternaria brassicicola* by a range of methods including gene disruption (34). These studies provide evidence for a differential and sequential induction of two classes of cutinolytic esterases. For instance, serine esterases with cutinolytic activities were expressed by conidia germinating on a host surface. However, the enzymes were not induced by surface wax or cutin monomers and were expressed only during initial contact of conidia with cutin on host surfaces free of wax and with cutin in aqueous suspensions. In contrast, contact with cutin had no immediate effect on the expression of CUTAB1, a gene encoding two cutinase isozymes with crucial functions in saprophytic utilization of cutin. The presence of cutin or prolonged exposure to cutin was required for the induction of CUTAB1 expression. The differential induction of cutinolytic esterases indicates a sequential recognition of cutin as a barrier to be penetrated and utilized as a carbon source in the saprophytic stages (35).

VI. ROLE OF ENZYMES IN PLANT–MICROBE INTERACTIONS

The most immediate role of enzymes occurs during attachment of both pathogenic and saprophytic microorganisms to leaf surfaces. Initial attachment of spores of active cells is a passive process that is reversible. However, with time the process becomes active, including the formation of adhesion pads and production of esterases and cutinases that lead to firm adhesion by erosion of the host cuticle and waxes (36). However, our knowledge of the attachment process and the enzymes involved is sparse (37). More is known of the plant response to microbial attachment and potential invasion, which typically, in many healthy resistant plants, results in a defense response in the form of enzyme production. This includes the production of elevated levels of plant enzymes, such as chitinases, and glucanses, that have the potential to degrade the cell wall components of fungal pathogens (19). In addition, less specific antimicrobials are produced by some plants, including active oxygen species, that cause the oxidation of proteins of both bacterial and viral pathogens (38). This response has been studied extensively in some strains such as *Botrytis cinerea*, the less vigorous strains of which have been found to produce a greater enzymatic response to pathogens (20). The presence of several plant enzymes has been correlated to resistance to microbial attack, including those due to peroxidase, oxidase, and polyphenol oxidase production (38). One approach suggested as a means of controlling microbial pathogenesis is to introduce genes for α-amylase production that reduce disease by pathogenic pseudomonads (39). Some pathogens, such as strains of *Erwinia chrysanthemi*, counteract this plant response by generating resistant forms that produce peptide methionine sulfoxide reductase, which repairs damage caused by protein oxidation (40).

Assuming the invading microbial pathogen has overcome the natural plant defense mechanisms, including production of enzymes that are antimicrobial, the organisms must physically penetrate the outer cuticular wax layer before a successful pathogenic interaction can become established. Jones (41) divides the early interaction of fungal spores of pathogens with host plants into five stages: (1) initial passive attachment, (2) active induced attachment, (3) spore differentiation and germination, (4) development of appressoria, and (5) penetration of the host substratum.

At least three of these stages (active attachment, development of appressoria, and penetration) involve the production of enzymes. As indicated earlier, the active attachment and penetration of the host substratum certainly involve a range of enzymes, including cutinases and lipases. Their role in the initial stages of plant host attack has been demonstrated when enzyme activity has been inhibited by antibodies (42). Similarly, treatment of spores of the parasitic pathogen *Uromyces viciae-fabae* with the serine esterase inhibitor diisopropyl fluorophosphate prevented formation of the adhesion pad on the broad bean host. This finding also implicates strongly the role of esterase and cutinase in adhesion (36). The list of enzymes implicated in the more advanced stages of pathogen attack is also extensive; it includes polygalacturonase, phenoloxidases, β-galactosidase. A considerable body of information concerns these enzymes. For instance, many hydrolytic enzymes are constitutive, whereas cellulases are almost entirely inducible (43). It has been shown that the activity of microbial enzymes, such as pectate lyases from pathogenic *Erwinia carotovora*, also induces a plant defense response.

VII. EXPLOITATION OF MICROBIAL ENZYMES FROM PLANT SURFACES

The repertoire of enzymes possessed by microorganisms that colonize and proliferate on the aerial parts of plants is extensive and in many respects distinguishable from those that occur in other less extreme habitats. The enzymes must be able to function in environmental extremes such as temperature and moisture that are also prone to sudden fluctuations.

The plant tissues (waxes, cutin, and cellulose) on which the microbial colonizers have to survive and from which they derive nutrients are typically chemically complex and resistant to degradation. Furthermore, even when conditions favor microbial growth, this leads to proliferation and intra- and interspecies competition. Nevertheless, despite what appear to be extremely unfavorable conditions for microbial growth and activity, the aerial parts of plants are colonized by a broad range of microorganisms. It follows that these microbes must produce enzymes that are adapted to function in extreme conditions and be effective against recalcitrant substrates such as plant polysaccharides. With the enormous array of selective forces acting on microbial communities that colonize the aerial parts of the plant, it is not surprising that they are a rich source of organisms and enzymes with potential in biotechnological exploitation. We describe later a few instances of microbial strains that originate from the aerial parts of the plant that have been or are currently being developed for a broad range of biotechnological exploitation. These applications include controlling plant diseases, producing plant products of increased dietary value, and degrading anthropogenic materials.

A. Biocontrol

Examination of the aerial parts of most plants reveals that microbial colonizers tend to occur in ''hot spots'' and that cell densities can be high. This inevitably leads to intense intra- and interspecies competition and a range of survival mechanisms, involving antibiotic and enzyme production, that confer competitive advantage. As a consequence of such interactions, plant surfaces are known to be a rich source of microorganisms that are inhibitory to phytopathogens. The exploitation of microbial enzymes for controlling phytopathogens has taken at least two routes. The most obvious has been direct inhibition of the pathogen with control agents. For instance, the biocontrol agent *Trichoderma hazianum* has been shown to inhibit the pathogen *Botrytis cinerea* by the production of a range of hydrolytic enzymes including cutin esterase, polygalacturonase, pectin methyl esterase, pectate, and carboxymethylcellulases. The degree of inhibition can be considerable (up to 80%) when compared to that of the control (43). An alternative approach in the exploitation of microbial enzymes in plant protection has been the use of hydrolytic enzymes to stimulate a defense response by the plant. In a 1998 study, purified cutinase, taken from one pathogen (*Venturia inaequalis*), was applied to leaves infected with the pathogen *Rhizoctonia solani*, and this process reduced the incidence of necrosis. The mechanisms of disease inhibition are still not known, but it is unlikely to be due to the direct effects of the enzyme since this is normally associated with cutin degradation of the plant. Consequently, it seems that the protective effect is most likely to be due to stimulation of a plant defensive response (44). This represents an avenue of research that reflects the broad range of possibilities for exploiting microbial enzymes for more effective biocontrol of disease causing organisms.

B. Animal Feed Quality

Many of the enzymes produced by microorganisms, specifically to gain access to the plant inner tissues, have potential in processes aimed at improving some aspect of the plant cell chemical characteristics that leads to improved nutritional value. For instance, pectinase, derived from plant surface colonizers, has been used to increase the concentration of lactic acid in silage and thus the dietary value as an animal feed (45). Similarly, cellulolytic, hemicellulolytic, and lignolytic enzymes from *Pleurotus* species have been used effectively to convert banana pseudostem biomass to more edible, protein-rich animal feed (46). The same approach has been used with a white-rot fungus, *Lentinula edodes*, that produces phytase, endocellullase, and polyphenol oxidase. These enzymes have been used to modify the total and insoluble protein of wheat bran, resulting in increased dietary value (47).

C. Fiber and Paper Production

There is considerable interest in the potential for developing a commercial process for enzymatic retting (i.e., separation of plant fibers from woody tissues) of flax. The most encouraging results have come from *Rhizomucor pusillus*, which has been demonstrated to grow well on flax stems. Investigations have highlighted the significant role of pectinases in the process (48).

The potential of white-rot fungi, such as *Trametes versicolor*, for bleaching and delignifying unprocessed kraft pulp for paper manufacture has generated considerable research interest. A broad range of enzymes have been implicated in the delignification process, including lignin peroxidases, manganese peroxidases, and laccases (49), although the first enzyme reported to facilitate pulp bleaching was xylanse (50), which now is used in several papermills. However, for enzymes such as laccase to be effective it is necessary to add a mediator such as 2,2′-azinobis (3-ethylbenzthiazoline-6-sulfonate) (ABTS), which extends the substrate range of laccase to include nonphenolic subunits (51). A natural laccase mediator, 3-hydroxyanthranilic acid, which is produced by the white-rot fungus *Pycnoporus cinnabarinus* (52), has been shown to increase the range of substrates on which laccases can act. Delignification of kraft pulp by using enzymes is now considered to be a commercially viable alternative to chemical bleaching. The use of enzymes for delignification has the additional advantage of requiring much lower pressure and temperature (49).

D. Decolorization and Degradation

Phenoloxidases, laccases, and peroxidases, all enzymes implicated in the delignification of wood pulp, also have great potential in decolorization of dyes in wastes produced by the printing and textile industries. These enzymes have been studied intensively and decolorization has been found to be associated particularly with laccase activity (53). These enzymes now have been isolated and are being exploited in decolorization in packed-bed bioreactors (54).

Another biotechnological development is the exploitation of enzymes from phytopathogenic bacteria that, coincidentally, degrade plastics such as polycaprolactone (PCL). For example, the pathogenic fungus *Fusarium solani* secretes cutinase, which hydrolyzes cutin, the polyester structural component of plant cuticle. Studies performed in 1998 show that the same strain grows on PCL, producing a cutinase that functions as a PCL depolymerase (55).

The potential of cutinase is being studied in the removal of fatty substances in laundry washes. This enzyme and other lipolytic forms occur commonly on plant surfaces and have been shown to be very effective at improving removal of triglyceride oil immobilized on test cloths (56). The authors conclude that the potential market for such applications is enormous.

The preceding brief description serves to provide some indication of the potential of enzymes derived from microorganisms that colonize plant surfaces. The list is far from exhaustive since the range of applications is growing rapidly. Some of the most exciting papers concerning novel application of microbial enzymes have been published since 1998 (53,55,56). Perhaps one of the most novel recent applications of microbial enzymes is the use of *Fusarium oxysporium* as a mycoherbicide to remove leaves from coca plants and thus control and destroy illegal cocaine crops (57). The authors describe a commercial preparation of cell wall–degrading enzymes that stimulates the level of ethylene production by the leaves, which leads to loss of leaves.

The important benefits offered by the approaches outlined are that they require relatively low input technologies and therefore offer sustainable approaches available to third world countries (e.g., renewable and local resources, low energy and investment).

VIII. MOLECULAR METHODS FOR STUDYING MICROBIAL INTERACTIONS, ENZYME PRODUCTION, AND ACTIVITY ON PLANT SURFACES

The best route to improved microbial exploitation is an understanding of the molecular biological characteristics of the microorganisms that produce potentially valuable enzymes. A range of methods has become available to improve understanding of enzyme activity on the aerial parts of plants, particularly those involved in pathogenicity. These employ established molecular techniques to define gene function and to detect ecologically significant genes associated with habitat niche. Molecular approaches have provided the means of in situ study of microbes, for example, the ability to apply reporter genes to study function, such as sensors for the detection of iron or sugar availability in the phytosphere (58,59). These investigations are highly relevant since they are the first to provide reliable data on the mechanistic interaction of bacteria and their plant habitat. They advance previous observational studies with quantitative assessments for what are presumed important components of the microbial communities. As methodologies develop, it is feasible that a series of organismal reporters or DNA/RNA ''chips'' will become available for the evaluation of specific gene expression in natural environments. Biosensor and molecular approaches now allow the determination of the role and impact of particular genes and the enzymes and biosynthetic pathways they direct. This includes the transformation of commensal or saprophytic colonizers with presumed virulence genes, isolated from phytopathogens, to produce recombinants with phytopathogenic traits (31). With gene disruption or knockout strategies, other functional transformation studies have been successfully undertaken (60–63).

Other techniques also are available but have been less widely applied. These include the use of specific antisera for immunodetection in the suppression/blocking of functional activities in plants, which rely on the application of antibodies raised against particular enzymes or antigens (42,64). Immunodetection methods have confirmed the role of many cellular components in pathogenicity and, in some instances, reduced infectious lesions. Furthermore, antibodies combined with genetic modification to introduce reporter genes for cellular monitoring provide rapid and reliable in situ measurements of microbe distribution and activity (65). Specific examples include the targeting of genes responsible for the coding of novel enzyme activity to determine activity or regulated expression of certain genes. A good example of this approach is the use of *lacZ* (β-galactosidase). In the presence of the expressed enzyme an indicator substrate, X-gal (5-bromo-4-chloro-3indolyl-β-D-galactopyranodisidase) turns blue. In combination, extracts can report on overall activity, on isolated colonies on agar plates containing indicator turn blue (66). Alternatively, individual transcriptionally active cells can be seen under epifluorescence microscopy when appropriate fluorescent substrates are used. Reporter genes have been adopted to demonstrate the interaction and association of cellular enzymatic activity and nutrient availability of colonizing populations of phyllosphere bacteria (58).

Molecular tools and the use of marker genes have been applied widely (67–69) in microbial ecology. When adopted as promoter probes, they have formed the basis for investigations into niche- and habitat-regulated gene expression. Studies of the fate and activity of introduced inocula also have benefited from these technologies. For example, the field release of genetically modified natural isolates has allowed the persistence, activity, and survival of phytosphere communities to be followed (66,70). In particular the influence of seasonal and local niche-specific conditions on the dispersal, through the phytosphere, of a fluorescent pseudomonad strain after seed inoculation has been demonstrated (71). Among the most intriguing observations that have emerged from monitoring released microbial populations in the phytosphere are the abundance and importance of the horizontal gene pool. The use of marked strains has demonstrated that in many habitats, plasmids are abundant and that self-transmissible forms are also common (72). Although the role of these plasmids has not been fully resolved, there is no doubt that they provide genotypic and phenotypic plasticity to the host bacteria by the transfer of local adaptive traits for survival (73). It is highly likely that they carry novel enzymatic pathways that enhance the fitness of the recipient bacteria. Indeed the horizontal gene pool is one of the predominant mechanisms for bacterial adaptation and evolution through the exchange of genetic information (74).

Molecular approaches allow not only identification of specific pathways but also in situ tracking of the natural hosts and their mobile elements back to the plant. These investigations reveal not only that there is a diversity of extrachromosomal elements (75) but also that large conjugative (i.e., transfer-proficient) plasmids carry accessory traits that contribute to host fitness (73,74). Although the exact nature of these adaptive plasmid-borne traits has yet to be resolved, it is highly relevant that an extension to the gene pool and available phenotype of host bacteria be carried on extrachromosomal elements. Many of these plasmids carry traits for specific tasks as the acquisition of these plasmids and selection for plasmid-carrying host always coincide with the maturing of the crop. But how do we extend our understanding?

Perhaps one of the most limiting factors in studying microbial ecological characteristics of the phyllosphere (and many other habitats) has been the inability to link the diversity recorded with the actual function of individual cells, populations, and indeed communities, except at the most crude level. This is less of a challenge for specific pathogens, as disease typicality correlates with clonal proliferation of the causative agent. The study of saprophytes or commencals is more difficult. As indicated, molecular methods have provided tools based generally on gene displacement or knockout using site-directed homologous recombination or random insertion of transposons and subsequent isolation and testing of mutants where the loss or enhancement of a function can be identified (67). The utility of transposon mutagenesis is dependent on the choice of selectable markers or reporter genes that can be used as promoter probes to create random insertion mutants (68). Limitations to the system are that mutant libraries have to be isolated in vitro and interrogated against a set of devised assays. Nonetheless these approaches are efficient, report on insertional inactivation, and identify the regulated expression when used with fusion-promoter probes (58). However, their use in the evaluation of environmentally regulated gene expression has been restricted by the lack of tools for the rapid identification of mutants in the library that circumvents the need for preselection on indicator plates.

Identification of Ecologically Significant Genes: In Vivo Expression Technology

Molecular tools for gene fusion technology are of considerable value and were originally described for the positive selection of fragments of the phytopathogen *Xanthomonas campestris*, which carried regions associated with the expression of genes only under certain conditions. This was achieved by randomly cloning fragments into a broad-host-range plasmid vector carrying a promoterless chloramphenicol-resistance gene (76). Once it was reintroduced to the environment the selective pressure of the antibiotic was applied and only those constructs that carried promoters induced under the imposed condition, expressed chloramphenicol resistance, and, therefore, survived for characterization. Such an approach was further refined and applied to the study of *Salmonella* sp. infection in the mouse model; from it the term *in vivo expression technology* (IVET) was coined (77). In 1996 a development of this system was described for the study of *Pseudomonas aeruginosa* infection (78) and for the determination of the mechanistic basis of phytosphere colonization by *Pseudomonas fluorescens* SBW25 (79). The essence of these approaches is in situ selection. The strategy is based on the screening libraries for the presence of promoters that drive the expression of an essential gene that has been deleted from the chromosome (Fig. 2).

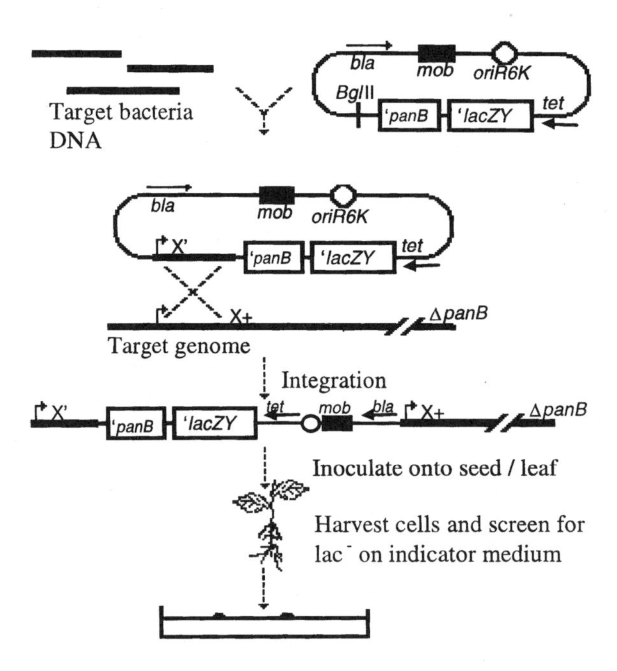

Figure 2 In vivo expression technology (IVET) strategy for the positive selection of phytosphere induced genes. (After Ref. 80.)

In the case of *P. fluorescens*, a deletion derivative that lacked *pan*B, which encodes pantothenate, was constructed. *pan*B deletion derivatives failed to colonize the rhizosphere (79). After the construction of a genomic library in a vector that carries promoterless *pan*B, integration into the chromosome occurs and mutants that synthesize pantothenate from promoters that are constitutive or habitat- (stimulus-) specific are selected. From these libraries, rescue mutants are isolated and genes associated with in situ activity characterized. Typically these promoters and associated genes are unique and not identified by the more classical approaches (80). Despite the elegance of these methods, they suffer from the limitation that for each organism to be studied a specific knockout mutant for the selected essential gene may have to be generated. This can be complex and is time-consuming.

In vivo expression technology (IVET) is essentially a promoter-trapping system that selects plant-induced (*pli*) genes through their ability to drive the expression of a gene that is itself essential for survival on plants. Random chromosomal fusions are generated upstream of a promoterless ''in vivo–selected'' marker in an autonomously replicating plasmid, that is then integrated into the chromosome of the chosen host bacterium. The technology is based upon a bacterial strain, with a conditional lethal mutation in a biosynthetic gene, that is involved in the synthesis of an essential growth factor. Provided that exogenous levels of the growth factor are negligible, then growth on plants cannot occur unless the missing gene is introduced into the cell. Introduction into the mutant strain of a promoterless, but otherwise functional copy of the biosynthetic gene to which random fragments of wild-type deoxyribonucleic acid (DNA) have been cloned (in place of the promoter) provides a means of selecting for promoters that are active on plant surfaces. Selection of such promoters (and corresponding genes) is performed simply by recovery of bacteria that have grown on plants; growth on plants can occur only if the function of the mutated biosynthetic gene has been restored. This can only occur when transcription of the promoterless biosynthetic gene is initiated via promoters responsive to plant signals. A subsequent screening step involving a second promoterless marker (*lacZ*) facilitates differentiation between those promoters active solely on plants from those active both on plants and in vitro.

A phytosphere IVET selection strategy based on the water-soluble vitamin pantothenate, which is a crucial component of the Gram-negative bacterial cell wall, has been developed (78). Because *pan* is not present at significant levels on plant surfaces *pan*-deletion mutants are unable to colonize plants. Complementation of *pan* synthesis with the vector insertion (Fig. 2) provides a highly sensitive method for the selection of plant-induced genes. This approach is particularly relevant to the study of the response of an individual strain in the natural environment. For example, the nature and variety of induced genes promoted in the presence of specific plant exudates (i.e., nutrients including amino acids or sugars) or in response to specific changes in plant physiological characteristics can be evaluated.

It is important to note that a duplication of the cloned fragment occurs during the integration process such that one promoter drives the in vivo–selected marker, while the other drives a wild-type copy of the gene or operon defined by the cloned fragment. The net result is that no genes in the target organism's genome are functionally interrupted by the integration. The pool of target bacteria carrying the fusion vector promoter probe IVET library is then introduced directly to the selective environment. Growth (and colonization) can occur only when the *pan*B lesion is genetically complemented as a result of a plant-induced (or plant-environment-induced) promoter's activating transcription of the promoterless *pan*B gene. This is because the gene product, pantothenate, has been deleted from the chromosome of the target bacteria. In order to select those promoters induced specifically during colonization of the plant (and not in vitro), bacteria are recovered from the phytosphere and plated on minimal lactose indicator medium supplemented with pantothenate and tetracylcline. This enables pIVET-PAN fusions to be recovered from nonsterile environments. Blue colonies, which display $lacZY^+$ phenotype, are ignored (because the promoters remained active in vitro); those that are $lacZY^-$, and therefore likely to contain fusions to plasmid genes activated on plants, should be further characterized. Sequence analysis of the insert region and comparison to databases may provide putative identity.

IX. CONCLUSIONS

The availability of sensitive assays and molecular methods provides greater opportunities for in situ activity assessments of microorganisms associated with the aerial parts of plants. It is, therefore, possible to identify chromosomal genes, and genes from the horizontal gene pool (e.g., plasmids), that are ecologically relevant and functional in the environment and define those local plant factors that regulate bacterial gene (enzyme) expression. Emerging technologies based on prokaryotic genomics that allow extremely detailed analysis of microbial transcription are of particular relevance. These methods, already applicable to laboratory cultures, are being developed for studies in situ (81). It is likely that the leaf will be a primary target, because of accessibility and relative (at least to soil) low microbial diversity. One of the first investigations will be the screening of communities for mRNA transcripts for niche-specific enzymes, perhaps even those of many microorganisms that as yet cannot be grown in the laboratory (82). These developments will inevitably improve understanding of the molecular basis of enzyme activity and microbial ecological characteristics. They also will lead to the isolation of genes for perhaps novel enzyme production, and other fitness traits that can be exploited by transferring these genes to other microbes more amenable to manipulation and better able to persist in the target habitat. The phylloplane must be a rich source of such enzymes, but they remain to be discovered.

Without question, we are entering an exciting period for microbial ecology. However, further technical developments are needed to resolve the genetic basis of the molecular and physiological interactions that occur between microbes and their environment. Indeed, little is known of the link between diversity and function, let alone the mechanistic nature of how bacteria communicate with their neighbors and regulate their immediate environment. Local sensing of nutrient availability and the regulated response of the cell are fundamental processes that require coordination of the individual organism and the community as a whole. By combining the significant amount of microbiological data amassed by observation and manipulation of populations and communities in the phyllosphere with data on genetic diversity and gene regulation, a more comprehensive insight of cellular enzymatics will be possible. These combined data sets will allow prescribed intervention and manipulation of the environment, either by direct means or by the introduction of genetically modified strains, so that optimal conditions for plant productivity through nutrification (biofertilization) and disease suppression (biocontrol) can be achieved. The improvement and use of plant growth-promoting bacteria are major goals for the biotechnology industry because considerable environmental benefits can be realized through phytoremediation and in the use of plants as cell factories.

REFERENCES

1. DC Clarke. Co-evolution between plants and pathogens of their aerial tissues. In: CE Morris, PC Nicot, C Nguyen, eds. Aerial Plant Surface Microbiology. New York: Plenum Press, 1996, pp 91–102.
2. ND Hallam, BE Juniper. The anatomy of the leaf surface. In: TF Preece, Dickinson CH, eds. Ecology of Leaf Surface. London: Academic Press 1970, pp 3–37.
3. PJ Holloway. Structure and histochemistry of plant cuticular membranes: an overview. In: DF Cutler, KL Alvin, CE Price, eds. The Plant Cuticle. London: Academic Press 1982, pp 1–32.
4. J Schonherr. Water permeability of isolated cuticular membranes: The effect of cuticular waxes on diffusion of water. Planta 131:159–164, 1976.
5. M Ferrandon, A Chamel. Foliar uptake and translocation of iron, zinc, manganese: Influence of chelating agents. Planta Physiol Biochem 25:713–722, 1989.
6. D Steinmuller, M Tevini. Action of ultraviolet radiation (UV-B) upon cuticular waxes in some crop plants. Planta 164:557–564, 1985.
7. J Schonherr. Resistance of plant surfaces to water loss: Transport properties of cutin, suberin, and associated lipids. In: OL Lange, Nobel PS, Osmond CB, Ziegler H, eds. Encyclopedia of Plant Physiology. Heidelberg: Springer-Berlin, 1982, pp 153–179.
8. M Reiderer, Schonherr J. Accumulation and transport of 2,4-dichlorophenoxy (2,4-D) acetic acid in plant cuticles. I. Sorption in the cuticular membrane and its components. Ecotox Environ Safety 8:236–247, 1985.
9. IP Thompson, MJ Bailey, JS Fenlon, TR Fermor, AK Lilley, JM Lynch, PJ McCormack, MP McQuilken, KJ Purdy, PB Rainey, Whipps JM. Quantitative and qualitative seasonal changes in the microbial community from the phyllosphere of sugar beet (*Beta vulgaris*). Plant Soil 150:177–191, 1993.
10. M-A Jacques. The effect of leaf age and position on the dynamics of microbial populations on aerial plant surfaces: In: CE Morris, P Nicot, C Nguyen-The, eds. Aerial Plant Surface Microbiology. New York: Plenum, 1996, pp 103–123.
11. HB Tukey. The leaching of substances from plant. Annu Rev Plant Physiol 21:305–324, 1970.
12. TJ Purnell, TF Preece. Effects of foliar washing on infection of leaves of swede (*Brassica napus*) by *Erysiphe cruciferarum*. Physiol Plant Pathol 1:123–132, 1971.
13. A Hatanaka. The biogeneration of green odour by green leaves. Phytochemistry 34:1201–1218, 1993.
14. OS Peries. Studies on strawberry mildew caused by *Spaerotheca macularis*: Host parasite relationships on foliage of strawberry varieties. Annu Rev Appl Biol 50:225–233, 1962.
15. JA Hargreaves, GA Brown, PJ Holloway. The structural and chemical characterisation of the leaf surface of *Lupinus albus* L. in relation to the distribution of antifungal compounds. In: DF Cutler, KL Alvin, CE Price, eds. The Plant Cuticle. London: Academic Press, 1962, pp 331–340.
16. LL Kinkel, M Wilson, SE Lindow. Effect of sampling scale on the assessment of epiphytic bacterial populations. Microb Ecol 29:283–289, 1995.
17. SS Hirano, CD Upper. Ecology and epidemiology of foliar bacterial plant pathogens. Annu Rev Phytopathol 28:243–269, 1990.
18. DG Cooper, JE Zajic. Surface active compounds from microorganism. Adv Appl Microbiol 26:229–253, 1980.
19. K Joseph. Molecular aspects of plant responses to pathogens. Acta Physiol Planta 19:551–559, 1997.
20. JP Derckel, F Bailleul, S Manteau, JC Audran, B Haye, B Lambert, L Legendre. Differential induction of grapevine defences by two strains of *Botrytis cinerea*. Phytopathology 89:197–203, 1999.
21. RS Sanguran, GP Lodhi, YP Luthra. Changes in total phenols and some oxidative enzymes

in cowpea leaves infected with yellow mosaic virus. Acta Phtyopathol Entomol Hungarica 31:191–197, 1996.
22. U Bonas, G Van den Ackerveken. Bacterial *hrp* and avirulence genes are key determinanats in plant-pathogen interactions. In: CE Morris, PC Nicot, C Nguyen-The, eds. Microbial attachment to plant aerial surfaces. Aerial Plant Surface Microbiology. New York: Plenum Press, 1996, pp 59–72.
23. CH Dickinson, B Austin, M Goodfellow. Quantitative and qualitative studies of phylloplane bacteria from *Lolium perenne*. J Gen Microbiol 91:157–166, 1975.
24. GL Ercolani. *Pseudomonas savastanoi* and other bacteria colonising the surface of olive leaves in the field. J Gen Microbiol 109:245–257, 1978.
25. CH Dickinson. Adaptation of microorganism to climatic conditions affecting plant surfaces. In: NJ Fokkema, J vanden Heuval, eds. Microbiology of the Phyllosphere. Cambridge: Cambridge University Press, 1986, pp 77–100.
26. AJ Dik. Interactions among fungicides, pathogenes, yeasts, and nutrients in the phyllosphere. In: JH Andrew, SS Hirano, eds. Micobial Ecology of Leaves. New York: Springer-Verlag, 1991, pp 412–429.
27. GL Ercolani. Distribution of epiphytic bacteria on olive leaves and the influence of leaf age and sampling time. Microbiol Ecol 21:6–48, 1975.
28. SS Hirano, CD Upper. Ecology and epidemiology of foliar bacterial pathogens. Annu Rev Phytopathol 21:243–269, 1983.
29. RJ Ellis, Thompson IP, MJ Bailey. Temporal fluctuations in the pseudomonad population associated with sugar beet leaves. FEMS Microbiol Ecol 28:345–356, 1999.
30. PB Rainey, MJ Bailey, IP Thompson. Phenotypic and genotypic diversity of fluorescent pseudomonads isolated from field grown sugar bee. Microbiology 140:2315–2331, 1995.
31. MB Dickman, GK Podila, PE Kolattukudy. Insertion of cutinase gene into a wound pathogen enables it to infect intact host. Nature 342:446–448, 1989.
32. RJ St Leger, L Joshi, DW Roberts. Adaptation of proteases and carbohydrases of saprophytic, phytopathogenic and entomopathogenic fungi to the requirements of their ecological niches. Microbiology 143:1983–1992, 1997.
33. W Knogge. Fungal infection of plants. Plant Cell 8:1711–1722, 1996.
34. W Koller, C Yao, F Trial, DM Parker. Role of cutinases in the invasion of plants. Can J Bot 73:1109–1118, 1995.
35. CY Fan, W Koller. Diversity of cutinases from plant pathogenic fungi: Differential and sequential expression of cutinolytic esterases by *Alternaria brassicicola*. FEMS Microbiol Lett 158: 33–38, 1998.
36. H Deising, RL Nicholson, M Hang, RJ Howard, K Mendgen. Adhesion pad formation and the involvement of cutinase and esterase in the attachment of urediniospores to the host cuticle. Plant Cell 4:1011–1111, 1992.
37. M Romantschuk, E Roine, K Bjorklof, T Ojanen, E-L Nurmiaho-Lassila, K Haahtela. Microbial attachment to plant aerial surfaces. In: CE Morris, PC Nicot, C Nguyen-The, eds. Aerial Plant Surface Microbiology. New York: Plenum Press, 1996, pp 43–58.
38. JN Sharma, JL Kaul. Biochemical nature of resistance in apple to *Venturia inaequalis* causing scab oxidative enzymes. Int J Trop Plant Dis 14:173–176, 1996.
39. Y Huang, RO Nordeen, M Di, LD Owens, JH McBeath. Expression of an engineered cecropin gene cassette in trangenic tobacco plants confers disease resistance to *Pseudomonas syringae* pv. *tabaci*. Phytopathol 87:494–499, 1997.
40. HM El, JP Chanbost, D Expert, GF Van, F Barras. The minimal gene set member msrA, encoding peptide methionine sulfoxide reductase, is a virulence determinant of the plant pathogen *Erwinia chrysanthemi*. Proc Natl Acad Sci USA 96:887–892, 1999.
41. EBG Jones. Fungal adhesions. Mycol Res 9:961–981, 1994.
42. P Commenil, L Belingheri, B Dehorter. Antilipase antibodies prevent infection of tomato leaves by *Botrytis cinerea*. Physiol Mol Plant Pathol 52:1–14, 1998.

43. A Kapat, G Zimand, Y Elad. Effect of two isolates of *Tricoderma harzianum* on the activity of hydrolytic enzymes produced by *Botrytis cinerea*. Physiol Mol Plant Pathol 52:127–137, 1998.
44. D Parker, W Koeller. Cutinase and other lipolytic estrases protect bean leaves from infection by *Rhizoctonia solani*. Mol Plant Microb Interact 6:514–522, 1998.
45. AC Sheppard, Maslanka M, Quinn D, King L. Additives containing bacteria with enzymes for alfafa silage. J Diary Sci 78:565–572, 1995.
46. M Ghosh, R Mukherjee, B Nandi. Production of extracellular enzymes by two *Pleurotus* species using banana pseudostem biomass. Acta Biotechnol 18:243–254, 1998.
47. GL Di, E Patroni, GB Quaglia. Improving the nutritional value of wheat bran by a white-rot fungus. Int J Food Sci Technol 32:513–519, 1997.
48. G Henriksson, ED Akin, D Slomczynski, LKK Erikkson. Production of highly efficient enzymes for flax retting by *Rhizomucor pusillus*. J Biotechnol 68:115–123, 1999.
49. FS Archibald, R Bourbonnais, L Jurasek, MG Paice, ID Reid. Kraft pulp bleaching and delignification by *Trametes versicolor*. J Biotechnol 53:215–236, 1997.
50. L Viikari, M Ranua, A Kantelinen, J Sunquist, M Linko. Bleaching with enzymes. Proceedings of the Third International Conference on Biotechnology in the Pulp Industry, Stockholm, STFI, Sweden, 1996, pp 67–69.
51. R Bourbonnais, MG Paice, B Freiermuth, E Bodie, S Borneman. Reactivities of various mediators and laccases with kraft pulp and lignin model compounds. Appl Environ Microbiol 63: 4627–4632, 1997.
52. C Eggert, T Ulrike, EC Eriksson. The liginolytic system of the white rot fungus *Pycnoporus cinnabarinus*: Purification and characterisation of the laccase. Appl Environ Microbiol 62: 1151–1158, 1996.
53. E Rodriguez, AM Pickard, RD Vazquez. Industrial dye decolorization by laccases from ligninolytic fungi. Curr Microbiol 38:27–32, 1999.
54. K Schliephake, GT Lonergan. Laccase variation during dye decolouration in a 200 L packed-bed bioreactor. Biotech Letts 18:881–886, 1996.
55. CA Murphy, JA Cameron, SJ Huang, RT Vinopal. A second polycaprolactone depolymerase from *Fusarium*, a lipase distinct from cutinase. Appl Microbiol Biotechnol 50:692–696, 1998.
56. JAC Flipsen, MCA Appel, HTMWM Van der Hijden, CT Verrips. Mechanisms of removal of immobilised triacylglycerol by lipolytic enzymes in a sequential laundry wash process. Enzyme Microb Technol 23:274–280, 1998.
57. BA Bailey, JC Jennings, JD Anderson. Sensitivity of coca (*Erythroxylum coca* var. *coca*) to ethylene and fungal proteins. Weed Sci 45:716–721, 1997.
58. JE Loper, SE Lindow. A biological sensor for iron available to bacteria in their habitats on plant surfaces. Appl Environ Microbiol 60:1934–1941, 1994.
59. CH Jaeger III, SE Lindow, W Miller, E Clark, MK Firestone. Mapping of sugar and amino acids in soil and roots with biosensors of sucrose and tryptophan. Appl Environ Microbiol 65:2685–2690, 1999.
60. PE Kolattudy, D Li, CS Hwang, MA Flaishman. Host signals in fungal gene expression involved in penetration into the host. Can J Bot 73:1160–1169, 1995.
61. W Koeller, Y Chenglin, F Trial, DM Parker. Role of cutinase in the invasion of plants. Can J Bot 73:1109–1118, 1995.
62. JAL van Kan, JW van't Klooster, CAM Wagemekers, DCT Dees, CJB van der Vlugt-Bergmans. Cutinase A of *Botrytis cinerea* is expressed, but not essential, during penetration of gerbera and tomato. Mol Plant Mic Interact 10:30, 38, 1995.
63. IA van Gemeren, A Beijersbergen, CAMJJ van den Hondel, CT Verrips. Expression and secretion of defined cutinase variants by Aspergillus awamori. Appl Environ Microbiol 64:2794–2799, 1998.
64. M Wilson, SS Hirano, SE Lindow. Location and survival of leaf-associated bacteria in relation

to pathogenicity and potential for growth within the leaf. Appl Environ Microbiol 65:1435–1443, 1999.
65. TM Timms-Wilson, RJ Ellis, MJ Bailey. Immuno-capture differential display method (IDDM) for the detection of environmentally induced promoters in rhizobacteria. J Microbiol Method 41:77–84, 2000.
66. MJ Bailey, AK Lilley, IP Thompson, RB Rainey, RJ Ellis. Site directed chromosomal marking of a fluorescent pseudomonad isolated from the phytosphere of sugar beet: Stability and potential for marker gene transfer. Mol Ecol 4:755–764, 1995.
67. SE Lindow. The use of reporter genes in the study of microbial ecology. Mol Ecol 4:555–566, 1995.
68. JI Prosser. Molecular marker systems for detection of genetically engineered microorganisms. Microbiology 140:5–17, 1994.
69. JJ Jansson, JD Van Elsas, MJ Bailey. Tracking Genetically Engineered Microorganisms: Method Development from Microcosms to the Field. Texas: RG Landes Company, 1999.
70. IP Thompson, AL Lilley, RJ Ellis, PA Bramwell, MJ Bailey. Survival, colonisation and dispersal of genetically modified *Pseudomonas fluorescens* SBW25 in the phytosphere of field grown sugar beet. Nature Bio/Tech 13:1493–1497, 1995.
71. AK Lilley, RS Hails, JS Cory, MJ Bailey. The dispersal and establishment of pseudomonad populations in the phyllosphere of sugar beet by phytophagous caterpillars. FEMS Microbiol Ecol 24:151–158, 1997.
72. AK Lilley, Bailey MJ. The acquisition of indigenous plasmids by a genetically marked pseudomonad population colonising the phytosphere of sugar beet is related to local environmental conditions. Appl Environ Microbiol 63:1577–1583, 1997.
73. AK Lilley, MJ Bailey. Impact of pQBR103 acquisition and carriage on the phytosphere fitness of *Pseudomonas fluorescens* SBW25: burden and benefit. Appl Environ Microbiol 63:1584–1587, 1997.
74. MJ Bailey, AK Lilley, JD Diaper. Gene transfer in the phyllosphere. In: CE Morris, P Nicot, C Nguyen-The, eds. Microbiology of Aerial Plant Surfaces. New York: Plenum, 1996, pp 103–123.
75. KE Ashleford, MJ Day, MJ Bailey, AK Lilley, JC Fry. *In situ* population dynamics of bacterial viruses in a terrestrial environment. Appl Environ Microbiol 65:169–174, 1999.
76. AE Osbourn, CF Barber, MJ Daniels. Identification of plant induced genes of the bacterial pathogen *Xanthomonas campestris* pathovar *campestris* using a promoter probe plasmid. EMBO J 6:323–328, 1987.
77. MJ Mahan, JM Slauch, JJ Mekalanos. Selection of bacterial virulence genes specifically induced in host tissue. Science 259:686–688, 1993.
78. J Wang, A Mushegian, A Lory, S Jin. Large scale isolation of candidate virulence genes of *Pseudomonas aeruginosa* by in vivo selection. Proc Natl Acad Sci USA 93:10434–10439, 1996.
79. GM Preston, B Haubold, PB Rainey. Bacterial genomics and adaptation to life on plants: Implications for the evolution of pathogenicity and symbiosis. Curr Opin Microbiol 1:589–597, 1998.
80. PB Rainey. Adaptation of *Pseudomonas fluorescens* to the plant rhizosphere. Environ Microbiol 1:243–258, 1999.
81. A Dellagi, PR Birch, J Heilbronn, GD Lyon, IK Toth. CDNA-AFLP analysis of differential gene expression in the prokaryotic plant pathogen *Erwinia carotovora*. Microbiology 146: 165–171, 2000.
82. L Diatchenko, L-FC Lau, AP Campbell et al. Suppression subtractive hybridisation: a method for generating differentially regulated or tissue specific cDNA probes and libraries. Proc Natl Acad Sci USA 93:6025–6030, 1996.

7

Microbial Enzymes in the Biocontrol of Plant Pathogens and Pests

Leonid Chernin and Ilan Chet
The Hebrew University of Jerusalem, Rehovot, Israel

I. INTRODUCTION

Despite many achievements in modern agriculture, food crop production continues to be plagued by disease-causing pathogens and pests. In many cases, chemical pesticides effectively protect plants from these pathogens. However, public concerns about harmful effects of chemical pesticides on the environment and human health have prompted a search for safer, environmentally friendly control alternatives (1–3). One promising approach is biological control that uses microorganisms capable of attacking or suppressing pathogens and pests in order to reduce disease injury. Biological control of plant pathogens offers a potential means of overcoming ecological problems induced by pesticides. It is an ecological approach based on the natural interactions of organisms with the use of one or more biological organisms to control the pathogen. Generally, biological control uses specific microorganisms that attack or interfere with specific pathogens and pests. Because of their specificity, different microbial biocontrol agents typically are needed to control different pathogens and pests, or the same ones in different environments.

Agriculture benefits, and is dependent on, the resident communities of microorganisms for naturally occurring biological control, but additional benefits can be achieved by introducing specific ones when and where they are needed (4–9). Many agrochemical and biotechnological companies throughout the world are increasing their interest and investment in the biological control of plant diseases and pests. For plant pathogens alone, the current list of microbial antagonists available for use in commercial disease biocontrol includes around 40 preparations (9–11). These are all based on the practical application of seven species of bacteria (*Agrobacterium radiobacter*, *Bacillus subtilis*, *Burkholderia cepacia*, *Pseudomonas fluorescens*, *Pseudomonas syringae*, *Streptomyces griseoviridis*, *Streptomyces lydicus*) and more than 10 species of fungi (*Ampelomyces quisqualis*, *Candida oleophila*, *Coniothyrium minitans*, *Fusarium oxysporum*, *Gliocladium virens*, *Phlebia gigantea*, *Pythium oligandrum*, *Trichoderma harzianum*, and other *Trichoderma* species). The current market for biological agents is estimated at only $500 million, which is about 1% of the world's total output for crop protection. The largest share of this market involves biopesticides marketed for insect control (mainly products based on *Bacillus thuringiensis* strains that produce a protein toxin with strong insecticidal activity), and these bioinsecticides represent around 4.5% of the world's insecticide sales. Other agents used for biocontrol exist on a much smaller scale commercially. However, the biopesticides market is expected to grow over the next 10 years at a rate of 10% to 15% per annum, vs. 1% to 2% for chemical pesticides (12).

Several modes of action have been identified in microbial biocontrol agents, no two of which are mutually exclusive. Biological control may be achieved by both direct and indirect strategies. Indirect strategies include the use of organic soil amendments and composts, which enhance the activity of indigenous microbial antagonists against a specific pathogen (13), and the use of indirect modes of the microbial-biocontrol-agent action. These include two main mechanisms. One is cross-protection, which involves the activation of physical and chemical self-defense responses (induced resistance) within the host plant against a particular pathogen by prior inoculation of the plant with a nonvirulent strain of that pathogen, resulting in partial or complete resistance to a variety of diseases in several types of plants (14,15). The other is plant growth promotion by root-colonizing bacteria and fungi that are able to stimulate plant growth and development; some of these also are capable of inducing resistance (16–18).

The direct approach involves the introduction of specific microbial antagonists into the soil or plant material. These antagonists need to proliferate and establish themselves in the appropriate ecological niche in order to be active against a pathogen or a pest. A beneficial organism used to protect plants is referred to as a biological control agent (BCA) or, often, as an antagonist, because it interferes with the target organisms that damage the plant. Antagonists generally are naturally occurring, mostly soil microorganisms with some trait or characteristic that enables them to interfere with pathogen or pest growth, survival, infection, or plant attack. Usually they have little effect on other soil organisms, leaving the natural biological characteristics of the ecosystem more balanced and intact than would a broad-spectrum chemical pesticide. Some BCAs have been modified genetically to enhance their biocontrol capabilities or other desirable characteristics.

There are four general direct mechanisms of biological control of plant diseases. The first is competition with the pathogen for limited resources such as nutrients or space. Antagonists capable of more efficiently utilizing essential resources (e.g., carbon, nitrogen, volatile organic materials, plant residues, iron, microelements) effectively compete with the pathogen for an ecological niche and colonization of the rhizosphere and/or phyllosphere, leaving the pathogen less able to grow in the soil or to colonize the plant. Many plant pathogens require exogenous nutrients to germinate, then penetrate and infect host tissue successfully. Therefore, competition for limiting nutritional factors, mainly carbon, nitrogen, and iron, may result in the biological control of plant pathogens (19,20).

The second mechanism is antibiosis, which is the inhibition or destruction of the pathogen by a metabolic product of the antagonist. That is, the antagonist produces some compound that is toxic or inhibitory to the pathogen, resulting in destruction of the latter's propagules or suppression of its activity. Antibiosis is restricted for the most part to those interactions that involve low-molecular-weight diffusible compounds (e.g., antibiotics or siderophores) produced by a microorganism that inhibit the growth of another microorganism (21–26). However, this definition excludes proteins or enzymes that kill the target organism. Hence, Baker and Griffin (19) extended its scope to ''inhibition or destruction of an organism by the metabolic production of another,'' thereby including small toxic molecules, and volatile and lytic enzymes. The impact of antibiosis on biological control under greenhouse and field conditions is still uncertain. Even in cases in which anti fungal metabolite production by an agent reduces disease, other mechanisms also may be operating.

Hypovirulence is another mechanism that reduces virulence in some pathogenic strains. Some natural- or laboratory-source hypovirulent strains were able to reduce the effect of the virulent ones. Hypovirulent strains of *Cryphonectria parasitica*, *Fusarium* spp., *Rhizoctonia solani*, *Sclerotinia homoeocarpa*, and others have been used as biocontrol agents of chestnut blight, wilt, rots, and other fungal diseases caused by the wild type of these pathogens. Some of these hypovirulent strains contain a single cytoplasmic element of double-stranded ribonucleic acid (dsRNA), which can be introduced into virulent strains by deoxyribonucleic acid– (DNA)-mediated transformation. This may be considered a specialized form of cross-protection that is limited to the control of only established compatible strains (27–29).

The fourth mechanism is predation/parasitism, which occurs when the BCA feeds directly on or inside the pathogen. In this case, the antagonist is a predator or parasite of the pathogen. When one fungus feeds on another fungus, generally it is called mycoparasitism. This process results in the direct destruction of pathogen propagules or structures (30–35).

All known BCAs utilize one or more of these general indirect or direct mechanisms. At the product level, this includes the production of antibiotics, siderophores, and cell wall lytic enzymes, and the production of substances that promote plant growth. Additionally, successful colonization of the root surface is considered a key property of prospective antagonists (9). The most effective BCAs use two or three different mechanisms. Antagonists also can be combined to provide multiple mechanisms of action against one or more pathogens. An understanding of this mechanism of action is important because it provides a wealth of information that can be useful in determining how to maintain, enhance, and implement this form of biological control.

Numerous comprehensive reviews on specialized topics, as well as proceedings and books describing the biocontrol activities of different microorganisms against plant pathogens and pests in laboratories, greenhouses, and the field, appeared in the late 1990s (9,10,34,36–41). However, the biological control of plant diseases is not as well established as biocontrol of insects in commercial agriculture. The latter has been a successful approach for decades and continues to be a rapidly developing area of research. In this chapter, we limit our discussion to enzymatic mechanisms of microbial control of plant pathogens and pests.

II. THE ROLE OF FUNGAL ENZYMES IN THE BIOLOGICAL CONTROL OF PLANT DISEASES

A. *Gliocladium* and *Trichoderma* Species Systems

The fungus *Gliocladium virens* Miller, Giddens and Foster (=*Trichoderma virens*, Miller, Giddens, Foster, and von Ark) is a common soil saprophyte and one of the most promising and studied fungal biocontrol agents. It originally was isolated from a sclerotium of the plant pathogenic fungus *Sclerotinia minor* and then was found to be active against several fungal plant pathogens. *Trichoderma*, a genus of hyphomycetes that is an anamorphic Hypocreaceae (class Ascomycetes), also is common in the environment, especially in soils. Many *Gliocladium* and *Trichoderma* spp. isolates obtained from natural habitats have been used in biocontrol trials against several soil-borne plant pathogenic fungi under both greenhouse and field conditions. In particular, isolates of *G. virens*, *G. roseum*, *T. viride*, *T. harzianum* Rafai, and *T. hamatum* have been reported to be antagonists of phytopathogenic fungi, including *Botrytis cinerea*, *Fusarium* spp., *Phytophthora cactorum*, *Pythium ultimum*, *Pythium aphanidermatum*, *Rhizoctonia solani*, *Sclerotinia sclerotiorum*, and *Sclerotium rolfsii*. These cause soil-borne and foliage diseases in a wide variety of economically important crops in a range of environmental conditions.

The antagonists kill the host by direct hyphal contact, causing the affected cells to collapse or disintegrate; vegetative hyphae of all species have been found susceptible. The biological and ecological characteristics and potential of these closely related genera for the biological control of plant pathogens have been reviewed extensively (4,9,31,34,35,42–48).

Among the biocontrol mechanisms proposed for *Gliocladium* and *Trichoderma* spp. are competition, antibiosis, and mycoparasitism. The last mechanism is based mainly on the activity of lytic exoenzymes (chitinases, glucanases, cellulases, and proteases) responsible for partial degradation of the host cell wall. Barnett and Binder (30) divide mycoparasitism into necrotrophic (destructive) parasitism, which results in death and destruction of the host fungus, and biotrophic (balanced) parasitism, in which the development of the parasite is favored by a living host structure. The sequential events involved in mycoparasitism have been described in several comprehensive reviews (31–35). Briefly, mycoparasitism is a complex process that involves ''recognition'' of the host, positive chemotropic growth, attachment, and de novo synthesis of a set of cell-wall-degrading enzymes that aid the parasite in penetrating the host and completing its destruction. Lectins, the sugar-binding proteins or glycoproteins of nonimmune origin, which agglutinate cells and are involved in interactions between the cell surface components and its extracellular environment, have been shown to play a role in the recognition and contact between necrotrophic mycoparasites of *Gliocladium* and *Trichoderma* spp. and soil-borne pathogenic fungi. This contact, in turn, initiates a signal transduction cascade toward the second, most important step of mycoparasitism, the induction of lytic enzymes able to degrade fungal cell walls.

Most fungi attacked by *Gliocladium* and *Trichoderma* spp. have cell walls that contain chitin as a structural backbone and laminarin (β-1,3-glucan) as a filling material. The other minor cell wall components are proteins and lipids. The ability to produce lytic enzymes has been shown to be a crucial property of these and other mycoparasitic fungi. Several contemporary reviews discuss the role of, in particular, chitinolytic enzymes of *Trichoderma* spp. in fungal mycoparasitism and biocontrol activity (33,49–51). In the last few years, the enzymatic patterns of various strains of *Trichoderma* and *Gliocladium* spp. have been determined, the corresponding genes cloned, and their products characterized. Some of these enzymes have been studied in more detail, with the goal of understanding their role in fungal biocontrol activity and principles of their expression regulation. In general, fungal cell-wall-degrading enzymes produced by *G. virens* and *Trichoderma* spp. are strong inhibitors of spore germination and hyphal elongation in a number of phytopathogenic fungi. The excretion of lytic enzymes enables *Trichoderma* spp. to degrade the target fungal cell wall and utilize its nutrients (52–55).

A considerable amount of recent research has been devoted to studying the individual lytic systems produced by *Trichoderma* spp. Most of the studies on the expression and regulation of these lytic enzymes have been performed in liquid cultures supplemented with different C sources (e.g., chitin, glucose, β-1,4-linked *N*-acetylglucosamine [GlcNAc], fungal cell walls) and their antifungal effects determined in vitro. These growth conditions facilitated the identification of the lytic enzymes induced in *Trichoderma* spp. to hydrolyze the polymers constituting the fungal cell walls. However, they did not reflect the exact conditions existing during the antagonistic interactions between *Trichoderma* spp. and its hosts. Thus, using *T. harzianum–R. solani* and *T. harzianum–S. rolfsii* interactions as model systems, Elad et al. (52) revealed lysed sites and penetration holes in the hyphae of the host fungus caused by the antagonist's attachment and coiling around it (Fig. 1). In the presence of the protein synthesis inhibitor cycloheximide, antagonism was prevented and enzymatic activity reduced. These observations suggested that the lytic enzymes whose synthesis de novo was induced as a result of early stages of interaction with the target phytopathogen excreted by *Trichoderma* spp. degrade *R. solani* and *S. rolfsii* cell walls at the interaction sites. According to more recent data obtained by electron microscopy of the interaction between *T. harzianum* and the arbuscular mycorrhizal fungus *Glomus intraradices*, chitinolytic degradation was seen only in areas adjacent to the sites of *Trichoderma* spp. penetration. The interaction between *T. harzianum* and *G. intraradices* involves the following events: (i) recognition and local penetration of the antagonist into mycorrhizal spores, (ii) active proliferation of antagonist cells in mycorrhizal hyphae, and (iii) release of the antagonist through moribund hyphal cells (56).

1. Chitinolytic Enzymes

Chitin, an unbranched insoluble homopolymer consisting of GlcNAc units, is the second (after cellulose) most common biodegradable polysaccharide in nature, being the main structural component of cell walls of most fungi and arthropods (insects, nematodes, and other invertebrates) including many agricultural pests (57–59). Many species of bacteria, streptomycetes and other actinomycetes, fungi, and plants produce chitinolytic enzymes that catalyze the hydrolysis of chitin. Chitinases produced by various microbes differ considerably in their molecular masses, high-temperature optima, and degrees of stability, probably because of glycosylation; they generally are active in a rather wide pH range. In recent years, soil-borne microorganisms that produce chitinases have become considered as potential biocontrol agents against fungal pathogens, insects, and nematodes that

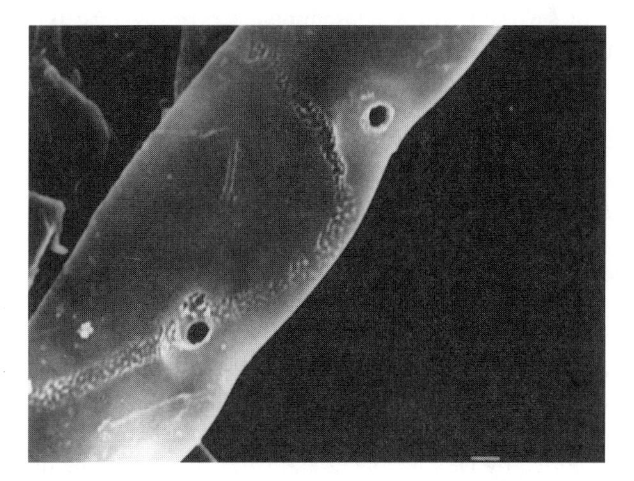

Figure 1 Scanning electron micrograph of *Trichiderma* spp. hyphae interacting with those of *S. rolfsii*. Hypha of *S. rolfsii*, from which a coiling hypha of *T. harzianum* was removed, showing digested zone with penetration sites caused by the antagonists (×5, 500). (From Ref. 52.)

cause diseases and damage in agricultural crops. Chitinases also play an important physiological and ecological role in ecosystems as recyclers of chitin, by generating C and N sources. Some producers of chitinases, including *Trichoderma* spp., are also sources of mycolytic enzyme preparations (51,59,60).

Chitinolytic enzymes are defined as enzymes that cleave a bond between the C1 and C4 of two consecutive GlcNAc units. On the basis of amino acid sequence similarities, all chitinases have been grouped into families 18, 19, and 20, under the main class of glycosyl hydrolases. Most of the microbial chitinases belong to family 18 (61,62). Even within the same family, chitinases show widely differing properties with respect to substrate specificity, reaction specificity, and pH optimum. The chitinolytic enzymes are divided into three principal types depending on their action on chitin substrates. According to the nomenclature suggested by Sahai and Manocha (59), endochitinases (EC 3.2.1.14) are defined as enzymes catalyzing the random hydrolysis of 1,4-β linkages of GlcNAc at internal sites over the entire length of the chitin microfibril. The products of the reaction are soluble, low-molecular-mass multimers of GlcNAc such as chitotetraose, chitotriose, and diacetylchitobiose. Exochitinases, also termed chitobiosidases or chitin-1,4-β-chitobiosidases (63), catalyze the progressive release of diacetylchitobiose units in a stepwise fashion as the sole product from the chitin chains, such that no monosaccharides or oligosaccharides are formed.

The third type of chitinolytic enzyme is chitobiase also termed as hexosaminidase (EC 3.2.1.52) or *N*-acetyl-β-1,4-D-glucosaminidase (EC 3.2.1.30) belongs to family 20 and also acts in exo splitting mode on diacetylchitobiose and higher analogs of chitin, including chitotriose and chitotetraose, to produce GlcNAc monomers. Rapid and specific methods have been developed for detection and quantitative assays of *N*-acetyl-β-glucosaminidase, chitobiosidase, and endochitinase in solutions using *p*-nitrophenyl-*N*-acetyl-β-D-glucosaminide, *p*-nitrophenyl-β-D-*N*,*N*′-diacetylchitotriose, and *p*-nitrophenyl-β-D-*N*,*N*′,*N*″-triacetylchitotriose or colloidal chitin as substrates, respectively (64). Procedures also are described for the direct assay of these three enzymes after their separation by sodium dodecyl sulfate (SDS)-polyacrylamide gel electrophoresis (PAGE) in which the enzymes are visualized as fluorescent bands by using an agarose overlay containing 4-methyl-umbelliferyl derivatives of *N*-acetyl-β-D-glucosaminide, β-D-*N*,*N*′-diacetylchitobioside, or β-D-*N*,*N*,*N*″-triacetylchitotriose, respectively (65).

A set of chitinolytic enzymes secreted by various strains of *T. harzianum* (e.g., TM, T-Y, 39.1, CECT 2413, P1 = *T. atroviride*), when grown on chitin as the sole C source, consists of *N*-acetylglucosaminidases, endochitinases, and exochitinases (chitobiosidases). In total, 10 separated chitinolytic enzymes were listed by Lorito (50); only one step in the microparasitic process of *T. harzianum*, which is the dissolution of the cell wall of the target fungus by enzyme activity, may involve more than 20 separate genes and gene products synergistic one to another (Table 1). Two *N*-acetylglucosaminidases with apparent molecular masses of 102 to 118 kD (depending on the isolate and the procedure used) and 72 to 73 kD (=NAG1) have been described by Ulhoa and Peberdy (66), Lorito et al. (67), and Haran et al. (68). The 102-kD enzyme (CHIT102) is the only chitinase of *T. harzianum* to be expressed constitutively when the fungus is grown with glucose instead of chitin as the sole C source (69). Four endochitinases—CHIT31, CHIT33, CHIT52, and CHIT42 (=ECH42)—have been reported by De La Cruz et al. (70), Ulhoa and Peberdy (66), Harman et al. (63), and Haran et al. (68). Additionally, a glycosylated chitobiosidase of 40 kD is secreted by strain P1 when grown on crab-shell chitin as the sole C source (63), and a 28-kD exochitinase releasing GlcNAc only was purified from the culture filtrate

Table 1 Examples of Lytic Enzymes Produced by Mycoparasitic Fungi which May Be Involved in Disease Biocontrol

Enzyme	Molecular mass (kDa)	Encoding gene	Fungus/strain	Reference
N-Acetylglucosaminidase (EC 3.2.1.30)	102–118	ND	*Trichoderma harzianum* (TM, 39.1)	(66, 68)
N-Acetylglucosaminidase (EC 3.2.1.30)	72–73	*nag1*	*T. harzianum* (TM, P1)	(67, 68, 88)
Endochitinase (EC 3.2.1.14)	52	ND	*T. harzianum* (TM)	(68)
Endochitinase (EC 3.2.1.14)	41–42	*ech42*	*T. harzianum* (39.1, P1, CEST2413); *G. virens* (41)	(63, 70, 78, 79, 84, 106)
Exochitinase (chitibiosidase)	40	ND	*T. harzianum* (P1)	(63)
Endochitinase (EC 3.2.1.14)	37	ND	*T. harzianum* (CEST 2413, TM)	68, 70)
Endochitinase (EC 3.2.1.14)	33	*chit33*	*T. harzianum* (CEST 2413, TM)	(68, 70)
Proteinase	31	*prb1*	*T. harzianum*	(55)
β-1,3-endoglucanase (EC 3.2.1.6; EC 3.2.1.39)	78	*bgn13.1*	*T. harzianum* (CECT2413)	(109)
β-1,3-endoglucanase	17	ND	*T. harzianum* (CECT2413)	(113)
β-1,3-endoglucanase	36	ND	*T. harzianum* (39.1)	(110)
β-1,3-exoglucanase (EC 3.2.1.58)	77–110	*lam1.3*	*T. harzianum* (P1, T-Y, IMI1206040)	(67, 111, 112)
β-1,6-endoglucanase	43	ND	*T. harzianum* (CECT2413)	(117, 118)
β-1,4-endoglucanase	51	*egl1*	*T. longibrachiarum*	(290)
β-1,3-exoglucanase	84	*exgA*	*Ampelomyces quisqualis*	(141)
Endochitinase	40	ND	*Fusarium chlamydosporum*	(130)
β-1,3-glucanase	ND	ND	*Trametes versicolor, Pleurotus eryngii*	(131)
β-1,3-glucanase, β-1,6-glucanase, chitinase	ND	ND	*Penicillium purpurogenum*	(132)
β-1,3-glucanase	ND	ND	*Tilletiopsis spp.*	(136)

of strain *T. harzianum* T198. This particular enzyme displayed activity on a wide array of chitin substrates of more than two GlcNAc units in length (71).

Lorito et al. (72,73) studied the antifungal activities of a 42-kD endochitinase and a 40-kD chitobiosidase from *T. harzianum* strain P1 in bioassays against nine different fungal species. Both spore germination and germ-tube elongation were inhibited in all chitin-containing fungi. The degree of inhibition was proportional to the level of chitin in the cell wall of the target fungus. Combining the two enzymes resulted in a synergistic increase in antifungal activity. A variety of synergistic interactions have been found when different enzymes were combined or associated with biotic or abiotic antifungal agents.

The levels of inhibition obtained by using enzyme combinations were, in some cases, comparable with those of commercial fungicides. Moreover, the antifungal interaction between enzymes and common fungicides allowed up to 200-fold reductions in the required chemical doses. These two enzymes, separately or in combination, substantially improved the antifungal ability of a biocontrol strain of *Enterobacter cloacae* (74). In an in vitro bioassay, different classes of cell-wall-degrading enzymes (glucan 1,3-β-glucosidase [EC 3.2.1.58], *N*-acetyl-β-glucosaminidase, endochitinase, and chitin 1,4-β-chitobiosidase) produced by *T. harzianum* and *G. virens* inhibited spore germination of *B. cinerea*. The addition of any chitinolytic or glucanolytic enzyme to the reaction mixture synergistically enhanced the antifungal properties of five different fungitoxic compounds against *B. cinerea* (73). Some of the combinations showed a high level of synergism, suggesting that the interaction between membrane-affecting compounds and cell-wall-degrading enzymes could be involved in biocontrol processes and plant self-defense mechanisms (75). A correlation between high capacity to produce chitinolytic enzymes and the superior biocontrol potential of the mycoparasitic fungi was also reported by Lima et al. (76). In general, chitinolytic enzymes from *Trichoderma* spp. appeared to be more effective in vitro against a number of fungal plant pathogens than were similar enzymes from plants or bacteria (72).

The *ech42* chitinase gene was shown to be highly conserved within the genus *Trichoderma* (77) and its product, the 42-kD chitinase, is believed to be one of the most crucial for mycoparasitic interactions between *Trichoderma* spp. and target pathogens. A similar endochitinase was purified from *G. virens* (78). Carsolio et al. (79) cloned and characterized *ech42* (previously named *ThEn42*) encoding a 42-kD endochitinase in the biocontrol strain *T. harzianum* IMI206040. Expression of the complementary deoxyribonucleic acid (cDNA) clone in *Escherichia coli* produced bacteria with chitinase activity. This chitinase displayed lytic activity on *B. cinerea* cell walls in vitro. The *ech42* gene was assigned to a double-chromosomal band (chromosome V or VI) upon electrophoretic separation and Southern analysis of the chromosomes. Expression of *ech42* was strongly enhanced during direct interaction of the mycoparasite with a phytopathogenic fungus when confronted in vitro and when it was grown in minimal medium containing chitin as sole C source. Similarly, light-induced sporulation resulted in high levels of transcript, suggesting developmental regulation of the gene. *T. virens* strains in which the 42-kD chitinase gene was disrupted or constitutively overexpressed were constructed through genetic transformation. The resulting transformants were stable and showed patterns similar to those of the wild-type strain with respect to growth rate, sporulation, antibiotic production, colonization efficiency on cotton roots, and growth/survival in soil. However, biocontrol activities of the ''disrupted'' and constitutively overexpressed strains were significantly decreased and enhanced, respectively, against cotton seedling disease incited by *R. solani* when compared with those of the parental strain (80).

However, several recently reported experiments have put into question the role of CHIT42 endochitinase as the only key enzyme in mycoparasitism. The biocontrol strain *T. harzianum* P1, recently attributed to *T. atroviride* (81), was genetically modified by targeted disruption of the single-copy *ech42* gene. A mutant, lacking the 42-kD endochitinase but retaining the ability to produce other chitinolytic and glucanolytic enzymes of this strain expressed during mycoparasitic activity, was unable to clear a medium containing colloidal chitin but grew and sporulated similarly to the wild type. In vitro antifungal activity of the *ech42*-disruptant culture filtrate against *B. cinerea* and *R. solani* was reduced by about 40% relative to that of the wild type, but its activity in protecting against *P.*

ultimum and *R. solani* in biocontrol experiments was the same or even better than that of strain P1. In contrast, the mutant's antagonism against *B. cinerea* on bean leaves was significantly reduced compared with that of strain P1. These results indicate that the antagonistic interaction between strain P1 and various fungal hosts is based on different mechanisms (82).

Corresponding results were obtained with several transgenic *T. harzianum* strains carrying multiple copies of *ech42*, and the corresponding gene disruptants were constructed. The level of extracellular endochitinase activity when *T. harzianum* was grown under inductive conditions increased up to 42-fold in multicopy strains relative to that of the wild type, whereas gene disruptants exhibited practically no activity. However, no major differences in the efficacies of the strains generated as biocontrol agents against *R. solani* or *S. rolfsii* were observed in greenhouse experiments (83). One possible explanation for these results is that other enzymes of *Trichoderma's* chitinolytic system are sufficient to control these fungal phytopathogens and that the lack of a certain protein can be compensated for by altering the levels of other proteins with similar activity. In view of the results showing efficient synergism between different chitinolytic enzymes produced by the same *Trichoderma* sp. isolate, it is not surprising that overexpression of one of these enzymes does not necessarily lead to an increase in biocontrol activity. Moreover, to achieve the highest level of antagonism toward target pathogens, a combination of several enzymes gives a better effect than the overproduction of only one of them.

Several groups have reported cloning genes *ech42* (79,84–86), *chit33* (87), and *nag1* (88). Very little is known, however, about the regulation of these genes and the roles of the corresponding enzymes in fungi during mycoparasitism. Generally, products of chitin degradation are thought to induce chitinolytic enzyme expression, and easily metabolizable C sources serve as repressors (59,89,90). Fungal cell walls, colloidal chitin, and C starvation have been shown to be inducers of the cloned chitinase genes (79,84,87,88,91).

To study the regulation of chitinolytic enzyme synthesis during the *Trichoderma* sp.–host mycoparasitic interaction, more specific confrontation assays (dual culture) on plates were developed (53,69,92). The differential expression of chitinolytic enzymes during the interaction of *T. harzianum* with *S. rolfsii* and the role of fungal–fungal recognition in this process were studied by Inbar and Chet (92). A change in the chitinolytic enzyme profile was detected during the interaction between the fungi grown in dual culture on synthetic medium. Before contact with one another, both fungi contained a protein with constitutive 1,4-β-*N*-acetylglucosaminidase activity. As early as 12 h after contact, the chitinolytic activity in *S. rolfsii* disappeared, while that in *T. harzianum* (a protein with a molecular mass of 102 kD, CHIT102) greatly increased. After 24 h of interaction, the activity of CHIT102 diminished concomitantly with the appearance of a 73-kD 1,4-β-*N*-acetylglucosaminidase, which became clear and strong at 48 h. This phenomenon did not occur if the *S. rolfsii* mycelium was autoclaved prior to incubation with *T. harzianum*, suggesting its dependence on vital elements from the host. Cycloheximide inhibited this phenomenon, indicating that de novo synthesis of enzymes takes place in *Trichoderma* spp. during these stages of the parasitism. A biomimetic system based on the binding of a purified surface lectin from the host *S. rolfsii* to nylon fibers was used to dissect the effect of recognition. An increase in CHIT102 activity was detected, suggesting that the induction of chitinolytic enzymes in *Trichoderma* sp. is an early event that is elicited by the recognition signal (i.e., lectin–carbohydrate interactions). Experiments with *T. harzianum* and the host lectin–covered nylon threads indicated that mere physical contact with the host triggers both the mycoparasitism-specific coiling of *Trichoderma* sp. hyphae around the host and chitinase formation (32,92). It is postulated that recognition is the first step in a cascade of antagonistic events that trigger the parasitic response in *Trichoderma* spp.

These observations were extended by Haran et al. (69), who showed that the expression of the various *N*-acetylglucosaminidases and endochitinases during mycoparasitism can be regulated in a very specific and finely tuned manner that is affected by the host. When strain *T. harzianum* T-Y antagonized *S. rolfsii*, the *N*-acetylglucosaminidase CHIT102 was the first to be induced. As early as 12 h after contact, its activity diminished, and the other *N*-acetylglucosaminidase, CHIT73, was expressed at high levels. However, when *T. harzianum* antagonized *R. solani*, the chitinase expression patterns differed considerably. Twelve hours after contact, CHIT 102 activity was elevated, and the activities of three additional endochitinases, at 52 kD (CHIT 52), 42 kD (CHIT 42), and 33 kD (CHIT 33), were detected. As the antagonistic interaction proceeded, CHIT102 activity decreased, whereas the activities of the endochitinases gradually increased.

Similarly, Carsolio et al. (79) detected the induction of *ech42* gene transcription only 24 h after contact of *T. harzianum* with *B. cinerea*. These data suggested that chitinase formation takes place during the later stages of the host–mycoparasite interaction, for example, to *T. harzianum* in penetration of the host hyphae. Therefore, chitinase induction generally has been regarded as a consequence of, rather than a prerequisite for, mycoparasitism. Krishnamurthy et al. (93) reported that differential induction of chitinase isoforms in vitro might depend on C sources in the growth medium. Nevertheless, in vivo the differential expression of *T. harzianum* chitinases may influence the overall antagonistic ability of the fungus against a specific host.

The specific and unique role of the 102-kD enzyme in triggering the expression of other chitinolytic enzymes was questioned by Zeilinger et al. (94). To monitor chitinase expression during mycoparasitism of strain *T. harzianum* P1 (=*T. atroviride*) in situ, strains were constructed containing fusions of the green fluorescent protein to the 5′-regulatory sequences of the *Trichoderma nag1* and *ech42* genes. Confronting these strains with *R. solani* led to induction of gene expression before or after physical contact in the cases of genes *ech42* and *nag1*, respectively. Separating the two fungi abolished *ech42* expression, indicating that macromolecules are involved in its precontact activation. No *ech42* expression was triggered by culture filtrates of *R. solani* or placement of *T. harzianum* on plates previously colonized by *R. solani*. Instead, high expression occurred upon incubation of *T. harzianum* with the supernatant of *R. solani* cell walls digested with culture filtrates or purified CHIT42. The results indicate that *ech42* is expressed before contact of *T. harzianum* with *R. solani* and its induction is triggered by soluble chitooligosaccharides produced by constitutive activity of CHIT42 and/or other chitinolytic enzymes. Therefore, *ech42* expression, in contrast to that of *nag1*, is a relatively early event, taking place prior to physical hyphal contact of the fungus with its host (*R. solani*). This indicates that this enzyme could be involved in the very early stages of the mycoparasitic process. Furthermore, the involvement of chitinase activity in the induction of *ech42* gene expression pre contact has been demonstrated by the effect of the chitinase inhibitor allosamidin, an actinomycete-derived metabolite. Expression of the 73-kD exochitinase *nag1* gene was observed only after contact of *Trichoderma* spp. with its host and was most active during overgrowth of *R. solani*. Therefore, different mechanisms of induction may occur for *ech42* and *nag1*, and *nag1* gene expression and may depend on products generated by CHIT42 activity. The results support the earlier suggestion by Lora and associates (95) that constitutive chitinases may partially degrade the cell walls of the host, thereby generating oligosaccharides containing GlcNAc that may act in turn as elicitors for the general antifungal response of *Trichoderma* sp. Although Zeilinger et al. (94) did not determine the number or expression patterns of other chitinase genes during this process, the ability of *R. solani* cell walls to induce *ech42* expression clearly was shown. The authors suggested that low constitutive activity of CHIT42 or some other chitinase triggers the induction of *ech42* when the host is at close range. A major role for CHIT42 in the induction process is implied by the fact that it generated the most strongly inducing mixture from *R. solani* cell walls. However the authors did not exclude the possibility that other chitinases, e.g., the 102-kD *N*-acetyl-β-D-glucosaminidase or CHIT33, as shown previously by Haran et al. (69) and Garcia et al. (84), respectively, also may be produced constitutively and act in a similar manner. This implies that chitinolytic enzymes not only are involved in the destruction of the host cell wall but also may play a role during the initial stages of mycoparasitism.

Cortes et al. (96) also studied whether physical contact between the mycoparasite and its host is necessary to induce expression of the *Trichoderma* sp. hydrolytic enzymes during the parasitic response. Dual cultures of *Trichoderma* sp. and a host, with and without contact, were used to study the mycoparasitic response in *Trichoderma* spp. Northern analysis showed a high level of expression of genes encoding a proteinase (*prb1*) and an endochitinase (*ech42*) in dual cultures, even when contact with the host was prevented by cellophane membranes. Neither gene was induced during the interaction of *Trichoderma* sp. with lectin-coated nylon fibers, even through the latter do induce hyphal coiling and appressorium formation (92). Therefore, the signal involved in triggering the production of these hydrolytic enzymes is independent of the recognition mediated by this lectin–carbohydrate interaction. The results showed that induction of *prb1* and *ech42* is contact-independent, and a diffusible molecule produced by the host is the signal that triggers expression of both genes in vivo. Furthermore, a molecule that is resistant to heat and protease treatment, obtained from *R. solani* cell walls, induced expression of both genes. Thus, this molecule is involved in regulating the expression of hydrolytic enzymes during mycoparasitism by *T. harzianum* (96). The antagonism observed in dual cultures, however, is not necessarily correlated with the fungus's chitinolytic activity. Thus, similarities as well as variations were observed in the abilities of various isolates of *G. virens* and *Trichoderma longibrachiatum* to invade the test pathogens *R. solani*, *S. rolfsii*, and *P. aphanidermatum* in dual culture. Although all the isolates produced enhanced levels of lytic enzymes, no correlation was observed between this attribute and the hyperparasitic potential of the various isolates in dual culture (97). Therefore, the relevance and role of enzymes and toxic metabolite(s) of these mycoparasitic fungi in their antagonism toward plant pathogens can vary among independent isolates and should be reassessed for each individual case. Moreover, the ability of lytic enzymes to provide biocontrol depends on both the type of plant being protected and the fungal pathogen. Thus, chitinase production does not appear to play a major role in protecting wood against fungal strains (98). Further characterization of the full chitinolytic system of *Trichoderma* sp. at the gene level should clarify which singular of these enzymes is really responsible for precontact gene expression. This, in turn, will help in understanding the relevance of this mechanism to biocontrol.

Studies on the regulation of *ech42* and *nag1* gene expression have been reported by Lorito et al. (99) and Mach and colleagues (81). Competition experiments, using oligonucleotides containing functional and nonfunctional consensus sites for binding of the C catabolite repressor Cre1, provided evidence that the complex from nonmycoparasitic mycelia involves the binding of Cre1 to both fragments of the *ech42* promoter. The presence of two and three consensus sites for the binding of Cre1 in the two *ech42* promoter fragments used is consistent with these findings. In contrast, formation of the protein–DNA complex from mycoparasitic mycelia is unaffected by the addition of the competing oligonucleotides and hence does not involve Cre1. The addition of equal amounts of protein of cell-free extracts from nonmycoparasitic mycelia converted the mycoparasitic DNA–protein complex into a nonmycoparasitic complex. The addition of purified Cre1 :: glutathione *S*-transferase protein to mycoparasitic cell-free extracts produced the same effect. These findings suggest that *ech42* expression in *T. harzianum* is regulated by (i) binding of Cre1 to two single sites in the *ech42* promoter, (ii) binding of a ''mycoparasitic'' protein–protein complex to the *ech42* promoter near the Cre1 binding sites, and (iii) functional inactivation of Cre1 upon mycoparasitic interaction to allow formation of the mycoparasitic protein–DNA complex (99,100). Using a reporter system based on the *Aspergillus niger* glucose oxidase *goxA* gene, Mach et al. (81) showed *ech42* gene expression during growth on fungal (*B. cinerea*) cell walls or after prolonged C starvation, independent of the use of glucose or glycerol as a C source, suggesting that relief of C catabolite repression is not involved in induction during starvation. In addition, *ech42* gene transcription was triggered by physiological stresses, such as low temperature, high osmotic pressure, or addition of ethanol. This corresponds to the finding that the *ech42* promoter contains four copies of a putative stress-response element CCCCT, also found in yeasts. The *nag1* gene expression was triggered by growth on chitin, GlcNAc, and the cell walls of *B. cinerea* used as a C source but, in contrast to *ech42*, also by a number of the chitin degradation products (chitooligomers) when added to mycelia pregrown on different C sources. The application of new techniques for examining the activities of the mycoparasite (fusion[s] of *ech42* or *nag1* with novel reporter genes such as green fluorescent protein or *A. niger goxA*) offers the possibility of revealing for the first time that (i) *ech42* transcription is induced before *Trichoderma* sp. physically contacts its host (94) and (ii) different regulatory signals are involved in triggering the expression of the 42-kD endochitinase and the 73-kD *N*-acetyl-β-D-glucosaminidase. This last enzyme revealed high similarity to *N*-acetyl-glucosaminidases from other eukaryotes, such as *Candida albicans*, and invertebrate and vertebrate animal tissues; the greatest similarity was to the corresponding gene from the silkworm (88).

The pattern of chitinolytic enzymes production can be an important marker for *Trichoderma* sp. strain identification and classification. The identification of *Trichoderma* sp. strains is important for their application as biocontrol agents. Schikler et al. (101) used a two-dimensional analysis in which extracellular proteins of *T. harzianum* strains T-35, Y, and TM were separated first according to their isoelectric point and then according to their molecular mass. Chitinase activities were detected in situ after the second separation. Each of the three strains exhibited a unique pattern of three to five different chitinases (one or two *N*-acetyl-β-glucosaminidases, and two or four endochitinases). These unique profiles can be used to differentiate among strains within this species, a requirement for specific biocontrol applications. Random amplification of polymorphic DNA (RAPD) was applied to characterize 34 strains of seven species of *Trichoderma*, including *T. hamatum*, *T. harzianum*, and *T. viride* isolated from various fungal sources. The RAPD patterns of *T. viride* strains were highly variable; isolates of *T. harzianum* proved to be more uniform; *T. hamatum* demonstrated remarkable intraspecific divergence. These three types comprised certain pairs of strains that have become promising participants in a strain-improving program since their strong genetic affinities offer good chances for combining their contrasting biocontrol traits (102).

2. Glucanases

β-1,3-glucan, or laminarin, is a polymer of D-glucose in a β-1,3 configuration, arranged as helical coils. Fungal cell walls contain more than 60% laminarin. Whereas chitin is arranged in regularly ordered layers, laminarin fibrils are arranged in an amorphic manner. There are chemical bonds between the laminarin and chitin, and together they form a complex net of glucan and GlcNAc oligomers (103). Laminarin is hydrolyzed mainly by β-1,3-glucanases, also known as laminarinases. These enzymes, described in fungi, bacteria, actinomycetes, algae, mollusks, and higher plants, are further classified as exo- and endo-β-glucanases. Exo-β-1,3-glucanases (β-1,3-glucan glucanohydrolase, [EC 3.2.1.58]) hydrolyze laminarin by sequentially cleaving glucose residues from the nonreducing ends of polymers or oligomers. Consequently, the sole hydrolysis products are glucose monomers. Endo-β-1,3-glucanases (β-1,3-glucan glucanohydrolase [EC 3.2.1.6 or EC 3.2.1.39]) cleave β-1,3 linkages at random sites along the polysaccharide chain, releasing smaller oligosaccharides. Both enzyme types are necessary for the full digestion of laminarin (104). These enzymes have several functions in fungi including nutrition in saprotropism, mobilization of β-glucans under conditions of C- and energy-source exhaustion, and a physiological role in morphogenetic processes during fungal development and differentiation (105).

Glucanases have been suggested as another group of key enzymes involved in the mycoparasitism of *Gliocladium* and *Trichoderma* spp. against fungal plant pathogens (Table 1). The substrate of these enzymes, β-1,3-glucan, is one of the major components of fungal cell walls along with chitin. Aside from the β-1,3-glucanases, the *Trichoderma* spp. also produce β-1,6-glucanases under specific growth conditions, and these enzymes hydrolyze minor structure polymers of fungal cells walls, β-1,6-glucans, which are thought to play an important role in the antagonistic action of *Trichoderma* spp. against a wide range of fungal plant pathogens (53). However, similarly to chitinases, glucanases are produced by *Trichoderma* sp. when it is grown in the presence of not only isolated fungal cell walls but chitin as well (106,107). Isolated plasma membranes of *B. cinerea* provide useful tools to study synergism between cell-wall-hydrolytic chitinases and glucanases of *T. harzianum* during the antagonism with phytopathogenic fungi. The data obtained in this system showed that cell wall synthesis is a major target of mycoparasitic antagonism by *T. harzianum*. Inhibition of the resynthesis of the host cell wall β-glucans sustained the disruptive action of β-glucanases and enhanced fungicidal activity. Therefore, cell wall turnover was considered a major target of mycoparasitic antagonism (100).

Large interstrain and interspecies differences exist in the production levels of both the laminarinase and chitinase enzymes by *Trichoderma* sp. isolates. Total activities of the enzymes were greater when isolates were cultured in malt medium, but specific chitinase and laminarinase activities were higher under low-nutrient conditions. Glucose appears to inhibit the formation of all of the inducible β-1,3-glucanases and chitinase, although this effect was not common to all *Trichoderma* sp. isolates for the latter enzyme (108). Similarly to chitinolytic enzyme production, the same strain of *Trichoderma* sp. can produce several extracellular β-1,3-glucanases. *T. harzianum* CECT 2413 was shown to produce at least three extracellular β-1,3-glucanases. The most basic 78-kD extracellular enzyme, named *BGN13.1*, was expressed when either fungal cell wall polymers or autoclaved mycelia from different fungi were used as the C source. The enzyme is specific for β-1,3 linkages and has an endolytic mode of action.

Sequence comparison shows that this β-1,3-glucanase, first described for filamentous fungi, belongs to a family different from that of its previously described bacterial, yeast, and plant counterparts. BGN13.1 hydrolyzes yeast and fungal cell walls; it is repressed by glucose and induced by either fungal cell wall polymers or autoclaved yeast cells and mycelia. A gene encoding the BGN13.1 endo-β-1,3-glucanase has been cloned and sequenced. Its structural analysis suggests that the enzyme contains a hydrophobic leader peptide that may be cleaved by an endoproteinase (109). A 36-kD endo-β-1,3-glucanase, purified from *T. harzianum* 39.1, was active toward glucans containing β-1,3-linkages and hydrolyzed laminarin to form oligosaccharides (110). At least seven extracellular β-1,3-glucanases ranging from 60 to 80 kD were produced by strain *T. harzianum* IMI206040 upon induction with laminarin or a soluble β-1,3-glucan or in the presence of different glucose polymers and fungal cell walls. The level of secreted β-1,3-glucanase activity was proportional to the amount of glucan present in the inducer. The properties of this complex group of enzymes suggest they have different roles in host cell wall lysis during mycoparasitism (111).

A novel 110-kD extracellular β-1,3-exoglucanase, LAM1.3, was purified from *T. harzianum* strain T-Y grown with laminarin. The corresponding gene, *lam1.3*, was cloned and the deduced amino acid sequence of the LAM1.3 enzyme showed high homology to EXG1, a β-1,3-exoglucanase of the phytopathogenic fungus *Cochliobolus carbonum*, and lower homology to BGN13.1 (112). Further studies of the β-1,3-glucanase system of *T. harzianum* strain T-Y revealed at least five different enzymes with molecular masses of 30 to 200 kD. In contrast to other β-1,3-glucanases, whose production is repressed by glucose and induced by a variety of polysaccharides as sole C source (109,111–113), the largest enzyme, Gβ-1,3-200, was the most abudant when strain T-Y was grown with no C source and was repressed by GlcNAc or malic acid (114). β-1,3-glucanases in *T. harzianum* are found in the periplasm, bound to cell walls, or secreted into the growth medium (115), and regulation of the enzymes' expression is considered a key step in β-glucan biodegradation and consequently in mycoparasitism.

Total β-1,3-glucanase activity has been found to be induced by different polysaccharides or by fungal cell walls and repressed by high glucose concentrations (109,113). Moreover, different fungal cell walls have been shown to induce different levels of β-1,3-glucanase activity and different enzyme patterns were observed when *T. harzianum* was grown on different C-source-containing media (109,111). The interaction between *T. harzianum* and the soil-borne plant pathogen *P. ultimum* (which is exceptional in that the cell walls contain β-[1,3]-[1,6]-D-glucans and cellulose instead of chitin as major structural components) and studied by electron microscopy and gold cytochemistry, revealed marked alteration of the β-1,3-glucan component of the *Pythium* sp. cell wall. This suggested that β-1,3-glucanases played a key role in the process (116). By specific detection of their activity in gels, different *Trichoderma* sp. strains grown under different growth conditions excreted the β-1,6-glucanase isozymes (107,116–118).

Despite considerable evidence that *Trichoderma* spp. produce chitinolytic enzymes and glucanases in vitro, much less is known about what happens in vivo under natural conditions, and no definitive evidence has shown the presence or activity of chitinases or endoglucanase in the rhizosphere (the zone immediately adjacent to the plant root) associated with a soil-borne fungal pathogen. Most studies have been performed on plates or in liquid cultures supplemented with different C sources, and these theories have not been fully studied in vivo. de Soglio et al. (119) detected chitobiosidase, endochitinase, endo-β-1-3-glucanase, and *N*-acetylglucosaminidase simultaneously in the roots of soybean seedlings and in cell-free culture filtrates of *T. harzianum* isolate Th008. With the exception of that of endochitinase, activity of these enzymes also was associated with *R. solani* isolate 2B-12, causal agent of soybean root rot. In greenhouse experiments, soybean seeds inoculated with *T. harzianum* Th008 were planted in a soil mixture infested with *R. solani* 2B-12. Fifteen days after emergence, the rhizosphere was assayed for chitinolytic enzymes and endoglucanase. Only the *N*-acetylglucosaminidase and endochitinase activities in the rhizosphere samples were significantly elevated above those of the controls. It was determined that *T. harzianum* Th008 was the source of the endochitinase in the rhizosphere. The results indicated that the probable source of the detectable endochitinase activity in rhizosphere extracts is the biocontrol agent rather than soybean root or the pathogen. A positive correlation was found between disease index and total protein (milligrams per gram [mg/g] soil) in rhizosphere samples and in *N*-acetylglucosaminidase activity in rhizosphere extracts. This finding suggests the release of *N*-acetylglucosaminidase into the rhizosphere results from a response of root cells to the pathogen.

3. Cellulases

Trichoderma spp. produce enzymes in the cellulolytic (exo- and endo-β-1-4-glucanase, β-1-4-glucosidase) and hemicellulolytic (especially xylanase and β-xylosidase) complexes that are effective in degrading natural lignocelluloses. Chitin and β-(1,3)-glucan are the two major structural components of many plant pathogenic fungi, except oomycetes, which contain cellulose in their cell wall and have no appreciable levels of chitin. Therefore, the biological control of such economically important plant pathogenic oomycetes as *Pythium* spp. can be provided by a biocontrol agent able to produce cellulases (Table 1).

Cellulose, a linear, essentially insoluble β-1,4-glucosidically linked homopolymer of about 8,000 to 12,000 glucose units, is used as an energy source by numerous diverse microorganisms, including fungi and bacteria, which produce cellulases. Among the best-characterized of these systems are the inducible cellulases of the saprophytic fungus *Trichoderma reesei* (=*T. longibrachiarum*), which include 1,4-β-D-glucan cellobiohydrolases (EC 3.2.1.91), endo-1,4-β-D-glucanases (EC 3.2.1.4), and 1,4-β-D-glucosidases (EC 3.2.1.21) (120).

There have been indications that endo-1,3-β-glucanase (EC 3.2.1.6) and endo-1,4-β-D-glucanase activity of *T. harzianum* isolate T3 is induced in sphagnum peat moss cultivations and dual culture experiments by the presence of *P. ultimum*. Further, *P. ultimum* stimulated the germination of *Trichoderma* sp. conidia. Low concentrations of purified 17-kD endo-1,3-β-glucanase and 40- and 45-kD cellulases were able to inhibit the germination of encysted zoospores and elongation of germ tubes of a plant-pathogenic *Pythium* sp. isolate. A strong synergistic effect was observed on the inhibition of cyst germination by a combination of endo-1,3-β-glucanase and fungicide (Fongarid). Finally, in a time-course study of colonization of the rhizosphere of cucumber seedlings, the active fungal mycelial biomass of a GUS-transformant of *T. harzianum* isolate T3 increased over 4 weeks. *Trichoderma* sp. appeared to colonize healthy roots only superficially, whereas the mucilage of the root hairs and of distal parts of the wounded areas or broken parts of the roots was extensively colonized (113).

The interaction between *T. harzianum* and *P. ultimum* has been studied by electron microscopy and further investigated by gold cytochemistry. Early contact between the two fungi was accompanied by the abnormal deposition of a cellulose-enriched material at sites of potential antagonist penetration. The antagonist displayed the ability to penetrate this barrier, indicating that cellulolytic enzymes had been produced. However, the presence of cellulose in the walls of severely damaged *Pythium* sp. hyphae indicated that cellulolytic enzymes were not the only critical factors involved in the antagonistic process. The marked alteration of the β-1,3-glucan component of the *Pythium* sp. cell wall suggested that β-1,3-glucanases played a key role in the process (118).

4. Proteases

Protease production is common in microorganisms, including fungi, among which *Trichoderma* spp. are well-known producers (121). Protease activity of *T. harzianum* can be induced by autoclaved mycelia, a fungal cell wall preparation, or chitin; however, the induction does not occur in the presence of glucose and increases when the liquid culture medium contains organic nitrogen sources (122). Rodriguez-Kabana et al. (123) provided evidence that *T. viride* proteolytic activity is involved in the biocontrol of *S. rolfsii*. A gene, *prb1*, of *T. harzianum* IMI 206040 was cloned and its product was biochemically characterized as a 31-kD basic serine proteinase (Prb1) (55). That was the first report of cloning a mycoparasitism-related gene (Table 1). This protease was suggested to provide the mycoparasite with nutrients, since it was involved in the degradation of pathogen cell walls and membranes and release of the proteins from the lysed pathogen (124). Strong expression of this protease was observed during mycoparasitic interactions with *R. solani* (125).

Foliage diseases have been some of the most difficult to control with biological agents because of the severe environment on the leaf surface. Until recently, most research on the biological control of aerial plant diseases was focused on the control of bacterial pathogens (10). The last decade, however, has seen increased activity in the development of biocontrol agents for foliar fungal pathogens. The strain *T. harziaum* T39, known as an efficient biocontrol agent of *B. cinerea*, which causes gray mold, a foliage disease of grapes and some other crops, was found to produce protease in liquid culture medium and directly on the surface of bean leaves. On the latter surface, the protease obtained from liquid culture medium of *T. harzianum* isolates resulted in a 56% to 100% reduction in disease severity. The hydrolytic enzymes endo- and exopolygalacturonase produced by *B. cinerea* were shown to be targets of the proteolytic activity secreted by strain T39. Since T39 was found to be a poor producer of chitinase and β-1,3-glucanase in vitro and these enzymes were not detected on leaves treated with T39, protease is suggested to be the key enzyme involved in biocontrol of *B. cinerea* by this *T. harzianum* isolate (126).

Other observations, however, have brought the role of proteases in *Trichoderma* sp. strain biocontrol activity into question. Methods for measuring protease activity from fungi based on the use of four chromogenic substrates were developed by Mischke (127). Digestion of azoalbumin, a water-soluble substrate, resulted in a level of dye release closely proportional to enzyme activity. Water-insoluble substrates were advantageous for time-course studies, and azocoll was more sensitive to digestion and easier to handle than powder azure. The optimal pH was 7 for measurements of extracellular protease activity from the *Trichoderma* sp. strains. The addition of calcium or serine protease inhibitors did not affect crude protease activity. The optimized protocol was used to demonstrate that the specific activity of proteases produced by the strains of *Trichoderma* sp. tested is not correlated to their known biocontrol ability.

B. Lytic Enzymes Involved in the Biocontrol Activity of Other Fungi

Besides *Trichoderma* spp., several other fungi exhibit the role of cell-wall-lytic enzymes in biocontrol activity (Table 1). One example is *Mucarales* sp., which can suppress *Fusarium oxysporum* f.sp. *lycopersici* via degradation of the fungal cell wall (128,129). A 40-kD endochitinase was purified from the culture filtrate of *Fusarium chlamydosporum*. The purified chitinase inhibited the germination of *Puccinia arachidis* uredospores and also lysed the walls of uredospores and germ tubes. Results from these experiments indicated that *F. chlamydosporum* chitinase plays an important role in the biocontrol of groundnut rust (130). The β-1,3-glucanase activity of two soil-borne fungal biocontrol agents, *Trametes versicolor* and *Pleurotus eryngii*, was shown to contribute to the degradation of the hyphal cell walls of *F. oxysporum* f.sp. *lycopersici* race 2, containing glucan as the principal component of its cell walls. The lack of cellulase and xylanase activities (acting on plant cell wall polysaccharides) in *T. versicolor* suggests this species to be a better alternative for the potential control of diseases caused by *Fusarium* spp. (131). β-1,3-glucanase, β-1,6-glucanase, and chitinase were shown responsible for biocontrol activity of the fungus *Penicillium purpurogenum* against the plant pathogens *Monilinia laxa* and *F. oxysporum* f.sp. *lycopersici* on peach and tomato. These lytic activities were inducible by cell walls and live mycelium of *M. laxa* but not of *F. oxysporum* f.sp. *lycopersici*, whereas crude enzyme preparations produced lysis of hyphae and spores of both these fungal pathogens. A relationship was found between the severity of the lytic effects on the fungi mycelia in vitro and the decrease in disease incidence caused by these pathogens in vivo (132). Similarly, correlation analyses between the extracellular enzymatic activities of different isolates of *Talaromyces flavus* and their ability to antagonize *S. rolfsii* indicated that mycoparasitism by *T. flavus* and biological control of *S. rolfsii* were related to the former's chitinase activity (133).

Transposon mutagenesis and subsequent in vivo assays have shown that the biocontrol ability of a *Stenotrophomonas maltophilia* strain against *P. ultimum* is mediated by chitinase and protease production (134). In a dual culture with *R. solani*, the mycoparasitic fungus *Schizophyllum commune* markedly enhanced production of endo-β-1,3(4)-glucanase compared with that of cultures of the mycoparasite alone (135). β-1,3-glucanase of *Tilletiopsis* spp. was shown to be responsible for biocontrol of powdery mildew by this yeast (136).

Ampelomyces quisqualis Ces. has been reported as a biotrophic mycoparasite and biocontrol agent of many fungi that cause powdery mildew (137,138). The anatomical characteristics of the mycoparasitic interaction between the fungus and its hosts, the Erysiphales, have been studied intensively (139); however, the enzymatic basis of *A. quisqualis* mycoparasitism is less clear. In vitro, the fungus constitutively produces several extracellular enzymes, among them a β-1,3-glucanase (140). Very recently, the *exgA* gene encoding an 84-kD endo-1,3-glucanase in strain *A. quisqualis* 10, a very efficient biocontrol agent of powdery mildew, was isolated and sequenced (141). The predicted polypeptide deduced from *exgA* showed 46%, 42%, and 30% identity to amino acid sequences of exo-β-1,3-glucanases produced by *T. harzianum* and *C. carbonum*, and of *T. harzianum* BGN13.1 endo-β-1,3-glucanase, respectively. All of these glucanases have a putative hydrophobic leader sequence of 33, 35, and 48 amino acids for *T. harzianum* endo-β-1,3-glucanase, *T. harzianum* exo-β-1,3-glucanase, and *A. quisqualis* exo-β-1,3-glucanase, respectively. These leader sequences end with the amino acids Lys–Arg and can be cleaved by an endoprotease (109,141). *exgA* was shown to be expressed during the late stages of mycoparasitism when the mycoparasite forms pycnidia, and transcription was induced by fungal cell wall components. The crude preparation of EXGA from *A. quisqualis* was able to lyse cell walls of *Sphaerotheca fusca*, a causative agent of powdery mildew (141). The differences in modes of mycoparasitism between *Trichoderma* spp. and *A. quisqualis*, considered necrotrophic and biotrophic mycoparasites, respectively, can be explained partially by the differt patterns of lytic enzymes produced by the fungi.

The role of cellulolytic enzymes in fungal mycoparasitism was shown by light and electron microscopic studies of interactions between the mycoparasitic oomycete *Pythium oligandrum* and the plant pathogenic oomycete *P. ultimum* (142). Localization of the host-wall cellulose component showed that cellulose was altered at potential penetration sites. At least two distinct mechanisms were suggested to be involved in the process of oomycete and fungal attack by *P. oligandrum*: (i) mycoparasitism, mediated by intimate hyphal interactions, and (ii) antibiosis, with alteration of the host hyphae prior to contact with the antagonist. However, the possibility that the antagonistic process relies on the dual action of antibiotics and hydrolytic enzymes appears plausible (142).

More evidence of the role of cellulolytic enzymes in fungal antagonism was obtained by studying the mode of action of a species of the antagonistic fungus *Microsphaeropsis* against *Venturia inaequalis*. Cytological observations indicated that the antagonistic interaction between the two fungi is likely to involve a sequence of events, including (i) attachment and local penetration of *Microsphaeropsis* sp. into *V. inaequalis* hyphae, (ii) induction of host structural response at sites of potential antagonist entry, (iii) alteration of host cytoplasm, and (iv) active multiplication of antagonistic cells in pathogen hyphae, leading to host-cell breakdown and release of the antagonist. The use of gold-complexed β-1,4-exoglucanase and a wheat germ agglutinin/ovomucoid gold complex to localize cellulosic β-1,4-glucan and chitin monomers, respectively, resulted in regular labeling of *V. inaequalis* cell walls. This finding supports other studies refuting the classification of ascomycetes as solely a glucan-chitin group. At an advanced stage of parasitism, the labeling pattern of cellulose and chitin, which clearly showed that the level of integrity of these compounds was affected, suggested the production of cellulolytic and chitinolytic enzymes by *Microsphaeropsis* sp. Wall appositions formed in *V. inaequalis* in response to antagonist attack contained both cellulose and chitin. However, penetration of this newly formed material was frequently successful (143).

In some cases, the interaction between cell-wall lytic enzymes produced by the antagonist and the pathogen can help the latter overcome the antagonist's attack. Cell-free culture filtrates of *R. solani* isolate 2B-12, causal agent of soybean (*Glycine max*) root rot, inhibited the growth of the biocontrol agent soil-borne *T. harzianum* isolate Th008 and the rhizosphere-competent bacterium *Bacillus megaterium* strain B153-2-2. The pathogen secretes endoproteinase, exochitinase, glucanase, and phospholipase, all of which potentially are detrimental to the cell wall/membrane integrity of the biocontrol agents. Compared to *R. solani* 2B-12, the *T. harzianum* isolate produced more extracellular endochitinase and endoproteinase, both of which can disrupt the cell wall and membrane structure of *R. solani* (144). Metabolites produced by *R. solani* and *P. ultimum* strains may reduce the density of *Trichoderma* sp. strain mycelial growth and the production of antagonist conidia on agar media (145). Similarly two isolates of *T. harzianum* (T39, a biocontrol agent, and NCIM 1185) reduced the level of hydrolytic enzymes produced by *B. cinerea* both in vitro and in vivo and inhibited infection caused by *B. cinerea* (146).

C. Involvement of Fungal Enzymes in Induced Resistance

To protect themselves against diseases, plants have defense mechanisms known as induced systemic resistance (ISR) that can be induced, prior to disease development, by pathogens, nonpathogens, and certain chemical compounds (147,148). The general plant's defense response consists of induction and accumulation of low-molecular-weight proteins, called pathogenesis-related (PR) proteins, and depositions of structural polymers such as callose and lignin. Acidic PR proteins, including acidic β-1,3-glucanases and chitinases, act against fungal and bacterial pathogens at an early stage of the infection process; basic β-1,3-glucanases and chitinases may interact with pathogens at a later stage of infection (149). Another group of enzymes includes peroxidases, which play a key role in the plant resistance process, since they are involved in synthesis of phenolic compounds and formation of structural barriers (150). Evidence was presented that *T. harzianum* strain 39 may participate in induced plant defense against foliage disease caused by *B. cinerea* (151). Significant increase of the activity of the most widely recognized PR proteins, chitinase, β-1,3-glucanase, cellulase, and peroxidase, was observed in cucumber roots treated by *T. harzianum* strain T-203. The expressed chitinase isozymes were derived from both the plant defense system and the fungus. Two proteins with apparent molecular weights of 102 and 73 kD were classified as exochitinases related to the mycoparasitic system that consists of six known chitinase isozymes of *T. harzianum*. A protein with an apparent molecular weight of 33 kD has been suggested as being of plant origin. All of these hydrolytic activities reached their maxima at 72 h after inoculation, indicating the activation of a general defense response in the plant (152).

Besides cell-wall lytic enzymes, a few examples showing the involvement of other enzymes in the biocontrol activity of *Trichoderma* sp. and other fungal antagonists of plant pathogens have been found. A xylanase produced by *T. viride* has induced defense responses, including ethylene biosynthesis and necrosis, in *Nicotiana tabacum* cv. Xanthi leaves. The sensitivity of the leaves to xylanase and ethylene was influenced by tissue age: young leaves were relatively insensitive to both; mature leaves were relatively insensitive to xylanase but became very sensitive to xylanase after treatment with ethylene; senescing leaves were more sensitive to xylanase than were young or mature leaves. A second ethylene treatment of tobacco plants, after loss of the effects of the initial treatment, restored the enhanced sensitivity of the tissues to xylanase. The continual presence of ethylene was required to maintain its effects, and the timing of the induction and subsequent loss of ethylene's effects were closely coordinated at the molecular and whole tissue levels (153). Glucose-oxidase activity may play a role in the antagonism of *T. flavus* against *V. dahliae* by retarding germination and hyphal growth and melanizing newly formed microsclerotia (133).

III. BACTERIAL ENZYMES IN THE BIOCONTROL OF PLANT PATHOGENS AND PESTS

A. Lytic Enzymes of Soil-Borne and Rhizospheric Bacteria in Plant-Pathogen Biocontrol

Chitinase activity has been found in a wide variety of bacteria (59). Bacteria produce chitinase to digest chitin, primarily to utilize it as a C and energy source. The ability to produce lytic enzymes is a widely distributed property of soil, marine, and rhizosphere bacteria. Many of these are potential biocontrol agents of chitin-containing plant pathogens. The list of such bacterial antagonists includes *Aeromonas caviae* (154), *Chromobacterium violaceum* (155,156), *Enterobacter agglomerans* (157), *Paenibacillus* sp. and *Streptomyces* sp. (158), *Pseudomonas fluorescens* (159,160), *Pseudomonas stutzeri* (161), *Serratia marcescens* (162,163), *Serratia liquefaciens* (164), and *Serratia plymuthica* (164,165) (Fig. 2, Table 2). Considerable interest has been focused on the role and production of cell-wall-degrading enzymes in bacteria and the ability of chitinolytic bacteria to protect plants against diseases and pests. Antifungal properties of chitinolytic soil bacteria

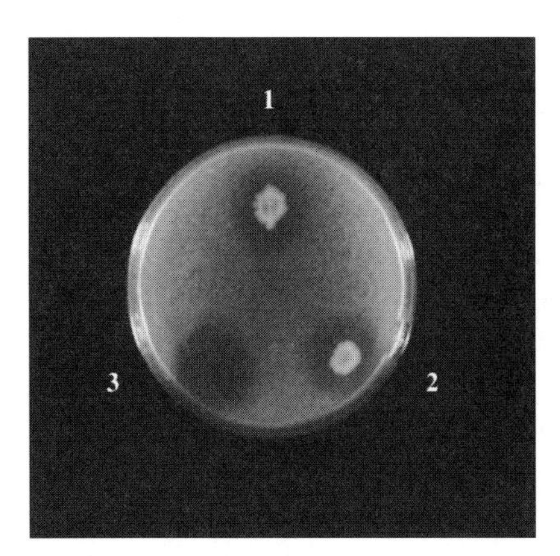

Figure 2 Clearing zones of colloidal chitin formed by chitinases produced by chitinolytic strains *E. agglomerans* (1), *A. caviae* (2), and *S. marcescens* (3).

may enable them to compete successfully with fungi for chitin. Moreover, the production of chitinase may be part of a lytic system that enables the bacteria to use living hyphae rather than chitin as the actual growth substrate since chitin is an important constituent of most fungal cell walls.

A strain of *S. marcescens*, isolated from the rhizosphere of plants grown in soil infested with *S. rolfsii* Sacc., was found to be an effective biocontrol agent under greenhouse conditions against this pathogen and *R. solani* Kuhn. A chitinase(s) produced by the bacterium caused degradation of *S. rolfsii* hyphae in vitro, which provides evidence that this enzyme has a role in biocontrol (163). *S. marcescens* was shown to produce several chitinolytic enzymes, including endochitinases of 58 kD (ChiA), 54 kD (ChiB), 48 kD (C1), 36 kD (C2) and 22 kD and a 94-kD chitobiase (166–170,306). The structural genes encoding some of these enzymes have been cloned and characterized (162,171,172). *S. marcescens* mutants in which *chiA* had been inactivated were used to prove the importance of the ChiA chitinase for biocontrol activity toward *Fusarium* sp. on pea seedlings (162). Shapira et al. (173) demonstrated the involvement of *S. marcescens* ChiA in the control of *S. rolfsii* via genetic engineering: the enzyme produced by an *E. coli* strain carrying the *chiA* gene of *S. marcescens* cloned under the control of a strong and regulated promoter caused rapid and extensive bursting of the pathogenic fungus's hyphal tip. A recombinant *E. coli* expressing the *chiA* gene from *S. marcescens* was effective in reducing disease incidence caused by *S. rolfsii* and *R. solani*. In addition to *S. marcescens*, other *Serratia* species have been found to be efficient biocontrol agents. Strains of *Serratia* spp. have been isolated from the rhizosphere of oilseed rape. The percentage of *Serratia* sp. in this microenvironment was determined to be 12.4% of the total antifungal bacteria. *S. liquefaciens*, *S. plymuthica*, and *S. rubidaea* also were found. All of the isolates showed antifungal activity against different phytopathogenic fungi in vitro, albeit at different efficiencies. The antifungal mechanisms of 18 selected strains were investigated. The direct antifungal effect may be based on antibiosis and the production of lytic enzymes (chitinases and β-1,3-glucanases). Potent siderophores are secreted by the strains to improve iron availability. No strain was able to produce cyanide. In addition, most of the strains secrete the plant growth hormone indole acetic acid (IAA), which can directly promote root growth. The mechanisms were specific for each isolate (164). Other strains of *S.*

Table 2 Examples of Lytic Enzymes Produced by Bacterial Biocontrol Agents.

Producer	Enzyme	Mol. masse, kDa	Encoding gene	Reference
Aeromonas caviae	Endochitinase	94	chiA	(154, 174)
Bacillus cereus	Chitobiosidase	36	ND	(194)
Chromobacterium violaceum	Endochitinase	52	ND	(155)
–''–	Endochitinase	37	ND	(155)
Enterobacter agglomerans	Endochitinase	58	chiA	(157, 176)
–''–	β-N-acetylglucosaminidase	89	ND	(157)
–''–	β-N-acetylglucosaminidase	67	ND	–''–
–''–	Chitobiosidase	50	ND	–''–
E. asburiae,	Cellulase	ND	ND	(195)
Kurthia zopfii	Chitinase	72	ChiSH-1	(215)
Pseudomonas fluorescens	Cellulase, Endochitinase, β-1,3-glucanase	ND	ND	(159, 160, 184, 195)
P. stutzeri	Endochitinase, β-1,3-glucanase	ND	ND	(161)
P. cepacia	β-1,3-glucanase	ND	ND	(183)
Serratia marcescens	Endochitinase	58	chiA	(162, 166–172)
–''–	Endochitinase	54	chiB	–''–
–''–	β-N-acetylglucosaminidase	98	ND	–''–
Serratia plymuthica	β-N-acetylglucosaminidase	ND	ND	(164, 165)
–''–	Endochitinase		ND	(164, 165)
–''–	Chitobiosidase		ND	(164, 165)
Streptomyces coelicolor, S. halstedi	Chitinase	ND	ND	(185)
S. lydicus	Chitinase	ND	ND	(213)
S. violaceusniger	Chitinase, β-1,3-glucanase	ND	ND	(214)
Xanthomonas maltophilia	Endochitinase	ND	ND	(187)

plymuthica isolated from the rhizosphere of oilseed rape showed antifungal activity against the phytopathogenic fungus *V. dahliae* var. *longisporum* in vitro. One of these isolates, C48, produced several chitinolytic enzymes (one *N*-acetyl-β-D-glucosaminidase, one chitobiosidase, and one endochitinase) but no antifungal antibiotics, siderophores, or glucanases. A C48 mutant, deficient in chitinolytic activity, not only lost inhibitory activity on plates but was unable to protect oilseed rape from *Verticillium* sp. wilt. Therefore, the chitinolytic activity was suggested to be exclusively responsible for strain C48's antifungal activity (165).

A chitinolytic strain of *A. caviae*, isolated from roots of healthy bean plants growing in soil artificially infested with *S. rolfsii*, was able to control *R. solani* and *F. oxysporum* f.sp. *vasinfectum* in cotton and *S. rolfsii* in beans under greenhouse conditions (154). The strain produced an extracellular ca. 94-kD chitinase with a high degree of similarity to

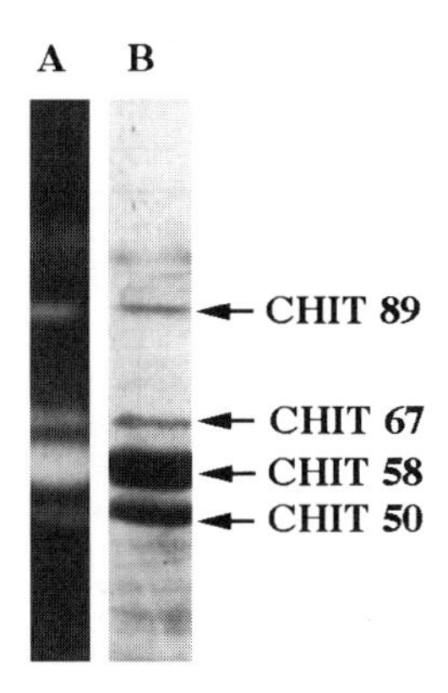

Figure 3 Detection of chitinolytic activity (A) and Coomassie blue staining (B) of extracellular proteins produced by an *E. agglomerans* strain grown on minimal media with chitin, after separation by SDS-PAGE. Chitinolytic activity was detected with the 4-methylumbelliferyl-β-D-*N,N′*-diacetylchitobioside (4-MU-$(GlcNAc)_2$). The bands on lane B corresponding to chitinolytic enzymes visible on lane A are indicated by arrows.

the ChiA endochitinase of *S. marcescens* (174). A soil-borne chitinolytic *E. agglomerans* strain IC1270 was found to be a strong antagonist of about 30 species of plant-pathogenic bacteria and fungi in vitro and an efficient biocontrol agent of several diseases caused by soil-borne fungal pathogens (157,175). The strain produced and excreted a complex of chitinolytic enzymes consisting of two *N*-acetyl-β-D-glucosaminidases with apparent molecular masses of 89 and 67 kD and a 58-kD endochitinase. Additionally, a 50-kD chitobiosidase was observed in two other strains of *E. agglomerans* tested in this work (157). The chitinolytic activity was induced when the strains were grown in the presence of colloidal chitin as the sole C source; the observed chitinolytic enzymes seemed to be the most abundant proteins secreted by the bacteria under this condition (Fig. 3). The *chiA* gene of the 58-kD endochitinase was cloned from strain IC1270 in *E. coli*. The nucleotide sequences of this gene showed an 86.8% identity with the corresponding gene *chiA* of *S. marcescens*. A database search revealed that the deduced Chia_Entag protein amino acid sequence was 87.7%, 71.9%, 52.2%, and 32.2% identical to Chia_Serma, Chia_Aerca from *A. caviae*, Chia_Altso from an *Alteromonas* sp., and Chi1_Bacci from *Bacillus circulans*, respectively. These comparisons suggest that the levels of diversity among various chitinases correlate with the evolutionary distances between the bacteria that produce them. Thus, the chitinases of *S. marcescens* and *E. agglomerans* (both Enterobacteriaceae) are closer to those of *A. caviae* (the Vibrionaceae family) than to those of *Alteromonas* sp. (a group of aerobic marine bacteria) or those of the Gram-positive *Bacillus circulans*. The antifungal activity of the endochitinase secreted by strain IC1270 has been demonstrated in vitro by inhibition of *F. oxysporum* spore germination. The ChiA_Entag-producing *E. coli* strain decreased the disease incidence of root rot caused by *R. solani* on cotton under greenhouse conditions (176).

In addition to its chitinolytic activity, the strain IC1270 produces an antibiotic pyrrolnitrin {3-chloro-4-(2′-nitro-3′-chlorophenyl)pyrrole} with a wide range of activity against many phytopathogenic bacteria and fungi in vitro (177). This antibiotic also was shown to be important to biocontrol activity of several *Pseudomonas* and *Serratia* spp. rhizosphere strains (164,178,179). However, the Tn5 mutants of strain IC1270, one of which is deficient in chitinolytic enzyme production but still possesses antibiotic activity, and the other of which is deficient in both of these activities, were equally unable to protect cotton against root rot caused by *R. solani* (157). These observations raised doubts as to whether pyrrolnitrin can be considered the main compound responsible for biocontrol activity of this *E. agglomerans* strain toward *R. solani* in the rhizosphere or whether it needs to be combined with cell-wall lytic enzymes to provide the host strain with biocontrol capacity. The mode of activity of pyrrolnitrin is not yet completely understood, although direct interference of pyrrolnitrin or its synthetic derivatives with fungal plasma membranes has been demonstrated (180,181). On the basis of these data, the ability of *E. agglomerans* IC1270 to produce pyrrolnitrin in combination with chitinases would be advantageous in attacking fungal phytopathogens.

Secreted chitinolytic activity of soil-borne *C. violaceum* C-61 has been shown to be important for this strain's ability to suppress damping off of cucumber and eggplant caused by *R. solani* (155). Tn5 mutants that cannot produce two of the four chitinase isoforms are unable to inhibit mycelial growth of *R. solani* on plates, and their ability to suppress the disease was much lower than that of the parental strain. Production of six chitinolytic enzymes in another *C. violaceum* strain, ATCC31532, was shown to be controlled by a two-component quorum-sensing mechanism (156).

Rhizosphere pseudomonads are receiving increasing attention as protectors of plants against soil-borne fungal pathogens (6,9,182). Many of these strains have been defined by Kloepper and coworkers (16,18) as plant-growth-promoting bacteria. Enzymatic activities important for bacterial biocontrol capacity occasionally have been reported among *Pseudomonas* spp. strains, but compared to the extensive work on these enzymes in other bacteria and fungi, very little has been done to explore these enzymes' role in the biocontrol provided by the producer strain. Lim et al. (161) presented probably the first piece of evidence that *Pseudomonas* sp. strains can produce cell-wall lytic enzymes important for the bacterium's biocontrol activity. *P. stutzeri* strain YPL-1 isolated from the rhizosphere of ginseng was found to produce β-1,3-glucanase (laminarase) and chitinase activities. These extracellular lytic enzymes markedly inhibited mycelial growth and also caused lysis of *F. solani* mycelia and germ tubes. Abnormal hyphal swelling and retreat were caused by the lysing agents from *P. stutzeri* YPL-1, and a penetration hole was formed on the hyphae in the region of interaction with the bacterium; the walls of this region were lysed rapidly, causing leakage of protoplasm. In several biochemical tests with culture filtrates of *P. stutzeri* YPL-1 and in mutational analyses of antifungal activities of reinforced or defective mutants, the authors found that the bacterium's anti–*F. solani* mechanism may involve a lytic enzyme rather than a toxic substance or antibiotic. Since that report, several groups have succeeded in isolating lytic-enzyme-producing bacteria with biocontrol activity. A β-1,3-glucanase-producing strain of *Pseudomonas cepacia*, isolated on a synthetic medium with laminarin as sole C source, significantly decreased the incidence of diseases caused by *R. solani*, *S. rolfsii*, and *P. ultimum*. The biocontrol ability of this *Pseudomonas* sp. strain was correlated with the induction of the β-1, 3-glucanase by different fungal cell walls in synthetic medium (183). Strain PF-21 of *P. fluorescens*, isolated from the rhizosphere of rice and producing chitinase and β-1,3-glucanase, was found to be very effective in inhibiting the growth of *R. solani* in vitro and in controlling rice sheath blight under greenhouse conditions. A significant relationship between the antagonistic activity of *P. fluorescens* and its level of chitinase production was observed (184).

In fact, chitinolytic pseudomonads are distributed widely in the environment: between 0.01% and 0.5% of the total aerobic counts isolated from airtight stored cereal grain were chitinolytic bacteria (185). Among them Gram-negative bacteria, mainly Pseudomonadaceae, constituted approximately 80% of the chitinolytic population. However, only 4% of the chitinolytic isolates exibited antagonism toward fungi (185). Several chitinolytic respesentatives of the Pseudomonadaceae family (*Pseudomonas* spp. and *Xanthomonas* spp.) with wide ranging antifungal activity were described by Andreeva and coworkers (186). A chitinolytic strain of *X. maltophilia* was shown to suppress *Magnaporthe poae*, the causal agent of summer patch on Kentucky bluegrass, efficiently in growth chamber studies (187). An endochitinase constitutively produced in low-glucose medium by several *P. fluorescens* strains was suggested as an antagonistic mechanism toward *R. solani* (160). To understand better the relationship between chitinolytic and antifungal properties of bacteria that occur naturally in soils, i.e., without artificial selection, three inner dune sites along the Dutch coast, two of which were lime-poor and one lime-rich, were selected as a natural source of chitinolytic bacteria. These bacteria constituted up to 5.7% of the total amount of culturable bacteria of these dune sites. Among them, *Pseudomonas* spp. were the most abundant at the lime-poor sites, whereas *Xanthomonas* and *Cytophaga* spp. were important at the lime-rich site. The percentage of bacterial isolates that were antagonistic to fungal dune strains (*Chaetomium globosum*, *Fusarium culmorum*, *F. oxysporum*, *Idriella* [*Microdochium*] *bolleyi*, *Mucor hiemalis*, *Phoma exigua*, *Ulocladium* sp.) was considerably higher for chitinolytic strains than for nonchitinolytic ones. However, in many cases the inhibition of fungal growth was not accompanied by bacterial chitinase production, indicating that other cell-wall-degrading enzymes (β-1,3-glucanase and protease) and/or antibiotics may also be involved in the antagonistic activities of chitinolytic bacteria against fungi (188).

B. Biocontrol Potential of Lytic-Enzyme-Producing Bacterial Endophytes

The term endophytic is applied to bacteria living inside a plant without causing any visible symptoms. The best-characterized microbial endophytes are nonpathogenic fungi, for which much compelling evidence of plant/microbe mutualism has been provided. Some endophytic fungi are thought to produce compounds that render plant tissues less attractive to herbivores, whereas other strains may increase host plant drought resistance. In return, fungal endophytes are thought to benefit from the comparatively nutrient-rich, buffered environment inside plants (189). However, endophytic fungi constitute only part of the nonpathogenic microflora found naturally inside plant tissues. Bacterial populations exceeding 1×10^7 colony forming units (cfu) g^{-1} plant matter have been reported within tissues of various plant species. Despite their discovery more than four decades ago, bacterial endophytes are much less known than are their fungal counterparts (190).

Compared to use of soil-borne and rhizospheric bacteria, only a few indications support the possibility of using endophytic bacteria as biocontrol agents. Even less is known about the role of endophytic exoenzymes in bacterial antagonism to plant pathogens. Nevertheless, data obtained with plant species of agricultural and horticultural importance indicate that some endophytic bacterial strains stimulate host plant growth by acting as biocontrol agents, either through direct antagonism of microbial pathogens or through induction of systemic resistance to disease-causing organisms. Other endophytic bacterial strains may protect crops from plant-parasitic nematodes and insects (191). Regardless of the mechanism(s) involved, bacterial endophytes appear to be part of a special type of mutualistic plant/microorganism symbiosis that warrants further study. Evidence has been presented that plants can be protected from pathogens by manipulating these naturally occurring microorganisms, and the potential of endophytes as biocontrol agents has been explored (192,193). Endophytic *Bacillus cereus* strain 65 isolated from *Sinapis* sp. was found to excrete a 36-kD chitobiosidase that exibited antifungal activity in a *F. oxysporum* spore-germination bioassay (194). The ability to produce cellulases that cause hydrolysis of wall-bound cellulose near bacterial cells was described for a systemic cotton-plant-colonizing bacterium, *Enterobacter asburiae* JM22, and a cortical root–colonizing *P. fluorescens* 89B-61, a plant-growth-promoting strain with biocontrol potential against various pathogens (195).

C. Genetic Systems for Regulating the Production of Enzymes and Secondary Metabolites Involved in Biocontrol Activity of Gram-Negative Bacteria

In many Gram-negative bacteria, including plant-growth-promoting pseudomonads, three types of control elements are involved in the production of some secondary metabolites and enzymes that are synthesized at the end of exponential growth or during the stationary phase and are involved in biological control. These are (i) two-component global regulatory systems that mediate transduction of environmental signals into the cells, (ii) sigma-factor-mediated transcription by RNA polymerase, and (iii) a diffusible *N*-acyl-homoserine lactone (*N*-acyl-HSL) quorum sensing signals.

Signaling pathways involve a two-component design consisting of a transmembrane sensor kinase (designated *LemA*, *ApdA*, or *GacS*) and a cognate cytoplasmic response regulator protein (GacA). The sensor kinase, when activated by a signal, phosphorylates its own conserved histidine residue, which then serves as a histidine protein kinase (HPK), a phosphoryl donor to an aspartate in the response-regulator protein. Two-component systems seem to be a common way for bacteria to sense and respond to their environment: when triggered by some environmental signals, the sensor phosphorylates the regulator. The phosphorylated regulator functions as a transcriptional activator of target genes (159,196–198). The genes *gacA*, encoding the response regulator (159,196), and *apdA* (also called *lemA*, *repA*, *pheN*, or *gacS*) (198,199), encoding the cognate sensor kinase, are highly conserved among *Pseudomonas* spp. When mutations in *gacA* and *apdA* occur, a similar pleiotropic phenotype develops, but production of several antibiotics, an extracellular protease(s), and a tryptophan side-chain oxidase disappears (198). In vitro, all of these compounds are synthesized at the end of exponential growth or during the stationary phase. In response to starvation or upon entry into the stationary phase, gram-negative bacteria undergo a process of differentiation that leads to the development of a cellular state with markedly enhanced tolerance to a variety of individual stresses.

Besides the *gacA-apdA* system of global regulation, the stationary-phase sigma factor σ^S (σ^{38}), encoded by the *rpoS* gene, plays a critical role as a regulator of the production of secondary metabolites responsible for the biocontrol potential of *P. fluorescens* (200). The two-component regulatory system and σ^S interact or operate through independent regulatory circuits; however, the GacA-ApdA system influences σ^S accumulation and the stress response of stationary-phase cells of *Pseudomonas* spp. (201). The importance of another sigma factor, σ^D (σ^{70}) encoded by the *rpoD* gene, for the production control of a number of secondary metabolites and biocontrol activity was demonstrated in *P. fluo-*

rescens strain CHAO (202). Amplification of σ^{70} enhances the production of some antibiotics and improves the protection of cucumber against damping off caused by *P. ultimum* under gnotobiotic conditions. The relative amounts of σ^{38} and σ^{70} may be particularly important in the stationary phase when the cellular levels of both sigma factors are controlled by sophisticated regulatory mechanisms: the σ^{38}/σ^{70} ratio rises and many σ^{38}-dependent genes are expressed (203).

Quorum-sensing control, a cell-density-dependent phenomenon mediated by intercellular communication and typically regulated by *N*-acyl-HSL signaling molecules (AHLs), has been established as a key feature in the regulation of exoenzyme production in many gram-negative bacteria. These signal molecules play a regulatory role in a multitude of characteristics, including extracellular enzyme production (204–206). *N*-acyl-HSL-mediated cross-interaction between isogenic bacterial populations occurs in the rhizosphere. *P. aureofaciens* strain 30-84, isolated from the rhizosphere of wheat, produces antifungal phenazine (Phz) antibiotics that inhibit a wide range of bacteria and fungi in vitro and are responsible for the bacterium's ability to suppress take-all disease caused by the ascomycete fungus *Gaeumannomyces graminis* var. *tritici* (6). Studies of the genetic control of Phz production and regulation in this strain have provided evidence that (i) *N*-acyl-HSLs produced by one population influence gene expression in a second population in the rhizosphere, (ii) *N*-acyl-HSL production is required for Phz expression in roots, and (iii) *N*-acyl-HSLs serve as a regulatory signal in nature (207). In *Pseudomonas* sp. the AHL-mediated quorum-sensing response may be controlled by the GacA/ApdA global regulation system (208). In 1998 *N*-acyl-HSL signals were shown to be produced not only by *Pseudomonas* spp. but by many other Gram-negative plant-associated bacteria as well (209). Still very little is known about the role of quorum sensing in the regulation of enzymatic activity important for the bacteria's antagonism to plant pathogens. In 1998 the production of six chitinolytic enzymes in *C. violaceum* strain ATCC31532 was shown to be controlled by AHL's signal molecules (156). Results from in vitro experiments show that *C. violaceum* ATCC31532 can suppress growth of the fungal phytopathogens *P. aphanidermatum* and *R. solani*. In the AHL-deficient mutant, this is related to supplementation of AHL to the growth medium (M. Winson and L. Chernin, unpublished observation). Although *C. violaceum* usually constitutes only a minor component of the total microflora found in soil and water, some strains isolated from rhizospheric soil of maize and used for the inoculation of maize seeds were found to increase plant yield significantly (210). This could be the result of antagonism to other soil-borne bacterial and fungal plant pathogens.

Most of the data on the role of global regulation pathways in bacterial antagonist biocontrol activity have been obtained by studying *Pseudomonas* sp. strains that produce various antibiotics and other secondary metabolites able to suppress mostly plant pathogenic fungi. The involvement of lytic enzymes in the biocontrol efficacy of *Pseudomonas* sp. strains is much less clear than the role of antibiotics. In root-colonizing *P. fluorescens* BL915, able to protect cotton seedlings against *R. solani*, the expression of uncharacterized chitinolytic activity was shown to be regulated by the two-component system (159). Cloning of the *gacA* regulatory region from strain BL915 in certain heterologous soil isolates of *P. fluorescens* was found to stimulate the expression of otherwise latent chitinase genes (159), indicating that global regulation by two-component regulators may be a common feature of the regulation of chitinase expression. However, to date, this probably is the only study of regulatory pathways involved in the production of lytic enzymes in biocontrol strains of *Pseudomonas* spp. Other data merely demonstrate that lytic enzymes produced by several soil-borne or rhizospheric pseudomonads play a role in the strains' efficacy at protecting crops against pathogens. It is worth noting that since, in *T. harzianum*, the synthesis of both hydrolytic enzymes (chitinase, β-1,3-glucanase, and protease) and peptaibol antibiotics is triggered by the same environmental signal (211), the existence of a type of global regulation mechanism can be predicted in this fungus as well.

An interesting extension of the role enzymes may play in the biocontrol capacity of a bacterium was suggested by 1998 work by Dekkers et al. (212). The authors studied a mutant of an efficient root-colonizing biocontrol strain of *P. fluorescens* that is impaired in competitive root-tip colonization of a number of grown crops under gnotobiotic conditions and in potting soil. A DNA fragment that was able to complement the mutation for colonization possessed an open reading frame (ORF) that was a homolog of *xer*C in *E. coli* and the *sss* gene in *P. aeruginosa*. Both these genes encode proteins that belong to the lambda integrase family of site-specific recombinases that play a role in phase variation caused by DNA rearrangements. The authors suggested a relationship between the processes of root colonization and genetic rearrangement, known to be involved in the generation of different phenotypes, thereby allowing a bacterial population to occupy various habitats. This work is the first to show the importance of phase variation in microbe–plant interactions.

D. Lytic Enzymes in the Biocontrol of Plant Pathogens by Gram-Positive Bacteria

Gram-positive chitinolytic bacteria, predominantly of the *Streptomyces* and *Bacillus* spp. groups, that exhibited biocontrol potential also were isolated (Table 2). Among these, isolates of *Streptomyces coelicolor* and *S. halstedii* inhibited growth or a broad range of fungi (185). Extracellular chitinase from culture filtrates of *S. lydicus* WYEC108, a broad-spectrum antifungal biocontrol agent, was characterized and purified. Activity was induced by GlcNAc or *N,N′*-diacetylchitobiose ($[GlcNac]_2$) and repressed by glucose, xylose, arabinose, raffinose, and carboxymethyl cellulose. Strong catabolite repression of the chitinase was observed. Addition of pectin, laminarin, starch, or β-glucan to the chitin-containing medium, however, increased chitinase production. Low constitutive levels of the enzyme were observed when cultures were grown with both simple and complex C substrates. Strong chitinase production was obtained when 1% colloidal chitin was present in the medium as a growth substrate; however, further enhancement was achieved when cells were grown in a medium containing colloidal chitin supplemented with certain fungal cell wall preparations, in particular those from *Pythium* or *Aphanomyces* species. The chitinase appears to play a role in the antifungal activities of an *S. lydicus* strain. Crude fungal cell walls were lysed by partially purified chitinase. The authors suggested that whereas *S. lydicus* also produces one or more antifungal antibiotics, its chitinase probably plays a significant role in the in vivo antifungal biocontrol activity of this rhizosphere-colonizing actinomycete (213).

Another ascomycete, strain YCED-9 of *Streptomyces violaceusniger*, antagonistic to many different classes of plant-pathogenic fungi, produces several antifungal secondary metabolites, as well as chitinase and β-1,3-glucanase under induction by colloidal chitin and laminarin, respectively. Fungal cell walls induced the production of both enzymes. A strong in vitro antagonism toward pathogenic isolates of *Pythium infestans* suggested that strain YCED-9 has potential for biological control of diseases caused by this fungus (214). A chitin-degrading strain of *S. anulatus* has been utilized for the control of *Fusarium* sp. wilt of tomato and strawberry. The antagonistic effect of the bacterium was attributed to hydrolytic degradation of fungal cell walls by their chitinolytic enzymes. Another gram-positive bacterium isolated from chitin-amended field soil and identified as *Kurthia zopfii* produces at least three major extracellular chitinases of ~72, 58, and 44 kD. The gene encoding the 72-kD chitinase, designated as *SH-1*, has been cloned in *E. coli*. The SH-1 chitinase was effective in digesting the appressoria or secondary hyphae of *Sphaerotheca* sp. powdery mildew pathogens of barley, and significant suppression of the disease was achieved on leaves sprayed with *E. coli* cells carrying the cloned chitinase gene (215).

Many groups have reported successful combinations of biological control agents, e.g., mixtures of fungi, mixtures of fungi and bacteria, and mixtures of bacteria, in improving biocontrol (216). In particular, a combination of chitinolytic and antibiotic activities can improve significantly the bacterial antagonistic biocontrol capacity. In a model experiment by Lorito and associates (74), nonchitinolytic biocontrol strains of *E. cloacae* and two chitinolytic enzymes from *T. harzianum* isolate P1 were combined and tested for antifungal activity in bioassays. Inhibitory effects on spore germination and germ-tube elongation of *B. cinerea*, *F. solani*, and *Uncinula necator* were synergistically increased by mixing fungal enzymes with cells of *E. cloacae*.

Sung and Chung (217) used chitinase-producing *Streptomyces* spp. and *B. cereus* strains in combination with pyrrolnitrin-producing *P. fluorescens* and *Bulkhoderia (Pseudomonas) cepacia* isolates: they had a synergistic effect on the suppression of rice sheath blight. The combination of these strains produced the same combined mechanism of biocontrol activity responsible for the antagonistic effect of strains such as *E. agglomerans* IC1270 (157) or *S. plymuthica* (165) strains, both of which possess simultaneously chitinolytic and antibiotic activities. Two biocontrol strains, *P. fluorescens* F113 and the nonfluorescent *Stenotrophomonas (Xanthomonas) maltophilia* W81, protect sugar beet from *Pythium*-mediated damping off through production of the antifungal antibiotic 2,4-diacetylphloroglucinol and extracellular protease activity, respectively. In a mixture, these two strains improve the level of protection compared to when each strain is used separately. In a field experiment, the only inoculation treatment capable of conferring effective protection of sugar beet was that in which W81 and F113 were coinoculated, and this treatment proved equivalent to the use of chemical fungicides. In another mixture of biocontrol agents, the combined use of a phloroglucinol-producing *P. fluorescens* and a proteolytic *S. maltophilia* improved protection of sugar beet against *Pythium*-mediated damping off (218).

IV. MICROBIAL ENZYMES IN THE BIOCONTROL OF POSTHARVEST DISEASES

Microbial control of postharvest diseases of fresh fruits and vegetables has been developed over the past decade. Effective antagonistic microorganisms have been isolated and efficiently applied as biocontrol agents of postharvest pathogens. Antibiosis, competition for nutrients and space, induction of host resistance, and direct interactions between the antagonist and the pathogen have been described as the main mechanisms of postharvest biocontrol agents (219,220). Nevertheless, their modes of action are still poorly understood in comparison to those of microbes surviving as biocontrol agents of crops at preharvest stage.

With few exceptions, most of the early reports on antagonists of postharvest diseases dealt with antibiotic-producing antagonists. The role of enzymes in the activity of postharvest biocontrol agents is much less understood. However, the assumption that practical application of antibiotic-producing microbes may not be approved readily, especially in fresh fruits, has stimulated a search for antagonists that suppress postharvest pathogens by other means. The nonantibiotic strain US-7 of the yeast *Pichia guilliermondii* (=*Candida oleophila*), which has exhibited efficacy against several postharvest diseases under various environmental conditions, was developed as the efficient biocontrol agent Aspire. This isolate protects apples from the postharvest fruit-rotting fungi *B. cinera* and *Pencillium expansum*. Culture supernatants from *P. guilliermondii* yielded two- to fivefold more β-1-3-glucanase activity than those from a yeast lacking biocontrol activity. Data indicate that tenacious attachment, along with secretion of cell-wall-degrading enzymes, may play a role in the biocontrol activity of this yeast antagonist (221). In 1998 the exo-β-1,3-glucanase activity of *Pichia anomala* strain K, a yeast antagonistic to *B. cinerea* on postharvest apples, was studied in a synthetic medium supplemented with laminarin, a cell-wall preparation of *B. cinerea*, or glucose. The highest enzyme activity was detected in culture media containing the *B. cinerea* cell wall preparation as the sole C source, whereas the lowest activity was observed in culture media supplemented with glucose. An exo-β-1,3-glucanase, designated *Exoglcl*, was purified to homogeneity from culture filtrates of strain K containing a cell wall preparation. The molecular mass of Exoglcl was estimated at less than 15 kD. In vitro, Exoglcl showed a stronger inhibitory effect on germ-tube growth than on conidial germination of *B. cinerea* and caused morphological changes such as cytoplasm leakage and cell swelling. Exo-β-1,3-glucanase activity was detected on apples treated with strain K and was similar to that of Exoglcl in terms of activity on a native gel. Moreover, the addition of a cell wall preparation of *B. cinerea* to a suspension of *P. anomala* stimulated both in situ exo-β-1,3-glucanase activity and protective activity against the pathogen, strengthening the hypothesis that exo-β-1,3-glucanase activity is one of the mechanisms of action involved in the suppression of *B. cinerea* by *P. anomala* strain K (222). The yeasts *Rhodotorula glutinis* (isolate LS-11) and *Cryptococcus laurentii* (isolate LS-28), showing different levels of antagonistic activity against a range of postharvest pathogens, were examined for their possible mode(s) of action in order to highlight the reasons for the higher activity of isolate LS-28. In vitro, the latter isolate was able to produce significantly higher levels of extracellular β-1,3-glucanase activity than was LS-11 when grown in the presence of hyphal cell walls of the pathogens *P. expansum* and *B. cinerea* as sole C sources. Antibiosis did not appear to be involved in the activity of either antagonist (223). Cellulase has been shown to act as an enzymatic elicitor, capable of enhancing resistance of grapefruit against green mold decay caused by *Penicillium digitarum*, the main postharvest pathogen of citrus fruits (224).

V. ENZYMES OF ENTOMOPATHOGENIC FUNGI AND BACTERIA FOR THE BIOCONTROL OF INSECT PESTS

Bioinsecticides are used for the control of many insect pests as an environmentally acceptable alternative to chemical insecticides. The agrochemical industry recently introduced highly insect-pest-specific insecticides of microbial origin, with modes of action that are targeted to a variety of insect species. Most of them are based on the use of *Bacillus thuringiensis* and *Bacillus sphaericus*, both of which produce insecticidal crystal protein toxin (Cry); of *Streptomyces* spp. producing avermectins, the macrocyclic lactone antibiotics; and of actinomycetes belonging to *Saccharopolyspora spinosa* species, which produces a family of new, unique macrolides called spinosyns (40). Considerable interest in chitinolytic enzymes for the biocontrol of pests (insects and nematodes) has been stimulated by many reports showing their involvement in defense against plant pathogens. During the last decade, much evidence has been obtained that chitinases of fungi, bacteria, and even insects themselves can be considered for potential use as biopesticides (59,90,225).

A. Exoenzymes of Fungal Mycoinsecticides

Entomopathogenic fungi (mycoinsecticides) are gaining increasing attention as environmentally friendly insect control agents. Although over 750 species have been reported to infect insects, few have received serious consideration as potential commercial candidates. The entomopathogenic deuteromycetes fungi *Beauveria bassiana* (Balsamo) Vuillemin and *Metarhizium anisopliae* (Metchnikoff) Sorokin appear to have the broadest potential as viable insect-control agents (226,227). Lytic enzymes are important components of these fungi's insecticidal activity.

Insect chitin, which is surrounded by a protein matrix, forms 30% (w/w) of the cuticle and functions as a barrier against entomopathogenic fungi whose mode of invasion is transcuticular. Chitin is a metabolic target of selective pest control agents. The production of chitinolytic enzymes, intended to penetrate the host cuticle, has been documented for several entomopathogenic fungi, including *Aspergillus flavus* (228,229), *B. bassiana* (229–231), *M. anisopliae* (232,233), *Nomuraea rileyi* (234,235), *Paecilomyces farinosus* (236), *Verticillium lecanii* (237), *Zoophthora* spp. (238), and *Entomophthora* spp. (239). Infection by these fungi is facilitated by the secretion of extracellular proteases (230,240–243). Strains of *Verticillium lecanii* and *V. indicum* able to infect insects are present in divergent groups in the consensus tree, suggesting that the ability to infect insects may have evolved independently many times. The insect and mushroom pathogens and several nematode pathogens (e.g., *V. chlamydosporium*) were distinguishable from the plant pathogenic *Verticillium* species in their ability to produce chitinases (244). Proteolytic enzymes appear early, in large quantities, during infection, assisting in initial host penetration and in later processes (245). These enzymes expose the chitin fibrils, which become accessible to the action of chitinases, which in turn degrade chitin to chains of glucose-*N*-acetyl of variable lengths. The exochitinase (*N*-acetyl-β-D-glucosaminidase) releases GlcNAc monomers from the nonreducing end of the chain (230,246).

Each of the *Metarhizium*, *Beauveria*, and *Aspergillus* species assayed for chitinolytic activity produced a heterogeneous collection of chitinolytic enzymes, regulated by a chitin inducer–repressor system. The number of isozymes was a result of posttranslational modifications, as glycosylation imparts a wide range of molecular weights (229,232). Ground chitin was found to be the most effective inducer of chitinase activity in *M. anisopliae*, colloidal chitin and GlcNAc were less efficient (90). In contrast, GlcNAc induced chitinase and chitobiase activities better than colloidal chitin in all fungi tested. The most pronounced difference was with *Hersutella necatrix*, because only GlcNAc, but not colloidal chitin, served as an inducer for this fungus. Additionally, *H. necatrix* produced its chitobiase in noninducing media, and this basal level was enhanced by GlcNAc. A constitutive *N*-acetyl-glucosaminidase of around 110 kD was reported in *M. anisopliae* (90). A novel 60-kD chitinase that has both endo- and exochitinase activity was detected in extracellular culture fluids of the entomopathogenic fungus *M. anisopliae* (ATCC 20500) grown in liquid medium containing chitin as a sole C source (233). This molecular mass is different from values of 33, 43.5, and 45 kD for endochitinases and 110 kD for an exochitinase (*N*-acetylglucosaminidase) from *M. anisopliae* ME-1 published previously (90,229,232).

Several studies have attempted to correlate chitinase activity of entomopathogenic fungi with their infectivity, an important trait in the fungus's ability to adapt to the different cuticle environments provided by different host insects (234,247). However, the relevance of chitinase production to fungal pathogenicity is still unclear. Studies with *B. bassiana* and *N. rileyi* have indicated that chitinolytic activity is partially correlated with virulence toward lepidopteran larvae and the penetration of their cuticle (228,234,235,248). Activity of *N*-acetyl-glucosaminidase was 15–18 times higher in the virulent than the avirulent isolates (230). The expression of chitinolytic enzymes is probably a specific response to contact with a cuticular surface (220). On the other hand, a chitinase-deficient mutant of *V. lecanii* was just as pathogenic to an aphid as the wild type (237).

Extracellular proteases produced by *B. bassiana* and *M. anisopliae* solubilize cuticle proteins, thus assisting penetration and providing nutrients for further growth. The penetration of entomopathogenic fungi through the intact arthropod cuticle appears to involve a combination of mechanical force and enzymatic degradation (247,249,250). The relative contributions of these components probably depend on the structure and composition of the cuticle encountered as well as on the available enzymes. In certain entomopathogenic fungi, enzymes are secreted in an orderly sequence, i.e., esterase and proteolytic enzymes (chemoelastase Pr1, aminopeptidase, and carboxypeptidase) first, followed by chitobiase, chitinase, and lipase (242). Both the fungus and the host insect have developed mechanisms involved in the host–parasitic interaction of these organisms; the insects release inhibitors against the fungal protease, influencing their susceptibility to the pathogen (251); while the protease released by the entomopathogenic fungi may cause suppression of cellular immune responses in infected insects (252).

The fungus *Myrothecium verrucaria* produces high activity of extracellular insect-cuticle-degrading enzymes, chitinases, proteinases, and lipases. Both first (I) and fourth (IV) instar larvae of a mosquito, *Aedes aegypti*, a vector of yellow fever and dengue, were susceptible to crude culture filtrate. The supplementation of purified *M. verrucaria* endochitinase with commercial lipase decreased the time lethality (LT_{50}) from 48 h to 24 h for instar I and from 120 h to 96 h for instar IV. These data indicate that the cuticle-degrading enzyme complex of *M. verrucaria* has good potential for the control of mosquitoes (253).

More than 50 species in the genus *Hirsutella* are known to infect a wide range of invertebrates: an acaropathogenic member of this genus even has been used as a biological control agent of mites (254). Six isolates of *Hirsutella thompsonii*, as well as one each of *H. necatrix* and *H. kirchneri*, were assayed for chitinolytic activity (255). Two isolates of *H. thompsonii* (#414, #255) and the *H. necatrix* were able to produce and excrete chitinolytic enzymes. A chitobiase of >205 kD was excreted by all fungi and a chitobiase of 112 kD only by isolate 414. An endochitinase of 162 kD was excreted by isolate 414, and two endochitinases, of 66 and 38 kD, by isolate 255. Both *H. thompsonii* isolates produced chitinolytic enzymes only under inductive conditions, in the presence of colloidal chitin as the sole C source. *H. necatrix* produced a chitobiase constitutively when grown in the presence of glucose. In addition to chitinolytic enzymes, the *H. thompsonii* isolates excreted proteolytic activities, including Pr1-like chymoelastase, as well as α-esterase and α-amylase activities. *H. necatrix* was unable to use casein, milk powder, or elastin as the sole C source. The *H. necatrix* isolate was unable to produce Pr1 protein and differed in the secretion pattern of chitinases and some proteolytic enzymes. This corroborates data on the acaropathogenicity of these fungi against the carmine spider mite (*Tetranychus cinnabarinus*): isolates 414 and 255 were more infective than *H. necatrix*. However, the ability of *H. necatrix* to kill mites, at least to a certain extent, indicates that in this fungus, the other enzymatic activities can partially substitute for Pr1 and chitinase deficiencies. Further studies on the functional relationship between these enzymes may engender a rational and cost-effective approach to selecting *Hirsutella* sp. isolates with greater virulence to pestiferous mites.

There are almost no reports of entire gene sequences coding for chitinolytic enzymes from entomopathogenic fungi, even though these enzymes act synergistically with proteolytic enzymes to solubilize the insect cuticle during the key step of host penetration, and this capacity has considerable importance in the biological control of some insect pests. Bogo et al. (256) reported the complete nucleotide sequence and analysis of the chromosomal and full-length cDNA copies of the regulated gene (*chit1*) encoding one of the chitinases produced by the biocontrol agent *M. anisopliae*. The *chit1* gene is interrupted by three short typical fungal introns and has a 1521-bp ORF that encodes a protein of 423 amino acids with a stretch of 35 amino acid residues displaying characteristics of a signal peptide. The deduced sequence of the mature protein predicts a molecular mass of 42-kD. Southern analysis of the genomic DNA indicates a single copy of *chit1* in the *M. anisopliae* genome.

B. Exoenzymes of Bacterial Insecticides

The most famous bacterium used for the biocontrol of various pests is *B. thuringiensis*, a Gram-positive spore-forming microorganism that produces, during the stationary and/or sporulation phase, parasporal inclusions composed of crystal (Cry) proteins that are toxic to a wide variety of insect species (257). However, bacterial chitinolytic activity also has been shown to play a substantial role in biocontrol activity against pests in several ways: (i) by isolating chitinolytic *B. thuringiensis* strains and comparing their activity against pests with that of the parental strain, (ii) by adding chitinolytic enzymes from other sources to *B. thuringiensis* with the aim of increasing its toxicity, (iii) by introducing cloned chitinase genes into a *B. thuringiensis* strain, and (iv) by directly using another chitinolytic bacterium as a potential biocontrol agent of insects.

Sampson and Gooday (258) used the first of these approaches to investigate whether the ability to produce chitinolytic enzymes would further improve the bioinsecticidal activity of *B. thuringiensis*. Strains of both *B. thuringiensis* subsp. *israelensis* IPS78 and *B. thuringiensis* subsp. *aizawai* HD133 secrete exochitinases when grown in a medium containing chitin: they were shown to be efficient against larvae of the midge *Culicoides nubeculosus* and caterpillars of the cotton leafworm *Spodoptera littoralis*, respectively. Inhibition of the bacteria's chitinolytic activities by allosamidin, a specific chitinase inhibitor, led to a significant decrease in the bacteria's insecticidal activity, demonstrating a role for bacterial chitinases in the attack on insects. Contrary to this inhibitory effect, adding chitinase A from *S. marcescens* considerably increased the bacteria's pathogenicity, confirming previous observations with different systems of the potentiation of *B. thuringiensis* entomopathogenesis by exogenous chitinases. The authors suggested that the most likely purpose of *B. thuringiensis* endogenous chitinases is to weaken the insects' peritrophic membranes, allowing more ready access of the bacterial toxins to the gut epithelia. Consequently, addition of exogenous chitinases increases this effect.

Synergism between chitinases of different origins and *B. thuringiensis* was employed to enhance the activity of the bacterium, and many authors reported increased potency of this and other entomopathogenic microorganisms as biocontrol agents (e.g., 259, 260). In an attempt to increase the insecticidal effect of the δ-endotoxin crystal protein CryIC on the relatively Cry-insensitive larvae of *S. littoralis*, Regev et al. (259) used a combination of CryIC and endochitinase from *S. marcescens*. On its own, the endochitinase, even at very low concentrations, perforated the larval midgut peritrophic membrane. When it was applied together with low concentrations of CryIC, a synergistic toxic effect was obtained. In the absence of chitinase, much more CryIC per milliliter was required to obtain maximal reduction in larval weight, and seven-fold less CryIC caused a similar toxic effect in the presence of endochitinase. Thus, a combination of the Cry protein and an endochitinase could result in effective insect control in transgenic systems in which the Cry protein is not expressed in a crystalline form. Esposito et al. (260) reported that crude chitinase preparations from a *Bacillus circulans* strain enhance the toxicity of *B. thuringiensis* subsp. *kurstaki* to diamondback moth larvae, and Tantimavanich and associates (261) obtained similar results by combining chitinase from a high-producing strain of *Bacillus licheniformis* with *B. thuringiensis* subsp. *aizawai* against the pest *Spodoptera exigua*. These and similar results obtained by other authors (225) suggested the usefulness of introducing chitinase-encoding genes into *B. thuringiensis* to increase the toxicity of the bacterium toward target insects. Thus, chitinase-encoding genes from chitinolytic strains of *Aeromonas hydrophila*, *Pseudomonas maltophilia*, and *B. lichenformis* were cloned and then introduced into several *B. thuringiensis* subspecies. Selected transformants stably maintained and expressed cloned chitinase genes, implying that potentially these strains could be used as effective biopesticides (260–262).

Other chitinolytic bacteria not belonging to *B. thuringiensis* also have been found to be efficient against insects. Bacteria from the genus *Serratia* currently are under development, or have been developed, as control agents for several insect pests. For example, *S. enthomophila* has been developed as the commercial pesticide Invade for the control of grass grub, *Costelytra zealandica* (263). *S. marcescens* and *S. proteamaculans* have shown potential as control agents of Diptera (264). It is worth adding that not only microbial but also insect chitinases, isolated from the tobacco hornworm, *Manduca sexta*, and several other insect species have started to be used in pest biocontrol. Transgenic plants that express hornworm chitinase constitutively have been developed and found to exhibit host-plant resistance. A transformed entomopathogenic virus that produces the enzyme displayed enhanced insecticidal activity. Insect chitinase and its gene are now available for biopesticidal applications in integrated pest management programs (225).

VI. ENZYMES OF FUNGI AND BACTERIA IN THE BIOCONTROL OF NEMATODES

Plant-parasitic nematodes are major agronomic pests. The economic loss caused by nematodes is estimated at U.S. $100 billion worldwide (265); the nematocide market share is currently estimated at $500 million, which is only 1/40 of the world pesticide market (266). Bacteria are omnipresent and destroy nematodes in virtually all soils because of their constant association in the rhizosphere. Parasitoid endospore-forming bacteria, such as *Pasteuria penetrans* destroy nematodes by parasitic action, the nonparasitic rhizobacteria reduce nematode populations by colonizing the rhizosphere of the host plant. The large number of rhizobacteria genera known to reduce nematode populations include

Agrobacterium, *Alcaligenes*, *Bacillus*, *Clostridium*, *Desulfovibrio*, *Pseudomonas*, *Serratia*, and *Streptomyces*. Application of some of these bacteria has shown very promising results (267). Combined use of *P. penetrans* with rhizosphere bacteria inhibited reproduction of the root-knot nematode *Meloidogyne incognita* and increased the attachment of *P. penetrans* endospores on the nematodes in vitro; this capacity could contribute greatly to the efficiency of nematode biocontrol with *P. penetrans* (268).

The role of chitinases and other lytic enzymes in nematode biocontrol now has become a matter of intensive investigation. Higher chitinase activity and early induction of specific chitinase isozymes in plants have been shown to be associated with resistance to root-knot nematode (269,270). Purified commercial chitinase inhibits egg hatch of the potato cyst nematode *Globodera rostochiensis* (Ro1) in vitro by up to 70% relative to that of an untreated control. A screening strategy was developed by Cronin et al. (271) to isolate chitinase-producing bacteria from a soil with no documented history of damage by potato cyst nematodes in the preceding 30 years that has been cropped with potato. About 4% of the bacterial isolates were found to be chitinase-positive. All the chitinase-producing bacteria tested in vitro could reduce the hatch of *G. rostochiensis* eggs, some by up to 90% relative to controls. Two of the most efficient strains belonged to *Stenotrophomonas maltophilia* and a *Chromobacterium* sp. Both isolates also reduced the ability of *G. rostochiensis* to hatch in soil microcosms planted with potato seed tubers (271). The inhibition of *G. rostochiensis* egg hatch by chitinase-producing bacteria was suggested as a biocontrol strategy for the defense of potato crops against potato cyst nematodes. Another approach to control plant-parasitic nematodes is to amend soil with chitin to stimulate chitinolytic activity. Sarathchandra et al. (272) demonstrated that such a treatment increases the populations of bacteria and fungi 13-fold and 2.5-fold, respectively. Simultaneously, the cyst nematode of white clover was significantly reduced in chitin-amended soil, possibly because of increased levels of chitinase produced by rhizosphere microorganisms. Two other plant-parasitic nematodes, *Pratylenchus* and *Tylenchus* were also reduced in ryegrass roots and in soil as a result of the chitin amendment. However, the total number of free-living nematodes increased 5.4-fold in amended soil, indicating this is a safe ecological soil treatment. The nematocidic potential of chitinolytic bacteria was demonstrated with strain 20M^T isolated from soil in Israel. The strain was described as *Pseudomonas chitinolytica* (273) and then attributed to *Telluria chitinolytica* sp. nov.—strictly aerobic, rod-shaped, bacteria that are active polysaccharide degraders (274). Among the various products formed by *B. thuringiensis* Berliner, chitinase and exotoxin seem to play an important role in the killing of eggs, egg masses, and juveniles of *M. incognita* and in the reduction of gall formation on pea roots (275).

The nematode's surface comprises a multilayered cuticle, which consists mainly of collagen proteins. Collagenolytic activity was shown therefore to play a role in the antagonism of some fungi and bacteria toward plant-pathogenic nematodes. Soil amendment by collagen sharply decreased the root-galling index of tomatoes inoculated with *Meloidogyne javanica* (276). This led to the enrichment of collagenolytic microorganisms, which drastically reduced the number of galls caused by *M. javanica*. The effect was even more pronounced when the amendment was supplemented with the collagenolytic fungus *Cunninghamella elegans*. This fungus has been shown to possess collagenolytic, elastolytic, keratinolytic, and nonspecific proteolytic activities when grown on collagen media, but only chitinolytic activity when grown on chitin media (276). In the 1990s a novel 42.8-kD collagenolytic/proteolytic enzyme secreted by *B. cereus* was shown to be able to digest collagens extracted from intact cuticles of second-stage juveniles of the root-knot nematode *M. javanica*, suggesting damage to the nematode cuticles in vivo (277).

Bonants et al. (278) purified a 33.5-kD protease from a culture filtrate of the nematode-egg-parasitic fungus *Paecilomyces lilacinus* infecting eggs of the root-knot nematode *Meloidogyne* spp.; this protease was very similar to several subtilisin-like serine proteases. The protease quantitatively bound to nematode eggs, and eggs incubated with the purified protease eventually floated. Incubation of the purified protease with nematode eggs significantly influenced their development: immature eggs were highly susceptible to the protease treatments, whereas those containing a juvenile were more resistant. In addition, hatched larvae were not visibly affected by the protease. It can be suggested that the serine protease plays a role in penetration of the fungus through the eggshell of nematodes. Another extracellular 35-kD protease, which belongs to the family of serine proteases, is produced by the nematode-trapping fungus *Arthrobotrys oligospora*. The enzyme immobilized the free-living nematode *Panagrellus redivivus*, suggesting that the enzyme(s) that hydrolyzes proteins of the purified cuticle is involved in the infection of nematodes (279). A potentially similar 33-kD protease, designated *VCP1*, was secreted by the nematophagous fungus *Verticillium chlamydosporium* and *V. lecanii*, pathogens of nematodes and insects, respectively, but not plant-pathogenic *Verticillium* species. Purified protease hydrolyzed proteins in situ from the outer layer of the eggshell of the host nematode *M. incognita* and exposed its chitin layer (280).

VII. MOLECULAR APPROACHES FOR THE IMPROVEMENT AND CREATION OF NEW BIOCONTROL AGENTS

A. Manipulation of Lytic Enzyme Systems Involved in Biocontrol Agents of Plant Pathogens

1. Manipulation of Fungal Systems

There are a number of ways in which chitinase can be used as a biocontrol agent against plant-pathogenic fungi. The free enzyme can be introduced into the irrigation water or incorporated into the seed coating to protect germinating seedlings. However, the activity of free enzymes in the soil may be short-lived. Research has been directed toward producing transgenic plants expressing a cloned chitinase gene. Another alternative is to introduce genetically engineered rhizospheric bacteria or fungi able to express and secrete chitinase into the soil.

Trichoderma spp. strains currently are the most attractive targets for further improvement of their biocontrol capacity. Preparations based on several isolates, belonging mainly to *G. virens*, *T. harzianum*, *T. hamatum*, and *T. viride*, already are being used as biocontrol agents in the soil and phyllosphere (31,44), but the challenge of improving existing biocontrol strains, in order to make plant-disease biocontrol a more effective and economical process, remains. No correlation has been obtained between field biocontrol effectiveness and lytic enzyme production by different strains; this may be because antibiotic metabolites also are involved in the action. One target is the production of *Trichoderma* sp. strains with elevated levels of degradative and lytic exocellular enzymes that also produce antibiotics and grow rapidly in soil and rhizospheric environments (5). This goal can be achieved only if we acquire detailed knowledge of the mechanisms underlying the mycoparasitic fungus's interaction and at the same time develop better molecular tools for genetic manipulations of *Gliocladium* and *Trichoderma* spp. strains. The current progress in cloning genes encoding lytic enzymes, characterizing their products, and elucidating their individual roles in the mycoparasitic activity of *T. harzianum* and other species of *Trichoderma* or *Gliocladium* has opened the way to improving the biocontrol capacity of these fungi.

Two different strategies have been used to increase the chitinase activity of one of the most potent biocontrol species, *T. harzianum*. Haran et al. (281) introduced the chitinase gene of *S. marcescens* into the genome of *T. harzianum* under the control of a constitutive 35S promoter. Margolles-Clark et al. (91) constructed a transformation vector that contained the coding region of the 42-kD endochitinase gene (ThEn-42) of *T. harzianum* under the control of a cellulase promoter, *cbh1* from *T. reesei*, and achieved a 10-fold increase in chitinase activity in most of the transformants. The 42-kD endochitinase encoding gene of *T. harzianum* has been shown to be triggered in mycoparasitic interactions (79). Transformants of *T. hamatum* with about a fivefold increase in chitinase activity were obtained when the 42-kD-endochitinase-encoding gene, designated *Tham-ch*, was reintroduced in one additional copy into the host strain genome under the control of its own regulatory elements (86). These authors suggested that duplication of this highly conserved gene appears to be a potential means of improving the biocontrol capability of *Trichoderma* species.

Limon et al. (282) isolated and characterized transformants of *T. harzianum* strain CECT 2413 that overexpressed a 33-kD chitinase (Chit33). Strain CECT 2413 was co-transformed with the *amdS* gene and its own *chit33* gene under the control of a constitutive promoter from *T. reesei*. There was no correlation between the number of integrated copies and the level of expression of the *chit33* gene in the transformants. When transformants were grown in glucose, their extracellular chitinase activity was up to 200-fold greater than that of the wild type, whereas in chitin, the activities of both the transformants and the wild type are similar. Under both conditions, the transformants were more effective in inhibiting the growth of *R. solani* as compared with that of the wild type. Similar results were obtained when culture supernatants from the transformants and the wild type were tested against *R. solani* (282).

Probably the most impressive evidence of the biocontrol potential of the *Trichoderma* sp. chitinolytic enzymes was presented by Lorito et al. (283), who very successfully improved disease resistance by inserting the *ech42* gene into tobacco and potato plants. High expression levels of the fungal gene were obtained in different plant tissues, but there was no visible effect on plant growth or development. Substantial differences in 42-kD endochitinase activity were detected among transformants. Selected transgenic lines were highly tolerant of or completely resistant to the foliar pathogens *Alternaria alternata*, *A. solani*, and *B. cinerea* and the soil-borne pathogen *R. solani*. The high level and broad spectrum of resistance obtained with a single chitinase gene from *Trichoderma* sp. overcame the limited efficacy of transgenic expression in plants by chitinase genes from plants and bacteria. Several transgenic apple lines expressing endochotinase Ech42 were shown more resistant to apple scab caused by the fungus *Venturia inaequalis* than nontransformed apple cultivar (284). The results of these works extend previous approaches to the use of chitinase and β-1,3-glucanase genes from a range of sources to construct transgenic plants with enhanced resistance to the disease (285–287).

Proteinases also can be overexpressed in *Trichoderma* spp. with the aim of improving the isolate's biocontrol activity. Flores et al. (288) attempted to do just that by increasing the copy number of the basic proteinase gene *prb1* in *T. harzianum*. The transformants were stable and carried from 2 to 10 copies of the gene. High levels of expression of *prb1* during the fungus–fungus interaction were detected when *T. harzianum* and *R. solani* were confronted in vitro. In liquid cultures, the proteinase was induced by cell walls of *R. solani*. However, analysis of different transformants indicated that high *prb1* messenger RNA (mRNA) production does not always result in high Prb1 protein production. The successful overexpression of the *prb1* gene driven by its own promoter led the authors to test the effectiveness of the transformants as biocontrol agents. Under greenhouse conditions, incorporation of *T. harzianum* transformants into pathogen-infested soil reduced the disease caused by *R. solani* in cotton plants up to fivefold as compared with that of the wild-type strain. It is worth noting, however, that the best protection was provided by a strain that produced only an intermediate level of proteinase protein. The authors suggested that extreme levels of the proteinase may degrade other enzymes that play a role in the mycoparasitic process. These results indicated that the introduction of multiple copies of *prb1* improves the biocontrol activity of *T. harzianum* and showed the importance of the proteinase Prb1 in biological control. The conclusion that Prb1 may be involved specifically in mycoparasitism now has found additional support in studies of other *T. harzianum* transformants with high alkaline proteinase activity. One of these transformant strains produced an elevated, constitutive *prb1* mRNA level during mycoparasitic interactions with *R. solani* (125).

In all of the reports mentioned, the *Trichoderma* strains were genetically improved to serve as better biocontrol agents of fungi that contain chitin and β-glucan as major cell-wall components. A similar approach was approved for obtaining *Trichoderma* strains with high activity against plant-pathogenic oomycetes, such as *Pythium* spp., containing cellulose as the main cell-wall component (289). The difference in this case, however, was that the *Trichoderma* strains were designed to overproduce cellulytic enzymes instead of chitinolytic ones. Hypercellulolytic strains of *T. longibrachiatum* were obtained by transformation with a multicopy hybrid plasmid bearing the *egl1* gene encoding the cellulose-degrading enzyme β-1,4-endoglucanase in this fungus (290). The transformants were tested for their ability to reduce *Pythium* sp. damping off on cucumber seedings and were shown to be significantly more effective in controlling this disease than was wild-type strain (disease incidence was reduced from 60% to 28%) (291). Transformants of *T. longibrachiatum* with extra copies of the *egl1* gene were studied for mitotic stability, endoglucanase production, and biocontrol activity against *P. ultimum* on cucumber seedlings. The transformants showed a significantly higher level of expression of the *egl1* gene than in the wild type under both inductive and noninductive growth conditions. Transformants with the *egl1* gene under the control of a constitutive promoter showed the highest enzymatic activity. Both the endoglucanase activity and the transforming sequences were stable under nonselective conditions. When applied to cucumber seeds sown in *P. ultimum*–infested soil, *T. longibrachiatum* transformants with increased inducible or constitutive *egl1* expression were generally more suppressive than was the wild-type strain (292). These results prove the involvement of cellulase activity in the biocontrol of *Pythium* spp. by *T. longibrachiatum*.

2. *Manipulation of Bacterial Systems*

Biocontrol activity of bacteria can be enhanced by genetic manipulations of regulatory systems responsible for antifungal enzyme expression, e.g., by genetically engineering them to overexpress several different antiphytopathogenic traits that can act synergistically. In engineering these strains, it is essential to ensure that the normal functioning of the bacterium is not impaired, i.e., that there is no problem with metabolic load (293). Genetic improvement is of vital importance for the enhancement of the biological control capability of BCAs, and even for upgrading of their adaptability to different stresses. At present, a widely used approach is to improve useful antagonists genetically by introducing foreign target genes, e.g., genes with biocontrol potential such as chitinase gene, β-1,3-glucanase gene, and even both genes together. For instance, several important rhizobacteria, such as *Rhizobium meliloti* and *Pseudomonas putida*, which are excellent root colonizers, lack the ability to synthesize chitinolytic enzymes. Hence, it is assumed that introducing the chitinase gene into these bacteria will enable them to provide protection against plant-pathogenic fungi. The cloned chitinase gene from *S. marcescens* has been expressed in *Pseudomonas* spp. and the plant symbiont *R. meliloti*. The modified *Pseudomonas* sp. strain controlled the pathogens *F. oxysporum* f.sp. *redolens* and *Gauemannomyces graminis* var. *tritici* (294,295). The antifungal activity of the transgenic *Rhizobium* sp. during symbiosis on alfalfa roots was verified by lysis of *R. solani* hyphal tips treated with cell-free nodule extracts (296). In another work, the *chiA* endochitinase gene of *S. marcescens* inserted into a Tn7 transposon and transferred to a *P. fluorescens* strain provided the new host with the capacity to control *S. rolfsii* and *R. solani* under greenhouse conditions (297). Cheng et al. (298) constructed a genetically improved *B. subtilis* strain B908 with a transformed chitinase gene. Wang (299) ligated the heterologous chitinase gene and β-1,3-glucanase gene and transformed them into the effective biocontrol agent B908, producing obviously increased antifungal activity. Several other systems utilizing gram-positive BCAs provide models for future biocontrol methods (300).

The current knowledge of regulatory mechanisms of various types of antifungal substance expression may help in the construction of strains with enhanced biocontrol activity. Manipulation of regulatory systems responsible for the production of lytic enzyme and antibiotics resulted in significant improvement of the bacterial biocontrol potential. The advantages of this approach were demonstrated by increasing of the doses of genes encoding the GacA-GacS system of global regulation or sigma factors of transcription in some biocontrol strains of *P. fluorescens* and other gram-negative bacteria (196,198,202,301–304). Antifungal activity of bacteria can be also enhanced by treatment with exogenous quorum sensing signal molecules (304). This finding supports the idea that quorum-sensing signaling by AHLs may contribute to the success of a bacterium in competition with plant pathogens and other rhizosphere inhabitants under natural conditions (207).

B. Manipulations of Enzymes Involved in the Biocontrol of Pests

A key aim of recent studies has been to increase the killing speed to improve the commercial efficacy of mycoinsecticides as biocontrol agents. This may be achieved by adding insecticidal genes to the fungus, an approach considered to have enormous potential for the improvement of biological pesticides. St. Leger et al. (305) described an approach for developing a genetically improved entomopathogenic fungus. Additional copies of the gene encoding a regulated cuticle-degrading protease (Pr1) from *M. anisopliae* were inserted into the genome of *M. anisopliae* such that Pr1 was constitutively overproduced in the hemolymph of *Manduca sexta*, activating the prophenoloxidase system. The combined toxic effects of Pr1 and the reaction products of phenoloxidase caused larvae challenged with the engineered fungus to exhibit a 25% reduction in time of death and 40%

reduced food consumption relative to infections by the wild-type fungus. In addition, infected insects were rapidly melanized, and the resulting cadavers were poor substrates for fungal sporulation. Thus, environmental persistence of the genetically engineered fungus is reduced, thereby providing biological containment.

VIII. CONCLUSIONS

Evidence has been provided that exoenzymes produced by many fungi and bacteria are strictly involved in the microbes' antagonism toward plant pathogens and pests. These exoenzymes, possessing chitinolytic, glucanolytic, cellulolytic, or proteolytic activities, can be used individually or in combination to provide an enzymatic basis for a number of processes, which ultimately lead to a biocontrol effect. Among these processes, one of the most intensively studied is the mycoparasitism seen in *Trichoderma* spp., *Gliocladium* spp., and some other fungal antagonists of plant-pathogenic fungi. To achieve efficient biocontrol, the collaborative interaction of the enzymes and secondary metabolites (particularly antibiotics) usually is required. Mycoparasitism is a specific mechanism of some fungal antagonists' biocontrol activity, the efficiency of which very much depends on their ability to produce lytic enzymes. However, many other antagonistic organisms serve as efficient biocontrol agents of plant pathogens that cause soil-borne, foliar, and postharvest diseases, mainly through the production of exoenzymes invading the pathogen. The same is true for fungi and bacteria with strong bioinsecticidal and nematocidal activities. Enthomopathogenic fungi secreting lytic enzymes are gaining more attention as environmentally friendly insect control agents. The ability to produce lytic enzymes enhances the biocontrol potential of *B. thuringiensis*, for which worldwide sales have overtaken those of any other biopesticide products.

Until recently, research on biological control has focused on the identification of potential organisms and the study of their mechanisms of competition and antagonism. However, the possibilities are promising for improving biocontrol activity by distinct microbial antagonists by using genetic engineering techniques. Molecular biology, especially molecular genetics of microorganisms, and genetic engineering technology have helped identify the modes of action of many biocontrol agents and create a strategy for their further improvement. Data obtained recently with transgenic *Trichoderma* spp. and several bacterial strains transformed to produce increased amounts of specific enzymes are encouraging, especially because the results of assays under natural environmental conditions have demonstrated the enhanced biocontrol capacity of the engineered organisms. Continued studies on mode of expression and regulation of antibiotics, siderophores, and enzymes effective against plant pathogens and pests will extend significantly the potential for developing biocontrol agents with increased production of these compounds.

Future research should focus on the development of strains expressing "multigene" combinations while preserving the intrinsic vigor and ecological competence of microbial antagonists. A definition of environmental signals that interact with the microbial recognition and regulatory complexes, such as two-component global regulation and quorum-sensing systems, will help elucidate the competitive mechanisms in pathogen suppression to benefit fully from soil and rhizosphere biocontrol microorganisms. Further greenhouse and field experiments should be conducted to investigate the biological control potential of the microorganisms. Once transgenic strains capable of producing highly efficient synergistic combinations of enzymes are obtained, a higher level of plant-pathogen and pest control should become possible. Transgenic BCAs therefore offer the potential of substantially reducing the amount of chemical fungicides required to produce crops protected from diseases and pests. A combination of transgenic BCAs and transgenic plants resistant to pathogens and pests would appear to yield a very environmentally friendly and efficient strategy of plant protection as we begin the third millennium—when chemistry will meet ecology.

ACKNOWLEDGMENTS

This work was in part supported under Grant No. CA16-012, U.S.–Israel Cooperative Development Research Program, Economic Growth, U.S. Agency for International Development, and by Israel Ministry of Agriculture, Grant No 823-0126-98.

REFERENCES

1. J Katan, JE DeVay, eds. Soil Solarization. Boca Raton, FL: CRC Press, 1991.
2. MA De Waard, SG Georgopoulos, DW Hollomon, H Ishii, P Leroux, NN Ragsdale, FJ Schwinn. Chemical control of plant diseases: Problems and prospects. Annu Rev Phytopathol 31:403–421, 1993.
3. KA Powell, AR Jutsum. Technical and commercial aspects of biocontrol products. Pesticide Sci 37:315–321, 1993.
4. KF Baker. Evolving concepts of biological control of plant pathogens. Annu Rev Phytopathol 26:67–85, 1987.
5. JM Lynch. Biological control of plant diseases: achievements and prospects. In: Brighton Crop Protection Conference: Pests and Diseases. Brighton, 1988, pp 587–595.
6. DM Weller. Biological control of soilborne plant pathogens in the rhizosphere with bacteria. Annu Rev Phytopathol 26:379–407, 1988.
7. I Chet, ed. Biotechnology in Plant Disease Control. New York: John Wiley & Sons, 1993.
8. RJ Cook. Making greater use of introduced micro-organisms for biological control of plant pathogens. Annu Rev Phytopathol 31:53–80, 1993.
9. JM Whipps. Developments in the biological control of soil-borne plant pathogens. Adv Bot Res 26:1–134, 1997.
10. M Wilson, PA Backman. Biological control of plant pathogens. In: FR Ruberson, ed. Hard Book of Pest Management. New York: Marcel Dekker, 1998, pp 309–335.
11. JM Whipps. Microbial interactions and biocontrol in the rhizosphere. J Exp Bot 52:487–511, 2001.
12. JJ Menn, FR Hall. Biopesticides. Present status and future prospects. In: FR Hall, JJ Menn, eds. Biopesticides: Use and Delivery. Totowa, NJ: Humana Press Inc., 1999, pp 1–10.
13. HAJ Hoitink, MJ Boehm, Y Hadar. Mechanisms of suppression of soilborne plant pathogens in compost-amendment substrates. In: HAJ Hoitink, HM Keener, eds. Science and Engineering of Composting: Design, Environmental, Microbiological and Utilization Aspects. Worthington, OH: Renaissance, 1993, pp 601–621.
14. XS Ye, N Strobel, J Kuc. Induced systemic resistance (ISR): Activation of natural defense mechanisms for plant disease control as part of integrated pest management (IPM). In: R Reuveni, ed. Novel Approaches to Integrated Pest Management. Boca Raton, FL: CRC Press, 1995, pp 95–113.
15. S Tuzun, JW Kloepper. Potential application of plant growth-promoting rhizobacteria to in-

duce systemic disease resistance. In: R Reuveni, ed. Novel Approaches to Integrated Pest Management. Boca Raton, FL: CRC Press, 1995, pp 115–127.
16. JW Kloepper, R Lifshitz, RM Zablotowicz. Free-living bacterial inocula for enhancing crop productivity. Trends Biotech 7:39–43, 1989.
17. BR Glick. The enhancement of plant growth by free-living bacteria. Can J Microbiol 41: 109–117, 1995.
18. S Tuzun, E Bent. The role of hydrolytic enzymes in multigenic and microbially-induced resistance in plants. In: AA Agrawal, S Tuzin, E Bent, eds. Induced plant defenses against pathogens and herbivores. St Paul, MN: APS Press, 1999, pp 95–115.
19. R Baker, GJ Griffin. Molecular strategies for biological control of fungal plant pathogens. In: R Reuveni, ed. Novel Approaches to Integrated Pest Management. Boca Raton, FL: Lewis Publishers, CRC Press, 1995, pp 153–182.
20. C Alabouvette, B Schippers, P Lemanceau, PAHM Bakker. Biological control of Fusarium wilts: Toward development of commercial products. In: GJ Boland, LD Kuykendall, eds. Plant-Microbe Interactions and Biological Control. New York: Marcel Dekker, 1998, pp 15–36.
21. J Handelsman, JL Parke. Mechanism in biocontrol of soilborne plant pathogens. In: T Kosuge, EW Nester, eds. Plant Microbe Interactions. Vol. 3. New York: McGraw-Hill, 1989, pp 27–61.
22. JE Loper, SF Lindow. Roles of competition and antibiosis in suppression of plant diseases by bacterial biological control agents. In: RD Lumsden, JL Vaughn, eds. Pest Management: Biologically Based Technologies. Washington, DC: American Chemical Society, 1993, pp 144–155.
23. DM Weller, LS Thomashow. Use of rhizobacteria for biocontrol. Curr Opin Biotech 4:306–311, 1993.
24. LL Barton, BC Hemming, eds. Iron Chelation in Plants and Soil Microorganisms. New York: Academic Press, 1993.
25. V Braun, K Hantke, W Köster. Bacterial iron transport: Mechanisms, genetics and regulation. In: A Sigel, H Sigel, eds. Metal Ions in Biological Systems. Vol. 35. Iron Transport and Storage in Microorganisms, Plants, and Animals. New York: Marcel Dekker, 1998, pp 67–145.
26. SA Leong, G Winkelmann. Molecular biology in iron transport in fungi. In: A Sigel, H Sigel, eds. Metal Ions in Biological Systems. Vol. 35. Iron Transport and Storage in Microorganisms, Plants, and Animals. New York: Marcel Dekker, 1998, pp 147–186.
27. LJ Herr. Biological control of *Rhizoctonia solani* by binucleate *Rhizoctonia* spp. and hypovirulent R. solani agents. Crop Protect 14:179–186, 1995.
28. O Kilic, GJ Griffin. Effect of dsRNA-containing and dsRNA-free hypovirulent isolates of *Fusarium oxysporum* on severity of *Fusarium* seedling disease of soybean in naturally infested soil. Plant Soil 201:125–135, 1998.
29. T Zhou, GJ Boland. Suppression of dollar spot by hypovirulent isolates of Sclerotinia homoeocarpa. Phytopathology 88:788–794, 1998.
30. HL Barnett, FL Binder. The fungal host-parasite relationship. Annu Rev Phytopathol 11: 273–292, 1973.
31. I Chet. Mycoparasitism—recognition, physiology and ecology. In: R Baker, P Dunn, eds. New Directions in Biological Control: Alternatives for Suppressing Agricultural Pests and Diseases. New York: Alan R. Liss, 1990, pp 725–783.
32. J Inbar, I Chet. Lectins and biocontrol. CRB Crit Rev Biotechnol 17:1–20, 1997.
33. S Haran, N Benhamou, I Chet. Mycoparasitism and lytic enzymes. In: GE Harman, CK Kubicek, eds. *Trichoderma* and *Gliocladium*. Vol. 2. London: Taylor and Francis, 1998, pp 153–172.
34. A Herrera Estrella, I Chet. Biocontrol of bacteria and phytopathogenic fungi. In: A Altman, ed. Agricultural Biotechnology. New York: Marcel Dekker, 1998, pp 263–282.

35. A Tronsmo, LG Hjeljord. Biological control with *Trichoderma* species. In: GJ Boland, LD Kuykendall, eds. Plant-Microbe Interactions and Biological Control. New York: Marcel Dekker, 1998, pp 111–126.
36. WH Tang, RJ Cook, A Rovira, eds. Advances in Biological Control of Plant Diseases. Proceedings of the International Workshop on Biological Control of Plant Diseases, Beijing, China Agricultural University Press, 1996.
37. ZK Punja. Comparative efficacy of bacteria, fungi, and yeasts as biological control agents for diseases of vegetative crops. Can J Plant Pathol 19:315–323, 1997.
38. GJ Boland, LD Kuykendall, eds. Plant-Microbe Interactions and Biological Control. New York: Marcel Dekker, 1998.
39. GE Harman, CK Kubicek, eds. *Trichoderma* and *Gliocladium*. Vol. 2. London: Taylor and Francis, 1998.
40. FR Hall, JJ Menn, eds. Biopesticides: Use and Delivery. Totowa, NJ: Humana Press, 1999.
41. FN Martin, JE Loper. Soilborne plant diseases caused by *Pythium* spp: Ecology, epidemiology, and prospects for biological control. Crit Rev Plant Sci 18:111–181, 1999.
42. GC Papavizas. *Trichoderma* and *Gliocladium*: Biology, ecology and potential for biocontrol. Annu Rev Phytopathol 23:23–54, 1985.
43. I Chet. *Trichoderma*: application, mode of action, and potential as a biocontrol agent of soilborne plant pathogenic fungi. In: I Chet, ed. Innovative Approaches to Plant Disease Control, New York: John Wiley & Sons, 1987, pp 137–160.
44. Y Elad, I Chet. Practical approaches for biocontrol implementation. In: R Reuven, ed. Novel Approaches to Integrated Pest Management. Boca Raton, FL: CRC Press, 1995, pp 323–338.
45. GJ Samuels. *Trichoderma*: A review of biology and systematics of the genus. Mycol Res 100:923–935, 1996.
46. E Esposito, N da Silva. Systematics and environmental application of the genus *Trichoderma*. Crit Rev Microbiol 24:89–98, 1998.
47. GE Harman, T Björjman. Potential and existing uses of *Trichoderma* and *Gliocladium* for plant disease control and plant growth enhancement. In: GE Harman, CK Kubicek, eds. *Trichoderma* and *Gliocladium*. Vol. 2. London: Taylor and Francis, 1998, pp 229–265.
48. KP Hebbar, RD Lumsden. Biological control of seedlings diseases. In: FR Hall, JJ Menn, eds. Biopesticides: Use and Delivery. Totowa, NJ: Humana Press, 1999, pp 103–116.
49. S Haran, H Schickler, I Chet. Molecular mechanisms of lytic enzymes involved in the biocontrol activity of *Trichoderma harzianum*. Microbiology 142:2321–2331, 1996.
50. M Lorito. Chitinolytic enzymes and their genes. In: GE Harman, CP Kubicek, eds. *Trichoderma* and *Gliocladium*. Vol. 2. London: Taylor and Francis, 1998, pp. 73–99.
51. MS Manocha, V Govindsamy. Chitinolytic enzymes of fungi and their involvement in biocontrol of plant pathogens. In: GJ Boland, LD Kuykendall, eds. Plant-Microbe Interactions and Biological Control. New York: Marcel Dekker, 1998, pp 309–327.
52. Y Elad, I Chet, P Boyle, Y Henis. Parasitism of *Trichoderma* spp. on *Rhizoctonia solani* and *Sclerotium rolfsii*—scanning electron microscopy and fluorescent microscopy. Phytopathology 73:85–88, 1983.
53. Y Elad, I Chet, Y Henis. Degradation of plant pathogenic fungi by *Trichoderma harzianum*. Can J Microbiol 28:719–725, 1982.
54. CJ Ridout, JR Coley-Smith, JM Lynch. Fractionation of extracellular enzymes from a mycoparasitic isolate of *Trichoderma harzianum*. Enzyme Microb Technol 10:180–187, 1988.
55. RA Geremia, GH Goldman, D Jacobs, W Ardiles, SB Vila, M Van Montagu, A Herrera-Estrella. Molecular characterization of the proteinase-encoding gene, prb1, related to mycoparasitism by Trichoderma harzianum. Mol Microbiol 8:603–613, 1993.
56. A Rousseau, N Benhamou, I Chet, Y Piche. Mycoparasitism of the extramatrical phase of *Glomus intraradices* by *Trichoderma harzianum*. Phytopathology 86:434–443, 1996.
57. G Gooday. The ecology of chitin degradation. Microbial Ecol 10:387–431, 1990.
58. I Havukkala. Chitinolytic enzymes and plant pests. In: LL Ilag, AK Raymundo, eds. Biotech-

nology in the Philippines Towards the Year 2000. Proceedings of the Second Asia-Pacific Biotechnology Congress, Los Banos, SEARCA, University of the Philippines, 1991, pp 127–140.

59. AS Sahai, MS Manocha. Chitinases of fungi and plants: Their involvement in morphogenesis and host-parasite interaction. FEMS Microbiol Rev 11:317–338, 1993.
60. PA Felse, T Panda. Regulation and cloning of microbial chitinase genes. Appl Microbiol Biotechnol 51:141–151, 1999.
61. A Perrakis, KS Wilson, I Chet, AB Oppenheim, CE Vorgias. Phylogenetic relationships of chitinases. In: RAA Muzzarelli, ed. Chitin Enzymology. Ancona: European Chitin Society, 1993, pp 217–232.
62. R Cohen-Kupiec, I Chet. The molecular biology of chitin digestion. Curr Opin Biotechnol 9:270–277, 1998.
63. GE Harman, CK Hayes, M Lorito, RM Broadway, A DiPietro, C Peterbauer, A Tronsmo. Chitinolytic enzymes of *Trichoderma harzianum*: Purification of chitobiosidase and endochitinase. Phytopathology 83:313–318, 1993.
64. WS Roberts, CP Seletrennikoff. Plant and bacterial chitinase differs in antifungal activity. J Gen Microbiol 134:169–176, 1988.
65. A Tronsmo, GE Harman. Detection and quantification of N-acetyl-b-D-glucosaminidase, chitobiosidase and endochitinase in solutions and in gels. Anal Biochem 208:74–79, 1993.
66. CJ Ulhoa, JF Peberdy. Regulation of chitinase synthesis in *Trichoderma harzianum*. J Gen Microbiol 137:2163–2169, 1991.
67. M Lorito, CK Hayes, A Di Pietro, SL Woo, GE Harman. Purification, characterization, and synergistic activity of a glucan 1,3-β-glucosidase and an N-acetyl-β-glucosaminidase from Trichoderma harzianum. Phytopathology 84:398–405, 1994.
68. S Haran, H Schickler, A Oppenheim, I Chet. New components of the chitinolytic system of Trichoderma harzianum. Mycol Res 99:441–446, 1995.
69. S Haran, H Schickler, A Oppenheim, I Chet. Differential expression of *Trichoderma harzianum* chitinases during mycoparasitism. Phytopathology 86:980–985, 1996.
70. J De La Cruz, A Hidalgo-Gallego, JM Lora, T Benitez, JA Pintor-Toro, A Llobell. Isolation and characterization of three chitinases from *Trichoderma harzianum*. Eur J Biochem 206: 859–867, 1992.
71. EE Deane, JM Whipps, JM Lynch, JF Peberdy. The purification and characterization of a *Trichoderma harzianum* exochitinase. Biochim Biophys Acta Protein Structure Mol Enzymol 1383:101–110, 1998.
72. M Lorito, GE Harman, CK Hayes, RM Broadway, A Tronsmo, SL Woo, A Di Pietro. Chitinolytic enzymes produced by *Trichoderma harzianum*: antifungal activity of purified endochitinase and chitobiosidase. Phytopathology 83:302–307, 1993.
73. M Lorito, CK Peterbauer, CK Hayes, GE Harman. Synergistic interaction between fungal cell-wall degrading enzymes and different antifungal compounds on spore germination. Microbiology 140:623–629, 1994.
74. M Lorito, A Di Pietro, CK Hayes, SL Woo, GE Harman. Antifungal, synergistic interaction between chitinolytic enzymes from *Trichoderma harzianum* and Enterobacter cloacae. Phytopathology 83:721–728, 1993.
75. M Lorito, SL Woo, M D'Ambrosio, GE Harman, CK Hayes, CP Kubicek, F Scala. Synergistic interaction between cell wall degrading enzymes and membrane affecting compounds. Mol Plant Microbe Interact 9:206–213, 1996.
76. LHC Lima, CJ Ulhoa, AP Fernandes, CR Felix. Purification of a chitinase from *Trichoderma* sp. and its action on *Sclerotium rolfsii* and *Rhizoctonia solani* cell walls. J Gen Appl Microbiol 43:31–37, 1997.
77. C Fekete, T Weszely, L Hornok. Assignment of a PCR-amplified chitinase sequence cloned from Trichoderma hamatum to resolved chromosomes of potential biocontrol species of Trichoderma. FEMS Microbiol Lett 145:385–391, 1996.

78. A Di Pietro, M Lorito, CK Hayes, RM Broadway, GE Harman. Endochitinase from *Gliocladium virens*: Isolation, characterization, and synergistic antifungal activity in combination with gliotoxin. Phytopathology 83:308–313, 1993.
79. C Carsolio, A Gutierrez, B Jimenez, M van Montagu, A Herrera-Estrella. Characterization of ech-42, a *Trichoderma harzianum* endochitinase gene expressed during mycoparasitism. Proc Natl Acad Sci USA 91:10903–10907, 1994.
80. JM Baek, CR Howell, CM Kenerley. The role of an extracellular chitinase from *Trichoderma virens* Gv29-8 in the biocontrol of *Rhizoctonia solani*. Curr Genet 35:41–50, 1999.
81. RL Mach, CK Peterbauer, K Payer, S Jaksits, SL Woo, S Zeilinger, CM Kullnig, M Lorito, CP Kubicek. Expression of two major chitinase genes of *Trichoderma atroviride* (*T. harzianum* P1) is triggered by different regulatory signals. Appl Environ Microbiol 65:1858–1863, 1999.
82. SL Woo, B Donzelli, F Scala, RL Mach, GE Harman, CP Kubicek, G Del Sorbo, M Lorito. Disruption of ech42 (endochitinase-encoding) gene affects biocontrol activity in *Trichoderma harzianum* strain P1. Mol Plant Microbe Interact 12:419–429, 1999.
83. C Carsolio, N Benhamou, S Haran, C Cortes, A Gutierrez, I Chet, A Herrera-Estrella. Role of the *Trichoderma harzianum* endochitinase gene, ech42, in mycoparasitism. Appl Environ Microbiol 65:929–935, 1999.
84. I Garcia, JM Lora, J de la Cruz, T Benitez, A Lobell, JA Pintor-Torro. Cloning and characterization of a chitinase (chit42) cDNA from the mycoparasitic fungus *Trichoderma harzianum*. Curr Genet 27:83–89, 1994.
85. CK Hayes, S Klemsdal, M Lorito, A Di Pietro, C Peterbauer, JP Nakas, A Tronsmo, GE Harman. Isolation and sequence of an endochitinase gene from a cDNA library of *Trichoderma harzianum*. Gene 138:143–148, 1994.
86. G Giczey, Z Kerenyi, G Dallmann, L Hornok. Homologous transformation of *Trichoderma hamatum* with an endochitinase encoding gene, resulting in increased levels of chitinase activity. FEMS Microbiol Lett 165:247–252, 1998.
87. MC Limon, JM Lora, I Garcia, J de la Cruz, A Llobell, T Benitez, JA Pintor- Toro. Primary structure and expression pattern of the 33-kDa chitinase gene from the mycoparasitic fungus Trichoderma harzianum. Curr Genet 28:478–483, 1995.
88. CK Peterbauer, M Lorito, CK Hayes, GE Harman, CP Kubicek. Molecular cloning and expression of nag1, a gene encoding N-acetyl-β-glucosaminidase from *Trichoderma harzianum*. Curr Genet 30:325–331, 1996.
89. RJ Smith, EA Grula. Chitinase is an inducible enzyme in Beauveria bassiana. J Invert Pathol 42:319–326, 1983.
90. RJ St. Leger, RM Cooper, AK Charnley. Cuticle-degrading enzymes of entomopathogenic fungi: Cuticle degradation in vitro by enzymes from entomopathogens. J Invert Pathol 47: 167–177, 1986.
91. E Margolles-Clark, CK Hayes, GE Harman, M Penttila. Improved production of *Trichoderma harzianum* endochitinase by expression in Trichoderma reesei. Appl Environ Microbiol 62: 2145–2151, 1996.
92. J Inbar, I Chet. The role of recognition in the induction of specific chitinases during mycoparasitism by *Trichoderma harzianum*. Microbiology 141:2823–2829, 1995.
93. J Krishnamurthy, R Samiyappan, P Vidhyasekaran, S Nakkeeran, E Rajeswari, JAJ Raja, P Balasubramanian. Efficacy of *Trichoderma* chitinases against *Rhizoctonia solani*, the rice sheath blights pathogen. J Biosciences 24:207–213, 1999.
94. S Zeilinger, C Galhaup, K Payer, SL Woo, RL Mach, C Fekete, M Lorito, CP Kubicek. Chitinase gene expression during mycoparasitic interaction of *Trichoderma harzianum* with its host. Fungal Genet Biol 26:131–140, 1999.
95. JM Lora, J de la Cruz, T Benitez, A Llobell, JA Pintor-Toro. Molecular characterization and heterologous expression of an endo-β-1,6-glucanase gene from the mycoparasitic fungus *Trichoderma harzianum*. Mol Gen Genet 242:461–466, 1994.

96. C Cortes, A Gutierrez, V Olmedo, J Inbar, I Chet, A Herrera-Estrella. The expression of genes involved in parasitism by *Trichoderma harzianum* is triggered by a diffusible factor. Mol Gen Genet 260:218–225, 1998.
97. S Sreenivasaprasad, K Manibhushanrao. Antagonistic potential of *Gliocladium virens* and Trichoderma longibrachiatum to phytopathogenic fungi. Mycopathologia 109:19–26, 1990.
98. J Liu, JJ Morrell. Effect of biocontrol inoculum growth conditions on subsequent chitinase and protease levels in wood exposed to biocontrols and stain fungi. Material Organismen 31:265–279, 1997.
99. M Lorito, RL Mach, P Sposato, J Strauss, CK Peterbauer, CP Kubicek. Mycoparasitic interaction relieves binding of Cre1 C catabolite repressor protein to promoter sequence of ech-42 (endochitinase-encoding) gene of *Trichoderma harzianum.* Proc Natl Acad Sci USA 93: 14868–14872, 1996.
100. M Lorito, V Farkas, S Rebuffat, B Bodo, CP Kubicek. Cell wall synthesis is a major target of mycoparasitic antagonism by *Trichoderma harzianum.* J Bacteriol 178:6382–6385, 1996.
101. H Schickler, BC Danin Gehali, S Haran, I Chet. Electrophoretic characterization of chitinases as a tool for the identification of *Trichoderma harzianum* strains. Mycol Res 102:373–377, 1998.
102. G Turoczi, C Fekete, Z Kerenyi, R Nagy, A Pomazi, L Hornok. Biological and molecular characterisation of potential biocontrol strains of *Trichoderma.* J Basic Microbiol 36:63–72, 1996.
103. RCW Berkeley, GW Gooday, DC Ellwood, eds. Microbial Polysaccharides and Polysaccharases. London: Academic Press, 1979.
104. SM Pitson, RJ Seviour, BM McDougall. Noncellulolytic fungal β-glucanases: Their physiology and regulation. Enzyme Microb Technol 15:178–192, 1993.
105. T Kamada. Stipe elongation in fruit bodies. In: JCH Wessels, F Meinhardt, eds. Mycota. Vol.1. Berlin, Heidelberg: Springer-Verlag, 1994, pp 367–376.
106. CJ Ulhoa, JF Peberdy. Purification and some properties of the extracellular chitinase produced by *Trichoderma harzianum.* Enzyme Microb Technol 14:236–240, 1992.
107. J de la Cruz, M Rey, JM Lora, A Hidalgo-Gallego, F Dominguez, JA Pintor-Toro, A Llobell, T Benitez. C source control on β-glucanases, chitobiase and chitinase from *Trichoderma harzianum.* Arch Microbiol 159:316–322, 1993.
108. A Bruce, U Srinivasan, HJ Staines, TL Highley. Chitinase and laminarinase production in liquid culture by *Trichoderma* spp. and their role in biocontrol of wood decay fungi. Int Biodeterior Biodegrad 35:337–353, 1995.
109. J de la Cruz, JA Pintor-Toro, T Benitez, A Llobell, and LC Romero. A novel endo-β-1,3-glucanase, BGN13.1, involved in the mycoparasitrism of *Trichoderma harzianum.* J Bacteriol 177:6937–6945, 1995.
110. EF Noronha, CJ Ulhoa. Purification and characterization of an endo-β-1,3-glucanase from *Trichoderma harzianum.* Can J Microbiol 42:1039–1044, 1996.
111. S Vazquez-Garciduenas, CA Leal-Morales, A Herrera-Estrella. Analysis of the β-1,2-glucanolytic system of the biocontrol agent *Trichoderma harzianum.* Appl Environ Microbiol 64: 1442–1446, 1998.
112. R Cohen-Kupiec, KE Broglie, D Friesem, RM Broglie, I Chet. Molecular characterization of a novel β-1,3-exoglucanase related to mycoparasitism of *Trichoderma harzianum.* Gene 226:147–154, 1999.
113. C Thrane, A Tronsmo, DF Jensen. Endo-1,3-β-glucanase and cellulase from *Trichoderma harzianum*: Purification and partial characterization, induction of and biological activity against plant pathogenic *Pythium* spp. Eur J Plant Pathol 103:331–344, 1997.
114. O Ramot, R Cohen-Kupiec, I Chet. Regulation of β-1,3-glucanase by carbon starvation in the mycoparasite Trichoderma harzianum. Mycol Res 104:412–420, 2000.

115. M Göhl, R Srinivas, W Dammertz, MR Udupa, T Panda. Localization of β-1,3-glucanase in Trichoderma harzianum. Bioprocess Engin 19:237–241, 1998.
116. N Benhamou, I Chet. Cellular and molecular mechanisms involved in the interaction between *Trichoderma harzianum* and *Pythium ultimum*. Appl Environ Microbiol 63:2095–2099, 1997.
117. J de la Cruz, JA Pintor-Toro, T Benitez, A Llobell. Purification and characterization of an endo-β-1,6-glucanase from *Trichoderma harzianum* that is related to its mycoparasitism. J Bacteriol 177:1864–1871, 1995.
118. A Soler, J de la Cruz, A Llobell. Detection of β-1,6-glucanase isozymes from *Trichoderma* strains in sodium dodecyl sulphate polyacrylamide gel electrophoresis and isoelectrofocusing gels. J Microbiol Methods 35:245–251, 1999.
119. FK de Soglio, BL Bertagnolli, JB Sinclair, GY Yu, DM Eastburn. Production of chitinolytic enzymes and endoglucanase in the soybean rhizosphere in the presence of *Trichoderma harzianum* and *Rhizoctonia solani*. Biol Control 12:111–117, 1998.
120. B Seiboth, S Hakola, RI Mach, PL Suominen, CP Kubicek. Role of four major cellulases in triggering of cellulase gene expression by cellulose in *Trichoderam reesei*. J Bacteriol 179:5318–5320, 1997.
121. D Haab, K Hagspiel, K Szakmary, CP Kubicek. Formation of the extracellular protease from *Trichoderma reesei* QM 9414 involved in cellulose degradation. J Biotechnol 16:187–198, 1990.
122. A Kapat, SK Rakshit, T Panda. Optimization of carbon and nitrogen sources in the medium and environmental factors for enhanced production of chitinase by *Trichoderma harzianum*. Bioprocess Eng 15:13–20, 1996.
123. R Rodriguez-Kabana, WD Kelley, EA Curl. Proteolytic activity of *Trichoderma viride* in mixed culture with Sclerotium rolfsii in soil. Can J Microbiol 24:478–490, 1978.
124. GH Goldman, C Hayes, GE Harman. Molecular and cellular biology of biocontrol by *Trichoderma* spp. Trends Biotech 12:478–482, 1994.
125. MHS Goldman, GH Goldman. 1998. *Trichoderma harzianum* transformant has high extracellular alkaline proteinase expression during specific mycoparasitic interactions. Gen Mol Biol 21:329–333, 1998.
126. Y Elad, A Kapat. The role of Trichoderma harzianum protease in the biocontrol of *Botrytis cinerea*. Eur J Plant Pathol 105:177–189, 1999.
127. S Mischke. Evaluation of chromogenic substrates for measurement of protease production by biocontrol strains of *Trichoderma*. Microbios 87:175–183, 1996.
128. F Santamaria, OM Nuero, C Alfonso, A Prieto, JA Leal, F Reyes. Biochemical studies on the cell wall degradation of *Fusarium oxysporum* f.sp. *lycopersici* race 2 by lytic enzymes from *Mucorales* for its biocontrol. Lett Appl Microbiol 18:152–155, 1994.
129. F Santamaria, OM Nuero, C Alfonso, A Prieto, JA Leal, F Reyes. Cell wall degradation of *Fusarium oxysporum* f.sp. lycopersici race 2 by lytic enzymes from different *Fusarium* species for its biocontrol. Lett Appl Microbiol 20:385–390, 1995.
130. N Mathivanan, V Kabilan, K Murugesan. Purification, characterization, and antifungal activity of chitinase from *Fusarium chlamydosporum*, a mycoparasite to groundnut rust, *Puccinia arachidis*. Can J Microbiol 44:646–651, 1998.
131. FJ Ruiz-Duenas, MJ Martinez. Enzymatic activities of *Trametes versicolor* and *Pleurotus eryngii* implicated in biocontrol of *Fusarium oxysporum* f.sp. *lycopersici*. Curr Microbiol 32:151–155, 1996.
132. I Larena, P Melgarejo. Biological control of *Monilinia laxa* and *Fusarium oxysporum* f.sp. *lycopersici* by a lytic enzyme-producing *Penicillium purpurogenum*. Biol Control 6:361–367, 1996.
133. L Madi, T Katan, J Katan, Y Henis. Biological control of *Sclerotium rolfsii* and *Verticillium dahliae* by talar *Talaromyces flavus* is mediated by different mechanisms. Phytopathology 87:1054–1060, 1997.

134. C Dunne, I Delany, A Fenton, F Ogara. Mechanisms involved in biocontrol by microbial inoculants. Agronomie 16:721–729, 1996.
135. SC Chiu, SS Tzean. Glucanolytic enzyme production by *Schizophyllum commune* Fr. during mycoparasitism. Physiol Mol Plant Pathol 46:83–94, 1995.
136. EJ Urquhart, JS Menzies, ZK Punja. Growth and biological control activity of *Tilletiopsis* species against powdery mildew (*Sphaerotheca fuliginea*) on greenhouse cucumber. Phytopathology 84:341–351, 1994.
137. A Sztejnberg, S Galper, S Mazar, N Lisker. *Ampelomyces quisqualis* for biological and integrated control of powdery mildews in Israel. J Phytopathol 124:285–295, 1989.
138. R Hofstein, A Chapple. Commercial development of biofungicides. In: FR Hall, JJ Menn, eds. Biopesticides: Use and Delivery. Totowa, NJ: Humana Press, 1999, pp 77–102.
139. Y Hashioka, Y Nakai. Ultrastructure of picnidial development and mycoparasitism of *Ampelomyces quisqualis* parasitic on *Erysiphales*. Trans Mycol Soc Jpn 21:329–338, 1980.
140. WD Philipp. Extracellular enzymes and nutritional physiology of *Ampelomyces quisqualis* Ces., hyperparasite of powdery mildew, in vitro. Phytopathol Z 114:274–283, 1985.
141. Y Rotem, O Yarden, A Szteinberg. The mycoparasite *Ampelomyces quisqualis* expresses exgA encoding an exo-β-1,3-glucanase in culture and during mycoparasitism. Phytopathology 89:631–638, 1999.
142. N Benhamou, P Rey, K Picard, Y Tirilly. Ultrastructural and cytochemical aspects of the interaction between the mycoparasite *Pythium oligandrum* and soilborne plant pathogens. Phytopathology 89:506–517, 1999.
143. M Benyagoub, N Benhamou, O Carisse. Cytochemical investigation of the antagonistic interaction between a *Microsphaeropsis* sp. (isolate P130A) and *Venturia inaequalis*. Phytopathology 88:605–613, 1998.
144. BL Bertagnolli, FK Dal Soglio, JB Sinclair. Extracellular enzyme profiles of the fungal pathogen Rhizoctonia solani isolate 2B-12 and of two antagonists, *Bacillus megaterium* strain B153-2-2 and *Trichoderma harzianum* isolate Th008. I. Possible correlations with inhibition of growth and biocontrol. Physiol Mol Plant Pathol 48:145–160, 1996.
145. JA Lewis, RD Lumsden. Do pathogenic fungi have the potential to inhibit biocontrol fungi? J Phytopathology 143:585–588, 1995.
146. A Kapat, G Zimand, Y Elad. Effect of two isolates of *Trichoderma harzianum* on the activity of hydrolytic enzymes produced by *Botrytis cinerea*. Physiol Mol Plant Pathol 52:127–137, 1998.
147. NR Madamanchi, J Kuc. Induced systemic resistance in plants. In: GT Cole, HC Hoch, eds. The Fungal Spore and Disease Initiation in Plants and Animals. New York: Plenum Press, 1991, pp 347–362.
148. E Kombrink, IE Somssich. Defence response of plants to pathogens. Adv Bot Res 21:1–34.
149. LC Van Loon, PAHM Bakker, CMJ Pieterse. Systemic resistance induced by rhizosphere bacteria. Annu Rev Phytopathol 36:453–483, 1998.
150. RF Dalisay, JA Kuc. Persistence of induced resistance and enhanced peroxidase and chitinase activities in cucumber plants. Physiol Mol Plant Pathol 47:315–327, 1995.
151. G De Meyer, J Bigirimana, Y Elad, M Hofte. Induced systemic resistance in *Trichoderma harzianum* T39 biocontrol of *Botrytis cinerea*. Eur J Plant Pathol 104:279–286, 1998.
152. I Yedidia, N Benhamou, I Chet. Induction of defense responses in cucumber plants (Cucumis sativis L.) by the biocontrol agent *Trichoderma harzianum*. Appl Environ Microbiol 65: 1061–1070, 1999.
153. BA Bailey, A Avni, JD Anderson. The influence of ethylene and tissue age on the sensitivity of *Xanthi* tobacco leaves to a *Trichoderma viride* xylanase. Plant Cell Physiol 36:1669–1676, 1995.
154. J Inbar, I Chet. Evidence that chitinase produced by *Aeromonas caviae* is involved in the biological control of soil-borne plant pathogens by this bacterium. Soil Biol Biochem 23: 973–978, 1991.

155. SK Park, HY Lee, KC Kim. Role of chitinase produced by *Chromobacterium violaceum* in suppression of *Rhizoctonia* damping-off. Korean J Plant Pathol 11:304–311, 1995.
156. LS Chernin, MK Winson, JM Thompson, S Haran, BW Bycroft, I Chet, P Williams, GSAB Stewart. Chitinolytic activity in *Chromobacterium violaceum*: substrate analysis and regulation by quorum sensing. J Bacteriol 180:4435–4441, 1998.
157. LS Chernin, ZF Ismailov, S Haran, I Chet. Chitinolytic *Enterobacter agglomerans* antagonistic to fungal plant pathogens. Appl Environ Microbiol 61:1720–1726, 1995.
158. PP Singh, YC Shin, CS Park, YR Chung. Biological control of *Fusarium* wilt of cucumber by chitinolytic bacteria. Phytopathology 89:92–99, 1999.
159. TD Gaffney, ST Lam, J Ligon, K Gates, A Frazelle, J Di Maio, S Hill, S Goodwin, N Torkewitz, AM Allshouse, HJ Kempt, JO Becker. Global regulation of expression of antifungal factors by a *Pseudomonas fluorescens* biological-control strain. Mol Plant Microbe Interact 7:455–463, 1994.
160. MN Nielsen, J Sorensen, J Fels, HC Pedersen. Secondary metabolite- and endochitinase-dependent anatagonism toward plant-pathogenic microfungi of *Pseudomonas fluorescens* isolates from sugar beet rhizosphere. Appl Environ Microbiol 64:3563–3569, 1998.
161. H Lim, Y Kim, S Kim. *Pseudomonas stutzeri* YPL-1 genetic transformation and antifungal mechanism against *Fusarium solani*, an agent of plant root rot. Appl Environ Microbiol 57: 510–516, 1991.
162. JDG Jones, KL Grady, TV Suslow, JR Bedbrook. Isolation and characterization of genes encoding two chitinase enzymes from *Serratia marcescens*. EMBO J 5:467–473, 1986.
163. A Ordentlich, Y Elad, I Chet. The role of chitinase of *Serratia marcescens* in biocontrol of *Sclerotium rolfsii*. Phytopathology 78:84–88, 1988.
164. C Kalbe, P Marten, G Berg. Strains of the genus *Serratia* as beneficial rhizobacteria of oilseed rape with antifungal properties. Microbiol Res 151:433–439, 1996.
165. J Frankowski, G Berg, H Bahl. Mechanisms involved in the antifungal activity of the rhizobacterium *Serratia plymuthica*. In: B Duffy, U Rosenberger, G Defago, eds. Biological Control of Fungal and Bacterial Plant Pathogens: Molecular Approaches in Biological Control. IOBC Bull 21:45–50, 1998.
166. J Monreal, ET Rees. The chitinase of *Serratia marcescens*. Can J Microbiol 15:689–696, 1969.
167. RL Roberts, E Calib. *Serratia marcescens* chitinase: one step purification and use for determination of chitin. Anal Biochem 127:402–412, 1982.
168. CE Vorgias, I Tews, A Perrakis, KS Wilson, AB Oppenheim. Purification and characterization of the recombinant chitin degrading enzymes, chitinase A and chitobiase from *Serratia marcescens*. In: RAA Muzzarelli, ed. Chitin Enzymology. Ancona: European Chitin Society, 1993, pp 417–422.
169. MB Brurberg, IF Nes, VGH Eijsink. Comparative studies of chitinases A and B from *Serratia marcescens*. Microbiology 142:1581–1589, 1996.
170. SW Gal, JY Choi, CY Kim, YH Cheong, YJ Choi, JD Bahk, SY Lee, MJ Cho. Isolation and characterization of the 54-kDa and 22-kDa chitinase genes of *Serratia marcescens* KCTC2172. FEMS Microbiol Lett 151:197–204, 1997.
171. RL Fuchs, SA McPharson, DJ Drabos. Cloning of *Serratia marcescens* gene encoding chitinase. Appl Environ Microbiol 51:504–509, 1986.
172. H Kless, Y Sitrit, I Chet, A Oppenheim. Cloning of the gene coding for chitobiase of *Serratia marcescens*. Mol Gen Genet 217:471–473, 1989.
173. R Shapira, A Ordentlich, I Chet, AB Oppenheim. Control of plant diseases by chitinase expressed from cloned DNA in *Escherichia coli*. Phytopathology 79:1246–1249, 1989.
174. Y Sitrit, CE Vorgias, I Chet, AB Oppenheim. Cloning and primary structure of the chiA gene from *Aeromonas caviae*. J Bacteriol 177:4187–4189, 1995.
175. LS Chernin, IA Khmel, ZF Ismailov, NB Lemanova, TA Sorokina, ID Avdienko, MI Ovadis, OA Grischenko, VN Pozdnyakov, MA Terentyev, VA Lipasova, LK Miroshnikova, MV Kovalchuk, IN Stekhin. Soilborne bacteria with wide host range of antagonistic activity

against phytopathobacteria and fungi promising for biological control of plant diseases. In: KG Skryabin, ed. Plant Biotechnology and Molecular Biology. Moscow: Pushchino Research Center, 1993, pp 76–86.

176. LS Chernin, L De La Fuenta, V Sobolev, S Haran, CE Vorgias, AB Oppenheim, I Chet. Molecular cloning, structural analysis, and expression in *Escherichia coli* of a chitinase gene from *Enterobacter agglomerans*. Appl Environ Microbiol 63:834–839, 1997.
177. LS Chernin, A Brandis, ZF Ismailov, I Chet. Pyrrolnitrin production by an *Enterobacter agglomerans* strain with a broad spectrum of antagonistic activity towards fungal and bacterial phytopathogens. Curr Microbiol 32:201–212, 1996.
178. Y Homma, Z Sato, F Hirayama, K Konno, H Shirahama, T Suzuki. Production of antibiotics by *Pseudomonas cepacia* as an agent for biological control of soilborne plant pathogens. Soil Biol Biochem 21:723–728, 1989.
179. KH van Pee, JM Ligon. Biosynthesis of pyrrolnitrin and other phenylpyrrole derivatives by bacteria. Nat Prod Rep 17:157–164, 2000.
180. M Nose, K Arima. On the mode of action of a new antifungal antibiotic, pyrrolnitrin (*Pseudomonas pyrrocinia*). J Antibiot 22:135–143, 1969.
181. ABK Jespers, LC Davidse, MA de Waard. Interference of the phenylpyrrole fungicide fenpriclonil with membranes and membrane function. Pestic Sci 40:133–140, 1994.
182. C Keel, G Defago. Interactions between beneficial soil bacteria and root pathogens: Mechanisms and ecological impact. In: AC Gande, VK Brown, eds. Multitrophic Interactions in Terrestrial Systems. Oxford: Blackwell Science, 1997, pp 27–47.
183. M Fridlender, J Inbar, I Chet. Biological control of soilborne plant pathogens by a β-1,3 glucanase-producing *Pseudomonas cepacia*. Soil Biol Biochem 25:1211–1221, 1993.
184. R Velazhahan, R Samiyappan, P Vidhyasekaran. Relationship between antagonistic activities of *Pseudomonas fluorescens* isolates against *Rhizoctonia solani* and their production of lytic enzymes. J Plant Dis Protect 106:244–250, 1999.
185. E Frandberg, J Schnurer. Antifungal activity of chitinolytic bacteria isolated from airtight stored cereal grain. Can J Microbiol 44:121–127, 1998.
186. NB Andreeva, TA Sorokina, IA Khmel. Chitinolytic activity of pigmented *Pseudomonas* and *Xanthomonas* bacteria. Microbios 87:53–57, 1996.
187. DY Kobayashi, M Guglielmoni, BB Clarke. Isolation of the chitinolytic bacteria *Xanthomonas maltophilia* and *Serratia marcescens* as biological-control agents for summer patch disease of turfgrass. Soil Biol Biochem 27:1479–1487, 1995.
188. W De Boer, PJAK Gunnewiek, P Lafeber, JD Janse, BE Spit, JW Woldendorp. Anti-fungal properties of chitinolytic dune soil bacteria. Soil Biol Biochem 30:193–203, 1998.
189. K Sivasithamparam. Root cortex—the final frontier for the biocontrol of root-rot with fungal antagonists: a case study on a sterile red fungus. Annu Rev Phytopathol 36:439–452, 1998.
190. CP Chanway. Bacterial endophytes: Ecological and practical implications. Sydowia 50:149–170, 1998.
191. J Hallmann, A Quadt-Hallmann, R Rodriguez-Kabana, JW Kloepper. Interactions between *Meloidogyne incognita* and endophytic bacteria in cotton and cucumber. Soil Biol Biochem 30:925–937, 1998.
192. CR Bell, GA Dickie, JWYF Chan. Variable response of bacteria isolated from grapevine xylem to control grape crown gall disease in plants. Am J Enol Viticult 46:499–508, 1995.
193. C Chen, EM Bauske, G Musson, R Rodriguez-Kabana, JW Kloepper. Biological control of Fusarium wilt of cotton by use of endophytic bacteria. Biol Control 5:83–91, 1995.
194. S Pleban, L Chernin, I Chet. Chitinolytic activity of an endophytic strain of *Bacillus cereus*. Lett Appl Microbiol 25:284–288, 1997.
195. A Quadt-Hallmann, N Benhamou, JW Kloepper. Bacterial endophytes in cotton: Mechanisms of entering the plant. Can J Microbiol 43:577–582, 1997.
196. J Laville, C Voisard, C Keel, M Maurhofer, G Defago, and D Haas. Global control in *Pseu-*

domonas fluorescens mediating antibiotic synthesis and suppression of black root rot of tobacco. Proc Natl Acad Sci USA 89:1562–1566, 1992.

197. JJ Rich, TG Kinscherf, T Kitten, DK Willis. Genetic evidence that the gacA gene encodes the cognate response regulator for the lemA sensor in *Pseudomonas syringae*. J Bacteriol 176:7468–7475, 1994.
198. N Corbell, JE Loper. A global regulator of secondary metabolite production in *Pseudomonas fluorescens* Pf-5. J Bacteriol 177:6230–6236, 1995.
199. T Kitten, TG Kinscherf, JL McEvoy, DK Willis. A newly identified regulator is required for virulence and toxin production in *Pseudomonas syringae*. Mol Microbiol 28:917–929, 1998.
200. A Sarnaguet, J Kraus, MD Henkels, AM Muehichen, JE Loper. The sigma factor σ^s affects antibiotic production and biological control activity of *Pseudomonas fluorescens*. Proc Natl Acad Sci USA 92:12255–12259, 1995.
201. CA Whistler, NA Corbell, A Sarniguet, W Ream, JE Loper. The two-component regulators GacA and GacS influence accumulation of the stationary-phase sigma factor σ^s and the stress response in *Pseudomonas fluorescens* Pf-5. J Bacteriol 180:6635–6641, 1998.
202. U Schnider, C Keel, C Blumer, J Troxler, G Defago, D Haas. Amplification of the houskeeping sigma factor in *Pseudomonas fluorescens* CHAO enhances antibiotic production and improves biocontrol abilities. J Bacteriol 177:5387–5392, 1995.
203. PC Loewen, R Hengge-Aronis. The role of the sigma factor σ^s (KatF) in bacterial global regulation. Annu Rev Microbiol 48:53–80, 1994.
204. P Greenberg. Quorum sensing in Gram-negative bacteria: acylhomoserine lactone signalling and cell-cell communication. In: R England, G Hobbs, N Bainton, DM Roberts, eds. Microbial Signalling and Communication. Cambridge: Cambridge University Press, 1999, pp 71–84.
205. S Swift, JA Downie, NA Whitehead, AML Barnard, GPC Salmond, P Williams. Quorum sensing as a population-density-dependent determinant of bacterial physiology. Adv Microb Physiology 45:199–270, 2001.
206. H Withers, S Swift, P Williams. Quorum sensing as an integral component of gene regulatory networks in Gram-negative bacteria. Curr Opin Microbiol 4:186–193, 2001.
207. LS Pierson III, DW Wood, EA Pierson. Homoserine lactone-mediated gene regulation in plant-associated bacteria. Annu Rev Phytopathol 36:207–225, 1998.
208. C Reimmann, M Beyeler, A Latifi, H Winteler, M Foglino, A Lazdunski, D Haas. The global activator GacA of *Pseudomonas aeruginosa* PAO positively controls the production of the autoinducer *N*-butyryl-homoserine lactone and the formation of the virulence factors pyocyanin, cyanide, and lipase. Mol Microbiol 24:309–319, 1997.
209. C Cha, P Gao, Y-C Chen, PD Shaw, SK Farrand. Production of acyl-homoserine lactone quorum-sensing signals by Gram-negative plant-associated bacteria. Mol Plant Microbe Interact 11:1119–1129, 1998.
210. A Hussain, V Vancura. Formation of biologically active substances by rhizosphere bacteria and their effect on plant growth. Folia Microbiol 15:468–478, 1970.
211. M Schirmböck, M Lorito, YL Wang, CK Hayes, I Arisan-Atac, F Scala, GE Harman, CP Kubicek. Parallel formation and synergism of hydrolytic enzymes and peptaibol antibiotics, molecular mechanisms involved in the antagonistic action of *Trichoderma harzianum* against phytopathogenic fungi. Appl Environ Microbiol 60:4364–4370, 1994.
212. LC Dekkers, CC Phoelich, L van der Fits, BJ Lugtenberg. A site-specific recombinase is required for competitive root colonization by *Pseudomonas fluorescens* WCS365. Proc Natl Acad Sci USA 95:7051–7056, 1998.
213. B Mahadevan, DL Crawford. Properties of the chitinase of the antifungal biocontrol agent *Streptomyces lydicus* WYEC108. Enzyme Microb Technol 20:489–493, 1997.
214. SR TrejoEstrada, A Paszczynski, DL Crawford. Antibiotics and enzymes produced by the biocontrol agent *Streptomyces violaceusniger* YCED-9. J Ind Microbiol Biotechnol 21:81–90, 1998.

215. H Toyoda, S Ikeda, Y Matsuda, S Ouchi. A molecular strategy for biological control of powdery mildew pathogens by chitinase gene chiSH1 cloned from Gram-positive bacterium Kurthia zopfii. 2nd International Symposium on Chitin Enzymology, Senigallia, AN, Italy, 1996, pp 110–111.
216. GS Raupach, JW Kloepper. Mixtures of plant growth-promoting rhizobacteria enhance biological control of multiple cucumber pathogens. Phytopathology 88:1158–1164, 1998.
217. KC Sung, YR Chung. Enhanced suppression of rice sheath blight using combination of bacteria which produce chitinases or antibiotics. In: A Ogoshi, K Kobayashi, Y Homma, F Kodama, N Kondo, S Akino, eds. Plant Growth-Promoting Rhizobacteria—Present Status and Future Prospects. Proc. 4th Int. Workshop on Plant Growth-Promotin Rhizobacteria, Sapporo, Japan, Nakanishi Printing, 1997, pp 370–372.
218. CY Dunne, Y Moenne-Loccoz, J McCarthy, P Higgins, J Powell, DN Dowling, F O'Gara. Combining proteolytic and phloroglucinol-producing bacteria for improved biocontrol of *Pythium*-mediated damping-off of sugar beet. Plant Pathol 47:299–307, 1998.
219. E Chalutz, S Droby. Biological control of post-harvest disease. In: GJ Boland, LD Kuykendall, eds. Plant-Microbe Interactions and Biological Control. New York: Marcel Dekker, 1998, pp 157–170.
220. WJ Janisiewicz. Biocontrol of postharvest diseases of temperate fruits: Challenges and opportunities. In: GJ Boland, LD Kuykendall, eds. Plant-Microbe Interactions and Biological Control. New York: Marcel Dekker, 1998, pp 171–198.
221. M Wisniewski, C Biles, S Droby, R McLaughlin, C Wilson, E Chalutz. Mode of action of the postharvest biocontrol yeast, *Pichia guilliermondii*. I. Characterization of attachment to *Botrytis cinerea*. Physiol Mol Plant Pathol 39:245–258, 1991.
222. MH Jijakli, P Lepoivre. Characterization of an exo-β-1,3-glucanase produced by *Pichia anomala* strain K, antagonist of *Botrytis cinerea* on apples. Phytopathology 88:335–343, 1998.
223. R Castoria, F De Curtis, G Lima, V De Cicco. β-1,3-glucanase activity of two saprophytic yeasts and possible mode of action as biocontrol agents against postharvest diseases. Postharvest Biol Technol 12:293–300, 1997.
224. R Porat, V Vinocur, B Weiss, L Cohen, A Daus, S Droby. Effects of various elicitors on resistance of grapefruit against *Penicillium digitatum*. Abstracts of XIV International Plant Protection Congress (IPPC), Jerusalem, 1999, p 80.
225. KJ Kramer, S Muthukrishnan. Insect chitinases: Molecular biology and potential use as biopesticides. Insect Biochem Mol Biol 27:887–900, 1997.
226. DW Roberts, AE Hajek. Entomopathogenic fungi as bioinsecticides. In: GF Leatham, ed. Frontiers in Industrial Mycology. New York: Chapman & Hall, 1992, pp 144–159.
227. SP Wraight, RI Carruthers. Production, delivery, and use of mycopesticides for control of insect pests on field crops. In: FR Hall, JJ Menn, eds. Biopesticides: Use and Delivery. Totowa, NJ: Humana Press, 1999, pp 233–269.
228. T Yanagita. The formaldehyde resistance of *Aspergillus* fungi attacking silkworm larvae. 4. The relationship between pathogenicity to silkworm larvae and chitinase activity of *Aspergillus flavus* and *A. oryzae*. J Seric Sci Jpn 49:440–445, 1980.
229. RJ St Leger, RC Staples, DW Roberts. Entomopathogenic isolates of *Metarhizium anisopliae*, *Beauveria bassiana*, *Aspergillus flavus* produce multiple extracellular chitinase isozymes. J Invert Pathol 61:81–84, 1993.
230. MJ Bidochka, KI Tong, GG Khachatourians. Partial purification and characterization of two extracellular *N*-acetyl-D-glucosaminidases produced by the entomopathogenic fungus *Beauveria bassiana*. Can J Microbiol 39:41–45, 1993.
231. I Havukkala, C Mitamura, S Hara, K Hirayae, Y Nishizawa, T Hibi. Induction and purification of *Beauveria bassiana* chitinolytic enzymes. J Invertebr Pathol 61:97–102, 1993.
232. RJ St Leger, M Goettel, DW Roberts, RC Staples. Prepenetration events during infection of host cuticle by Metarhizium anisopliae. J Invertebr Pathol 58:168–179, 1991.
233. SC Kang, S Park, DG Lee. Purification and characterization of a novel chitinase from the

entomopathogenic fungus, *Metarhizium anisopliae*. J Invertebr Pathol 73:276–281, 1999.
234. GN El-Sayed, TA Coudron, CM Ignoffo. Chitinolytic activity and virulence associated with native and mutant isolates of an entomopathogenic fungus *Nomuraea rileyi*. J Invertebr Pathol 54:394–403, 1989.
235. GN El-Sayed, CM Ignoffo, TD Leathers, SC Gupta. Insect cuticule and yeast extract effects on germination, growth and production of hydrolytic enzymes by *Nomuraea rileyi*. Mycopathologia 122:143–147, 1993.
236. S Harney, P Widden. Physiological properties of the entomopathogenic hyphomycete *Paecilomyces farinosus* in relation to its role in the forest ecosystem. Can J Bot 69:1–5, 1991.
237. CW Jackson, JB Heale, RA Hall. Traits associated with virulence to the aphid *Macrosiuphoniella sanborni* in eighteen isolates of *Verticillium lecanii*. Ann Appl Biol 106:39–48, 1985.
238. MJ Urbanczyk, A Zabza, S Balazy, W Peczynska-Czoch. Laboratory culture media and enzyme activity of some entomopathogenic fungi of Zoophthora (Zygomycetes:Entomophthoraceae). J Invertebr Pathol 59:250–257, 1992.
239. BP Gabriel. Enzymatic activities of some entomophthorous fungi. J Invertebr Pathol 11:70–81, 1968.
240. JH Willis. Cuticular protein: The neglected component. Arch Insect Biochem Physiol 6:203–215, 1987.
241. RJ St Leger, PK Durrands, AK Charnley, RM Cooper. Role of extracellular chymoelastase in the virulence of *Metarhizium anisopliae for Manduca sexta*. J Invertebr Pathol 52:285–293, 1988.
242. AK Charnley, RJ St Leger. The role of cuticle-degrading enzymes in fungal pathogenesis in insects. In: GT Cole, MC Hoch, eds. The Fungal Spore and Disease Initiation in Plants and Animals. New York: Plenum Press, 1991, pp 267–289.
243. MJ Bidochka, GG Khachatourians. Identification of *Beauveria bassiana* extracellular protease as a virulence factor in pathogenicity toward the migratory grasshopper *Melanoplus sanguinipies*. J Invertebr Pathol 56:362–370, 1990.
244. MJ Bidochka, RJ St Leger, A Stuart, K Gowanlock. Nuclear rDNA phylogeny in the fungal genus *Verticillium* and its relationship to insect and plant virulence, extracellular proteases and carbohydrases. Microbiology 145:955–963, 1999.
245. GN El-Sayed, CM Ignoffo, TD Leathers, SC Gupta. Effects of cuticule source and concentration on expression of hydrolytic enzymes by an entomopathogenic fungus *Nomuraea rileyi*. Mycopathologia 122:149–152, 1993.
246. TA Coudron, MJ Kroha, CM Ignoffo. Levels of chitinolytic activity during development of three entomopathogenic fungi. Comp Biochem Physiol B 79:339–348, 1984.
247. GG Khachatourians. Physiology and genetics of entomopathogenic fungi. In: DK Arora, KG Mukerji, E Drouhet, eds. Handbook of Applied Mycology. Vol. 2. New York: Marcel Dekker, 1991, pp. 613–663.
248. C Bajan, S Kalalova, K Kimitowa, A Samsinakova, M Wojciechowska. The relationship between infectious activities of entomophagous fungi and their production of enzymes. Bull Acad Pol Sci Cl 2 Ser Sci Biol 27:963–968, 1979.
249. RI Samuels, IC Paterson. Cuticle degrading proteases from insect moulting fluid and culture filtrates of entomopathogenic fungi. Comp Biochem Physiol 110B:661–669, 1995.
250. MJ Bidochka, RJ St Leger, DW Roberts. Mechanisms of deuteromycete fungal infections in grasshoppers and locusts: An overview. Memoirs Entomol Soc Can 171:213–224, 1997.
251. A Vilcinskas, M Wedde. Inhibition of *Beauveria bassiana* proteases and fungal development by inducible protease inhibitors in the haemolymph of *Galleria mellonella* larvae. Biocontrol Sci Technol 7:591–601, 1997.
252. J Griesch, A Vilcinskas. Proteases released by entomopathogenic fungi impair phagocytic

activity, attachment and spreading of plasmatocytes isolated from haemolymph of the greater wax moth *Galleria mellonella*. Biocontrol Sci Technol 8:517–531, 1998.
253. ES Mendonsa, PH Vartak, JU Rao, MV Deshpande. An enzyme from *Myrothecium verrucaria* that degrades insect cuticles for biocontrol of *Aedes aegypti* mosquito. Biotech Lett 18: 373–376, 1996.
254. CW McCoy. Pathogens of eriophyoid mites. In EE Lindquist, SW Sabelis, J Bruin, eds. Eriophyoid Mites, Their Biology, Natural Enemies and Control. Amsterdam: Elsevier, 1996, pp 481–490.
255. L Chernin, A Gafni, R Moses-Koch, U Gerson, A Szteinberg. Chitinolytic activity of the acaropathogenic fungi *Hirsutella thompsonii* and *Hirsutella necatrix*. Can J Microbiol 43: 440–446, 1997.
256. MR Bogo, CA Rota, H Pinto, M Ocampos, CT Correa, MH Vainstein, A Schrank. A chitinase encoding gene (chit1 gene) from the entomopathogen *Metarhizium anisopliae*: Isolation and characterization of genomic and full-length cDNA. Curr Microbiol 37:221–225, 1998.
257. JA Baum, TB Johnson, BC Carlton. Bacillus thuringiensis: Natural and recombinant biopesticide products. In: FR Hall, JJ Menn, eds. Biopesticides: Use and Delivery. Totowa, NJ: Humana Press, 1999, pp 189–209.
258. MN Sampson, GW Gooday. Involvement of chitinases of *Bacillus thuringiensis* during pathogenesis in insects. Microbiology 144:2189–2194, 1998.
259. A Regev, M Keller, N Strizhov, B Sneh, E Prudovsky, I Chet, I Ginzberg, Z KonczKalman, C Knocz, J Schell, A Zilberstein. Synergistic activity of a *Bacillus thuringiensis* delta-endotoxin and a bacterial endochitinase against *Spodoptera littoralis* larvae. Appl Environ Microbiol 62:3581–3586, 1996.
260. E Esposito, M da Silva, C Wiwat, M Lertcanawanichakul, P Siwayapram, S Pantuwatana, A Bhumiratana. Expression of chitinase-encoding genes from *Aeromonas hydrophila* and *Pseudomonas maltophilia* in *Bacillus thuringiensis* subsp. *israelensis*. Gene 179:119–126, 1996.
261. S Tantimavanich, S Pantuwatana, A Bhumiratana, W Panbangred. Cloning of a chitinase gene into *Bacillus thuringiensis* subsp. *aizawai* for enhanced insecticidal activity. J Gen Appl Microbiol 43:341–347, 1997.
262. W Chanpen, L Monton, S Patcharapron, P Somak, B Amaret. Expression of chitinase encoding genes from *Aeromonas hydrophila* and *Pseudomonas maltophila* in *Bacillus thuringiensis* sub sp. *israelensis*. Gene 179:119–126, 1997.
263. TA Jackson, JF Pearson, M O'Callaghan, HK Mahanty, MJ Willocks. Pathogen to product—development of *Serratia entomophila* (*Enterobacteriacea*) as a commercial biological control agent for the New Zealand grass grub (*Costelytra zealandica0*). In: TA Jackson, TR Glare, eds. Use of Pathogens in Searab Pest Management. Andover, England: Intercept, 1992, pp 191–198.
264. M O'Callaghan, MT Garnham, TL Nelson, D Baird, TA Jackson. The pathogenicity of *Serratia* strains to *Lucia sericata* (*Diptera*:*Calliphoridae*). J Invertebr Pathol 68:22–27, 1996.
265. D Nordmeyer. The search for novel nematocidal compounds. In: FJ Gommers, WT Maas, eds. Nematology from Molecules to Ecosystems. Invergowrie, Scotland: European Society of Nematologists, 1992, pp 281–293.
266. B Zechendorf. Biopesticides as integrated pest management strategy in agriculture. In: R Reuveni, ed. Novel Approaches to Integrated Pest Management. Boca Raton, FL: Lewis, 1995, pp 231–260.
267. ZA Siddiqui, I Mahmood. Role of bacteria in the management of plant parasitic nematodes: A review. Biores Technol 69:167–179, 1999.
268. R Duponnois, AM Ba, T Mateille. Beneficial effects of *Enterobacter cloacae* and *Pseudomonas mendocina* for biocontrol of *Meloidogyne incognita* with the endospore-forming bacterium *Pasteuria penetrans*. Nematology 1:95–101, 1999.
269. J Qiu, J Hallmann, N Kokalis-Burelle, DB Weaver, R Rodriguez-Kabana, S Tuzun. Activity

and differential induction of chitinase isozymes in soybean cultivars resistant or susceptible to root-knot nematodes. J Nematol 29:523–530, 1997.

270. S Rahimi, DJ Wright, RN Perry. Identification and localisation of chitinases induced in the roots of potato plants infected with the potato cyst nematode *Globodera pallida*. Fund Appl Nematol 21:705–713, 1998.
271. D Cronin, Y Moenne-Loccoz, C Dunne, F O'Gara. Inhibition of egg hatch of the potato cyst nematode *Globodera rostochiensis* by chitinase-producing bacteria. Eur J Plant Pathol 103: 433–440, 1997.
272. SU Sarathchandra, RN Watson, NR Cox, ME Di Menna, JA Brown, G Burch, FJ Neville. Effects of chitin amendment of soil on microorganisms, nematodes, and growth of white clover (*Trifolium repens* L.) and perennial ryegrass (*Lolium perenne* L.). Biol Fertil Soils 22:221–226, 1996.
273. Y Spiegel, E Cohen, S Galper, E Sharon, I Chet. Evaluation of a newly isolated bacterium, *Pseudomonas chitinolytica* sp. nov., for controlling the root-knot nematode *Meloidogyne javanica*. Biocontrol Sci Technol 1:115–125, 1991.
274. JP Bowman, LI Sly, AC Hayward, Y Spiegel, E Stackebrandt. *Telluria mixta* (*Pseudomonas mixta* Bowman, Sly, and Hayward 1988) gen. nov., comb. Nov., and *Telluria chitinolytica* sp. nov., soil-dwelling organisms which actively degrade polysaccharides. Int J System Bacteriol 43:120–124, 1993.
275. VPS Chahal, PPK Chahal. Control of *Meloidogyne incognita* with *Bacillus thuringiensis*. Dev Plant Soil Sci 45:677–680, 1991.
276. S Galper, E Cohn, Y Spiegel, I Chet. A collagenolytic fungus, *Cunninghamella elegans*, for biological control of plant parasitic nematodes. J Nematol 23:269–274, 1991.
277. S Sela, H Schickler, I Chet, Y Spiegel. Purification and characterization of a *Bacillus cereus* collagenolytic/proteolytic enzyme and its effect on *Meloidogyne javanica* cuticular proteins. Eur J Plant Pathol 104:59–67, 1998.
278. PJM Bonants, PFL Fitters, H Thijs, E den Belder, C Waalwijk, JWDM Henfling. A basic serine protease from *Paecilomyces lilacinus* with biological activity against *Meloidogyne hapla* eggs. Microbiology 141:775–784, 1995.
279. A Tunlid, S Rosen, B Ek, L Rask. Purification and characterization of an extracellular serine protease from the nematode-trapping fungus *Arthrobotrys oligospora*. Microbiology 140: 1687–1695, 1994.
280. R Segers, TM Butt, BR Kerry, JF Peberdy. The nematophagous fungus *Verticillium chlamydosporium* produces a chymoelastase-like protease which hydrolyses host nematode proteins in situ. Microbiology 140:2715–2723, 1994.
281. S Haran, H Schickler, S Pe'er, S Logemann, A Oppenheim, I Chet. Increased constitutive chitinase activity in transformed *Trichoderma harzianum*. Biol Control 3:101–108, 1993.
282. MC Limon, JA Pintor-Toro, T Benitez. Increased antifungal activity of *Trichoderma harzianum* transformants that overexpress a 33-kDa chitinase. Phytopathology 89:254–261, 1999.
283. M Lorito, SL Woo, I Garcia, G Colucci, GE Harman, JA Pintor-Toro, E Filippone, S Muccifora, CB Lawrence, A Zoina, S Tuzun, F Scala. Genes from mycoparasitic fungi as a source for improving plant resistance to fungal pathogens. Proc Natl Acad Sci USA 95:7860–7865, 1998.
284. JP Bolar, JL Norelli, K-W Wong, CK Hayes, GE Harman, HS Aldwinckle. Expression of endochitinase from *Trichoderma harzianum* in transgenic apple increases resistance to apple scab and reduce vigor. Phytopathology 90:72–77, 2000.
285. K Brogle, I Chet, M Holliday, R Cressman, P Biddle, S Knowlton, CJ Mauvais, R Broglie. Transgenic plants with enhanced resistance to the fungal pathogen *Rhizoctonia solani*. Science 254:1194–1197, 1991.
286. G Jach, S Logenmann, G Wolf, A Oppenheim, I Chet, J Schell, J Logemann. Expression of

a bacterial chitinase leads to improved resistance of transgenic tobacco plants against fungal infection. Biopractice 1:33–41, 1992.

287. Q Zhou, EA Maher, S Masoud, RA Dixon, CJ Lamb. Enhanced protection against fungal attack by constitutive co-expression of chitinase and glucanase genes in transgenic tobacco. Biotechnology 12:807–812, 1994.
288. A Flores, I Chet, A Herrera-Estrella. Improved biocontrol activity of *Trichoderma harzianum* by over-expression of the proteinase-encoding gene prbl. Curr Genet 31:30–37, 1997.
289. S Bartincki-Garcia. Cell wall chemistry, morphogenesis, and taxonomy of fungi. Annu Rev Microbiol 22:87–108, 1968.
290. R Gonzalez, D Ramon, Ja Perezgonzalez. Cloning, sequence-analysis and yeast expression of the egl1 gene from *Trichoderma longibrachiatum*. Appl Microbiol Biotechnol 38:370–375, 1992.
291. Q Migheli, O Friad, D Ramon-Vidal, L Gonzalez Candelas. Hypercellulolytic transformants of *Trichoderma longibrachiatum* are active in reducing *Pythium* damping-off on cucumber. In: MJ Daniels, ed. Advances in Molecular Genetics of Plant-Microbe Interactions. Dordrecht: Kluwer, 1994, pp 395–398.
292. Q Migheli, L Gonzalez Candelas, L Dealessi, A Camponogara, D Ramon Vidal. Transformants of *Trichoderma longibrachiatum* overexpressing the β-1,4-endoglucanase gene egl1 show enhanced biocontrol of *Pythium ultimum* on cucumber. Phytopathology 88:673–677, 1998.
293. BR Glick, Y Bashan. Genetic manipulation of plant growth-promoting bacteria to enhance biocontrol of phytopathogens. Biotechnol Adv 15:353–378, 1997.
294. L Sundheim, AR Poplawsky, AH Ellingboe. Molecular cloning of two chitinase genes from *Serratia marcescens* and their expression in Pseudomonas species. Physiol Mol Plant Pathol 33:483–491, 1988.
295. L Sundheim. Effect of chitinase encoding genes in biocontrol *Pseudomonas* spp. In: EC Tjamos, GC Papavizas, RJ Cook, eds. Biological Control of Plant Diseases: Progress and Challenges for the Future. New York: Plenum Press, 1991, pp 331–333.
296. Y Sitrit, Z Barak, Y Kapulnik, AB Oppenheim, I Chet. Expression of a *Serratia marcescens* chitinase gene in *Rhizobium meliloti* during symbiosis on alfalfa roots. Mol Plant Microb Interact 6:293–298, 1993.
297. S Koby, H Schickler, I Chet, A Oppenheim. The chitinase encoding Tn7-based chiA gene endows *Pseudomonas fluorescens* with the capacity to control plant pathogens in soil. Gene 147:81–83, 1994.
298. AL Cheng, YM Wang, WH Tang. The transformation and expression chitinase gene in *Bacillus subtilis* B-908. Acta Phytopathol Sinica 26:204–206, 1996.
299. YM Wang. Cloning of chitinase gene and β-1,3-glucanase gene and expression in *Bacillus subtilis* B-908. PhD Thesis, China Agricultural University, Beijing, China, 1997.
300. EA Emmert, J Handelsman. Biocontrol of plant disease: A (Gram-) positive perspective. FEMS Microbiol Lett 171:1–9, 1999.
301. M Maurhofer, C Keel, D Haas, G Defago. Influence of plant species on disease suppression by *Pseudomonas fluorescens* strain CHAO with enhanced antibiotic production. Plant Pathol 44:40–50, 1995.
302. JM Ligon, ST Lam, TD Gaffney, DS Hill, PE Hammer, N Torkewitz. Biocontrol: genetic modification for enhanced antifungal activity. In: G Stacey, B Mullin, PM Gresshoff, eds. Biology of Plant-Microbe Interactions. St. Paul, MN: International Society for Molecular Plant-Microbe Interactions, 1996, pp 457–462.
303. S Niemann, C Keel, A Puhler, W Selbitschka. Biocontrol strain *Pseudomonas fluorescens* CHAO and its genetically modified derivative with enhanced biocontrol capability exert comparable effects on the structure of a *Sinorhizobium meliloti* population in gnotobiotic systems. Biol Fertil Soils 25:240–244, 1997.
304. L Chernin, L Zhou, M Ovadis, Z Ismailov, I Chet. Enhancement of a soilborne *Enterobacter*

strain's chitinolytic activity by heterologous global regulation and quorum-sensing systems. Proceedings of International Conference on Enzymes in the Environment, Granada, 1999, p 41.

305. R St. Leger, L Joshi, MJ Bidochka, DW Roberts. Construction of an improved mycoinsecticide overexpressing a toxic protease. Proc Natl Acad Sci USA 93:6349–6354, 1996.
306. K Suzuki, M Taiyoji, N Sugawara, N Nikaidou, B Henrissat, T Watanabe. The third chitinase gene (*chiC*) of *Serratia marcescens* 2170 and the relationship of its product to other bacterial chitinases. Biochem J 343:587–596, 1999.

8

Microbiology and Enzymology of Carbon and Nitrogen Cycling

Robert L. Tate III
Rutgers University, New Brunswick, New Jersey

I. INTRODUCTION

The title of this chapter brings to mind the diversity and essentiality to living systems of processes associated with the biogeochemical cycles involving nitrogen and carbon. A quick perusal of any basic biochemistry text suggests a nearly endless array of metabolic enzymes that catalyze the reactions necessary for energy transformation and cell replication and survival. Indeed, on a larger scale, ecosystem stability and sustainability (terms frequently linked to native and managed systems, respectively) rely nearly in toto on a foundation of a functional microbial community, including the complexities of intermediary metabolism of the diverse soil microbial population. Fortunately, in analyzing the status of current research relating to this topic, a relatively limited number of nitrogen and carbon catabolic enzymes have served as indicators of the metabolic status or activity of the soil biological community.

Justification of studies of carbon and nitrogen cycling enzymes has frequently been linked to agricultural systems, but associations with soil management in general, as well as reclamation concerns in particular, are becoming more common. With the need to ameliorate the impact of past anthropogenic intrusion into terrestrial systems through appropriate management as well as the desire to preclude or minimize future damage to soil systems (i.e., enhance our capacity to discern proper soil system stewardship), it would be ideal if a clear understanding of carbon- and nitrogen-based processes were attainable. Carbon and nitrogen cycling not only are essential processes for the maintaining, transformation, and flux of essential elements and energy in the biosphere, but are also crucial to management and reduction of the impact of many organic and some inorganic pollutants. Worldwide implications of soil-based carbon and nitrogen processes are exemplified by their impact on global greenhouse and ozone depleting gas production and consumption. For example, selection of cultivation methods can have a significant impact on carbon dioxide production from microbial respiration as well as reduction of atmospheric carbon dioxide loading (35,71). Similarly, quantities of nitrous oxide (both a greenhouse and an ozone-depleting gas) evolved from terrestrial systems are affected by fertilizer use (nitrification and denitrification effects) as well as by protection and creation of wetlands (4,7,39,57,85,122).

It is against this backdrop of the major environmental relevance of the enzymes of nitrogen and carbon cycling processes that this chapter is presented. The utility of soil enzyme activities as indicators of soil quality and in monitoring of the effects of soil pollution is presented elsewhere (14,34,60,116,131) and in Chapters 15, 16, and 17. The general objective of this chapter is to highlight the current status of our understanding of soil carbon and nitrogen processes and the properties of the soil system that controls activity of the enzymes catalyzing these nitrogen and carbon transformations.

II. NITROGEN AND CARBON TRANSFORMATION PROCESSES

A. General Metabolic Considerations

Enzymes associated with carbon and nitrogen transformations are central to cellular growth and energy processes. Thus, it is logical to conclude that any enzyme involved in cellular metabolism must be present in soil. Furthermore, quantities of the enzyme present in a particular soil and the reaction kinetics should reflect the basic metabolic properties of all cell systems. However, the utility of assessing quantities of carbon and nitrogen transformation enzymes in soil for describing overall system function is more complicated. The environment within which the enzymatic transformations occur is a complex array of sand, silt, and clay particles intermixed with a diverse array of organic substances. Some of the organic matter is readily available to and transformed by soil enzymes, but a significant portion is intrinsically more resistant to biodegradation because of its chemical structure. Additionally, substrates generally expected to be more ephemeral may exhibit extended longevity that is due to their physical location within the soil matrix (1,122). Further complications in interpreting or predicting biodecomposition kinetics may arise from the limited water solubility of the potential substrate. Frequently, only a small portion of the organic complex in soil is water-soluble (121). Because of the necessity of conversion of the water-insoluble energy resources to a form that can enter the cell and be metabolized, the cell must produce enzymes that function outside the confines of the cellular membrane—beyond the relatively safe environment of the cell. Thus, our concept of carbon and nitrogen transformation in soil must include an evaluation of the sorptive (e.g., clay interactions), physically adverse (e.g., temperature and moisture variations), and chemically limiting (e.g., pH, water-soluble heavy metals) extracellular environment.

The emphasis of this presentation is on the current status of basic enzyme studies involved in carbon and nitrogen transformations in soil. A number of excellent reviews (17,33,36,113) that are available on this topic are useful when considering its historical context. More current examples of the types of reactions studied in soil, considerations of the implications of the physical structure of the soil ecosystem on enzyme activities, and future research needs are examined herein.

B. Specific Enzymatic Activities

Although an interminable array of enzymes involved with carbon and nitrogen metabolism could be evaluated in soil and associated ecosystems, only a limited number of enzymic activities are commonly studied. Many of the enzymes are those generally found to exist and to express their catalytic activities extracellularly, such as cellulase. Others, such as urease, are found to catalyze reactions both within and outside cells. Ideally, enzyme activities selected as indicators of soil fertility or soil quality should be easily quantified and vary with ecosystem type, condition, or degree of human intervention.

Until the more recent era most soil-based research has been directed toward meeting agricultural needs (34,36). Therefore, to a large degree, the historical evaluation of enzyme activity in soil has been concentrated on quantification of cropping and management effects on activities involved with biogeochemical cycles (e.g., recycling of plant biomass nutrients and nitrogen fixation) or more directly agriculturally pertinent enzyme activities, such as urease. As can be concluded from the investigations cited later, the commonly studied activities involving nitrogen transformations have been associated with ammonium generation (amidases and urease), hydrolysis of proteins (proteases), nitrogen fixation (nitrogenases), and loss of nitrogen from soil ecosystems (nitrogen oxide reductases). Similarly, activities involving carbonaceous substances have included those associated with hydroxylation of aromatic rings (e.g., polyphenyl oxidases, laccases), leading ultimately to either mineralization or humification of the parent compounds; hydrolysis of polysaccharides (e.g., amylases, cellulases, xylanases); and a variety of lipases and esterases; plus the indicator of respiratory activity, dehydrogenase. These enzymatic activities have proved useful for assessment of more general ecological concerns, such as organic matter transformations in native soil systems, as well as of the effect of human intervention. For example, in the latter arena, any of the general carbon or nitrogen catabolism enzymes (e.g., cellulase, hydrolases, dehydrogenase) is useful in assessing impacts of recycling waste organic matter (e.g., composts, sludges) through soil ecosystems, whereas polyphenyl oxidases and laccase activity assessments are commonly linked to decomposition and humification of aromatic ring–containing xenobiotic chemicals.

III. ENZYME ASSAYS AND THEIR EFFECT ON DATA INTERPRETATION

The two primary questions that must be addressed in assessing carbon and nitrogen metabolic processes in soil are, How can the activity be quantified and what is an appropriate assay method? and How does the activity vary both in a relatively defined system in the test tube as well as in the more complex, heterogeneous environment of the soil? A primary property that is intimately linked to the latter question is the kinetics of the reaction. Although a sound assay method based on a clear understanding of the specific reactants and the reaction kinetics of the individual enzyme is essential to provide reliable data, the effect of soil particulates on the reaction properties must also be understood. Responses to either of the preceding questions are nontrivial when considering a soil ecosystem, especially when dealing with those enzymatic activities most closely linked with cellular energy and nutrient management.

The characteristics of the enzyme reaction must clearly be understood (i.e., reaction substrates, products, optimal conditions, and activity curves), but more important are the properties of the environment (extra- or intracellular) within which the enzyme functions. Concerns with the possibility of changes in enzymatic activity during sample collection, storage, and analysis are particularly acute when evaluating those activities associated with carbon and nitrogen transformations. A common general objective of enzyme studies is to estimate the quantity of enzymic activity expressed in the native soil site. Thus, changes in activity due to synthesis of new enzyme; either in the reaction mixture or in the soil sample itself prior to quantification, must be prevented. As is documented in the following section, there is a delicate balance between the amount of enzymatic activity and enzyme molecules, cellular metabolic state, and substrate (or inducer) level. A slight change in the soil physical structure can result in a significant change in quantity of enzyme in a soil sample in an assay mixture as a result of induction of new activity or enzyme repression. This variation in enzyme activity could result from liberation of substrate, inducers, or inhibitors from the soil matrix by disruption of the soil structure. Thus, appropriate design of a study of soil enzymes must include an appreciation of not only the basic traits of the enzyme reaction itself but also the ways that the soil properties may alter the measured activity.

Presentation of specific assay methods for the various enzymes commonly quantified in soil samples can be found in Chapter 21. Nonetheless, consideration of some of the general factors associated with the physical status of the enzyme and the state of the cells producing the enzymes is essential, because both affect the quantity of enzyme detected in the environmental sample and the kinetics of the reaction. Ultimately, the objective of any assessment of enzymatic activity is to relate the amount of activity to properties or conditions of the site from which the sample was collected. Furthermore, current questions relating to appropriate soil stewardship necessitate sufficient understanding of the variability of enzyme activity with soil properties to allow prediction of the relationship between enzyme and changes in ecosystem conditions, anthropogenically generated or other.

A. Soil Sample Collection and Data Interpretation

Although, as indicated, considerable effort is expended to assure the accurate measurement of enzyme activity in a reaction mixture, experimental objectives are usually directed at elucidating the activity expressed in a particular soil site. The two values are not necessarily equivalent. As soon as a soil sample is collected, the environmental parameters determining the amount of enzyme present and the proportion of that enzyme that is active are altered (121,122). Two examples of changes that can affect the metabolic status of the enzyme-producing cells are soil oxygen tension and the availability of the carbon and energy source. Oxygen concentration in soil is generally controlled by its diffusion rate from the atmosphere above the soil into the soil matrix as well as the rate of its consumption. This supply/consumption relationship can result in anaerobic microsites within the larger soil aggregates. Disruption of soil aggregates through the mechanics of soil sample collection (as well as by the common practice of sieving the soil in preparation for enzyme assays) alters this distribution of aerobic and anaerobic microsites and affects microbial metabolism accordingly. Additionally, much of the native soil organic matter is physically protected from access by microbes and their enzymes. That is, the organic material is physically occluded within soil aggregates, trapped in soil nanopores, or sorbed onto particle surfaces (1,98,122). Thus, the simple act of collecting a soil sample alters its physical state and likely increases the accessibility of the soil organic matter to enzymes and microbes. As a consequence, induction of new enzymatic activities and augmentation of existing activity through microbial replication may occur. Thus, an altered microbial community and its associated enzyme activities are necessarily created by the simple act of sample collection. At least a minimal change in soil enzyme activity, particularly that central to the metabolism of the microbial cell, by sample collection and storage is inevitable. However, it must be noted that the quantities of immobilized (stabilized) extracellular enzymes are not likely to be greatly changed by this process.

Soil sampling procedures may also affect the activity associated with mineralization of xenobiotic contaminants. This is especially true in aged, contaminated soils where the accessibility of organics is often reduced by sequestration (1). During the aging processes (i.e., as the interval between input of the contaminant and sampling), the xenobiotic substances and their metabolites become distributed among soil micro-, macro-, and nanopores in free and sorbed states. That portion of the chemical retained in interstitial waters of macro- and some micropores is most available for interaction with soil microbes and their enzymes. Therefore, equilibrium solution concentrations dependent upon the sequestering or sorption of the chemical pollutant can be altered by soil sampling and manipulation when the equilibrium is altered through disruption of soil structure and redistribution of soil water (122).

Each of these alterations of enzyme activity due to sample collection reflects on the validity of extrapolating the activity detected in the laboratory to that expressed in situ. In each case, the changes may be reduced or minimized by lessening destruction of soil structure during sampling and storing the soil sample under conditions that minimize the potential for microbial growth and enzyme synthesis (commonly at 4°C).

Generally, assay procedures for soil enzymes are designed to prevent increases in microbial numbers in enzyme levels during the assay. Thus many assay protocols recommend the use of growth inhibitors (e.g., toluene, mercaptoethanol, sodium azide, radiation sterilization, antibiotics) or utilization of an assay time that is insufficient for microbial growth and production of significant quantities of de novo synthesized enzyme (16, 75,113).

Although it is reasonable to assume that the level of activity expressed in a freshly collected soil sample is optimized for the in situ conditions, these conditions are not static. Therefore, induction of new enzyme activity can occur when soil conditions change. Examples of evidence supporting the conclusion that enzyme induction occurs readily in soil are provided by studies of L-histidine ammonia lyase (19) and nitrogen oxide reductases (114). Burton and McGill (19) found an increase in L-histidine ammonia lyase activity in soil in the absence of microbial growth when specific inducers of the enzyme were added to soil samples. An additional example of the importance of enzyme induction in soil is provided by studies of denitrification rates. Quantification of the kinetics for appearance of new enzyme, albeit from enzyme induction or microbial growth, has been used to estimate nitrous oxide reductase activity in field soils. A common means of estimating denitrification is to inhibit nitrous oxide reductase activity with acetylene. Thus, all of the nitrate denitrified accumulates in the reaction vessel as nitrous oxide. Smith and Tiedje (114) observed three-phase reaction kinetics for this process when quantifying nitrous oxide production in freshly collected soil samples incubated under controlled conditions in the laboratory. In the first few hours nitrous oxide production results from the activity of preexisting enzyme. This is followed by a transition period that results from the production of new enzyme by induction of enzyme synthesis in preexisting cells. In the third phase new enzyme activity results from the increased enzyme levels provided by an increased population density of active denitrifiers. Because of its critical nature in estimating native nitrous oxide reductase enzyme levels in soil samples the duration of the initial phase of the process has been evaluated by several investigators. It is reasonable to assume that the duration of each of the three phases varies with ecosystem type and status. Differences in the metabolic status of the denitrifier population vary (e.g., inactive as a result of the presence of O_2 or already maximized through optimal conditions—therefore no further growth or induction of the population or activity may occur), as would the occurrence of indirect inhibitors of the denitrification process (e.g., excessively high or low pH or temperature) that limit or slow growth and enzyme induction. Luo and associates (79) recommend an incubation period of not longer than 5 h at 20°C for estimating preexisting denitrification activity in soils. Similarly, Dendooven and Anderson (27) found that de novo synthesis of nitrite reductase started in their system 5 hours after imposition of anoxic conditions and that of nitrous oxide reductase after 16 hours. Other procedures that are useful in evaluating preexisting or indigenous nitrogen–oxide reductase activity in native soil samples include gamma sterilization of the soil (75) and incorporation of chloramphenicol into the assay mixture (28–30,94). The assumption in all of these studies is that the initial rate observed in the incubated fresh soil sample represents that enzyme's presence in the soil prior to collection from the field. The activity is still considered to be a potential activity in that it is likely that the nitrogen oxide substrate does not exist at saturating concentrations in the field site.

B. Relationship of Laboratory Enzyme Activity to Enzyme Expression in Field Soils

Of concern when relating the laboratory generated data to field situations is the fact that the laboratory assessments are based on maximizing the interaction between enzyme and substrate. Thus, something approaching total activity is usually assessed in the laboratory, whereas in the field the interaction of the enzyme and its substrate may be reduced as a result of a variety of soil properties affecting the efficiency of interaction of the substrate and enzyme molecules. In other words, a portion of the enzyme molecules existing in the field soil may not be actively engaged in catalyzing their requisite reaction or may be transforming the substrate at a suboptimal rate. Therefore, the enzyme activity expressed in the laboratory assay must be assumed to be maximal (given the defined conditions of the assay) until demonstrated otherwise. Enzyme activities measured in the laboratory are "potential" activities.

C. Control of Expression of Enzymes in Soil Microsites

Two forms of interaction between the enzyme and its physical environment can delineate enzyme function within a soil microsite: occlusion within a living cell, cell debris, or even a soil aggregate and sorption or binding to soil minerals or non-water-soluble organic substances. Thus, manipulation of a soil sample that disturbs native associations (e.g., disrupts aggregates or fractures cells) or alters the equilibrium between sorption and desorption results in reaction rates that differ from those of the native environment.

As was described in detail by Burns (17,18), enzymes exist in a variety of states in soil: that is, in growing or nongrowing microbial cells, cell debris, associated with clay minerals or soil organic matter, and soluble in the aqueous phase as free enzyme or enzyme–substrate complexes. Most commonly quantified soil enzyme components assayed are the activities contained within living cells, bound to soil organic matter, or soluble in the soil interstitial water. Additionally, soil enzyme may be associated with clay minerals or occluded within soil aggregates. The consideration of the inclusion of enzymes within soil aggregates is rarely taken into account because enzyme activities are usually measured in soil suspensions in the laboratory. This practice ensures maximum rates of enzyme–substrate interaction and adheres to basic enzyme assay principles when total enzyme within a system is considered. A future concern may be to evaluate in greater detail the impact of heterogeneity in location within the soil system on the portion of the enzyme activity that is expressed in situ. To reiterate, these distribution considerations related to enzymes of nitrogen and carbon cycling in soil affect the total quantity of activity expressed as well as the rate of the reactions—thereby controlling overall ecosystem and population dynamics.

The general spatial variability of microbes, enzymes, and their activities in soil has long been appreciated (92). This variation in activity is accentuated in the enzymes associated with carbon and nitrogen metabolism because of their strong linkage with inputs of readily metabolized fixed carbon resources. Thus, these enzymatic activities are highest in regions of native biomass production or inputs (e.g., rhizosphere) or in soils receiving organic wastes (e.g., composts or biosolids) and generally correlate significantly with levels of native soil organic matter (10,113).

Macrosite variability is of interest in assessing general ecosystem nutrient dynamics, but considerations at a microsite level may be more useful in determining the means and kinetics of reactions associated with organic pollutant decomposition or the effects of management decisions relating to the sustaining or improving of soil quality. From the foregoing, it could be concluded that increases in soil aggregation would result in a decline in soil carbon and nitrogen transformations. Generally, this is not observed (121). In fact, management of soil in a manner that increases soil aggregate formation usually results in a stimulation of the microbial and enzymatic activity associated with carbon and nitrogen metabolism. For example, Kandeler and Murer (68) noted that increased soil aggregation in a conventionally tilled agricultural soil converted to grassland was accompanied by significant increases in dehydrogenase, protease, and xylanase activities. Conversely, returning the soil ecosystem to conventional agricultural management caused a decline in the elevated enzymatic activities.

The distribution of enzymes involved in carbon and nitrogen transformation within the soil profile and aggregates reflects a central dogma of soil enzymology: that activities of carbon and nitrogen metabolizing enzymes measured in a soil sample correlate with levels of soil organic matter and readily available organic matter. For example, activities of xylanase, invertase, and protease have been found to be stimulated in the detritosphere (the soil litter interphase) (67). In another study, xylanase and invertase levels were elevated in the soil particle-size fraction (>200-μm fraction) containing the decomposing maize straw (117,118). Association of individual enzymes with specific size fractions relates in part to their interaction with fresh organic matter and the degree to which the activity is linked to humic acids (i.e., humic acid stabilized enzymes) (95,118). These relationships between organic matter sources and enzyme activities support a hypothesis that any soil management procedures that encourage the maintenance or development of soil aggregates optimize plant biomass production. Therefore, since the primary source of energy for the soil microbial community and substrates for the associated enzymes is the carbon fixed by the plants, the soil microbial biomass and their associated intra- and extracellular activities are in turn optimized by the improved soil management.

D. Distribution of Enzymes in Soil and Enzyme Kinetic Parameters

Reactions catalyzed by enzymes in soils, including those complexed to clays or organic matter, can be anticipated to follow Michaelis–Menten kinetics. In the case of humic– and clay–enzymes complexes, any divergence in reaction properties is a result of impairment of enzyme–substrate interactions due to alteration of the basic conformation of the enzyme protein when in the sorbed state (clay micelles) or covalently bound to soil humic acid. Additionally, it must be remembered that in a multicomponent system the activity quantified could result from the summation of a number of enzyme types in various locations that catalyze the same reaction with different kinetic constants. Therefore, the resultant kinetic parameters are an average of all forms and states of the enzyme molecules and their compliance with Michaelis–Menten kinetics may be at times coincidental.

A primary consequence of the physical location of enzyme molecules in soil is its impact on the probability of substrate–enzyme interactions (i.e., free diffusion) as well as the potential for the induction of enzyme synthesis (19). Sorption of extracellular enzymes with clay and/or humic substances can also alter the efficiency (K_m) of the reaction. Thus, both the total enzyme as reflected in the V_{max} and the efficiency of the transformation are environmentally controlled. Analysis of the kinetics of a reaction can be used to show occurrence of multiple forms and states of an enzyme in soil (see Ref. 122 for discussion of use of Eadie Scatchard plots in this analysis). An example of use of enzyme kinetic parameters to demonstrate occurrence of specific isoenzymes in soil is a study of urease (21) in which K_m values were shown to vary between 0.5 and 1.3 M depending on soil type and pH. Thus, enzyme kinetic patterns observed in soil may reflect properties of the reaction in situ but not the kinetics of a purified enzyme. Therefore, enzyme properties assessed in the complex soil sample are described as apparent reaction kinetic parameters.

Hope and Burns (63) developed a method of assessing extracellular enzyme diffusion in soil that also reveals the variable affinities of enzymes with specific clay minerals—thereby adding a consideration of soil type to any evaluation of enzyme activity variation in soil ecosystems. These workers studied the variable effect of bentonite (high surface area and high cation exchange capacity) and kaolinite (low surface area and low cation exchange capacity) on diffusion of endoglucanase and β-D-glucosidase in soil. The kaolinite had no effect (i.e., binding) on enzyme diffusion, whereas the bentonite significantly reduced (bound the enzyme) mobility. Thus, these studies showed that movement of an enzyme molecule from the vicinity of the cell synthesizing it is environmentally controlled. Concern over the effect of this high affinity of some clay molecules for the enzyme proteins on enzyme kinetics was also discussed, especially in the context of extracellular enzyme efficiency.

Clay interactions with or effects on biological systems and products (e.g., enzymes) are frequently discussed as if clay properties are relatively uniform. The example noted previously involving kaolinite and bentonite already suggests that there is considerable disparity in properties of different clay minerals. Because of the high variability in clay mineral quantities and types among various soil types (12), generalizations regarding the role of clay in expression of activities of the enzymes of the carbon and nitrogen cycles are difficult, if not impossible, to derive. Among the foremost of the variable properties of clay minerals affecting interactions with soil enzymes are their surface area and cation exchange capacity. The effects of these properties and their variation with clay type and enzyme have been documented with a variety of enzymes (104,121,122). Examples of this analysis of nitrogen and carbon metabolism associated enzymes are provided by studies of urease and invertase in the early 1990s. Gianfreda and coworkers (50) examined the interaction of invertase (β-fructosidase) with montmorillonite, aluminum hydroxide, and aluminum hydroxide–montmorillonite complexes. The sorption of invertase varied with pH of the reaction mixture and the specific clay mineral, most sorption was detected with montmorillonite and least with aluminum hydroxide. Sorption reduced the enzyme activity in general, the proportion of enzyme activity lost due to sorption varied with pH and clay type. Invertase was stabilized by association with the clay surface in that resistance to heat was increased in the sorbed state. In a similar study with urease (51), using the same clay minerals, the heat stability of the sorbed enzyme was reduced, as were the Michaelis constants, V_{max}, and K_m. Lai and Tabatabai (74), in an evaluation of the sorption of jackbean urease on kaolinite and montmorillonite, found that the K_m values of the sorbed enzyme were similar to that of free enzyme.

E. Enzyme Binding to Soil Humus

It has long been appreciated that stable extracellular enzymes occurring in soil are usually covalently linked to humic acid (20,86). For example, Nannipieri and colleagues (86) fractionated urease and proteolytic activity into a variety of molecular weight fractions. The number of molecular weight peaks varied with specific enzymes: that is, the enzyme was fractionated on the basis of the size of humic acid molecules with which it was associated. As with the sorption of enzymes to clay, the enzyme kinetic parameters are altered by any resulting occlusion of the enzyme's active site by the humic acid molecule or by conformational alterations to the enzyme structure due to changing molecular forces induced by the covalent linkage between the two macromolecules (121).

Interactions between macromolecules and enzyme proteins may also alter enzyme properties in a soil system. For example, the effect of binding of enzymes to polysaccharides on enzymatic activity was evaluated with urease purified from *Bacillus pasteurii* immobilized on calcium polygalacturonate: a model for mucigel (24). It was noted that the adsorption parameters of the enzyme and polysaccharide varied with sodium chloride concentration of the reaction mixture, suggesting that the nature of the interaction between the enzyme and the polysaccharide involved electrostatic associations rather than covalent linkages. Variation in the kinetic parameters and stability of the enzyme were used to assess the effect of association of the sugar polymer on the accessibility of the enzyme and its conformation. The bound extracellular enzyme exhibited increased stability with time and to heat denaturation compared to the soluble enzyme. The similarity of the V_{max} values between the two enzyme forms suggests that little conformational change in the enzyme structure had occurred but accessibility of the enzyme to its substrate was altered, as indicated by variation in the K_m value.

Similar studies have evaluated the effect of humification of proteins on their enzymatic activities. Examples using enzyme involvement in carbon and nitrogen transformations include the effect of bonding of β-D-glucosidase to a phenolic copolymer of L-tyrosine, pyrogallol, or resorcinol (108) and of linking of urease to tannic acid (49,52). Sarkar and Burns (108) found that their copolymers had several properties in common with those of native soil humic acid–enzyme complexes (E_4/E_6 ratios; carbon, hydrogen, nitrogen, and sulfur ratios; and infrared [IR] spectra). A lowering of the efficiency by polymerization was shown by both reduced V_{max} and increased K_m values. Association of the copolymer with bentonite resulted in a complex that was resistant to protease and much more stable than the native enzyme. Similarly, Gianfreda and associates (52) found that inclusion of ferric ions and aluminum hydroxide species with tannic acid in forming organomineral urease complexes resulted in maintenance of the conformational and enzymatic properties of the free enzyme.

These latter studies demonstrate how one enzyme commonly found in an extracellular state (urease) and one more closely related to the microbial cell (β-D-glucosidase) can be stabilized with soil organomineral complexes with activity levels sustained at a level that allows the enzyme to continue to contribute to ecosystem carbon and nitrogen dynamics. Considering that the long-term and heat stability of the enzymes were generally enhanced in the bound and/or mineral associated forms, it could be reasonably concluded that the total impact of the enzyme on overall ecosystem function was extended by the interactions with the abiotic soil constituents. This dissociation of enzyme activities from living cell is a factor that must be considered in assessing the utility of assessing soil enzymatic activities as an indicator of soil status or quality.

The foregoing analysis of the properties of enzyme reactions leads to the conclusion that a number of questions need to be addressed before applying basic enzyme kinetics analyses to studies of soil ecosystem function, such as environmental impact or soil quality assessments, or ultimately before predicting and measuring the outcome of soil ecosystem management decisions. Soil enzyme activities generally can be anticipated to follow Michaelis kinetics, especially intracellular enzymatic activities linked to essential metabolic processes. Interactions of extracellular enzymes with soil components result in variation in anticipated reaction rates or efficiencies. Additionally, should the physical conditions of the soil dictate, enzyme or substrate accessibility and enzyme reaction kinetics may deviate in toto from that anticipated. This latter situation can occur when the parameter quantified is the sum of several enzymatic reactions (e.g., dehydrogenase activity, carbon dioxide evolution from complex organic matter substrates) or when the enzyme accessibility is controlled by processes such as diffusion or by the surface area of non-water-soluble substances. In the latter situation, dissolution or solubilization of the substrate may be a nonenzymatic process (122). Even though at the site of action product generation may reflect Michaelis–Menten kinetics, the kinetics of the reaction for the ecosystem in toto would reflect diffusional limitations or other kinetic parameters. Furthermore, it is much more difficult to quantify individual enzyme reactions in a soil. As a result of its great heterogeneity (compared with the environment of a test tube with a defined mixture of reactants), the reaction kinetics occurring in the field may reflect enzyme induction and microbial growth rates, altered reaction kinetics (due to sorption and humification of the enzymes), as well as variations in the degree of saturation of the enzyme molecules with substrate and the distribution of the catalysts and reactions within the soil matrix.

These considerations may be particularly important for data interpretation in regard to xenobiotic chemical mineralization since it is not uncommon to use radio-labeled carbon dioxide evolution from labeled parent compound to assess mineralization capacity. Thus, the kinetics for the overall mineralization processes as estimated by radio-labeled carbon dioxide production from the parent compound in soil reflects the totality of the processes occurring between entry of the parent molecule into the ecosystem and generation of the mineralization products. Therefore, a variety of zero-order (substrate levels are generally saturating the available enzymes) and first-, second- or mixed-order models are used to describe these enzyme reactions in soil. In all cases, the reactions approximate the product of the interaction of the enzyme between its substrate and environment. More detailed analyses of this aspect of soil enzymology are found elsewhere (2,122).

IV. CARBON AND NITROGEN TRANSFORMATION ENZYMATIC ACTIVITY IN FIELD SITUATIONS

A consideration of the status of studies of carbon and nitrogen transformations in soil could be initiated by the question, Why study soil enzymes? In attempting to answer this question, five general areas of endeavor, some more novel, others strongly linked to the history of soil enzyme research, are apparent. These are (i) determination of the basic levels of various activities in native ecosystems—including methods development (i.e., basic enzyme studies); (ii) provision of an understanding of an essential soil process (e.g., denitrification, nitrogen fixation, humification); (iii) elucidation of basic soil ecosystem properties; (iv) quantification of the impact of management on soil ecosystem function; and (v) assessment of the impact of anthropogenic activities on a system (i.e., pollution, climate change).

A. Basic Enzyme Studies

Reviewed under the topic of basic enzyme studies are research aimed at optimizing the techniques needed for the study of soil enzymatic activity as well as the application of these methods to answer some basic soil science or ecological questions. A large variety of enzyme activities associated with carbon and nitrogen metabolism have been evaluated in soil. These range from the general activity measured by dehydrogenase, through a variety of hydrolases and amidases, to a number of activities directly related to the mineralization of xenobiotic compounds. Typically such studies involve the optimization of a specific assay procedure then the application of the procedure to the study of a limited number of soils from one or more different ecosystem types. A general objective of this type of research is commonly to determine specific environmental factors (e.g., pH, moisture, organic carbon or nitrogen, total soil organic matter) that delimit the extent of the reaction.

One major soil biological property of interest in environmental studies is the general respiration rate of the biological community. Many such studies are based on the quantification of carbon dioxide evolution. If the objective of an assessment of carbon dioxide fluxes from a soil system is to estimate organic carbon mineralization rates, field measurements of carbon dioxide fluxes would necessarily be an overestimate. This is because the total carbon dioxide flux from a soil reflects not only microbial respiration but also plant and animal respiration plus any carbon dioxide produced through chemical processes (e.g., limestone dissolution). A logical resolution to this appears to be to collect soil samples and sieve the soil to remove material that contributes a significant portion of the excess carbon dioxide. The assumption underlying this procedure is that any remaining nonprimary decomposers in the soil sample would contribute minimal quantities of carbon dioxide to the total evolved. This is a reasonable assumption, but as described earlier other problems are created by sieving the soil sample. In this example, sieving of the soil increases the quantities of soil organic matter available to the microbial cell (121,122) and gives rise to respired carbon dioxide in excess of that likely to be generated in the field.

Development of an enzyme-based assay that would be less sensitive to soil sample manipulation than is the assessment of carbon dioxide flux is desirable. As is shown in the research cited later, dehydrogenase activity has served frequently in that capacity. Dehydrogenase is a nonspecific assay in that it represents the activity of several different enzymes (122). An electron acceptor, generally triphenyltetrazolium chloride (TTC), is added to soil as a terminal electron acceptor. The colorless TTC is reduced to triphenylformazan (TPF) by cellular respiratory enzymes. TPF, a red product, is quantified spectrophotometrically. The quantity of triphenylformazan yielded in the assay is proportional to microbial respiration and in some cases to microbial biomass (122). Because of its general utility for estimating soil respiratory activity under a variety of conditions, the general procedure for quantifying dehydrogenase activity is consistently being revised to augment its effectiveness (47,132,126). Since a reasonably nonspecific activity is quantified, a variety of electron acceptors have been evaluated as substitutes for TTC. Also, even the "best" enzyme assays have limitations. For dehydrogenase, substances capable of competing with the TTC in the electron transfer necessarily decrease the quantity of dehydrogenase activity detected. Potential inhibitors for this activity include nitrate, nitrite, ferric iron, and copper (13,22,127). In spite of these limitations, dehydrogenase has proved to be a useful enzyme for quantifying general microbial respiration in a variety of soil ecosystems, including those where inhibitors might be anticipated to present complications for data interpretation. Examples of the latter situations include flooded soils (88,93), drained and flooded organic soils (119,120), and metal contaminated soils (70).

When evaluating the utility of the activities of soil enzymes as indicators or descriptors of soil ecosystem properties, it is necessary to ask, What enzymatic activity (or activities) would be most characteristic of the system of interest? The response to this query could appear daunting considering the vast variety of enzymes associated with intermediary metabolic transformations and related metabolic processes (e.g., toxicant inactivation, methylation of metals) available for selection. The task of characterization of a soil ecosystem is simplified by the general biochemical principle that the enzymes involved with intermediary metabolic processes are essential for the existence and function of the microbial cells. Thus to ensure balanced growth and activity of the microbial cell, its enzymic activities must be coordinated. Increases and decreases in their activities are correlated. Quantities of key metabolic enzymes can be assumed to reflect variation in levels of related enzymic activities. Note that the occurrence of extracellular pools of stabilized enzyme does not negate this assumption, but it may obscure some of the expected activity relationships if high levels of stabilized enzyme exist in the system. Changes in activity should result in significant increases or decreases in the enzymatic activity to provide a useful indication of soil biological status. Additionally, should there be sufficient stabilized pools of an enzymatic activity that changes in production (intra- or extracellular) are insignificant, then the ecosystem is defined by the stabilized portion of the enzyme activity and therefore useful data of another nature are provided by the enzyme activity assessment.

Is there evidence that such a coordination carbon and nitrogen transformation enzyme activities occurs in soil? If this query can be answered affirmatively, then soil systems could be characterized by quantification of a few specific enzymes rather than assessment of a battery of such activities. A number of studies have shown direct correlation among a variety of microbial metabolic activities in soil. For example, in a study of activities of 11 soil enzymes, alkaline phosphatase, amidase, α-glucosidase, and dehydrogenase highly correlated (1% level) with carbon dioxide evolution in glucose-amended soils, and phosphodiesterase, arylsulfatase, invertase, α-galactosidase, and catalase correlated at the 5% level (42). Of the carbon and nitrogen cycle enzymes, only urease activity did not correlate with the carbon dioxide evolution. Frankenberger and Tabatabai (46) found that L-glutaminase activity correlated with amidase, urease, and L-asparaginase activities in their soil samples. Similarly, in a study of enzyme activities in rhizosphere soils, Tate and coworkers (123,124) found direct relationships among a variety of carbon and nitrogen enzymatic activities in pitch pine (*Pinus rigida* Miller) rhizosphere soils. Therefore, it appears that a few enzyme activities may be studied with a reasonable assumption that they represent a more general trend in overall cellular metabolic processes.

Once the enzymes are selected for study, the next question of concern is, How does its activity vary in natural or managed soil ecosystems? A vast number of studies are recorded in the literature in which a specific enzyme activity is quantified in several soils sampled from different ecosystem types—with the objective of determining the activity range for the enzyme. It is reasonable to assume that the soil microbes are adapted to their environment, but it does not necessarily follow that the enzyme activities expressed by them are occurring under optimal conditions in situ in the field or in the assay mixture. Even in the latter situation, it is reasonable to conclude that existence of metals, heterogeneous distribution of substrates (although this may be overcome by substrate saturation and agitation), and variation of soil physical and chemical properties at the microsite wherein the enzyme resides could reduce the activity below that which would be expressed by purified enzymes in a totally defined reaction mixture. Field soil parameters commonly evaluated by varying the laboratory assay mixture to test for effects on soil enzymic activity are temperature and pH (43,89,120,130), moisture (81,108), organic carbon and nitrogen (45,87,128,134), soil salinity (41), and seasonal changes (129,133). Each of these soil properties has a reasonably predictable impact on soil enzymatic activity. Carbon and nitrogen metabolism enzyme activities generally correlate with total soil organic carbon or readily metabolized organic matter levels and are inhibited by high salt or metal concentrations (121). The assay pH provides an interesting data set in that although each individual enzyme has a specific optimal pH, different forms (isoenzymes) of the same enzyme exist with variant pH optima. The most commonly studied examples of this type of enzyme are acid and alkaline phosphatases. Since each of these general soil properties is highly variable among soil and ecosystem types, it is reasonable to respond to the question with a conclusion that soil enzyme activity assessments can provide a valid indication of ecosystem properties.

B. Soil Process Evaluation

Some enzyme activities, such as those of nitrogenase and nitrous oxide reductase, are useful for the estimation of nitrogen cycling rates in complex soil ecosystems. Both are exclusively intracellular enzymes. The results of studies of these enzymes are commonly presented as if a single enzyme following Michaelis–Menten kinetics were being evaluated. However, unless the specific substrate for the enzyme activity in question is added to the reaction mixture, the kinetic parameters measured are the result of interaction of all enzymes in the reaction sequence. For nitrous oxide production by denitrification frequently nitrate is added to the reaction mixture and nitrous oxide production assessed. Enzymes preceding nitrous oxide reductase in the reaction sequence include nitrate reductase and nitrite reductase, and thus, the kinetic parameters for the process would be a combination of at least three enzymes. Similarly, activity of nitrogenase, which is itself a complex "multiple" enzyme, reflects not only the kinetics of the enzyme transformations but also the accessibility of the enzyme to its primary substrate, whether dinitrogen or acetylene (122).

As noted previously and elsewhere (15), acetylene inhibition of nitrous oxide reductase either in the presence of chloramphenicol or through use of limited incubation times has been used to estimate total denitrification in a variety of ecosystems. Acetylene reduction by nitrogenase, although not necessarily stoichiometrically equivalent to reduction of dinitrogen (122), allows ready demonstration of nitrogen fixation potential in soil ecosystems ranging from flooded to cold desert soils, native forest to agricultural and grassland soils (6,15,78,80,99).

Another example of a major native soil process affecting fate of carbon in soil is humification. This process is instrumental in increasing soil organic carbon pools, thereby potentially reducing greenhouse gas production and removing potentially toxic xenobiotic chemicals from soil interstitial waters through covalent linkage to humic acid. Both abiotic and biotic agents in soil catalyze these processes. Enzymatic contribution to the process involves formation of free radicals by two common soil enzymes, laccase and tyrosinase. The potential role of these enzymes in the humification of anilinic and phenolic compounds and reduction of their bioavilability with the passage of time (aging) is sufficient reason to investigate their function in soil further (8,26,90,125).

C. Elucidation of Basic Ecosystem Properties

Current research about the environmental variability of enzyme activity and how it relates to properties of the soil system is commonly associated with maintaining a highly productive ecosystem and optimal soil quality. Enzymes whose activities are selected for analysis often include proteases, urease, dehydrogenase, and cellulases. Examples of such studies include evaluations of enzyme activity variability within a reasonably uniform ecosystem, such as a pastureland (106); in a native system subject to some degree of environmental perturbation, such as wetland soils (58); forest soils in general (37); forests experiencing wild fires (62); forest degradation and regeneration (64); agricultural soils (3); and other soils in a Mediterranean climate (48). In each situation, the enzymatic activity was highly variable and influenced by soil properties such as pH, moisture, and organic matter content and major changes in aboveground productivity (e.g., crop management) and climatic variation. A general conclusion from these studies is that increased aboveground productivity commonly results in stimulation of the soil enzymatic activities.

D. Soil Enzymes and Soil Management

The activity of carbon and nitrogen cycle enzymes in soil has been used to assess soil ecosystem adaptation to anthropogenic intervention and the effects of reclamation management decisions. Of major interest are questions relating to the effect of elevated atmospheric carbon dioxide loading on equilibrium levels of soil organic matter. It is generally noted that increasing the amount of carbon dioxide available to plants for photosynthesis increases biomass production (104). Any increase in plant biomass production is likely to stimulate soil microbial activity through the increased availability of a carbon and energy, thereby resulting in stimulation of carbon and nitrogen cycling enzymes. For example, Ross et al. (106) documented enhanced soil respiration and invertase activity in a ryegrass–white clover system incubated under elevated levels of carbon dioxide. In this situation, the enhanced input of plant biomass resulted from the stimulation of plant growth by the augmented atmosphere loading of carbon dioxide. Data interpretation was complicated in this study by the large spatial variability of the activities, underscoring a need to understand better the spatial variation of enzymatic activities in soils. Similar studies evaluating soil enzymatic responses within the boundaries defined to a certain degree by anthropogenic intrusion include variation of enzymatic activities in abandoned agricultural soils (91), alteration of plant cover (9), fertilization and agrochemical inputs into pasturelands and agricultural soils (72,105), enzyme differences between constructed and native wetlands (38), and variation due to tillage type (111).

A current major concern impacting general soil enzymatic activity is the need to use soil ecosystems for recycling the carbon in a variety of societal organic waste products (e.g., recycled biosolids from composting and the wastewater purification process). A priori conclusions regarding this activity are that since the general soil microbial community is carbon limited (122) and since the activities of the enzymes associated with carbon and nitrogen transformations are commonly directly linked to levels of microbial activity, then any amendment of a fixed-carbon resource would result in an increase in the relevant enzyme activities. Of perhaps greater interest is the potential to use duration and/or magnitude of change in the enzyme activities to predict positive and negative changes in the system (in terms of soil quality, ecosystem stability and function, as well as environmental risk). These applications of enzyme activity data are currently developing, but considerable quantification of variation of enzyme activity with waste–organic material amendment has been conducted, for example, composted municipal solid wastes (56,97), sewage sludges (11,23,61), manure and dairy effluent (59,65,135), petroleum (44), and agricultural residues (31,32,40). As these data are collected and collated, a more clear picture of their utility in assessing ecosystem sustainability, resilience, and environmental risk will emerge.

E. Pollution Impact Assessment

Enzymic activities in soil have been used frequently to estimate biological process rates in a number of contaminated soil systems and to document the results of reclamation management. The effect of metal contamination on soil enzymatic activity is an excellent example of such studies. It has long been known that a number of cations, such as many of the heavy metals commonly found in soils, through their association with proteins (e.g., interaction with the sulfur of sulfide and sulfhydral groups), denature the protein and through related mechanisms inhibit cellular respiration (122). Thus, it is logical that sufficient levels of such metals could occur in soil interstitial water to affect expression of such enzymes. Since these cations are also normal substituents of soil minerals, it is also reasonable to assume that soil microbes are capable of adapting to the presence of these substances in their environment. Therefore, it could be concluded that the primary concern for management of sites with metal contamination is limited to those with excessive levels of such contamination, as are noted around metal smelters. But as the limited examples provided in the following discussion document, even low levels of such metals can alter the equilibrium level of soil enzymatic activity. A variety of enzymes involved in essential carbon and nitrogen transformations, including dehydrogenase, have been assessed in metal contaminated soils in both laboratory and field studies. The metals have a varying impact on enzyme activity, depending not simply on their total concentration in the soil but rather on capacity to interact with the enzyme protein. Thus, the direct impact of the cations is affected by soil pH, soil organic matter levels, and interactions with other soil minerals and organic matter (5,12,28,55,66,73). In a detailed evaluation of 13 soil enzymes involved with carbon, nitrogen, phosphorus, and sulfur cycling enzymes, Kandeler et al. (66) found significant decreases in critical soil enzyme levels, even at soil metal loading levels approximating current European Community limits. A clear example demonstrating that even low levels of metals can alter general metabolic enzyme activity (including a wide variety of extracellular and intracellular enzymatic activities) was provided by an examination of the effect of copper concentrations on carbohydrases, proteases, lipases, and phosphatases in organic soils receiving low levels of copper amendment (82–84). In all cases, the enzyme activities were reduced in the presence of the copper.

As a result of the complexities of the interaction between soil enzymes and soluble cations and our current lack of understanding of the extent to which soil enzyme levels can be reduced before an environmental impact is observed, it is difficult to assess the importance of changes in levels of specific enzymes. This consideration is particularly acute because metal contamination rarely occurs in isolation: that is, generally more than one toxic metal is present in soil interstitial water, and other factors such as reduction of pH due to elevated sulfate levels also affect soil enzymic activity (69). Thus, assessment of a combination of soil metabolic and enzyme activity levels is necessary for monitoring of soil pollution by metals (15). The emerging use of soil nitrogen and carbon cycling activities in environmental quality assessment is also exemplified by studies of soils receiving agricultural chemicals, including pesticides (53,54,76,102), smoke exposure (77), ozone and acid rain (101,103), and fly ash (96).

The question associated with such studies is not whether the enzymatic activity will be reduced by the introduction of toxicants into soil (negative correlation between various carbon and nitrogen enzyme activities and concentrations of inhibitors are commonly reported), rather, it is whether the amount of reduction in enzyme activity and its duration are meaningful with regard to ecosystem sustainability and its ability to recover after appropriate reclamation management.

V. CONCLUSIONS

There is a rich history of soil enzyme research in the assessment of soil carbon and nitrogen transformation processes, and levels of these activities in native and anthropogenically impacted soils are well described. This review has documented a continuation of what could be considered to be a historical propensity to apply well established enzyme assay procedures to quantifying the activities in a variety of the world's soils. From such studies, a more complete understanding of the effect of soil properties and anthropogenic activities on enzyme expression has emerged. Unfortunately, as we read the list of conclusions of such research, we frequently see the results of data analysis reduced to an almost generic statement that the soil enzyme activity of interest is correlated with changes in soil pH, moisture, and organic matter levels (natural or laboratory induced) or that the contaminant of choice (various metals and organic compounds) inhibits the enzyme activity. What is missing is a direct link (or even a statement regarding the real meaning) between the relationship of the levels of expression of the enzymes and overall ecosystem function. That is, the reader is commonly left with the question, So what? The challenge of soil enzyme research in general, and of studies of carbon and nitrogen enzymes specifically, is to develop models that allow responses to such questions as, How much of a decline in soil enzymatic activity can occur before the quality or sustainability of the total ecosystem is affected? Does an increase in enzyme activity due to reclamation management indicate improvement in total ecosystem?, or even Which enzymatic activities should be quantified to provide an indication of an impact of management, pollution, or reclamation on general soil microbial activity, redundancy, and resilience? Therein lies the challenge to soil enzymologists.

REFERENCES

1. M Alexander. How toxic are toxic chemicals in soil. Environ Sci Technol 29:2713–2717, 1995.
2. M Alexander. Biodegradation and Bioremediation. 2nd. ed. New York: Academic Press, 1999.

3. L Badalucco, PJ Kuikman, P Nannipieri. Protease and deaminase activities in wheat rhizosphere and their relation to bacterial and protozoan populations. Biol Fertil Soils 23:99–104, 1996.
4. BC Ball, GW Horgan, H Clayton, JP Parker. Spatial variability of nitrous oxide fluxes and controlling soil and topographic properties. J Environ Qual 26:1399–1409, 1997.
5. RD Bardgett, TW Speir, DJ Ross, GW Yeates, HA Kettles. Impact of pasture contamination by copper, chromium and arsenic timber preservatives on soil microbial properties and nematodes. Biol Fertil Soils 18:71–79, 1994.
6. J Belnap. Soil surface disturbances in cold deserts: Effects on nitrogenase activity in cyanobacterial-lichen crusts. Biol Fertil Soils 23:362–367, 1996.
7. AM Blackmer, JM Bremner. Inhibitory effect of nitrate on reduction of N_2O to N_2 by soil microorganisms. Soil Biol Biochem 10:187–191, 1978.
8. J-M Bollag. Decontaminating soil with enzymes: An in situ method using phenolic and anilinic compounds. Environ Sci Technol 26:1876–1881, 1992.
9. H Bolton Jr, JL Smith, SO Link. Soil microbial biomass and activity of a disturbed and undisturbed shrub-steppe ecosystem. Soil Biol Biochem 25:545–552, 1993.
10. M Bonmati, B Ceccanti, P Nanniperi. Spatial variability of phosphatase, urease, protease, organic carbon and total nitrogen in soil. Soil Biol Biochem 23:391–396, 1991.
11. M Bonmati, M Pujola, J Sana, M Soliva, MT Felipo, M Garau, B Ceccanti, P Nanipieri. Chemical properties, populations of nitrite oxidizers, urease and phosphatase activities in sewage sludge-amended soils. Plant Soil 84:79–91, 1985.
12. NC Brady, RR Weil. The Nature and Properties of Soils. 12th ed. Upper Saddle River, NJ: Prentice Hall, 1999.
13. JM Bremner, MA Tabatabai. Effects of some inorganic substances on TTC assay of dehydrogenase activity in soil. Soil Biol Biochem 5:385–386, 1973.
14. PC Brookes. The potential of microbiological properties as indicators in soil pollution monitoring. In: R Schuline, A Desaules, R Webster, B von Steiger, eds. Soil Monitoring: Early Detection and Surveying of Soil Contamination and Degredation. Basel, Switzerland: Birkhauser Verlag AG, 1993, pp 229–264.
15. PC Brookes. 1995. The use of microbial parameters in monitoring soil pollution by heavy metals. Biol Fertil Soils 19:269–279, 1995.
16. KA Brown. Biochemical activities in peat sterilized by gamma-irradiation. Soil Biol Biochem 13:469–474, 1981.
17. RG Burns. Soil enzymology. Sci Prog 64:275–285, 1977.
18. RG Burns. Enzyme activity in soil: Location and possible role in microbial ecology. Soil Biol Biochem 14:423–427, 1982.
19. DL Burton, WB McGill. Inductive and repressive effects of carbon and nitrogen on L-histidine ammonia-lyase activity in black chernozemic soil. Soil Biol Biochem 23:939–946, 1991.
20. MD Busto, M Perez-Mateos. Extraction of humic-β-glucosidase fractions from soil. Biol Fertil Soils 20:77–82, 1995.
21. ML Cabrera, DE Kissel, BR Bock. Urea hydrolysis in soil: Effects of urea concentration and soil pH. Soil Biol Biochem 23:1121–1124, 1991.
22. K Chander, PC Brookes. Is the dehydrogenase assay invalid as a method to estimate microbial activity in copper-contaminated soils? Soil Biol Biochem 23:909–915, 1991.
23. K Chander, PC Brookes, SA Harding. Microbial biomass dynamics following addition of metal-enriched sewage sludges to a sandy loam. Soil Biol Biochem 27:1409–1421, 1995.
24. S Ciurli, C Marzadori, S Benini, S Deiana, C Gessa. Urease from the soil bacterium *Bacillus pasteurii*: Immobilization on Ca-polygalacturonate. Soil Biol Biochem 28:811–817, 1996.
25. GH Dar. Effects of cadmium and sewage-sludge on soil microbial biomass and enzyme activities. Bioresource Technol 56:141–145, 1996.
26. J Dec, J-M Bollag. Determination of covalent and noncovalent binding interactions between xenobiotic chemicals and soil. Soil Sci 162:858–874, 1997.

27. L Dendooven, JM Anderson. Dynamics of reduction of enzymes involved in the denitrification in pasture soil. Soil Biol Biochem 26:1501–1506, 1994.
28. L Dendooven, JM Anderson. Maintenance of denitrification potential in pasture soil following anaerobic events. Soil Biol Biochem 27:1251–1260, 1995.
29. L Dendooven, E Pemberton, JM Anderson. Denitrification potential and reduction enzyme dynamics in a Norway spruce plantation. Soil Biol Biochem 28:151–157, 1996.
30. L Dendooven, P Splatt, JM Anderson. Denitrification in permanent pasture soil as affected by different forms of C substrate. Soil Biol Biochem 28:141–149, 1996.
31. SP Deng, MA Tabatabai. Effect of tillage and residue management on enzyme activities in soils. I. Amidohydrolases. Biol Fertil Soils 22:202–207, 1996.
32. SP Deng, MA Tabatabai. Effect of tillage and residue management on enzyme activities in soils. II. Glycosidases. Biol Fertil Soils 22:208–213, 1996.
33. RP Dick. A review: Long-term effects of agricultural systems on soil biochemical and microbial parameters. Agric Ecosyst Environ 40:25–36, 1992.
34. RP Dick. Soil enzyme activities as indicators of soil quality. In: JW Doran, DC Coleman, DF Bezdicek, BA Stewart, eds. Defining Soil Quality for a Sustainable Environment. Madison, WI: Soil Science Society of America, 1994, pp 107–124.
35. WA Dick. Influence of long-term tillage and crop rotation combinations on soil enzyme activities. Soil Sci Soc Am J 48:569–574, 1984.
36. WA Dick, MA Tabatabai, Significance and potential uses of soil enzymes. In: FB Metting Jr, ed. Soil Microbial Ecology: Applications in Agricultural and Environmental Management. New York: Marcel Dekker, 1992, pp. 95–225.
37. O Dilly, JC Munch. Microbial biomass content, basal respiration and enzyme activities during the course of decomposition of leaf litter in a black alder (*Alnus glutinosa*(L.) Gaertn.) forest. Soil Biol Biochem 28:1073–1081, 1996.
38. CP Duncan, PM Groffman. Comparing microbial parameters in natural and constructed wetlands. J Environ Qual 23:298–305, 1994.
39. JM Duxbury, DR Bouldin, RE Terry, RL Tate III. Emissions of nitrous oxide from soils. Nature 298:462–464, 1982.
40. AMK Falih, M Wainwright. Microbial and enzyme activity in soils amended with a natural source of easily available carbon. Biol Fertil Soils 21:177–183, 1996.
41. WT Frankenberger Jr, FT Bingham. Influence of salinity on soil enzyme activities. Soil Sci Soc Am J 46:1173–1177, 1982.
42. WT Frankenberger Jr, WA Dick. Relationships between enzyme activities and microbial growth and activity indices in soil. Soil Soc Am J 47:945–951, 1983.
43. WT Frankenberger Jr, JB Johanson. Effect of pH on enzyme stability in soils. Soil Biol Biochem 14:433–437, 1982.
44. WT Frankenberger Jr, JB Johanson. Influence of crude oil and refined petroleum products on soil dehydrogenase activity. J Environ Qual 11:602–607, 1982.
45. WT Frankenberger Jr, JB Johanson. Distribution of L-histidine ammonia-lyase activity in soils. Soil Sci 136:347–353, 1983.
46. WT Frankenberger Jr, MA Tabatabai. Factors affecting L-glutaminase activity in soils. Soil Biol Biochem 23:875–879, 1991.
47. JK Friedel, K Molter, WR Fischer. Comparison and improvement of methods for determining soil dehydrogenase activity by using triphenyltetrazolium chloride and iodonitrotetrazolium chloride. Biol Fertil Soils 18:291–296, 1994.
48. C Garcia, T Hernandez, F Costa. Microbial activity in soils under Mediterranean environmental conditions. Soil Biol Biochem 26:1185–1191, 1994.
49. L Gianfreda, A De Cristofaro, MA Rao, A Violante. Kinetic behavior of synthetic organo- and organo-mineral-urease complexes. Soil Sci Soc Am J 59:811–815, 1995.
50. L Gianfreda, MA Rao, A Violante. Invertase (β-fructosidase): Effects of montmorillonite,

Al-hydroxide and Al(OH)x-montmorillonite complex on activity and kinetic properties. Soil Biol Biochem 23:581–587, 1991.

51. L Gianfreda, MA Rao, A Violante. Adsorption, actiity and kinetic properties of urease on montmorillonite, aluminum hydroxide and Al(OH)x montmorillonite complexes. Soil Biol Biochem 24:51–58, 1992.
52. L Gianfreda, MA Rao, A Violante. Formation and activity of urease-tannate complexes affected by aluminum, iron, and manganese. Soil Sci Soc Am J 59:805–810, 1995.
53. L Gianfreda, F Sannino, N Ortega, P Nanipieri. Activity of free and immobilized urease in soil: Effects of pesticides. Soil Biol Biochem 26:777–784, 1994.
54. L Gianfreda, F Sannino, A Violante. Pesticide effects on the activity of free, immobilized and soil invertase. Soil Biol Biochem 27:1201–1208, 1995.
55. PL Giusquiani, G Gigliotti, D Businelli. Long-term effects of heavy metals from composted municipal waste on some enzyme activities in a cultivated soil. Biol Fertil Soils 17:257–262, 1994.
56. PL Giusquiani, M Pagliai, G Gigliotti, D Businelli, A Benetti. Urban waste compost: Effects on physical, chemical, and biochemical soil properties. J Environ Qual 24:175–182, 1995.
57. LL Goodroad, DR Keeney. Nitrous oxide emissions from forest, marsh, and prairie ecosystems. J Environ Qual 13:448–452, 1984.
58. PM Groffman, GC Hanson, E Kiviat, G Stevens. Variation in microbial biomass and activity in four different wetland types. Soil Sci Soc Am J 60:622–629, 1996.
59. A Hadas, L Kautsky, R Portnoy. Mineralization of composted manure and microbial dynamics in soil as affected by long-term nitrogen management. Soil Biol Biochem 28:733–738, 1996.
60. JJ Halvorson, JL Smith, RI Papendick. Integration of multiple soil parameters to evaluate soil quality: A field example. Biol Fertil Soils 21:207–214, 1996.
61. H Hattori. Microbial activities in soil amended with sewage sludges. Soil Sci Plant Nutr 34: 221–232, 1988.
62. T Hernandez, C Garcia, I Reinhardt. Short-term effect of wildfire on the chemical, biochemical, and microbiological properties of Mediterranean pine forest soils. Biol Fertil Soils 25: 109–116, 1997.
63. CFA Hope, RG Burns. The barrier-ring plate technique for studying extracellular enzyme diffusion and microbial growth in model soil environments. J Gen Microbiol 131:1237–1243, 1985.
64. DK Jha, GD Sharma, RR Mishra. Soil microbial population numbers and enzyme activities in relation to altitude and forest degradation. Soil Biol Biochem 24:761–767, 1992.
65. E Kandeler, G Eder, M Sobotik. Microbial biomass, N mineralization, and the activities of various enzymes in relation to nitrate leaching and rood distribution in a slurry-amended grassland. Biol Fertil Soils 18:7–12, 1994.
66. E Kandeler, C Kampichler, O Horak. Influence of heavy metals on the functional diversity of soil microbial communities. Biol Fertil Soils 23:299–306, 1996.
67. E Kandeler, J Luxhoi, D Tscherko, J Magid. Xylanase, invertase and protease at the soil-litter interface of a loamy sand. Soil Biol Biochem 31:1171–1179, 1999.
68. E Kandeler, E Murer. Aggregate stability and soil microbial processes in a soil with different cultivation. Geoderma 56:503–513, 1993.
69. JJ Kelly, MM Häggblom, RL Tate III. Changes in soil microbial communities over time resulting from one time application of zinc: A laboratory microcosm study. Soil Biol Biochem 31:1455–1465, 1999.
70. JJ Kelly, RL Tate III. Effects of heavy metal contamination and remediation on soil microbial communities in the vicinity of a zinc smelter. J Environ Qual 27:609–617, 1998.
71. JS Kern, MG Johnson. Conservation tillage impacts on natural soil and atmospheric carbon levels. Soil Sci Soc Am J 57:200–210, 1993.

72. MJ Kirchner, AG Wollum II, LD King. Soil microbial populations and activities in reduced chemical input agroecosystems. Soil Sci Soc Am J 57:1289–1295, 1993.
73. P Lahdesmaki, R Piispanen. Soil enzymology: Role of protective colloid systems in the preservation of exoenzyme activities in soil. Soil Biol Biochem 24:1173–1177, 1992.
74. CM Lai, MA Tabatabai. Kinetic parameters of immobilized urease. Soil Biol Biochem 24: 225–228, 1992.
75. R Lensi, C Lescure, C Steinberg, JM Savoie, G Faurie. Dynamics of residual enzyme activities, denitrification potential, and physico-chemical properties in a gamma-sterilized soil. Soil Biol Biochem 23:367–373, 1991.
76. G Lethbridge, AT Bull, RG Burns. Effects of pesticides on 1,3-β-glucanase and urease activities in soil in the presence and absence of fertilisers, lime and organic materials. Pestic Sci 12:147–155, 1981.
77. SW Li, JK Fredrickson, MW Ligotke, P van Voris, JE Rogers. Influence of smoke exposure on soil enzyme activities and nitrification. Biol Fertil Soils 6:341–346, 1988.
78. C Limmer, HL Drake. Non-symbiotic N_2 fixation in acidic and pH-neutral forest soils: Aerobic and anaerobic differentials. Soil Biol Biochem 28:177–183, 1996.
79. J Luo, RE White, PR Ball, RW Tillman. Measuring denitrification activity in soils under pasture: Optimizing conditions for the short-term denitrification enzyme assay and effects of soil storage on denitrification activity. Soil Biol Biochem 28:409–417, 1996.
80. J Maggs, RK Hewett. Nitrogenase activity (C_2H_2 reduction) in the forest floor of a *Pinus elliotii* plantation following superphosphate addition and prescribed burning. Forest Ecol Management 14:91–101, 1986.
81. HMA Magid, MA Tabatabai. Amide nitrogen transformation in waterlogged soil. J Agric Sci 116:281–285, 1991.
82. SP Mathur, JI MacDougall, M McGrath. Levels of activities of some carbohydrases, protease, lipase, and phosphatase in organic soils of differing copper content. Soil Sci 129:376–385, 1980.
83. SP Mathur, CM Preston. The effect of residual fertilizer copper on ammonification, nitrification and proteolytic population in soil organic soils. Can J Soil Sci 61:445–450, 1981.
84. SP Mathur, RB Sanderson. The partial inactivation of degradative soil enzymes by residual fertilizer copper in Histosols. Soil Sci Soc Am J 44:750–755, 1980.
85. AR Mosier, GL Hutchinson. Nitrous oxide emissions from cropped fields. J Environ Qual 10:169–173, 1981.
86. P Nannipieri, B Ceccanti, D Bianchi, M Bonmati. Fractionation of hydrolase-humus complexes by gel chromatography. Biol Fertil Soils 1:25–29, 1985.
87. P Nannipieri, A Gelsomino, M Felici. Method to determine guaiacol oxidase activity in soil. Soil Sci Soc Am J 55:1347–1352, 1991.
88. M Okazaki, E Hirata, K Tensho. TTC reduction in submerged soils. Soil Sci Plant Nutr 29: 489–497, 1983.
89. P O'Toole, MA Morgan. Thermal stabilities of urease enzymes in some Irish soils. Soil Biol Biochem 16:471–474, 1984.
90. S Pal, J-M Bollag, PM Huang. Role of abiotic and biotic catalysts in the transformation of phenolic compounds through oxidative coupling reactions. Soil Biol Biochem 26:813–820, 1994.
91. SK Pancholy, EL Rice. Soil enzymes in relation to old field succession: Amalase, cellulase, invertase, dehydrogenase and urease. Soil Sci Soc Am Proc 37:47–50, 1973.
92. TB Parkin. Spatial variability of microbial processes in soil—a review. J Environ Qual 22: 409–417, 1993.
93. FR Pedrazzini, KL McKee. Effect of flooding on activities of soil dehydrogenases and alcohol dehydrognease in rice (*Oryza sativa* L.) roots. Soil Sci Plant Nutr 30:359–366, 1984.
94. M Pell, B Stenberg, J Stenström, L Torstensson. Potential denitrification activity assay in soil—with or without chloramphenicol. Soil Biol Biochem 28:393–398, 1996.

95. M Perez Mateos, S Gonzalez Carcedo. Effect of fractionation on location of enzyme activities in soil structural units. Biol Fertil Soils 1:153–159, 1985.
96. JR Pitchtel, JM Hayes. Influence of fly ash on soil microbial activity and populations. J Environ Qual 19:593–597, 1990.
97. JC Rad, M Navarro-Gonzalea, S Gonzalez-Carcedo. Characterization of proteases extracted from a compost of municipal solid wastes. Geomicrobiol J 13:45–56, 1995.
98. M Radosevich, SJ Traina, OH Tuovinen. Atrazine mineralization in laboratory-aged soil microcosms inoculated with S-triazine-degrading bacteria. J Environ Qual 26:206–214, 1997.
99. R Rai, RP Singh. Effect of salt stress on interaction between lentil (*Lens culinaris*) genotypes and *Rhizobium* spp. strains: Symbiotic N_2 fixation in normal and sodic soils. Biol Fertil Soils 29:187–195, 1999.
100. GB Reddy, A Faza, R Bennett Jr. Activity of enzymes in rhizosphere and non-rhizosphere soils amended with sludge. Soil Biol Biochem 19:203–205, 1987.
101. GB Reddy, RA Reinert, G Eason. Enzymatic changes in the rhizosphere of loblolly pine exposed to ozone and acid rain. Soil Biol Biochem 23:1115–1119, 1991.
102. GB Reddy, RA Reinert, G Eason. Loblolly pine needle nutrient and soil enzyme activity as influenced by ozone and acid rain chemistry. Soil Biol Biochem 27:1059–1064, 1995.
103. KN Reddy, RM Zablotowicz, MA Locke. Chlorimuron adsorption, desorption and degradation in soils from conventional tillage and no-tillage systems. J Environ Qual 24:760–767, 1995.
104. C Rosenzweig, D Hillel. Climate Change and the Global Harvest. New York: Oxford University Press, 1998.
105. DJ Ross, TW Speir, HA Kettles, AD Mackay. Soil microbial biomass, C and N mineralization and enzyme activities in a hill pasture: Influence of season and slow-release P and S fertilizers. Soil Biol Biochem 27:1431–1443, 1995.
106. DJ Ross, KR Tate, PCD Newton. Elevated CO_2 and temperature effects on soil carbon and nitrogen cycling in ryegrass/white clover turfs of an Endoaquept soil. Plant Soil 176:37–49, 1995.
107. P Ruggiero, VM Radogna. Inhibition of soil humus-laccase complexes by some phenoxyacetic and s-triazine herbicides. Soil Biol Biochem 17:309–312, 1985.
108. JM Sarkar, RG Burns. Synthesis and properties of β-D-glucosidase-phenolic copolymers as analogues of soil humic-enzyme complexes. Soil Biol Biochem 16:619–625, 1984.
109. KL Sahrawat. Effects of temperature and moisture on urease activity in semi-arid tropical soils. Plant Soil 78:401–408, 1984.
110. ZN Senwo, MA Tabatabai. Aspartase activity in soils. Soil Sci Soc Am J 60:1416–1422, 1996.
111. P Sequi, G Cercignani, M de Nobili, M Pagliai. A positive trend among two soil enzyme activities and a range of soil porosity under zero and conventional tillage. Soil Biol Biochem 17:255–256, 1985.
112. C Serra-Wittling, S Houot, E Barriuso. Soil enzymatic responses to addition of municipal solid-waste compost. Biol Fertil Soils 20:226–236, 1985.
113. JJ Skujins. Enzymes in soil. In AD McLaren, GA Peterson, eds. Soil Biochemistry. Vol. 1. New York: Marcel Dekker, 1967, pp 371–414.
114. MS Smith, JM Tiedje. Phases of denitrification following oxygen depletion in soil. Soil Biol Biochem 11:261–267, 1979.
115. TW Speir, DJ Ross, VA Orchard. Spatial variability of biochemical properties in a taxonomically-uniform soil under grazed pasture. Soil Biol Biochem 16:153–160, 1984.
116. ML Staben, DF Bezdicek, JL Smith, MR Fauci. Assessment of soil quality in conservation reserve program and wheat-fallow soils. Soil Sci Soc Am J 61:124–130, 1997.
117. M Stemmer, MH Gerzabek, E Kandeler. Organic matter and enzyme activity in particle-size fraction of soils obtained after low-energy sonication. Soil Biol Biochem 30:9–17, 1998.

118. M Stemmer, MH Gerzabek, E Kandeler. Invertase and xylanase activity of bulk soil and particle-size fractions during maize straw decomposition. Soil Biol Biochem 31:9–18, 1999.
119. RL Tate III. Effect of flooding on microbial activities in organic soil: Carbon metabolism. Soil Sci 128:267–273, 1979.
120. RL Tate III. Effect of several environmental parameters on carbon metabolism in histosols. Microbial Ecol 5:329–336, 1980.
121. RL Tate III. Soil Organic Matter: Biological and Ecological Effects. New York: John Wiley & Sons, 2000.
122. RL Tate III. Soil Microbiology. 2nd ed. New York: John Wiley & Sons, 2000.
123. RL Tate III, RW Parmelee, JG Ehrenfeld, L O'Reilly. Enzymatic and microbial interactions in response to pitch pine root growth. Soil Sci Soc Am J 55:998–1004, 1991.
124. RL Tate III, RW Parmelee, JG Ehrenfeld, L O'Reilly. Nitrogen mineralization: Root and microbial interactions in pitch pine microcosms. Soil Sci Soc Am J 55:1004–1008, 1991.
125. K Tatsumi, A Freyer, RD Minard, J-M Bollag. Enzymatic coupling of chloroanilines with syringic acid, vanillic acid and protocatechuic acid. Soil Biol Biochem 26:735–742, 1994.
126. JT Trevors. Dehydrogenase activity in soil: A comparison between the INT and TTC assay. Soil Biol Biochem 16:673–674, 1984.
127. JT Trevors. Effect of substrate concentration, inorganic nitrogen, O_2 concentration, temperature and pH on dehydrogenase activity in soil. Plant Soil 77:285–293, 1984.
128. H Ueno, K Miyashita, Y Sawada, Y Oba. Assay of chitinase and N-acetylglucosaminidase activity in forest soils with 4-methylumbelliferyl derivatives. Z Pflanzen Bodenk 154:171–175, 1991.
129. E Vardavakis. Seasonal fluctionations of soil microfungi in correlation with some soil enzyme activities and VA mycorrhizae associated with certain plants of a typical calcixeroll soil in Greece. Mycologia 82:715–726, 1990.
130. D Vaughan, RE Malcolm. Influence of assay conditions on invertase activity in different Scottish soils. Plant Soil 80:285–292, 1984.
131. S Visser, D Parkinson. Soil biological criteria as indicators of soil quality: Soil microorganisms. Am J Alt Agric 7:33–37, 1992.
132. M von Mersi, F Schiner. An improved and accurate method for determining the dehydrogenase activity of soils with iodonitrotetrazolium chloride. Biol Fertil Soils 11:216–220, 1991.
133. K Watanabe, K Hayano. Seasonal variation in extracted proteases and relationship to overall soil protease and exchangeable ammonia in paddy soils. Biol Fertil Soils 21:89–94, 1996.
134. SG Wirth, GA Wolf. Micro-plate colourimetric assay for endo-acting cellulase, xylanase, chitinase, 1,3-β-glucanase and amylase extracted from forest sol horizons. Soil Biol Biochem 24:511–519, 1992.
135. M Zaman, HJ Di, KC Cameron, CM Frampton. Gross nitrogen mineralization and nitrification rates and their relationships to enzyme activities and the soil microbial biomass in soils treated with dairy shed effluent and ammonium fertilizer at different water potentials. Biol Fertil Soils 29:178–186, 1999.

9

Enzyme and Microbial Dynamics of Litter Decomposition

Robert L. Sinsabaugh
University of Toledo, Toledo, Ohio

Margaret M. Carreiro
University of Louisville, Louisville, Kentucky

Sergio Alvarez
Universidad Autónoma de Madrid, Madrid, Spain

I. INTRODUCTION

The decomposition of plant litter may be the biosphere's most complex ecological process in that it involves the interactions of a large number of taxa, spanning much of the range of biotic diversity. Because of the complexity, nearly all efforts to model plant litter decomposition have approached the problem from an ecosystem perspective: predicting mass loss (the emergent biotic process) from litter composition and physical conditions—the abiotic template (26,53). Technological developments of the past two decades have removed many impediments to the study of microbial dynamics in natural systems; however, many of these tools have not been applied extensively to the study of microdecomposer communities; this is beginning to change (22,75). Consequently, fundamental information on structural and functional patterns across systems, a requisite for development of general models, is lacking. The significance of this gap is apparent in the context of global change. The effects of atmospheric carbon enrichment, nitrogen deposition, climate alteration, and other anthropogenic processes on ecosystems cannot be predicted without decomposition models grounded in biotic process (51,52).

The biotic process of decomposition spans three levels of organization: biochemical, organismal, and community. At the biochemical level, the topics of interest are the structure of plant fiber and the enzymological characteristics of degradation. The enzymological features are complex. For the polysaccharides at least one enzyme is required for every combination of monomer, linkage, and secondary structure (69). For lignin and other aromatic molecules, the process is principally oxidative; the enzymes have lower specificity, but the full range of enzymes and oxidants involved has not been defined (2,25). At the organismal level, the questions focus on the regulation of enzyme expression and the kinetics of growth. Studies of model organisms suggest a common pattern for the regula-

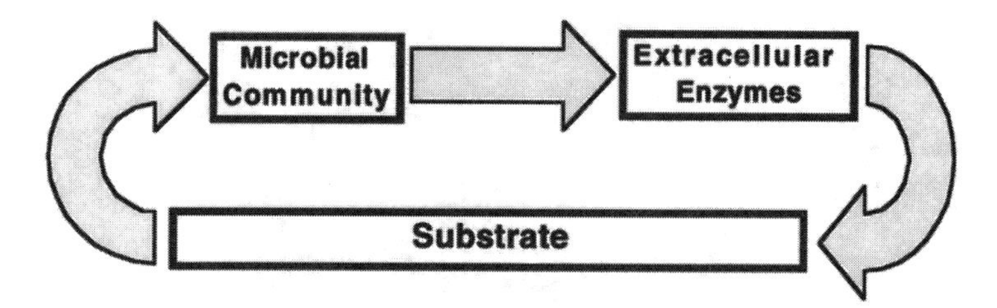

Figure 1 The decomposition process presented as a successional loop. The diagram emphasizes the dynamic interactions among microdecomposers, extracellular enzymes, and substrate and highlights the role of extracellular enzymes as the rate-controlling agents of decomposition.

tion of extracellular enzyme expression that proceeds from environmental induction, to derepression of transcription, translational expression, and then transcriptional repression, if enzymatic reaction products exceed metabolic needs (70). This system optimizes the allocation of resources to the production of enzymes that once deployed outside the cell are subject only to environmental regulation (5). At the community level, the subjects include metabolism, structure, competition, succession, and diversity. Aside from patterns of biomass and respiration, and in some cases fungal succession (19), there is probably less known at this level than the others.

The first step toward integrating microbial decomposition with traditional ecosystem perspectives is to integrate these three levels into a useful representation. This can be done by considering the decomposition process as a successional loop (Fig. 1). The substrate selects the microbial community, which produces extracellular enzymes that degrade and modify the substrate, which in turn, drives community succession. In this model, extracellular enzymes link substrate composition and microbial community metabolism. This central role, plus substrate specificity and ease of assay, make enzyme kinetics a powerful tool for investigating the functional diversity of decomposers and the mechanisms that link environmental disturbance to ecosystem responses (60,61,72).

In this chapter, we review the literature on enzyme activities and decomposition, propose some metrics for comparative analyses, and present conceptual models for future research. Our review is limited to studies of natural systems that include quantitative data on the activities of one or more extracellular enzymes in relation to mass loss from a cohort of particulate plant detritus. Studies in which enzyme activities are measured in relation to microbial biomass, production, or respiration are also included if they are directly related to plant litter decomposition. Information on the enzymatic capabilities of individual taxa is not presented. Assay methodologies are not described except when relevant for cross-study comparisons.

II. ENZYMES OF INTEREST

The enzymes of interest in decomposition studies are generally those that break down the principal components of plant fiber (cellulose, hemicellulose, pectin, lignin) into soluble units that either are directly assimilated by microorganisms or enter the dissolved organic matter pool. Because the main constituents of plant fiber do not contain nitrogen (N) or phosphorus (P), extracellular enzyme systems involved in the acquisition and recycling of N and P for microbial growth are also of interest. The most commonly measured classes of enzymes are cellulases, hemicellulases (xylanases, mannanases), pectinases, (poly) phenol oxidases, peroxidases, chitinases, peptidases, ureases, and phosphatases. Each of these functional classes includes multiple forms of the enzymes, whose structure, kinetics, and deployment may vary considerably across taxa. In general, a different enzyme is required for each type of linkage and each type of monomer; also enzymes that act on the interior linkages of polymers (endoenzymes) are usually distinct from those that attack free ends (exoenzymes). This complexity creates a hierarchy of synergistic interactions: individual enzymes are components of multienzyme systems that collectively degrade specific polymers, multiple systems of enzymes degrade the matrix of polymers that constitute plant cell walls, and diverse microbial taxa deploy enzyme systems that interact to effect decomposition. The biochemical characteristics of these enzyme systems have been reviewed extensively (3–5,9,14,20,21,25,28,34,41,47,56,58,69,83,84).

A. Aquatic Systems

The literature on enzyme dynamics in relation to litter decomposition is not extensive. The first study, by Sinsabaugh and associates (64), reported patterns of cellulase activity (β-1,4-exoglucanase, β-1,4-endoglucanase, β-glucosidase) for senescent *Cornus florida* (flowering dogwood), *Acer rubrum* (red maple), and *Quercus prinus* (chestnut oak) leaves decomposing in a woodland stream. They observed that each enzyme showed a distinct temporal pattern and that the ratio of endoglucanase to exoglucanase activities increased through time and with initial lignin content. Chamier and Dixon (8) found that pectinolytic enzymes produced by hyphomycetes, the principal fungal decomposers of plant litter in aquatic systems, were important components of the decomposition process. Tanaka (76,77) studied decomposition of senescent leaves from the reed *Phragmites communis* in a coastal saline lake. The microbial community, initially dominated by bacteria, became fungus-dominant after a few months. Both groups produced cellulolytic and xylanolytic enzymes whose activities were correlated with mass loss.

Sinsabaugh and Linkins (68) collected particulate organic matter (POM) from depositional areas of a boreal river and examined the distribution of enzyme activities in relation to particle size and composition. This descriptive study was followed by two others (74,85) that included analyses of structural similarity in POM-associated microdecomposer communities. These studies showed that POM generally becomes more recalcitrant with decreasing particle size and that carbohydrase activities tend to decrease while oxidative activities increase. Fungi become scarce as POM size decreases below 1 mm and microbial community diversity steadily increases as size approaches 0.1 mm.

To determine how particle comminution and the shift from a fungus-dominated to a bacteria-dominated decomposer community affect decomposition, POM was collected from a woodland stream and sorted into three ranges (1–4, 0.25–1, 0.063–0.25 mm), which were dried, placed in litter bags, and returned to the stream (73). Mass loss and the activities of seven enzymes involved in lignocellulose and chitin degradation were followed. Natural POM accumulations were also collected, size-sorted, and assayed. When the enzyme activities were integrated over time and regressed against mass loss, it appeared that the decomposition of particles <1 mm-was less efficient (i.e., lower mass loss increment per unit of enzyme activity) than the decomposition of POM > 1 mm by factors of 1.5 to 7. In addition, enzyme activities associated with the POM confined in litter bags were generally lower than those associated with in situ POM, suggesting that the litter bag technique was underestimating in situ turnover rates. Estimates of in situ turnover rates, generated from regression models relating enzyme activities and mass loss, were up to twice that of confined POM. This approach was applied by Jackson et al. (29) to study the spatial and temporal dynamics of POM turnover in a *Typha* sp. (cattail) marsh. Three size ranges of POM were collected and placed in litter bags at two sites. In situ enzyme activities were monitored along transects across the marsh. Size-specific relationships between enzyme activity and mass loss generated from the litter bag data were used to estimate instantaneous mass loss rates across the spatial grid.

Sinsabaugh and Findlay (65) collected four size ranges of POM from a *Typha* sp. wetland, a *Trapa* sp. (water chestnut) wetland, and two channel sites along the Hudson River estuary and assayed for lignocellulose and chitin-degrading enzyme activities; bacterial and fungal biomass and productivity were estimated. Bacterial biomass and productivity increased as particle size declined, whereas fungal biomass decreased. Using estimates of POM turnover rate based on enzyme activities, they calculated that production efficiency, i.e., production rate/decomposition rate, ranged from 1% to 30%; thus most of the soluble products of decomposition were exported as dissolved organic matter (DOM) rather than metabolized in situ.

Denward et al. (15) used microcosms containing *Phragmites australis* to assess the effects of solar radiation on decomposition. Compared to that of shaded controls, bacterial abundance increased relative to that of fungi. β-Glucosidase activity also increased, shifting the α-glucosidase/β-glucosidase ratio from >1 to <1.

Other studies in aquatic systems have taken a comparative ecosystem approach. Kok and Van der Velde (36) placed litter bags containing fragments of senescent water lily leaves (*Nymphaea alba*) in alkaline (pH ca. 8) and acidic (pH ca. 5) freshwater ponds. The contents of each bag were analyzed for mass loss, cellulase activity, and xylanase activity. In a parallel study, they followed the decomposition of *Nymphaea alba* leaf disks in six freshwater microcosms with pH values from 4.0 to 8.0. Litter from each microcosm was analyzed for mass loss and the activities of cellulase, xylanase, polygalacturonase (pectinase), and pectin lyase. Corroborating the work of Chamier and Dixon (8), they concluded that the pH dependence of pectinolytic activity was a critical factor underlying differences in mass loss rates with system pH.

The decomposition of *Liriodendron tulipifera* (tulip poplar) wood was studied in a small mountain stream from which new litter inputs were excluded (78). The activities of phosphatase and five lignocellulose-degrading enzymes were followed along with fungal biomass and breakdown rates. Compared to those in a reference stream, fungal biomass, enzyme activities, and breakdown rates were higher in the litter-excluded stream. Differences in the ratios of phosphatase to carbohydrase and phenol oxidase to carbohydrase between the systems suggested that the increased decomposition activity was the result of higher nitrogen and phosphorus availability, a finding confirmed by water chemical analyses.

Raviraja et al. (54) looked at hyphomycete diversity in relation to enzyme activities and mass loss at organically polluted river sites, using two litter types: *Ficus benghalensis* and *Eucalyptus globulus*. Although diversity was strongly depressed compared to that of unpolluted sites, mass loss rates and enzyme (cellulase, amylase, xylanase, pectinase) activities did not differ.

Alvarez et al. (2) examined POM turnover in two ephemeral wetlands. Toro pond was surrounded by a *Pinus pinea* forest and had a littoral belt of *Juncus* and *Scirpus* spp. Oro pond was surrounded by a eucalyptus plantation and had a disturbed littoral zone. Both ponds dried completely during the summer. Two size ranges of POM (>1 mm and 0.063–0.5 mm) were collected from each site and placed in litter bags. In using the approach of Sinsabaugh et al. (73) and Jackson et al. (29), confined and in situ POM samples were assayed monthly for β-glucosidase, β-*N*-acetylglucosaminidase, β-xylosidase, phenol oxidase, and alkaline phosphatase. When regressions of integrated enzyme activity and mass loss were compared, it appeared that decomposition of coarse particles was about 20 times more efficient than that of fine particles at the Toro site and about 10 times more efficient at the Oro site. Differences between sites were attributed to differences in organic matter quality and to the lower pH of Oro pond. At both sites, enzyme activities measured on material confined in litter bags were lower than those measured in situ.

B. Terrestrial Systems

Like those in aquatic systems, the terrestrial studies can be roughly classified into those dealing with fine-scale questions, such as the relationship between enzyme activities and litter composition or microbial dynamics and those that take a larger-scale comparative ecosystem approach. Linkins et al. (42) followed the decomposition of senescent *Cornus florida* (flowering dogwood), *Quercus prinus* (white oak), and *Acer rubrum* (red maple) leaves in a deciduous woodland in southwest Virginia. This step was followed by a microcosm study using the same litter types (43). They found that activity levels varied with litter type, that cellulose disappearance and mass loss were correlated with cellulase activities, and that cellulolytic activity declined sharply as the lignocellulose index (LCI) approached 0.7. (LCI is the fraction of acid-insoluble material in the residual plant fiber: [lignin + humus]/[lignocellulose + humus]).

Zak et al. (88) examined fungal diversity, lignocellulase activities, and mass loss rates on mesquite sticks incorporated into the middens of desert wood rats. The dominant fungal taxa and fungal diversity varied with moisture availability, but these structural changes did not correlate with enzyme activity patterns or mass loss in these arid systems.

Litter decomposition in a suburban forest in relation to N deposition has been studied (6). Senescent leaves of *Quercus rubra* (red oak), *Acer rubrum* (red maple), and *Cornus florida* (flowering dogwood) were placed on forest floor plots that were sprayed monthly with distilled water or with NH_4NO_3 solution at dose rates equivalent to 2 or 8 g N $m^{-2}y^{-1}$. Mass loss responses to N amendment varied with the lignin content of the litter. Dogwood, a fast-decomposing, low-lignin litter, decomposed up to 25% faster than did the control plots. Maple, intermediate in lignin content, decomposed slightly faster at the lower N deposition rate and slightly slower at the higher rate. Mass loss rates for heavily lignified oak litter declined by up to 25%. Fungal biomass increased for all litter types (40% maple, 32% dogwood, 15% oak). Cellulolytic activity, measured by assays for β-glucosidase, cellobiohydrolase, and endoglucanase, increased with N deposition for all litter types; ligninolytic activity, measured by assays for phenol oxidase and peroxidase, varied with the lignin content of the litter. With added N, oxidative activity increased on dogwood litter, decreased on oak litter, and stayed about the same on maple litter. For all litter types, phosphatase activity increased with N deposition, indicating higher P demand. For dogwood, the activities of peptidase and chitinase, enzymes involved in N acquisition, were repressed by added N; for maple and oak, these activities increased. The results suggested that white rot fungi, which produce ligninases in response to low N availability, were displaced by supplemental N, slowing the decomposition of recalcitrant litter.

Henriksen and Breland (27) also focused on the role of N in the decomposition process. Using a microcosm system of wheat straw and soil, they found that carbon mineralization, fungal biomass, and activities of cellulolytic and hemicellulolytic enzymes decreased with N availability.

In the area of comparative ecosystem studies, Sinsabaugh et al. (62,63) followed mass loss, N and P immobilization, and activity of 11 types of extracellular enzymes for birch sticks (*Betula papyfera*) decomposing at eight upland, riparian, and lotic sites over a first-order watershed. Mass loss rates among sites varied by a factor of 5 and were correlated with lignocellulase activities. In contrast, relationships between mass loss and activities of acid phosphatase and β-1,4-*N*-acetylglucosaminidase varied widely among sites. These relationships along with analyses of the N and P content of the sticks suggested that differences in mass loss rates among sites were tied to differences in nutrient availability.

In another experiment, litter bags containing senescent leaves of *Ageratum conizoides* and *Mallotus philippinensis* were placed on the floor of a young tropical forest site in northeast India (38). Other litter bags containing leaves of *Holarrhena antidysenterica* and *Vitex glabrata* were placed at a mature tropical forest site. At higher-elevation subtropical sites, litter bags containing *Pinus kesiya* and *Myrica esculenta* leaves were placed in a young forest and bags containing *Pinus kesiya* and *Alnus nepalensis* leaves were placed in a mature forest. Samples were analyzed for mass loss, bacterial and fungal numbers, cellulose content, N content, soluble sugar content, and activities of cellulase, amylase, and invertase. Cellulase and amylase activities were correlated with microbial numbers. Invertase activity correlated with soluble sugar content. Enzyme activities and mass loss rates were higher at the lower elevation sites but were not related to stand age. In a similar study, the decomposition of *Pinus kesiya* and *Alnus nepalensis* at a disturbed roadside forest site was compared with that at an undisturbed site (30). Again cellulase and amylase activities were correlated with microbial numbers, whereas invertase activity was linked to soluble sugars.

Dilly and Munch (18) studied enzyme activities and microbial respiration for *Alnus glutinosa* (black alder) leaves decomposing at wet and dry sites within a fen forest. Mass loss rates were more than twice as fast at the wet site. Microbial biomass and respiration decreased over time (16 to 2.3 μmol $g^{-1}h^{-1}$), but the efficiency of C utilization increased. These trends were paralleled by decreasing β-glucosidase activity and increasing protease activity.

III. COMPARATIVE ANALYSES

In the context of the successional loop model (Fig. 1), there are three dimensions for comparing studies of enzymatic decomposition: enzyme activity and litter composition, enzyme activity and mass loss rate, enzyme activity and community composition. A fourth dimension is large-scale patterns in relation to ecosystem type or disturbance type. These comparisons are external to the model but integrate enzymatic decomposition into large-scale perspectives. At present, there are too few studies in any of these areas to support much beyond inference, and in any case, comparisons are generally complicated by methodological diversity (72).

A. Enzyme Activity and Litter Composition

It is clear that enzyme activities vary with litter composition. Patterns are most easily seen when different types of litter decompose in the same environment. The patterns arise from both biotic and abiotic processes. The biotic processes are substrate selection of microdecomposer populations and physiological regulation of enzyme secretion. In addition, litter-specific activity patterns to some degree reflect physicochemical processes of adsorption and stabilization (5,66), which are functions of the architecture and composition of the litter. The relative contribution of biotic and abiotic processes to enzyme activity patterns across litter types is probably a function of the structural resistance of the enzyme to inhibition, denaturation, and proteolysis. If the turnover time for a particular enzyme activity is longer than that for microbial populations, sorption processes may contribute to litter-specific patterns; if it is lower, then organismal processes predominate.

Enzymes like α-glucosidase and invertase that process soluble saccharides appear to be in the latter category. Their activities are correlated with the soluble saccharide content of the litter (38,30). Activities generally peak early in the decomposition process, then decline markedly; activities tend to be higher on fast-decomposing litter (Fig. 2). β-

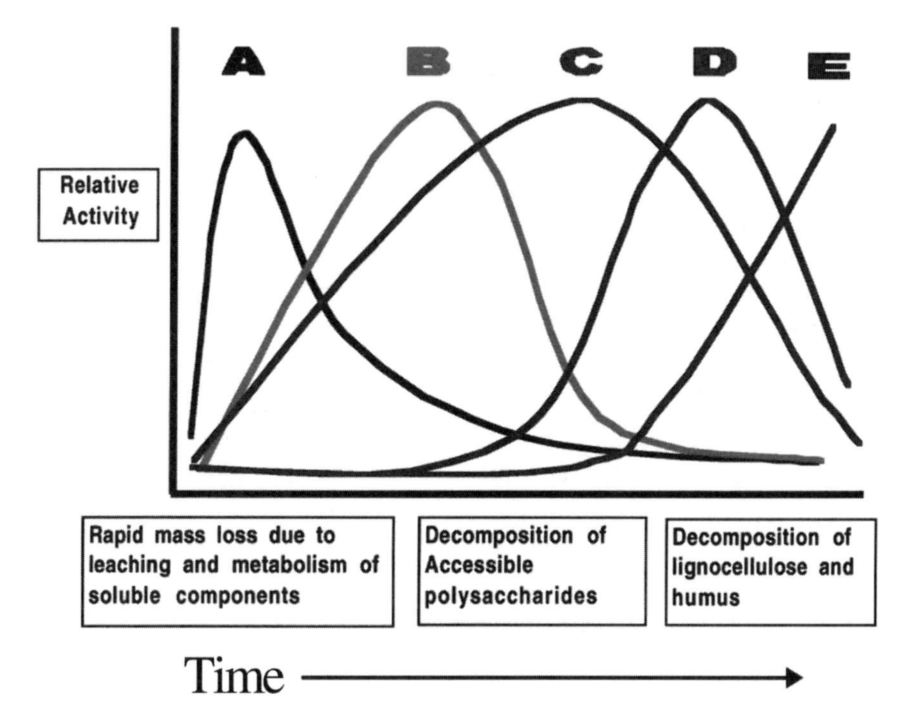

Figure 2 Hypothetical distribution of relative enzyme activities with time for a decomposing cohort of herbaceous plant litter. The patterns reflect general trends reported in the literature. (A), invertase, α-glucosidase; (B), β-1,4-exoglucanase (exocellulase), (C), β-1,4-endoglucanase (endocellulase), (D), (poly)phenol oxidase; (E), peroxidase.

Glucosidase activity also tends to be highest during the early stages of decomposition, but because of its role in cellulolysis, activity remains significant even at high mass loss values (18,64). The activities of the other cellulases, β-1,4-endoglucanase and β-1,4,-exoglucanase, increase more slowly and generally peak about midway (40–80% mass loss) through decomposition; early peaks are associated with heavily lignified litter and later peaks with labile litter (42,62). As accessible cellulose disappears, the ratio of endoglucanase to exoglucanase tends to increase, at least partly as a result of differential sorption (66). (Poly)phenol oxidase activity tends to increase with lignin-humus content (68), but activities can also be relatively high early in decomposition for litters that have high tannin contents (6). In heavily humified material peroxidase activities predominate. These trends appear to be general: they occur in both aquatic and terrestrial systems; they apply to individual litter types decomposing through time, as well as to differences among litters of varying initial composition; and they can be observed along gradients of decreasing particle size. The generalities suggest that enzyme activities may be used to make inferences about organic matter quality across environmental templates. Ratios of β-1,4-endoglucanase to β-1,4-exoglucanase activity (64) or cellulolytic: ligninolytic activity (65) have been suggested for this purpose. However, environmental fluctuations that alter temperature, moisture, and nutrient availability also alter enzyme activities and may obscure or overwhelm patterns linked to litter quality.

B. Enzyme Activities and Mass Loss

Correlating enzyme activities with litter mass loss provides information on the mechanics of decomposition. The activities of several enzymes have been correlated with the rate of disappearance of specific litter constituents or with mass loss in general. This information can be used in various ways. One is to model enzymatic decomposition in relation to temperature and water potential (48,72). Another is to use statistical models to estimate instantaneous mass loss rates from enzyme activities. This second approach has been applied to provide estimates of organic matter turnover in heterogeneous systems (65) and to estimate the turnover rate of fine particulate organic matter, which appears to be underestimated by the traditional litter bag method (2,29). The third use is to describe the efficiency and functional diversity of enzymatic decomposition. For these comparisons, turnover activities are a useful measurement.

Turnover activities are calculated from models of mass loss as a function of cumulative enzyme activity, analogous to traditional models that describe litter decay over time as a first-order function of residual mass (Fig. 3). Cumulative enzyme activity is expressed in units of activity-days, which are calculated by integrating the area under a curve of enzyme activity vs. time. A linear regression, LN (cumulative activity-days) vs. time, generates a first-order rate constant called the *apparent enzymatic efficiency* with units of activity-day^{-1}. Inverting this rate constant produces an estimate of litter turnover expressed in units of activity-days. Implicit in this model is the premise, supported by field data (72), that mass loss per unit of enzyme activity decreases through time because of the increasing recalcitrance of the residual material.

Like the traditional mass loss constant, and its inverse turnover time, these turnover activities provide a basis for comparison across sites, treatments, and litter types. These comparisons convey a sense of the quantity and type of "work" a microbial community has to do to decompose a cohort of litter and how this work is linked to substrate heterogeneity and enzyme synergisms. They also provide a way to "map" the functional diversity

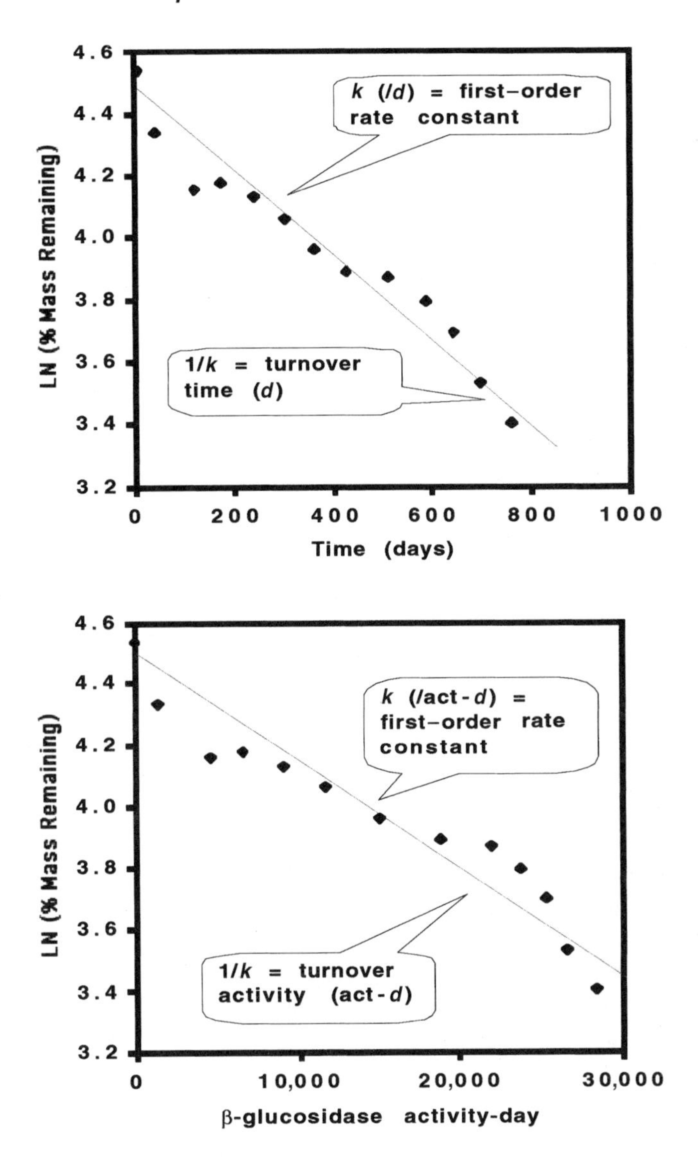

Figure 3 Comparison between turnover time and turnover activity for a litter cohort. In this example, β-glucosidase activity was assayed during the decomposition of *Acer rubrum* litter at a forest site. The upper graph shows a traditional first-order exponential decay curve with cumulative mass loss plotted as a function of time. The slope of the linear regression is a rate constant (k) with units of day^{-1}; $1/k$ is the turnover time for the litter in days. The lower graph shows mass loss as a function of cumulative β-glucosidase activity. Cumulative activity is calculated by integrating the area under the curve of β-glucosidase activity vs. time and is expressed in units of activity-days. The slope of the regression is a first-order rate constant (k) with units of activity-day^{-1}; $1/k$ is the β-glucosidase turnover activity for the litter.

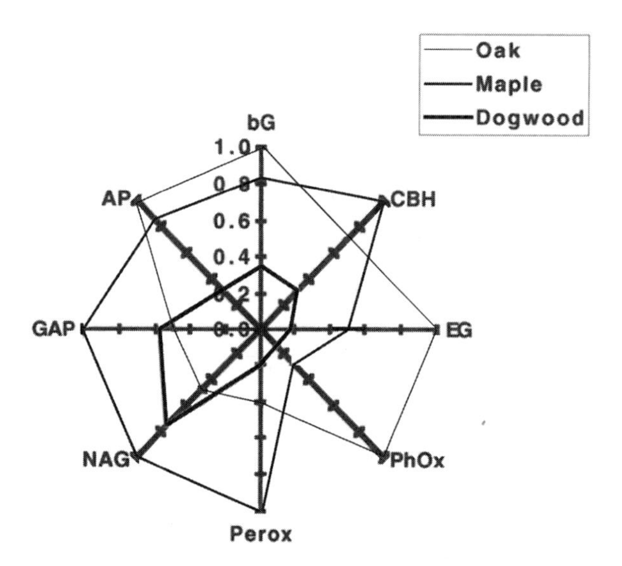

Figure 4 Functional profiles of *Cornus florida* (flowering dogwood), *Acer rubrum* (red maple), and *Quercus borealis* (red oak) leaf litter decomposition based on turnover activities. A comparison of relative turnover activities shows that the enzymatic decomposition of dogwood leaves was more efficient than that of maple and oak. Oak leaves required the most phenol oxidase activity, whereas maple required the most peroxidase activity. Phosphorus acquisition activity was highest for oak; nitrogen acquisition activity was highest for maple. G, β-glucosidase; CBH, cellobiohydrolase; EG, β-1,4-endoglucanase; PhOx, phenol oxidase; Perox, peroxidase; NAG, β-1,4-N-acetylglucosaminidase; GAP, glycine amino peptidase; AP, acid phosphatase. (Data from Ref. 6b).

of decomposition. One example (6) shows that the decomposition of flowering dogwood leaves was accomplished with much less enzyme activity than that needed to turn over red maple and red oak litter and that extensive phenol oxidase activity was needed to decompose oak leaves, which have a lot of lignin, whereas a lot of peroxidase activity was required for decomposing maple leaves, which have a lot of nonlignin phenols (Fig. 4). Other studies suggest that the apparent enzymatic efficiency of decomposition declines with particle size (2,29), coinciding with the transition from fungal to bacterial dominance but also with increasing humification.

Turnover activities can also be calculated for enzymes that are not directly involved in the decomposition of major litter components. For enzymes such as phosphatase, urease, peptidase, and chitinase, turnover activities are measures of relative effort directed toward obtaining N and P from organic sources. Such comparisons have proved useful in understanding differences in decomposition rates among systems (63,78). Even within the same system, the enzymatic effort directed toward the acquisition of organic N and P varies among litter types (Fig. 4). A major constraint on the value of turnover activities is that direct comparisons across studies cannot be made unless the same assay methodology was used.

C. Enzyme Activity and Community Composition

In some decomposition studies, microbial numbers, biomass, or respiration has been linked with extracellular enzyme activities, but there have been few attempts to link community composition or biodiversity with enzymatic process. Maire and associates (46) examined the relationships among soil respiration, microbial diversity (using phospholipid fatty acid analysis), and activities of xylanase, laminarinase, phosphatase, urease, and chitinase. They found a correspondence between functional diversity and structural diversity, both peaking in spring. Others (44,45) found that differences in breakdown rates and cellulolytic activity between permanent and temporary stream sites were associated with differences in fungal diversity and bacterial biomass. However, Raviraja et al. (54) and Zak et al. (88) found no relationships among fungal diversity, enzyme activities, and mass loss.

Pollution gradients may be good systems for investigating such relationships. Kandeler et al. (31) reported that heavy metal contamination decreased microbial biomass and functional diversity in soils. C-acquiring enzymes (cellulase, xylanase, β-glucosidase) were the least affected, phosphatase and sulfatase the most affected; N-acquiring enzymes (urease) were intermediate. Another study of heavy metal contamination in grassland soils (39) showed that reductions in microbial biomass and substrate-induced respiration paralleled 10- to 50-fold reductions in extracellular enzyme activities. β-Glucosidase activity was the most depressed, phosphatase and endocellulase activities were the least; reduction in β-*N*-acetylglucosaminidase activity was intermediate.

From a systematic perspective, many fungal taxa can be classified as sugar fungi or brown rot, soft rot, and white rot decomposers whose extracellular enzyme complements have been characterized to varying degrees (57). Fungal succession on major categories of litter has been well studied. Within a particular habitat, the dominant populations vary more or less predictably through time, selected by their substrate utilization capabilities, their tolerance of inhibitory phenolic compounds, and their effective ranges of temperature, water potential, and nitrogen availability (17,19,23,55). The result is a dynamic community that at any particular time is dominated by a relatively small number of populations, whose identity varies with litter type and habitat. Thus the taxonomic diversity of fungal communities may be viewed as either low or high, depending on the spatial and temporal scales under consideration (86).

From an ecosystem perspective, it seems likely that the prominence of ligninase-producing basidiomycetes (35) in terrestrial systems and pectinase-producing hyphomycetes (7,89) in aquatic systems probably affects the functional profile and efficiency of enzymatic decomposition. Such differences have not been explicitly described but have the potential to influence the quantity of carbon metabolized in situ, and the quantity exported, as well as the form. Differences in efficiency probably exist between bacterially and fungally dominated systems as well. Within bacterial systems, the functional diversity and efficiency of enzymatic decomposition probably vary with the distribution of oxygen and other electron acceptors. For detritus with a high concentration of aromatic residues, decomposition may all but shut down in the absence of oxygenases and peroxidases. In experimentally manipulated wetland soils, McLatchey and Reddy (49) found microbial biomass and mineralization of C, N, and P decreased with redox potential; phosphatase, protease, and β-glucosidase activities also declined; phenol oxidase was detectable only under aerobic conditions.

Much remains to be learned about the significance of functional and structural diversity in the decomposition process. Understanding these mechanics—the relationships among microbial community composition, enzyme activity, and litter breakdown—is a requisite for better understanding of ecosystem function.

D. Enzyme Activities and Ecosystems

Outside the context of the successional loop, decomposition studies focus on regulation by climate and nutrient availability. Freeze–thaw and wet–dry events affect the temporal pattern of decomposition across systems but also influence enzyme activities and the decomposability of litter (10,11,59,67,79,81). The role of nutrient availability is generally assessed by using amendment studies. An alternative approach is to examine the relative distribution of enzyme activities directed toward the acquisition of C, N, and P.

Sinsabaugh and Moorhead (70,71) proposed a model for the distribution of extracellular enzyme activities in relation to mass loss, litter composition, and nutrient availability. The model, called *Microbial Allocation of Resources among Community Indicator Enzymes* (MARCIE), is based on the observation that the production of extracellular enzymes is often controlled by induction/repression mechanisms tied to substrate availability. At the community level, this type of regulation resembles an optimal resource allocation strategy for maximizing microbial productivity. The model links litter mass loss with the activities of C-acquiring enzymes (e.g., cellulases, hemicellulases), which are constrained by effort directed toward the acquisition of N and P. In this model, ratios of N acquisition activity (e.g., chitinase, urease, peptidase) to C-acquistion activity and of P-acquisition activity (e.g., phosphatase) to C-acquisition activity become indicators of relative nutrient availability. The general utility of the model remains to be established, but it has provided insight into nutrient regulation of decomposition rates in at least two studies (63,78) and has potential application in the area of global change research.

Evidence that decomposer communities respond quickly to global change disturbances is accumulating (12,24,50,87,90). Körner and Arnone (37) and Dhillion et al. (16), working in tropical and mediterannean systems, respectively, reported increased activities of several soil enzymes in response to atmospheric CO_2 enrichment, presumably an effect of fine-root C priming. Others (6) found that the effects of N deposition on decomposition were litter-specific because of the connections among N availability, ligninase production, and distribution of white rot fungi. The effects of atmospheric CO_2 accumulation, warming, and N deposition on decomposition have the potential to alter soil carbon storage in ways that may accelerate or mitigate global climate change. At present, these changes are both difficult to predict and difficult to measure directly. The pool sizes are large. The addition of nutrients that alter the composition of the microbial community may reduce decomposition and increase soil organic matter storage even as soil respiration increases. For these questions, enzyme responses interpreted in the context of the successional loop and MARCIE model may be the most sensitive indicators of the magnitude and direction of change.

IV. CONCLUSIONS

Extracellular enzymes directly mediate organic matter breakdown. They link microbial community organization, litter composition, and environmental conditions. They are also the most readily monitored components of the decomposition process. This combination means that models centered on enzyme dynamics have the potential for wide application in ecological research.

The simplest models relate mass loss to enzyme activities. The potential applications are monitoring of organic matter pools in heterogeneous systems (65) and conversion of spatial patterns of enzyme activity (13,32) into estimates of decomposition rate. Because this is an empirical approach, models must be developed for each system. The models may incorporate the activities of several enzymes, but it is also possible that one to two ''critical activities'' may have adequate predictive power in some systems. Because systems with homogeneous litter quality and stable environmental conditions are likely to produce the most powerful models, this approach may work better for aquatic systems than for terrestrial ones.

Resource allocation models like MARCIE are valuable for investigating nutrient controls on decomposition, reducing the need for resource intensive amendment studies. Because they are constructed from the microbial community perspective, they can resolve fine-scale patterns with respect to time, space, or litter type. As information on the regulation of extracellular enzyme expression accumulates, the utility of this approach may increase.

At present no models link functional diversity with decomposition. Functional diversity may affect the efficiency of decomposition, which in turn is likely to influence the growth of microdecomposers, rate of particle comminution, quantity and form of carbon exported as dissolved organic matter, and nutrient availability for plant growth. Ligninases are a clear case: loss of activity as a result of insufficient oxygen or excess nitrogen can markedly reduce the enzymatic efficiency of decomposition. Studies focusing on microbial production dynamics in relation to enzyme activities and dissolved organic matter production have not been done in terrestrial systems. The subject gets a lot of attention in aquatic systems (1,33,65,81,82), though not in the context of plant litter decomposition. Research in this area has the potential to improve understanding of food web organization and carbon transduction in heterotrophic systems.

Simulation models can address mechanical questions of decomposition and community metabolism that are not amenable to direct measurement. Models that recreate patterns of litter composition, mass loss, and enzyme dynamics observed in field studies identify critical parameters and conceptual gaps and provide predictive hypotheses (51). Some fundamental questions such as the relationship between microbial biomass turnover and extracellular enzyme turnover and the number of C or N atoms assimilated for each C or N atom deployed in extracellular enzymes may be difficult to address outside the realm of modeling but clearly affect the movement of detrital carbon into the food web.

Over the past four decades, most of the research on enzymatic decomposition has focused on assay methodology and comparative description of varied systems. The challenge for the next decade is to move enzymic analysis toward the realm of application. To achieve this, more research is needed on functional diversity and the efficiency of decomposition, and on the regulation of enzyme expression and activity. As this information develops, enzyme assays will become increasingly useful tools for monitoring and understanding ecosystem function.

REFERENCES

1. M Agis, J Iriberri, M Unanue. Bacterial colonization and ectoenzymatic activity in phytoplankton-derived model particles. Part II. Cleavage and uptake of carbohydrates. Microb Ecol 36: 66–74, 1998.
2. S Alvarez, MC Guerrero. Enzymatic activities associated with decomposition of particulate organic matter in two shallow ponds. Soil Biol and Biochem 32:1941–1951.
3. RA Blanchette. Delignification by wood-decay fungi. Annu Rev Phytopathol 29:381–398, 1991.

4. P Biely. Microbial xylanolytic systems. Trends Biotechnol 3:286–290, 1985.
5. RG Burns. Extracellular enzyme-substrate interactions in soil. In: JH Slater, R Whittenbury, JWT Wimpenny, eds. Microbes in Their Natural Environments. Cambridge: Cambridge University Press, 1983, pp 249–298.
6. MM Carreiro, RL Sinsabaugh, DA Repert, DF Parkhurst. Microbial enzyme shifts explain litter decay responses to simulated nitrogen deposition. Ecology
7. A-C Chamier. Cell-wall degrading enzymes of aquatic hyphomycetes: A review. Linnean Soc 91:67–81, 1985.
8. A-C Chamier, PA Dixon. Pectinases in leaf degradation by aquatic hyphomycetes: The enzymes and leaf maceration. J Gen Microbiol 128:2469–2483, 1982.
9. RJ Chróst. Environmental control of the synthesis and activity of aquatic microbial ectoenzymes. In: RJ Chróst, ed. Microbial Enzymes in Aquatic Environments. New York: Springer-Verlag, 1991, pp 25–59.
10. JS Clein, JP Schimel. Reduction in microbial activity in birch litter due to drying and rewetting events. Soil Biol Biochem 26:403–406, 1994.
11. JS Clein, JP Schimel. Microbial activity of tundra and taiga soils at sub-zero temperatures. Soil Biol Biochem 27:1231–1234, 1995.
12. MF Cotrufo, P Ineson, AP Rowland. Decomposition of tree leaf litters grown under elevated CO_2: Effect of litter quality. Plant Soil 163:121–130, 1994.
13. KLM Decker, REJ Boerner, SJ Morris. Scale-dependent patterns of soil enzyme activity in a forested landscape. Can J Forest Res 1998.
14. RFH Dekker. Biodegradation of hemicelluloses. In: T Higuchi, ed. Biosynthesis and Biodegradation of Wood Components. New York: Academic Press, 1985, pp 505–533.
15. CMT Denward, H Edling, LJ Tranvik. Effects of solar radiation on bacterial and fungal density on aquatic plant detritus. Freshwater Biol 41:575–582, 1999.
16. SS Dhillion, J Roy, M Abrams. Assessing the impact of elevated CO_2 on soil microbial activity in a Mediterranean model ecosystem. Plant Soil 165:1966.
17. J Dighton. Nutrient cycling by saprotrophic fungi in terrestrial habitats. In: DT Wicklow, B Söderström, eds. The Mycota. Vol. IV. Environmental and Microbial Relationships. Berlin: Springer-Verlag, 1997, pp 271–280.
18. O Dilly, J-C Munch. Microbial biomass content, basal respiration and enzyme activities during the course of decomposition of leaf litter in a black alder (Alnus gluinosa (L.) Gaertn.) forest. Soil Biol Biochem 28:1073–1081, 1996.
19. NJ Dix, J Webster. Fungal Ecology. London: Chapman & Hall, 1995.
20. K-E Eriksson, RA Blanchette, P Ander. Microbial and enzymatic degradation of wood components. Berlin: Springer-Verlag, 1990.
21. K-E Eriksson, TM Wood. Biodegradation of cellulose. In: T Higuchi, ed. Biosynthesis and Biodegradation of Wood Components. New York: Academic Press, 1985, pp 469–503.
22. A Felske, A Wolterink, R Van Lis, ADL Akkermans. Phylogeny of the main bacterial 16S rRNA sequences in Drentse A grassland soils (The Netherlands). Appl Environ Microbiol 64: 871–879, 1998.
23. MO Gessner, K Suberkropp, E Chauvet. Decomposition of plant litter by fungi in marine and freshwater ecosystems. In: DT Wicklow, B Söderström, eds. The Mycota. Vol. IV. Environmental and Microbial Relationships. Berlin: Springer-Verlag, 1997, pp 303–322.
24. B Griffiths, K Ritz, N Ebblewhite, E Paterson, K Kilham. Rye grass rhizosphere microbial community structure under elevated carbon dioxide concentrations with observations on wheat rhizosphere. Soil Biol Biochem 30:315–321, 1998.
25. KE Hammel. Fungal degradation of lignin. In: G Cadisch, KE Giller, eds. Driven by Nature: Plant Litter Quality and Decomposition. Wallingford: CAB International, 1997, pp 33–46.
26. OW Heal, JM Anderson, MJ Swift. Plant litter quality and decomposition: An historical overview. In: G Cadisch, KE Giller, eds. Driven by Nature: Plant Litter Quality and Decomposition. Wallingford: CAB International, 1997, pp 3–32.

27. TM Henriksen, TA Breland. Nitrogen availability effects on carbon mineralization, fungal and bacterial growth, and enzyme activities during decomposition of wheat straw in soil. Soil Biol Biochem 31:1121–1134, 1999.
28. T Higuchi. Lignin biochemistry: Biosynthesis and biodegradation. Wood Sci Technol 24:23–63, 1990.
29. C Jackson, C Foreman, RL Sinsabaugh. Microbial enzyme activities as indicators of organic matter processing rates in a Lake Erie coastal wetland. Freshwater Biol 34:329–342, 1995.
30. SR Joshi, RR Mishra, GD Sharma. Mirobial enzyme activities related to litter decomposition near a highway in a sub-tropical forest of north east India. Soil Biol Biochem 25:1763–1770, 1993.
31. E Kandeler, C Kampichler, O Horak. Influence of heavy metals on the functional diversity of soil microbial communities. Biol Fertil Soils 23:299–306, 1996.
32. H Kang, C Freeman, D Lee, WJ Mitch. Enzyme activities in constructed wetlands: Implication for water quality amelioration. Hydrobiologia 368:231–235, 1998.
33. M Karner, GJ Herndl. Extracellular enzymatic activity and secondary production in free-living and marine-snow-associated bacteria. Mar Biol 113:341–347, 1992.
34. TK Kirk, RL Farrell. Enzymatic ''combustion'': The microbial degradation of lignin. Annu Rev Microbiol 41:465–505, 1987.
35. TK Kirk, M Shimada. Lignin biodegradation: The microorganisms involved, and the physiology and biochemistry of degradation by white-rot fungi. In: T Higuchi, ed. Biosynthesis and Biodegradation of Wood Components. New York: Academic Press, 1985, pp 579–605.
36. CJ Kok, G Van der Velde. The influence of selected water quality parameters on the decay rate and exoenzymatic activity of detritus of *Nymphaea alba* L. floating leaf blades in laboratory experiments. Oecologia 88:311–316, 1991.
37. C Körner, JA Arnone. Responses to elevated carbon dioxide in artificial tropical ecosystems. Science 257:1672–1675, 1992.
38. S Kshattriya, GD Sharma, RR Mishra. Enzyme activities related to litter decomposition in forests of different age and altitude in North East India. Soil Biol Biochem 24:265–270, 1992.
39. RG Kuperman, MM Carreiro. Soil heavy metal concentrations, microbial biomass and enzyme activities in a contaminated grassland ecosystem. Soil Biol Biochem 29:179–190, 1997.
40. P Lahdesmaki, R Piispanen. Degradation products and hydrolytic enzyme activities in the soil humification process. Soil Biol Biochem 20:287–292, 1988.
41. LG Ljungdahl, K-E Eriksson. Ecology of microbial cellulose degradation. Adv Microb Ecol 8:237–299, 1985.
42. AE Linkins, RL Sinsabaugh, CM McClaugherty, JM Melillo. Comparison of cellulase activity on decomposing leaves in a hardwood forest and woodland stream. Soil Biol Biochem 22: 423–425, 1990.
43. AE Linkins, RL Sinsabaugh, CM McClaugherty, JM Melillo. Cellulase activity on decomposing leaf litter in microcosms. Plant Soil 123:17–25, 1990.
44. A Maamri, E Chauvet, H Chergui, F Gourbiere, E Pattee. Microbial dynamics on decaying leaves in a temporary Moroccan river. I. Fungi. Arch Hydrobiol 144(1):41–59, 1998.
45. A Maamri, E Pattee, X Gayte, H Chergul. Microbial dynamics on decaying leaves in a temporary Moroccan river. II. Bacteria. Arch Hydrobiol 144:157–175, 1999.
46. N Maire, D Borcard, E Laczko, W Matthey. Organic matter cycling in grassland soils of the Swiss Jura mountains: Biodiversity and strategies of the living communities. Soil Biol Biochem 31:1281–1293, 1999.
47. WL Marsden, PP Gray. Enzymatic hydrolysis of cellulose in lignocellulosic materials. CRC Crit Rev Biotechnol 3:235–276, 1986.
48. CA McClaugherty, AE Linkins. Temperature response of extracellular enzymes in two forest soils. Soil Biol Biochem 22:29–34, 1990.
49. GP McLatchey, KR Reddy. Regulation of organic matter decomposition and nutrient release in a wetland soil. J Environ Qual 27:1268–1274, 1998.

50. L Marilley, U Hartwig, M Aragno. Influence of an elevated atmospheric CO2 content on soil and rhizosphere bacterial communities beneath *Lolium perenne* and *Trifolium repens* under field conditions. Microb Ecol 38:39–49, 1999.
51. DL Moorhead, RL Sinsabaugh. Simulated patterns of litter decay predict patterns of extracellular enzyme activities. Applied Soil Ecology.
52. DL Moorhead, RL Sinsabaugh, AE Linkins, JF Reynolds. Decomposition processes: Modelling approaches and applications. Sci Total Environ 183:137–149, 1996.
53. K Paustian, G Ågren, E Bosatta. Modelling litter quality effects on decomposition and soil organic matter dynamics. In: G Cadisch, KE Giller, eds. Driven by Nature: Plant Litter Quality and Decomposition. Wallingford: CAB International, 1997, pp 313–336.
54. NS Raviraja, KR Sridhar, F Bärlocher. Breakdown of *Ficus* and *Eucalyptus* leaves in an organically polluted river in India: Fungal diversity and ecological functions. Freshwater Biol 39: 537–545, 1998.
55. ADM Rayner. Fungi, a vital component of ecosystem function in woodland. In: D Allsop, RR Colwell, DL Hawksworth, eds. Microbial Diversity and Ecosystem Function. New York: CAB International, 1995, pp 231–254.
56. ADM Rayner, L Boddy. Fungal Decomposition of Wood. Its Biology and Ecology. Chichester: John Wiley, 1988.
57. K Ruel, F Barnoud. Degradation of wood by microorganisms. In: T Higuchi, ed. Biosynthesis and Biodegradation of Wood Components. New York: Academic Press, 1985, pp 441–467.
58. T Sakai, T Sakamoto, J Hallaert, EJ Vandamme. Pectin, pectinase, and protopectinase: Production, properties, and applications. Adv Appl Microbiol 39:213–294, 1993.
59. JP Schimel, JS Clein. Microbial response to freeze-thaw cycles in tundra and taiga soils. Soil Biol Biochem 28:1061–1066, 1996.
60. RL Sinsabaugh. Enzymic analysis of microbial pattern and process. Biol Fertil Soils 17:69–74, 1994.
61. RL Sinsabaugh, RK Antibus, AE Linkins. An enzymic approach to the analysis of microbial activity during plant litter decomposition. Agric Ecosystems Environ 34:43–54, 1991.
62. RL Sinsabaugh, RK Antibus, AE Linkins, CA McClaugherty, L Rayburn, D Repert, T Weiland. Wood decomposition over a first-order watershed: Mass loss as a function of lignocellulase activity. Soil Biol Biochem 24:743–749, 1992.
63. RL Sinsabaugh, RK Antibus, AE Linkins, L Rayburn, D Repert, T Weiland. Wood decomposition: Nitrogen and phosphorus dynamics in relation to extracellular enzyme activity. Ecology 74:1586–1593, 1993.
64. RL Sinsabaugh, EF Benfield, AE Linkins. Cellulase actvity associated with the decomposition of leaf litter in a woodland stream. Oikos 36:184–190, 1981.
65. RL Sinsabaugh, S Findlay. Microbial production, enzyme activity and carbon turnover in surface sediments of the Hudson River Estuary. Microb Ecol 30:127–141, 1995.
66. RL Sinsabaugh, AE Linkins. Adsorption of cellulase components by leaf litter. Soil Biol Biochem 20:927–932, 1988.
67. RL Sinsabaugh, AE Linkins. Natural disturbance and the activity of *Trichoderma viride* cellulase complexes. Soil Biol Biochem 21:835–839, 1989.
68. RL Sinsabaugh, AE Linkins. Enzymic and chemical analysis of particulate organic matter from a boreal river. Freshwater Biol 23:301–309, 1990.
69. RL Sinsabaugh, M Liptak. Enzymatic conversion of plant biomass. In: B Soderstrom, DT Wicklow, eds. The Mycota. Vol. 4. Environmental and Microbial Relationships. Berlin: Springer-Verlag, 1997, pp 347–357.
70. RL Sinsabaugh, DL Moorhead. Resource allocation to extracellular enzyme production: A model for nitrogen and phosphorus control of litter decomposition. Soil Biol Biochem 26: 1305–1311, 1994.
71. RL Sinsabaugh, DL Moorhead. Synthesis of litter quality and enzyme approaches to decompo-

sition modeling. In: G Cadisch, K Giller, eds. Driven by Nature: Plant Litter Quality and Decomposition. London: CAB International, 1997, pp 363–375.
72. RL Sinsabaugh, DL Moorhead, AE Linkins. The enzymic basis of plant litter decomposition: Emergence of an ecological process. Appl Soil Ecol 1:97–111, 1994.
73. RL Sinsabaugh, M Osgood, S Findlay. Enzymatic models for estimating decomposition rates of particulate detritus. J North Am Benthological Soc 13:160–169, 1994.
74. RL Sinsabaugh, T Weiland, AE Linkins. Enzymic and molecular analysis of microbial communities associated with lotic particulate organic matter. Freshwater Biol 28:393–404, 1992.
75. E Smit, P Leeflang, B Glandorf, JD van Elsas, K Wernars. Analysis of fungal diversity in the wheat rhizosphere by sequencing of cloned PCR-amplified genes encoding 18S rRNA and temperature gradient gel electrophoresis. Appl Environ Microbiol 65:2614–2621, 1999.
76. Y Tanaka. Microbial decomposition of reed (*Phragmites communis*) leaves in a saline lake. Hydrobiologia 220:119–129, 1991.
77. Y Tanaka. Activities and properties of cellulase and xylanase associated with *Phragmites* leaf litter in a seawater lake. Hydrobiologia 262:65–75, 1993.
78. JL Tank, JR Webster, EF Benfield, RL Sinsabaugh. Effect of leaf litter exclusion on microbial enzyme activity associated with wood biofilms in streams. North Am Benthological Soc 17: 95–103, 1998.
79. BR Taylor, D Parkinson. Does repeated wetting and drying accelerate decay of leaf litter? Soil Biol Biochem 20:647–656, 1988.
80. BR Taylor, D Parkinson. Does repeated freezing and thawing accelerate decay of leaf litter? Soil Biol Biochem 20:657–665, 1988.
81. M Uchida. Enzyme activities of marine bacteria involved in Laminaria-thallus decomposition and the resulting sugar release. Mar Biol 123:639–644, 1995.
82. M Unanue, JM Arrieta, I Azua. Bacterial colonization and ectoenzymatic activity in phytoplankton-derived model particles: Cleavage of peptides and uptake of amino acids. Microb Ecol 35:136–146, 1998.
83. L Viikari, M Tenkanen, J Buchert, M Rättö, M Bailey, M Siika-aho, M Linko. Hemicellulases for industrial applications. In: JN Saddler, ed. Bioconversion of Forest and Agricultural Plant Residues. Wallingford: CAB International, 1993, pp. 131–182.
84. KKY Wong, LUL Tan, JN Saddler. Multiplicity of β-1,4-xylanase in microorganisms: functions and applications. Microbiol Reviews 52:305–317, 1988.
85. P Yeager, RL Sinsabaugh. Microbial diversity along a sediment particle size gradient. Aquat Ecol 32:281–289, 1998.
86. JC Zak, SC Rabatin. Organization and description of fungal communities. In: DT Wicklow, B Söderström, eds. The Mycota. Vol. IV. Environmental and Microbial Relationships. Berlin: Springer-Verlag, 1997, pp 33–46.
87. D Zak, D Ringelberg, K Pregitzer, D Randlett, D White, P Curtis. Soil microbial communities beneath *Populus grandidentata* Michx. grown under elevated atmospheric CO2. Ecol Appl 6:257–262, 1996.
88. JC Zak, RL Sinsabaugh, W MacKay. Windows of opportunity in desert ecosystems: their implications to fungal community development. Can J Bot 73:S1407–S1414, 1995.
89. J Zemek, L Marvanova, L Kuniak, A Kadlecikova. Hydrolytic enzymes in aquatic hyphomycetes. Folia Microbiol 30:363–372, 1985.
90. G Zogg, D Zak, D Ringelberg, N MacDonald, K Pregitzer, D White. Compositional and functional shifts in microbial communities due to soil warming. Soil Sci Soc Am J 61:475–481, 1997.

10

Fungal Communities, Succession, Enzymes, and Decomposition

Annelise H. Kjøller and Sten Struwe
University of Copenhagen, Copenhagen, Denmark

I. INTRODUCTION

Fungi are essential for nutrient mobilization, storage, and release during decomposition of plant material in terrestrial ecosystems. Saprophytic microfungi are the least visible group of fungi in soil but are, nevertheless, key decomposers of the massive amounts of leaves, stalks, and other plant parts deposited on and in the ground each year. Because of their hyphal growth pattern, production of vegetative spores, specific survival strategies, and capacity to produce a variety of enzymes important in decomposition processes, these fungi are ubiquitous and respond rapidly to the addition of new substrates.

During the decomposition of plant material the composition of the fungal community changes, a process referred to as *microbial succession* (1). This succession can be viewed as changes in taxonomic diversity, and if the role of the fungal population is known, then functional diversity can also be considered. Although some individual species of fungi are capable of producing many different enzymes, communities that comprise different fungi usually contribute collectively to the decomposition of physically and chemically complex substrates such as leaves (2). Fungal communities vary in species composition from site to site, reflecting fungal versatility and functional resilience and thereby ensuring efficient decomposition and mobilization of nutrients in most environments.

Microfungi are able to degrade virtually all of the organic compounds generated by primary production in the various ecosystems of the world. Moreover, they are also able to degrade xenobiotic compounds, many of which are comparatively new to the environment (3,4). Bacteria also produce a large variety of enzymes in the environment, and an understanding of the interaction between fungi and bacteria is important to comprehension of the decomposition process. In most soils, the fungal biomass corresponds to or exceeds the bacterial component such that fungi play the major role, especially in the initial stages of cellulose, lignin, and chitin decomposition (5). Hyphal growth enables fungi to grow toward and through dense organic material and to grow from one source to another through a depleted zone. Because of the comparatively slow decomposition rate of hyphae, which is due to their high cell wall chitin content, nutrients are immobilized in fungal biomass for a longer period than in bacteria (6). The fungal biomass therefore comprises an important soil nutrient pool. Because fungal growth is affected by tillage, fertilization, fungicides, and the removal of plant biomass, fungal biomass in undisturbed, uncultivated soil is normally higher than in cultivated, agricultural soil. As a consequence of the lower fungal biomass, the sustainability (i.e., organic matter content, soil structure, resilience to impacts) of the soil diminishes (6).

II. THE FUNGAL COMMUNITY

The regulation and secretion of fungal extracellular enzymes in pure culture, in vitro, are not reviewed in this chapter; this topic is discussed in detail elsewhere (7,8). Although the fungal enzymes and their principal substrates are well defined, comparatively little is known about their regulation in nature. Some enzymes are induced in the presence of substrates and products, whereas others are regulated by repression/derepression. However, few studies have investigated this recognized but undoubtedly complex situation. One of the reasons for the involvement of the whole fungal community in decomposition could be regulatory factors that do not specifically favor one strain but rather stimulate several fungi to utilize the substrate or components of it. The most important degradative extracellular enzymes produced by fungi are the proteases, amylases, pectinases, cellulases, ligninases, and xylanases, although enzymes such as chitinases, cutinases, phytases, and phosphatases also play a role.

A. Principal Groups of Soil Microfungi

There are two principal taxonomic groups of microfungi active in the decomposition process in litter and soil: the Zygomycetes and the Deuteromycetes. These have various morphological and enzymatic traits that enable them to grow and proliferate on diverse substrates. Examples of functional groups of fungi are shown in Table 1.

Mucorales is the largest group of the Zygomycetes, encompassing such important genera as *Mucor*, *Rhizopus*, *Absidia*, and *Mortierella*. These all have fast-growing mycelia, are devoid of hyphal septa, and exhibit various kinds of vegetative sporangiospores. Some species also produce sexual spores, which are often thick-walled and able to survive under adverse environmental conditions. Members of the Mucorales are unable to degrade polysaccharides such as cellulose and lignin, but they rapidly penetrate organic material and utilize soluble sugars in competition with bacteria during the initial phases of decomposi-

Table 1 Examples of Functional Genera of Fungi in the Key Taxonomic Groups

	Zygomycetes	Ascomycetes	Basidiomycetes	Deuteromycetes
Soil fungi	*Mortierella*	*Peziza*	*Agaricus*	*Trichoderma*
Litter fungi	*Mucor*	*Chaetomium*	*Mycena*	*Cladosporium*
Wood fungi		*Xylaria*	*Trametes*	imperfect stages[a]
Mycorrhizal fungi	*Endogone*	*Tuber*	*Cantharellus*	
Pathogenic fungi	*Enthomophthora*	*Erysiphe*	*Ustilago*	*Verticillium*

[a] Imperfect stages of genera of Ascomycetes and Basidiomycetes.

tion (9). Species of the large genus *Mortierella* have different capacities and are specialized for chitin degradation, producing enzymes such as β-*N*-acetylglucosaminidase (10).

The Deuteromycetes comprise a very large (approximately 17,000 species) and heterogeneous group of filamentous fungi—the hyphomycetes. These only reproduce vegetatively and hence are traditionally referred to as *Fungi Imperfecti.* When a sexual phase is known, the taxon should be classified among the Ascomycetes (or Basidiomycetes), but for identification purposes, the vegetative stages are also included in the Deuteromycetes. The Deuteromycetes are enzymatically extremely versatile and hence often play the major role in fungal degradation of organic matter. Moreover, many species produce a large number of conidia and are fast-growing, thereby spreading rapidly throughout the environment and germinating when conditions are optimal. Many strains are unable to form conidia and remain sterile, and such sterile mycelia may account for up to half of the strains isolated from a site. The Deuteromycetes are known to be capable of producing enzymes important for the decay processes (7). Some pathogenic imperfect fungi are also saprophytic and produce a variety of enzymes, primarily those necessary for penetrating insect cuticles (e.g., chitinase and protease) and cellulase for decomposing plant material. An example of such a versatile fungus is *Paecilomyces farinosus* (11).

Two other fungal groups are also important in the decomposition processes. The first group, the Ascomycetes, includes genera that produce both vegetative conidia and sexual spores (e.g., *Penicillium* and *Aspergillus* spp.) as well as a large group of yeasts common in certain soils and fruits with a high sugar content (9). The second group, the Basidiomycetes, include the wood-decaying fungi with large groups of soft rot, brown rot, and white rot fungi producing lignocellulose-degrading enzymes (12). An important functional group of Basidiomycetes are the ectomycorrhizal fungi, which are in direct mycelial contact with the roots of trees, bushes, and herbs in terrestrial ecosystems (12).

B. Fungal Biomass

Determination of fungal biomass is important for estimating the organic C pool in fungal hyphae; for comparing fungal biomass in different soils and horizons; for determining the effects of pollution and changes in climate and land use; and for using biomass data in decomposition models (13). The mycelium is often well hidden in soil aggregates and is not easily accessible. As a result, many different approaches have been employed to determine fungal biomass, including microscopy, cultivation, substrate utilization, and analyses of structural components. Determination of fungal biomass in litter and soil is usually based on the estimation of fluorescein diacetate (FDA) or cell components such as ergosterol and phospholipid fatty acids (PLFA), procedures that are fully described elsewhere (14,15). A physiological method much used for determining total soil microbial biomass is substrate-induced respiration (SIR) (16), and by combining SIR with antibiotic inhibition the contributions of the bacterial and fungal biomass can be separated (17). Although the selective inhibition technique has the potential to be the most precise means of measuring the active fungal biomass, and many attempts to improve the procedure have been reported, it nevertheless remains very difficult to obtain reliable, reproducible results, especially when using soil samples with a high organic matter content (18–23).

An integrated experiment to demonstrate fungal and bacterial competition was carried out by Hu and van Bruggen (24), who investigated microbial dynamics during the decomposition of cellulose-amended soil. Measuring respiration in combination with the selective inhibition technique, they showed that the fungal population played the major role in cellulose decomposition since the bacterial respiration was very low during the 30-day experimental period. Fungal respiration peaked within 10 days when the bacteria (and fungi) had depleted the easily available C; after 10 days the fungi initiated cellulose degradation.

A recently developed technique based on the production of the fungal enzyme chitinase, has been employed to estimate fungal presence and activity in soil and litter. Adding a fluorogenic substrate analog, 4-methylumbelliferyl *N*-acetyl-β-D-glucosamide (MUF), to the sample allows *N*-acetylglucosaminidase (NAGase) activity to be measured. Laboratory experiments have shown that the NAGase activity is significantly correlated with both the ergosterol content and the fungal PLFA (25,26), thus confirming that fungi are the predominant source of the activity. This method was used to compare the spatial and temporal changes in fungi and fungal enzyme activity during decomposition of maize litter in two agricultural soils from the northern temperate and the southern Mediterranean zones (27). Chitinase activity was determined by the MUF technique on six sampling occasions during one year (25). The level of enzyme activity differed between the two soils; activity was considerably lower and the lag time before production of enzymes longer in the Mediterranean zone soil than in the temperate zone soil. Moreover, enzyme activity was considerably lower in bulk soil than in the "residuesphere" (i.e., the interface between soil and plant residues).

Fungal–bacterial interaction during the decomposition of beech leaves was demonstrated by Møller and associates (26), who showed that the chitinase (*N*-acetylglucosaminidase) activity was fungal in origin and that bacteria made only a marginal contribution to chitin degradation despite the fact that the bacterial community (as revealed by the Biolog method) exhibited high functional diversity. A significant correlation exists between chitinase activity and both exo- and endocellulase activity, possibly indicating that both enzymes are mainly fungal in origin.

The validity of chitinase activity as a measure of fungal biomass was substantiated in a study of the influence of fungal–bacterial interaction on the bacterial conjugation rate in the residuesphere (28). The aim was to determine whether the residuesphere is a hot spot for conjugal gene transfer and whether fungal colonization of the leaves affects conjugation efficiency. In a microcosm experiment with soil and barley straw precolonized by soil fungi, chitinase activity increased after 17 days whereas the number of transconjugants decreased. The activity of chitinase and *N*-acetylglucosidase as measured by the MUF technique decreased with depth in four different Japanese forest soil profiles (29). It was concluded that the higher levels of these enzymes in the upper part of the profile could be due to the presence of fungi (chitin in the cell walls) and arthropods (chitin in the exoskeleton) serving as substrates.

Enzyme determination using MUF substrates is a highly sensitive technique and the enzymes can be measured in nanomolar concentrations and under in situ conditions. Other MUF substrates have also been used to measure various enzymes in soils and sediments, including cellulases, peptidases, and glucosidases (30–32).

III. INFLUENCE OF RESOURCE QUALITY ON FUNGAL ACTIVITY

The two main plant compounds, cellulose and lignin, are degraded by both bacteria and fungi but the literature on fungal enzymes states that the Basidiomycetes play the major role (33–35). Bacteria are generally unable to degrade lignin completely. Even the Actinomycetes, which exhibit mycelial growth, do not have the same lignin-degrading capacity as fungi and do not appear to play a significant part in lignin degradation. Many genera of saprophytic microfungi, which colonize leaf litter during decomposition and operate in the different soil horizons, degrade cellulose and lignin compounds to different degrees. However, this group of microfungi tends to be ignored in many of the reviews of cellulose and lignin degradation.

The lignin content markedly affects the decomposition rate of both leaf and needle litter types; lignin concentration and living fungal biomass are inversely related (36). This indicates that fungal growth during colonization is repressed by lignin and that decomposition rates in humified litter are very low. Entry and Backman (37) also argued that the lignin content of organic matter is a major factor controlling decomposition of organic matter in terrestrial ecosystems. In experiments involving the addition of C and N to forest soils they found that as the C and N concentration increased, so did cellulose and lignin degradation and the active fungal biomass. Fungal biomass correlated with both cellulose and lignin degradation, indicating the importance of the fungal population in the decomposition processes. It was concluded that the cellulose/lignin/N ratio more accurately predicts the rate of organic matter decomposition (and hence substrate quality) than overall C/N ratios.

It is not possible to test microfungi for lignin degradation ability directly. Alternative substrates have been introduced over the years; these include vanilin, indulin, ferrulic acid, and, most importantly, ^{14}C-labeled synthetic lignins. Various fungal enzymes are involved in lignin degradation, including lignin peroxidase, manganese peroxidase, polyphenol oxidases, and especially laccase (34,38–43).

As fungi or other microorganisms capable of attacking humic acid or gallic acid are also able to degrade lignin (44), the effect of these two compounds on microfungi combined with determination of their degradation ability may be used as an indicator of lignin degradation. Gallic acid has been shown to inhibit the growth of fungi isolated from litter and soil from temperate forests. Radial growth of the frequently isolated microfungi (e.g., species of *Cladosporium*, *Aureobasidium*, *Epicoccum*, *Alternaria*, and *Ulocladium*) was restricted on agar containing gallic acid as the sole carbon source as compared with growth on medium devoid of gallic acid. Some *Penicillium* species producing polyphenol oxidase were able to grow in the presence of gallic acid and may be the only fungi able to tolerate gallic acid in the environment (45).

In a study of deciduous forest litter, Rai et al. (46) reported marked inhibition of *Curvularia*, *Cladosporium*, and *Myrothecium* spp. in cultures containing gallic acid. Although most of the isolated strains of these genera are able to produce polyphenol oxidase, only *Penicillium*, *Aspergillus*, and *Trichoderma* spp. were not inhibited and were able to utilize gallic acid as a source of carbon and energy.

In a study of the humic acids–degrading efficiency of fungi and bacteria, Gramass et al. (47) investigated the growth of 36 fungi and 9 bacteria isolated from soil and plant material, including wood-degrading and soil-inhabiting saprophytic Basidiomycetes, ectomycorrhizal fungi, and soil-borne microfungi and bacteria. The wood-degrading Basidiomycetes decomposed humic acid at twice the rate of other groups of fungi, whereas the bacteria exhibited little humic acid degradation.

Decomposition of beech leaves has been investigated by Rihani and associates (48). Pure cultures of two white rot fungal strains (Basidiomycetes), isolated from beech soil and litter, were able to use pectin, cellulose, lignin substitutes, and phenols as their sole carbon source in pure cultures. Thus, when the two strains were inoculated separately on sterile fresh leaves, cellulose, lignin, and phenol degradation was initiated immediately. Fourteen days later, when 20% of the cellulose had been degraded, the rate of lignin degradation increased. Decomposition was rapid during the first month but virtually ceased after four months.

Low resource quality and adverse environmental conditions (e.g., low water availability) result in low decomposition rates. This has been examined by incubating pine needles in litter bags in a southern Italian pine forest (49). Both the C/N ratio and the lignin content of the litter were high. Measurement of biological parameters such as CO_2 evolution and fungal biomass over a three-year period revealed a significant positive correlation between respiration rate and moisture content of the litter. There was no obvious relationship between fungal biomass and other measured parameters (i.e., litter mass loss, lignin content, and nitrogen content). It was concluded that since the litter was very dry for most of the year, an autochthonous fungal flora had developed that was able to degrade these litter types under adverse conditions albeit at a low rate.

The examples of interactions between substrates and fungal groups mentioned and the influence of different concentrations of substrates illustrate the complex and dynamic processes involved in litter decomposition. In the next section the successional stages of decomposition are discussed in the context of enzyme activity.

IV. FUNGAL POPULATIONS AND ENZYME ACTIVITY

Numerous studies on fungal succession have been published, many of which discuss the identification of fungi at different stages of decomposition (1). However, the emphasis is usually on taxonomic identity rather than on enzymatic diversity. Those genera most frequently mentioned in connection with early colonization of the organic debris in the temperate zone are *Alternaria*, *Aureobasidium*, *Cladosporium*, and *Epicoccum*. In her review, Frankland (1) concluded, "Let us ecologists not neglect to study in greater depth more of the star performers in fungal succession, on which the maintenance of entire ecosystems may depend." In this context "star performers" encompass the important enzyme producers and hence the key decomposers.

The link between taxonomic and functional diversity in the fungal population has been discussed in reviews by Miller (50) and Zak and Visser (51), both of which emphasize the importance of succession studies. The relationship between fungal succession and the enzymatic potential of the fungi has been observed during decomposition of forest litter, e.g., of alder (2,52–57) and beech (54–57) (see Table 2).

On beech leaves, fungal species of the genera *Aureobasidium*, *Cladosporium*, *Epicoccum*, and *Alternaria* appear first, although *Mucor*, *Phoma*, and *Acremonium* are often early colonizers (Table 3). *Acremonium* spp. isolates attack cellulose and chitin as well as gallic acid, although the main role of these fungi in the initial phases of decomposition is to degrade pectin and starch. The second wave of degraders varies in different litters, consisting of a wider variety of genera (e.g., *Cylindrocarpon*, *Phialophora*, *Phoma*, and *Phomopsis*). These fungi are versatile with regard to enzyme production and secrete cellulases, polygalacturonases, xylanases, lipases, and proteases. A third group of degraders, which come into play when the litter is almost completely decomposed, chiefly consists of cellulose-degrading fungi but also includes lignin and chitin degraders of genera such as *Trichoderma*, *Penicillium*, *Fusarium*, *Acremonium*, and *Mortierella*. In the later stages

Table 2 The Most Frequent Microfungal Genera in Alder, Ash, and Beech Litter Able to Utilize Pectin, Cellulose, Chitin, and Gallic Acid

	Alder	Ash	Beech Year 1 new	Beech Year 2 old
Pectin	*Phoma*	*Phomopsis*	*Acremonium*	*Trichoderma*
	Cladosporium	*Phoma*	*Sterile mycelia* dark	*Sterile myelia* hyaline
	Cylindrocarpon	*Epicoccum*	*Aureobasidium*	*Mortierella*
			Heteroconium	*Chrysosporium*
			Cladosporium	*Penicillium*
Cellulose	*Phoma*	*Phoma*	*Acremonium*	*Trichoderma*
	Verticillium	*Cylindrocarpon*	*Heteroconium*	*Acremonium*
	Cylindrocarpon		*Phialophora*	*Mortierella*
Chitin	*Mortierella*	*Phoma*	*Acremonium*	*Mortierella*
	Verticillium			*Trichoderma*
	Aureobasidium			*Penicillium*
Gallic acid	*Cladosporium*	*Phoma*		
	Cylindrocarpon	*Phomopsis*	nd[a]	nd[a]
		Cylindrocarpon		

[a] nd, not determined.
Source: Refs. 2, 53, and 54.

of decomposition, *Mortierella* spp. strains constitute 28% of the isolates, all of which are able to degrade chitin, whereas only a few also attacked pectin and cellulose. *Mortierella* spp. isolates have been tested for the production of hydrolytic enzymes by Terashita and associates (58), who reported that 18 isolates were able to produce acid protease, β-1,3-glucanase and chitinase, whereas cellulase was produced by a smaller number only.

The applicability of laboratory findings to events in the environment depends on how reliably the environmental conditions are simulated in the model and culture studies. Moreover, as isolation procedures for fungi and enzyme assays differ among studies, the findings of different research groups are not always directly comparable. However, the methods used in the cases discussed later concerning forest litter decomposition are almost identical, thereby allowing valid comparisons to be made.

In each of the studies the fungi were isolated by blending soil or litter in water and washing the soil particles to remove conidia. Growing hyphae remained attached to the particles, which were placed on appropriate agar plates and incubated at 10°C or 15°C until growth of the fungal strains was sufficient to allow identification. Although the choice of medium varied, soil fungi, unlike bacteria, are able to grow on both complex and very dilute (oligotrophic) media. Temperature significantly influences enzyme production in the environment, but this influence is difficult to study in situ and most of our knowledge stems from applied studies of enzyme production in the laboratory. Flanagan and Scarborough (44) reported that a fungal strain isolated from an arctic soil and grown on cellulose or pectin as the carbon source produced cellulase at low temperature (4°–5°C), whereas pectinase production was optimal at much higher temperatures (15°–25°C).

Table 3 Microfungal Succession and Substrate Utilization Pattern During Decomposition of Beech Litter over an 18-Month Peroid[a]

Fungal genera	Fresh Leaves			Months 3			6			13			16			18		
Acremonium spp.	Pe	Ce	Ch	Pe	Ce		Pe		Ch	Pe	Ce		Pe					
Cladosporium cl.	Pe																	
Sterile mycelia black	Pe																	
Sterile mycelia black	Pe																	
Phialophora sp.	Pe	Ce																
Sterile mycelia grey	Pe																	
Sterile mycelia dark grey				Pe			Pe		Ch									
Aureobasdium pullulans				Pe			Pe		Ch									
Heteroconium sp.				Pe	Ce													
Cladosporium herbarum				Pe			Pe											
Pseudofusarium sp.				Pe														
Sterile mycelia grey				Pe														
Sterile mycelia brown								Ce										
Sclerotia																		
Trichoderma viride										Pe	Ce	Ch					Ce	Ch
Mortierella spp.										Pe	Ce	Ch			Ch	Pe	Ce	Ch
Penicillium spp.														Ce		Pe	Ce	Ch
Sterile mycelia hyaline													Pe					
Chrysosporium sp.													Pe					
Mortierella vinaceae													Pe					

[a] Pe, pectin; Ce, Cellulose; Ch, chitin.
Source: Ref. 54.

A. Forest Litter Decomposition

The literature on decomposition encompasses numerous studies on many different types of forest litter from all over the world. Both recent and older reports concentrate on temperate forest litter (mostly from deciduous forests); tropical litter is only rarely included. Research on fungal activity and carbon sequestration in relation to the high decomposition rate in tropical rain forest should thus be accorded high priority in future studies.

Alder litter was investigated by Rosenbrock et al. (52), who showed that fungal amylase, polygalacturonidase, cellulase, xylanase, pectinase, protease, and laccase were produced at the beginning of the decomposition period. A high proportion of the fungal isolates produced amylase (80–100%) and polygalacturonase (50–95%) throughout the first year of decomposition, whereas the percentages of fungi producing cellulase, xylanase, pectinlyase, protease, and lipase decreased with time. Pectinase and protease were only produced by approximately half of the isolated strains. Laccase activity was restricted to only 2–6% of the isolates and occurred sporadically throughout the year. After the initial dominance of *Mucor*, *Alternaria*, and *Epicoccum* spp. these fungi were displaced by a number of different *Fusarium* species.

The potential of fungi to produce a large range of various enzymes during the initial stages of litter decay was also observed in our own decomposition studies of alder, ash, and beech litter. In the beech litter study (54), fungal strains were isolated and tested on pectin, cellulose, and chitin. Three months after litter fall, 90% of the isolates were recorded as pectinase producers, e.g., *Aureobasidium* and *Cladosporium* spp; *Heteroconium* and *Acremonium* spp. were able to degrade both pectin and cellulose. After 10 months the proportion of pectinase-positive fungi had decreased to 40%, whereas the proportion of cellulose-decomposing fungi had increased from 20% to 60%, the latter mainly accounted for by various sterile mycelia. After 18 months the active fungal flora was dominated by *Mortierella*, *Penicillium*, and *Trichoderma*, which degrade both cellulose and chitin. At this stage a single fungal strain could be highly versatile, able to attack more than one polymer (and its lower-molecular-mass oligomers). This study thus demonstrates the occurrence of taxonomic and functional succession during decomposition of beech litter, as the fungal flora change composition and functional role as the substrate resource is depleted.

Fungal succession and decay of beech litter were investigated in a transect/transplant experiment in four European beech forests in the (CANIF) project (57,59). Fungal activity was measured as endo- and exocellulase activity (endo 1,4-β-glucanase/exo cellobiohydrolase) using a MUF substrate, 4-methylumbelliferyl β-D-lactoside (27). Although the MUF technique does not distinguish between fungal and bacterial cellulase activity, Møller and coworkers (26) showed that the cellulase activity measured by the MUF technique is mainly fungal in origin with very few bacteria active. Moreover, Miller et al. (25) showed that MUF cellulase activity correlated with ergosterol and fungal PLFA, Cotrufo and colleagues (59) reported a correlation of cellulase activity with decomposition rate (litter weight loss). Thus the MUF cellulase activity measured probably reflects the activity of live fungi colonizing the litter. In the CANIF project, fungal strains were isolated and identified and the Simpson diversity index was calculated (60). In the transect experiment, samples of leaves from an Italian beech forest were placed in the litter layer of beech forests in France, Germany, and Denmark. In the transplant experiment, beech leaves from these three forests were placed together with the local litter in the Italian beech forest. By incubating identical litter types in different climatic zones and by placing litter of different origin in the same climate, interesting decay patterns and biodiversity changes were revealed. A linear regression model of mass loss as a function of cumulative cellulase activity for pooled data from all sites (both transplant and transect) revealed a significant correlation between the two sets of data (59).

When the three types of ''foreign'' litter were placed at the southern site in Italy, both cellulase activity and fungal diversity were lower than in the native litter layer. The Italian litter had the highest cellulase activity but the lowest fungal diversity, thus indicating that the fungal population was adapted to the local climate and soil. Another interesting finding was that when the Italian litter was placed along the transect in France, Germany, and Denmark, the rate of cellulase activity increased to much higher levels than when incubated in Italy, especially during the second year of decomposition. When the litter was placed in a less adverse climate, decomposition proceeded at a higher rate. At the Danish site, for example, decomposition was twice as fast as in Italy. Fungal diversity was high during the first months but diminished with time, while the cellulase activity remained high. Key functions are undertaken by different fungi at different sites and stages of decomposition, thus indicating the occurrence of functional substitution. The most frequent fungi on the Italian beech litter were *Chalara* species, which initially constituted 40% of the population but disappeared rapidly after the first eight months of decomposition. *Cladosporium* and *Aureobasidium* spp. were present during the entire period, whereas *Chalara* sp. was replaced by cellulase-producing fungi such as *Penicillium*, *Acremonium*, and *Alternaria* spp., and at the Danish site also by *Trichoderma* sp. At no time was it possible to correlate fungal diversity with decomposition rate as measured by cellulase activity.

In two significant papers, Andrén et al. (61,62) discussed biodiversity and species redundancy among litter decomposers and the influence of soil microorganisms on ecosystem-level processes. Some of the hypotheses put forward in Andrén (61) are relevant to the CANIF data, for example, the hypothesis ''If diversity is important, there should be a positive correlation between diversity and decomposition rate. . . .'' When closely examining the preceding findings it is apparent that fungal diversity was low in all four types of litter when ''foreign'' beech litter was incubated in Italy. On each sampling occasion during the two-year study period the transplant experiment also revealed low cellulase activity. In the transect experiment, however, in which Italian litter was placed in France, Germany, and Denmark, a different picture emerged. Thus the cellulase activity increased at all sites during the incubation period, and the highest level of activity was reached during the second year of decomposition. Simultaneously, fungal diversity was initially high but decreased to very low levels toward the end of the decomposition period, lower than in the transplant experiment. As a consequence, fungal diversity and decomposition activity were inversely correlated. The difference in the results of the two experiments may be attributable to a number of factors. For example, decomposition of cellulose seems to proceed well under conditions of low diversity.

Another hypothesis put forward by Andrén (61) was, ''If a particular organism group controls decomposition, it should be possible to relate its dynamics to decomposition rate. . . .'' This was demonstrated for the fungal community in the preceding experiments. If the fungal populations are removed from the litter or their growth is inhibited from the beginning of the decomposition process, however, the decomposition proceeds extremely slowly and is solely due to bacterial cellulase activity (26).

A third hypothesis proposed by Andrén (61) can be summarized as follows: ''During decomposition there is a succession of fungi adapted to changes in substrate quality but the decomposition rate may nevertheless remain constant.'' The question here is whether a change in the succession observed in the two experiments mentioned will affect the litter decay rate. There is a marked succession of fungi, but would a change in the composition of the fungal flora at a certain stage affect the decomposition rate? This is difficult to answer since it is not easy to manipulate natural systems and exclude specific members of the fungal succession.

The paradox of the apparent simplicity of ecosystem process control and the high diversity of soil organisms (invertebrates, bacteria, and microfungi) has been discussed (62). Although the most simple decomposition models operate without including diversity, the microbial component, expressed, for example, as microbial biomass, may still be able to predict the turnover of organic matter. Most decomposition models include components such as temperature, moisture, and resource quality, and since these variables have direct effects on microbial growth and activity, they are also important for decomposition. When a more detailed view is necessary, enzyme production seems to be a useful tool. Fungal enzyme production is essential to decomposition, and it is thus important to study the response of individual fungi to environmental changes.

B. Decomposition of Crop Residues

Postharvest decomposition of crop litter plays an import role in returning nutrients to agricultural soils. Much research has been undertaken to determine the effect of land use changes, management practices, and resource quality on litter decomposition. Some examples of effects on the fungal community and enzyme production are examined later in order to highlight the role played by microfungi in the sustainability of agricultural soils. As previously discussed, the fungal community of agricultural soil is under stress due to the management procedures employed in modern agriculture, and fungal biomass is consequently much lower than in natural soils (6). It was stated that agricultural practice especially would affect the fungal biomass. Since fungal hyphae and fungal production of polysaccharides are essential for soil stability, consequences could include less stable soil aggregates.

The effect of reduced soil management on fungal activity in agricultural soil was investigated in laboratory experiments in which maize litter was either placed on top of or incorporated into the soil to simulate reduced and a traditional soil management practice, respectively (27). Cellulase activity and chitinase activity in the bulk soil were both low. When maize litter was incorporated into the soil, enzyme activities increased. When the litter was placed on top of the soil, the level of activity was consistently higher. Moreover, fungal isolation frequency was also higher (i.e., more fungi per soil particle) and the fungal communities were more diverse. However, after one year, the total degree of mineralization of maize litter was the same irrespective of whether the litter had been placed on top of the soil or incorporated into it. The fungal community on maize litter was initially dominated by members of the Mucorales, e.g., species of *Mucor*, *Mortierella*, and including *Rhizopus*, genera that are only occasionally found in soil. After three months the Mucorales were replaced by cellulolytic fungi such as *Fusarium*, *Acremonium*, and *Penicillium* (27).

During decomposition of maize litter in soil, protease, xylanase, and invertase activities were two orders of magnitude higher than in control soil (63). These enzymes are not specific for fungi, however, and the authors made no attempt to distinguish between fungal and bacterial activities. Some crop litter components decompose very slowly. The decomposition rates of different components of wheat litter (internodes and leaves) in the soil have been compared in an eight-month incubation study by Robinson and colleagues (64). A high lignin content (10%) of the initial dry matter and a high C/N ratio (approximately 100) resulted in a low decomposition rate. At the end of the experiment the lignin and cellulose contents were still high in the internodes, whereas the leaves had decomposed almost completely. The fungal genera acting on the two components were cellulose-decomposing fungi such as *Trichoderma* and *Cladorrhinum* spp. In the internodes these two were supplemented with other cellulose decomposers such as *Fusarium*, *Phoma*, and *Penicillium* spp. On the leaves, in contrast, they were supplemented by genera capable of degrading both cellulose and lignin, e.g., *Epicoccum* and *Cladosporium* spp. Ligninolytic Basidiomycetes can be expected to appear much later to complete the decomposition of the internodes.

In similar experiments, Bowen and Harper (65,66) examined the succession of saprophytic microfungi on decomposing wheat straw in agricultural soil during a one-year period together with the cellulase- and lignin-degrading ability of the fungi. The first colonizers were *Mucor* and *Cladosporium* spp. The number of *Mucor* spp. isolates decreased after the first months of decomposition, whereas *Cladosporium* spp. remained abundant. Other fungi played an important role after the first month, e.g., *Penicillium* and particularly *Fusarium* spp. *Chaetomium* was abundant during the first months but was subsequently functionally replaced by *Trichoderma* spp. The fungi that became more abundant as decomposition progressed were tested for their ability to degrade cellulose, lignin, and phenolic acids; it turned out that they were only able to degrade cellulose. Later-appearing Basidiomycetes, which included a *Typhola* sp., were able to degrade both cellulose and lignin, although the lignin decomposition rate differed among the individual fungi. It was argued that these Basidiomycetes with multiple degradation abilities degrade the straw more efficiently than the strong cellulolytic, nonlignolytic, filamentous fungi and hence may play an important role in agricultural soil sustainability. It was also demonstrated that mixed communities of cellulose- and lignin-degrading fungi almost always exhibited higher rates of decomposition than single strains of efficient degraders.

The effect of nitrogen availability on enzyme activity during decomposition of wheat straw in soil has been examined in a two-month study by Henriksen and Breland (67). The carbon mineralization rate was reduced in straw-amended soil having a low N content (less than 1.2% of straw dry weight). Moreover, when the fungal biomass decreased, exocellulase, endocellulase, and hemicellulase activity also decreased. These findings demonstrate the need for available N to improve enzyme production and decomposition of recalcitrant substrates.

The two methodological approaches used, i.e., the testing of strains and the extraction of enzymes, provide complementary information on enzyme production by emphasizing the potential of the living hyphae and the sum of past and present activities respectively. The use of MUF-linked substrates allows work to be undertaken with small samples.

V. ENVIRONMENTAL IMPACT ON ENZYME ACTIVITY

Many changes in the physical and chemical environment influence fungal activity in the soil. The following examples from the recent literature highlight specific cases. Both environmental and anthropogenic stresses are important. Other chapters of the present volume (see Chapters 17–20) examine pertinent aspects of this such as pesticides, xenobiotics, heavy metals, and various other pollutants.

Environmental conditions and specific stress factors can markedly affect microbial enzyme activity in the soil. For example, freezing and thawing enhance bacterial and fungal phosphatase, urease, xylanase, and cellulase activity, thereby accelerating decomposition compared with that during continuous snow cover (68).

Cellulase activity is a key element of decomposition, and the determination of this predominantly fungal enzyme is essential in both general decomposition studies and in studies of the effect of stress factors. The impact of industrial pollution on cellulase activity has been investigated in forest humus in northern Finland by two methods (69). The use of cellulose strips inserted into the soil proved to be much less efficient at detecting differences in cellulase activity than traditional, chemical analyses in the laboratory. Only the latter analyses were sufficiently sensitive to demonstrate pollution-induced changes. Cellulase activity correlated well with basal respiration, decreasing significantly with increasing pollution.

An effect of pollution on enzyme activity has also been observed in a study of cellulase, amylase, and invertase activity in litter from coniferous and deciduous trees near a busy highway in northeast India (70). The enzyme activities were much higher in the litter at a site 500 m away from the highway than at a site immediately beside the highway, where decomposition of the polluted litter close to the highway decrease. Cellulase and amylase activities were significantly correlated with the number of fungi and bacteria present.

The antibiotic tylosin has been shown to stimulate soil fungal biomass in soil as measured as chitinase activity by the MUF technique. Thus chitinase activity was higher after 10 days of exposure, and although the level decreased somewhat after 20 days, it remained higher than in untreated soil (71). Mercury at various concentrations had no significant effect on chitinase activity; only a minor decrease was observed at high mercury concentrations (511 μg Hg g^{-1} dry soil) (72).

The activity of a number of soil enzymes has been studied in a grassland soil contaminated with various heavy metals, including As, Cd, Cr, Cu, Ni, Pb, and Zn (73). Total microbial biomass, fungal biomass, and bacterial biomass were also determined. Both biomass and enzyme activity were inversely proportional to the level of contamination, and there was a high degree of correlation between enzyme activities and both SIR and fungal length. The respiration rate and cellulolytic activity of some cellulose-decomposing fungi isolated from salinized Egyptian soils were found to decrease at increasing salinity (74). Although fungi are normally considered to be tolerant of high salt concentrations, this study indicates that the decomposition rate of organic matter is lower in salinized soil. Salinization is a global phenomenon of increasing extent as a result of the drier climate in some areas and especially the impact of human activities.

The effect of the increasing atmospheric concentration of CO_2 has also been focused on in recent years. In a study of the effect of three-year exposure to elevated CO_2 levels on the activity of some of the enzymes essential to the decomposition process, Moorhead and Linkins (75) found that cellulase activity increased in ectomycorrhizae in an arctic tussock soil but decreased in the surrounding soil. They concluded that the decrease in cellulase activity would reduce cellulose turnover by 45%, leading to "a substantial increase in activities associated with nutrient acquisition by plants and microorganisms, a reduction in litter cellulose decay and an increase in soil mineral nutrient size." Dhillion and associates (76) reported an increase in saprophytic fungal hyphal length and cellulase activity in a Mediterranean soil under conditions of elevated CO_2, thus suggesting stimulation of organic matter decomposition.

The impact of enhanced ultraviolet-B (UV-B) radiation on the decomposition of plant material has been investigated by Gehrke et al. (77), who reported that the lignin content of the plant material decreased with time but at a lower rate than in nonirradiated material. The total microbial respiration rate decreased after exposure to UV-B radiation, but only transiently. Changes in the fungal community indicate that some genera were sensitive to the radiation (*Mucor hiemale* and *Truncatella truncata*), and others were indifferent (*Penicillium brevicompactum*). Although UV-B radiation may influence fungal communities, insufficient information is available to allow any conclusion to be drawn concerning the implications for decomposition.

VI. CONCLUSIONS

Decomposition was extensively examined by Swift et al. (5). Selected parts of their text focusing on litter quality, the influence of environmental factors, and mathematical modeling were updated by the same authors (78). Although decomposer organisms are not specifically examined by the authors, their conclusions pose three important questions: First, to what extent do decomposer organisms adapt to reductions in diversity by increasing their functional niche? Second, do general relationships exist between species diversity and decomposer function? Third, are there definable levels of diversity at which decomposer processes change significantly? Some of these questions are addressed in the recent literature.

Numerous recent environmental studies have provided evidence concerning fungal enzymes and their regulation, promoting a much more complete understanding of the role of fungal consortia and communities in the decomposition process. The various functional groups of fungi are active in ecosystem niches. The hyphomycetes in particular produce important enzymes able to degrade the plant polymers. Mycologists have traditionally analyzed these fungi on the basis of their morphological characteristics and appearance in a decomposer sequence. Now new methods allow a more detailed analysis of enzyme production and activity. The capacity of many fungi to perform the same hydrolyses on a particular substrate appears to be a less efficient way of organizing the decomposition process. Given the important role of decomposition in nutrient cycling, this overlap in functionality should be viewed as a necessity to ensure decomposition under all circumstances.

In a global change context, it is of great concern whether the dynamic balance among the different functions or enzymes that constitute the decomposition process will continue under changed climatic and soil conditions and changes in land use. It could be hypothesized that a warmer climate will not be accompanied by an increase in cellulose decomposition as great as the increase in lignin decomposition and that the composition and quality of soil organic matter would consequently change. The decomposition of less common compounds could become more susceptible to changes with loss or addition of certain fungal activities. Analysis of all the elements of the decomposition process is therefore important for an overall assessment of the effects of global change.

Considerably more is known about bacterial enzymes than about fungal enzymes, and although the functional diversity of bacteria is greater than that of fungi, fungi are nevertheless most important for the decomposition of plant material. When a greater number of molecular methods have been more widely applied for studying fungal gene activity and in situ determination of fungal enzyme synthesis, it will be possible to clarify the interactions among enzymes and substrates, and among fungi and other organisms. This approach is already used in plant pathology and will undoubtedly form the basis for many new decomposition studies. Most previous research on molecular microbial ecology has been carried out with soil and rhizosphere bacteria, often with a view to protecting plants against fungal attack. Extending this type of study to litter fungi would provide useful information on the regulation of enzyme synthesis and interaction with other organisms. Besides enhancing our knowledge of ecosystem function, it would also pave the way for new applications in agriculture and industry.

REFERENCES

1. JC Frankland. Fungal succession—unravelling the unpredictable. Mycol Res 102:1–15, 1998.
2. A Kjøller, S Struwe. Microfungi of decomposing red alder leaves and their substrate utilization. Soil Biol Biochem 12:425–431, 1980.
3. UD Singh, N Sethunathan, K Raghu. Fungal degradation of pesticides. Handbook Appl Mycol Soil Plants 1:541–558, 1991.
4. S Kremer, H Anke. Fungi in bioremediation. In: T Anke, ed. Fungal Biotechnology. Weinheim: Chapman & Kall, 1997, pp 276–290.
5. MJ Swift, OW Heal, JM Anderson. Decomposition in terrestrial ecosystems. Oxford: Blackwell, 1979, pp 1–372.
6. PD Stahl, TB Parkin, M Christensen. Fungal presence in paired cultivated and uncultivated soils in central Iowa, USA. Biol Fertil Soils 29:92–97, 1999.
7. WM Fogarty, CT Kelly. Microbial enzymes and biotechnology. 2nd ed. London: Elsevier Applied Science, 1990, pp 1–472.
8. DB Finkelstein, C Ball. Biotechnology of Filamentous Fungi: Technology and Products. Boston: Butterworth-Heinemann, 1992, pp 1–520.
9. MJ Carlile, SC Watkinson. The Fungi. London: Academic Press, 1996, pp 1–482.
10. A Hodge, IJ Alexander, GW Gooday. Chitinolytic enzymes of pathogenic and ectomycorrhizal fungi. Mycol Res 99:935–941, 1995.
11. S Harney, P Widden. Physiological properties of the entomopathogenic hyphomycete Paecilomyces farinosus in relation to its role in the forest ecosystem. Can J Bot 69:1–5, 1991.
12. JW Deacon. Modern Mycology. 3rd. ed. Oxford: Blackwell, 1997, pp 1–303.
13. DS Powlson. The soil microbial biomass: Before, beyond and back. In: K Ritz, J Dighton, KE Giller, eds. Beyond the Biomass: Compositional and Functional Analysis of Soil Microbial Communities. Chicester: John Wiley & Sons, 1994, pp 3–20.
14. JC Frankland, J Dighton, L Boddy. Methods for studying fungi in soil and forest litter. In: R Grigorova, JR Norris, eds. Methods in Microbiology. London: Academic Press, 1990, pp 343–403.
15. A Tunlid, DC White. Biochemical analysis of biomass, community structure, nutritional status, and metabolic activity of microbial communities in soil. Soil Biochem 7:229–262, 1992.
16. JPE Anderson, KH Domsch. A physiological method for the quantitative measurement of microbial biomass in soils. Soil Biol Biochem 10:215–221, 1978.
17. JPE Anderson, KH Domsch. Measuring of bacterial and fungal contributions to respiration of selected agricultural and forest soils. Can J Microbiol 21:314–322, 1975.
18. K Sakamoto, Y Oba. Effect of fungal to bacterial biomass ratio on the relationship between CO_2 evolution and total soil microbial biomass. Biol Fertil Soils 17:39–44, 1994.
19. J Alphei, M Bonkowski, S Scheu. Application of the selective inhibition method to determine bacterial:fungal ratios in three beechwood soils rich in carbon—Optimization of inhibitor concentrations. Biol Fertil Soils 19:173–176, 1995.

20. CK Johnson, MF Vigil, KG Doxtader, WE Beard. Measuring bacterial and fungal substrate-induced respiration in dry soils. Soil Biol Biochem 28:427–432, 1996.
21. H Velvis. Evaluation of the selective respiratory inhibition method for measuring the ratio of fungal: bacterial activity in acid agricultural soils. Biol Fertil Soils 25:342–360, 1997.
22. HH Schomberg, JL Steiner. Estimating crop residue decomposition coefficients using substrate-induced respiration. Soil Biol Biochem 29:1089–1097, 1997.
23. Q Zhang, JC Zak. Potential physiological activities of fungi and bacteria in relation to plant litter decomposition along a gap size gradient in a natural subtropical forest. Microb Ecol 35: 172–179, 1998.
24. S Hu, AHC van Bruggen. Microbial dynamics associated with multiphasic decomposition of ^{14}C-labeled cellulose in soil. Microb Ecol 33:134–143, 1997.
25. M Miller, A Palojärvi, A Rangger, M Reeslev, A Kjøller. The use of fluorogenic substrates to measure fungal presence and activity in soil. Appl Environ Microbiol 64:613–617, 1998.
26. J Møller, M Miller, A Kjøller. Fungal-bacterial interaction on beech leaves: Influence on decomposition and dissolved organic carbon quality. Soil Biol Biochem 31:367–374, 1999.
27. M Miller. An enzymatic approach in soil microbial ecology, with special emphasis on fungal presence and activity. PhD dissertation, University of Copenhagen, 1998.
28. G Sengeløv. Influence of fungal-bacterial interactions on bacterial conjugation in the residuesphere. FEMS Microbiol Ecol 31:39–45, 2000.
29. H Ueno, K Miyashita, Y Sawada, Y Oba. Assay of chitinase and N-acetylglucosaminidase activity in forest soils with 4-methylumbelliferyl derivatives. Z Pflanzenernähr Bodenk 154: 171–175, 1991.
30. HTS Boschker, TE Cappenberg. A sensitive method for using 4- methylumbelliferyl-β-cellobiose as a substrate to measure (1,4)-β-glucanase activity in sediments. Appl Environ Microbiol 60:3592–3596, 1994.
31. C Freeman, G Liska, NJ Ostle, SE Jones, MA Lock. The use of fluorogenic substrates for measuring enzyme activity in peatlands. Plant Soil 175:147–152, 1995.
32. C Freeman, GB Nevison, S Hughes, B Reynolds, J Hudson. Enzymic involvement in the biogeochemical responses of a Welsh peatland to a rainfall enhancement manipulation. Biol Fertil Soils 27:173–178, 1998.
33. P Markham, MJ Bazin. Decomposition of cellulose by fungi. In: DK Arona, B Rai, KG Mukerji, GR Knudsen, eds. Handbook of Applied Mycology: Soil and Plants. Vol. 1. New York: Marcel Dekker, 1991, pp 379–424.
34. JA Buswell. Fungal degradation of lignin. In: DK Arona, B Rai, KG Mukerji, GR Knudsen, eds. Handbook of Applied Mycology: Soil and Plants. Vol. 1. New York: Marcel Dekker, 1991, pp 425–480.
35. KE Hammel. Fungal degradation of lignin. In: G Cadish, KE Giller, eds. Driven by nature: Plant litter quality and decomposition. Oxon: CAB International, 1997, pp 33–46.
36. B Berg. FDA-active fungal mycelium and lignin concentrations in some needle and leaf litter types. Scand J For Res 6:451–462, 1991.
37. JA Entry, CB Backman. Influence of carbon and nitrogen on cellulose and lignin degradation in forest soils. Can J For Res 25:1231–1236, 1995.
38. S Criquet, S Tagger, G Vogt, G Iacazio, J Le Petit. Laccase activity of forest litter. Soil Biol Biochem 31:1239–1244, 1999.
39. C Chasseur. Étude de la dynamique fongique dans le processus de décomposition de la litière de Fagus sylvatica, pour 2 forets de Bassin de Mons (Belgique) partage du substrat au sein des groupes successionnels. Belg Bot 125:16–28, 1992.
40. A Leonowicz, J-M Bollag. Laccases in soil and the feasibility of their extraction. Soil Biol Biochem 19:237–242, 1987.
41. ML Niku-Paavola, L Raaska, M Itävaara. Detection of white-rot fungi by a nontoxic stain. Mycol Res 94:27–31, 1990.

42. PJA Howard, CH Robinson. The use of correspondence analysis in studies of successions of soil organisms. Pedobiologia 39:518–527, 1995.
43. I Nioh, T Isobe, M Osada. Microbial biomass and some biochemical characteristics of a strongly acid tea field soil. Soil Sci Plant Nutr 39:617–626, 1993.
44. PW Flanagan, A Scarborough. Physiological groups of decomposer fungi on tundra plant remains. In: AJ Holding, QW Heal, SF Maclean Jr, PW Flanagan eds. Soil Organisms and Decomposition in Tundra. Fairbanks: University of Alaska, 1974, pp 159–181.
45. NJ Dix. Inhibition of fungi by gallic acid in relation to growth on leaves and litter. Trans Br Mycol Soc 73:329–336, 1979.
46. B Rai, RS Upadhyay, AK Srivastava. Utilization of cellulose and gallic acid by litter inhabiting fungi and its possible implication in litter decomposition of tropical deciduous forest. Pedobiologia 32:157–165, 1988.
47. G. Gramass, D Ziegenhagen, S. Sorge. Degradation of soil humic extract by wood- and soil-associated fungi, bacteria, and commercial enzymes. Microb Ecol 37:140–151, 1999.
48. M Rihani, E Kiffer, B Botton. Decomposition of beech leaf litter by microflora and mesofauna. I. In vitro action of white-rot fungi on beech leaves and foliar components. Eur J Soil Biol 31:57–66, 1995.
49. A Fioretto, A Musacchio, G Andolfi, V De Santo. Decomposition dynamics of litters of various pine species in a Corsican pine forest. Soil Biol Biochem 30:721–727, 1998.
50. S Miller. Functional diversity in fungi. Can J Bot 73(Suppl):50–57, 1995.
51. JC Zak, S Visser. An appraisal of soil fungal biodiversity: The crossroads between taxonomic and functional biodiversity. Biodiversity Conservations 5:169–183, 1996.
52. P Rosenbrock, F Buscot, JC Munch. Fungal succession and changes in the fungal degradation potential during the initial stage of litter decomposition in a black alder forest (Alnus glutinosa (L.) Gaertn.). Eur J Soil Biol 31:1–11, 1995.
53. A Kjøller, S Struwe. Functional groups of microfungi on decomposing ash litter. Pedobiologia 30:151–159, 1987.
54. A Kjøller, S Struwe. Decomposition of beech litter: A comparison of fungi isolated on nutrient rich and nutrient poor media. Trans Mycol Soc Japan 31:5–16, 1990.
55. A Kjøller, S Struwe. Functional groups of microfungi in decomposition. In: GC Carroll, DT Wicklow, eds. The Fungal Community. New York: Marcel Dekker, 1992, pp. 619–630.
56. A Kjøller, S Struwe. Analysis of fungal communities on decomposing beech litter. In: RK Ritz, J Dighton, KE Giller, eds. Beyond the biomass: Compositional and functional analysis of soil microbial communities. Chicester: John Wiley & Sons, 1994, pp. 191–199.
57. A Kjøller, M Miller, S Struwe, V Wolters, A Pflug. Diversity and role of microorganisms. In: E Schulze, ed. Carbon and Nitrogen Cycling in European Forest Ecosystems: Ecological Studies. Vol. 142. Heidelberg: Springer Verlag, 2000, pp. 382–402.
58. T Terashita, T Sakai, K Yoshikawa, J Shishiyama. Hydrolytic enzymes produced by the genus Mortierella. Trans Mycol Soc Jpn 34:487–494, 1993.
59. MF Cotrufo, M Miller, B Zeller. Litter decomposition In: E Schulze, ed. Carbon and Nitrogen Cycling in European Forest Ecosystems: Ecological Studies. Vol. 142. Heidelberg: Springer Verlag, 2000, pp 276–296.
60. AE Magurran. Ecological Diversity and Its Measurement. Princeton, NJ: Princeton University Press, 1988, pp 1–179.
61. O Andrén, J Bengtsson, M Clarholm. Biodiversity and species redundancy among litter decomposers. In: HP Collins, GP Robertson, MJ Klug, eds. The Significance and Regulation of Soil Biodiversity. Netherlands: Klüwer Academic, 1995, pp 141–151.
62. O Andrén, L Brussard, M Clarholm. Soil organism influence on ecosystem-level processes—bypassing the ecological hierarchy? Applied Soil Ecology 11:177–188, 1999.
63. E Kandeler, J Luxhøi, DV Tscherko, J Magid. Xylanase, invertase and protease at the soil-litter interface of a loamy sand. Soil Biol Biochem 31:1171–1179, 1999.

64. CH Robinson, J Dighton, JC Frankland, JD Roberts. Fungal communities on decaying wheat straw of different resource qualities. Soil Biol Biochem 26:1053–1058, 1994.
65. RM Bowen, SHT Harper. Fungal populations on wheat straw decomposing in arable soils. Mycol Res 93:47–54, 1989.
66. RM Bowen, SHT Harper. Decomposition of wheat straw and related compounds by fungi isolated from straw in arable soils. Soil Biol Biochem 22:393–399, 1990.
67. TM Henriksen, TA Breland. Nitrogen availability effects on carbon mineralization, fungal and bacterial growth, and enzyme activities during decomposition of wheat straw in soil. Soil Biol Biochem 31:1121–1134, 1999.
68. F Schinner. Litter decomposition, CO_2-release and enzyme activities in a snow bed and on a windswept ridge in an alpine environment. Oecologia (Berlin) 59:288–291, 1983.
69. R Ohtonen, P Lähdesmäki, AM Markkola. Cellulase activity in forest humus along an industrial pollution gradient in Oulu, Northern Finland. Soil Biol Biochem 26:97–101, 1994.
70. SR Joshi, GD Sharma, RR Mishra. Microbial enzyme activities related to litter decomposition near a highway in a sub-tropical forest of North East India. Soil Biol Biochem 25:1763–1770, 1993.
71. AK Müller, K Westergaard, S Christensen, SJ Sørensen. The effect of long-term mercury pollution on the soil microbial community. FEMS Microbiol Ecol 36:11–19, 2001.
72. AK Müller, K Westergaard, S Christensen, SJ Sørensen. The diversity, structure and function of soil microbial communities exposed to different disturbances. OIKOS 2001.
73. RG Kuperman, MM Carreiro. Soil heavy metal concentrations, microbial biomass and enzyme activities in a contaminated grassland ecosystem. Soil Biol Biochem 29:179–190, 1997.
74. RAM Badran. Cellulolytic activity of some cellulose-decomposing fungi in salinized soils. Acta Mycol 29:245–251, 1994.
75. DL Moorhead, AE Linkins. Elevated CO_2 alters belowground exoenzyme activities in tussock tundra. Plant Soil 189:321–329, 1997.
76. SS Dhillion, J Roy, M Abrams. Assessing the impact of elevated CO_2 on soil microbial activity in a Mediterranean model ecosystem. Plant Soil 187:333–342, 1996.
77. C Gehrke, U Johanson, TV Callaghan, D Chadwick, CH Robinson. The impact of enhanced ultraviolet-B radiation on litter quality and decomposition processes in Vaccinium leaves from the Subarctic. Oikos 72:213–222, 1995.
78. OW Heal, JM Anderson, MJ Swift. Plant litter quality and decomposition: A historical overview. In: G Cadish, KE Giller, eds. Driven by Nature: Plant Litter Quality and Decomposition. Oxon: CAB International, 1997, pp 3–30.

11

Enzyme Adsorption on Soil Mineral Surfaces and Consequences for the Catalytic Activity

Hervé Quiquampoix
Institut National de la Recherche Agronomique, Montpellier, France

Sylvie Servagent-Noinville and Marie-Hélène Baron
Centre National de la Recherche Scientifique, Université Paris VI, Thiais, France

I. INTRODUCTION

Soil enzymes are either actively secreted by living microorganisms and plant roots or released after the death of soil biota by cell lysis. One class of enzymes, the hydrolases, have a very important role in the biogeochemical cycles of major elements (C, N, P, and S) since their substrates in soil are mainly in a polymerized form. Usually microorganisms (and plant roots) cannot take up macromolecules directly from the external medium. There are few exceptions to this, such as the uptake of fragments of deoxyribonucleic acid (DNA) or plasmids, that lead to transformation in bacteria. In general, the membrane transport systems are specific and recognize the universal biological monomers, such as amino acids, sugars, and nucleotides; low number oligomers, such as cellobiose or maltose; or mineral ionic groups that can be released by enzymatic hydrolysis of organic molecules, such as orthophosphate and sulfate. Extracellular enzymes perform three main functions in soil: (1) they reach substrates in pores with dimensions roughly 100 times smaller than those of bacteria; (2) they hydrolyze these substrates and make them soluble and consequently able to diffuse back to the microorganisms or plant roots; and at the same time (3) they transform polymers into monomers or oligomers that can be recognized and taken up by membrane transport systems to undergo intracellular metabolism.

As enzymes are proteins, they share with this class of macromolecules their strong affinity with interfaces. Proteins are potentially flexible polypeptide chains, even if they have a stable folded configuration in solution, with individual amino acids whose lateral chains have various physicochemical properties: hydrophilic or hydrophobic; negatively, neutrally, or positively charged. From a thermodynamic point of view, these properties give rise to both enthalpic (related to intermolecular forces) and entropic (related to the spatial arrangement of the molecules) contributions to the interactions with surfaces. The strong and often largely irreversible adsorption of enzymes on the mineral phase of the soil has important consequences, not only for their mobility, but also for their survival and catalytic activity.

The most well-known effect of the adsorption of enzymes on negatively charged surfaces, such as clay minerals, is a shift of their optimal catalytic activity towards a higher pH range (1–4). A second general effect, which can be extended to all proteins, is that the maximal adsorption is observed near the isoelectric point (i.e.p.) of the enzyme. McLaren et al. described these two properties in the 1950s (5–7). However, until the beginning of the 1990s, no effort, supported by independent structural study of the adsorbed enzymes or proteins by modern physical methods, was made to propose a mechanism explaining both observations.

II. NATURE OF THE DRIVING FORCES LEADING TO ADSORPTION OF ENZYMES ON SURFACES

Studies on the quantity of protein adsorbed on surfaces cannot be separated from the study of their conformation on these surfaces. The reason is that a modification toward a more disordered structure contributes to the driving forces of adsorption, since it increases the entropy of the system and thus decreases the Gibbs energy. The modification of conformation also can have an effect on the maximal quantity of protein adsorbed, since conformational changes may affect the area occupied by each single protein on the surface.

The spontaneous adsorption of proteins at constant temperature and pressure leads to a decrease of the Gibbs energy of the system, according to the second law of thermodynamics (8–13). The Gibbs energy, G, depends on enthalpy, H, which is a measure of the potential energy (energy that has to be supplied to separate the molecular constituents from one another), and entropy, S, which is related to the disorder of the system:

$$\Delta_{ads} G = \Delta_{ads} H - T\Delta_{ads} S < 0$$

where T is the absolute temperature and Δ_{ads} is the change in the thermodynamic functions resulting from adsorption.

Some difficulties arise in the analysis of these processes because enthalpic effects, related to intermolecular forces, and entropic effects, related to the spatial arrangements of molecules, are not totally independent. Intermolecular forces influence the distribution of molecules, and the potential energy also is also dependent on the molecular structure of the system.

A. Enthalpic Effects

1. *Coulombic Interactions*

The electrical charge of proteins results from the ionization of the carboxylic, tyrosyl, amine, imine, and imidazole groups of the side chains of some amino acids. The electrical charge of mineral surfaces can result from pH-independent isomorphic substitutions in the crystal lattice, as in some clays (basal surfaces of illite or montmorillonite), or from pH-dependent ionization of hydroxyls (edge sites of clays, oxyhydroxides). Coulombic forces are very strong and long-range intermolecular forces. They can be screened by the ions in the solution. As all electrical charges have to be compensated by an equal number of electrical charges of the opposite sign, a diffuse double layer is established around the macromolecules and mineral surfaces. The electrostatic interactions between proteins and surfaces thus can be analyzed as an overlap of their electrical double layers (8,10). The electrostatic part of the Gibbs energy is given by the isothermal and isobaric work of charging the electrical double layer:

$$G_{el} = \int_0^{\sigma} \phi(\sigma) d\sigma$$

where ϕ is the variable electrostatic potential and σ is the variable surface density during the charging process.

2. Lifshitz–van der Waals Interactions

Contrary to the Coulombic interactions, van der Waals forces act on all molecules, even if they are electrically neutral. They are short-range forces and are composed of three different components. The main component are the dispersion (or London) forces, which originate from the instantaneous dipolar moment resulting from the fluctuation of the electrons around the nuclear protons. The electric field created induces, by polarization, a dipole moment in nearby molecules, which, in turn, creates an instantaneous attractive interaction. The two other components are the induction (or Debye) forces, related to the interaction between a polar molecule and a nonpolar molecule, and the orientation (or Keesom) forces, related to the interaction between two polar molecules.

B. Entropic Effects

1. Hydrophobic Interactions

The stability of proteins in solution results mainly from the shielding of amino acids with a hydrophobic side chain in the core of the protein from contact with water. It is due to the hydrophobic effect that causes water molecules around a nonpolar group to establish more hydrogen bonds among themselves than around a polar group. This process maximizes the mutual association of water molecules by hydrogen bonds and results in an increased order of the surrounding water, and thus a favorable decrease in entropy of the system.

Sometimes hydrophobic interactions are involved in interactions of proteins with hydrophilic mineral surfaces. An example is given by the higher affinity of a hydrophobic methylated derivative of bovine serum albumin (BSA) for the hydrophilic montmorillonite surface than the native, less hydrophobic BSA (14). The explanation is that the adsorption of proteins is accompanied by the exchange of charge-compensating cations on the clay, which are the true hydrophilic centers, leaving a hydrophobic siloxane layer (15). Thus, the montmorillonite surface presents hydrophilic properties for molecules whose adsorption does not result in the removal of charge-compensating cations, and hydrophobic properties for molecules that replace the charge-compensating cations.

2. Modifications in Protein Molecular Structure

The entropic contribution to adsorption also can result from a modification of the conformation of the protein. This phenomenon is related to an increase of the rotational freedom of the peptide bonds engaged in secondary structures, such α helices and β sheets. The ordered secondary structures are an important part of the densely packed hydrophobic core of proteins. After adsorption, internal hydrophobic amino acids can reach more external positions in contact with the surface, since the amino acids remain shielded from contact with the water molecules of the surrounding solvent phase. If a decrease of internal ordered secondary structures accompanies this process, it results in an increase of conformational entropy. The gain of conformational entropy, S_{conf}, can be calculated from the assumption that four different conformations are possible for peptide units in random structures as compared with only one in α helices and β sheets:

$$\Delta_{ads} S_{conf} = R \ln 4^n$$

where R is the molar gas constant and n is the number of peptide units involved in the transfer from an ordered secondary structure to a random secondary structure (9,10).

III. EXPERIMENTAL EVIDENCE FOR pH-DEPENDENT CHANGES IN THE STRUCTURE OR ORIENTATION OF ADSORBED PROTEINS

The previous thermodynamical considerations indicate that modifications in conformation are an important parameter to consider in the adsorption of proteins, since the entropic effect related to these structural changes is itself a factor of adsorption. These three-dimensional changes affect the catalytic activity of the adsorbed enzymes.

A major difficulty in the evaluation of the extent of such events is that no experimental method allows the direct measurement of the conformation of proteins in an adsorbed state. Only two methods are suitable for the determination of the tertiary structure of the proteins, and neither can be employed when proteins are adsorbed. One method, X-ray diffraction, necessitates the preparation of protein crystals, which is impossible for adsorbed proteins. The other, nuclear magnetic resonance (NMR) spectroscopy, is confined to molecules with a sufficiently high tumbling rate to obtain spectra with narrow linewidth peaks, a condition not compatible with the adsorption on a surface of larger dimension than the protein itself since even a small adsorbed molecule experiences the slower rotational movement of the mineral colloid.

Without dramatic advances in solid-state NMR spectroscopy, information on the conformation can be deduced only from lower levels of structural information than the tertiary structure, such as the secondary structure and the specific interfacial area occupied by adsorbed proteins. Two spectroscopical approaches that permit such investigations are now discussed.

A. Study of the Interfacial Area of Protein-Surface Contact by Nuclear Magnetic Resonance Spectroscopy

The study of adsorption isotherms of proteins on clay mineral surfaces has been disappointing with regard to the interpretation of the adsorbed enzyme activity. The idea that can be advanced is that the quantity of protein adsorbed is by itself insufficient to describe a complex phenomenon that can involve at least four other parameters: (1) the orientation of the protein on the surface (this is important as proteins are rarely perfect spheres and are more often described as ellipsoids with a long and a large axis; thus an end-on adsorption involves a higher quantity of protein adsorbed than a side-on adsorption); (2) a possible unfolding of the protein on the surface changing the interfacial area between individual protein and surface and the quantity of protein adsorbed at saturation; (3) the surface coverage at saturation, which could be less than 100% for packing reasons, if the adsorption is irreversible; and (4) the possibility of a multilayer adsorption. Thus, it is always possible to find several explanations to interpret a protein adsorption isotherm, with no experimental evidence available to choose among them. The advantage of the NMR method is that it simultaneously gives the quantity of adsorbed protein, the surface coverage of the solid by the protein, and the monolayer or multilayer mode of adsorption (16). Only knowledge of these three factors allows a possible unfolding of the proteins on the clay surfaces to be detected and quantified.

1. Nuclear Magnetic Resonance Detection of the Exchange of a Paramagnetic Cation on Protein Adsorption on Clays

The principle of the method (16) is based on the fact that the adsorption of proteins on clays causes the release of charge-compensating cations (7,17). It also uses the sensitivity of the relaxation times T_1 and T_2 of nuclear spins to paramagnetic cations in NMR spectroscopy (18,19).

A small quantity (between 3 and 20 μM depending on the pH) of a paramagnetic cation, Mn^{2+}, is added to a sodium-saturated montmorillonite suspension (1 g L^{-1}) with a 10-mM concentration of orthophosphate. The suspension is studied by ^{31}P NMR spectroscopy. An interesting phenomenon is observed: (1) the Mn^{2+} cations that are adsorbed on the clay surface do not interact at all with the orthophosphate, as shown by the comparison between the clay suspension and supernatant after removal of the clay by centrifugation; and (2) the Mn^{2+} cations in solution interact with the orthophosphate, leading to a linear increase of the linewidth at half height, $\Delta\nu_{1/2}$, of the orthophosphate peak on the NMR spectrum. This last effect is the result of the paramagnetic contribution to the decrease of the spin–spin relaxation time, T_2, of the orthophosphate signal. When a given quantity of protein is introduced into this suspension, it disturbs the equilibrium between the paramagnetic Mn^{2+} adsorbed on the clay surface and that in solution. The analysis of the resulting linewidth of the orthophosphosphate signal gives the quantity of cations exchanged on adsorption.

With a 300-MHz NMR spectrometer, the measurement takes a few minutes; with a 500-MHz spectrometer, 1 min is sufficient (even less if higher concentrations of orthophosphate are used). As no centrifugation is required with this method, this short time of signal acquisition is compatible with kinetic studies. The results are expressed as $\Delta\nu_P$, which is the difference between $\Delta\nu_{1/2}$ in the system with paramagnetic cations and $\Delta\nu_{1/2}$ in a control of the same composition, (but without paramagnetic cations) divided by the concentration of paramagnetic cations. The surface coverage of the clay by the protein can be deduced from the fraction of Mn^{2+} released. The knowledge of both the quantity of protein adsorbed and the surface coverage of the solid allows the calculation of the interfacial area of contact between a single protein molecule and the clay surface at different pH and ionic strengths.

2. Conformational Changes on Adsorption of a Soft Protein, Bovine Serum Albumin

a. Description of the Progressive Surface Coverage of the Clay Figure 1 shows the evolution of $\Delta\nu_p$, i.e., the release of the paramagnetic cation Mn^{2+}, when the total quantity of bovine serum albumin (BSA) introduced in the clay suspension increases, and at a pH corresponding to the i.e.p. of the BSA (pH 4.7). The increase is linear, followed by a plateau. The plateau corresponds to the saturation of the montmorillonite surface, as shown by the comparison with the measurement, by UV adsorption (A_{279} nm), of the BSA in the supernatant solution after centrifugation. The linear increase of $\Delta\nu_p$ before the pla-

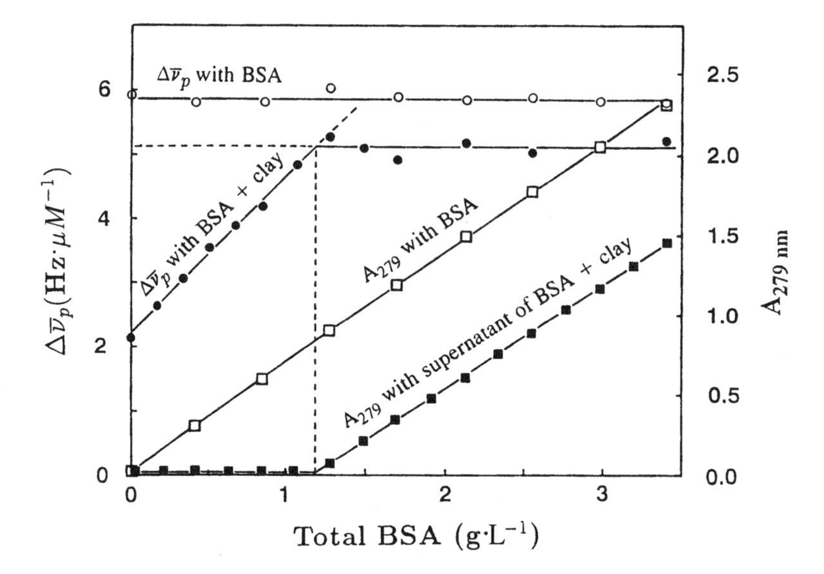

Figure 1 Effect of the addition of bovine serum albumin on the release of Mn^{2+}, as detected by its line-broadening effect $\Delta\nu p$ on orthophosphate by NMR, and on the UV absorption A_{279} nm of the protein. When present, the montmorillonite suspension is at 1 g dm^{-3}, pH 4.65. (Adapted from Ref. 16.)

teau indicates that the adsorbed proteins always have the same interfacial area of contact with the clay surface, whatever the surface coverage. No change, from a side-on to an end-on state, or from an unfolded to a more native state, resulting from an increase of lateral repulsions with packing can be invoked. If there were such a change in the mode of adsorption, the amount of paramagnetic cation released per unit mass of protein would be greater at low surface coverage than at high surface coverage, and this would be seen as a convexity rather than a linearity of the curve.

b. Monolayer Mode of Adsorption The comparison between the Mn^{2+} exchange data and the depletion data in Figure 1 shows also that the maximal adsorption of BSA corresponds to a monolayer. Indeed, only the contact of the protein with the clay surface can lead to the exchange of the charge-compensating cations. A second layer would involve a protein–protein contact, with no release of Mn^{2+}. Thus, the occurrence of the breaks in both cation exchange and protein depletion curves at the same protein concentration is compatible only with a monolayer of protein on the clay surface.

c. Maximum of Adsorption at the i.e.p. The data, such as those presented in Fig. 1, have been collected over a large pH range. They all have the same general aspect; only two parameters vary. Fig. 2 shows the evolution with pH of these two parameters: the plateau amount of BSA adsorbed on montmorillonite, measured by either NMR or the depletion method, and the maximal fraction of Mn^{2+} that is displaced. As often is observed, the maximal amount of protein adsorbed occurs near the i.e.p. of the protein, which is 4.7 for BSA. Several hypotheses have been advanced to explain this phenomenon. One class of hypotheses is based on the same (symmetrical) mechanisms above and below the i.e.p. to explain the decrease of adsorption. They can be based on the effect of lateral electrostatic repulsions between the adsorbed proteins, which increase as the pH is more

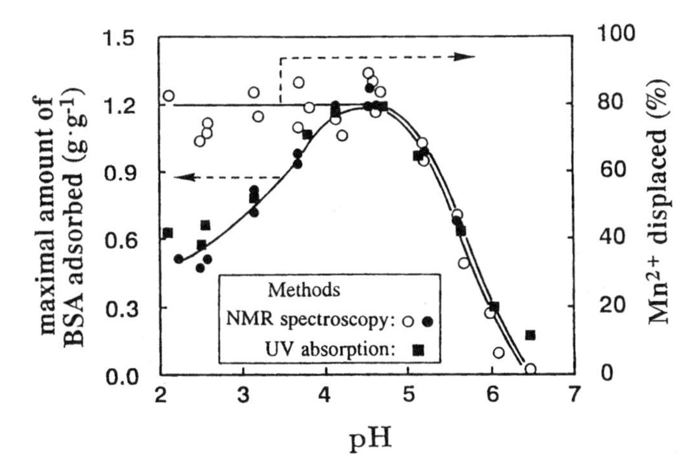

Figure 2 Effect of pH on the maximal amount of bovine serum albumin adsorbed on montmorillonite and on the clay surface coverage followed by the release of Mn^{2+} on protein adsorption. (Adapted from Ref. 16.)

distant from the i.e.p. (20–22). Alternatively, they can be based on a decrease in the structural stability of the protein when the net electric charge increases, leading to an unfolding of the protein (7,8,10). In addition, different (asymmetrical) mechanisms can be invoked above and below the i.e.p., and the comparison of the NMR exchange data and the protein adsorption data shows that such an asymmetrical mechanism is involved in the adsorption of BSA on montmorillonite, as explained later.

d. Repulsive Electrostatic Interactions and Protein Unfolding Figure 2 shows that, above the i.e.p., the maximal amount of BSA adsorbed and the fraction of Mn^{2+} exchanged decrease in exactly the same proportion. This good correlation between the quantity of protein adsorbed and the surface coverage of the clay surface supports a mechanism based on an increase of the electrostatic repulsion between the protein, whose net negative charge increases with pH above the i.e.p., and the montmorillonite, which carries a permanent negative charge. Figure 2 also shows that, below the i.e.p., the maximal amount of BSA adsorbed decreases, but the fraction of Mn^{2+} exchanged remains nearly constant. Below the i.e.p. of the BSA (pH 4.7) there is no important variation in the number of positively charged side chains of the protein because the pKa of histidine (His), lysine (Lys), and arginine (Arg) is approximately pH 7, 10, and 12, respectively. The constant proportion of the cation exchanged by the BSA, despite a decreasing quantity adsorbed, can be explained only by an unfolding of the protein on the clay surface, moving more positively charged side chains of His, Lys, and Arg to near the surface. The increase of the specific interfacial area of the protein resulting from this unfolding is compatible with a smaller quantity adsorbed at constant surface coverage. It can be calculated from the data reported in Fig. 2 that the interfacial surface area occupied by a molecule of BSA on montmorillonite is 60 nm^2 at pH 4.5 and increases to 120 nm^2 at pH 3.0. Again, electrostatic interactions appear to be the main driving force, but here they are attractive since the protein becomes more positively charged as the pH decreases below the i.e.p. and the montmorillonite remains negatively charged. BSA is a soft protein since, even at the i.e.p., the interfacial area of 60 nm^2 is higher than would be expected from the X-ray structure of the analogous human serum albumin, which has the shape of an equilateral triangle with sides of 8 nm and a depth of 3 nm (23,24). If no modification of conformation had occured on adsorption at the i.e.p., the interfacial area of contact should have been 28 nm^2 for a side-on adsorption. This is half of the value obtained by NMR at pH 4.5.

B. Study of the Modification in Secondary Structures by Fourier Transform Infrared Spectroscopy

Although the determination of the tertiary structure of adsorbed proteins is impossible with the present state of scientific knowledge, the study of changes in the repartition of the different secondary structures on adsorption is possible. This determination allows the deduction of the occurrence of a modified conformation, and it also allows direct calculation of the component of the Gibbs energy of adsorption that is related to these structural changes. Circular dichroism and infrared spectroscopy can be used to investigate the secondary structure of proteins. Circular dichroism has, nevertheless, limitations in turbid suspensions because of light-scattering effects and can be applied only to particles with a size below 30 nm (25–27). Fourier transform infrared spectroscopy (FTIR) does not have this disadvantage and can be applied to more turbid suspensions of clays of a greater size.

1. Fourier Transform Infrared Spectral Analysis

Transmission FTIR spectra in the 1800- to 1500-cm^{-1} region give information on the protonation state, the secondary structure, and the solvation of the protein. All samples were prepared in ^2H_2O medium in order to shift the spectral absorption domain of water molecules, bound to the polypeptide backbone of the studied protein, out of the Amide I and II spectral range. A phosphate buffer ($Na^2H_2PO_4$) was used at a final concentration of 0.055 mol L^{-1} in ^2H_2O. Several p^2H values in the 4–12 range were obtained by adding ^2HCl or NaO^2H (11). The state of the protein in solution is obtained from the spectral difference between the protein in solution and the corresponding buffer; the spectral difference between the solid protein–clay mixture and the corresponding clay suspension reveals the state of the adsorbed protein. Spectral decomposition could be achieved by second-derivative, curvature analysis, or self-deconvolution procedures. The same number of principal components of the overall spectrum range (1500–1800 cm^{-1}) was obtained at similar wavenumbers (± 1 cm^{-1} or less). A least-square iterative curve-fitting program (Levenberg–Marquardt) was applied to fit the overall spectrum with the found number of principal components. The fixed parameters (frequency, IR band profile) for any spectral decomposition allows a comparative quantitative analysis of intensity changes for each component from one spectrum to another. Examples of initial difference spectra and spectral decomposition of BSA in solution and adsorbed on montmorillonite are reproduced in Fig. 3.

The assignments specific for the Amide I′/I region are deduced from the literature and our own experiments on model amides, polypeptides, and proteins (28–40). The area of each Amide I component is expressed as a percentage of the sum of the areas of all Amide I components. Intensities (percentage peptide CO) are used to deduce the proportion of peptide units involved in the various solvated structural domains of the polypeptide backbone.

The solvation parameter is given by the percentage of N^2H. The level of exchange at a given time depends on the rate at which water molecules gain access to internal

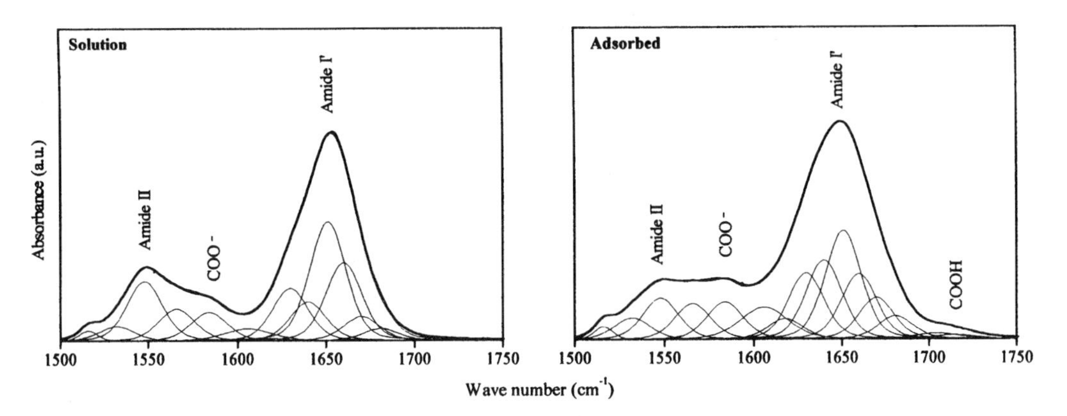

Figure 3 FTIR-vibrational absorption spectra (1750–1500 cm^{-1}) and computed decomposition of spectral profiles for BSA in solution (left) or adsorbed on montmorillonite (right) at $p^2H = 5.6$. (Adapted from Ref. 43.)

peptide groups in the protein core (28–32,41). The protonation parameter is given by COO^- fractions (percentage) deduced from measurements of the area of the $\nu(CO)_{COOH}$ absorption for the remaining COOH species (with respect to the overall Amide I intensity at a given p^2H). These relative areas are expressed with respect to the corresponding relative areas (percentage) obtained at low p^2H when Asp and Glu side chains are all fully protonated (100% COOH).

2. *Conformational Changes on Adsorption of a Soft Protein, Bovine Serum Albumin*

Adsorption on montmorillonite surfaces implies pH-dependent changes in BSA solvation, unfolding of helical domains, as well as changes in hydration and self-associated domains (42,43).

a. BSA Solvation Figure 4 shows the adsorption effects induced by the negatively charged montmorillonite surface on the BSA solvation with p^2H, protein concentration, and time. The adsorption effects already are established after 10 min; the relative intensity is simply more pronounced at 2 hours. For acidic p^2H, the weaker exchange after adsorption suggests that the electronegative surface protects some domain of the protein. In contrast, in the i.e.p. range, adsorption increases the NH/N^2H exchange, and at higher p^2H, the rate of water diffusion is no longer influenced by adsorption.

b. BSA Protonation BSA adsorption on montmorillonite leads to the protonation of the ionizable carboxylic groups of the protein, aspartic acid (Asp), and glutamic acid (Glu), at least to p^2H 6.5 (Fig. 5). Various reasons may explain such a shift of the apparent pK_a of Asp and Glu. In the primary structure of BSA, some Asp and Glu side chains are adjacent to R^+ functions. Embedded among positively charged side chains interacting with the electronegative clay or embedded in self-associated domains, external Asp and Glu side chains are assumed to become indifferent to buffer. Moreover, the electronegative charge of the clay surface could favor a protonation of the Asp and Glu carboxylates to decrease the coulombic repulsion between the protein and the surface, as observed by titration on other systems (8,10).

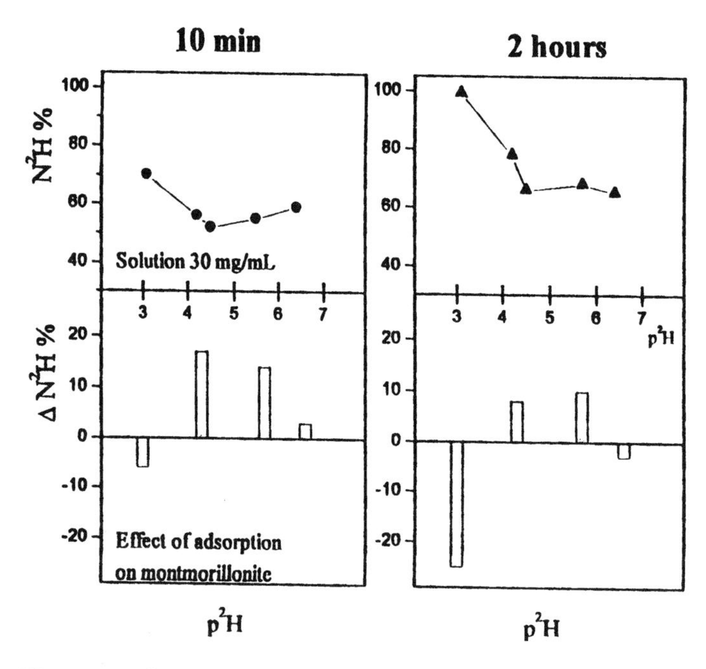

Figure 4 p^2H-Dependent NH/N^2H exchange (expressed as $N^2H\%$) of BSA in ^2H_2O at 10 min and 2 h. Δ $N^2H\%$ represents the change in protein solvation for BSA adsorbed on montmorillonite with respect to the solution. (Adapted from Ref. 43.)

c. Helix Unfolding Montmorillonite induces important unfolding of helical domains of BSA (Fig. 6). After adsorption on montmorillonite, the largely p^2H-independent external helix unfolding is related to new orientations for the Lys^+ and Arg^+ side chains forced close to the negative clay surface. In contrast, unfolding in internal and packed helices is largely p^2H-dependent. This change could be related to the disruption of some

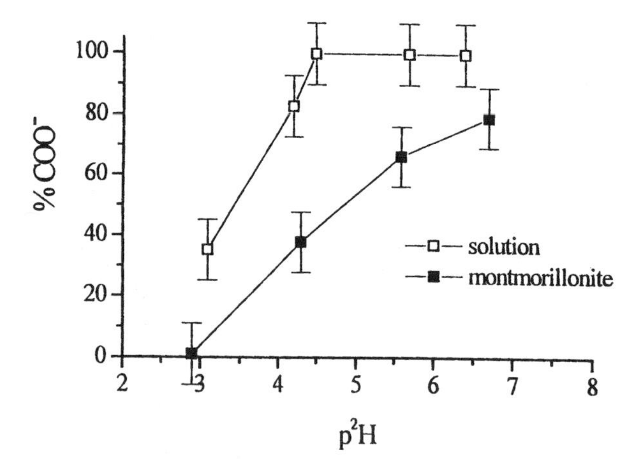

Figure 5 Effect of BSA adsorption on montmorillonite on the p^2H-dependent Asp and Glu deprotonation. (Adapted from Ref. 43.)

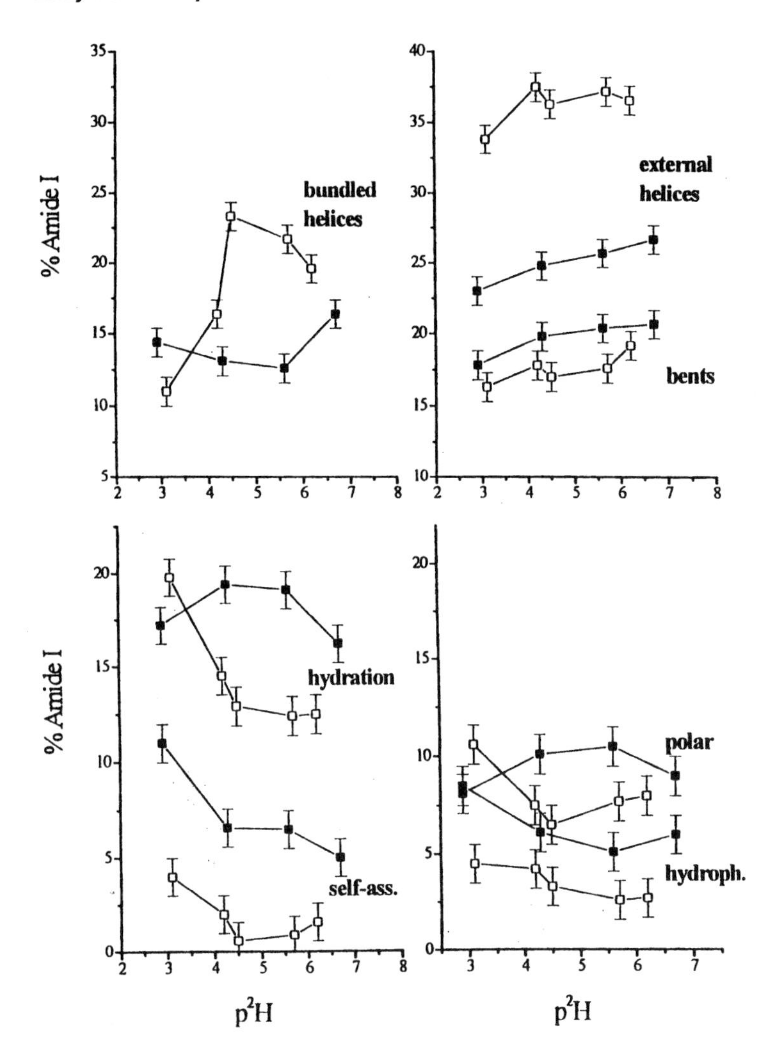

Figure 6 Effect of BSA adsorption on montmorillonite on the p²H-dependent secondary structures; open symbols, BSA in solution; closed symbols, BSA adsorbed on montmorillonite. Abbreviations are for the environment of the CO peptide groups (free CO in polar or hydrophobic environments; H-bonded CO in bundled or external helices, in bents or in protein self-association; hydrated CO). (Adapted from Ref. 43.)

$(Asp^-/Glu^-)–His^+$ salt bridges that enhance helix formation in solution. Adsorption involves protonated His^+, as proved by the recovery of helices when all His side chains become deprotonated. At low p²H, the protein molecules adsorbed on montmorillonite spread over the entire mineral surface, as is in agreement with the instability of BSA structure in solution. It should be noted that the p²H-dependent profiles for bundled helices follow those observed for the decreases of the Amide II bands for both adsorbed and solution states. Unfolding of bundled/internal helical domains increases water diffusion inside the core of the protein (Fig. 6).

d. Hydrated and Self-Associated Domains Helix unfolding increases the amount of self-association, free polar CO, and hydrated peptide CO. Among the peptide units that are unfolded, those embedded in hydrophobic regions probably are responsible for the increase in protein self-association; the others in polar environments should become hydrated. On average, adsorption on montmorillonite entails a large degree of protein self-association (Fig. 6). The level of protein self-association on montmorillonite is even more important at low p^2H.

e. Entropy of Conformation Changes The loss of ordered secondary structures, such as internal and external α helices, is compensated by the increase of unordered structures such as hydrated and self-associated domains. As previously emphasized, this represents an important contribution of the entropy of conformation to the adsorption process, since a loss of approximatively 20% of ordered secondary structures near the i.e.p. represents for BSA an entropic contribution of -400 kJ mol^{-1} to the decrease of Gibbs energy accompanying adsorption. Thus, BSA can be considered as a soft protein, as the study of the variation of the interfacial area on adsorption by NMR spectroscopy already has shown.

3. *Orientation Changes on Adsorption of the Hard Protein, Bovine Pancreatic α-Chymotrypsin*

a. Enzyme Activity The adsorption of chymotrypsin on montmorillonite results in a complete inhibition below pH 7, a progressive recovery of the activity from pH 7 to pH 9, and an activity quite similar to that observed in solution above pH 9 (Fig. 7) (42).

b. Preservation of the Secondary Structure on Adsorption Figure 8 shows that the adsorption of α-chymotrypsin on montmorillonite has only a very small effect on the secondary structure of this protein (32). Only 10 to 20 peptide units in peripheral β sheets are lost on adsorption below pH 7, and this value decreases above this pH. The contrast with the low structural stability of the secondary structure of BSA is thus well marked. These weak structural changes should not perturb the structure of α-chymotrypsin at the level of the enzymatic site. The transmission-FTIR analysis of solutions also confirms that the optimal catalytic activity near pH 8 results from the convergence of several parameters: (1) the deprotonation of the carboxylic side chains; (2) the deprotonation of two His side chains, increasing both protein flexibility and hydration; and (3) a local β sheet folding that results from the formation of a salt bridge between the Ile^{+}-16 (isoleucine) end chain aminium group and the Asp-194 side chain carboxylate. For pH > 10, the

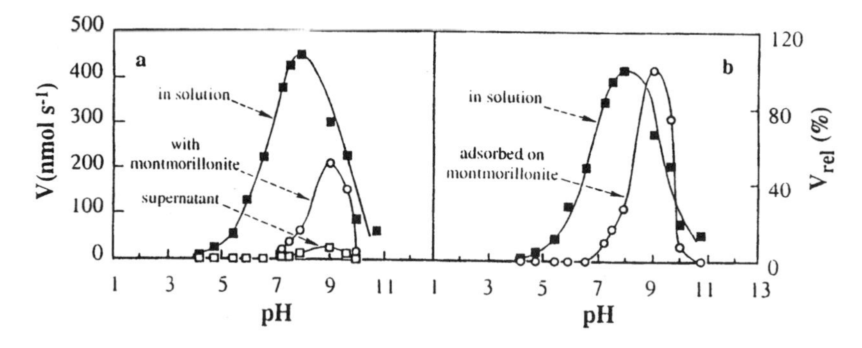

Figure 7 Catalytic activity of α-chymotrypsin in the presence of montmorillonite. (a) experimental data; (b) data normalized with respect to the maximal value. (Adapted from Ref. 42.)

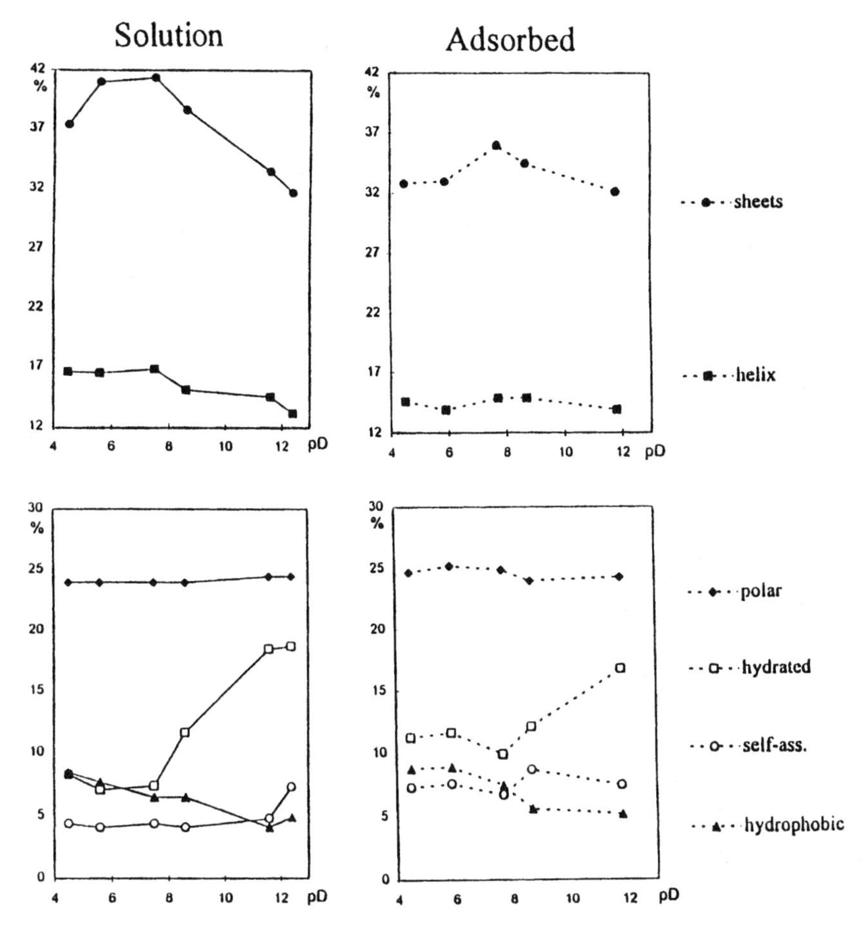

Figure 8 Secondary structure of α-chymotrypsin in solution or adsorbed on montmorillonite. Abbreviations as in Figure 6. (Adapted from Ref. 32.)

complete inhibition of the catalytic activity should result not only from peripheral secondary structure unfolding caused by external Lys or Tyr (tyrosine) deprotonations, but also from internal Tyr deprotonations entailing excessive internal hydration in the vicinity of the catalytic center.

c. Orientation of the Catalytic Site of the Enzyme The pH dependence of the adsorbed enzyme catalytic activity shows that electrostatic interactions are involved. Nevertheless, the dynamic structural transition of the protein, resulting from both flexibility and hydration that are slightly enhanced with respect to solution phase, cannot explain the inactivation of the enzyme in the 5–9 pH range by these weak structural changes alone. If the tertiary structure of α-chymotrypsin, as determined by X-ray diffraction studies, is taken as relatively invariant on adsorption and if the time dependence of the Amide II intensity is analyzed for varying pH, information on the pH dependence of the α-chymotrypsin orientation on the montmorillonite surface can be obtained (32). The kinetics of the NH/N^2H exchange measured by the Amide II intensity indicates which class of amino acids are protected from water contact. The analysis of the results shows that most of the inhibition would aries from a steric hindrance by the clay of the substrate access to the α-chymotrypsin catalytic site. This is due to an interaction involving positively charged His^{+}-40 and His^{+}-57 imidazole and Ala^{+}-149 (alanine) end chain aminium that control the initial specific recognition of the substrate by the enzyme. At pH higher than 8.5, when His-40, His-57, and Ala-149 are deprotonated, the enzyme is adsorbed with a different orientation, which allows a recovery of activity, similar to that measured in solution in the same pH range, since the catalytic site now is exposed to the solvent (Fig. 9). Even if adsorption on montmorillonite implies weak effects on the structure of α-chymotrypsin, which can be considered as a hard protein, whereas major ones are observed for BSA, a soft protein, a pH-dependent orientation effect nevertheless can affect the catalytic activity of α-chymotrypsin.

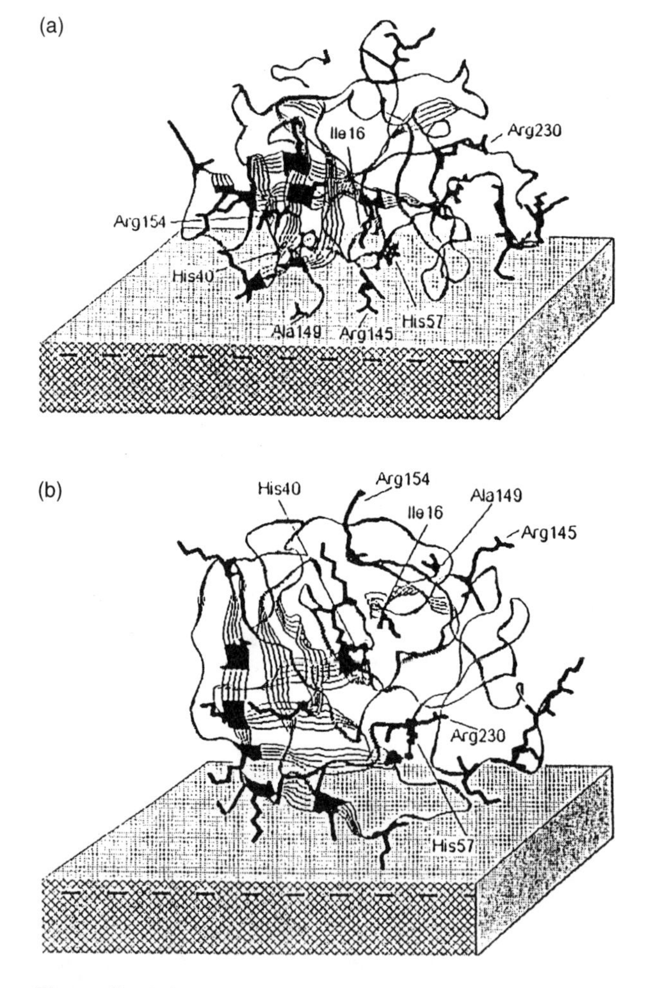

Figure 9 Schematic representation of an orientation for α-chymotrypsin adsorbed on montmorillonite below pH 7 (A) and above pH 8 (B). See text. (Adapted from Ref. 32.)

IV. EFFECT ON CATALYTIC ACTIVITY OF ENZYME ADSORPTION ON MINERAL SURFACES

A. Models of Interaction

Four different mechanisms currently are invoked to explain the modification of properties with pH between the free and adsorbed enzymes; they are based either on the activity of the protons or the substrates in the microenvironment of the enzyme (interfacial pH effects, diffusional effects) or on the state of the adsorbed enzyme itself (orientation effects, conformational effects).

1. Interfacial pH Effects

Although increasingly contested, this hypothesis is the most frequently invoked to explain the pH shift in the optimal catalytic activity when an enzyme is adsorbed on an electrically charged surface. It originates from the double diffuse layer theory and assumes that there is a difference in pH between the bulk of the solution (pH_b) and the liquid layer near the solid surface (pH_2) (6,44–46) due to the increased concentration of protons on electronegative surfaces such as clays. Thus, the active site of an adsorbed enzyme on a clay surface could be at a pH lower than that measured in the bulk of the solution, and the apparent pH_b for maximal activity of a bound enzyme could appear to be higher than that of a free enzyme.

2. Diffusion-Limited Reactions

The apparent enzyme activity could decrease as a result of limitations in the diffusion of substrates toward adsorbed enzymes (47–50). The concentration of substrate can be lower near the surface than in the bulk of the solution as the result of its consumption by the adsorbed enzyme. A concentration gradient is thus established. If diffusion is slow with respect to substrate consumption, a steady state is eventually established such that the rate of diffusion of the substrate in the unstirred layer equals the consumption rate of the substrate. It should be noted that as the rate of consumption of the substrate is the only factor that depends on the enzyme in this process, this phenomenon should not cause a shift in the pH of the optimal catalytic activity. The enzyme activity versus pH follows a bell-shaped curve, and thus, there are always two pH values, below and above the optimal pH for activity, where the enzyme activity is similar. Consequently, the decrease in catalytic activity, as a result of a limitation in diffusion of the substrate, should be the same. The expected effect should be symmetrical with respect to the optimal pH, and with no pH shift.

3. Orientation Effects

Adsorption of an enzyme with its active site facing the mineral surface obviously limits the access of the substrate to this site. Such a case has been described for the interaction of α-chymotrypsin with montmorillonite (32).

4. pH-Dependent Modifications of Conformation

In contrast to the three preceding models, which assume that the enzymes retain the same conformation in the adsorbed state and in solution, another model is based on a pH-dependent unfolding of the enzyme on the surface. The mechanism could be analogous to the modification of conformation observed on BSA adsorbed on montmorillonite (43).

B. Experimental Study of the Interaction of β-D-Glucosidases with Mineral Surfaces

1. Experimental Approach

The interaction between the β-D-glucosidases and the mineral surfaces at different pH values was studied by three procedures (Fig. 10). Procedure A is a measurement of the enzyme activity in the absence of adsorbent surfaces. Procedure B is a measurement of the enzyme activity in the presence of the adsorbent surface. Procedure C is designed to determine the catalytic activity of the nonadsorbed fraction in procedure B. In other words, it is a measurement of the enzyme activity in the supernatant solution after centrifugation of the enzyme-adsorbent surface suspension. If A, B, and C are the values of the catalytic activity measured by the respective procedures, two parameters well suited to a physico-chemical analysis of the interaction of enzymes with solid surfaces may be calculated (Fig. 11). The first is F, the proportion of nonadsorbed enzyme, which is inversely related to the affinity for the surface:

$$F = C/A$$

The second is the relative activity, R, a structure-related parameter that is defined by the ratio of the catalytic activity resulting from the enzyme fraction adsorbed ($B - C$) and that of an equal quantity of enzyme in solution ($A - C$):

$$R = (B - C)/(A - C)$$

2. Interactions with Montmorillonite

The β-D-glucosidase of *Aspergillus niger* (51) has an i.e.p. of 4.0. At a pH above the i.e.p., the nonadsorbed fraction, F, increases progressively from pH 4 until at pH 6 no enzyme is adsorbed (Fig. 11). This shows the electrostatic repulsions between the net

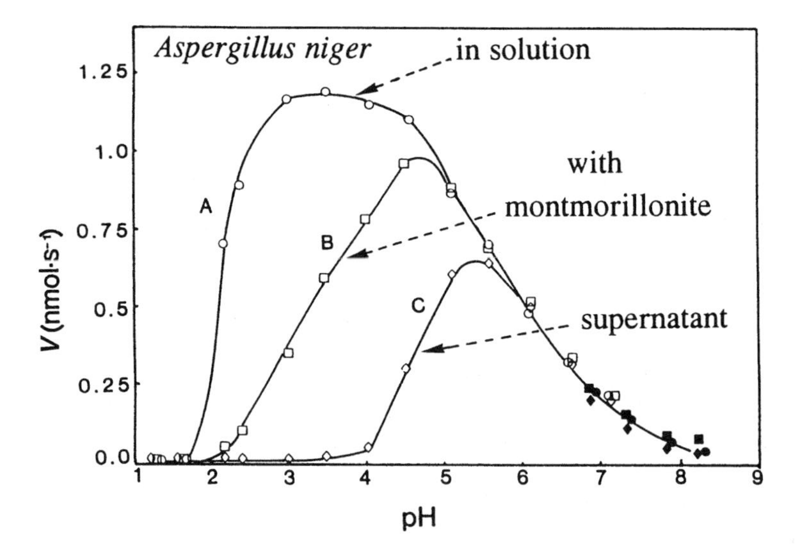

Figure 10 Effect of pH on the activity of *Aspergillus niger* β-D-glucosidase in solution (A); in the presence of montmorillonite (B), where the activities of both the free and the bound enzyme are measured; and in the supernatant (C), where the bound fraction has been eliminated by centrifugation. (Adapted from Ref. 51.)

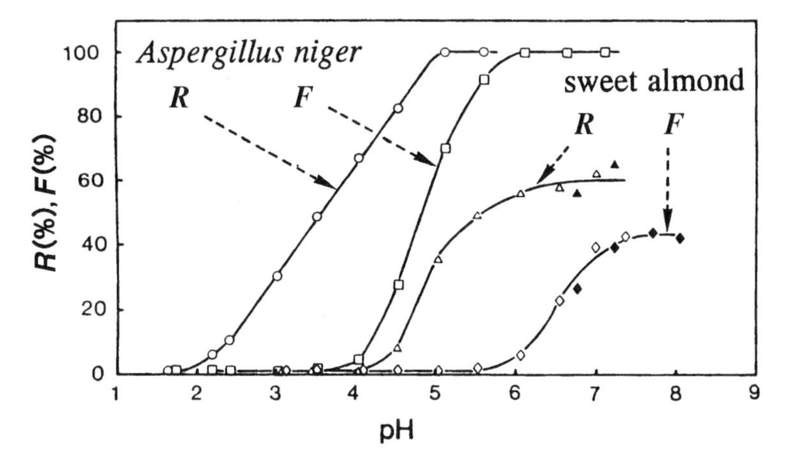

Figure 11 Effect of pH on the relative catalytic activity R in the adsorbed state and on the relative quantity F in the nonadsorbed state of two β-D-glucosidases from *Aspergillus niger* and sweet almond. (Adapted from Ref. 51.)

negatively charged enzyme and the electronegative clay surface. In contrast, when the pH decreases below the i.e.p., the relative activity, R, of the enzyme decreases. It can be the result of a progressive unfolding caused by the electrostatic attractions between the electronegative surface and the positively charged enzyme in this range of pH, as shown with BSA by the NMR and FTIR approach (16,43). Alternatively, it can be the result of a change of orientation of the enzyme with pH, as shown with α-chymotrypsin by the comparison of catalytic activity and FTIR data (32,42). Although, in our opinion, the modification of conformation hypothesis has the advantage that it always can explain an alkaline shift of the optimal catalytic activity of enzymes adsorbed on electronegative surface, it is not really possible to choose between these two hypotheses in this case.

3. Irreversible Effects on Adsorbed Enzyme Activity

The sweet almond β-D-glucosidase adsorption on montmorillonite presents characteristics that are more in favor of a modification of conformation (52). Figure 11 shows results qualitatively similar to the β-D-glucosidase of *A. niger*. But the results of another experiment can be explained only by structural changes of the sweet almond β-D-glucosidase. Indeed, irreversible effects influence the activity of this enzyme when adsorbed at a pH that is different from a subsequent pH of catalytic reaction (52). The sweet almond β-D-glucosidase first was incubated with montmorillonite at the pH of adsorption for 2 h (pH_{ads}). After 2 h, the pH was changed (pH_{react}), and the catalytic activity was measured (Fig. 12). For a given pH_{react}, the catalytic activity decreased when the pH_{ads} was decreased. At a pH_{ads} of 3.6, no catalytic activity is detected over the entire range of pH_{react}, despite complete stability of the enzyme in solution in this pH range. This behavior suggests two remarks. First, an irreversible effect would not be expected if surface pH was involved, as the 2 h of equilibration should be sufficient to reach a new repartition of the protons in the double diffuse layer. Thus, no lasting effect of pH_{ads} should have been observed. Second, irreversible effects are common with polymers adsorbed on surfaces and are explained by irreversible changes of conformation (8,10,53). If the enzyme is unfolded, the number of points of contact with the surface increases and the energy necessary for

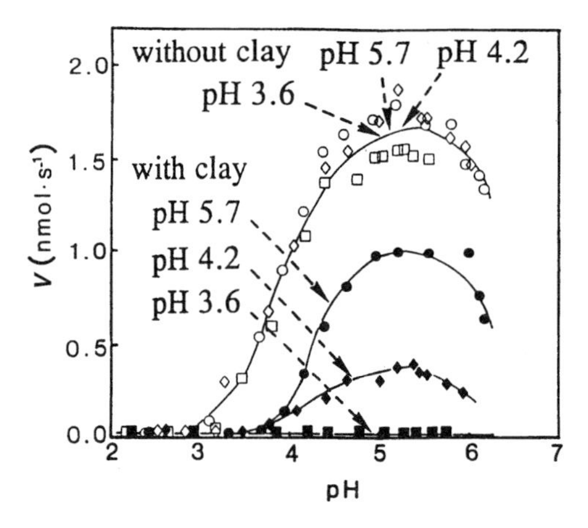

Figure 12 Effect of the pH of adsorption on montmorillonite (numbers on curves) on the subsequent pH profile of activity of sweet almond β-D-glucosidase when the pH is adjusted after the adsorption stage to give the pH of the catalytic reaction on the x axis. (Adapted from Ref. 52.)

the reversal of the unfolding of the adsorbed enzyme therefore would be greater than the thermal energy, kT, available to the system.

4. *Inadequacy of the Interfacial pH Hypothesis*

The irreversible effect of pH changes on adsorbed enzyme activity is not the only fact that mitigates against the interfacial pH interpretation of the pH shift of the catalytic activity. An important consequence of a local pH mechanism is that the catalytic activity of the adsorbed enzyme should be higher than that of the enzyme in solution in the alkaline pH range. The results obtained for α-chymotrypsin (42) show that it is not the case when absolute values are considered (Fig. 7a), contrary to conclusions drawn considering values normalized with respect to their maximum (Fig. 7b). In addition, even an increase in ionic strength increase cannot eliminate the pH shift effect (54). Finally, it should been pointed out that although the concentration of protons is different near an electrically charged surface and in the bulk of the solution, their electrochemical potentials must, nevertheless, be identical, since they are in thermodynamic equilibrium. Because the electrochemical potential determines the work available for a reaction, higher proton activity near a surface does not mean higher reactivity of these protons.

5. *Comparison of the Effect of Different Mineral Surfaces*

Although electrostatic interactions between surfaces and enzymes explain most of the observed conformational changes, other types of interaction sometimes need to be taken into account. Figure 13 compares the effects of talc, goethite, and montmorillonite on the relative catalytic activity of sweet almond β-D-glucosidase. The hydrophilic surface of goethite, which had a low electrical charge as a result of complexation of oxyanions under the conditions of the experiment (52), has the lower destabilizing effect. The hydrophobic talc, which also has no electrical charge on its basal surface, has a more pronounced destabilizing effect, confirming the intervention of hydrophobic interactions.

6. *Interaction with Organomineral Surfaces*

Organic coatings on clays can have a protective effect on the adsorbed enzyme activity (54). Figure 14 shows the effect of different natural and artificial clay–polymer complexes on the

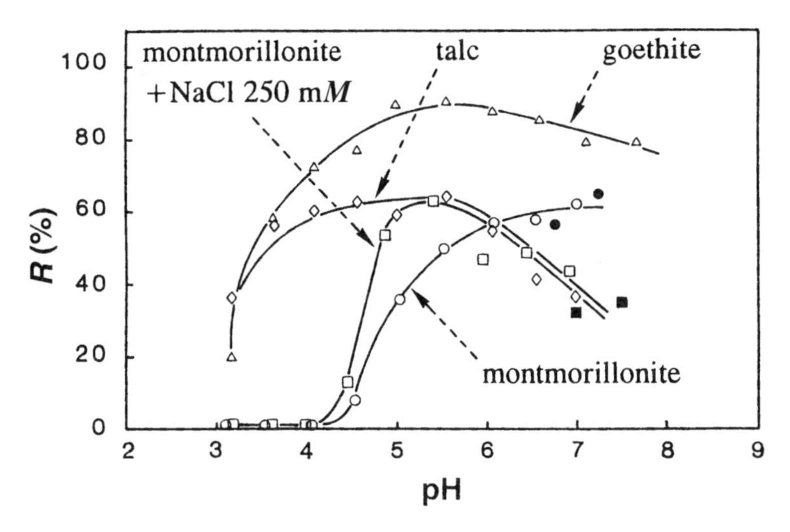

Figure 13 Effect of different mineral surfaces on the relative activity *R* of adsorbed sweet almond β-D-glucosidase. (Adapted from Ref. 52.)

relative catalytic activity of sweet almond β-D-glucosidase. A soil fraction rich in organic matter has a pronounced protective effect, at least above pH 4. A polyethylene glycol–montmorilonite complex has a similar effect. In this case, it could be explained by a protective effect of this hydrophilic polymer. The lack of protective effect below pH 4 could be due to an exchange mechanism between the positively charged enzyme at this pH and the neutral polymer. This mechanism is supported by the absence of inhibition of the β-D-glucosidase at pH 4 by a lysozyme–montmorillonite complex. In this case, it is the strong interaction of the lysozyme with the clay surface, the high i.e.p. of the protein, and its positive charge over the entire pH range studied that prevents exchange with the β-D-glucosidase.

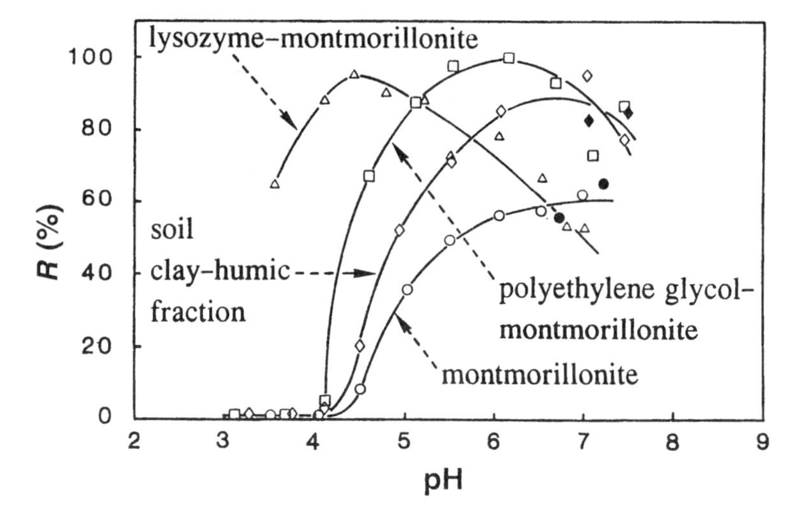

Figure 14 Effect of different organic coatings on mineral surfaces on the relative activity *R* of adsorbed sweet almond β-D-glucosidase. (Adapted from Ref. 54.)

IV. CONCLUSIONS

The adsorption of enzymes on mineral surfaces is a complex phenomenon that involves both enthalpic and entropic effects. An important and difficult challenge is the determination of possible changes in conformation of the adsorbed enzyme. NMR anf FTIR spectroscopies are useful tools to answer this question since they, respectively, give information on the interfacial area of the surface in contact with the protein and on the secondary structure of adsorbed proteins. It has been shown that both pH-dependent modification of conformation and pH-dependent orientation of the catalytic site of the enzyme can explain the alkaline pH shift of the enzyme activity on electronegative soil mineral surfaces. Soft proteins, such as BSA, are more prone to the first mechanism, whereas hard proteins, such as α-chymotrypsin, are more prone to the second. On the other hand, it has been shown, from experimental observations as well as for theoretical reasons, that a surface pH effect cannot explain adequately the shift in the pH of the optimal activity when an enzyme is adsorbed on clays. In addition to electrostatic forces, hydrophobic interactions are implied in the interaction of proteins with clays. An important aspect is the interplay of different driving forces in adsorption. For example, the hydrophobic interactions with clays can result from an electrostatic exchange of the hydrophilic counterions on the clay surface that leaves a hydrophobic siloxane surface. The rearrangement of the protein structure on the clay surface subsequently can be facilitated when hydrophobic amino acids come in contact with the hydrophobic siloxane layer and remain shielded from the water molecules of the solution. If this rearrangement is accompanied by a decrease of ordered secondary structures, it results in an additional increase in conformational entropy, lowering the Gibbs energy of the system. The combination of all these different subprocesses is responsible for the irreversible aspects of the modifications of conformation of enzymes on clay surfaces.

REFERENCES

1. AD McLaren. Concerning the pH dependence of enzyme reactions on cells, particulates and in solution. Science 125:697, 1957.
2. P Monsan, G Durand. Préparation d'invertase insolubilisée par fixation sur bentonite. FEBS Lett 16:39–42, 1971.
3. J Skujinš, A Pukite, AD McLaren. Adsorption and activity of chitinase on kaolinite. Soil Biol Biochem 6:179–182, 1974.
4. P Douzou, P Maurel. Ionic control of biochemical reactions. Trends Biochem Sci 2:14–17, 1977.
5. AD McLaren. The adsorption and reactions of enzymes and proteins on kaolinite. J Phys Chem 58:129–137, 1954.
6. AD McLaren, EF Estermann. Influence of pH on the activity of chymotrypsin at a solid-liquid interface. Arch Biochem Biophys 68:157–160, 1957.
7. AD McLaren, GH Peterson, I Barshad. The adsorption and reactions of enzymes and proteins on clay minerals. IV. Kaolinite and montmorillonite. Soil Sci Soc Am Proc 22:239–244, 1958.
8. W Norde. Adsorption of proteins from solution at the solid-liquid interface. Adv Colloid Interface Sci 25:267–340, 1986.
9. W Norde, J Lyklema. Why proteins prefer surfaces. J Biomater Sci Polymer Edn 2:183–202, 1991.
10. CA Haynes, W Norde. Globular proteins at solid/liquid interfaces. Colloids Surfaces B Biointerfaces 2:517–566, 1994.
11. W Norde, J Lyklema. Thermodynamics of protein adsorption: Theory with special reference

to the adsorption of human plasma albumin and bovine pancreas ribonuclease at polystyrene surfaces. J Colloid Interface Sci 71:350–366, 1979.
12. J Lyklema. Proteins at solid-liquid interfaces: A colloid-chemical review. Colloids Surfaces 10:33–42, 1984.
13. W Norde. Energy and entropy of protein adsorption. J Dispersion Sci Technol 13:363–377, 1992.
14. S Staunton, H Quiquampoix. Adsorption and conformation of bovine serum albumin on montmorillonite: Modification of the balance between hydrophobic and electrostatic interactions by protein methylation and pH variation. J Colloid Interface Sci 166:89–94, 1994.
15. P Chassin, C Jouany, H Quiquampoix. Measurement of the surface free energy of calcium-montmorillonite. Clay Miner 21:899–907, 1986.
16. H Quiquampoix, RG Ratcliffe. A ^{31}P NMR study of the adsorption of bovine serum albumin on montmorillonite using phosphate and the paramagnetic cation Mn^{2+}: modification of conformation with pH. J Colloid Interface Sci 148:343–352, 1992.
17. JT Albert, RD Harter. Adsorption of lysozyme and ovalbumin by clay: Effect of clay suspension pH and clay mineral type. Soil Sci 115:130–136, 1973.
18. H Quiquampoix, G Bacic, BC Loughman, RG Ratcliffe. Quantitative aspects of the ^{31}P-NMR detection of manganese in plant tissues. J Exp Bot 44:1809–1818, 1993.
19. H Quiquampoix, BC Loughman, RG Ratcliffe. A ^{31}P-NMR study of the uptake and compartmentation of manganese by maize roots. J Exp Bot 44:1819–1827, 1993.
20. TJ Su, JR Lu, BK Thomas, ZF Cui, J Penfold. The effect of solution pH on the structure of lysozyme layers adsorbed at the silica-water interface studied by neutron reflection. Langmuir 14:438–445, 1998.
21. S Duinhoven, R Poort, G Van der Voet, WGM Agterof, W Norde, J Lyklema. Driving forces for enzyme adsorption at solid-liquid interfaces. 1. The serine protease savinase. J Colloid Interface Sci 170:340–350, 1995.
22. S Duinhoven, R Poort, G Van der Voet, WGM Agterof, W Norde, J Lyklema. Driving forces for enzyme adsorption at solid-liquid interfaces. 2. The fungal lipase lipolase. J Colloid Interface Sci 170:351–357, 1995.
23. XM He, DC Carter. Atomic structure and chemistry of human serum albumin. Nature 358: 209–214, 1992.
24. DC Carter, JX Ho. Structure of serum albumin. Adv Protein Chem 45:155–203, 1994.
25. A Kondo, K Higashitani. Adsorption of model proteins with wide variation in molecular properties on colloidal particles. J Colloid Interface Sci 150:344–351, 1992.
26. A Kondo, F Murakami, K Higashitani. Circular dichroism studies on conformational changes in protein molecules upon adsorption on ultrafine polystyrene particles. Biotechnol Bioeng 40:889–894, 1992.
27. A Kondo, F Murakami, M Kawagoe, K Higashitani. Kinetic and circular dichroism of enzymes adsorbed on ultrafine silica particles. Appl Microbiol Biotechnol 39:726–731, 1993.
28. L Boulkanz, N Balcar, MH Baron. FTIR analysis for structural characterisation of albumin adsorbed on reversed phase support. Appl Spectrosc 49:1737–1746, 1995.
29. B De Collongue, B Sebille, MH Baron. Chromatography of the Interferon γ and the Analogue II : FTIR Analysis. Biospectros 2:101–111, 1996.
30. L Boulkanz, C Vidal-Madjar, N Balcar, MH Baron. Adsorption mechanism of human serum albumin on a reversed-phase support by kinetic, chromatographic and FTIR methods. J Colloid Interface Sci 188:58–67, 1997.
31. A Pantazaki, MH Baron, M Revault, C Vidal-Madjar. Characterisation of human serum albumin adsorbed on a porous anion-exchange support. J Colloid Interface Sci 207:324–331, 1998.
32. MH Baron, M Revault, S Servagent-Noinville, J Abadie, H Quiquampoix. Chymotrypsin adsorption on montmorillonite: Enzymatic activity and kinetics FTIR structural analysis. J Colloid Interface Sci 214:319–332, 1999.
33. JLR Arrondo, A Muga, J Castresana, FM Goni. Quantitative studies of the structure of proteins

in solution by Fourier transform infrared spectroscopy. Prog Biophys Mol Biol 59:23–56, 1993.
34. DM Byler, H Susi. Application of computerized infrared and Raman spectroscopy to conformation studies of casein and other food proteins. J Ind Microbiol 3:73–88, 1988.
35. D Byler, H Susi. Examination of the secondary structure of proteins by deconvolved FTIR spectra. Biopolymers 25:469–487, 1986.
36. WK Surewicz, HH Mantsch. New insight into protein secondary structure from resolution-enhanced infrared spectra. Biochim Biophys Acta 952:115–130, 1988.
37. F Fillaux, C de Loze. Spectroscopic study of monosubstituted amides. II. Rotation isomers in amides substituted by aliphatic side-chain models. Biopolymers 11:2063–2077, 1972.
38. C de Loze, MH Baron, F Fillaux. Interactions of the CONH group in solution. Interpretation of the infrared and Raman spectra in relationship to secondary structures of peptides and proteins. J Chimie Physique 75:631–647, 1978.
39. MH Baron, C de Loze, C Toniolo, GD Fasman. Structure in solution of protected homo-oligopeptides of L-valine, L-isoleucine and L-phenylalanine: An infrared absorption study. Biopolymers 17:2225–2239, 1978.
40. M Hollòsi, ZS Majer, AZ Rònai, A Magyar, K Medzihradszky, S Holly, A Perczel, GD Fasman. CD and FTIR spectroscopic studies of peptides. II. Detection of βturns in linear peptides. Biopolymers 34:117–185, 1994.
41. RB Gregory, A Rosenberg. Protein conformational dynamics measured by hydrogen isotope exchange techniques. Methods Enzymol 131:448–511, 1986.
42. H Quiquampoix, J Abadie, MH Baron, F Leprince, PT Matumoto- Pintro, RG Ratcliffe, S Staunton. Mechanisms and consequences of protein adsorption on soil mineral surfaces. In: TA Horbett, JL Brash, eds. Proteins at Interfaces II, ACS Symposium Series 602. Washington, DC: American Chemical Society, 1995, pp 321–333.
43. S Servagent-Noinville, M Revault, H Quiquampoix, MH Baron. Conformational changes of bovine serum albumin induced by adsorption on different clay surfaces: FTIR analysis. J Colloid Interface Sci 221: 273–283, 2000.
44. G Durand. Modification de l'activité de l'uréase en présence de bentonite. C R Acad Sci 259: 3397–3400, 1964.
45. RA Aliev, VS Gusev, DG Zvyagintsev. Influence of adsorbents on the optimum pH of catalase. Vestn Moskov Univ Serie 6 Biol Pochvoved 31:67–70, 1976.
46. P Douzou, GA Petsko. Proteins at work: ''Stop-action'' pictures at subzero temperatures. Adv Protein Chem 36:245–361, 1984.
47. J Konecny, W Voser. Effects of carrier morphology and buffer diffusion on the expression of enzymic activity. Biochim Biophys Acta 485:367–378, 1977.
48. M Arrio-Dupont, JJ Béchet. Diffusion limited kinetics of immobilised myosin ATPase. Biochimie 71:833–838, 1989.
49. V Bille, D Plainchamp, S Lavielle, G Chassaing, J Remacles. Effect of the microenvironment on the kinetic properties of immobilized enzymes. Eur J Biochem 180:41–47, 1989.
50. CA Ku, BB Lentrichia. Effects of diffusion limitation on binding of soluble ligand to avidin-coupled latex particles. J Colloid Interface Sci 132:578–584, 1989.
51. H Quiquampoix, P Chassin, RG Ratcliffe. Enzyme activity and cation exchange as tools for the study of the conformation of proteins adsorbed on mineral surfaces. Prog Colloid Polym Sci 79:59–63, 1989.
52. H Quiquampoix. A stepwise approach to the understanding of extracellular enzyme activity in soil. I. Effect of electrostatic interactions on the conformation of a β-D-glucosidase on different mineral surfaces. Biochimie 69:753–763, 1987.
53. J Israelachvili. Intermolecular and Surface Forces. 2nd ed. London: Academic Press, 1991.
54. H Quiquampoix. A stepwise approach to the understanding of extracellular enzyme activity in soil. II. Competitive effects on the adsorption of a β-D-glucosidase in mixed mineral or organo-mineral systems. Biochimie 69:765–771, 1987.

12

Microbes and Enzymes in Biofilms

Jana Jass
Umeå University, Umeå, Sweden

Sara K. Roberts
University of Illinois–Chicago, Chicago, Illinois

Hilary M. Lappin-Scott
Exeter University, Exeter, England

I. INTRODUCTION

Microbial enzymes and their activities have been studied primarily in pure liquid cultures under laboratory conditions. However, in natural environments microorganisms grow at interfaces as attached (sessile) mixed communities rather than as suspended planktonic populations (1). Studies of microbial enzymes in soil go some way to recognizing this, but data interpretation has often been difficult because the methodologies do not easily differentiate between enzymes associated with surface-attached populations and those loosely attached or free in the liquid phase. It is the aim of this chapter to discuss the biological characteristics of these sessile microbial populations with particular reference to their enzyme activities.

II. WHAT IS A BIOFILM?

Microorganisms attached to a surface are collectively referred to as a biofilm and are of current interest because they are different in their phenotype and physiological characteristics from the planktonic populations. Early research on biofilms was conducted in the 1920s and 1930s, primarily by Claude ZoBell (2–4), who was one of the first people to note that bacteria existed as what he termed attached films. Three of ZoBell's many observations (3,4) include that bacteria attach rapidly to surfaces; planktonic bacteria are not covered in ''sticky'' material, but sessile bacteria are; and these organisms, once associated with a surface, secrete a ''cementing'' substance. It is significant that many of the problems encountered by biofilm researchers today are the same as those from as long as 60 years ago; they include understanding the mechanisms underlying attachment, detachment (3), microbial interactions, population diversity, biofilm structure, and growth (5).

There have been many proposed definitions of a biofilm over the years (6,7), the most useful is given by Costerton and associates (8), who defined the biofilm as bacteria attached to surfaces and aggregated in a hydrated polymeric matrix of their own synthesis.

However, it is important to be aware that in many instances, particularly in natural environments, biofilms consist not only of bacteria but of fungi (9,10), yeasts (11), algae (12), and protozoa (13). Nonetheless, much of the published literature concentrates on bacterial biofilms, although with the increased emergence of infections (11) and problem associated with fungal biofilms (10) the importance of studying more complex mixed biofilms containing both prokaryotes and eukaryotes has recently been realized (14).

III. BIOFILM LIFE CYCLE

It is often easier to understand what a biofilm is in terms of the events that lead to its formation. Most studies of biofilms in natural environments have concentrated on events at solid–liquid interfaces, the colonization of a submerged abiotic surface is depicted in Fig. 1. Upon immersion of a nonbiological material, such as glass or silica, the surface becomes coated rapidly with a layer of proteinacous material called a conditioning layer (15–17). Ions and other nutrient sources accumulate at the interface, giving rise to higher microenvironment concentrations that will attract microorganisms from the nutrient- and energy-starved liquid phase to the surface (18). Bacteria, which are often the first colonizers, begin to synthesize copious amounts of exopolysaccharide (EPS) material initiated

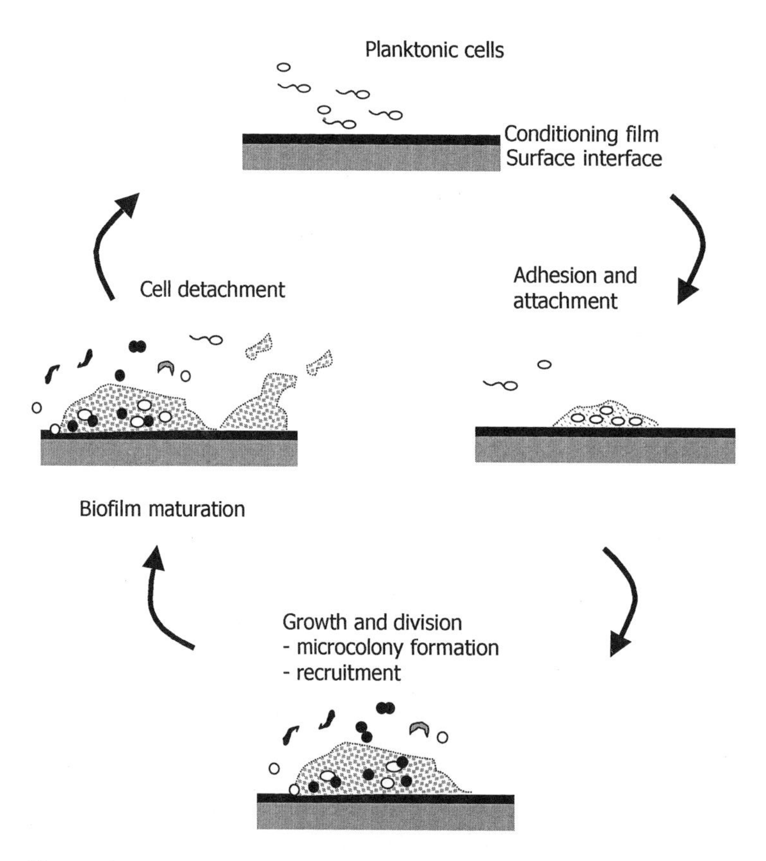

Figure 1 A schematic illustrating the life cycle of a biofilm.

upon contact with a surface (19,20). The microbial cells become embedded within this matrix, grow, and divide to form microcolonies. Other microorganisms present in the surrounding environment are recruited into the biofilm at all stages of biofilm development to form complex functioning communities (14,21,22). Bryers and Characklis (23) proposed a three-step colonization process that is widely accepted by many authors: initial biofilm formation, exponential accumulation of cells and biomass, and steady state. This pattern of colonization dictates a typical sigmoidal growth curve. Steady state is reached when attachment of cells is equal to detachment of cells as a consequence of such processes as predation, sloughing, and erosion.

Although biofilms are complex and dynamic and differ from environment to environment, they all have three primary common features. First, a biofilm is associated with an interface at which the cells accumulate. The solid–liquid interface is most frequently studied and well characterized, however, biofilms may also form at air–liquid (24,25), solid–air (26), and, in some cases; when a phase separation occurs, liquid–liquid interfaces. Second, a biofilm contains a number of microbial cells of one or more species at an interface. A single attached microorganism does not constitute a biofilm although opinions differ as to how dense the attached organisms must be to constitute, a biofilm (27). Third, the sessile microorganisms produce an extracellular polymer matrix within which they are embedded. This matrix, often composed of EPS synthesized by the bacteria, may contain materials and components trapped from their surrounding environment. For example, biofilms in natural water habitats contain particles of sediments and plant material trapped within the matrix. In addition, the EPS matrix is believed to be important in a variety of biofilm functions, which are discussed in the following sections. Studies have also shown that biofilm bacteria are more resistant to antimicrobial regimes than their planktonic counterparts (8). The exopolymer matrix may contribute to the increased resistance to antimicrobial agents by either ionically binding the compounds or physically reducing penetration of the agent through the structure, although other factors may be involved (28–30).

IV. FUNCTION OF BIOFILM STRUCTURE AND ARCHITECTURE

It would be naive to assume that a biofilm community is simply defined as microorganisms residing and growing at an interface. Microbes are, in fact, components of complex communities continuously responding to both their immediate microenvironment and their surrounding habitat. This is reflected in the range of biofilm structures: from thin layers of attached cells, as seen with monocultures of some Pseudomonads or smooth colony variants of *Vibrio cholerae* (31), to more complex forms of attached communities containing multiple species interacting with each other (22,32).

Biofilm structure (three-dimensional) and architecture (microbial organization) are strongly connected to the functions and survival of the microorganisms within. Research has shown that there are many conditions that contribute to biofilm architecture; these can be categorized as physical factors (i.e., flow rates, hydrodynamic forces, and viscosity), chemical factors (i.e., nutrient availability and EPS composition), and biological factors (i.e., competition and predation) (33). In practice it is difficult to separate the influences of these categories, as there is overlap between them. The combination of species specificity and physical, chemical, and biological factors influence biofilm structure in such a manner that it is virtually impossible (and probably unrealistic) to agree on a standard

Table 1 Factors That Influence Biofilm Structure

Factor	Examples of variables	Reference
Surface	Hydrophobicity	Bos et al. (15)
	Roughness	Lewandowski et al. (34)
	Electrochemical properties	Geesey et al. (35)
Hydrodynamics	Mass transfer	Lewandowski et al. (36)
	Flow rate/velocity	Stoodley et al. (37,38)
Nutrients	Concentration	Stoodley et al. (37,38)
	Mass transfer	Xu et al. (39)
	Availability	deBeer and Stoodley (40)
		Møller et al. (41)
Exopolymeric matrix	Exopolysaccharide production	Sutherland (42, 43)
		Skillman et al. (44)
Ecology	Consortia	Stoodley et al. (37)
	Predation	Caron (45)
		Rogers et al. (46)
	Cell signaling	Davies et al. (47)

Source: Adapted From Ref. 13.

biofilm model. In practice, different models are available for different growth conditions, based on a consensus of variables that influence biofilm architecture (Table 1). With advances in imaging technology, such as confocal scanning laser microscopy (48), real-time image capture (49), and fluorescent staining (41,50,51), our understanding of biofilm structure is increasing rapidly. Some researchers believe that biofilm structure and increased resistance to antimicrobial regimes are attributable to the production of chemical signals (52).

An increasing number of microorganisms, including bacteria and fungi, are found to produce a range of molecules that regulate their population density; these are called quorum sensing or cell–cell communication molecules (53). Many gram-negative bacteria produce N-acylhomoserine lactones (AHL-s) as sensor molecules (54); however, other substances have been implicated in signaling including 3-hydroxypalmatic acid methyl ester produced by the plant pathogen *Ralstonia solanacearum* (55). Gram-positive organisms (e.g. *streptomyces* spp.) produce different signal molecules such as small posttranslationally modified peptides or other compounds related to AHLs such as γ-buytrolactones (56). These small diffusible molecules accumulate at high cell densities within the biofilm and, at a critical concentration, activate a genetic response in the microorganisms. The response is not always restricted to the same species producing the sensing molecules; other bacterial species or even eukaryotic cells (fungi, plant, or animal cell cultures) may respond to these chemical signals (57,58). Davies et al. (47) reported that the quorum sensing system of *Pseudomonas aeruginosa* that affects biofilm formation is composed of a two-gene cascade systems, *lasR-lasI* and *rhlR-rhlI*. The *lasI* and *rhlI* gene products are involved in the synthesis of two different AHL molecules, *N*-(3-oxododecanoyl)-L-homoserine lactone and *N*-buytryl-L-homoserine lactone, respectively (47). The AHL molecules are required to activate the transcriptional regulators (products of *lasR* and *rhlR*) in a sequential order, where the gene product of *lasR* activates the *rhlR-rhlI* system and a number of virulence factors and secondary metabolites. Mutants lacking both *lasI* and *rhlI* or just *lasI* gene products were able to adhere to a glass surface but were not able to differentiate into thick multilayered biofilms. This system also regulates the expression of other factors (59), such as type IV pili in *P. aeruginosa* (twitching motion), which have also been found to influence the differentiation of adherent monolayers to thick biofilm structures (60). An increasing number of bacteria are being found to be associated with new density-dependent communication molecules, both in the laboratory and in situ (53,61). For example, the presence of AHLs was detected in naturally occuring aquatic biofilms on stones by introducing *Agrobacterium tumefaciens* A136 with a *lac*Z fusion as an indicator organism (61). However, it would be naïve to assume that adhesion and biofilm formation rest solely on the production of chemical signals (52,62). Other research has shown that although AHLs play an important role in the accumulation of cells on the surface and the formation of biofilms, the overall structure of biofilms growing in aqueous environments during the early stages of colonization is determined largely by the flow conditions (37,63).

There are two main delimiting factors that influence the structure of a biofilm in aqueous environments: flow rate and nutrient availability. Flow can be categorized as laminar or turbulent. Laminar flow is the smooth flow of fluid through a pipe or duct. In contrast, when flow becomes erratic and irregular it is described as turbulent. Lewandowski and Walser (64) found that the thickness of a mixed culture biofilm was at a maximum near the transition between laminar and turbulent flows. However, many different biofilm structures have been observed, often explained by examining the mass transfer properties of the bulk liquid. In a turbulent system there is good mixing of nutrients, and the bulk liquid comes into contact with large proportions of the biofilm where uptake of nutrients can take place. In comparison, under laminar flow conditions there is poor mixing of nutrients in the bulk liquid, limiting nutrient availability. Indeed, Lewandowski and Walser (64) hypothesized that there was an optimal flow rate below which biofilm accumulation was limited by mass transfer and above which biofilm accumulation was limited by continual cell detachment. Many of the recent confocal microscope studies have shown that a biofilm consists of microcolonies of bacteria in a dense EPS matrix with less dense interstitial voids or water channels (38,65,66). Using microelectrodes (50) it has been demonstrated that these interstitial voids contain greater concentrations of nutrients than the microcolonies and thus can act as transport channels for nutrients and the removal of byproducts, making them an essential structure in any biofilm (66). Others (67) have shown that there were fewer channels, which were less defined in a maturing biofilm. Reduction of these channels would decrease the mass transport characteristics within the bulk liquid phase, thereby controlling growth rate of the microbes within the biofilm due to reduced nutrient and, possibly, oxygen availability (68). In the laboratory under nutrient-rich conditions, bacterial monocultures may form thin layer biofilms; however, even these biofilms are not uniform in their structure. For example, thin layered biofilms produced by *P. aeruginosa* often contain bacteria distributed over a surface interdispersed with uncolonized regions (Figs. 2 and 3), and these spaces are as important to a biofilm as the regions containing the bacteria. Dalton et al. (69) showed that a marine bacterium, *Psychrobacter* sp, SW5, produced a tightly packed multilayered biofilm on a hydrophobic surface (silanized glass). In contrast, the biofilm formed on a hydrophilic surface (glass) was composed of multicellular chains arranged in a more open architecture with greater distances between the chains of bacteria. The more open biofilm structure may improve nutrient flux and availability; however, it may have a negative effect on other processes such as plasmid transfer, nutrient exchange, and effects of signaling molecules (70).

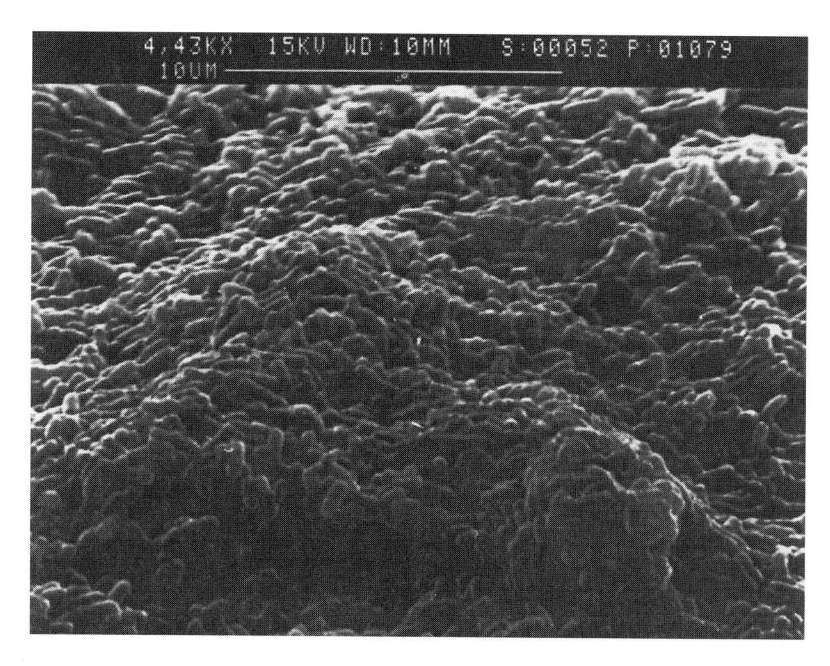

Figure 2 A scanning electron micrograph of a *P. aeruginosa* biofilm formed on a silastic surface over 48 h. This biofilm has thick regions visible here and areas that are only sparsely covered with cells.

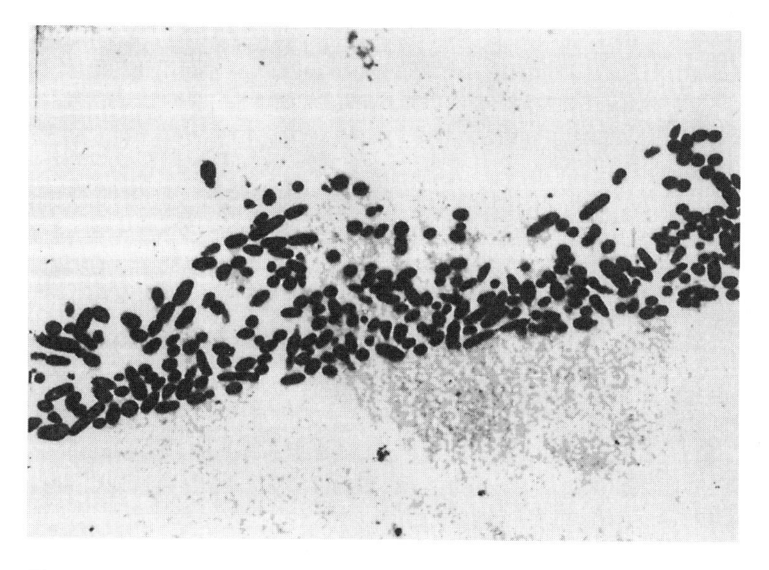

Figure 3 A transmission electron micrograph of a cross section of a *P. aeruginosa* biofilm on a silastic surface demonstrating the cell distribution and biofilm thickness.

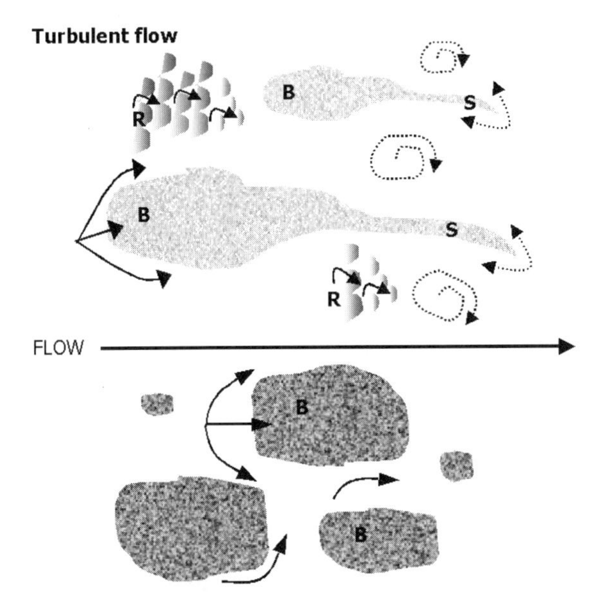

Figure 4 A schematic illustrating some of the variability in biofilm formation under different flow conditions. In aerial view: B, biofilm clusters, shading, biofilm thickness; S, streamer structures; R, ripples; dashed arrows, oscillation with flow; bold arrows, direction of flow around the channels and biofilm clusters. (Based on Ref. 38.)

Stoodley et al. (38) showed that a mixed culture biofilm grown under laminar flow conditions was ''patchy'' in that it consisted of rounded clusters of cells up to 100 μm in diameter separated by interstitial voids containing only a thin dispersion of single cells on the surface. Biofilms grown in turbulent flow conditions were also patchy but consisted of migratory ripple-like patches and elongated tapered colonies termed streamers, which oscillated in the direction of flow (38). Figure 4 is a schematic representation of the different structures under these flow conditions. In addition to flow dynamics, the biofilm structure was affected by changing nutrient conditions. When the glucose concentration was increased from 40 to 400 mg L^{-1}, there was a parallel increase in biofilm thickness from 30 to 130 μm over a 2-day period (38). However, 10 hours after the addition of glucose, migratory ripple-like structures had disappeared and the streamers became rounded to form larger porous structures. When the glucose concentration was reduced to the original concentration, the migratory ripple formation was again observed after 2 days. This may be indicative of the biofilm responding to a decrease in nutrient availability, thereby increasing its surface area and thus contact with the bulk fluid.

V. WHERE ARE BIOFILMS FOUND?

Biofilms are ubiquitous and may be beneficial or detrimental, depending on where they are found. Beneficial biofilms are those actively employed in processes such as wastewater and drinking water treatment (71). Slimy adherent microbial populations on the surface of rocks (trickling filter) or associated with a rotating biological contactor (biodisk) are used in the removal of the organic carbon during sewage treatment (72). Wastewater is passed over the surface containing the adherent microbial communities, formed of primarily slime producing *Zooglea ramigera* and other bacteria (72). The thick EPS matrix can retain a large number of other organisms to produce a consortium that is able to absorb and utilize the dissolved organic carbon present in the water. Similar systems have been used for biodegradation and remediation of industrial wastewaters. In the environment, natural selection favors microbial communities that can survive and grow by utilizing the waste as nutrients. However, this is often a slow process. Studies are under way into methods for increasing the population of biodegrading organisms at contaminated sites by enrichment techniques and immobilization of the organisms to substrata. Using immobilized communities in biofilms is more advantageous because higher concentrations of toxic compounds can be applied and they are less susceptible to washout under high flow or loading (73). In drinking water purification systems, sand filters containing microbial communities are used to remove potential pathogens by trapping them within the EPS matrix of the biofilm (72). In most instances, however, considerable problems are associated with biofilm growth or biofouling in industrial processes and cost industry a significant amount of money to develop control regimens (74). In water distribution systems, biofilms cause corrosion and degrade the quality of the water through microbial by-products. Biofilms may also harbor pathogens that put consumers and workers at risk. In the food and drink industry, biofilms cause contamination and spoilage of the product.

One of the most common occurrences of a biofilm community is dental plaque, which has been studied for nearly 300 years (75). Over 500 different microbial species have been identified in dental plaque (76). Whereas normal microbial flora can exist in the mouth without causing any problems, when pathogens are present there is a potential for periodontal disease. This biofilm exemplifies cooperation and coexistence in a complex microbial consortium in response to continual environmental changes. One example of this within the dental biofilm is the presence of the obligate anaerobe *Fusobacterium nucleatum*, which aggregates with both aerobes and anaerobes within a microbial population. The presence of this organism aids in the survival of obligate anaerobes by promoting aggregation in association with aerobes that remove oxygen from the immediate environment, thereby creating a localized anaerobic region. Bradshaw and associates (76) found that without *F. nucleatum* present in the microbial consortium, the anaerobic population was significantly decreased. Therefore, within this particular microbial community, bacteria interact with each other to create suitable microenvironments that support the growth of a diverse microbial population that often would not survive as monocultures in the same environment (76).

Other frequently studied biofilms are those found in aquatic habitats, including freshwater, groundwater, and marine environments, where the microorganisms are attached to abiotic or biotic surfaces (Fig. 5). These biofilms include a large number of bacteria and unicellular marine organisms. However, there are many other habitats that are currently being investigated with respect to microbial adhesion and biofilms, including soil particles (77,78), plant surfaces (25,79), and animal guts (80). Microbial adhesion and physiological processes are much more difficult to investigate in these habitats because of their diversity and range in conditions. These habitats are divided into aquatic and nonaquatic environments and are discussed separately.

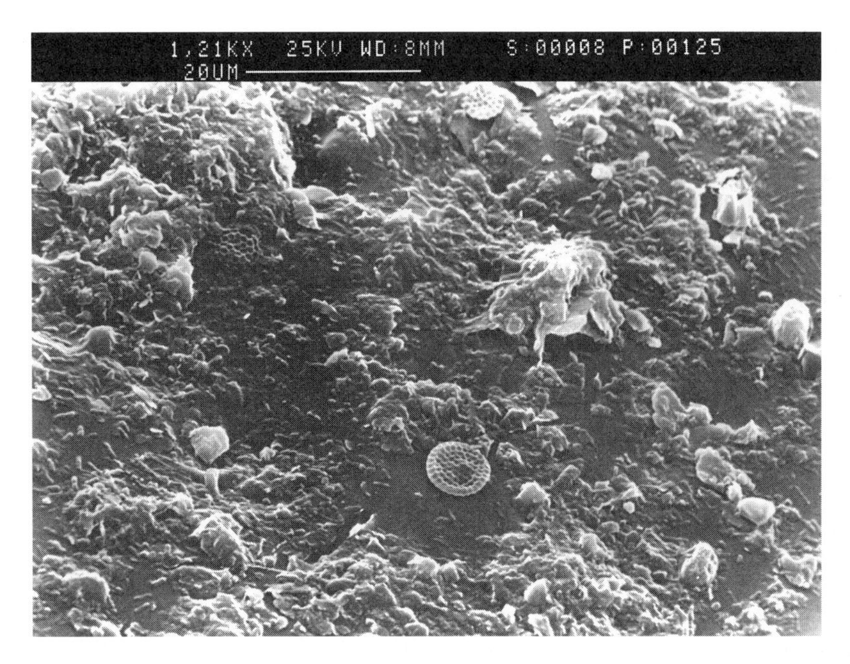

Figure 5 A scanning electron micrograph of a biofilm formed on a glass slide immersed in pond water. This multispecies biofilm demonstrates the diverse population, variable structure, and debris present within a natural biofilm.

VI. AQUATIC ENVIRONMENTS

In fresh alpine rivers, there are nearly 1000 times more bacteria attached to surfaces (square centimeters) than are present as planktonic cells (ml) (1,81). Biofilms composed of bacteria and algae have been found on sediments and rock surfaces in both freshwater and marine ecosystems. The organisms synthesize large amounts of exopolymer material, creating a complex matrix that aids in sediment cohesion and stability in intertidal sediments (82). In other instances, when the river is polluted and has high organic matter content, these biofilms may become so thick that they clog the river beds, creating drainage problems and stagnation (78).

Microorganisms in aquatic environments adhere to inorganic rocks and clay particles as well as biological/organic surfaces. Although at times biofilms are also found on living marine animals (83) and plants (84), their surfaces have mechanisms that resist microbial adhesion and often remain free of biofilms. In some cases, however, a biofilm on a plant or animal surface is in a symbiotic relationship whereby the microorganisms enhance the growth of the higher organisms. In the highly integrated rhizobia–legume symbiosis, biofilm formation is preceded by recognition and attachment of the microorganisms to the root surface. Root colonization is often multifunctional in that the organisms aid in nutrient acquisition and also provide a protective environment for the plant. For example, the colonization of mangrove roots is believed not only to help with nitrogen fixation and solublization of phosphorus but also to protect mangroves growing in saline or brackish waters (85).

Microbial mats are examples of thickly layered biofilms of photosynthetic microorganisms attached to rocks and sediment particles in aqueous habitats (25). They are often found under extreme environmental conditions. For example, in the vicinity of deep sea hydrothermal vents, microorganisms within biofilms survive extreme temperatures (86,87). Hot springs are another extreme habitat where both high temperatures and sulfide concentrations harbor mats containing layers primarily composed of Archaea, including sulfate-reducing purple bacteria (e.g., *Chloroflexis* spp., *Chromatium* spp., *Thiopedia roseopersicinia*) in association with cyanobacteria (25). Additional extreme environments where microbial mats may be found include hypersaline lakes (88), terrestrial deserts with cyclical drought and desiccation, soda lakes and acid thermal waters containing extreme pH conditions, and regions with high levels of ultraviolet (UV) irradiation (88). The microbial species that are found in these extreme environments are limited to primarily cyanobacteria (e.g., *Oscillatoria* and *Spirulina* spp.) and others such as *Desulfovibrio* spp., *Beggiatoa* spp., and *Thiovulum* spp., with differing and varying degrees of tolerance (89). Although mats are primarily composed of prokaryotes, other organisms, such as the eukaryotic *Cyanidium* sp., have been found at pH levels below 4.5 (89). Studies have shown that most of the organisms within a mat are often not physiologically adapted to the extreme environment but growth within layers of a thick biofilm helps them survive and find a suitable microniche (89). Microbial mats are a good example of the protective nature of biofilm growth and the method with which stratification can encourage nutrient availability and cycling (90).

Biofilms have been observed at other aquatic interfaces besides those at a solid–liquid interface. For example, in stagnant waters, biofilms are sometimes found at the air–liquid interface and are often seen as brown or green layers composed of algae and other aquatic microorganisms. Another example is the waxy type biofilm at the air–liquid interface formed from the rugose phenotype of *Vibrio cholerae* isolated from starvation medium (91). The interface between jet fuels and water can also harbor biofilm growth, such as the fungus *Cladosporium resinae* (92).

VII. NONAQUATIC ENVIRONMENTS

Although biofilms have often been studied in aquatic environments, more recent studies have shown that microorganisms within thick EPS matrices or biofilms are also found in nonaquatic environments such as the rhizosphere (Chapter 4), soil, and subsurface environments (93,94). One of the more complex environments is the soil ecosystem, with its many different particles and pore spaces (95). Microorganisms in the soil adhere to surfaces such as inorganic solid particles, humic matter, plant material (roots), and microfauna. Plants provide large amounts of carbon and other nutrients to encourage microbial growth in the vicinity of the roots, and, in turn, the microorganisms fix nitrogen, assist the plant in adsorption of nutrients from the soil, and protect the roots against pathogens. Another example of a nonaquatic biofilm is the colonization of the leaves of plants—the phyllosphere (96; Chapter 6). These biofilms consist of a diverse population of microorganisms, including gram-positive and gram-negative bacteria, yeasts, and filamentous fungi, supported within extensive exopolymer matrices (96,25).

The primary component of biofilms is the EPS matrix produced by the bacteria. In nonaquatic environments, the EPS matrix is of primary importance for microbial survival since they experience intermittent flux of nutrients and water. Roberson and Firestone

(93) reported that a soil *Pseudomonas* sp. produced more exopolysaccharide and less protein under low-water conditions than when growing in a water-rich environment. The polysaccharide adsorbs large amounts of water, thus reducing the rate of drying and protecting the cells from desiccation. This suggests that biofilms are important for the survival of microorganisms in the soil. Mucoid strains of *Escherichia coli*, *Acinetobacter calcoaceticus*, and *Erwinia stewartii* exhibited up to 35% greater survival under desiccating conditions compared with isogenic nonmucoid mutants (97). The EPS matrices not only are necessary for the microbes to survive low relative humidity and desiccation but may also have a role in plant microbe symbiosis and plant pathogenesis (98). These polysaccharides have also been shown to be virulence factors on plant pathogens such as *Pseudomonas*, *Erwinia*, and *Xanthomonas* spp., however, they are also important in the symbiosis of Rhizobia spp. (99). Both events depend on bacterial adhesion to surfaces and the formation of a temporary biofilm.

VIII. ENZYMES IN BIOFILMS

A number of different processes that occur within a microbial biofilm contribute to the creation of a heterogenous, dynamic environment. These processes, among many others, include cycling and exchange of nutrients, plasmid transfer, communication via chemical signals, and frequent deterioration of the surface (corrosion or degradation). The biofilm microorganisms respond by having different physiological characteristics, metabolic activity, and growth rates from those of unattached organisms growing outside the biofilm (e.g., soil aqueous phase, river water) (100).

Differences in the enzyme activities occurring within biofilms could account for some of the differences between biofilm and planktonic cells. Although there have been only a few reports on enzyme activities within biofilms, studies undertaken on samples from natural environments suggest that many biofilm processes would be mediated by enzymes. The bioremediation of pollutants by bacteria attached to soil particles, on the surface of biodegradable materials, and on rocks in rivers or trickling filters exemplifies enzymatic activity while bacteria are within biofilms. Potential substrates concentrated/precipitated at the surface or diffusing into the biofilm are accessible to the attached microorganisms in a way that is not possible for planktonic bacteria. Vetter and Deming (101) suggested that bacteria degrade both particulate and dissolved organic carbon by secreting extracellular enzymes, as they are too large for direct uptake into the cell. In natural aquatic environments it is difficult to imagine that these enzymes may have any substantial effects except when localized near the carbon source, such as in adherent cells and biofilms (101). This is supported by laboratory studies in which bacterial enzyme activity was important in colloidal organic matter degradation by a biofilm community in a reactor (102). This book identifies many processes in natural environments, and since biofilms are often found in those environments, the enzymatic processes described would therefore be relevant here.

Biofilms that accumulate on suspended particles in rivers, lakes, and marine systems are considered beneficial to their environments as they play an essential role of purification by removing suspended, settled, and dissolved organic material. Freeman et al. (103) described river biofilms as a trophic link between dissolved nutrients in the water column and the higher trophic levels of the ecosystem. Natural biofilms are able to biodegrade organic compounds and transform inorganic compounds as part of their natural metabolic pathway. It is assumed that in a natural environment, such as the soil habitat, bacteria and fungi do form biofilms and therefore the degradation of environmental pollutants occurs in a biofilm state. The role of bacterial–fungal biofilms in the degradation of environmental pollutants is not well understood. It is thought that the success of fungi in the role of biodegradation is mainly attributable to the production of extracellular enzymes. Manganese peroxidases and lignin peroxidases produced by *Phanerochaete chrysosporium* play a key role in the detoxification and decolorization of pulp bleach plant effluent (104). A 1993 report has shown that enzymes produced by fungi and bacteria would act simultaneously in the biodegradation of pesticides under field conditions (105). Levanon (105) described how the mineralization of alachlor and atrazine was mainly due to fungal activity and that the mineralization of carbofuran and malathion was mainly due to bacterial activity. The bacterial enzyme identified to degrade malathion was a carboxylesterase. Fungi release many hydrolytic enzymes into the soil, and these enzymes are capable of hydrolyzing pesticides. Degradation by fungal enzymes may be due to less specific enzymes, as in the case of lignin-degrading enzyme systems. However, the ability of fungi to degrade a wide range of pesticides is believed to be related to the structural similarity of lignin to the pesticides.

The degradation of inorganic minerals and the precipitation of toxic metals result from enzymatic activity of microorganisms within biofilms. The degradation of mining tailings by *Thiobacillus ferrooxidans* occurs when reduced iron and sulfur compounds in ores or wastes are converted to sulfuric acid and iron (III) (106). The oxidation of these metals is a chemical process enhanced by microbial biocatalysts, that is, enzymes that encourage the flow of electrons (107,108). With *Thiobacillus ferrooxidans*, these enzymes are believed to be located on the cell envelope; therefore, it is important that the minerals are closely associated with the bacterium (109). Although the details of the bacterial–mineral interaction are not fully understood, it is known that the bacterium colonizes the mineral with the aid of its lipopolysaccharide (110). Studies have shown that even in environments that solubilize metals, bacteria precipitate and accumulate metals within their matrix (110). This is a natural system for cleaning the surrounding environment of toxic metals so that other organisms and bacteria may survive.

Some surfaces act as both a colonization surface and a nutrient source. Often enzymes are required to degrade substances to a soluble and low-molecular-weight form that can be utilized by the bacterium as a carbon and nitrogen source. Enzymatic degradation of organic materials is often preceded by bacterial attachment and biofilm formation. A primary example is the hydrolysis of cellulose by anaerobic, thermophylic *Clostridium thermocellum* (111). This organism produces an extracellular protein complex called a cellulosome that has a dual purpose; it both binds to and degrades cellulose. The cellulosome produces catabolic enzymes that solubilize cellulose substances, primarily to the disaccharide cellobiose, which then can be taken up by the bacterium as a nutrient source (111). It is presumed that biofilm formation is required to position the bacterium close to the substrate, thereby concentrating the enzyme for hydrolytic activity. Similar studies have been undertaken on other cellulose-degrading microorganisms such as *Fibrobacter succinogenes* and *Ruminococcus albus* (112).

In many cases the microorganism uses the degraded compounds as a nutrient source; however, this process also liberates nutrients and ions for other bacteria in the biofilm community as well as for higher organisms such as plants or animals. For example, biofilms in the guts of ruminant animals help enzymatically degrade feed particles into substances that the animal is able to utilize (80). It is likely that each insoluble substance, such as cellulose, amylose, starch, and proteins, has its own biofilm population. The bacteria produce the appropriate suite of extracellular degradative enzymes that convert the substances into soluble nutrients available for the uptake by both bacteria and animal (80).

Many bacteria and fungi grow in association with a surface, and this process often results in the enzymic deterioration of that surface. The surface can therefore act as a substrate. For example, polyvinyl chloride (PVC) is largely composed of carbon, hydrogen, and chlorine, although these are unlikely to be available to a colonizing population; the added UV stabilizers and colourant may well influence attachment. A 1999 report (113) has shown that plasticizers increase the adhesion of the deteriogenic fungus *Aureobasidium pallulans* to PVC. The loss of plasticizers in PVC due to microbial degradation results in brittleness, shrinkage, and ultimate failure of the PVC in its intended application (113). The production of extracellular esterase by *A. pallulans* is thought to be instrumental the biodeterioration of PVC (114). *Stachybotrys chartarum (atra)* is a saprophytic green–black fungus that is found in many habitats and grows particularly well in high-cellulose material, damp wallpaper, fiber board, lint, and dust. It produces potent macrocyclic trichothecene toxins (satratoxins H, G, F; roridin E; verrucarin J; and trichoverrols A and B), and is associated with chronic health problems, and is implicated in ''sick building syndrom.''

Dental biofilms are highly complex consortia that cooperate via nutrient cycling. The aerobic or facultative organisms often aid the survival of anaerobic bacteria by utilizing the oxygen (115); thus the anaerobes are able to survive transient levels of oxygen. They also produce degrading enzymes that can catalyze the degradation of polysaccharides, glycoproteins, and complex macromolecules encountered in the oral cavity to smaller compounds usable by the microflora. For example, some of the enzymes known to be produced by dental microorganisms for the hydrolysis of polysaccharide components include β-galactosidase, β-*N*-acetylglucosaminidase, β-*N*-acetylgalactosaminidase, α- and β-mannosidase, and α-fucosidase (116). Other bacteria then produce proteolytic enzymes, such as glyprodiamino peptidase, which catalyzes the degradation of proteins into peptides and amino acids. The degradation of these complex molecules requires cooperative and synergistic enzymic interactions within the microbial community and, inevitably, bacterial diversity. In turn, the resultant nutrients released support a diverse range of organisms with different nutrient requirements (116).

Microorganisms are attracted to an inert interface as a result of the increased concentration of nutrients at that interface, including some ions and larger molecules such as amino acids and glycoproteins, which may not be easily transported into the cell. Dissolved organic matter, ions (e.g., NH_4^+) and metals that serve as energy sources for chemolithotrophs (e.g., Fe^{2+}, Mn^{2+}) may initially become adsorped at a surface and also by ionic and physical entrapment within the exopolymer. Both these processes concentrate nutrients in the proximity of the microorganisms. Extracellular enzymes produced by some microorganisms hydrolyze and degrade organic ions into soluble products available for their or other organisms' use. For example, an exoprotease producing *Pseudomonas* sp. strain S9 was used to determine adhesion by taking advantage of its degradation of ribulose-1,5-bisphosphate carboxylase adsorbed to the surface (117). In many cases, enzymes within biofilms are one of the normal consequence of microbial metabolism. The microenvironments within a biofilm may provide the conditions in which enzyme induction and secretion are stimulated. The activity of some of these enzymes outside the biofilm may be restricted as a result of the comparatively low concentrations of substrates, inducers, and other effector molecules outside the biofilm.

Biofilm maintenance or thickness may be controlled by specific enzymes produced by the bacterium in response to environmental stimuli. Boyd and Chakrabarty (118) discovered a *P. aeruginosa* strain that produces alginate lyase and degrades alginate. This may be a mechanism through which bacteria are detached from a biofilm. Similarly, *Streptococcus mutans* produces a protein called surface protein-releasing enzyme that catalyzes the release of bacteria from biofilms (119). This may be one method whereby organisms become dissociated from the biofilm environment when nutrients are depleted.

IX. CONCLUSIONS

Direct evidence for the production of specific enzymes within biofilms is sparse, possibly because of the limitations of research techniques. However, there is no doubt that extracellular enzymes, either secreted or arising from dead and lysing cells, are found in the biofilm matrix. Biofilms represent a dynamic and heterogeneous environment; therefore, localized (and low) concentrations of enzymes have been difficult to detect. With the current development of more sensitive microscopy methods, reporter gene technology, molecular biology, and nanotechnology, investigating single-cell and small population responses to environmental stimuli is possible. A number of useful reviews (120,121) and textbooks (122–124) have been published on current techniques to study microbial adhesion and biofilm formation in natural environments. This has provided a large resource for future studies into enzymes within biofilms. Indirect evidence suggests that there are a number of different processes that involve microbial enzymes and that they occur within biofilms but do not occur in planktonic cells. It is assumed that biofilm cells produce enzymes in response to environmental stimuli in order to maintain the integrity of the overall structure, composition, and activity of the biofilm.

This chapter has shown biofilms to have three primary functions with respect to enzyme production and activity. First, biofilm formation places the bacterium close to the substrate or the substratum where it is concentrated for which an enzyme may be specific. In this case, the bacterium may be producing enzymes still associated with the cell wall (i.e., mural enzymes) requiring attachment to the substrate for catalysis to occur. Second, a biofilm matrix may concentrate the enzyme so that it may reach concentrations at which the activity becomes relevant. Third, the biofilm may create an environment that induces the production of a specific enzyme.

REFERENCES

1. GG Geesey, R Mutch, JW Costerton, RB Green. Sessile bacteria: An important component of the microbial population in small mountain streams. Limnol Oceanogr 23:1214–1223, 1978.
2. CE ZoBell, EC Allen. Attachment of marine bacteria to submerged slides. Proc Soc Exp Biol Med 30:1409–1411, 1933.
3. CE ZoBell, EC Allen. The significance of marine bacteria in the fouling of submerged surfaces. J Bacteriol 29:239–251, 1935.
4. CE ZoBell. The effect of solid surfaces upon bacterial activity. J Bacteriol 46:39–56, 1935, 1943.

5. HM Lappin-Scott. Claude E. ZoBell-his life and contributions to biofilm microbiology. In: CB Bell, M Brylinsky, P Johnson-Green, eds. Proceedings of the 8th International Symposium on Microbial Ecology, Atlantic Canada Society for Microbiol Ecology, Halfax, 2000, pp 43–50.
6. WG Characklis, KC Marshall. Biofilms: A basis for an interdisciplinary approach. In: WG Characklis, KC Marshall, eds. Biofilms. New York: John Wiley & Sons, 1990, pp 3–15.
7. JW Costerton, Z Lewandowski, DE Caldwell, DR Korber, HM Lappin-Scott. Microbial biofilms. Annu Rev Microbiol 49:711–745, 1995.
8. JW Costerton, PS Stewart, EP Greenberg. Bacterial biofilms: A common cause of persistent infections. Science 284:1318–1322, 1999.
9. MV Jones. Fungal biofilms: Eradication of a common problem. In: J Wimpenny, P Handley, P Gilbert, HM Lappin-Scott, eds. The Life and Death of Biofilm. Cardiff: BioLine, 1995, pp 157–160.
10. JD Elvers, K Leeming, CP Moore, HM Lappin-Scott. Bacterial-fungal biofilms in flowing water photo-processing tanks. J Appl Microbiol 84:607–618, 1998.
11. KW Millsap, HC van der Mei, R Bos, HJ Busscher. Adhesive interactions between medically important yeasts and bacteria. FEMS Microbiol Rev 21:321–336, 1998.
12. KE Cooksey. Bacterial and algal interactions in biofilms. In: LF Melo et al., eds. Biofilms: Science and Technology. Netherlands: Kluwer Academic, 1992, pp 163–173.
13. J Rogers, CW Keevil. Immunogold and fluorescein labelling of *Legionella pneumophila* within an aquatic biofilm visualized by episcopic differential interference contrast microscopy. Appl Environ Microbiol 58:2326–2330, 1992.
14. P Gilbert, DG Allison. Dynamics in microbial communities: Alamarkian perspective. In: J Wimpenny, P Gilbert, J Walker,M Brading, R Bayston, eds. Biofilms: The good, the bad and the ugly. Cardiff: BioLine, 1999, pp 263–268.
15. R Bos, HC van der Mei, HJ Busscher. Physico-chemistry of initial microbial adhesive interaction: Its mechanisms and methods for study. FEMS Microbiol Rev 23:179–230, 1999.
16. HJ Busscher, R Bos, HC van der Mei. Initial microbial adhesion is a determinant for the strength of biofilm adhesion. FEMS Microbiol Lett 128:229–234, 1995.
17. DJ Bradshaw, PD Marsh, K Watson, K Schilling. The effect of the conditioning films on adhesion. In: J Wimpenny, P Handley, P Gilbert, HM Lappin-Scott, eds. The Life and Death of Biofilm. Cardiff: BioLine, 1995, pp 47–52.
18. PC Griffith, M Fletcher. Hydrolysis of protein and model dipeptide substrates by attached and non-attached marine Pseudomonas sp. Strain NCIMB 2021. Appl Environ Microbiol 57: 2186–2191, 1991.
19. P Vandevivere, DL Kirchman. Attachment stimulates exopolysaccharide synthesis by a bacterium. Appl Environ Microbiol 59:3280–3286, 1993.
20. DG Davies, AM Chakrabarty, GG Geesey. Exopolysaccharide production in biofilms: Substratum activation of alginate gene expression by *Pseudomonas aeruginosa.* Appl Environ Microbiol 59:1181–1186, 1993.
21. P Gilbert, DG Allison, AE Jacob, D Korber, G Wolfaardt, I Foley. Immigration of planktonic *Enterococcus faecalis* cells into mature *E. faecalis* biofilms In: J Wimpenny, P Handley, P Gilbert, HM Lappin-Scott, M Jones, eds. Biofilms Community Interactions and Control. Cardiff: BioLine, 1997, pp 133–141.
22. LC Skillman, IW Sutherland, MV Jones. Co-operative biofilm formation between two species of *Enterobacteriaceae.* In: J Wimpenny, P Handley, P Gilbert, HM Lappin-Scott, M Jones, eds. Biofilms Community Interactions and Control. Cardiff: BioLine, 1997, pp 119–127.
23. J Bryers, WG Characklis. Early fouling biofilm formation in a turbulent flow system: Overall kinetics. Water Res 15:483–491, 1981.
24. SN Wai, Y Mizunoe, S-I Yoshida. How *Vibrio cholerae* survive during starvation. FEMS Microbiol Lett 180:123–131, 1999.
25. TD Brock. Life at high temperatures. Yellowstone Association, Yellowstone National Park, 1994, p 9.

26. CE Morris, J-E Monier, M-A Jaques. Methods for observing microbial biofilms directly on leaf surfaces and recovering them for isolation of culturable microorganisms. Appl Environ Microbiol 63:1570–1576, 1997.
27. PS Handley. Is there a universal biofilm structure? J Wimpenny, P Handley, P Gilbert, HM Lappin-Scott, eds. The Life and Death of Biofilm. Cardiff: BioLine, 1995, pp 21–25.
28. P Gilbert, DG Allison. Biofilms and their resistance towards antimicrobial agents. In: HN Newman, M Wilson, eds. Dental plaque revisited: Oral biofilms in health and disease. Cardiff: BioLine, 1999, pp 125–143.
29. DG Alison, P Gilbert. Modification by surface association of antimicrobial susceptibility of bacterial populations. J Ind Microbiol 15:311–317, 1995.
30. WW Nichols. Biofilm permeability to antibacterial agents. In: J Wimpenny, W Nichols, D Stickler, HM Lappin-Scott, eds. Bacterial Biofilms in Medicine and Industry. Cardiff: BioLine, 1993, pp 141–149.
31. FH Yildiz, GK Schoolnik. *Vibrio cholerae* 01 El Tor: Identification of a gene cluster required for the rugose colony type, exopolysaccharide production, chlorine resistence, and biofilm formation. Proc Natl Acad Sci USA 96:4028–4033, 1999.
32. HM Dalton, PE March, KC Marshall. Community interaction in marine bacterial biofilms. In: J Wimpenny, P Handley, P Gilbert, HM Lappin-Scott, M Jones, eds. Biofilms community interactions and control. Cardiff: BioLine, 1997, pp 129–132.
33. MG Brading, J Jass, HM Lappin-Scott. Dynamics of bacterial biofilm formation. In: HM Lappin-Scott, JW Costerton, eds. Microbial biofilms. Cambridge: Cambridge University Press, 1995, pp 46–63.
34. Z Lewandowski, W Dickinson, W Lee. Electrochemical interactions of biofilms with metal surfaces. Water Sci Technol 36:295–302. 1997.
35. GG Geesey, RJ Gillis, R Avci, D Daly, M Hamilton, E Shope, G Harkin. The influence of surface features on bacterial colonisation and subsequent changes of 316L stainless steel. Corrosion Sci 38:73–95, 1996.
36. Z Lewandowski, P Stoodley. Flow induced vibrations, drag force and pressure drop in conduits covered with biofilm. Water Sci Technol 32:19–26, 1995.
37. P Stoodley, F Jørgensen, P Williams, HM Lappin-Scott. The role of hydrodynamics and AHL signalling molecules as determinants of the structure of *Ps. aeruginosa* biofilms. In: J Wimpenny, P Gilbert, J Walker, M Brading, R Bayston, eds. Biofilms: The Good, the Bad and the Ugly. Cardiff: BioLine, 1999, pp 223–230.
38. P Stoodley, I Dodds, J Boyle, HM Lappin-Scott. Influence of hydrodynamics and nutrients on biofilm structure. J Appl Microbiol 85:19S–28S, 1999.
39. KD Xu, P Stewart, F Xia, CT Huang, GA McFeters. Spatial physiological heterogeneity in *Pseudomonas aeruginosa* Biofilm is determined by oxygen availability. Appl Environ Microbiol 64:4035–4039, 1998.
40. D de Beer, P Stoodley. Relation between the structure of an aerobic biofilm and transport phenomena. Water Sci Technol 32:1–18, 1995.
41. S Møller, C Sternberg, JB Andersen, BB Christensen, JL Ramos, M Givskov, S Molin. *In situ* gene expression in mixed-culture biofilms: Evidence of metabolic interactions between community members. Appl Environ Microbiol 64:721–732, 1998.
42. I Sutherland. Biofilm matrix polymers. In: HN Newman, M Wilson, eds. Dental Plaque Revisited: Oral Biofilms in Health and Disease. Cardiff: BioLine, 1999, pp 49–62.
43. I Sutherland. Microbial biofilm exopolysaccharides—superglue or velcro? In: J Wimpenny, P Handley, P Gilbert, HM Lappin-Scott, M Jones, eds. Biofilms Community Interactions and Control. Cardiff: BioLine. 1997, pp 33–40.
44. LC Skillman, IW Sutherland, MV Jones. The role of exopolysaccharides in dual species biofilm development. J Appl Microbiol 85:S13–S18, 1999.
45. DA Caron. Grazing of attached bacteria by heterotrophic microflagellates. Microb Ecol 13: 203–218, 1987.

46. J Rogers, AB Dowsett, PJ Dennis, JV Lee, CW Keevil. Influence of plumbing materials on biofilm formation and growth of *Legionella pneumophila* in potable water systems. Appl Environ Microbiol 60:1842–1851, 1994.
47. DG Davies, MR Parsek, JP Pearson, BH Iglewski, JW Costerton, EP Greenberg. The involvement of cell-to-cell signals in the development of bacterial biofilm. Science 280:295–298, 1998.
48. JR Lawrence, DR Korber, BD Hoyle, JW Costerton, DE Caldwell. Optical sectioning of microbial biofilms. J Bacteriol 173:6558–6567, 1991.
49. P Stoodley, Z Lewandowski, JD Boyle, HM Lappin-Scott. Structural deformation of bacterial biofilms caused by short-term fluctuations in fluid shear: and in situ investigation of biofilm rheology. Biotech Bioeng 65:83–92, 1999.
50. C-T Huang, FP Yu, GA McFeters, PS Stewart. Nonuniform spatial patterns of respiratory activity within biofilms during disinfection. Appl Environ Microbiol 61:2252–2256, 1995.
51. PJ Robinson, JT Walker, CW Keevil, J Cole. Reporter genes and fluorescent probes for studying the colonisation of biofilms in a drinking water supply line by enteric bacteria. FEMS Microbiol Lett 129:183–188, 1995.
52. HG MacLehose, P Gilbert, D Allison. Homoserine lactones and biocide susceptibility: A cautionary note. In: J Wimpenny, P, Gilbert, J Walker, M Brading, R Bayston, eds. Biofilms: The Good, the Bad and the Ugly. Cardiff: BioLine, 1999, pp 231–236.
53. WC Fuqua, SC Winans, EP Greenberg. Quorum sensing in bacteria: The LuxR-LuxI family of cell density-responsive transcriptional regulators. J Bacteriol 176:269–275, 1994.
54. MTG Holden, SR Chhabra, R de Nys, P Stead, NJ Bainton, PJ Hill, M Manefield, N Kumar, M Labatte, D England, S Rice, M Givskov, GPC Salmond, GSAB Stewart, BW Bycroft, S Kjelleberg, P Williams. Quorum-sensing cross talk: isolation and chemical characterisation of cyclic dipeptides from *Pseudomonas aeruginosa* and other Gram-negative bacteria. Mol Microbiol 33:1254–1266, 1999.
55. AB Flavier, SJ Clough, MA Schell, TP Denny. Identification of 3-hydroxypalmitic acid methyl ester as a novel autoregulatory controlling virulence in *Ralstonia solanacearum*. Mol Microbiol 26:251–259, 1997.
56. S Horinouchi, T Beppu. A-factor as a microbial hormone that controls cellular differentiation and secondary metabolism in *Streptomyces griseus*. Mol Microbiol 12:859–864, 1994.
57. B Baker, P Zambryski, B Staskawicz, SP Dinesh-Kumar. Signalling in plant-microbe interactions. Science 276:726–733, 1997.
58. B Rodleas, JK Lithgow, F Wisniewski-Dye, A Hardman, A Wilkinson, A Economou, P Williams, JA Downie. Analysis of quorum-sensing-dependent control of rhizosphere-expressed (rhi) genes in *Rhizobium leguminosarum* bv. viciae. J Bacteriol 181:3816–3823, 1999.
59. A Glessner, RS Smith, BH Iglewski, JB Robinson. Roles of *Pseudomonas aeruginosa las* and *rhl* quorum sensing systems in control of twitching motility. J Bacteriol 181:1623–1629, 1999.
60. GA O'Toole, R Kolter. Flagellar and twitching motility are necessary for *Pseudomonas aeruginosa* biofilm development. Mol Microbiol 30:295–304, 1998.
61. RJC McClean, M Whieley, DJ Stickler, WC Fuqua. Evidence of autoinducer activity in naturally occurring biofilms. FEMS Microbiol Lett 154:259–263, 1997.
62. MJ Lynch, S Swift, D Kirke, CER Dodd, CW Keevil, G Stewart, P Williams. Investigation of quorum sensing in *Aeromonas hydrophilia* biofilms formed on stainless steel. In: J Wimpenny, P Gilbert, J Walker, M Brading, R Bayston, eds. Biofilms: The Good, the Bad and the Ugly. Cardiff: BioLine, 1999, pp 209–222.
63. MG Brading, J Boyle, HM Lappin-Scott. Biofilm formation in laminar flow using *Pseudomonas fluorescens* EX101. J Ind Microbiol 15:297–304, 1995.
64. Z Lewandowski, G Walser. Influence of hydrodynamics on biofilm accumulation. Environmental engineering proceedings, EE Div/ASCE, Reno, 1991, pp 619–624.

65. P Stoodley, JD Boyle, I Dodds, HM Lappin-Scott. Consensus model of biofilm structure. In: J Wimpenny, P Handley, P Gilbert, HM Lappin-Scott, M Jones, eds. Biofilms community interactions and control. Cardiff: BioLine, 1997, pp 1–10.
66. JW Costerton, Z Lewandowski, D deBeer, D Caldwell, D Korber, G James. Biofilms, the customized microniche. J Bacteriol 176:2137–2142, 1994.
67. SK Roberts, HM Lappin-Scott, K Leeming. The control of bacterial-fungal biofilms. In: J Wimpenny, P Gilbert, J Walker, M Brading, R Bayston, eds. Biofilms: the Good, the Bad and the Ugly. Cardiff: BioLine. 1999, pp 93–104.
68. WG Characklis, MH Turakhia, N Zelver. Transport and interfacial transport phenomena. In: WG Characklis, KC Marshall, eds. Biofilms. New York: John Wiley & Sons, 1990, pp. 265–340.
69. HM Dalton, LK Poulsen, P Halasz, ML Angles, AE Goodman, KC Marshall. Substratum-induced morphological changes in a marine bacterium and their relevance to biofilm structure. J Bacteriol 176:6900–6906, 1994.
70. M Hausner, S Wuertz. High rates of conjugation in bacterial biofilms as determined by quantitative in situ analysis. Appl Environ Microbiol 65:3710–3717, 1999.
71. J Boyle. Biofilms in natural waters. In: J Wimpenny, W Nichols, D Stickler, HM Lappin-Scott, eds. Bacterial Biofilms and Their Control in Medicine and Industry. Cardiff: BioLine, 1994, pp 37–40.
72. LM Prescott, JP Harley, DA Klein. Microbiology. 2nd ed. Oxford: Wm C Brown Communications, 1993, pp 832–838.
73. RC Wyndham, KJ Kennedy. Microbial consortia in industrial wastewater treatment. In: HM Lappin-Scott, JW Costerton, eds. Microbial Biofilms. Cambridge: Cambridge University Press, 1995, pp 183–195.
74. HM Lappin-Scott, JW Costerton. Bacterial biofilms and surface fouling. Biofouling 1:323–342, 1989.
75. SK Roberts, C Bass, M Brading, HM Lappin-Scott, P Stoodley. Biofilm formation and structure: What's new? In: HN Newman, M Wilson, eds. Dental Plaque Revisited: Oral Biofilms in Health and Disease. Cardiff: BioLine, 1999, pp 15–36.
76. DJ Bradshaw, PD Marsh, GK Watson, C Allison. Interspecies interactions in microbial communities. In: J Wimpenny, P Handley, P Gilbert, HM Lappin-Scott, M Jones, eds. Biofilms Community Interactions and Control. Cardiff: BioLine, 1997, pp 63–71.
77. Å Aakra, M Hesselsoe, LR Bakken. Surface attachment of ammonia-oxidizing bacteria in soil. Microb Ecol 39:222–235, 2000.
78. TJ Battin, D Sengschmitt. Linking sediment biofilms, hydrodynamics, and river bed clogging: evidence from a large river. Microb Ecol 37:185–196, 1999.
79. I Carmichael, IS Harper, MJ Coventry, PWJ Taylor, J Wan, MW Hickey. Bacterial colonisation and biofilm development on minimally processed vegetables. J Appl Microbiol 85:45s–51s, 1999.
80. K-J Cheng, TA McAllister, JW Costerton. Biofilms of the ruminant digestive tract. In: HM Lappin-Scott, JW Costerton, eds. Microbial Biofilms. Cambridge: Cambridge University Press, 1995, pp 221–232.
81. GG Geesey, WT Richardson, HG Yeomans, RT Irvin, JW Costerton. Microscopic examination of natural sessile bacterial populations from an alpine stream. Can J Microbiol 23:1733–1736, 1977.
82. ML Yallop, DM Paterson, P Wellsbury. Interrelationships between rates of microbial production, exopolymer production, microbial biomass, and sediment stability in biofilms of intertidal sediments. Microb Ecol 39:116–127, 2000.
83. J Jolly, HM Lappin-Scott, J Anderson, CD Clegg. Scanning electron microscopy of the gut microflora of two earthworms: *Lumbricus terrestris* and *Octolasion cyaneum*. Microbial Ecol 26:235–245, 1993.
84. M Givskov, R de Nys, M Manefield, L Gram, R Maximilien, L Eberl, S Molin, PD Steinberg,

S Kjelleberg. Eukaryotic interference with homoserine lactone mediated prokaryotic signalling. J Bacteriol 178:6618–6622, 1996.
85. ME Puente, G Holguin, BR Glick, Y Basham. Root-surface colonization of black mangrove seedlings by *Asperillium halopraeferens* and *Azospirillum brasilense* in seawater. FEMS Microbiol Ecol 29:283–292, 1999.
86. J Guezennec, O Ortega-Morales, G Raguenes, G Geesey. Bacterial colonization of artificial substrate in the vicinity of deep-sea hydrothermal vents. FEMS Microbiol Ecol 26:89–99, 1998.
87. D Wynn-Williams, C Elli-Evans, R Leakey. Microbial ecology in Antarctica. SGM Q November, 1992, pp 99–104.
88. C Kruschel, RW Castenholz. The effect of solar UV and visible irradiance on the vertical movement of cyanobacteria in microbial mats of hypersaline waters. FEMS Microbiol Ecol 27:53–72, 1998.
89. F Garcia-Pichel, M Mechling, RW Castenholz. Diel migration of microorganisms within a benthic, hypersaline mat community. Appl Environ Microbiol 60:1500–1511, 1994.
90. D Krekeler, A Teske, H Cypionka. Strategies of sulfate-reducing bacteria to escape oxygen stress in a cyanobacterial mat. FEMS Microbiol Ecol 25:89–96, 1998.
91. SN Wai, Y Mizunoe, A Takade, S-I Kawabata, S-I Yoshida. *Vibrio cholerae* 01 strain TSI-4 produces the exopolysaccahride material that determine colony morphology, stress resistance, and biofilm formation. Appl Environ Microbiol 64:3648–3655, 1998.
92. LM Prescott, JP Harley, DA Klein. Microbiology. 2nd ed. Oxford: Wm C Brown Communications, 1993, pp 906–907.
93. EB Roberson, MK Firestone. Relationship between desiccation and exopolysaccaride production in a soil *Pseudomonas* sp. Appl Environ Microbiol 58:1284–1291, 1992.
94. RC Foster. Polysaccharides in soil fabrics. Science 214:665–667, 1981.
95. RG Burns. Microbial and enzymic activities in soil. In: WG Characklis, PA Wilderer, eds. Structure and Function of Biofilms. New York: John Wiley & Sons, 1989, pp 333–349.
96. CE Morris, J-M Monier, M-A Jacques. A technique to quantify the population size and composition of biofilm components in communities of bacteria in the phyllosphere. Appl Environ Microbiol 64:4789–4795, 1998.
97. T Ophir, DL Gutmick. A role for exopolysaccharides in the protection of microorganisms from desiccation. Appl Environ Microbiol 60:740–745, 1994.
98. M Wilson, SE Lindow. Effect of phenotypic plasticity on epiphytic survival and colonization by *Pseudomonas syringae*. Appl Environ Microbiol 59:410–416, 1993.
99. JA Leigh, DL Coplin. Exopolysaccharide in plant-bacteria interaction. Annu Rev Microbiol 46:307–346, 1992.
100. C Freeman, MA Lock. [^{3}H] Thymidine incorporation as a measure of bacterial growth within intact river biofilms. Sci Total Environ 138:161–167, 1993.
101. YA Vetter, JW Deming. Growth rates of marine bacterial isolates on particulate organic substrates solubilized by freely released extracellular enzymes. Microb Ecol 37:86–94, 1999.
102. T Larsen, P Herremoes. Degradation mechanisms of colloidal organic matter in biofilm reactors. Wat Res 28:1443–1452, 1994.
103. C Freeman, PJ Chapman, K Gilman, MA Lock, B Reynolds, HS Wheater. Ion exchange mechanisms and the entrapment of nutrients by river biofilms. Hydrobiologia 297:61–65, 1995.
104. FC Michel, SB Dass, EA Grulke, CA Reddy. Role of manganese perioxidases and lignin perioxidases of *Phanerochaete chrysosporium* in the decolorization of kraft bleach plant effluent. Appl Environ Microbiol 57:2368–2375, 1991.
105. D Levanon. Roles of fungi and bacteria in the mineralization of the pesticides atrazine, alachlor, malathion and carbofuran in soil. Soil Biol Biochem 25:1097–1105, 1993.
106. F Baldi, T Clark, SS Pollack, GJ Olson. Leaching of pirites of various reactivities by *Thiobacillus ferrooxidans*. Appl Environ Microbiol 58:1853–1856, 1992.

107. PLAM Corstjens, JPM de Vrind, P Westbroek, EW de Vrind-de Jong. Enzymatic iron oxidation by *Leptothrix discophora*: Identification of an iron-oxidising protein. Appl Environ Microbiol 58:450–454, 1992.
108. G Southam, FG Ferris, TJ Beveridge. Mineralized bacterial biofilms in sulphide tailings and in acid mine drainage systems. In: HM Lappin-Scott, JW Costerton, eds. Microbial Biofilms. Cambridge: Cambridge University Press, 1995, pp 148–170.
109. G Southam, TJ Beveridge. Enumeration of Thiobacilli within pH-neutral and acidic mine tailings and their role in the development of secondary mineral soil. Appl Environ Microbiol 58:1904–1912, 1992.
110. G Southam, TJ Beveridge. Examination of lipopolysaccharide (o-antigen) populations of *Thiobacillus ferrooxidans* from two mine tailings. Appl Environ Microbiol 59:1283–1288, 1993.
111. EA Bayer, E Morag, Y Shoham, J Tormo, R Lamed. The cellulosome: A cell surface organelle for the adhesion to and degradation of cellulose. In: M Fletcher, ed. Bacterial Adhesion: Molecular and Ecological Diversity. New York. Wiley-Liss, 1996, pp 155–182.
112. J Gong, CW Forsberg. Factors affecting adhesion of *Fibrobacter succinogenes* subsp. Succinogenes S85 and adherence-defective mutants to cellulose. Appl Environ Microbiol 55: 3039–3044.
113. JS Webb, HC Van der Mei, M Nixon, IM Eastwood, M Greenhalgh, S Read, GD Robson, PS Handley. Plasticizers increase adhesion of the deteriogenic fungus *Aureobasidium pullulans* to polyvinyl chloride. Appl Environ Microbiol 65:3575–3581, 1999.
114. JS Webb, M Nixon, IM Eastwood, M Greenhalgh, GD Robson, PS Handley. Fungal colonisation and biodeterioration of plasticized polyvinyl chloride. Appl Environ Microbiol 66:3194–3200, 2000
115. DJ Bradshaw, PD Marsh, GK Watson, C Allison. Role of *Fusobacterium nucleatum* and coaggregation in anaerobe survival in planktonic and biofilm oral microbial communities during aeration. Infect Immun 66:4729–4732, 1998.
116. DJ Bradshaw, KA Holmer, PD Marsh, D Beighton. Metabolic cooperation in oral microbial communities during growth on mucin. Microbiol 140:3407–3412, 1994.
117. M-O Samuelsson, DL Kirchman. Degradation of adsorbed protein by attached bacteria in relation to surface hydrophobicity. Appl Environ Microbiol 56:3643–3648, 1991.
118. A Boyd, AM Chakrabarty. Role of alginate lyase in cell detachment of *Pseudomonas aeruginosa*. Appl Environ Microbiol 60:2355–2359, 1994.
119. SF Lee, YH Li, GH Bowden. Detachment of *Streptococcus mutans* biofilm cells by an endogenous enzymatic activity. Infect Immun 64:1035–1038, 1996.
120. JC Rayner, HM Lappin-Scott. Experimental biofilms and their applications on the study of environmental processes. In: C Edwards, ed. Environmental monitoring of bacteria. Clifton NJ: Humana Press, 1999 pp 279–306.
121. L Hall-Stoodley, JC Rayner, P Stoodley, HM Lappin-Scott. Establishment of experimental biofilms using the modified Robbins device and flow cells. In: C Edwards ed. Environmental monitoring of bacteria. Clifton, NJ: Humana Press, 1999 pp 307–319.
122. HA Yuehuei, RJ Friedman. eds. Handbook of bacterial adhesion. Clifton NJ: Humana Press, 2000.
123. H-C Flemming, U Szewzyk, T Griebe ed. Biofilms; Investigative methods and applications. Lancaster: Technomic, 2000.
124. R Doyle. Biofilms. Vol. 310. Methods in Enzymology, Biofilms. San Diego: Academic Press, 1999.

13

Search for and Discovery of Microbial Enzymes from Thermally Extreme Environments in the Ocean

Jody W. Deming and John A. Baross
University of Washington, Seattle, Washington

I. INTRODUCTION

The vast array of organic and inorganic compounds produced in and delivered to the ocean make it a haven for an extraordinary diversity of microorganisms that use a diverse array of catalytic agents in the synthesis, transformation, and degradation of these materials. The microbial production of enzymes functional at moderate thermal conditions in the marine environment is well known (see Chapter 4). Here we argue from observations, experimental results, and hypothetical scenarios that thermally extreme environments in the ocean offer new and continuing vistas for discovering enzymes of unique metabolic, ecological, and evolutionary significance, as well as for creating practical applications in the realms of biotechnology and bioremediation. We explore the physical features and, perhaps surprising, commonalities of very hot and very cold marine habitats, as well as the enzymes and producing microorganisms already known from these environments, to develop predictions of enzymes awaiting discovery. The potential for application of genomics, proteomics, and other forms of genetic access and manipulation as new search and discovery tools that can be made even more powerful with ecological insight is highlighted.

The thermal end members of marine habitats on this planet are submarine hydrothermal systems (1), including the virtually unexplored subsurface biosphere beneath their seafloor expressions (2–5), and subzero sea-ice systems (6), including their connections to permanently cold deep waters and sediments of polar regions (7). In considering these thermally extreme environments, we build upon the axiom that extreme temperatures, especially the sharp thermal gradients they create in the ocean, have provided powerful evolutionary forces to select for microbial enzymes with unique characteristics, unlike those of their moderate temperature counterparts. Inextricably related to the selective pressure of extreme temperature in these targeted environments are the elevated hydrostatic pressures at hydrothermal vents, which act to keep superheated fluids liquid in the deep sea (to temperatures above 400°C) (1), and the elevated salinities and other solutes in sea-ice formations, which act to keep supercooled fluids liquid even at the coldest of wintertime temperatures (to −35°C) (8,9). Just as organisms themselves must have suitable intracellular and membrane-bound enzymes to metabolize, replicate, and transcribe deoxyribonucleic acid (DNA) and grow under such combinations of extreme conditions, the extracellular transforming and degradative enzymes they release into their surroundings must be able to persist long enough to provide a useful nutritional return. Although our tendency to focus more on extracellular than intracellular enzymes in this chapter stems from the more abundant information available on them and the applied interests in them (the two are linked), it is also rooted in an ecological appreciation of their importance to the producing organisms in their natural settings (10).

II. TERMINOLOGY

A. The Microorganisms

Until recently, the search for microbial enzymes from extreme environments has invariably involved the producing organisms themselves, either in laboratory culture or in their native habitats. Terms used to describe the behavior of microorganisms at both ends of the temperature spectrum have undergone a series of revisions over the years, and sometimes apparent misuse, as the information base has increased and research emphases have changed (11,12). Here, as elsewhere (1,11), we consider hyperthermophilic microorganisms as those that grow optimally at a temperature of 80°C or higher and to a maximal temperature of at least 90°C; most are of marine origin, many of them isolated from the type of hydrothermal vent systems that we consider here. All are relatively new to science; they are differentiated from the more moderate thermophiles, known for decades, that have maximal growth temperatures between 55°C and 80°C (thermophilic eukaryotes, algae and fungi, have maximal temperatures of 50°C to 60°C). At the lower end of the temperature spectrum, we follow Morita's (13) definition of psychrophilic microorganisms as those that grow optimally at 15°C or lower and to a maximal temperature of 20°C. Psychrophiles are distributed worldwide in every type of cold environment, but those from marine environments have been the object of study for nearly a century (14). Each of these definitions, for hyperthermophiles and for psychrophiles, clearly emphasizes optimal activity at an extreme end of the temperature spectrum, coupled with the upper temperature limit for growth. Consistent terms emphasizing the lower-temperature limit for growth are generally missing from the literature.

The issue of a lower-temperature limit has contributed to some confusion in the literature, especially in the realm of psychrophily, because of the many organisms capable of growth (albeit slow) at near-freezing temperatures, even though they grow optimally above, and often well above, 20°C. Depending on author and research perspective, such organisms have been called psychrotrophic, psychrotolerant, or facultatively psychrophilic (12,13,15). Blurring the picture further is the recent use of psychrophilic to mean any organism capable of growth at near-freezing temperatures, regardless of its growth optimum. Some have argued that growth yield can be the more important variable and that yield is not always linked to growth rate (16). Here we follow our oceanographic perspectives and use psychrotolerant to refer to those organisms that can grow at near-freezing temperatures but most rapidly at approximate room temperature. This choice parallels the (high-pressure) deep-sea literature, in which barotolerant refers to an organism that can grow at elevated hydrostatic pressures but most rapidly at approximate room (atmospheric) pressure (17). We reserve the use of *psychrophilic* to refer only to an organism optimized for growth at low temperatures ($T_{opt} \leq 15°C$ and $T_{max} < 20°C$) (13).

Less confusion in terminology figures in the high-temperature literature, since the situation in the reverse direction does not appear to apply; that is, few if any organisms with optimal growth temperatures below 80°C can also grow at 90°C or higher. As organisms with increasingly higher temperature optima and maxima for growth (as a general rule, they increase in tandem) have been discovered, terms have easily accommodated the new information, from *thermophilic* ($T_{opt} \geq 55°C$ and $T_{max} = 80°C$) to *hyperthermophilic* ($T_{opt} \geq 80°C$ and $T_{max} \geq 90°C$) to *superthermophilic* ($T_{max} \geq 115°C$) (1,2). The term *extremely thermophilic*, used somewhat loosely in the past, has generally been retired in favor of the defined term *hyperthermophilic*. In the cold direction, the trend appears to be from *psychrophilic* ($T_{opt} \leq 15°C$ and $T_{min} = 0°C$) (13) to *extremely psychrophilic* ($T_{opt} \leq 5°C$ and $T_{min} = -5°C$); (18) to superpsychrophilic ($T_{min} < -5°C$) (9).

B. The Enzymes

A uniform approach to classifying enzymes on a similar thermal basis so far has eluded the research community. A common approach has been to refer to the enzyme according to the growth optimum of the organism that produces it, rather than a particular thermal feature of the enzyme itself. Enzymes produced by hyperthermophilic organisms have thus been called hyperthermophilic enzymes, whereas those produced by psychrophilic organisms have been called psychrophilic enzymes. This approach works reasonably well at the high end of the temperature scale, in the sense that most enzymes produced by hyperthermophiles also tend to be hyperthermophilic in their behavior, showing a temperature optimum for catalytic activity (Table 1) close to or greater than the T_{opt} for growth of the organism. The exceptions are almost always intracellular enzymes (four of five cases shown in Table 1).

The greater potential for confusion again emerges at the low end of the temperature scale. Reference to an enzyme under study as psychrophilic rarely means that the enzyme itself expresses optimal activity at a temperature of 15°C or lower, since so few enzymes with such a low T_{opt} for catalytic activity are known (Table 2). Depending on the perspective of the investigator, reference to an enzyme as psychrophilic can mean that it was produced by a psychrophilic organism, produced by a psychrotolerant organism, active at low temperatures (even if not optimally), or unstable at high temperatures (regardless of its thermal activity optimum). Some papers report T_{opt} for catalytic activity but not for organism growth, or vice versa, whereas others report maximal temperatures for enzyme stability (or enzyme stability at a temperature selected for reasons of convenience, not necessarily biological or ecological relevance) but not thermal optima (Table 2). We can find no examples of an enzyme from a psychrophile that has a T_{opt} for activity lower than the growth optimum of the producing organism (Table 2), in contrast to the situation for hyperthermophiles (Table 1).

The most common terms in use for enzymes from psychrophilic (or psychrotolerant) microorganisms are *cold-active* and *cold-adapted*, circumventing the terminology problem that stems from an emphasis on T_{opt} for catalytic activity and focusing instead on the ability of the enzyme to express significant activity at low temperatures, given a reference point for maximal activity at room temperature or higher. In light of the still limited information available on enzymes from psychrophiles (compared to hyperthermophiles), we adopt a similar approach in this chapter, at the same time underscoring the prediction,

Table 1 Examples of Enzymes from Hyperthermophilic Heterotrophic Microorganisms, All Isolated from Marine Hydrothermal Vents, Ordered by Strain (T_{opt} for Growth) and Thermal Activity Optimum

Organism (domain)	Growth T_{opt}(°C)	Enzyme, function	Enzyme activity T_{opt}(°C)	Enzyme half-life (time, °C)[a]	References
Thermotoga maritima (Bacteria)	80	4-α-Glucano-transferase, starch hydrolysis[b]	70	3 h @ 80	81
		β-Galactosidae, lactose hydrolysis[b]	80	N.A.	79
		Hydrogenase, hydrogen production	>90	2 h @ 95	147, 148
		Xylanase A, xylan hydrolysis[b]	92	45 min @ 90	149
		Xylanase B, xylan hydrolysis[b]	105	3 h @ 90	149
		Glucose isomerase, glucose to fructose	105–110	10 min @ 120	73
Thermotoga neopolitana (Bacteria)	80	Mannanase, mannan hydrolysis[b]	91	13 h @ 90 35 min @ 100	84
		Cellulase celA, cellulose hydrolysis[b]	95	N.A.	85
		Xylanase, xylan hydrolysis[b]	>100	N.A.	150
		α-Galactosidase, lactose hydrolysis[b]	100–103	9 h @ 85 2 h @ 90 3 min @ 100	84
		Cellulase celB, cellulose hydrolysis[b]	106	130 min @ 106 26 min @ 110	85

Thermococcus litoralis (Archaea)	85	DNA polymerase, DNA amplification	75	7 h @ 95	88, 151
		Amylopullulanase, starch hydrolysis[b]	115	N.A.	73
Pyrococcus furiosus (Archaea)	100	DNA polymerase, DNA amplification	>75	20 h @ 95	92, 152
		Protease, peptide bond hydrolysis	85	N.A.	153, 154
		Hydrogenase, hydrogen production	95	2 min @ 100	69, 147, 155
		α-Amylase, starch hydrolysis[b]	100	2 min @ 120	156, 157
		β-Glucosidase, cellobiose hydrolysis[b]	102–105	85 h @ 98	158
		Protease PfpI, peptide bond hydrolysis	105	N.A.	159
		β-Mannosidase, mannan hydrolysis[b]	105	60 h @ 90	160
		Invertase, sucrose inversion	105	48 min @ 95	161
		α-Glucosidase, maltose hydrolysis	110	48 min @ 98	158, 162
		Serine protease, peptide bond hydrolysis	115	33 min @ 98	163
		Amylopullulanase, starch hydrolysis[b]	125	12 min @ 125	73
Pyrococcus sp. strain ES4 (Archaea)	100	Amylopullulanase, starch hydrolysis[b]	118	10 min @ 120	74

[a] N.A., not available.
[b] Extracellular enzyme.

Table 2 Recent[a] Examples of Enzymes from Psychrophilic[b] Heterotrophic Marine Bacteria, Ordered by Strain (Environmental Source and T_{opt} for Growth) and Thermal Activity Optimum

Source	Organism	Growth T_{opt}(°C)	Enzyme	Enzyme activity T_{opt}(°C)	Enzyme half-life (time, °C)	References
Sea ice, Antarctic	*Vibrio* sp. strain ANT300	7	Triosephosphate isomerase	N.A.[c]	9 min @ 25	166
	Colwellia demingiae strains ACAM607, IC169	10–12	Trypsins[d,e]	12–14	N.A.[c]	105
			Phosphatases[d,e]	17–23		
			Proteases[d,e]	28–30		
	Cytophaga-like sp. strains IC164, IC166	N.A.	β-Galactosidase[d,e]	15	N.A.[c]	105
			Phosphatase[d,e]	19		
			Proteases[d,e]	20—27		
			α-Amylase[d,e]	25		
			Trypsin[d,e]	30		
	Pseudoalteromonas sp. strain IC000	N.A.	Trypsin[d,e]	22	N.A.[c]	105
			Protease[d,e]	29		
	Shewanella gelidimarina strain ACAM456	N.A.	β-Galactosidase[d,e]	24	N.A.[c]	105

Sediments, Arctic	*Aquaspirillum arcticum*	4	Malate dehydrogenase	N.A.[c]	10 min @ 55	121
	Colwellia sp. strain 34H	5–8	Protease[d,e]	20	N.A.[c]	20
Seawater, Antarctic	*Pseudomonas aeruginosa*	N.A.	Protease[e]	<25	2 min @ 45	23, 167
	Alteromonas haloplanctis	<15	α-Amylase[e]	27	10 min @ 50	16, 168
	Bacillus sp. strains TA39, TA41	N.A.	Subtilisin[e]	45	90 min @ 25 10 min @ 50	16, 130
Animal-associated, Antarctic (or N.A.)	*Psychrobacter immobilis*	<10	β-Lactamase[e]	35	5 min @ 50	169,170
			Lipase[e]	N.A.	N.A.[c]	171
	Shewanella sp.	N.A.	Phosphatase	30	10 min @ 50 5 min @ 60	172
	Flavobacterium balustinum strains P104–107	10–20	Proteases[e]	30–40	30 min @ 20 15 min @ 60	173
	Flavobacterium sp.	N.A.	β-Mannanase[e]	35	N.A.[c]	174
Deep sea	*Vibrio* sp. strain 5709	20	Protease[e]	40	20 min @ 40	120
	Vibrio sp. strain 5710	N.A.	Malate dehydrogenase	N.A.[c]	N.A.[c]	140
	Photobacterium sp. strain SS9	N.A.	Malate dehydrogenase	N.A.	N.A.	175

N.A., not available.

[a] Earlier work (e.g., 15,164,165), sometimes based on culture supernatants or partially purified protein preparations, reported similar thermal activity optima (in the range of 25°C–50°C) and thermostabilities (e.g., 10 min at 40°C–70°C; 164).

[b] Only the organisms listed from sea ice or sediments are strict psychrophiles (with both T_{opt} for growth ≤15°C and T_{max} ≤ 20°C); those from polar seawater, polar animals, or the deep sea have been called psychrophilic by various investigators, but are not or may not be psychrophiles as defined here.

[c] Indications of optimal activity shifted to lower temperature or of pronounced heat lability.

[d] Preparation may have included multiple isozymes.

[e] Extracellular enzyme.

as supported by information from 1999 and 2000 (Table 2) (19,20), that new discoveries will refocus attention on thermal activity optima that are indeed psychrophilic (≤15°C). In the realm of applications at either end of the temperature spectrum, however, neither activity optima nor thermal stability may be the essential enzyme feature: fidelity of amplification (in the case of DNA polymerases) or novelty of chemical transformation may take precedence. Ultimately, an understanding of the balance between activity optima and thermal stability must be achieved. Fortunately, this goal motivates much of the recent research on enzymes from both extremely hot and extremely cold marine environments.

III. BIOCHEMICAL CHALLENGES AT THERMAL EXTREMES

A. Common and Divergent Themes

The ability of an organism to grow or survive at an extreme temperature poses special physiological and biochemical challenges. Success depends upon both extrinsic and intrinsic factors: elevated hydrostatic pressure or solute concentration at high temperatures (as at deep-sea vents) and high salt or other solute concentration at low temperatures (as in sea-ice brines). These can extend the permissive temperature range by their effects on the liquid state of water (and on other molecules); intrinsic factors associated with uniquely evolved structural, catalytic, and informational macromolecules are essential. The stark contrast in levels of thermal energy inherent to very hot and very cold environments has led to divergent growth and survival strategies for hyperthermophiles and psychrophiles. In the face of very high thermal energy in superheated fluids, the successful hyperthermophile maintains metabolic integrity and control by virtue of remarkably heat-stable membranes, cell walls, and macromolecules, surviving supramaximal temperatures via unique ''heat-shock'' proteins that stabilize macromolecules at otherwise denaturing temperatures (21,22). In the face of very low thermal energy in subzero fluids, the successful psychrophile must contend with greatly reduced rates of physical (e.g., diffusional), physiological, and biochemical processes, maintaining adequate membrane fluidity simply to acquire nutrition from its surroundings.

Compared to extrinsic factors involved in growth and survival strategies at thermal extremes or to the intrinsic factors of structural lipids and informational macromolecules, less is known about the vast array of enzymes required by an organism to be successful at either end of the temperature spectrum. Some basic properties that emerge from a comparison of the extremes include that enzymes from psychrophiles have a lower free energy of activation than enzymes from thermophiles (23), in keeping with the disparate levels of thermal energy in their respective environments. By definition, cold-adapted enzymes have upper denaturation thresholds at relatively moderate temperatures (30°C–60°C) compared to hyperthermophilic enzymes, even though the same does not hold true in the reverse direction: only some hyperthermophilic enzymes are known to denature at cold temperatures (near room temperature or below) (24); most remain stable as the temperature drops. In spite of limited data, a relationship does appear to exist between the thermal optima for enzyme activity and the half-life or thermostability of the enzyme at supraoptimal temperatures: the higher the T_{opt} for activity the longer the lifetime at even higher temperatures (compare data for xylanases, for DNA polymerases, and for amylopullanases in Table 1 and for proteases in Table 2).

Although analyses of the most basic features of enzymes—their amino acid sequences—have yielded some insight into what makes an enzyme uniquely adapted to one thermal extreme or the other, the combination of this information with other biochemical and theoretical studies has been the most revealing (e.g., 25–27). For example, features of a successful hyperthermophilic enzyme can include increased compactness, stabilization of α helices, increased salt bridges and ion pairs for stabilizing secondary structures, or an increased number of hydrogen bonds, each toward retaining stability in the face of very high denaturing temperatures. The cold-adapted enzyme, in contrast, shows greater flexibility and less compaction, lacks salt bridges and ion pairs, and has a reduced number of hydrogen bonds, all toward retaining activity under very-low-energy near-freezing conditions. No organism, however, appears to have evolved a uniform strategy for stabilizing or allowing activity of all of its enzymes at a given extreme temperature. Instead, its suite of enzymes encompasses a range of unique combinations of molecular adapations that reflect the host of complex evolutionary and ecological factors, including acquisition of successful traits through genetic exchange in the environment (28), that define a contemporary microorganism.

A common theme for hyperthermophily and psychrophily, relating enzymes directly to the producing organism (and thus allowing at least some common terminology), is that the higher the T_{opt} for growth of the organism, the higher the T_{opt} for its enzymes: just as enzymes optimized for activity at the highest temperatures clearly derive from hyperthermophiles adapted to growth at the highest temperatures (Table 1), enzymes with the lowest thermal optima derive from psychrophiles with the lowest growth optima (Table 2). In fact, all known cold-adapted enzymes express thermal activity optima that fall above the T_{opt} for growth of the producing strain (Table 2). Although the same holds true in large part for hyperthermophilic enzymes, some (mostly intracellular) enzymes are optimized for activity below the T_{opt} for growth (Table 1), retaining only minimal stability (half-lives of a few minutes) as the maximal temperature for growth is approached (29).

B. Intracellular Versus Extracellular Enzymes

If activity optima for enzymes, whether from hyperthermophiles or psychrophiles, are examined according to general cellular location of the enzyme—intracellular (essential to metabolism, DNA processing, growth) versus extracellular (typically hydrolytic enzymes that act independently of the organism)—the T_{opt} for extracellular enzymes almost always falls above, and sometimes well above, the T_{opt} for growth (80% of the hyperthermophilic cases in Table 1; 100% of the psychrophilic cases in Table 2). Attempts to understand this locational discrepancy in thermal optima have been made by researchers studying psychrophilic and psychrotolerant bacteria. They have asked, What evolutionary pressure would select for extracellular enzymes optimized for activity at temperatures well above the T_{opt} for growth (25)? Would not an extracellular enzyme with greatest activity at the T_{opt} for growth be ideal—or better yet, at the in situ temperature of the environment, which in the case of psychrophiles is invariably lower than its growth optimum? Why should extracellular enzymes have evolved differently in thermal properties than intracellular enzymes?

The biochemical processes underlying enzyme activity versus stability at a given temperature have been proposed as a primary explanation for the phenomenon (16,23,25). Basically, an enzyme is least stable at the higher end of the temperature range over which it is active. For example, a psychrophile-derived extracellular enzyme optimized for activity at 30°C has a shorter half-life at that temperature than at lower ones. It is thus more stable at the growth optimum for the organism (≤15°C) and has its longest lifetime at the typical subzero temperatures encountered in a polar marine setting. As long as the enzyme retains enough activity at the lower temperatures, its longer lifetime can be seen as a benefit to the producing organism. What constitutes ''enough'' and ''benefit'' has been explored in a quantitative model of microbial foraging by extracellular enzymes under thermally moderate conditions (10); no similar quantitative analysis is available for thermal extremes (but see 9). The following consideration of enzyme foraging in light of the common physical features of our focal environments underscores the promise of this strategy for hyperthermophiles and psychrophiles and for a new generation of foraging models incorporating extreme temperatures.

IV. MICROBIAL FORAGING USING EXTRACELLULAR ENZYMES

A. General Features

The quantitative modeling work of Vetter and associates (10) addresses the specific use of extracellular enzymes as a bacterial foraging strategy in moderate-temperature microenvironments rich in particulate organic material (POM), of a size too large to pass the cell membrane without extracellular hydrolysis. The typical marine environment where this strategy is demonstrably advantageous involves an aggregation of POM-rich particles, either mineral grains with sorbed POM (as in marine sediments), or organic-rich detrital particles (as in aggregates sinking through the water column). Adequate pore space through which various solutes, from enzymes to POM hydrolysate, can diffuse is also essential (Fig. 1). Although not yet considered in a modeling context for their special features, both hydrothermal structures and sea ice are rich in interior colonizable surfaces, often laden with organic material in patches, and can be highly porous. They represent ideal settings for the use of extracellular enzymes as a foraging strategy and for recognition and improved dissection of the consequences of evolutionary pressure on enzyme adaptation at extreme temperatures.

B. Foraging in Hot Sulfide Structures

Actively venting sulfide structures on the seafloor are, by definition, composed of mineral grains, of variable composition depending on local chemical and thermal conditions for precipitation and deposition (30). They are colonized in their cooler portions by animals that produce organic, especially chitinous, structures and polymers that remain after the organism has sought new territory or succumbed to either a predator or thermochemical change in the habitat. The sulfide formations are clearly porous, often functioning as visible diffusers releasing cooled hydrothermal fluids into the surrounding seawater. Their interior portions are known to support abundant heterotrophic (and other) microbial populations (1,31,32), zoned phylogenetically (Bacteria versus Archaea; Fig. 2) according to permissive temperatures. The prediction from this combination of features alone is that POM foraging using extracellular enzymes is an important strategy for the growth and survival of heterotrophic microorganisms living within these structures. The additional fact that known hyperthermophilic heterotrophs release a wide variety of highly thermostable enzymes into culture media in the laboratory (Table 1) makes seafloor sulfide structures obvious sites for future exploration and discovery of new enzymes, especially extracellular hydrolytic enzymes. Although no direct environmental searches of enzyme activity in hydrothermal structures have yet been made to our knowledge, such an approach could

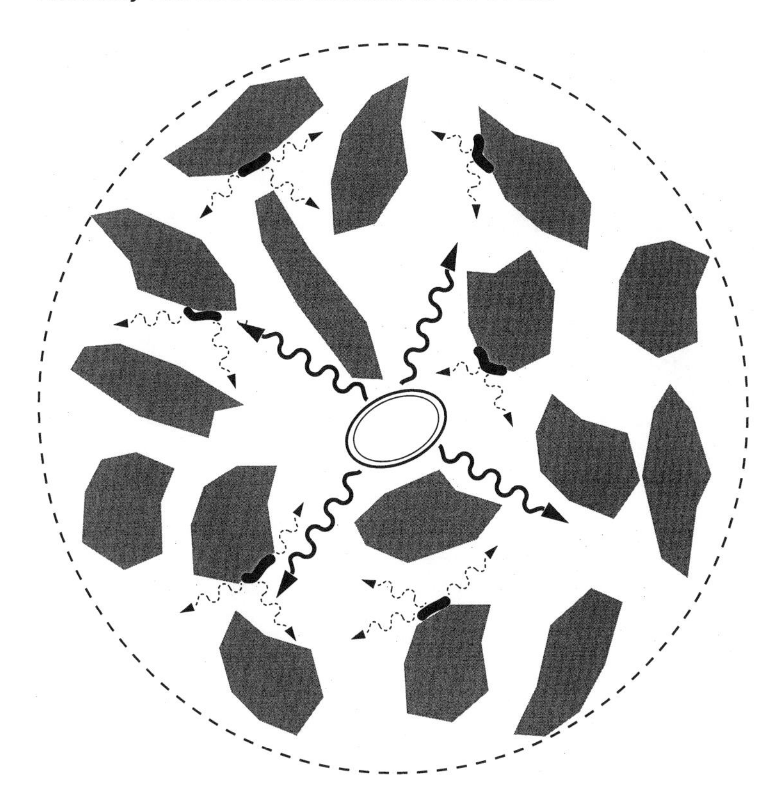

Figure 1 Schematic diagram of an aggregate of particles with an immobile enzyme-releasing bacterium at the center. The aggregate environment is composed of impermeable inorganic grains (shaded shapes), patches of organic material (black shapes, too large to cross the cell membrane) sorbed to the grains, and seawater-filled (or brine-filled, in sea ice) spaces through which solutes diffuse. Heavy arrows represent enzyme diffusing away from the organism; dashed arrows, hydrolysate diffusing away from the organic material, where it is produced enzymatically. (Modified from Ref. 10.)

prove fruitful, just as new microorganisms continue to emerge from direct examination of seafloor hydrothermal sites and effluents from the subsurface biosphere (5,32,33).

Special features to consider in the search for hyperthermophilic enzymes (or organisms) or in predictive modeling efforts in advance are the sharp temperature gradients, established across distances that measure in centimeters, in actively venting sulfide structures (Fig. 2); the other thermally linked gradients in chemical parameters, from pH to Eh (oxidation status) to a multitude of dissolved inorganic and organic species (1,28); and the influence of advection versus diffusion through pore spaces. Since hydrostatic pressure figures importantly at deep-sea hydrothermal vents, acting to keep superheated fluids in the liquid phase and to select for barophilic and barotolerant microorganisms (2), it must also be considered in the study of microbial enzymes from these environments (28,29,34). The critical feature of the lifetime of a given enzyme, especially when sorbed to mineral grains within an actively flushing vent structure, determines not only its detectability and utility as a foraging tool to the producing organism and its neighbors (10), perhaps in zones too hot to permit the organism itself (28), but also its performance immobilized under extreme conditions and thus its attraction to biotechnologists.

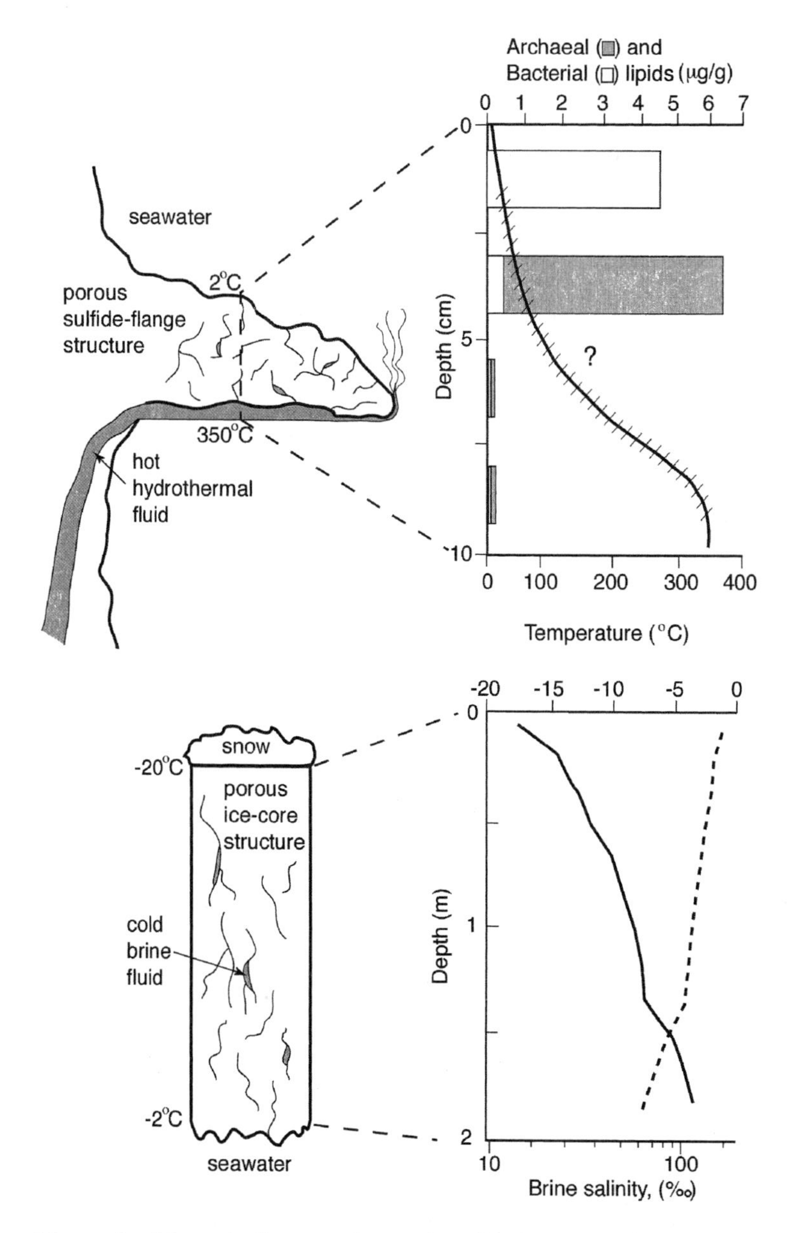

Figure 2 Schematic diagrams of examples of the hottest (seafloor hydrothermal sulfide structure) and coldest (wintertime Arctic sea ice) marine habitats, depicting common physical features of interior colonizable surface area, fluid-filled pore space, and sharp thermal gradients. Note that bacterial and archaeal zonations have been explored in sulfide structures (modified from Ref. 28), but not yet in sea ice (modified from Ref. 45), whereas fine-scale temperature and chemical properties of pore fluids are better known for sea ice (e.g., salinity gradients parallel thermal gradients inversely) than sulfide structures.

C. Foraging in Subzero Sea Ice

The three basic features of the enzyme foraging model of Vetter and coworkers (10) for particle aggregates (Fig. 1) also pertain to the other end of the temperature spectrum for microbial life and enzymatic activity epitomized by sea ice. Aggregates of mineral grains and other particles and precipitates (including microorganisms and salts) are known to concentrate within the fluid inclusions of sea ice (6), most notably in the Arctic, where seabed sediments entrain into coastal ice as it forms (35). These aggregates include POM-rich detrital particles (36) and large exopolymers (37) as a result of the autotrophic and heterotrophic communities that develop annually within the ice cover (38–40), as well as generally elevated levels of dissolved organic carbon (41,42) including enzymes (19,20). The sea-ice matrix is also highly porous, especially in summertime, flushing regularly with the tides or influence of waves while retaining particle aggregates and organisms within it (43,44). Even during wintertime (in the Arctic), when sea-ice temperatures can drop below −20°C (Fig. 2) to as low as −35°C, depending on snow cover and atmospheric conditions (8), interior movements of brine fluid through finely connected channels are possible on a scale relevant to bacteria and enzymes. This has been demonstrated by physical analyses of undisturbed ice sections using nuclear magnetic resonance (NMR) and transmission microscopy (45).

In contrast to research on hydrothermal structures, less information is available on the abundance or possible zonation, phylogenetic or otherwise (Fig. 2), of microorganisms in these coldest of wintertime sea-ice habitats (e.g., 18, 36). Only in 1999 was a nondestructive (nonwarming, nonmelting) method for studying microbial life in supercooled ice developed (36). Although extreme temperatures determine the solid phase of both hydrothermal structures (by controlling mineral precipitation reactions) and sea ice (by freezing water), only the hydrothermal structure remains intact for ready study at temperatures less extreme than those in situ. Sea-ice structure changes nonuniformly with every incremental change (up or down) in temperature, presenting special challenges to a postsampling evaluation of in situ microbial communities, products, or processes.

Nevertheless, the prediction from the three basic features (abundant attachment sites, organic material, and porosity) that enzyme foraging is an important microbial strategy for growth and survival in sea ice has been supported by direct environmental measurements in both wintertime (18) and summertime sea-ice samples (19,20). Not only have hydrolytic activities on substrate analogs for protein, chitin, and various carbohydrates been readily detected, but, where measured and compared across other subzero environments (Arctic seawater and sinking aggregates), the lowest thermal optima for enzyme activities were observed in multiyear sea ice (19). The optima were consistently psychrophilic, down to 10°C, compared to previous reports of 30°C–50°C (19, 20, and citations therein) (Table 2). In other words, the ice cover over the Arctic Ocean, which in some areas persists through a decade of winters (rarely if ever the case in Antarctic waters), clearly selects for cold-adapted and even strictly psychrophilic enzymes, as it does for psychrophilic organisms (discussed later), making it an obvious environment for continued search and discovery of new enzymes in this thermal class.

Special features to consider in a search for cold-adapted enzymes in sea ice resemble those for seafloor sulfide structures, albeit at subzero temperatures: sharp thermal gradients in wintertime ice (Fig. 2), linked salinity (and other chemical) gradients (Fig. 2), and the influence of advection versus diffusion. Elevated salinities, as well as concentrations of other solutes, are key to depressing the freezing point and maintaining fluid-filled pore spaces. In fact, physiological studies of sea-ice bacteria suggest that salinity (and pH) gradients may be as critical to potential microbial succession and zonation with the ice-brine matrix as the cardinal growth parameter of temperature (46). Unique to the cold end of the temperature spectrum exemplified by sea ice are the problems of increased viscosity as temperature drops and salinity rises (in wintertime) and of interior vertical mixing with tidal and wave action (in summertime). The lifetime of an extracellular enzyme sorbed to particle surfaces in situ under extremely cold, saline, and viscous conditions is as important a factor to the organisms in sea ice (and to those who study them) as it is in the case of extremely hot, chemically reduced, and nonviscous conditions within hydrothermal vent structures.

Indeed, enzyme lifetime or stability emerges as a critical factor in scenarios that help to explain why extracellular enzymes appear optimized for activity at temperatures well above what they experience in their natural settings, whether very hot or very cold. In an early ecological scenario for subzero marine sediments (47), absent specific information on enzyme activity relative to lifetime, both characteristics were assumed to be restricted at in situ temperature, such that luxury or excessive production of extracellular enzymes was invoked to account for sufficient hydrolytic return to support the microbial community clearly present in the environment. Indeed, some psychrophiles have since been shown in the laboratory to produce maximal amounts of extracellular enzymes at suboptimal growth temperatures (48), where they also appear to require elevated concentrations of dissolved organic matter for activity (49,50). However, we also now understand that an enzyme optimized for activity at a temperature well above the growth optimum for the producing organism (and thus the in situ temperature, in the case of psychrophiles) is more stable at that lower growth (or in situ) temperature than at its own T_{opt} for activity (see 16, 23, 25 for discussion at the biochemical–molecular level as to why a cold-adapted enzyme is relatively unstable in its optimally active state). The balance between activity and stability at the environmentally relevant temperature is thus also understood to determine the extracellular enzyme of greatest benefit to its producer (9,10,12,16,19,20,23, 25,51). We suggest that evolutionary pressure on microorganisms to feed competitively and thus survive in microenvironments rich in POM (10) (Fig. 1), but at temperatures suboptimal for growth, has selected for the production of extracellular enzymes with a balance between cell-free activity and lifetime that favors longevity at the in situ temperature and thus long-term return of hydrolysate to the organism and its neighbors. In contrast, membrane-bound and intracellular enzymes (which remain under cellular control and thus can be recycled and produced anew, as needed) with maximal catalytic activity at a given temperature help define that temperature as the optimum for growth.

This enzyme-based foraging and growth scenario is so far exemplified by a psychrophilic bacterium, *Colwellia* sp. strain 34H, enriched from near-freezing Arctic sediments (7) and later shown to be optimized for growth at 5°C–8°C (20). The organism produces extracellular proteases with unusually low thermal optima for activity (20°C) (Table 2). A crude preparation of these proteases from mass cultivation of the organism, one that still includes some intracellular proteases, expresses a lower thermal optimum for activity (13°C) (Fig. 3), as would be predicted for such an enzyme mix. The fraction of enzyme activity remaining in this preparation after a holding period at the environmentally relevant temperature of 0°C was significantly greater than that remaining at warmer temperatures more permissive of both growth and enzyme activity (Fig. 3). In other words, to survive and even grow in a permanently cold environment, an organism is well served both by extracellular enzymes adapted for maximal activity at temperatures well above in situ (and

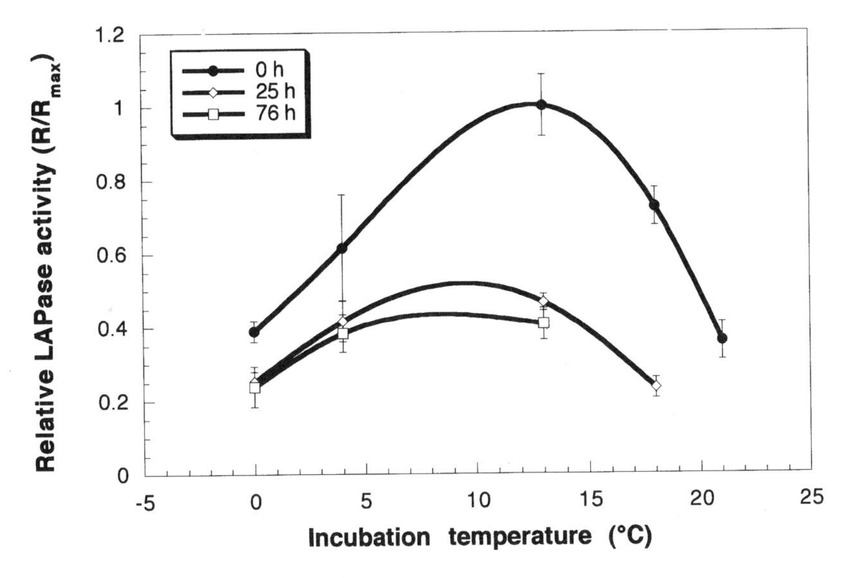

Figure 3 Activity of a crude preparation of proteases, specifically leucine-aminopeptidase (LAPase) activity from *Colwellia* sp. strain 34H (19) relative to maximum hydrolytic rate (R/R_{max}) over time at different incubation temperatures. Note LAPase activity decreasing to greater extent at higher temperatures (e.g., 59% loss after 76 h at 13°C, where loss is attributed to enzyme denaturation) than lower temperatures (e.g., 33% loss after 76 h at 0°C), as well as the shift in activity optimum from 13°C to 10°C to 8°C with increasing holding time. Solid lines, cubic spline curve fits; error bars, 95% confidence intervals for triplicate measurements at each temperature. (From A Huston and J Deming, unpublished.)

optimal growth) temperature, and thus well-designed at the molecular level for a long lifetime of activity (albeit at modest catalytic rate) at the environmentally relevant temperature, and by intracellular enzymes more closely optimized to in situ conditions and thus for metabolism and growth. A similar scenario, one not yet discussed in the literature, can be proposed for hyperthermophilic enzymes and their producing organisms, often found at growth-permissive temperatures along sharp thermal gradients within seafloor sulfide structures (Fig. 2). Some surprises in the hyperthermophilic scenario may be in store, however, given the fact that some intracellular enzymes from hyperthermophiles (e.g., DNA polymerases) express thermal activity optima below growth optima of the producing strains yet also retain significant activity even at extremely high temperatures (Table 1).

D. Future Foraging Scenarios

For both ends of the temperature spectrum, the described enzyme-based foraging and growth scenarios and new ones yet to be developed benefit from rigorous and innovative analysis of the characteristics of both intracellular and extracellular enzymes purified from model organisms; the research and technology communities may benefit from discovery of new enzymes or features of enzyme activity and thermostability in the process. Heeding an ecological perspective should continue to be useful, given that thermal optima apparent to researchers in short-term incubation experiments in the laboratory may not be synonymous with thermal optima relevant to an organism dependent (52) on a flux of dissolved compounds from enzymes already released and functioning over longer periods in the environment. For example, note the shift in thermal activity optima for cold-adapted proteases from 13°C to 10°C to 8°C as a function of increasing holding time (Fig. 3). We also suggest that the further exploratory study of microbial enzymes produced in environments characterized by sharp thermal gradients may yield enzymes with both high catalytic activity and long lifetimes at extreme temperatures (hot or cold), a combination of features that so far has been observed only as a result of genetic engineering (described later) and apparently not of evolutionary pressures in nature. The temporally and spatially fluctuating thermal gradients within sulfide structures and sea ice may have provided the necessary selective pressure.

V. STATUS OF THE SEARCH FOR HYPERTHERMOPHILIC MICROORGANISMS AND ENZYMES

A. Focus on Culturable Hyperthermophiles

Although the discovery of hyperthermophilic microorganisms at marine hydrothermal vents was reported in 1982 (53,54), their potentially exciting activities in situ have been studied by few and remain poorly constrained (1, 28). The in situ activities of enzymes that hyperthermophiles may release into their surroundings are completely unknown. This general lack of ecological information on the functioning of either hyperthermophilic organisms or enzymes in their natural settings stands in contrast to what is known about organisms and enzymes at the other end of the temperature spectrum (see Sec. V. B); marine psychrophiles have been known and studied for almost a century, much of the work ecologically motivated from the outset (14). Perhaps because of the immediate recognition of practical applications for new organisms functional at ever higher temperatures (11,55), research efforts now ongoing worldwide have focused heavily on organism and enzyme performance under controlled laboratory conditions, with specific biotechnological or industrial goals motivating the choice of organism, enzyme, or test conditions. The desire to achieve a fundamental understanding of the biochemical, metabolic, and genetic basis for hyperthermophily has often been presented as a better means to manipulate strains and their products in vitro for commercial purposes. However, the first whole-genome sequence for any organism, information of the most fundamental nature, was obtained for the deep-sea hyperthermophile *Methanococcus jannaschii* (56).

Although ecological considerations beg study and enzyme foraging scenarios for hyperthermophiles have not yet been formulated, the acquisition of culturable hyperthermophiles from marine hydrothermal vents now borders on routine. Current repositories of marine hyperthermophiles, virtually all of which are obligately anaerobic, include representatives of 25 genera (examples of which are shown in Fig. 4 in italics) and physiological processes as diverse as methanogenesis, iron oxidization, other forms of chemoautotrophy, sulfate reduction, and other forms of heterotrophy. With the exception of methanogenic genera, which contain a wide range of thermal classes of methanogens, all of these genera contain only hyperthermophilic organisms. As species have been added over time, culture collections have become dominated by heterotrophic hyperthermophiles in a limited number of genera of both Archaea and Bacteria. These organisms typically require complex organic compounds, including peptides or carbohydrates, to meet their carbon demands (21,57). For the initially exciting pace of discovery of new and diverse hyperthermophilic

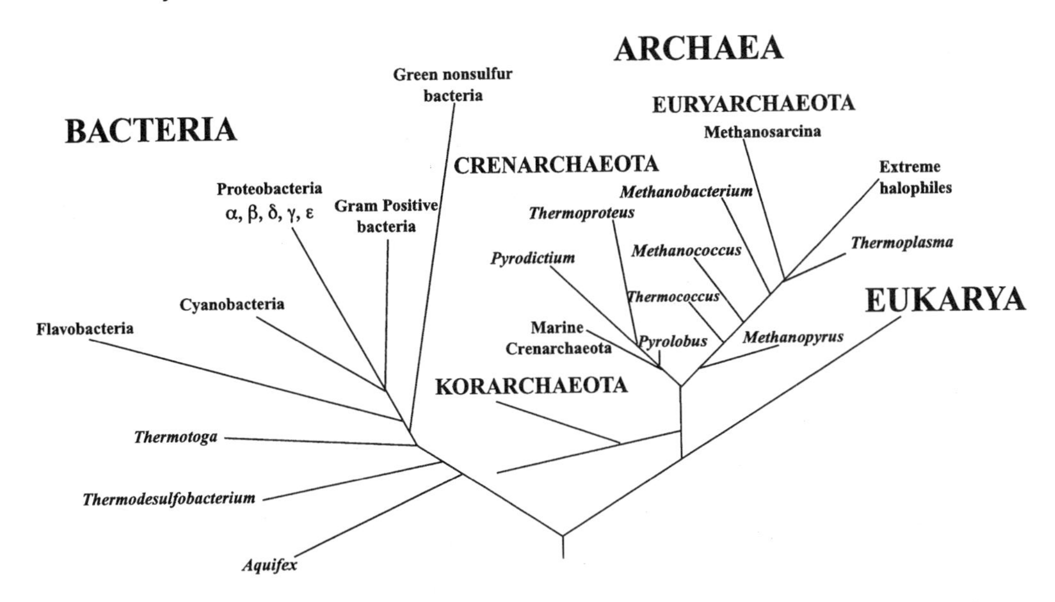

Figure 4 Universal phylogenetic tree based on 16S rRNA sequences, showing the three domains of Bacteria, Archaea, and Eukarya and featuring the hyperthermophilic genera (in italics), which fall within the prokaryotic domains of Bacteria and Archaea. Cultured psychrophiles fall within the bacterial divisions of Proteobacteria, Flavobacteria, and Gram-positive bacteria (see Fig. 5); the marine Crenarchaeota comprise a large group of uncultured presumptive psychrophiles within the Archaea. Distances were derived from numbers of mutations; the root, from sequences of the two subunits of the F1-ATPases and the translation elongation factors EF-1α (Tu) and EF-2 (G). (Modified from Ref. 65.)

organisms to continue, however, innovative sampling and culturing approaches beyond the now-standard heterotrophic sulfur-based media must be pursued. The continuing promise of discovery is evidenced by sampling strategies that access, at the seafloor, recent fluid emissions from the subsurface biosphere (2,4) and by inventive culturing media that yield new organisms and even the potential for novel metabolisms (33). For example, an organism isolated in 2000 from a seafloor eruption site on the Gorda Ridge in the Northeast Pacific Ocean likely represents a new genus and can function metabolically as a heterotroph, autotroph, and iron reducer (33).

Direct molecular approaches to assessing microbial diversity underscore the immense and still untapped diversity of hyperthermophiles. Analyses of small-subunit ribonucleic acid (RNA) sequences in environmental samples from terrestrial hot springs first revealed this untapped diversity in both domains of Bacteria and Archaea (58,59). In an early similar analysis of a submarine vent environment, specifically a sample of microbial mat from Loihi Seamount (60), sequence analyses of fewer than 50 clones appeared to reflect the untapped diversity in that habitat adequately. Subsequent analyses of the type of sulfidic structures that we have targeted here, with their sharp thermal gradients and higher end member values, have revealed a much greater degree of diversity, such that analysis of hundreds of clones can still be inadequate to the task (31,32,61). The presence of so many potentially unique and hyperthermophilic organisms necessarily indicates an equally untapped diversity of hyperthermophilic enzymes, awaiting discovery.

In the meantime, culture collections have provided rich depths to plumb: several well-studied heterotrophic hyperthermophiles have become the targets of concentrated searches for enzymes with unusual thermal or other properties. Of all of the known genera of hyperthermophiles, only three heterotrophic ones—*Pyrococcus* and *Thermococcus* of the Archaea and *Thermotoga* of the Bacteria—have yielded species that can be grown reproducibly to high cell densities in the laboratory, making them model organisms for enzyme studies in vitro. In fact, most of the physiological and enzymological studies have been carried out with three species (Table 1), *Pyrococcus furiosus* (62), *Thermococcus litoralis* (63), and *Thermotoga maritima* (64).

By far the largest number of hyperthermophilic species, and most of the characterized hyperthermophilic proteins, belong to the archaeal family of Thermococcales of the kingdom Euryarchaeota (65) (Fig. 4). This family is cosmopolitan in that its members have been isolated from all hydrothermal environments sampled so far. All are known to utilize carbohydrates by a glycolytic pathway that includes some unusual tungsten-containing enzymes (24,66); most have a growth requirement for amino acids and peptides, and for elemental sulfur (21). Tungsten concentrations, as well as levels of a wide variety of other metals, can be very high in portions of seafloor sulfide structures deposited at extreme temperatures (1,67), raising interesting evolutionary (and biotechnological) questions about metal-based enzymes and proteins in general. External sources of amino acids, proteins, and other organic compounds for hyperthermophiles in their native settings have been hypothesized (1), but not yet confirmed quantitatively.

For the two genera, *Pyrococcus* and *Thermococcus*, that constitute the family of Thermococcales, approximately 40 species have been described. They are routinely isolated from near-vent sites on the seafloor, from samples of originally hot sulfidic rocks and other structures, from alvinellid worms that colonize actively venting structures, and occasionally from samples of hot fluid emerging from such structures (1). Not only are these hyperthermophiles easy to grow, they are also hardy, surviving storage under low-temperature oxic conditions (68; J Baross, unpublished observations). Their ability to utilize a wide range of organic substrates is reflected in their production of a diverse array of hydrolytic enzymes (55,69,70) (Table 1); the oxidoreductases and dehydrogenases involved in their fermentative metabolisms are the enzymes with metal (tungsten) centers (24). Both the diversity and the properties of high-temperature enzymes can vary significantly among similar species (69). Enough information is now available to recognize that for a single organism the thermal properties of enzymes in the same functional class can also vary significantly (see proteases for *Pyrococcus furiosus* in Table 1). The latter finding opens new evolutionary and ecological scenarios for hyperthermophiles, e.g., the likelihood that thermal gradients fluctuating in time and space in the vent environment, and perhaps especially in the subsurface biosphere (1,28,33), may have selected for suites of isozymes that make an organism uniquely adapted to survive, and possibly grow, across a wider range of conditions than previously envisioned.

B. Focus on Commercially Important Enzymes

Apart from enzyme studies of some of the metalloproteins involved in hyperthermophilic metabolism mentioned, most of the studies of the best-known species of *Pyrococcus* and *Thermococcus* have focused on enzymes of commercial importance. Classes of hydrolytic enzymes have received particular emphasis; catalytic activity at high temperatures and extreme thermostability are frequently the properties of greatest interest. Hydrolytic enzymes are used by organisms to degrade peptides, complex carbohydrates, and lipids to low-molecular-weight components that are easily transported into the cell. The best studied of the hyperthermophilic hydrolytic enzymes are the proteases and the glycosyl hydrolases, known to be remarkably thermostable (Table 1) and to show very high rates of activity at their thermal optima. Their functions for the cell differ, however, with cellular location. Some hyperthermophilic proteases have been reported as intracellular, performing regulatory and catabolic functions, including degrading inactive proteins or activating others; others have been reported as periplasmic or membrane-associated, involved in degrading peptides for nutrition and growth. Those hydrolytic enzymes used extracellularly figure importantly in the development of enzyme foraging scenarios for hyperthermophiles. Differently located proteases have been studied from *Pyrococcus furiosus*: their thermal activity optima range from 85°C to 115°C (Table 1) (70), but with no obvious relationship between T_{opt} and intra- or extracellular location. Compared to *Escherichia coli*, however, from which 36 proteases have been identified (71), only a limited number of proteases have been obtained and studied from hyperthermophilic (or psychrophilic) organisms.

The glycosyl hydrolases from hyperthermophiles, which act on complex carbohydrates, are the most thermally stable enzymes yet characterized. When *P. furiosus* and *T. litoralis*, normally cultured in heterotrophic media containing peptides, are grown on maltose instead, they reach high cell yields and produce an extremely thermostable α-glucosidase with an activity optimum of 108°C and a half-life of 48 h at 98°C (72) (Table 1). The amylopullanases from *Pyrococcus* species ES4 and *P. furiosus* have the most thermophilic enzymes yet described. In hydrolyzing both the α-1,4 and α-1,6 linkages in starch, they show activity at temperatures above 125°C, where they are stabilized by the addition of Ca^{2+} (73,74). Glycosyl hydrolases in general may thus meet the starting premise for a successful extracellular forager when the in situ environmental temperature falls at or below the optimal growth temperature of the producing strain (e.g., at 100°C or below in a sulfide structure): their thermal activity optima are much higher, implying greater longevity at the in situ temperature. Even where enzyme lifetime is limited, at extreme points in the thermal gradient spanning a sulfide structure (Fig. 2), beneficial enzymic "work" for the producing organism or its neighbors is possible. Utilizable hydrolysate may be produced and returned via diffusion to milder locations along the gradient where the organisms reside.

Perhaps one of the more ecologically interesting discoveries about glycosyl hydrolases from hyperthermophiles is that *Thermococcus chitinophagus*, an archaeal hyperthermophile from deep-sea hydrothermal vents, can hydrolyze chitin (75). Subsequent analysis of the genome sequence for *Pyrococcus furiosus* revealed the presence of two chitinases (as well as other carbohydrate-hydrolyzing enzymes), even though this organism is not known to degrade chitin in culture (76). Fewer glycosyl hydrolases in general have been found in the *Pyrococcus horikoshii* genome (77), and only one glycosyl hydrolase was identified from the genome of the archaeal hyperthermophile *Archaeglobus fulgidus* (78). The abundance and characteristics of chitinases and other glycosyl hydrolases in hyperthermophilic Archaea clearly vary with species; that variation in turn may be related to their ecological niche. The major sources of chitin for hyperthermophiles in submarine hydrothermal vent environments are the tubes constructed by vestimentiferan tubeworms, which are known to become buried in hot sulfide structures during their formation, and by polychaetes that directly inhabit sulfide structures (1). The combination of *Thermococcus chitinophagous* or *Pyrococcus furiosus* as the organism, chitin as the target POM (Fig. 1), chitinases (or other glycosyl hydrolases) as the foraging tools, and sulfide structures as the environmental setting may provide an ideal start for an enzyme foraging analysis, as discussed earlier, at high temperatures. If the ecological niche for archaeal hyperthermophiles centers on chitin, then the otherwise curious absence of any known cellulose, lignin, or agar digesting enzymes from vent archaeal hyperthermophiles (70) may be explained, although future genome sequences may yet show the presence of unique glycosyl hydrolases not transcribed in conventional culturing media (see Sec. VII).

In contrast to hyperthermophilic Archaea, the hyperthermophilic Bacteria *Thermotoga maritima, T. litoralis*, and *T. neapolitana* are able to flourish on β-linked polysaccharides such as cellulose, xylan, and mannan, as evidenced by their production of thermostable cellulases, amylases, mannanases, galactosidases, and xylanases in culture (79–86) (see examples in Table 1). When cultured on galactomannan, for example, *Thermotoga neopolitana* produces an α-galactosidase active at 95°C–100°C, making it the most thermophilic α-galactosidase reported so far (87). In 1998 two cellulases were purified from *T. neapolitana* (85); they share only 60% amino acid sequence similarity and have different optima for pH and temperature (Table 1). These three *Thermotoga* species are also distinguished from the hyperthermophilic Archaea by their lower thermal optima for growth (Table 1) and their marine hydrothermal habitats, which are shallow and promixal to land. Common terrestrial sources for the β-linked polysaccharides may thus figure in the ecological niches of these hyperthermophiles, though no environmental chemical measurements to confirm or refute this idea are available.

The interest level in thermostable deoxyribonucleic acid (DNA) polymerases from hyperthermophiles, in addition to proteases and glycosyl hydrolases, is considerable. Whereas much of the incentive for purifying DNA polymerases from hyperthermophiles has been for biotechnological applications, researchers are also motivated to better understand DNA replication and repair at high temperatures and factors involved in stabilizing informational macromolecules at high temperature (88–91). At the present time a total of about 30 DNA polymerases have been characterized from archaeal and bacterial hyperthermophiles (88). The optimal temperatures of activity for the approximate 15 polymerases from hyperthermophilic Archaea fall between 70°C and 80°C, similar to the range for polymerases from thermophilic bacteria (88). For most of these hyperthermophilic species of Archaea, the temperature for optimal polymerase activity is about 20°C below that for optimal growth of the organism, whereas for the bacterial species examined, the enzyme optima are identical to or slightly higher than the growth optima (88). In spite of the discrepancy in thermal optima between archaeal species and their polymerases, the DNA polymerase from *P. furiosus* remains very stable at high temperatures, showing a half-life of 20 h at 95°C (92). The match between enzyme and growth optima may not be as important as continued stability of the enzyme at increasingly severe temperatures, pointing to the imperative of balance between activity and stability, as discussed earlier.

Alternatively, environmental factors determining optimal enzyme activity may not be fully appreciated by those studying hyperthermophilic enzymes. Hydrostatic pressures equivalent to the ocean depth of isolation of vent hyperthermophiles (where 22–45 Mpa is equivalent to the range of vent depths, 2200–4500 m) were observed experimentally to stabilize DNA polymerases from *Pyrococcus* strain ES4, *P. furiosus*, and *Thermus aquaticus* at denaturing temperatures by two to five times over those of 3-Mpa controls (29). Moderate pressures have also been shown to stabilize other enzymes from hyperthermophiles, though not equivalent enzymes from mesophiles (34). Studies combining the effects of hydrostatic pressure and high temperature are too few to draw general conclusions. The available data, however, make clear that considering only the factor of temperature, without accounting for the often ameliorating effects of elevated pressure, salinity, or other solute concentrations that uniquely characterize extreme thermal habitats in the ocean, can lead to inaccurate assessments of the evolution or relatedness of the relevant organisms and their enzymes. The effects of hydrostatic pressure on the activity or stability of extracellular hydrolases, so central to any microbial foraging scenario, are entirely unknown for hyperthermophiles. An increase in thermostability or an extension of the maximal temperature for activity would have important implications for the organisms that live within thermal gradients and close promixity to temperatures higher than their apparent (at atmospheric pressure) maximum for growth and survival (Fig. 2). Much work remains to be done.

VI. STATUS OF THE SEARCH FOR PSYCHROPHILIC MICROORGANISMS AND ENZYMES

A. Focus on Organisms and Enzymes in Their Native Habitats

In contrast to the study of hyperthermophiles and their enzymes, in which commercial interests are often at the forefront of the field, the study of cold-adapted organisms and enzymes in polar marine environments was first motivated by ecological interests. As a result, early notions of Arctic and Antarctic environments as biological deserts due to extreme conditions (never an issue for hydrothermal vents, where a wide variety of life forms obviously flourish) have given way to the opposite view. The characteristic subzero temperatures of polar marine habitats clearly do not preclude high rates of microbial production, metabolism, geochemical cycling, or even viral interactions (38–40,93–101). The involvement of cold-adapted enzymes in all of these processes is axiomatic and has often been measured in situ (discussed later), but the search for specific, cold-adapted intracellular or extracellular enzymes is still in an early stage. We have targeted sea ice, especially wintertime sea ice, in this chapter, believing that it holds great promise for discovery of novel psychrophilic enzymes and the organisms that produce them.

Microbial studies of wintertime sea ice are rare in the literature (18) and almost exclusive to Antarctic locales (see 36, 37 for Arctic work). Those available point to surprising heterotrophic bacterial production within the ice even during the coldest of winter months. Evidence suggests that the older the ice, the more selective it becomes for true or extreme psychrophiles (18). Although the phylogenetic diversity of wintertime sea-ice microorganisms is unknown, some unidentified isolates have been tested successfully for heterotrophic growth to temperatures as low as −5°C (18).

A similar selection process for heterotrophic psychrophiles appears to occur in polar sediments (102), where age of burial in the permanently cold (near-freezing) seabed would provide the selective force. Research on both sea ice and sediments, where psychrophiles dominate and organic materials concentrate, reveals substantial extracellular enzymatic activity and cold adaptation at the characteristic near-freezing or subzero in situ temperatures (18–20,51,103). Aggregates of organic material that sink through a subzero water column and connect sea ice to the seabed also provide at least temporary organic-rich sites for the selective flourishing of psychrophiles (7, A Huston and J Deming, unpublished data) and concentrated activities of their hydrolytic enzymes (19,20,51). Only a limited number of types of enzymes—primarily proteases, selected carbohydrases, and phosphatases—have been assessed in any of these extremely cold environments and then primarily to improve understanding of microbial growth and survival strategies and biogeochemical cycling. Again, directed search efforts for a wider range of cold-adapted enzymes of either academic or biotechnological value should be rewarding.

Unlike in the understudied regime of wintertime sea ice, substantial progress has been made, especially in the last decade, toward understanding thermal classes and phylogenetic diversity of microorganisms in seasonal sea ice, especially summertime ice that does not usually experience temperatures below −2°C. In spite of the more limited age and thermal severity of seasonal sea ice, organic-rich zones within it (e.g., harboring diatom blooms) nevertheless select for heterotrophic bacteria that are psychrophilic (104). Although microbial diversity across a variety of polar ecosystems can be described as broad, with representation from both the archaeal and bacterial domains (105), diversity appears to decrease when only cultured isolates from polar marine environments are considered (Fig. 5). Strains selected for this analysis do not exhaust the known genera (exam-

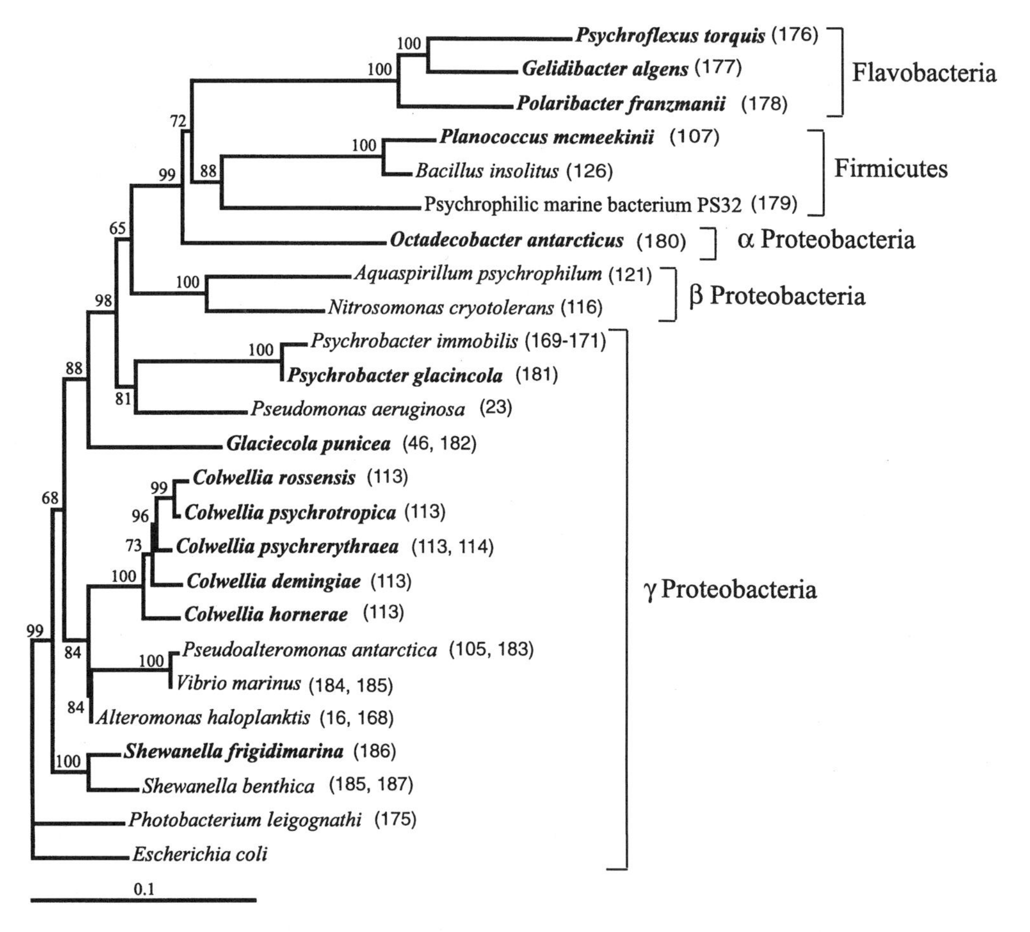

Figure 5 Phylogenetic relationships among 16S rRNA sequences retrieved from the GenBank database for representative psychrophiles from perennially cold marine environments—Arctic, Antarctic, and deep-sea locales; organisms derived from sea ice are indicated in bold; relevant citations, parenthetically. The neighbor-joining method was used to construct a 50% majority tree; percentages of 1000 bootstrap resamplings that supported the branching orders are shown near the relevant nodes. Scale bar corresponds to a 10% (or approximate 4-bp) difference in nucleotide sequence.

ples of missing genera include *Acinetobacter* (106), *Arthrobacter* (107,108), *Brachybacterium* (107), *Polaromonas* (109), and *Pseudomonas* [110,111]), but an attempt has been made to represent all of the larger groupings. Prokaryotes available in culture from sea ice are more limited in diversity, to heterotrophic bacteria from four phylogenetic groups—the α and γ subdivisions of the *Proteobacteria*, the Gram-positive branch, and the *Cytophaga–Flexibacter–Bacteroides* phylum (104,112) (Fig. 5). One cosmopolitan genus, *Colwellia* in the γ-Proteobacteria group (Fig. 5), with species isolated from Antarctic sea ice (113), Arctic sea ice (112), polar marine sediments (20), and the cold deep sea (114), appears to be unique in the microbial world in that all known members of the genus are psychrophilic. The genome of one of its members, *Colwellia* sp. strain 34H, already discussed earlier for its enzyme features, is being sequenced, so that the research community will soon have access to a complete set of genes from which to deduce much about the evolution and performance of enzymes at cold temperatures.

Since an unusual percentage of the total number of bacteria in sea ice (up to 87%) is often culturable on heterotrophic media (18,112), heterotrophy may be the rule in this cold environment. On the other hand, reported searches in sea ice have been limited to culturable heterotrophs. Often more than 99% of the total number of bacteria fail to grow on such media, suggesting that novel cold-adapted microorganisms, as well as enzymes, await discovery in seasonal sea ice more accessible to study than wintertime ice. The presence of vast numbers of uncultured archaeal cells in subzero Antarctic waters (100) and of ammonia-oxidizing bacteria of the β-*Proteobacteria* division in Arctic Ocean waters (115), the source fluids for sea-ice formation, supports this prediction. A cultured ammonia-oxidizing strain from Arctic waters has been shown to grow to temperatures as low as −5°C (116), suggesting adaptation to wintertime sea-ice conditions.

Although sea ice has been the focus of rewarding research on cold-adapted microorganisms and their enzymes in recent years, and wintertime sea ice with its extreme conditions holds new promise, polar marine sediments may yield maximum diversity of both psychrophilic microorganisms and enzymes. This prediction arises easily from what is already known about the diversity of microorganisms and enzymes in the characteristic zones and gradients of buried marine sediments across temperate latitudes (28,117), but also from direct experimentation in Arctic sediments using molecular techniques independent of culturing goals; i.e., development of 16S ribosomal DNA clone libraries (118). Whereas application of this approach to other extreme environments, such as a saltern (119) or an acidic microbial mat (60), has indicated that the screening of fewer than 50 bacterial clones was sufficient to detect the majority of taxa, even screening as many as 353 clones from Arctic sediments only partially covered the actual diversity present (118). To the extent that the relevant studies can be compared, polar marine sediments (118) and hot deep-sea sulfide structures (31,32,61) appear to support similarly high levels of microbial diversity. Although work with enzymes derived from a polar sedimentary source remains rare (20,120,121), the presence of high concentrations of a diversity of cold-adapted hydrolytic enzymes is clear (28,47,51,103,122). Working against confidence in sediments as the ultimate habitat for psychrophilic diversity is the evidence that cold-adapted deep-sea invertebrates may harbor considerable diversity in their guts (28); comparable molecular studies of polar invertebrate guts have not been undertaken. Nor has sea ice been probed by direct molecular analysis independent of culturing. Since sea ice in the Arctic entrains bacteria-laden sediments (as well as small invertebrates) and exposes them to colder temperatures than ever experienced at the seafloor, eventually releasing them for descent to the seabed, sea ice remains at the forefront for new searches.

B. Focus on Applications in Bioremediation and Biotechnology

An applied incentive for expanding the search for psychrophiles beyond the heterotrophs already available in culture is the finding that at low temperatures some autotrophs, especially ammonia oxidizers but also methanotrophs, can effectively degrade toxic organic contaminants such as trichloroethylene (123). In some low-temperature situations nonheterotrophs are better candidates for enhanced degradation through biostimulation, because providing gases (ammonia, methane) can be more feasible, affordable, and environmentally safe than introducing fertilizers or organic material to stimulate their heterotrophic counterparts. Polar marine ammonia oxidizers, present in significant numbers in the Arctic Ocean (115), await study from this perspective.

Even widely available psychrophilic heterotrophs, however, are understudied from the perspective of bioremediation, especially for the marine environment. The promise of new and useful by-products and processes appears to be great (105). For example, a psychrophilic strain of *Arthrobacter protophormiae* was shown to produce a biosurfactant when grown on *n*-hexadecane at 10°C. The surfactant was active over a wide range of temperatures (30°C–100°C) and pH (2–12) and effective in recovering 90% of residual oil from an oil-saturated sand column in the laboratory, attesting to its potential value in enhanced oil recovery (108). Hydrocarbon degradation by indigenous Antarctic bacteria has also been demonstrated, along with a possibly novel pathway for phenanthrene degradation, encouraging the concept of in situ bioremediation at very low temperatures (124). Even in sea ice, resident bacteria confronted with diesel fuel contamination rise to the challenge with greatly increased population size and hydrocarbon-utilizing activity (125).

As with hyperthermophilic research, most work to date on cold-adapted enzymes has centered on hydrolytic enzymes (proteases, amylases) or sometimes enzymes central to metabolism (malate dehydrogenase, citrate synthetase). Potentially novel enzymes of biotechnological or molecular research interest have been targeted more recently. For example, an unidentified Antarctic bacterium has yielded a novel restriction endonuclease with a low-temperature optimum of 15°C–20°C (126), and the psychrophile *Bacillus insolitus* has yielded a promoter with remarkably high transcription activity compared to that of mesophilic counterparts (127). In general, though, progress on enzymes that process informational macromolecules at low temperatures pales in comparison with work at high temperatures.

An alternative to searching for thermally unique enzymes in nature, either directly or from a cultured isolate, where they have evolved as a result of selective environmental forces, is to attempt to engineer an enzyme with the desired characteristics by genetic manipulation. Recent reports at both ends of the temperature spectrum suggest the promise of the approach, not only for engineering desired thermal characteristics but also for learning more about the structural basis for thermal adaptation at the fundamental molecular level. For example (128), use of an in vitro random mutagenesis and screening system has yielded a cold-adapted subtilisin with proteolytic activity 100% higher than that of the wild type at 10°C, an increase created by triple mutations, each at a significantly different location on the molecule (in a conserved region, in an unconserved region, and near the substrate binding area). In another directed evolution study (129), both the thermostability (at high temperatures) and the activity (at low temperatures) of the cold-adapted protease subtilisin S41 from the Antarctic *Bacillus* strain TA41 (130) were improved substantially and simultaneously, i.e., in the same mutant strain. Similar approaches using random mutagenesis have also been successful in improving thermal stability and pH tolerance or modifying catalytic activity and substrate specificity of enzymes performing at the high end of the temperature spectrum (131–134).

VII. CONCLUSIONS

Whether initially motivated by applied interests, ecological issues, or a desire to understand the fundamental basis for thermal adaptations, researchers exploring thermally extreme habitats in the ocean for novel microorganisms and enzymes have relied primarily upon the conventional approach of purifying enzymes from the biomass of culturable microorganisms. Since a useful number of both hyperthermophiles and psychrophiles have been amenable to culture and to high growth yields, the approach has yielded significant new knowledge at many levels of interest. The approach of improving features of various enzymes from known organisms via genetic engineering or directed evolution has also resulted in significant progress. More insight on various aspects of hyperthermophily and psychrophily is sure to come from a continued analysis of native and engineered traits of enzymes from strains in existing culture collections and from organisms yet to be isolated. Successful strategies for culturing new organisms will benefit from enzyme-based ecological analyses of the targeted habitats, especially the hot subsurface biosphere and wintertime sea ice, which have been sampled from a microbial perspective only rarely.

Adding a new dimension to the search and characterization of novel enzymes from cultured microorganisms is the accumulation of genome sequences. To date, sequencing of more than 50 bacterial and archaeal genomes has been completed or is in progress. Complete genome sequences exist for the hyperthermophilic Archaea *Methanococcus jannaschii*, *Pyrococcus horikoshii*, and *Archaeoglobus fulgidus*, and for the hyperthermophilic bacterium *Aquifex aeolicus*. Genome sequences from other hyperthermophiles are almost complete, including those of the heterotrophic Archaea *Pyrococcus furiosus*, *Pyrodictium abyssi*, and *Pyrobaculum aerophilum* and the heterotrophic bacterium *Thermotoga maritima*. Although no genome sequence data are yet available for psychrophiles, we are aware of at least one strictly psychrophilic bacterium in the pipeline, as mentioned earlier.

Genome sequence data can be useful in the discovery of a new enzyme provided that some sequence homology exists between the targeted sequenced enzyme and the open reading frames that encode for the same enzyme. This method for obtaining new enzymes is also dependent on expression of the genes in a suitable host that can reproduce a functional enzyme. Success at cloning and expression has so far been a relative measure, given the recurring evidence that the expression of genes from a hyperthermophile in a mesophilic host, such as *Escherichia coli*, yields recombinant proteins that differ from the native proteins (135–137). The differences have been ascribed to incorrect folding or assembly of the protein (138). For example, the half-life of native glutamate dehydrogenase from *Pyrococcus furiosus* is 10 h at 100°C, whereas the half-life of the recombinant protein measures only in minutes at the same temperature (135). In order to take full advantage of the potential for discovery of novel enzymes from genome sequences of either hyperthermophiles or psychrophiles, more efficient expression systems must be designed and verified. Suitable new host expression systems could derive from mesophiles, modified in some way to assemble enzymes correctly from hyperthermophiles or psychrophiles, or could involve the direct use of hyperthermophiles and psychrophiles. In the case of cold-adapted enzymes, the promise of successful overexpression of the gene coding for the target protein in a mesophilic host has been demonstrated. The heat-labile α-amylase from the Antarctic psychrophile *Alteromonas haloplanktis* was expressed in *Escherichia coli* at a compromise temperature of 18°C, which accommodated a reasonable growth rate for the vector and stability for the enzyme (139). In another study (140), a cold-adapted malate dehydrogenase from a deep-sea psychrophilic *Vibrio* sp. strain was also faithfully expressed in recombinant *E. coli* cells. However, expression of strictly psychrophilic enzymes likely requires development of an alternative host expression system.

One of the conundrums arising from the available genome sequences for microorganisms is that up to 50% of the open reading frames encode for unknown proteins. Moreover, some enzymes involved in metabolic pathways known to be present in the organism are absent in the genome, suggesting either that the pathways are incomplete or that enzymes with unknown sequences or low sequence homology are involved (141). Also detected are genes that encode for enzymes involved in pathways not observed to function in the organisms, such as the ribulose-bisphosphate carboxylase in *Methanococcus jannaschii* (56). Clearly, the challenges of functional genomics in the coming years will be to determine the catalytic or structural function of all of the genes for a given organism and to understand better the factors that control their expression. For hyperthermophiles and psychrophiles, the importance of environmental factors other than temperature—elevated hydrostatic pressures and salt or other solute concentrations—cannot be overestimated.

By far the highest diversity of enzymes lies hidden in the greater than 99% of uncultured microorganisms typically detected in environmental samples using small-subunit rRNA gene sequences. More than 40 phylogenetic divisions of Bacteria based on rRNA sequences have been estimated (142). (Fig. 4 does not attempt to represent all of these). For many of these divisions, few representative organisms have been isolated, yet many are likely to contain psychrophilic or psychrotolerant bacteria, since they include rRNA sequences from cold marine environments. Phylogenetic surprises have become a regular feature of environmental DNA analyses; they can be expected to reflect similar surprises in physiological and enzymological diversity. For example, archaeal sequences detected in great abundance in Antarctic waters and in association with a cold marine sponge (143,144) were detected in 1999 in fluids from hot subsurface marine environments (145). None of these marine Crenarchaeota (Fig. 4) is closely related to any cultured Archaea. Moreover, members of the archaeal kingdom Korarchaeota, also based entirely on environmental sequences (no known cultured members), have only been detected in hot environments (58). New molecular approaches are being developed to tap into this almost unlimited diversity of uncultured microorganisms specifically for novel enzymes and other bioactive compounds. The methods involve expression cloning of DNA from environmental samples and high throughput robotic screening (146).

This new generation of genome-based investigative tools clearly produce a heightened sense of excitement in the search and discovery of enzymes—from cloning and expression of targeted genes as an integrated component of functional genomics, to the direct measurement of amino acid sequences and their alignment to genome sequences using proteomics, to the construction of genomic expression libraries from environmental DNA. Combined with more traditional approaches, genomics and molecular ecology have expanded our horizons with new models and methods and the knowledge that a plethora of uncultured microorganisms and their enzymes from previously inaccessible or unexplored extreme environments await our perusal. The future for search and discovery of uniquely hyperthermophilic or psychrophilic enzymes is clearly very bright.

ACKNOWLEDGMENTS

Preparation of this chapter was supported by NOAA WA State Sea Grant and NSF awards to J. Deming to study psychrophiles and to J. Baross to study hyperthermophiles. We thank our students for their dedication to research and discovery in these areas, and especially Adrienne Huston, Julie Huber, and Shelly Carpenter for discussions and help in preparing the figures.

REFERENCES

1. JA Baross, JW Deming. Growth at high temperatures: Isolation and taxonomy, physiology, and ecology. In: DM Karl, ed. The Microbiology of Deep-Sea Hydrothermal Vents. Boca Raton, FL: CRC Press, 1995, pp 169–217.
2. JW Deming, JA Baross. Deep-sea smokers: Windows to a subsurface biosphere? Cosmochem Geochem Acta 57:3219–3230, 1993a.
3. JR Delaney, DS Kelley, MD Lilley, DA Butterfield, JA Baross, WSD Wilcock, RW Embley, M Summit. 1998. The quantum event of oceanic crustal accretion: impacts of diking at mid-ocean ridges. Science 281:222–230, 1998.
4. JF Holden, M Summit, JA Baross. Thermophilic and hyperthermophilic microorganisms in 3–30°C hydrothermal fluids following a deep-sea volcanic eruption. FEMS Microbiol Ecol 25:33–41, 1998.
5. M Summit, JA Baross. Thermophilic subseafloor microorganisms from the 1996 North Gorda Ridge eruption. Deep Sea Res II 45:2751–2766, 1998.
6. WF Weeks, SF Ackley. The growth, structure and properties of sea ice. In: N Untersteiner, ed. The Geophysics of Sea Ice (NATO ASI B146). Dordrecht: Martinus Nijhoff, 1986, pp 9–164.
7. JW Deming. Psychrophily in the deep sea. In: R Guerrero and C Pedros-Alio, eds. Proceedings of Sixth International Symposium on Microbial Ecology, Barcelona, 1993, pp 33–36.
8. GA Maykut. The surface heat and mass balance. In: N Untersteiner, ed. The Geophysics of Sea Ice (NATO ASI B146). Dordrecht: Martinus Nijhoff, 1986, pp 395–463.
9. JW Deming, AL Huston. An oceanographic perspective on microbial life at low temperatures with implications for polar ecology, biotechnology and astrobiology, In: J Seckbach, ed. Journey to Diverse Microbiol Worlds. Dordrecht, the Netherlands: Kluwer, 2000, pp 149–160.
10. Y-A Vetter, JW Deming, PA Jumars, BB Krieger-Brockett. A predictive model of bacterial foraging by means of freely-released extracellular enzymes. Microb Ecol 36:75–92, 1998.
11. MWW Adams, RM Kelly. Enzymes from microorganisms in extreme environments. Chemical Engineering News, Dec 18, 1995, pp 32–42.
12. JE Brenchley. Psychrophilic microorganisms and their cold-active enzymes. J Ind Microbiol 17:432–437, 1996.
13. RY Morita. Psychrophilic bacteria. Bacteriol Rev 39:144–167, 1975.
14. JA Baross, RY Morita. Microbial life at low temperatures: Ecological aspects. In: DJ Kushner, ed. Microbial Life in Extreme Environments. New York: Academic Press, 1978, pp 9–71.
15. JT Staley, JJ Gosink. Poles apart: Biodiversity and biogeography of sea-ice bacteria. Annu Rev Microbiol 53:189–215, 1999.
16. G Feller, E Narinx, JL Arpigny, M Aittaleb, E Baise, S Genicot, C Gerday. Enzymes from psychrophilic organisms. FEMS Microbiol Rev 18:189–202, 1996.
17. JW Deming. Ecological strategies of barophilic bacteria in the deep ocean. Microbiol Sci 3: 205–212, 1986.

18. E Helmke, H Weyland. Bacteria in sea ice and underlying water of the eastern Weddell Sea in midwinter. Mar Ecol Prog Ser 117:269–287, 1995.
19. AL Huston, JW Deming. Low-temperature activity optima for extracellular enzymes released by Arctic marine bacteria. Proceedings ASM Annual Meeting, Chicago, May 1999.
20. AL Huston, BB Krieger-Brockett, JW Deming. Remarkably low temperature optima for extracellular enzyme activity from Arctic bacteria and sea ice. Environ Microbiol 2:383–388, 2000.
21. JA Baross, JF Holden. Overview of hyperthermophiles and their heat-shock proteins. Adv Protein Chem 48:1–34, 1996.
22. JF Holden, MWW Adams, JA Baross. Heat-shock response in hyperthermophilic microorganisms. In: CR Bell, M Berlinsky, P Johnson-Green, eds. Stress Genes: Role in Physiological Ecology, Progress in Microbial Ecology. Proceedings of the 8th International Symposium on Microbial Ecology, Halifax, Canada, 2000, pp 663–670.
23. C Gerday, M Aittaleb, JL Arpigny, E Baise, J-P Chessa, G Garsoux, I Petrescu, G Feller. Psychrophilic enzymes: A thermodynamic challenge. Biochim Biophys Acta 1342:119–131, 1997.
24. MWW Adams, A Kletzin. Oxidoreductase-type enzymes and redox proteins involved in fermentative metabolisms of hyperthermophilic Archaea. Adv Protein Chem 48:101–180, 1996.
25. G Feller, C Gerday. Psychrophilic enzymes: Molecular basis of cold adaptation. Cell Mol Life Sci 53:830–841, 1997.
26. HC Helgeson. Thermodynamic prediction of the relative stabilities of hyperthermophilic enzymes. Chem Thermodynamics 301–312, 1999.
27. R Jaenicke, G Böhm. The stability of proteins in extreme environments. Curr Opinion Structural Biol 8:738–748, 1998.
28. JW Deming, JA Baross. Survival, dormancy and non-culturable cells in deep-sea environments. In: RR Colwell, DJ Grimes, eds. Non-Culturable Microorganisms in the Environment. Washington, DC: ASM Press, 2000, pp 147–197.
29. M Summit, B Scott, K Nielson, E Mathur, J Baross. Pressure enhances thermal stability of DNA polymerase from three thermophilic organisms. Extremophiles 2:339–345, 1998.
30. MK Tivey, RE McDuff. Mineral precipitation in the walls of black smoker chimneys: A quantitative model of transport and chemical reaction. J Geophys Res 95:12617, 1990.
31. HJM Harmsen, D Prieur, C Jeanthon. Distribution of microorganisms in deep-sea hydrothermal vent chimneys investigated by whole-cell hybridization and enrichment culture of thermophilic subpopulations. Appl Environ Microbiol 63:2876–2883, 1997.
32. MO Schrenk, DS Kelley, JA Baross. Attachment of hyperthermophilic microorganisms to mineral substrata: In-situ observations and subseafloor analogs. Proc Geol Soc Am 31:7, 1999.
33. M Summit. Ecology, physiology and phylogeny of thermophiles from mid-ocean ridge subseafloor environments. PhD dissertation, University of Washington, Seattle, WA, 2000.
34. PC Michels, D Hei, DS Clark. Pressure effects on enzyme activity and stability at high temperatures. Adv Prot Chem 48:341–376, 1996.
35. H Eicken, E Reimnitz, V Alexandrov, T Martin, H Kassens, T Viehoff. 1997. Sea-ice processes in the Laptev Sea and their importance for sediment transport. Cont Shelf Res 17: 205–233, 1997.
36. K Junge, C Krembs, J Deming, A Stierle, H Eicken. A microscopic approach to investigate bacteria under in-situ conditions in sea-ice samples. Ann Glaciol 33, 2001b (in press).
37. C Krembs, K Junge, J Deming, H Eicken. First observations on concentration and potential production and fate of organic polymers in winter sea ice from the Chukchi Sea. Proceedings of International Glaciology Society Symposium on Sea Ice and its Interactions with the Ocean, Atmosphere and Biosphere, Fairbanks, June 2000.
38. ST Kottmeier, CW Sullivan. Bacterial biomass and production in pack ice of Antarctic marginal ice edge zones. Deep Sea Res 37:1311–1330, 1990.

39. GA Palmisano, DL Garrison. Microorganisms in Antarctic sea ice. In: EI Friedmann, ed. Antarctic Microbiology. New York: Wiley-Liss, 1993, pp 167–219.
40. S Grossmann, GS Dieckmann. Bacterial standing stock, activity, and carbon production during formation and growth of sea ice in the Weddell Sea, Antarctica. Appl Environ Microbiol 60:2746–2753, 1994.
41. DN Thomas, RJ Lara, H Eicken, G Kattner, A Skoog. Dissolved organic matter in Arctic multi-year sea ice during winter: Major components and relationship to ice characteristics. Polar Biol 15:477–483, 1995.
42. DN Thomas, R Engbrodt, V Ginnelli, G Kattner, H Kennedy, C Hass, GS Dieckmann. Dissolved organic matter in Antarctic sea ice. Ann Glaciol 33, 2001 (in press).
43. R Gradinger, M Spindler, J Weissenberger. On the structure and development of Arctic pack ice communities in Fram Strait: A multivariate approach. Polar Biol 12:727–733.
44. C Krembs, R Gradinger, M Spindler. Implications of brine channel geometry and surface area for the interaction of sympagic organisms in Arctic sea ice. J Exp Mar Ecol 243:55–80, 2000a.
45. H Eicken, C Bock, R Wittig, H Miller, H-O Poertner. Nuclear magnetic resonance imaging of sea-ice pore fluids: Methods and thermal evolution of pore microstructure. Ann Glaciol 33, 2001 (in press).
46. DS Nichols, AR Greenhill, CT Shadbolt, T Ross, TA McMeekin. Physicochemical parameters for growth of the sea ice bacteria *Glaciecola punicea* ACAM 611 super(T) and *Gelidibacter* sp. strain ICI 158. Appl Environ Microbiol 65:3757–3760, 1999.
47. W Reichardt. Impact of the Antarctic benthic fauna on the enrichment of biopolymer-degrading psychrophilic bacteria. Microb Ecol 15:311–321, 1988.
48. G Feller, E Narinx, JL Arpigny, Z Zekhnini, J Swings, C Gerday. Temperature dependence of growth, enzyme secretion and activity of psychrophilic antarctic bacteria. Appl Microbiol Biotechnol 41:477–479, 1994.
49. LR Pomeroy, WJ Wiebe, D Deibel, RJ Thompson, GT Rowe, JD Pakulski. Bacterial responses to temperature and substrate concentration during the Newfoundland spring bloom. Mar Ecol Prog Ser 75:143–159, 1991.
50. WJ Wiebe, WM Sheldon Jr, LR Pomeroy. Bacterial growth in the cold: Evidence for an enhanced substrate requirement. Appl Environ Microbiol 58:359–364, 1992.
51. Y-A Vetter, JW Deming. Extracellular enzyme activity in the Arctic Northeast Water Polynya. Mar Ecol Prog Ser 114:23–34, 1994.
52. Y-A Vetter, JW Deming. Bacterial subsistence exclusively on particulate organic matter via the action of extracellular enzymes. Microb Ecol 37:86–94, 1999.
53. JA Baross, MD Lilley, LI Gordon. Is the CH_4, H_2, and CO venting from submarine hydrothermal systems produced by thermophilic bacteria? Nature 298:366–369, 1982.
54. KO Stetter. Ultrathin mycelia-forming organisms from submarine volcanic areas having an optimum growth temperature of 105°C. Nature 300:258–261, 1982.
55. MWW Adams, RM Kelly. Hyperthermophilic enzymes: Finding them and using them. Trends Biotechnol 16:329–332, 1998.
56. CJ Bult, et al. Complete genome sequence of the methanogenic archaeon, *Methanococcus jannaschii*. Science 273:1058–1073, 1996.
57. RM Kelly, MWW Adams. Metabolism in hyperthermophilic microorganisms. Antonie van Leeuvenhook 66:247–270, 1994.
58. SM Barns, RE Fundyga, MW Jeffries, NR Pace. Remarkable archaeal diversity detected in a Yellowstone National Park hot spring environment. Proc Natl Acad Sci USA 91:1609–1613, 1994.
59. MJ Ferris, SC Nold, NP Revsbeck, DM Ward. Population structure and physiological changes within a hot spring microbial mat community following disturbance. Appl Environ Microbiol 63:1367–1374, 1997.
60. CL Moyer, FC Dobbs, DM Karl. Estimation of diversity and community structure through

restriction fragment length polymorphism distribution analysis of bacterial 16S rRNA genes from a microbial mat at an active, hydrothermal vent system, Loihi Seamount, Hawaii. Appl Environ Microbiol 60:871–879, 1994.

61. K Takai, K Horikoshi. Genetic diversity of Archaea in deep-sea hydrothermal vent environments. Genetics 152:1285–1297, 1999.
62. G Fiala, KO Stetter. *Pyrococcus furiosus*, sp. nov., represents a novel genus of marine heterotrophic archaebacteria growing optimally at 100°C. Arch Microbiol 145:56–61, 1986.
63. A Neuner, HW Jannasch, S Belkin, KO Stetter. *Thermococcus litoralis* sp. nov.: A new species of extremely thermophilic marine archaebacteria. Arch Microbiol 153:205–207, 1990.
64. R Huber, TA Langworthy, H Konig, M Thomm, CR Woese, UB Sleytr, KO Stetter. *Thermotoga maritima* sp. nov. represents a new genus of unique extremely thermophilic eubacteria growing up to 90°C. Arch Microbiol 144:324–333, 1986.
65. CR Woese, O Kandler, ML Wheelis. Toward a natural system of organisms: Proposal for the domains Archaea, Bacteria, and Eukarya. Proc Natl Acad Sci USA 87:4576–4579, 1990.
66. MWW Adams. The biochemical diversity of life near and above 100°C in marine environments. J Appl Microbiol 85 (suppl S):108S–117S, 1999.
67. A Kletzin, MWW Adams. Tungsten in biology. FEMS Microbiol Rev 18:5–64, 1996.
68. HW Jannasch, CO Wirsen, SJ Molyneaux, TA Langworthy. Comparative physiological studies on hyperthermophilic archaea isolated from deep sea hot vents with emphasis on *Pyrococcus* strain GB-D. Appl Environ Microbiol 58:3472–3481, 1992.
69. MWW Adams. Enzymes and proteins from organisms that grow near and above 100°C. Annu Rev Microbiol 47:627–658, 1993.
70. MW Bauer, SB Halio, RM Kelly. Protease and glycosyl hydrolases from hyperthermophilic microorganisms. Adv Protein Chem 48:271–310, 1996.
71. MR Maurizi. Proteases and protein degradation in *Escherichia coli*. Experientia 48:178–201, 1992.
72. HR Costantino, SH Brown, RM Kelly. Purification and characterization of an α-glucosidase from a hyperthermophilic archaebacterium, *Pyrococcus furiosus*, exhibiting a temperature optimum of 105 to 115°C. J Bacteriol 172:3654–3660, 1990.
73. SH Brown, RM Kelly. Characterization of amylolytic enzymes, having both α-1,4 and α-1,6-hydrolytic activity, from the thermophilic archaea *Pyrococcus furiosus* and *Thermococcus litoralis*. Appl Environ Microbiol 59:2614–2621, 1993.
74. JW Schuliger, SH Brown, JA Baross, RM Kelly. Purification and characterization of a novel amylolytic enzyme from ES4, a marine hyperthermophilic Archaeum. Mol Mar Biol Biotech 2:76–87, 1993.
75. R Huber, J Stöhr, S Hohenhaus, R Rachel, S Burggraf, HW Jannasch, KO Stetter. *Thermococcus chitinophagus* sp. nov., a novel chitin-degrading, hyperthermophilic archaeum from a deep-sea hydrothermal vent environment. Arch Microbiol 164:255–264, 1995.
76. MW Bauer, LE Driskill, W Callen, MA Snead, EJ Mathur, RM Kelly. An endoglucanase, EglA, from the hyperthermophilic archaeon *Pyrococcus furiosus* hydrolyzes β-1,4 bonds in mixed linkage (1 → 3), (1 → 4)-β-D-glucans and cellulose. J Bacteriol 181:284–290, 1999.
77. Y Kawarabayasi, et al. Complete sequence and gene organization of the genome of a hyperthermophilic archaebacterium, *Pyrococcus horikoshii* OT3. DNA Res 5:147–155, 1998.
78. HP Klink, et al. The complete genome sequence of the hyperthermophilic, sulphate-reducing archaeon *Archaeoglobus fulgidus*. Nature 390:364–370, 1997.
79. J Gabelsberger, W Liebl, K-H Schleifer. Cloning and characterization of β-galactosidase and glucosidase hydrolyzing enzymes of *Thermotoga maritima*. FEMS Microbiol Lett 109:131–138, 1993.
80. K Bronnenmeier, A Kern, W Liebl, WL Studenbauer. Purification of *Thermotoga maritima* enzymes for the degradation of cellulosic materials. Appl Environ Microbiol 61:1399–1407, 1995.

81. W Liebl, R Feil, J Gadelsberger, J Kellermann, K-H Schleifer. Purification and characterization of a novel thermostable 4-α-glucanotransferase of *Thermotoga maritima* cloned in *Escherichia coli*. Eur J Biochem 207:81–88, 1992.
82. G Swaitek, JD Box, DA Yernool. Cloning and sequence analysis of a novel β-glucan-glucohydrolase A gene from *Thermotoga neapolitana*. Proceedings of the 96th General Meeting of the American Society for Microbiology, American Society for Microbiology, Washington, DC, 1996, p 367.
83. A Sunna, M Moracci, G Antranikian. Glycosyl hydrolases from hyperthermophilic microorganisms. Extremophiles 1:2–13, 1997.
84. GD Duffaud, CM McCutchin, P Leduc, KN Parker, RM Kelly. Purification and characterization of extremely thermostable β-mannosidase, and α-galactosidases from the hyperthermophilic eubacterium *Thermotoga neapolitana* 5068. Appl Environ Microbiol 63:169–177, 1997.
85. JD Box, DA Yernool, DE Eveleigh. Purification, characterization and molecular analysis of thermostable cellulases CelA and CelB from *Thermotoga neapolitana*. Appl Environ Microbiol 64:4774–4781, 1998.
86. MR King, DA Yernool, DE Eveleigh, BM Chassy. Thermostable α-galactosidase from *Thermotoga neapolitana*: Cloning, sequencing and expression. FEMS Microbiol Lett 163:37–42, 1998.
87. CM McCutchen, GD Duffaud, P Leduc, ARH Petersen, A Tayal, SA Kahn, RM Kelly. Characterization of extremely thermostable enzymatic breakers (α-1,6-galactosidases and β-1,4-mannanase) from the hyperthermophilic bacterium *Thermotoga neapolitana* 5068 for hydrolysis of guar gum. Biotechnol Bioeng 52:332–339, 1996.
88. FB Perler, S Kumar, H Kong. Thermostable DNA polymerases. Adv Protein Chem, 48:377–435, 1996.
89. C Hethke, A Bergerat, W Hausner, P Forterre, M Thomm. Cell-free transcription at 95 degrees: Thermostability of transcriptional components and DNA topology requirements of *Pyrococcus* transcription. Genetics 152:1325–1333, 1999.
90. P Forterre, CB de-la-Tour, H Philippe, M Duguet. Reverse gyrase from hyperthermophiles: probable transfer of thermoadaptation train from Archaea to Bacteria. Trends Genetics 16: 152–154, 2000.
91. H Myllykallio, P Lopez, P Lopez-Garcia, R Heilig, W Saurin, Y Zivanovic, H Philippe, P Forterre. Bacterial mode of replication with eukaryotic-like machinery in a hyperthermophilic archaeon. Science 288:2212–2215, 2000.
92. EJ Mathur, MWW Adams, WN Callen, JM Cline. The DNA polymerase gene from the hyperthermophilic archaebacterium, *Pyrococcus furiosus*, shows sequence homology with α-like DNA polymerases. Nucleic Acid Res 19:6952–6953, 1991.
93. RP Griffith, SS Hayasaka, TM McNamara, RY Morita. Relative microbial activity and bacterial concentrations in water and sediment samples taken in the Beaufort Sea. Can J Microbiol 24:1217–1226, 1978.
94. RE Smith, P Clement, GF Cota. Population dynamics of bacteria in Arctic sea ice. Microb Ecol 17:63–76, 1989.
95. R Maranger, DF Bird, SK Juniper. Viral and bacterial dynamics in Arctic sea ice during the spring algal bloom near Resolute, NWT, Canada. Mar Ecol Prog Ser 111:121–127.
96. GF Cota, LR Pomeroy, WG Harrison, EP Jones, F Peters, WM Sheldon Jr, TR Weingartner. Nutrients, primary production and microbial heterotrophy in the southeastern Chukchi Sea: Arctic summer nutrient depletion and heterotrophy. Mar Ecol Prog Ser 135:247–258, 1996.
97. GF Steward, DC Smith, F Azam. Abundance and production of bacteria and viruses in the Bering and Chukchi Seas. Mar Ecol Prog Ser 131:287–300, 1996.
98. PA Wheeler, M Gosselin, E Sherr, D Thibault, DL Kirchman, R Benner, TE Whitledge. Active cycling of organic carbon in the central Arctic Ocean. Nature 380:697–699, 1996.
99. J Rich, M Gosselin, E Sherr, B Sherr, DL Kirchman. High bacterial production, uptake and

concentrations of dissolved organic matter in the Central Arctic Ocean. Deep Sea Res II 44: 1645–1663, 1997.

100. AE Murray, CM Preston, R Massana, LT Taylor, A Blakis, K Wu, EF Delong. Seasonal and spatial variability of bacterial and archaeal assemblages in the coastal waters near Anvers Island, Antarctica. Appl Environ Microbiol 64:2585–2595.
101. K Sahm, C Knoblauch, R Amann. Phylogenetic affiliation and quantification of psychrophilic sulfate-reducing isolates in marine Arctic sediments. Appl Environ Microbiol 53:2332–2337, 1999.
102. E Helmke, H Weyland. Effect of temperature on extracellular enzymes occurring in permanently cold marine environments. Kieler Meeresforsch Sonderh 8:198–204, 1991.
103. C Arnosti. Rapid potential rates of extracellular enzymatic hydrolysis in Arctic sediments. Limnol Oceanogr 43:315–324, 1998.
104. JP Bowman, SA McCammon, MV Brown, DS Nichols, TA McMeekin. Diversity and association of psychrophilic bacteria in Antarctic sea ice. Appl Environ Microbiol 63:3068–3078, 1997b.
105. D Nichols, J Bowman, K Sanderson, CM Nichols, T Lewis, T McMeekin, PD Nichols. Developments with Antarctic microorganisms: Culture collections, bioactivity screening, taxonomy, PUFA production and cold-adapted enzymes. Curr Opin Biotechnol 10:240–246, 1999.
106. HM Alvarez, OH Pucci, A Steinbuechel. Lipid storage compounds in marine bacteria. Appl Microbiol Biotechnol 47:132–139, 1997.
107. K Junge, JJ Gosink, H-G Hoppe, JT Staley. *Arthrobacter, Brachybacterium* and *Planococcus* isolates identified from Antarctic sea ice brine and description of *Planococcus mcmeekinii*, sp. nov. Syst Appl Microbiol 21:306–314, 1998.
108. V Pruthi, SS Cameotra. Production and properties of a biosurfactant synthesized by *Arthrobacter protophormiae*: An Antarctic strain. World J Microbiol Biotechnol 13:137–139, 1997.
109. RL Irgens, JJ Gosink, JT Staley. *Polaromonas vacuolata* gen. nov., sp. nov., a psychrophilic, marine, gas vacuolate bacterium from Antarctica. Int J Syst Bacteriol 46:822–826, 1996.
110. JT Staley, WL Boyd. L-Serine dehydratase (deaminase) of psychrophiles and mesophiles from polar and temperate habitats. Can J Microbiol 13:1333–1342, 1966.
111. T Hoshino, K Ishizaki, T Sakamoto, H Kumeta, I Yumoto, H Matsuyama, S Ohgiya. Isolation of a *Pseudomonas* species from fish intestine that produces a protease active at low temperature. Lett Appl Microbiol 25:70–72, 1997.
112. K Junge, JF Imhoff, JT Staley, JW Deming. Phylogenetic diversity of numerically important bacteria cultured at subzero temperature from Arctic sea ice. Microb Ecol 2001 (submitted).
113. JP Bowman, JJ Gosink, SA McCammon, TE Lewis, DS Nichols, PD Nichols, JH Skerratt, JT Staley, TA McMeekin. *Colwellia demingiae* sp. nov., *Colwellia hornerae* sp. nov., *Colwellia rossensis* sp. nov. and *Colwellia psychrotropica* sp. nov.: Psychrophilic Antarctic species with the ability to synthesize docosahexaenoic acid (22:6w3). Int J Syst Bacteriol 48:1171–1180, 1998.
114. JW Deming, LK Somers, WL Straube, DG Swartz, and MT MacDonell. Isolation of an obligately barophilic bacterium and description of a new genus *Colwellia* gen. nov. Syst Appl Microbiol 10:152–160, 1988.
115. N Bano, JT Hollibaugh. Diversity and distribution of DNA sequences with affinity to ammonia-oxizing bacteria of the β subdivision of the class Proteobacteria in the Arctic Ocean. Appl Environ Microbiol 66:1960–1969, 2000.
116. RD Jones, RY Morita. Low-temperature growth and whole-cell kinetics of a marine ammonium-oxidizing bacterium, *Nitrosomonas cryotolerans* sp. Can J Microbiol 34:1122–1128, 1985.
117. JW Deming, JA Baross. The early diagenesis of organic matter: Microbial activity. In: MH Engel, SA Macko, eds. Organic Geochemistry. Vol 6. Topics in Geobiology. New York: Plenum Press, 1993, pp 119–144.

118. K Ravenschlag, K Sahm, J Pernthaler, R Amann. High bacterial diversity in permanently cold marine sediments. Appl Environ Microbiol 65:3982–3989, 1999.
119. AJ Martinez-Murcia, SG Acinas, F Rodriguez-Valera. Evaluation of prokaryotic diversity by restriction digestion of 16S rDNA directly amplified from hypersaline environments. FEMS Microb Ecol 17:247–256, 1995.
120. T Hamamoto, M Kaneda, T Kudo, K Horikoshi. Characterization of a protease from psychrophilic *Vibrio* sp. strain 5709. J Mar Biotechnol 2:219–222, 1995.
121. S-Y Kim, KY Hwang, S-H Kim, H-C Sung, YS Han, Y Cho. Structural basis for cold adaptation. J Biol Chem 274:11761–11767, 1999.
122. A Boetius, K Lochte. High proteolytic activities of deep-sea bacteria from oligotrophic polar sediments. Aquat Microb Ecol 48:269–276, 1996.
123. BN Moran, WJ Hickey. Trichloroethylene biodegradation by mesophilic and psychrophilic ammonia oxidizers and methanotrophs in groundwater microcosms. Appl Environ Microbiol 63:3866–3871, 1997.
124. JE Cavanaugh, PD Nichols, PD Franzmann, TA McMeekin. Hydrocarbon degradation by Antarctic coastal bacteria. Antarct Sci 10:386–397, 1998.
125. D Delille, A Brasseres, A Dessomoes. A seasonal variation of bacteria in sea ice contaminated by diesel fuel and crude oil. Microb Ecol 3:97–105, 1997.
126. M Kawalec, P Borsuk, S Piechula, PP Stepien. A novel restriction endonuclease UnbI, a neoschizomer of Sau96I from an unidentified psychrophilic bacterium from Antarctica is inhibited by phosphate ions. Acta Biochim Polonica 44:849–852, 1997.
127. HJ Park, SK Park, JS Jang, SH Lee, SM Byun. Molecular cloning of the promoters derived from *Bacillus insolitus* ATCC23299. Korean Biochem J 25:403–408, 1992.
128. S Taguchi, A Ozaki, H Momose. Engineering of a cold-adapted protease by sequential random mutagenesis and a screening system. Appl Environ Microbiol 64:492–495, 1998.
129. K Miyasaki, PL Wintrobe, RA Grayling, DN Rubingh, FH Arnold. Directed evolution study of temperature adaptation in a psychrophilic enzyme. J Mol Biol 297:1015–1026, 2000.
130. S Davail, G Feller, E Narinx, C Gerday. Cold adaptation of proteins: Purification, characterization, and sequence of the heat-labile subtilisin from the Antarctic psychrophile *Bacillus* TA41. J Biol Chem 26:17448–17453, 1994.
131. J-H Zhang, G Dawes, WPC Stemmer. Directed evolution of a fucosidase from a galactosidase by DNA shuffling and screening. Proc Natl Acad Sci USA 94:4504–4509, 1997.
132. FH Arnold. Enzyme engineering reaches the boiling point. Proc Natl Acad Sci USA 95: 2035–2036, 1998.
133. L Giver, A Gershenson, PO Freskgard, FH Arnold. Directed evolution of a thermostable esterase. Proc Natl Acad Sci USA 95:12809–12813, 1998.
134. RK Schopes. Longer is better—random elongation mutagenesis. Nature Biotechnol 17:12, 1999.
135. FT Robb, J-B Park, MWW Adams. Characterization of an extremely thermostable glutamate dehydrogenase: A key enzyme in the primary metabolism of the hyperthermophilic archaebacterium, *Pyrococcus furiosus*. Biochim Biophys Acta 1120:267–272, 1992.
136. JM Muir, RJM Russell, DW Hough, MJ Danson. Citrate synthase from the hyperthermophilic Archaeon, *Pyrococcus furiosus*. Protein Eng 8:583–592, 1995.
137. M Ghosh, A Grunden, R Weiss, MWW Adams. Characterization of the native and recombinant forms of an unusual cobalt-dependent proline dipeptidase (prolidase) from the hyperthermophilic archaeon *Pyrococcus furiosus*. J Bacteriol 180:4781–4789, 1998.
138. J DiRuggiero, FT Robb. Expression and in vitro assembly of recombinant glutamate dehydrogenase from the hyperthermophilic archaeon *Pyrococcus furiosus*. Appl Environ Microbio 61:159–164, 1995.
139. G Feller, O Le Bussy, C Gerday. Expression of psychrophilic genes in mesophilic hosts: Assessment of the folding state of a recombinant α-amlyase. Appl Environ Microbiol 64: 1163–1165, 1998.

140. M Ohkuma, K Ohtoko, N Takada, T Hamamoto, R Usami, T Kudo, K Horikoshi. Characterization of malate dehydrogenase from deep-sea psychrophilic *Vibrio* sp. strain no. 5710 and cloning of its gene. FEMS Microbiol Lett 137:247–252, 1996.
141. SJ Cordwell. Microbial genomes and missing enzymes: Redefining biochemical pathways. Arch Microbiol 172:269–279, 1999.
142. R Hugenholtz, BM Goebel, NR Pace. Impact of culture-independent studies on the emerging phylogenetic view of bacterial diversity. J Bacteriol 180:4765–4774, 1998.
143. E DeLong. Microbiology—archaeal means and extremes. Science 280:542–543, 1998.
144. C Schleper, EF DeLong, CM Preston, RA Feldman, KY Wu, RV Swanson. Genomic analysis reveals chromosomal variation in natural populations of the uncultured psychrophilic archaeon *Cenarchaeum symbiosum*. J Bacteriol 180:5003–5009, 1998.
145. JA Huber, JA Baross. Characterization and quantification of microorganisms in diffuse flow sites on Axial Seamount, Juan de Fuca Ridge. Proceedings of the NSF Hyperthermophile Symposium '99, Athens, GA, May 19–21, 1999, p 23.
146. DE Robertson, EJ Mathur, RV Swanson, BL Marrs, JM Short. The discovery of new biocatalysts from microbial diversity. SIM News 46:3–8, 1996.
147. MWW Adams. The metabolism of hydrogen by extremely thermophilic, sulfur-dependent bacteria. FEMS Microbiol Rev 75:219–238, 1990.
148. A Juszezak, S Aono, MWW Adams. The extremely thermophilic eubacterium, *Thermotoga maritima*, contains a novel iron-hydrogenase whose cellular activity is dependent upon tungsten. J Biol Chem 266:13834–13841, 1991.
149. C Winterhalter, W Liebl. Two extremely thermostable xylanases of the hyperthermophilic bacterium *Thermotoga maritima* MSB8. Appl Environ Microbiol 61:1810–1815, 1995.
150. ON Dakhova, NE Kurepina, VV Zverlov, VA Svetlichnyi, GA Velikodvorskaya. Cloning and expression in *Escherichia coli* of *Thermotoga neapolitana* genes coding for enzymes of carbohydrate substrate synthesis. Biochem Biophys Res Commun 194:1359–1364, 1993.
151. FB Perler, DG Comb, WE Jack, LS Moran, B Qiang, RB Kucera, J Benner, BE Slatko, DO Nwankwo, SK Hempstead, CKS Carlow, H Jannasch. Intervening sequences in an archaeal DNA polymerase gene. Proc Natl Acad Sci USA 89:5577–5581, 1992.
152. KS Lundberg, DD Shoemaker, JM Short, JA Sorge, MWW Adams, E Mathur. High fidelity amplification with a thermostable DNA polymerase isolated from *Pyrococcus furiosus*. Gene 108:1–6, 1991.
153. SB Halio, IL Blumentals, SA Short, BM Merrill, RM Kelly. Sequence, expression in *Escherichia coli* and analysis of the gene encoding a novel intracellular protease (PfpI) from the hyperthermophilic archaeon *Pyrococcus furiosus*. J Bacteriol 178:2605–2612, 1996.
154. SB Halio, MW Bauer, S Mukund, MWW Adams, RM Kelly. Purification and biochemical characterization of two functional forms of intracellular protease PfpI from the hyperthermophilic archaeon *Pyrococcus furiosus*. Appl Environ Microbiol 63:289–295, 1997.
155. FO Bryant, MWW Adams. Characterization of hydrogenase from the hyperthermophilic archaebacterium, *Pyrococcus furiosus*. J Biol Chem 264:5070–5079, 1989.
156. R Koch, P Zablowski, A Spreinat, G Antranikian. Extremely thermostable amylolytic enzyme from the archaebacterium *Pyrococcus furiosus*. FEMS Microbiol Lett 71:21–26, 1990.
157. KA Laderman, K Asada, T Uemori, H Mukai, Y Taguchi, I Kato, C Anfinsen. α-Amylase from the hyperthermophilic archaebacterium *Pyrococcus furiosus*. G B J Biol Chem 268: 24402–24407, 1993.
158. SWM Kengen, EJ Luesink, JM Stams, AJB Zehnder. Purification and characterization of an extremely thermostable β-glucosidase from the hyperthermophilic archaeon *Pyrococcus furiosus*. Eur J Biochem 213:305–312, 1993.
159. IL Blumentals, AS Robinson, RM Kelly. Characterization of sodium dodecyl sulfate-resistant proteolytic activity in the hyperthermophilic archaebacterium *Pyrococcus furiosus*. Appl Environ Microbiol 56:1992–1998, 1990.
160. MW Bauer, R Swanson, E Bylina, RM Kelly. Comparison of a β-glucosidase and a β-man-

nosidase from the hyperthermophilic archaeon *Pyrococcus furiosus*: Purification, characterization, gene cloning, and sequence analysis. J Biol Chem 271:23749–23755, 1996b.
161. HR Badr, KA Sims, MWW Adams. Purification and characterization of sucrose α-glucohydrolase (invertase) from the hyperthermophilic archaeon, *Pyrococcus furiosus*. Syst Appl Microbiol 17:1–6, 1994.
162. MW Bauer, LE Driskill, RM Kelly. Glycosyl hydrolases from hyperthermophilic microorganisms. Curr Opin Biotechnol 9:141–145, 1998.
163. R Eggen, A Geerling, J Watts, WM de Vos. Characterization of pyrolysin, a hyperthermoactive serine protease from the archaebacterium *Pyrococcus furiosus*. FEMS Microbiol Lett 71:17–20, 1990.
164. N Kato, T Nagasawa, S Adachi, Y Tani, K Ogata. Purification and properties of proteases from a marine-psychrophilic bacterium. Agric Biol Chem 36:1185–1192, 1972.
165. MS Weimer, RY Morita. Temperature and hydrostatic pressure effects on gelatinase activity of a *Vibrio* sp. and partially purified gelatinase. Zeitschrift Allg Mikrobiologiue 14:719–725, 1974.
166. E Adler, J Knowles. A thermolabile triosephosphate isomerase from the psychrophile *Vibrio* sp. strain ANT-300. Arch Biochem Biophys 321:137–139, 1995.
167. V Villeret, J-P Chessa, C Gerday, J van Beeumen. Preliminary crystal structure determination of the alkaline protease from the Antarctic psychrophile *Pseudomonas aeruginosa*. Protein Sci 6:2462–2464, 1997.
168. N Aghajari, G Feller, C Gerday, R Haser. Structures of the psychrophilic *Alteromonas haloplanctis* α-amylase give insights into cold adaptation at a molecular level. Structure 6: 1503–1516, 1998.
169. G Feller, P Sonnet, C Gerday. The β-lactamase secreted by the Antarctic psychrophile *Psychrobacter immobilis* A8. Appl Environ Microbiol 61:4474–4476, 1995.
170. G Feller, Z Zekhnini, J Lamotte-Brasseur, C Gerday. Enzymes from cold-adapted microorganisms: The class C β-lactamase from the Antarctic psychrophile *Psychrobacter immobilis* A5. Eur J Biochem 244:186–191, 1997.
171. J-L Arpigny, G Feller, C Gerday. Cloning, sequence and structural features of a lipase from the Antarctic facultative psychrophile *Psychrobacter immobilis* B10. Biochim Biophys Acta 1171:331–333, 1993.
172. H Tsuruta, ST Tsuneta, Y Ishida, K Watanabe, T Uno, Y Aizono. Purification and some characteristics of phosphatase of a psychrophile. J Biochem 123:219–225, 1998.
173. Y Morita, T Nakamura, Q Hasan, Y Murakami, K Yokoyama, E Tamiya. Cold-active enzymes from cold-adapted bacteria. J Am Oil Chem Soc 74:441–444, 1997.
174. MM Zakaria, M Ashiuchi, S Yamamoto, T Yagi. Optimization for beta-mannanase production of a psychrophilic bacterium *Flavobacterium* sp. Biosci Biotechnol Biochem 62:655–660, 1998.
175. TJ Welch, DH Bartlett. Cloning, sequencing and overexpression of the gene encoding malate dehydrogenase from the deep-sea bacterium *Photobacterium* species strain SS9. Biochim Biophys Acta 1350:41–46, 1997.
176. JP Bowman, SA McCammon, TE Lewis, JL Brown, PD Nichols, TA McMeekin. *Psychroflexus torquis* gen. nov., sp. nov., a psychrophilic bacterium from Antarctic Sea ice with the ability to form polyunsaturated fatty acids and the reclassification of *Flavobacterium gondwanense* Dobson, Franzmann 1993 as *Psychroflexus gondwanense* gen. nov., comb. nov. Microbiology 144:1601–1609, 1998.
177. JP Bowman, SA McCammon, MV Brown, DS Nichols, TA McMeekin. *Psychroserpens burtonensis* gen. nov., sp. nov., and *Gelidibacter algens* gen. nov., sp. nov., psychrophilic bacteria isolated from Antarctic lacustrine and sea ice habitats. Int J Syst Bacteriol 47:670–677, 1997c.
178. JJ Gosink, CR Woese, JT Staley. *Polaribacter* gen. nov., with three new species, *P. irgensii* sp. nov., *P. franzmannii* sp nov., and *P. filamentus* sp nov., gas vacuolate polar marine bacte-

ria of the *Cytophaga–Flavobacterium–Bacteroides* group and reclassification of *Flectobacillus glomeratus* as *Polaribacter glomeratus* comb. nov. Int J Syst Bacteriol 48:223–235, 1998.

179. R Borriss, A Sajidan, A-Z Farouk. Extracellular enzyme activities and 16S rRNA gene analysis of cold adapted sea water bacteria. Genbank, 1999.
180. JJ Gosink, RO Herwig, JT Staley. *Octadecobacter arcticus* gen. nov., sp. nov., and *O. antarcticus*, sp. nov., nonpigmented psychrophilic gas vacuolate bacteria from polar sea ice and water. Syst Appl Microbiol 20:356–365, 1997.
181. JP Bowman, DS Nichols, TA McMeekin. *Psychrobacter glacincola* sp. nov., a halotolerant, psychrophilic bacterium isolated from Antarctic sea ice. Syst Appl Microbiol 20:209–215, 1997.
182. JP Bowman, SA McCammon, JL Brown, TA McMeeking. *Glaciecola punicea* gen. nov., sp. nov. and *Glaciecola pallidula* gen. nov., sp nov.: Psychrophilic bacteria from Antarctic sea-ice habitats. Int J Syst Bacteriol 48:1205–1212, 1998a.
183. N Bozal, E Tudela, R Rossello-Mora, J Lalucat, J Guinea. *Pseuodoalteromonas antarctica* sp. nov., isolated from an Antarctic coastal environment. Int J Syst Bacteriol 47:345–351, 1997.
184. H Urakawa, T-K Kita, SE Steven, K Ohwada, RR Colwell. A proposal to transfer *Vibrio marinus* (Russell 1891) to a new genus *Moritella* gen. nov. as *Moritella marina* comb. nov. FEMS Microbiol Lett 165:373–378, 1998.
185. G Gautier, M Gautier, R Christen. Phylogenetic analysis of the genera *Alteromonas*, *Shewanella*, and *Moritella* using genes coding for small-subunit rRNA sequences and division of the genus *Alteromonas* into two genera, *Alteromonas* (emended) and *Pseuodoalteromonas* gen. nov., and proposal of twelve new species combinations. Int J Syst Bacteriol 45:755–761, 1995.
186. JP Bowman, SA McCammon, DS Nichols, JH Skerrat, SM Rea, PD Nichols, TA McMeeking. *Shewanella gelidimarina* sp. nov. and *Shewanella frigidimarina* sp. nov.: Novel species with the ability to produce eicosapentaenoic acid (20:5–3) and grow anaerobically by dissimilatory Fe(III) reduction. Int J Syst Bacteriol 47:1040–1047, 1997.
187. C Kato, L Li, Y Nogi, Y Nakamura, J Tamaoka, K Horikoshi. Extremely barophilic bacteria isolated form the Mariana Trench, Challenger Deep, at a depth of 11,000 meters. Appl Environ Microbiol 64:1510–1513, 1998.

14

Molecular Methods for Assessing and Manipulating the Diversity of Microbial Populations and Processes

Søren J. Sørensen, Anne Kirstine Müller, Lars H. Hansen, and Lasse Dam Rasmussen
University of Copenhagen, Copenhagen, Denmark

Julia R. de Lipthay
Geological Survey of Denmark and Greenland, Copenhagen, Denmark

Tamar Barkay
Cook College, Rutgers University, New Brunswick, New Jersey

I. INTRODUCTION

Because most soil bacteria cannot grow on standard laboratory media, a discrepancy of several orders of magnitude between direct microscopic- and viable-cell counts results (1). The reason for this discrepancy is one of the most important questions in microbial ecology (2). This difference most likely reflects the imperfections of our culturing techniques, although attempts to improve these techniques produce only a minimal quantitative change (3). In addition, some bacteria, while remaining viable, may lose the ability to grow on media on which they are routinely cultured in response to certain environmental stresses. This suggests that stress may induce a viable but nonculturable (VBNC) physiological state (4). This inability to culture most bacteria present in natural soils has, until recently, impaired studies of the relationships between the structure and function of soil microbial communities. One such relationship is flat between soil enzyme activities and the microbial populations that express the genes encoding for these enzymes. This shortcoming may now be remedied by the application of a rapidly growing number of molecular-based techniques that allow detection, enumeration, and characterization of microbial populations in natural environments but that do not depend on cultivation. This evolution has been facilitated by the studies of Carl Woese and coworkers, who introduced the concept of 16S ribosomal deoxyribonucleic acid–(rDNA)–based molecular phylogeny (5) and its application to the analysis of microbial communities in their natural habitats (6).

The introduction of specific detection techniques and the development of fingerprinting techniques for analysis of complex communities have provided the means for determination of the biodiversity of microbial communities without the bias of culturability. Yet, applying molecular techniques to the analysis of soil microbial communities is a big challenge requiring substantial method development before these techniques become generally applicable. Nevertheless, it is clear that we now have an opportunity to link the functional analysis of soils with community composition. This opens up the opportunities to address a new range of questions of the sort, Who is doing what, when, and why?, and for the first time will allow the coupling of enzyme activities to the physiological state of specific microbial populations within the soil environment. The first part of this chapter describes different molecular approaches for the analysis of microbial communities responsible for major enzyme activities in soil. In addition, these molecular approaches present opportunities for the introduction of new enzyme activities into soils by the deliberate release of recombinant bacteria. This strategy has a great potential in the bioremediation of disrupted and polluted environments and is the subject of the second part of this chapter. Although this review primarily deals with descriptions of bacteria in soil environments, the ideas are relevant to other microbes and to other environments.

II. A HOLISTIC APPROACH TO SOIL COMMUNITY DESCRIPTION

The composition of complex communities can be described by using a biomarker that is present in all bacteria but shows variation among taxa or functional groups. Several macromolecules, such as nucleic acid (ribonucleic acid [RNA] and DNA) and phospholipid fatty acid (PLFA), the major constituents of the membrane of all living cells except the *Archaea*, have been used frequently as biomarkers in environmental studies. Whole-community PLFA profiles are very useful in studies that define similarities or differences among microbial communities but give less information on the organisms accounting for these similarities or differences. In 1999 Zelles (7) reviewed the use of PLFA in the analysis of soil communities. Therefore, the discussion is focused on nucleic acid–based techniques.

A vast number of methods have been developed to analyze the DNA and RNA that are recovered directly from soil samples. The most detailed genetic information is obtained by sequencing the genes of interest, and one may argue that it is the most obvious method for investigating the heterogeneity of the community. Indeed, the construction of 16s rDNA clone libraries from DNA extracts of natural samples and the subsequent sequence analysis of these clones have revealed the genetic diversity of bacterial communities from many environments, including soil (8–11). These studies showed high genetic diversity and previously undescribed 16s rDNA sequences; in several only novel sequences were found. However, since this is a time-consuming and costly process, only a limited number of clones can be sequenced. In the studies mentioned 30–124 clones were sequenced, numbers that are probably too low to give an accurate overview of the genetic diversity of the microbial community. For example, Borneman and Triplett (9) found no duplicate sequences when investigating 100 clones obtained from Amazonian soil.

So, although DNA sequence data provide a suitable descriptor of the many unknown species in the environment, their use is not the method of choice when investigating the dynamics of microbial communities by trying to link enzyme activity in soil to community structure. Instead the use of genetic fingerprinting techniques combined with probe hybridization and sequencing of representative samples could be a better choice.

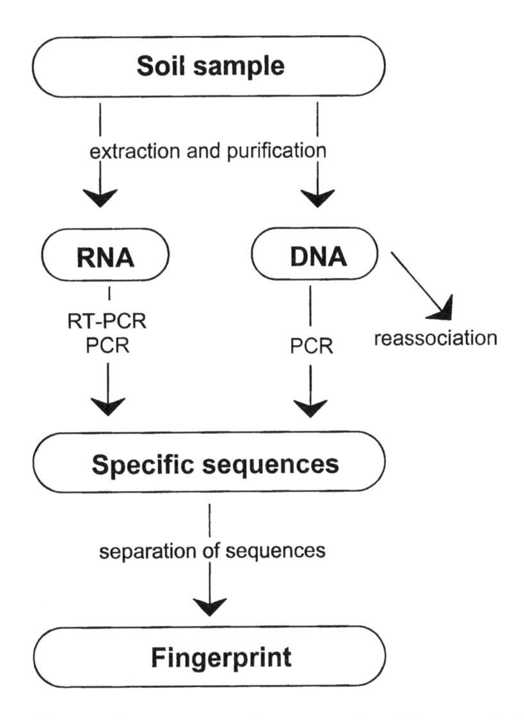

Figure 1 Diagram of steps employed in genetic fingerprinting of soil bacterial communities.

Genetic fingerprinting techniques (Fig. 1) involve extraction and purification of nucleic acids directly from the soil, although some investigators prefer to separate the cells from the environment prior to cell lysis (12–14). Extraction is followed by amplification of specific target sequences by using the polymerase chain reaction (PCR). When RNA is the molecule of interest, reverse transcriptase is used to transcribe the RNA sequence to complementary DNA (cDNA), which subsequently is used in PCR amplification. Examination of variations in the amplified target sequences is achieved by separation techniques such as denaturing gradient gel electrophoresis (DGGE), temperature gradient gel electrophoresis (TGGE), amplified ribosomal DNA restriction analysis (ARDRA), or restriction fragment length polymorphism (RFLP). The steps involved in fingerprinting analyses of soil microbial community structure and function are described in the following sections.

A. Nucleic Acids

Community analysis may use extracts of both community DNA and RNA from the soil. Genomic DNA is present in all bacteria, active as well as dormant, and in an extracellular form that is protected by adsorption to soil particles (15). Hence, a genetic fingerprint based on community DNA may overestimate the number of intact species present in the community at the time of sampling. On the other hand, the RNA content is generally presumed to be higher in active bacteria than in inactive bacteria. With pure cultures the number of ribosomes and the amount of rRNA are almost proportional to the growth rate of the organism (16). Therefore, RNA-based analysis may provide information on metabolically active subpopulations in the microbial community.

Teske et al. (17) found that DGGE profiles of 16S rDNA and 16S rRNA extracted from the same water column differed markedly. Further results from hybridization with group-specific probes (different groups of sulfate-reducing bacteria) suggested that certain strains played a more significant role in the community because of their activity rather than their abundance. Ribosomal RNA extraction procedures have successfully been applied to soil ecosystems (18,19). Thus, RNA-based analysis is a more appropriate choice in studies that link community structure to enzyme activity. However, recent studies showing rRNA persistance long after cell death questioned the use of rRNA in assessing metabolic activity of cells in natural samples (20).

B. Genes Used as Biomarkers

The most commonly used genetic marker in community analysis is the small subunit ribosomal RNA (16S rRNA) or the gene encoding for it (16S rDNA). This is a suitable genetic marker in investigations of diversity because each 16S rRNA/DNA nucleotide contains both highly conserved regions that are shared by all organisms and variable regions that are unique to specific organisms (or, at least, to closely related groups of organisms). This means that PCR primers with universal sequences can be designed to amplify DNA with species-/genus-specific sequences. It is also possible to analyze the genetic diversity of monophyletic groups of bacteria by designing primers specific to the group of interest, e.g., ammonia oxidizers or actinomycetes (21–23).

Since some functions have a polyphyletic distribution (i.e., they are present among distantly related taxa) genes specifying the function of interest must be used as the biomarker rather than rDNA genes (24,25). Henckel et al. (24), who investigated the methane oxidizing microbial community in rice field soil, targeted genes encoding the methane monooxygenase and methanol dehydrogenase enzymes. Likewise, genes coding for specific resistances such as resistance to mercury (26) have been used as targets in genetic fingerprinting techniques.

C. Fingerprinting Techniques

1. Restriction Fragment Patterns

Restriction fragment length polymorphism (RFLP), also known as *amplified rDNA restriction analysis* (ARDRA), targets sequence differences in species-/group-specific regions of 16S rDNA as reflected by variable number and locations of restriction enzyme recognition sites. Thus, restriction enzyme digests result in a specific number and size of DNA fragments that are distinguished after separation by gel electrophoresis. The more diverse the community is, the more elaborate are its RFLP patterns. Usually three to four different tetrameric enzymes are used to ensure a sufficient number of fragments for diversity analysis. The choice of restriction enzymes may greatly influence the results: e.g., 18 of the 23 correct clones found in a study of an anaerobic cyanide degrading consortium remained uncut even after treatment with four different enzymes (27). Either a mixture of PCR-amplified community 16S rDNA or clones derived by cloning this DNA may be analyzed. In both cases, PCR amplification of environmental DNA with 16S rDNA primers precedes RFLP analysis.

This method was used to show differences in genetic diversity of bacterial communities from extreme environments. In hypersaline ponds, RFLP performed with the 16S rDNA amplification products of the native community showed a decreased eubacterial genetic diversity with increasing salinity, whereas the reverse was true for Archaea (28). Structural changes in the microbial community in soil due to copper contamination were detected by Smit and associates (29) on the basis of ARDRA profiles of isolates, clones, and soil community DNA. However, a problem arises when using RFLP in the investigation of more complex microbial communities. In analyzing the amplification products of community DNA, different sequences result in a different number of DNA fragments, depending on how many restriction sites are present in a particular sequence. Presence of many bands does not necessarily reflect high diversity. Instead they may be due to the presence of many restriction enzyme sites in the amplified sequences. This problem is solved by the use of terminal restriction fragment length polymorphism (T-RFLP) (30), whereby the terminal end of the amplification product is labeled with a fluorescent marker during PCR. After digestion with restriction enzymes, only the terminal restriction fragment is detected, ensuring that each bacterium is represented by a single fluorescent fragment. This method has been used to analyze the effect of temperature on community structure of a methanogenic community by using Archaea-specific PCR primers (31) and of communities in mercury contaminated and uncontaminated soil (26). This method is still affected by the choice of the restriction enzyme as DNA lacking the specific site are not digested.

2. *Distinguishing Deoxyribonucleic Acid Molecules by Melting Characteristics*

Another approach for fingerprinting is the analysis of the PCR amplification products by denaturing gradient gel electrophoresis (DGGE). This technique, originally developed for the detection of point mutations (32–34), was more recently applied to studies of microbial genetic diversity (35). Unlike common electrophoresis, in which DNA fragments are separated by size differences, DGGE separates DNA fragments of the same length according to sequence heterogeneity that results in dissimilar melting properties. The complex mixture of amplified DNA is electrophoresed (at elevated temperature, usually 60°C) through an acrylamide gel that contains a linear gradient of denaturant concentrations (formamide and urea). DNA migrating through the gel partially melts at a specific denaturant concentration, depending on its primary sequence. This partial melting of the DNA results in branching of the molecule, thus lowering its mobility in the gel. To prevent complete melting a GC-clamp (an approximately 40-bases-long GC-rich sequence) is attached to the end of one of the PCR primers. At the denaturant concentration at which the 16S rDNA is completely denatured, the complex of single-stranded DNA and double-stranded GC-clamp is almost totally immobile, resulting in a DNA band at a location specific to this 16S rDNA sequence (Fig. 2). By using a GC-clamp it is possible to detect almost 100% of all possible sequence variations (33). Another approach used to create a continuous gradient in denaturing conditions is based on temperature in the so-called temperature gradient gel electrophoresis (TGGE). Since DGGE and TGGE are in principle the same no distinction between the two is made in the following discussion.

Specific fingerprints using DGGE/TGGE have been performed with many environmental samples (17,35–39), including soil (40–43). Furthermore, DGGE/TGGE is an excellent tool for monitoring changes in community diversity resulting from environmental disturbances. Diversity has been shown to change in agricultural soil by fumigation and pesticide treatments (40,43,44).

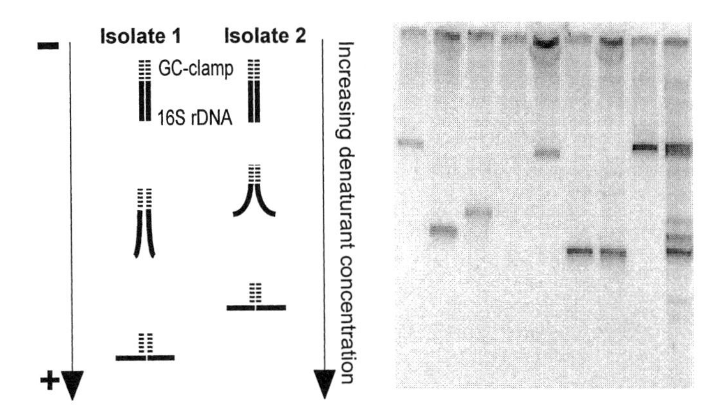

Figure 2 Theoretical and actual appearance of DGGE. Arrows indicate direction of migration in the gel.

One way by which DGGE/TGGE analysis may overestimate diversity is due to the presence of more than one 16S-rDNA gene in a single species. This has so far been reported only in the case of *Paenibacillus polymyxa*, when DGGE analysis revealed several bands (45). How this phenomenon affects the diversity analysis of microbial communities still needs to be investigated, but if it is a common phenomenon, diversity may be greatly overestimated by all fingerprinting techniques based on rRNA or rDNA. On the other hand, it is not possible to distinguish more than approximately 50–100 bands in DGGE/TGGE profiles. Therefore, this analysis does not identify all the species diversity in a complex community.

The sensitivity of the method was found to be limited to populations representing at least 1% of the total bacterial community (35). Newly developed dyes like SYBR-green and silver staining are more sensitive than ethidium bromide when bound to DNA and may improve the current detection limit. The number of bands that can be distinguished in a gradient gel may limit analysis of very complex bacterial communities in which high diversity is expected, e.g., soil communities. Nevertheless, changes in population composition in communities revealing more than 100 bands on DGGE gels have been reported (43).

3. Deoxyribonucleic Acid Reassociation

A very different approach to diversity analysis is the DNA reassociation technique described by Torsvik and colleagues (13,46). The heterogeneity of DNA extracted from soil samples was determined by reassociation kinetics, measured spectrophotometrically, after DNA denaturation. At a given DNA concentration, the time required for reassociation is proportional to the complexity of DNA. The method has been used to estimate the number of different species in soil communities by comparing the total genomic size of the mixed soil community DNA with that of the mean genome size of cultivated bacterial isolates from the same soil (13,46). This pioneering work estimated the presence of 4000 completely different bacterial genomes in 1 g of soil, equivalent to 13,000 species (13).

In 1998, the method was used to investigate the impact of environmental disturbances on community diversity (47). The genetic diversity of the microbial community was reduced 20-fold in CH_4-perturbed soil compared with that of communities in nondisturbed soil (38).

D. General Problems in Community Analysis

Although molecular approaches have facilitated inclusion of nonculturable microbes in community analysis, the question whether the results are representative of the total microbial community still needs to be addressed. The ideal analysis should of course reflect both the quantitative (''how many'') and qualitative (''who'') composition of the community. Each step in the molecular analysis could result in bias, making it difficult to interpret the results. The following section considers nucleic acid extraction and amplification as crucial steps in the accurate analysis of community diversity.

1. Extraction of Nucleic Acids

A number of methods to extract DNA from soil samples have been developed, including cycles of freezing and thawing, sonication, sodium dodecylsulphate (SDS) treatment, boiling, liquid nitrogen, and bead beating (14,48–51). Whereas some investigators have evaluated the quantity and quality of the extracted DNA by different methods (14,52), few have compared the relative compositions of the extracted gene pool. Our own experiments have shown that DGGE profiles from soil community DNA varied markedly, depending on the extraction method (Fig. 3). Two different extraction methods were used: a sonication method (representing a gentle treatment of the cells) and a harsher bead beating method. It is very likely that the sonication method mainly extracted fragile, thin-walled cells,

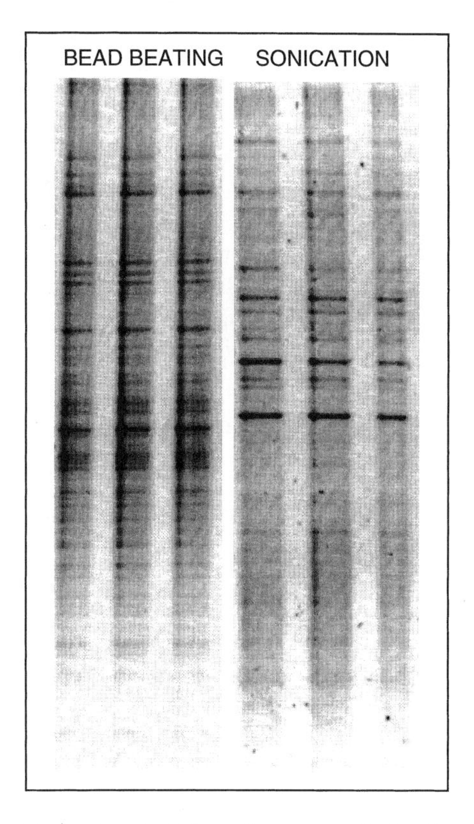

Figure 3 DGGE profiles of amplified 16S rDNA fragments extracted from a sandy loam soil by a bead beating method or by a sonication method.

leaving more sturdy cells intact. Bead beating methods, on the other hand, yielded DNA from more sturdy cells and spores (52), but loss of specific sequences could result from shearing of the DNA by the rather harsh procedure (53,54). The DGGE profiles differed both in the position and in the number of bands, clearly showing that our conclusion regarding ''who and how much is there'' depends on our DNA extraction procedure. Even when two DNA extraction protocols, both containing a bead-mill homogenization step, were compared, different DGGE profiles were observed (55). This study also concluded that the relative composition of the extracted gene pools varied with methods. Therefore, the suggestion that rDNA amplicons visualized by DGGE represent the dominant species in the community (35,42) needs to take into account bias introduced by the chosen DNA extraction methods.

2. *The Amplification of Specific Sequences*

The next crucial step in most analyses is the amplification of the target genes by PCR. When it is applied to community DNA several problems may result in a biased synthesis of amplification products (for detailed review see von Wintzingerode et al. [56]). The amplification efficiency has to be the same for all the sequences in the DNA mixture if the PCR products are to reflect the community composition. It is known that environmental samples may contain inhibitors of DNA amplification, e.g., phenolic compounds, humic acids, and heavy metals (for an extensive list see Table 3 in Wilson [57]). Amplification also depends on primer specificity and hybridization efficiency, template and primer concentration, and number of PCR cycles (58). Even when quantitative PCR is achieved, the number of target sequences before amplification could depend not only on the number of bacteria containing the specific sequence, but also on the number of gene copies in each cell (59), which for unculturable bacteria is still unknown.

The formation of chimeric DNA molecules and other PCR artifacts (60), as well as cross contamination (61,62), are additional problems. Even though in situ hybridization has shown that phylogenetic groups found by sequencing of clones in a 16S library were present in the environment (61), this is not always the case. Tanner et al. (62) showed that 16S rDNA sequences obtained in controls without DNA template corresponded to sequences found in environmental samples. This finding indicates that some sequences present in community DNA pools were contaminants of unknown origin. This cross contamination, if prevalent, may blur differences in genetic diversity among habitats.

E. Comparative Studies

Few investigators have compared community diversity analyses obtained by molecular approaches to those obtained by more traditional approaches, e.g., substrate utilization (40,44,63–66), or to typing according to colony morphological characteristics (65,66). We evaluated the effect of mercury on soil microbial communities by three different approaches: DGGE, colony morphological features, and sole carbon utilization patterns (BIOLOG). All methods were able to detect structural changes of the community in the presence of mercury. The effect of mercury on the diversity of the community (here expressed as number of types [Table 1]) was revealed by both DGGE and colony morphological analysis. DGGE was the most sensitive of the methods, showing a reduced number of bands not only in the most contaminated soil but also in the intermediate contaminated soil.

Torsvik et al. (13) reported agreement between the genetic diversity as described by DNA reassociation kinetics and by phenotypic diversity of isolated strains, and Engelen

Table 1 A Comparison Among Different Methods That Assess Community Diversity[a]

	Number of		
Soil	Morphotypes	DGGE bands	Substrates utilized
A	20.7 ± 1.9	57.3 ± 0.7[a]	15.0 ± 1.2
B	23.3 ± 2.2	53.3 ± 1.2[a]	13.3 ± 0.3
C	14.7 ± 0.9[b]	47.0 ± 1.2[a]	15.7 ± 1.2

[a] Numbers (mean ± standard error [SE]) of morphotypes represented by the morphological features of 50 randomly selected colonies, bands in the DGGE profiles, and substrates utilized in the BIOLOG Ecoplates for three soils with different levels of mercury contamination: soil A (7 μg Hg g^{-1} dw soil), soil B (28 μg Hg g^{-1} dw soil), and soil C (511 μg Hg g^{-1} dw soil).
[b] Significantly different from the others (t-test; $p < .05$). The t-test was only performed if analysis of variance (ANOVA) showed significant differences ($p < .05$).

and associates (40) showed correlation between pesticide-induced changes in the genetic and functional diversity in soils by DGGE and substrate utilization patterns. On the other hand Duineveld and colleagues (63) found similar DGGE profiles of communities from rhizosphere and bulk soil, where large differences in metabolic properties existed.

F. Ribonucleic Acid Hybridization

Together with the increase in known 16S and 23S rRNA sequences, new hybridization techniques for studying bacterial community structure have emerged. Two of the most promising of these techniques are fluorescent in situ hybridization (FISH) and rRNA slot- or dot-blot hybridization.

FISH can be used for detecting the abundance and distribution of specific bacteria at different phylogenetic levels. Using confocal scanning laser microscopy (CSLM) it is possible to visualisze the spatial distribution of the target organisms in complex environments. Enumeration of fluorescently marked species is achieved by submitting the hybridized cells to analysis in a fluorescence activated cell sorter (FACS) (67,68).

Numerous studies have examined biofilms, using the FISH technique to identify key species and their positions in environmental matrices (69–71). These include activated sludge (72–76), sediments (77,78), soils (79), and plant roots (80). The technique requires cell fixation by paraformaldehyde, dehydration with ethanol, and incubation with one or even several fluorescent oligonucleotide probes. The probes are specific to the target organisms' rRNA. Since rRNAs are present in copious amounts in the cell, hybridization results in fluorescent signals that are easily detected under the microscope or by FACS. Oligonucleotide probes can be designed to target different levels of phylogenetic specificity (i.e., kingdom-, genus-, and species-specific probes). For example, Logeman and associates (82) studied microbial diversity in a nitrifying reactor system using probes designed to hybridize with either all eubacteria, all *Cythophaga–Flexibacter–Bacterioides* spp. groups, or only nitrifiers in the *Nitrosomonas* sp. cluster.

rRNA slot- or dot-blot hybridization is a technique that, like FISH, relies on known sequences to examine the abundance of specific populations. The procedure is inexpensive compared to FISH as it does not require fluorescence microscopes; however, information about the spatial distribution of the microorganisms is lost. Total pools of rRNA isolated from natural samples or cultures are blotted onto a nylon membrane, and the rRNA is hybridized with both specific and more general oligonucleotide probes. By comparing signals that have been hybridized with a more specific probe to the signals from a more general probe, a relative abundance of the target bacterial species or genus can be calculated. This approach has been used to examine microbial communities in aquatic (74,81) and soil (79,83) samples. Hybridization can also be applied to PCR-amplified rDNA. In one study (81), 353 clones from a 16S rDNA clone library representing the community of permanently cold marine sediments were hybridized with group- and species-specific oligonucleotide probes. The study showed high bacterial diversity in the sediment and dominance of sulfate reducing bacteria, among them *Desulfotalea* spp. and other closely related species.

A limitation to rRNA hybridization is the fact that oligonucleotide probes can target only species for which sequences are known. However, the number of known sequences is increasing rapidly, and now comprises more than 7336 Bacteria and 324 Archaea for 16S rRNA (84). By combining, for example, DGGE, in which 16S rRNA from complex environments can be separated in a gel on the basis of sequence, with subcloning and sequencing, new sequences can be obtained and used to design oligonucleotide probes that are relevant to the environment in question.

G. How Do the Diversity and Structure of the Bacterial Community Influence the Function of the Soil System?

The use of molecular approaches to describe the composition of soil bacterial communities offers an opportunity to study the relationships among community diversity and the structure and function of soils. The ecological importance of these relationships cannot be overstated. A great many of the examples discussed reveal the enormous microbial diversity in soils, but only a few address diversity as it relates to the functioning of the ecosystem. If they do, they have mainly done so on the basis of functional diversity (85) as defined by carbon utilization patterns in BIOLOG microtiter plates. BIOLOG measures the *potential* of a fraction of the community, which does not necessarily represent the numerically dominant species (86), to grow on a particular substrate rather than the actual activity of the community (87). Furthermore, most of the test substrates have no special ecological relevance. Therefore, it is difficult to relate a change in a utilization profile to the presence or absence of specific enzyme activities in the soil. Theories have been proposed to explain how species diversity is related to ecosystem function (88). For example, it has been suggested that enhanced species diversity is beneficial for ecosystem functioning (89,90). Others have proposed that the properties of the system are more dependent on the functional abilities of some species than on the total number of different species (91–93). Studies have focused mainly on plants, and only recently has attention been given to soil microbial communities (94–96). The diversity of the soil microbial community is enormous even within a small area (such as a soil aggregate or a root surface), yet not all the bacteria contribute to the observed activity. Rather, a large proportion of the cells are inactive and become active only when conditions are favorable. Therefore, there is probably a large difference between the diversity of the potentially active bacteria and that of those that are actually active, and this is reflected in differences between the potential and actual activity in soils.

Numerous enzyme activities can be the focus when characterizing the function of the soil microbial community. It is very likely that highly specialised functions (connected only to a few species) or complex functions (depending on a consortia of organisms) are sensitive to lowered diversity, since they are dependent on the presence of particular species, whereas more general functions are not (97). Furthermore there may be a considerable redundancy of function between different bacterial species (98) and a high degree of adaptation to a changing environment (99), making at least some functions of the soil ecosystem more robust. Griffiths and coworkers (95) found decreased nitrification, potential denitrification, and methane oxidation in a soil with decreased microbial diversity, where more general functions were unaltered.

Another functional aspect of the soil ecosystem is stability as defined by both the capacity of the system to avoid species displacement after perturbation (resistance) and the ability of the system to return to the former state after perturbation (resilience) (100). The stability of the soil system has been hypothesized (94) and shown (95) to be related to the diversity of the microbial community. The community with the lowest diversity showed lower stability of the decomposition process of grass residues under applied perturbation (addition of $CuSO_4$) than communities with higher microbial diversity. Again it is very likely that diversity–stability relationships also depend on which process is examined.

Direct in situ detection of mRNA of functional genes can provide the link between enzymatic function in soil and the activity of the microbial community. Although this technique has been used in medical studies, it has so far, not been successfully applied to soil samples. Meanwhile, microbial community analysis using available techniques (as discussed here) combined with enzyme activity measurements can provide crucial information on the relationship between community structure and soil function.

III. COMMUNITY MANIPULATION

Molecular techniques for the manipulation of natural microbial communities and their activities have broad applications in environmental management. Because many natural and anthropogenic processes depend on the enzymatic activities of microbes, manipulation of the genetic potential of microbial communities could enhance and optimize these processes. This concept has been most closely examined with the degradation of industrial pollutants by microorganisms.

In just over a century, industrialization has led to a wide distribution of xenobiotic compounds in the environment. Because of their toxicity to humans and the risk they pose to the integrity of natural ecosystems, the removal of these compounds is of great benefit to society. A few processes contribute to this goal, among them (1) transport, (2) evaporation, (3) sorption, and (4) biodegradation. The first three, however, only alter the physical state or the location of the contaminants. In contrast, biodegradation is the primary process involved in the transformation and mineralization of xenobiotic compounds and, in the latter process, results in the elimination of the pollutant and its metabolites. Abiotic degradation may occur, but it is less common and often results in incomplete decontamination and sometimes the formation of more toxic chemical intermediates (101). Enhanced biodegradation of toxic xenobiotics could be achieved by in situ manipulation of degradative genes and their expression. For this approach to succeed, an understanding of how microorganisms evolve new catabolic capabilities and how they express enzymatic activities of degradative pathways is needed.

A. Microbial Adaptation to Degradation of Xenobiotic Compounds

The natural environment presents microorganisms with an enormous variety of substrates for growth and energy production, and consequently microbes have evolved enormous catabolic potential for the breakdown of organic molecules (102,103). Synthetic chemicals, however, are often substituted for by the addition of halogens, nitro-, sulfur-, or azo-groups that convert otherwise easily degraded substances to recalcitrant compounds (101). Because microbial enzymes do not recognize the substituted compound as a substrate, degradation is delayed or even eliminated. Typically a lag period of varying length ensues upon exposure of microbial communities to toxic contaminants. This lag phase might be due to the toxicity of the compound or to its nonavailability as a growth substrate. During this lag period changes in microbial community structure and enzymatic activities occur: i.e., the community acclimates to the presence of the contaminant. Acclimation facilitates the degradation of the contaminant and, thus, survival under the changed conditions. When it is challenged by repeated exposure to the same chemical a more rapid response occurs, as the microbial community now has been acclimated (see Fig. 4)(104,105). Many factors

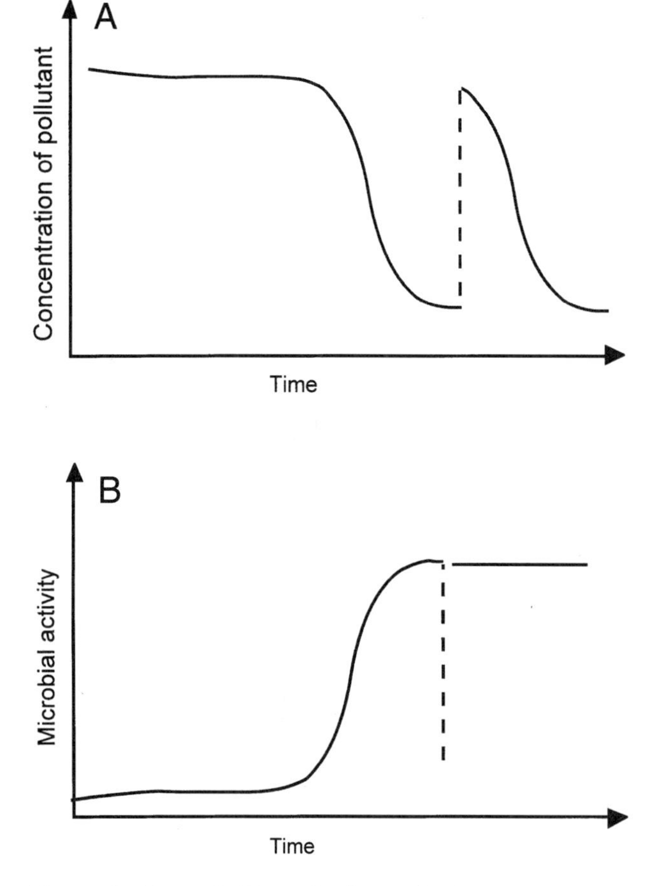

Figure 4 Fate and effects of pollutants in the environment. (A) Biodegradation of pollutant. (B) Effects of pollutants on microbial activities. Solid lines in A and B, response of the microbial community to the presence of the pollutant in question; dashed lines, second application of the pollutant.

affect the dynamics of the acclimation response, including chemical structure and concentration of the pollutant, presence of organic and inorganic nutrients, type and physiological state of the community, physical/chemical parameters of the environment (temperature, pH, redox potential, salinity), and biological factors such as predation and competition (104–109). The response depicted in Fig. 4 is a community level response, composed of numerous responses at the cellular level. Three mechanisms for microbial acclimation to degradation of xenobiotic compounds have been suggested (105,107,109,110): (1) enrichment of populations that carry the required degradative capabilities, (2) induction of enzymes involved in the uptake and turnover of the pollutant, and (3) genetic adaptation. The first two are manifested by previously existing subpopulations of degrading strains, whereas the process of genetic adaptation creates changes in the existing genetic pool of the microbial community and thus implies the evolution of new catabolic capabilities and an increase in the functional diversity of the microbial community.

B. Genetic Change in Acclimation

Little has been done to elucidate the molecular events that take place during genetic adaptation in situ, partly as a result of the extended time of evolutionary events. On the organism level, many studies have examined the genetics and the enzymatic capabilities of pure bacterial strains that degrade xenobiotic compounds (for review see, e.g., 109,111–115). The knowledge accumulated by these studies reveals different molecular processes for the acquisition of new enzymatic activities and for the formation of new catabolic pathways. A wide variety of processes cause genetic change and, consequently, altered metabolic functions. These can be divided into vertical and horizontal processes, as suggested by van der Meer (116). Vertical processes lead to the inheritance of accumulated mutations by daughter cells. These mutations can be single base-pair changes or larger changes in DNA sequence such as deletions and duplications. Horizontal processes involve an exchange of DNA at the intermolecular or the intercellular level. Intermolecular exchange results in DNA rearrangements within a single cell, e.g., between the chromosome and extrachromosomal elements. Intercellular processes involve exchange of DNA between the genomes of two different organisms, e.g., conjugation, transformation, and transduction. In bacterial conjugation plasmid DNA is transferred between two cells that are in physical contact. In transformation, competent cells take up naked DNA molecules, and in transduction, DNA transfer is mediated by bacteriophages. Numerous studies have demonstrated the occurrence of horizontal gene exchange in soil environments (117–119), where intercellular exchange is more prevalent than vertical processes and thus more significantly contributes to the evolution of new degradative capabilities (115). This is supported by the location of a large number of catabolic genes on plasmids (120,121), as well as by the higher incidence of plasmid DNA among bacteria isolated from polluted environments (122,123). Natural environments are characterized by the scarcity of nutrients, which results in extended periods of arrested growth of indigenous microbes. As it is now known that DNA rearrangement events (e.g., transposition) occur during stationary phase (124,125), natural microbial populations should participate in genetic change under in situ conditions. This may give rise to a large and flexible gene pool that facilitates genetic plasticity in the microbial community. The presence of specific conditions (e.g., xenobiotic contamination) selects for beneficial phenotypes, ensuring the stable inheritance of specific (e.g., degradative) genes in the community. The role of adaptive mutations in evolution, the increased incidence of mutations in the presence of a potentially useful substrate, although tempting as a possible mechanism of genetic adaptation, is still being debated (115).

C. Bioremediation

The adaptation concept has been most often examined with regard to the degradation of recalcitrant contaminants (104,126–128). The underlying premise of attempts to exploit genetic adaptation in environmental management is that if adaptation could be accelerated or even initiated by human intervention, more effective rehabilitation of contaminated environments could be achieved. As genetic adaptation may take a long time, acceleration of this process by the addition of catabolic genes to the microbial genetic pool of contaminated sites has been attempted. Three strategies have been employed: (1) the introduction of microorganisms harboring genes for complete degradation pathways (129–131), (2) the in situ transfer of catabolic genes to indigenous microorganisms (132,133), and (3) the expansion of the indigenous community's catabolic range by the complementation of an existing pathway by transfer of genes encoding additional enzymatic functions from added donors (134–136). Each of these approaches has advantages and disadvantages and has met with mixed success (Table 2).

1. *Application of Degrading Strains and the Construction of New Catabolic Capabilities*

The course of microbial adaptation is dependent on, among other factors, the chemical structure of the pollutant. It should be noted that some recently introduced xenobiotics are readily degraded; it is as if the appropriate catabolic enzymes already exist in the environment (115). Thus, it seems that the physiological versatility of natural microbial communities has evolved catabolic pathways for the degradation of substances not previously encountered in the environment. For nondegraded substances, new enzymatic activities and catabolic pathways have to evolve. In due time, during which changes in the genetic capacity and metabolic functions of natural microbial communities take place, even highly recalcitrant compounds are degraded. This approach has been named *natural attenuation*. However, the time for this evolution may be prohibitively long, and more proactive approaches are needed. For example, new catabolic capabilities could be generated by facilitating genetic adaptation. This can be done by either a horizontal or a vertical expansion of existing catabolic capabilities.

In horizontal expansion, the substrate profile of an existing pathway is broadened, so that analogs of the natural compounds are metabolized. The bases of this strategy are (1) the existence of isofunctional routes for degradation of structurally related compounds, (2) the existence of broad substrate specificity enzymes, and (3) the ability to alter the specificity of proteins for their substrates/effectors by mutagenesis (137). An example of this strategy is the expansion of the TOL encoded pathway for degradation of alkylbenzoates. Although specifying a pathway for the degradation of various alkylbenzoates, hosts carrying the TOL (pWWO) plasmid, originating in *Pseudomonas putida*, do not degrade 4-ethylbenzoate. By selection of mutants in three independent steps the substrate range of the catabolic pathway was broadened to include 4-ethylbenzoate (137).

In vertical expansion of degradative capabilities, the existing pathway is used as a base to which additional enzymes that extend the pathway are added (137). Structurally diverse aromatic compounds are degraded by a number of converging pathways that lead to the formation of dihydroxylated aromatic key intermediates (Fig. 5). Depending on the

Table 2 Strategies for Genetic Manipulations of Soil Microbial Communities to Stimulate Biodegradation

Strategy	Advantages	Disadvantages
Application of degrading strains	Strains carry complete biodegradative pathways	May not survive in the environment because of inability to invade established microbial community in contaminated site
Application of donor strains carrying the biodegradative pathway on a conjugative plasmid, resulting in transfer of catabolic genes to indigenous bacteria	Maintenance of catabolic genes is enhanced by transfer to indigenous bacteria	Effect depends on transfer and expression of large number of genes Expression in new hosts may be impossible or less than optimal
Expansion of the substrate range of pathways in indigenous bacteria by complementation with small number of catabolic genes originating in added donor strains	Effect depends on the transfer and expression of a small number of genes in strains that already express a catabolic function	Expression of new enzymatic activities depends on regulation of the transferred genes, resulting in possible obstacles of expression in new bacterial hosts Effect depends on number of potential recipients in indigenous community

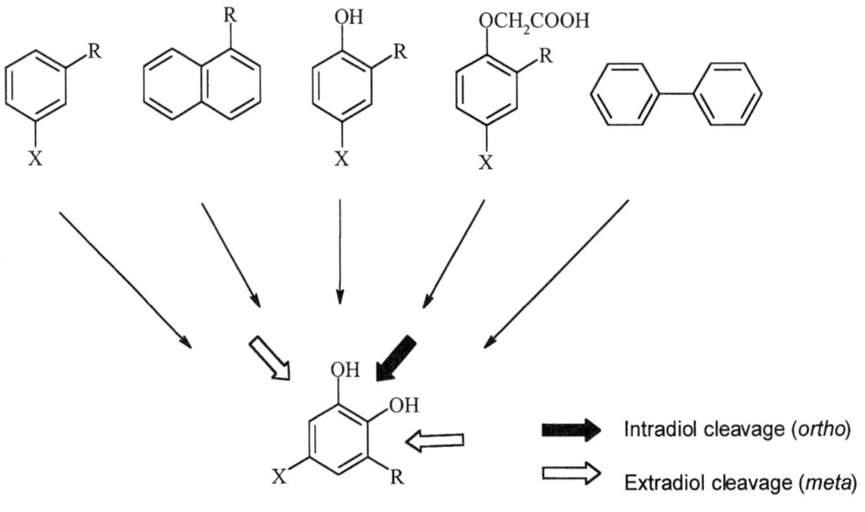

Figure 5 Formation of dihydroxylated aromatic key intermediates (illustrated as a substituted catechol) in the metabolic channeling of aromatic compounds by converging pathways. After ring cleavage the products are converted to intermediates of the tricarboxylic acid cycle by either the *ortho* or the *meta* cleavage pathway. R and X, alkyl and halogen substituents, respectively.

type of substitution, a key intermediate is converted to substrates of the tricarboxylic acid cycle by either an intradiol (or *ortho*) cleavage, general for halogenated substances, or an extradiol (or *meta*) cleavage, general for alkylated substances, of the aromatic ring. Probably evolution has ensured that aromatic substances are exclusively degraded by either an *ortho* or a *meta* cleavage pathway, as channeling of a substance through the wrong pathway can result in formation of toxic intermediates or dead-end products (138–140). For example, the extradiol cleavage of 3-chlorocatechol forms an acylchloride, which irreversibly inhibits the activity of catechol 2,3-dioxygenase (the *meta* ring cleavage enzyme). It was shown that autoxidation of the accumulated 3-chlorocatechol produced toxic intermediates (138,140). Pieper et al. (140) found that intradiol ring cleavage of methylated phenols leads to accumulation of dead-end products that do not serve as substrates for the next enzyme in the *ortho* cleavage pathway.

Strains with novel enzymatic capabilities prepared by the approaches described, or by the isolation from enrichment cultures, may be added to polluted environments to enhance biodegradation. The advantage of this approach is that the added strains carry the genetic information needed for the expression of the complete catabolic pathway. This approach has been applied for the degradation of 1,2,4-trichlorobenzene (129), 4-ethylbenzoate (141), atrazine (130,142) phenoxyacetic acid (143), 3-chlorobenzoate (144), 3-phenoxybenzoic acid (145), and 2,4-dichlorophenoxyacetic acid (145). Although all these studies report a decline in the concentration of the contaminants, it is very clear that success depends on a variety of factors. For example, Tchelet et al. (129) reported an effect of environmental matrix, as application in soil columns was successful whereas no effect could be observed in sewage sludge microcosms. The survival in soil of the 4-ethylbenzoate recombinant strain, described previously, depended on the nature of the aromatic contaminant (141), and removal of atrazine by a degrading bacterial consortium only took place after several applications of the herbicide (142). It seems though that selection of the appropriate degrader might be critical because Struthers and colleagues (130) successfully reduced the concentration of atrazine in soil by using a strain of *Agrobacterium radiobacter*. In addition the concentration of the contaminant seems to be crucial, too. Short et al. (143) demonstrated that the presence of substrate was essential for the survival of a phenoxyacetic acid degrading inoculant, and Daane and Häggblom (131) showed that the encapsulation of the degrading strains in structures that protect them from the toxicity of the contaminant enhanced their performance. Thus, although the activity of the applied strains has been optimized in the lab, they may not survive or express their degradative activity under field conditions. Furthermore, concerns with the release of recombinant strains may limit applications even if this approach has been fully successful.

2. *Transfer of Catabolic Genes to the Indigenous Floroa*

The strategy of in situ transfer of catabolic genes to the indigenous populations of microorganisms exploits the fact that these organisms are well adapted to the environment in question as well as to the constantly changing conditions that prevail in nature. The use of this strategy has potential to ensure more efficient maintenance as well as dispersal of the catabolic genes. Because gene transfer is stimulated in environments with high biomass and availability of growth substrates, a role for gene transfer in the degradation of contaminants in bioreactors (146) and activated sludge (147) has been proposed.

Top et al. (132) showed that transfer of two 2,4-dichlorophenoxy acetic acid (2,4-D) degradation plasmids (that were unrelated to the ''prototype'' 2,4-D plasmid, pJP4, or to each other) enhanced the degradation of 2,4-D in soil. In the presence of 2,4-D the plasmids were transferred to indigenous organisms in the soil to create a large biomass of degraders; in the absence of 2,4-D fewer transconjugants were detected. Transfer of pJP4 from its native host, *Ralstonia eutropha* JMP134, to a *Variovorax paradoxus* recipient, although quite frequent on agar plates, was dramatically reduced in sterile soil and even more so in unsterile soils (148). However, when the same donor was added to unsterile soil that was supplemented with 1000 μg 2,4-D g^{-1} soil, a large number of indigenous transconjugants arose and the rate of 2,4-D disappearance increased in inoculated relative to noninoculated soils (133). Because the experimental design employed by these studies did not eliminate donor strains, the role of gene transfer in the bioaugmentation of 2,4-D cannot be discerned (i.e., 2,4-D degradation could have been caused by donors that were enriched by the availability of the substrate). Furthermore, these studies did not compare the survival of the donor strains with the survival of the 2,4-D degrading genes. Such an analysis would allow evaluation of the premise that gene transfer to indigenous bacteria facilitates the maintenance of the catabolic capability in the community.

3. *Expansion of the Substrate Range of Indigenous Bacteria by Complementation with Plasmid-Borne Catabolic Genes*

The dissemination of catabolic genes to the indigenous flora could be improved if only a small number of essential genes were transferred to complement previously existing catabolic capabilities. As a result, an expanded degradative profile of the microbial community would emerge (134,136). Using this approach, Barkay et al. (134) showed that resistance to the organomercury compound phenylmercury acetate was established in transconjugants after filter matings between donors carrying *merB* (the gene encoding organomercurial lyase) on conjugal plasmids and bacterial strains from freshwater samples. To create a pool of potential recipients the microbial community had been acclimated to inorganic mercury and thus enriched for populations containing *merA* (the gene encoding mercuric reductase) prior to gene transfer (149). Thus, a complementation of *merA* resulted in expansion of the range of mercurial substrates that were degraded and volatilized by the mer system of aquatic strains. However, when this approach was attempted in microcosms that simulated an estuarine environment, no transconjugants were selected (150), possibly because of low transfer efficiency due to the scarcity of potential recipients among mercury-resistant strains in the microcosms. Thus, complementation is dependent on the particular microbial population of the environment in question as well as on the number of potential recipients.

Furthermore, even if transfer occurs, proper expression of the newly created catabolic pathway is not assured. An example of this was presented by de Lipthay et al. (136), who showed that transfer of the mobilizable plasmid pKJS32 (specifying a 2,4-dichlorophenoxyacetic acid/2-oxoglutarate dioxygenase, the product of the *tfdA* gene [151]) to phenol degrading strains resulted in expression of a new catabolic pathway for the degradation of phenoxyacetic acid, in transconjugant strains using the *ortho*, but not the *meta*, cleavage pathway for phenol degradation. By application of a similar approach in soil microcosms, de Lipthay et al. (152) showed that conjugal transfer of the *tfdA* gene resulted in establishment of the pathway for phenoxyacetic acid degradation in indigenous transconjugant strains (Fig. 6).

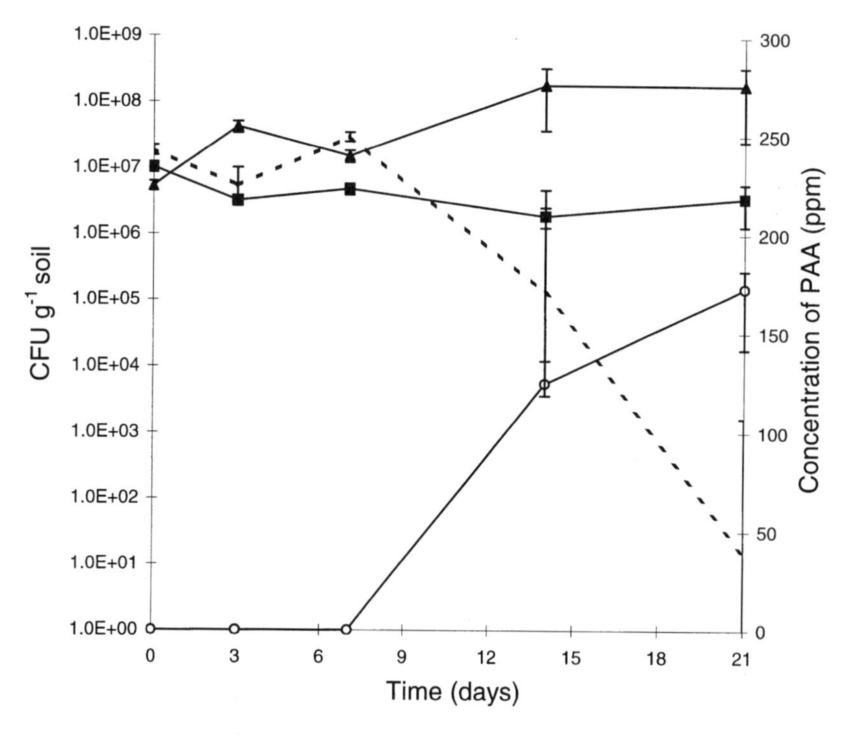

Figure 6 Establishment of new catabolic capabilities in transconjugant bacteria in soil by conjugal transfer of the plasmid pRO103 harboring the *tfdA* gene, encoding a 2,4-dichlorophenoxy acetic acid (2,4-D) dioxygenase. The recipient strain, *Ralstonia eutropha* AEO106 (▲), degrades phenol by an *ortho* cleavage pathway, and by complementation with the *tfdA* gene, transconjugant strains (○) acquire the ability to degrade phenoxyacetic acid (PAA), as shown by the dashed line. The donor strain, *Escherichia coli* HB101/pRO103 (■), is unable to degrade PAA because of its auxotrophic nature solid lines, bacterial counts. (From Ref. 152.)

A few studies have examined the significance of gene transfer as an adaptive mechanism in degradation of xenobiotic compounds in natural environments (e.g., 132,135,150,153). All pointed to the significance of selection for the efficiency and establishment of transconjugant organisms. Fulthorpe and Wyndham (153) found increased evolution of 3-chlorobenzoate degrading indigenous bacteria in lake microcosms exposed to 3-chlorobenzoate and inoculated with a donor carrying the chlorobenzoate-catabolic transposon Tn*5271*. Likewise, Ravatn and coworkers (135) reported increased frequency of catabolic genes in activated-sludge microcosms under selective conditions. In this study the chlorocatechol degradative genes, *clc*, were transferred from *Pseudomonas* sp. B13 to *Pseudomonas putida* F1. Strain F1 is able to degrade toluene, and, by acquisition of the *clc* genes, a new catabolic pathway allowing F1 to degrade chlorinated benzenes was established.

The strategy of complementing existing catabolic pathways by the in situ transfer of catabolic genes to indigenous populations, although valid, is still at the very initial stages of development, requiring additional experimentation and problem solving before its true potential can be evaluated.

IV. CONCLUSIONS

Contemporary molecular techniques provide the ecologist with tools to describe the dominating species in the soil microbial community and to identify the active members of the community under various conditions. Furthermore, they give the opportunity to introduce new genetic traits and thereby change the functional potential of the microbial community. All these features have been applied successfully to soil communities. There is an overwhelming amount of literature on the evolution of new methods, but only a few studies have applied these methods to fundamental ecological questions. It is a great challenge for ecologists in the new millennium to try to close the gap in knowledge on the structural–functional dynamics within the soil environment. A combined approach, giving a functional description of soil systems using enzyme activity measurements and a structural description of the communities present in the soil (using the molecular techniques that are discussed in this chapter), will certainly narrow this gap substantially. This knowledge can in turn be used to introduce novel catabolic capabilities into the soil environment.

REFERENCE

1. L Bakken. Culturable and nonculturable bacteria in soil. In: JD van Elsas, JT Trevors, EMH Wellington, ed. Modern Soil Microbiology. New York: Marcel Dekker, 1997, pp 47–62.
2. T Hattori, H Mitsui, H Haga, N Wakao, S Shikano, K Gorlach, Y Kasahara, A El Beltagy, R Hattori. Advances in soil microbial ecology and the biodiversity. Antonie van Leeuwenhoek 72:21–28, 1997.
3. JD Oliver. Formation of viable but nonculturable cells. In: S Kjelleberg, ed. Starvation in Bacteria. New York: Plenum Press, 1993, pp 239–271.
4. DB Roszak, RR Colwell. Metabolic activity of bacterial cells enumerated by direct viable count. Appl Environ Microbiol 53:2889–2893, 1987.
5. C Woese. Bacterial evolution. Microbiol Rev 51:221–271, 1987.

6. NR Pace. A molecular view of microbial diversity and the biosphere. Science 276:734–740, 1997.
7. L Zelles. Fatty acid patterns of phospholipids and lipopolysaccharides in the characterisation of microbial communities in soil: A review. Biol Fertil Soils 29:111–129, 1999.
8. J Borneman, PW Skroch, KM O'Sullivan, JA Palus, NG Rumjanek, JL Jansen, J Nienhuis, EW Triplett. Molecular microbial diversity of an agricultural soil in Wisconsin. Appl Environ Microbiol 62:1935–1943, 1996.
9. J Borneman, EW Triplett. Molecular microbial diversity in soils from eastern Amazonia: Evidence for unusual microorganisms and microbial population shifts associated with deforestation. Appl Environ Microbiol 63:2647–2653, 1997.
10. R Grosskopf, S Stubner, W Liesack. Novel euryarchaeotal lineages detected on rice roots and in the anoxic bulk soil of flooded rice microcosms. Appl Environ Microbiol 64:4983–4989, 1998.
11. W Liesack, E Stackebrandt. Occurrence of novel groups of the domain Bacteria as revealed by analysis of genetic material isolated from an Australian terrestrial environment. J Bacteriol 174:5072–5078, 1992.
12. WE Holben, JK Jansson, BK Chelm, JM Tiedje. DNA probe method for the detection of specific microorganisms in the soil bacterial community. Appl Environ Microbiol 54:703–711, 1988.
13. V Torsvik, J Goksøyr, FL Daae. High diversity in DNA of soil bacteria. Appl Environ Microbiol 56:782–787, 1990.
14. RJ Steffan, J Goksøyr, AK Bej, RM Atlas. Recovery of DNA from soils and sediments. Appl Environ Microbiol 54:2908–2915, 1988.
15. JD van Elsas, GF Duarte, AS Rosado, K Smalla. Microbiological and molecular methods for monitoring microbial inoculants and their effects in the soil environment. J Microbiol Methods 32:133–154, 1998.
16. R Wagner. The regulation of ribosomal RNA synthesis and bacterial cell growth. Arch Microbiol 161:100–106, 1994.
17. A Teske, C Wawer, G Muyzer, NB Ramsing. Distribution of sulfate-reducing bacteria in a stratified fjord (Mariager Fjord, Denmark) as evaluated by most-probable-number counts and denaturing gradient gel electrophoresis of PCR-amplified ribosomal DNA fragments. Appl Environ Microbiol 62:1405–1415, 1996.
18. A Felske, B Engelen, U Nubel, H Backhaus. Direct ribosome isolation from soil to extract bacterial rRNA for community analysis. Appl Environ Microbiol 62:4162–4167, 1996.
19. A Felske, ADL Akkermans, WM de Vos. Quantification of 16 S rRNA in complex bacterial communities by multiple competitive reverse transcription-PCR in temperature gradient gel electrophoresis fingerprints. Appl Environ Microbiol 64:4581–4587, 1998.
20. AM Prescott, CR Fricker. Use of PNA oligonucleotides for the in situ detection of Escherichia coli in water. Mol Cell Probes 13:261–268, 1999.
21. H Heuer, M Krsek, P Baker, K Smalla, EMH Wellington. Analysis of Actinomycete communities by specific amplification of genes encoding 16S rRNA and gel-electrophoretic separation in denaturing gradients. Appl Environ Microbiol 63:3233–3241, 1997.
22. MA Bruns, JR Stephen, GA Kowalchuk, JI Prosser, EA Paul. Comparative diversity of ammonium oxidizer 16 S rRNA gene sequences in native, tilled and successional soils. Appl Environ Microbiol 65:2994–3000, 1999.
23. GA Kowalchuk, JR Stephen, W de Boer, JI Prosser, TM Embley, JW Woldendorp. Analysis of ammonia-oxidizing bacteria of the b subdivision of the class Proteobacteria in coastal sand dunes by denaturing gradient gel electrophoresis and sequencing of PCR-amplified 16S ribosomal DNA fragments. Appl Environ Microbiol 63:1489–1497, 1997.
24. T Henckel, M Friedrich, R Conrad. Molecular analyses of the methane-oxidizing community in rice field soil by targeting the genes of the 16 S rRNA, particulate methane monooxygenase and methanol dehydrogenase. Appl Environ Microbiol 65:1980–1990, 1999.

25. AS Rosado, GF Duarte, L Seldin, JD van Elsas. Genetic diversity of *nifH* gene sequences in *Paenibacillus azotofixans* strains and soil samples analyzed by denaturing gradient gel electrophoresis of pCR-amplified gene fragments. Appl Environ Microbiol 64:2770–2779, 1998.
26. D Bruce Kenneth. Analysis of mer gene subclasses within bacterial communities in soils and sediments resolved by fluorescent-PCR-restriction fragment length polymorphism profiling. Appl Environ Microbiol 63:4914–4919, 1997.
27. TB Britschgi, RD Fallon. PCR-amplification of mixed 16S rRNA genes from an anaerobic, cyanide-degrading consortium. FEMS Microbiol Ecol 13:225–232, 1994.
28. AJ Martínez-Murcia, SG Acinas, F Rodriguez-Valera. Evaluation of prokaryotic diversity by restrictase digestion of 16S rDNA directly amplified hypersaline environments. FEMS Microbiol Ecol 17:247–256, 1995.
29. E Smit, P Leeflang, K Wernars. Detection of shifts in microbial community structure and diversity in soil caused by copper contamination using amplified ribosomal DNA restriction analisys. FEMS Microbiol Ecol 23:249–261, 1997.
30. W Liu, TL Marsh, H Cheng, LJ Forney. Characterization of microbial diversity by determining terminal restriction fragment length polymorphisms of genes encoding 16 S rRNA. Appl Environ Microbiol 63:4516–4522, 1997.
31. KJ Chin, T Lukow, R Conrad. Effect of temperature on structure and function of the methanogenic archaeal community in an anoxic rice field soil. Appl Environ Microbiol 65:2341–2349, 1999.
32. NF Cariello, TR Skopek. Mutational analysis using denaturing gradient gel electrophoresis and PCR. Mutat Res 288:103–112, 1993.
33. RM Myers, SG Fischer, LS Lerman, T Maniatis. Nearly all single base substitutions in DNA fragments joined to a GC-clamp can be detected dy denaturing gradient gel electrophoresis. Nucleic Acid Res 13:3131–3145, 1985.
34. RM Myers, T Maniatis, LS Lerman. Detection and localozation of single base changes dy denaturing gradient gel electrophoresis. Methods Enzymol 155:501–527, 1987.
35. G Muyzer, EC de Waal, AG Uitterlinder. Profiling of complex microbial populations by denaturing gradient gel electrophoresis analysis of polymerase chain reaction-amplified genes coding for 16S rRNA. Appl Environ Microbiol 59:695–700, 1993.
36. U Nübel, F Garcia-pichel, G Muyzer. PCR primers to amplify 16S rRNA genes from Cyanobacteria. Appl Environ Microbiol 63:3327–3332, 1997.
37. S Rölleke, G Muyzer, C Wawer, G Wanner, W Lubitz. Indetification of bacteria in a biodegraded wall painting by denaturing gradient gel electrophoresis of PCR-amplified gene fragments coding for 16S rRNA. Appl Environ Microbiol 62:2059–2065, 1996.
38. L Øvreås, L Forney, FL Daae, V Torsvik. Distribution of bacterioplankton in meromictic Lake Sælenvannet, as determined by denaturing gradient gel electrophoresis of PCR-amplified gene fragments coding for 16S rRNA. Appl Environ Microbiol 63:3367–3373, 1997.
39. CM Santegoed, SC Nold, DM Ward. Denaturing gradient gel electrophoresis used to monitor the enrichment culture of aerobic chemoorganotrophic bacteria from a hot spring cyanobacterial mat. Appl Environ Microbiol 62:3922–3928, 1996.
40. B Engelen, K Meinken, F von Wintzingerode, H Heuer, H-P Malkomes, H Backhaus. Monitoring impact of pesticide treatment on bacterial soil communities by metabolic and genetic fingerprinting in addition to conventional testing procedures. Appl Environ Microbiol 64: 2814–2821, 1998.
41. A Felske, A Wolterink, R van Lis, ADL Akkermans. Phylogeny of the main bacterial 16S rRNA sequences in Drentse A grassland soils (The Netherlands). Appl Environ Microbiol 64:871–879, 1998.
42. H Heuer, K Smalla. Application of denaturing gradient gel electrophoresis and temperature gradient gel electrophoresis for studying soil microbial communities. In: JD van Elsas, JT

Trevors, EMH Wellington, eds. Modern Soil Microbiology. New York: Maccel Dekker, 1997, pp 353–373.

43. L Øvreås, S Jensen, FL Daae, V Torsvik. Microbial community changes in a perturbed agricultural soil investigated by molecular and physiological approaches. Appl Environ Microbiol 64:2739–2742, 1998.
44. SE Fantroussi, HL Verschuere, W Verstraete, EM Top. Effect of herbicides on soil microbial communities estimated by analysis of 16 S rRNA gene fingerprinting and community level physiological profiles. Appl Environ Microbiol 65:982–988, 1999.
45. U Nübel, B Engelen, A Felske, J Snaidr, A Wieshuber, RI Amann, W Ludwig, H Backhaus. Sequence heterogeneities of genes encoding 16S rRNA in Paenibacillus polymyxa detected by temperature gradient gel electrophoresis. J Bacteriol 178:5636–5643, 1996.
46. V Torsvik, K Salte, R Sørheim, J Goksøyr. Comparison of phenotypic diversity and DNA heterogeneity in a population of soil bacteria. Appl Environ Microbiol 56:776–781, 1990.
47. V Torsvik, FL Daae, R Sandaa, L Øvreås. Novel techniques for analysing microbial diversity in natural and pertubated environments. J Biotech 64:53–62, 1998.
48. LA Porteous, JL Armstrong, RJ Seidler, LS Watrud. An effective method to extract DNA from environmental samples for polymerase chain reaction amplification and DNA fingerprint analysis. Curr Microbiol 29:301–307, 1994.
49. PA Rochelle, JC Fry, RJ Parkes, AJ Weightman. DNA extraction for 16S rRNA gene analysis to determine genetic diversity in deep sediment communities. FEMS Microbiol Lett 100: 59–66, 1992.
50. Y-L Tsai, BH Olson. Rapid method for direct extraction of DNA from soil and sediments. Appl Environ Microbiol 57:1070–1074, 1991.
51. T Volossiouk, EJ Robb, RN Nazar. Direct DNA extraction for PCR-mediated assays of soil organisms. Appl Environ Microbiol 61:3972–3976, 1995.
52. MI Moré, JB Herrick, MC Silva, WC Ghiorse, EL Madsen. Quantitative cell lysis of indigenous microorganisms and rapid extraction og microbial DNA from sediment. Appl Environ Microbiol 60:1572–1580, 1994.
53. A Orgam, GS Sayler, T Barkay. The extraction and purification of microbial DNA from sediments. J Microbiol Methods 7:57–66, 1987.
54. K Smalla, N Cresswell, LC Mendonca-Hagler, A Wolters, JDv Elsas. Rapid DNA extraction protocol from soils for polymerase chain reaction-mediated amplification. J Appl Bacteriol 74:78–85, 1993.
55. W Liesack, PH Janssen, FA Rainey, NL Ward-Rainey, E Stackebrandt. Microbial diversity in soil: The need for a combined approach using molecular and cultivation techniques. In: JD Elsas, JT Trevors, EMH Wellington, eds. Modern Soil Microbiology. New York: Maccel Dekker, 1997.
56. F von Wintzingerode, UB Gobel, E Stackebrandt. Determination of microbial diversity in environmental samples: pitfalls of PCR-based rRNA analysis. FEMS Microbiol Rev 21:213–29, 1997.
57. IG Wilson. Inhibition and facilitation of nucleic acid amplification. Appl Environ Microbiol 63:3741–3751, 1997.
58. MT Suzuki, SJ Giovannoni. Bias caused by template annealing in the amplification of mixtures of 16S rRNA genes by PCR. Appl Environ Microbiol 62:625–630, 1996.
59. V Farelly, FA Rainey, E Stackebrandt. Effect of genome size and rrn gene copy number on pcr amplification of 16S rRNA from a mixture of bacterial species. Appl Environ Microbiol 61:2798–2801, 1995.
60. W Liesack, H Weyland, E Stackebrandt. Potential risks of gene amplification by PCR as determined by 16 S rDNA of a mixed-culture of strict barophilic bacteria. Microbial Ecol 21:191–198, 1991.
61. R Amann, J Snaidr, M Wagner, W Ludwig, K-H Schleifer. In situ visualization of high genetic diversity in a natural microbial community. J Bacteriol 178:3496–3500, 1996.

62. MA Tanner, BM Goebel, MA Dojka, NR Pace. Specific ribosomal DNA sequences from diverse environmental settings correlate with experimental contaminants. Appl Environ Microbiol 64:3110–3113, 1998.
63. BM Duineveld, AS Rosado, JDv Elsas, JA Veen. Analysis of the dynamics of bacterial communities in the rhizosphere of the chrysanthemum via denaturing gradient gel electrophoresis and substrate utilization patterns. Appl Environ Microbiol 64:4950–4957, 1998.
64. LD Rasmussen, SJ Sørensen. Effects of mercury contamination on the culturable heterotrophic, functional and genetic diversity of the bacterial community in soil. Antonie Van Leeuwenhoek J Microbiol.
65. LD Rasmussen, SJ Sørensen, RR Turner, T Barkay. Application of a mer-lux biosensor for estimating bioavailable mercury in soil and its utility in relating the response of soil microbial communities to bioavailable mercury. Soil Biol Biochem.
66. A Müller, K Westergaard, SJ Sørensen. The effect of long term mercury pollution on the soil microbial community.
67. JC Thomas, M Desrosiers, Y Stpierre, P Lirette, JG Bisaillon, R Beaudet, R Villemur. Quantitative flow cytometric detection of specific microorganisms in soil samples using rRNA targeted fluorescent probes and ethidium bromide. Cytometry 27:224–232, 1997.
68. G Wallner, B Fuchs, S Spring, W Beisker, R Amann. Flow sorting of microorganisms for molecular analysis. Appl Environ Microbiol 63:4223–4231, 1997.
69. N Araki, A Ohashi, I Machdar, H Harada. Behaviors of nitrifiers in a novel biofilm reactor employing hanging sponge-cubes as attachment site. Water Sci Technol 39:23–31, 1999.
70. LW Clapp, JM Regan, F Ali, JD Newman, JK Park, DR Noguera. Activity, structure, and stratification of membrane-attached methanotrophic biofilms cometabolically degrading trichloroethylene. Water Sci Technol 39:153–161, 1999.
71. W Manz, K Wendt-Potthoff, TR Neu, U Szewzyk, JR Lawrence. Phylogenetic composition, spatial structure, and dynamics of lotic bacterial biofilms investigated by fluorescent in situ hybridization and confocal laser scanning microscopy. Microb Ecol 37:225–237, 1999.
72. S Rocheleau, CW Greer, JR Lawrence, C Cantin, L Laramee, SR Guiot. Differentiation of Methanosaeta concilii and Methanosarcina barkeri in anaerobic mesophilic granular sludge by fluorescent in situ hybridization and confocal scanning laser microscopy. Appl Environ Microbiol 65:2222–2229, 1999.
73. PH Nielsen, K Andreasen, N Lee, M Wagner. Use of microautoradiography and fluorescent in situ hybridization for characterization of microbial activity in activated sludge. Water Sci Technol 39:1–9, 1999.
74. BE Rittmann, CS Laspidou, J Flax, DA Stahl, V Urbain, H Harduin, JJ van der Waarde, B Geurkink, MJC Henssen, H Brouwer, A Klapwijk, M Wetterauw. Molecular and modeling analyses of the structure and function of nitrifying activated sludge. Water Sci Technol 39: 51–59, 1999.
75. S Juretschko, G Timmermann, M Schmid, KH Schleifer, A Pommereningroser, HP Koops, M Wagner. Combined molecular and conventional analyses of nitrifying bacterium diversity in activated sludge: nitrosococcus mobilis and nitrospira-like bacteria as dominant populations. Appl Environ Microbiol 64:3042–3051, 1998.
76. M Wagner, DR Noguera, S Juretschko, G Rath, HP Koops, KH Schleifer. Combining fluorescent in situ hybridization (fish) with cultivation and mathematical modeling to study population structure and function of ammonia-oxidizing bacteria in activated sludge. Water Sci Technol 37:441–449, 1998.
77. KZ Falz, C Holliger, R Grosskopf, W Liesack, AN Nozhevnikova, B Muller, B Wehrli, D Hahn. Vertical distribution of methanogens in the anoxic sediment of Rotsee (Switzerland). Appl Environ Microbiol 65:2402–2408, 1999.
78. KL Straub, BEE Buchholz-Cleven. Enumeration and detection of anaerobic ferrous iron-oxidizing, nitrate-reducing bacteria from diverse European sediments. Appl Environ Microbiol 64:4846–4856, 1998.

79. A Chatzinotas, RA Sandaa, W Schonhuber, R Amann, FL Daae, V Torsvik, J Zeyer, D Hahn. Analysis of broad-scale differences in microbial community composition of two pristine forest soils. Syst Appl Microbiol 21:579–587, 1998.
80. R Großkopf, PH Janssen, W Liesack. Diversity and structure of the methanogenic community in anoxic rice paddy soil microcosms as examined by cultivation and direct 16S rRNA gene sequence retrieval. Appl Environ Microbiol 64:960–969, 1998.
81. K Ravenschlag, K Sahm, J Pernthaler, R Amann. High bacterial diversity in permanently cold marine sediments. Appl Environ Microbiol 65:3982–3989, 1999.
82. S Logemann, J Schantl, S Bijvank, M Van Loosdrecht, JG Kuenen, M Jetten. Molecular microbial diversity in a nitrifying reactor system without sludge retention. FEMS Microbiol Ecol 27:239–249, 1998.
83. RA Sandaa, O Enger, V Torsvik. Abundance and diversity of Archaea in heavy-metal-contaminated soils. Appl Environ Microbiol 65:3293–3297, 1999.
84. YV de Peer, E Robbrecht, S de Hoog, A Caers, P De Rijk, R De Wachter. Database on the structure of small subunit ribosomal RNA. Nucleic Acids Res 27:179–183, 1999.
85. JC Zak, MR Willig, DL Moorhead, HG Wildman. Functional diversity of microbial communities: A quantitative approach. Soil Biol Biochem 26:1101–1108, 1994.
86. K Smalla, U Wachtendorf, H Heuer, W-T Liu, L Forney. Analysis of BIOLOG GN substrate utilisation patterns by microbial communities. Appl Environ Microbiol 64:1220–1225, 1998.
87. A Winding, NB Hendriksen. Biolog substrate utilization assay for metabolic fingerprints of soil bacteria: Incubation effects. In: H Insam, A Rangger, ed. Microbial Communities, Functional Versus Structural Approaches. Berlin Heidelberg: Springer-Verlag, 1997, pp 195–205.
88. JH Lawton. What do species do in ecosystems? Oikos 71:367–374. 1994.
89. S Naeem, LJ Thompson, SP Lawler, JH Lawton, RM Woodfin. Empirical evidence that declining species diversity may alter the performance of terrestrial ecosystems. Philos Trans R Soc London 347:249–262, 1995.
90. D Tilman, D Wedin, J Knops. Productivity and sustainability influenced by biodiversity in grassland ecosystems. Nature 379:718–720, 1996.
91. DA Wardle, KI Bonner, KS Nicholson. Biodiversity and plant litter: Experimental evidence which does not support the view that enhanced species richness improves ecosystem function. Oikos 79:247–258, 1997.
92. D Tilman, J Knops, D Wedin, P Reich, M Ritche, E Siemann. The influence of functional diversity and composition on ecosystem processes. Science 277:1300–1302, 1997.
93. DU Hooper, PM Viousek. The effects of plant composition on ecosystem processes. Science 277:1302–1305, 1997.
94. BS Griffiths, K Ritz, RE Wheatley. Relationship between functional diversity and genetic diversity in complex microbial communities. In: H Insam, A Rangger, ed. Microbial Communities, Functional Versus Structural Approaches. Springer-Verlag, Berlin, 1997, pp 1–9.
95. BS Griffiths, K Ritz, RD Bardgett, R Cook, S Christensen, F Ekelund, SJ Sørensen, E Bååth, J Bloem, P De Ruiter, J Dolfin, B Nicolardot. Stability of soil ecosystem processes following the experimental manipulation of soil microbial community diversity. Soil Biol Biochem (in press).
96. J Mikola, H Setälä. Relating species diversity to ecosystem functioning: Mecanistic backgrounds and experimental approach with a decomposer food web. Oikos 83:180–194, 1998.
97. V Wolters. Methods and approaches in studying functional implications of biodiversity in soil. In: V Wolters, ed. Functional Implications of Biodiversity in Soil. European Commission. Brussels, Belgium, 1997, pp 17–21.
98. FS Chapin, BH Walker, RJ Hobbs, DU Hooper, JH Lawton, OE Sala, D Tilman. Biotic control over the function of ecosystems. Science 277:500–504, 1997.
99. LB Finlay, SC Maberly, JI Cooper. Microbial diversity and ecosystem function. Oikos 80: 220–225, 1997.
100. M Begon, JL Harper, CR Townsend. Ecology. 3rd ed. Blackwell Science, Cambridge, 1996.

101. M Alexander. Biodegradation of chemicals of environmental concern. Science 211:132–138, 1981.
102. DT Gibson. The microbial oxidation of aromatic hydrocarbons. Crit Rev Microbiol 1:199–223, 1971.
103. H-J Knackmuss. Basic knowledge and perspectives of bioelimination of xenobiotic compounds. J Biotechnol 51:287–295, 1996.
104. JC Spain, PH Pritchard, AW Bourquin. Effects of adaptation on biodegradation rates in sediment/water cores from estaurine and freshwater environments. Appl Environ Microbiol 40:726–734, 1980.
105. T Barkay, H Pritchard. Adaptation of aquatic microbial communities to pollutant stress. Microbiol Sci 5:165–169, 1988.
106. CM Aelion, CM Swindoll, FK Pfaender. Adaptation to and biodegradation of xenobiotic compounds by microbial communities from a pristine aquifer. Appl Environ Microbiol 53: 2212–2217, 1987.
107. JC Spain, PA van Veld. Adaptation of natural microbial communities to degradation of xenobiotic compounds: Effects of concentration, exposure time, inoculum, and chemical structure. Appl Environ Microbiol 45:428–535, 1983.
108. CM Swindoll, CM Aelion, FK Pfaender. Influence of inorganic and organic nutrients on aerobic biodegradation and on the adaptation response of subsurface microbial communities. Appl Environ Microbiol 54:212–217, 1988.
109. JR van der Meer, WM de Vos, S Harayama, AJB Zehnder. Molecular mechanisms of genetic adaptation to xenobiotic compounds. Microbiol Rev 56:677–694, 1992.
110. D Ghosal, I-S You, DK Chatterjee, AM Chakrabarty. Microbial degradation of halogenated compounds. Science 228:135–142, 1985.
111. S Dagley. Biochemistry of aromatic hydrocarbon degradation in Pseudomonads. In: JR Sokatch, LN Ornston, ed. The Bacteria. Vol. 10. Academic Press, Orlando, 1986, pp 527–555.
112. LCM Commandeur, JR Parsons. Degradation of halogenated aromatic compounds. Biodegr 1:207–220, 1990.
113. M Schlömann. Evolution of chlorocatechol catabolic pathways. Biodegr 5:301–321, 1994.
114. PA Williams, JR Sayers. The evolution of pathways for aromatic hydrocarbon oxidation in Pseudomonas. Biodegr 5:195–217, 1994.
115. JR van der Meer. Evolution of novel metabolic pathways for the degradation of chloroaromatic compounds. Antonie van Leeuwenhoek 71:159–178, 1997.
116. JR van der Meer. Genetic adaptation of bacteria to chlorinated aromatic compounds. FEMS Microbiol Rev 15:239–249, 1994.
117. N Cresswell, EMH Wellington. Detection of genetic exchange in the terrestrial environment. In: EMH Wellington, JD van Elsas, ed. Genetic Interactions Among Microorganisms in the Natural Environment. Oxford: Pergamon Press, 1992, pp 59–82.
118. SJ Sørensen, LE Jensen. Transfer of plasmid RP4 in bulk soil and barley seedling microcosms. Antonie van Leeuwenhoek J Microbiol 73:69–77, 1998.
119. SJ Sørensen, T Schyberg, R Rønn. Predation by protozoa on *Escherichia coli* K12 in soil and transfer of resistance plasmid RP4 to indigenous bacteria in soil. Appl Soil Ecol 11:79–90, 1999.
120. B Frantz, AM Chakrabarty. Degradative plasmids in Pseudomonas. In: JR Sokatch, LN Ornston, ed. The Bacteria. Vol. 10. Academic Press, Orlando, 1986, pp 295–323.
121. GS Sayler, SW Hooper, AC Layton, JMH King. Catabolic plasmids of environmental and ecological significance. Microb Ecol 19:1–20, 1990.
122. JIA Campbell, CS Jacobsen, J Sørensen. Species variation and plasmid incidence among fluorescent Pseudomonas strains isolated from agricultural and industrial soils. FEMS Microbiol Ecol 18:51–62, 1995.
123. LD Rasmussen, SJ Sørensen. The effect of longterm exposure to mercury on the bacterial community in marine sediment. Curr Microbiol 36:291–297, 1998.

124. MM Zambrano, DA Siegele, M Almirón, A Tormo, R Kolter. Microbial competition: Escherichia coli mutants that take over stationary phase cultures. Science 259:1757–1760, 1993.
125. W Arber, T Naas, M Blot. Generation of genetic diversity by DNA rearrangements in resting bacteria. FEMS Microbiol Ecol 15:5–14, 1994.
126. ST Kellogg, DK Chatterjee, AM Chakrabarty. Plasmid-assisted molecular breeding: New technique for enhanced biodegradation of persistent toxic chemicals. Science 214:1133–1135, 1981.
127. L Kröckel, DD Focht. Construction of chlorobenzene-utilizing recombinants by progenitive manifestation of a rare event. Appl Environ Microbiol 53:2470–2475, 1987.
128. JO Ka, WE Holben, JM Tiedje. Analysis of competition in soil among 2,4-dichlorophenoxyacetic acid-degrading bacteria. Appl Environ Microbiol 60:1121–1128, 1994.
129. R Tchelet, R Meckenstock, P Steinle, JR van der Meer. Population dynamics of an introduced bacterium degrading chlorinated benzenes in a soil column and in sewage sludge. Biodegr 10:113–125, 1999.
130. JK Struthers, K Jayachandran, TB Moorman. Biodegradation of atrazine by Agrobacterium radiobacter J14a and use of this strain in bioremediation of contaminated soil. Appl Environ Microbiol 64:3368–3375, 1998.
131. LL Daane, MM Häggblom. Earthworm egg capsules as vectors for the environmental introduction of biodegradative bacteria. Appl Environ Microbiol 65:2376–2381, 1999.
132. EM Top, P van Daele, N de Saeyer, LJ Forney. Enhancement of 2,4-dichlorophenoxyacetic acid (2,4-D) in soil by dissemination of catabolic plasmids. Antonie van Leeuwenhoek 73: 87–94, 1998.
133. GD DiGiovanni, JW Nielson, IL Pepper, NA Sinclair. Gene transfer of Alcaligenes eutrophus JMP34 plasmid pJP4 to indigenous soil recipients. Appl Environ Microbiol 62:2521–2526, 1996.
134. T Barkay, C Liebert, M Gillman. Conjugal transfer to aquatic bacteria detected by the generation of a new phenotype. Appl Environ Microbiol 59:807–814, 1993.
135. R Ravatn, AJB Zehnder, JR van der Meer. Low-frequency horizontal transfer of an element containing the chlorocatechol degradation genes from Pseudomonas sp. strain B13 to Pseudomonas putida F1 and to the indigenous bacteria in laboratory-scale activated-sludge microcosms. Appl Environ Microbiol 64:2126–2132, 1998.
136. JR de Lipthay, T Barkay, J Vekova, SJ Sørensen. Utilization of phenoxyacetic acid, by strains using either the ortho or meta cleavage of catechol during phenol degradation, after conjugal transfer of tfdA, the gene encoding a 2,4-dichlorophenoxyacetic acid/2-oxoglutarate dioxygenase. Appl Microbiol Biotechnol 51:207–214, 1999.
137. JL Ramos, KN Timmis. Experimental evolution of catabolic pathways of bacteria. Microbiol Sci 4:228–237, 1987.
138. I Bartels, H-J Knackmuss, W Reineke. Suicide inactivation of catechol 2,3-dioxygenase from Pseudomonas putida mt-2 by 3-halocatechols. Appl Environ Microbiol 47:500–505, 1984.
139. F Rojo, DH Pieper, K-H Engesser, H-J Knackmuss, KN Timmis. Assemblage of ortho cleavage route for simultaneous degradation of chloro- and methylaromatics. Science 238:1395–1398, 1987.
140. DH Pieper, K Stadler-Fritzsche, H-J Knackmuss, KN Timmis. Formation of dimethylmuconolactones from dimethylphenols by Alcaligenes eutrophus JMP134. Appl Environ Microbiol 61:2159–2165, 1995.
141. JL Ramos, E Duque, MI Ramos-Gonzalez. Survival in soils of an herbicide-resistant Pseudomonas putida strain bearing a recombinant TOL plasmid. Appl Environ Microbiol 57: 260–266, 1991.
142. DA Newcombe, DE Crowley. Bioremediation of atrazine-contaminated soil by repeated applications of atrazine-degrading bacteria. Appl Microbiol Biotechnol 51:877–872, 1999.
143. KA Short, RJ King, RJ Seidler, RH Olson. Biodegradation of phenoxyacetic acid in soil by

Pseudomonas putida PP0301 (pRO103), a constitutive degrader of 2,4-dichlorophenoxyacetate. Mol Ecol 1:89–94, 1992.

144. RN Pertsova, F Kunc, LA Golovleva. Degradation of 3-chlorobenzoate in soil by pseudomonads carrying biodegradative plasmids. Folia Micrbiol (Praha) 29:242–247, 1984.
145. RU Halden, SM Tepp, BG Halden, DF Dwyer. Degradation of 3-phenoxybenzoic acid in soil by Pseudomonas pseudoalcaligenes POB310 (pPOB) and two modified Pseudomonas strains. Appl Environ Microbiol 65:3345–3349, 1999.
146. BE Rittmann, BF Smets, DA Stahl. The role of genes in biological processes. Environ Sci Technol 24:23–29, 1990.
147. H van Limbergen, EM Top, W Verstraete. Bioaugmentation in activated sludge: Current features and future perspectives. Appl Microbiol Biotechnol 50:16–23, 1998.
148. JW Nielson, KL Josephason, IL Pepper, RB Arnold, GD Di Giovanni, NA Sinclair. Frequency of horizontal gene transfer of a large catabolic plasmid (pJP4) in soil. Appl Environ Microbiol 60:4053–4058, 1994.
149. T Barkay. Adaptation of aquatic microbial communities to Hg2+ stress. Appl Environ Microbiol 53:2725–2732, 1987.
150. T Barkay, N Kroer, LD Rasmussen, SJ Sørensen. Conjugal transfer at natural population densities in a microcosm simulating an estuarine environment. FEMS Microbiol Ecol 16: 43–54, 1995.
151. WR Streber, KN Timmis, MH Zenk. Analysis, cloning, and high-level expression of 2,4-dichlorophenoxyacetate monooxygenase gene tfdA of Alcaligenes eutrophus JMP134. J Bacteriol 169:2950–2955, 1987.
152. J Radnoti de Lipthay, T Barkay, SJ Sørensen. Enhanced degradation of phenoxyacetic acid in soil by horizontal transfer of the tfdA gene encoding a 2,4-dichlorophenoxyacetic acid dioxygenase. FEMS Microbiology Ecology 35:75–84.
153. RR Fulthorpe, RC Wyndham. Involvement of a chlorobenzoate-catabolic transposon, Tn5271, in community adaptation to chlorobiphenyl, chloroaniline, and 2,4-dichlorophenoxyacetic acid in a freshwater ecosystem. Appl Environ Microbiol 58:314–325, 1992.

15

Bioindicators and Sensors of Soil Health and the Application of Geostatistics

Ken Killham
University of Aberdeen, Aberdeen, Scotland

William J. Staddon
Eastern Kentucky University, Richmond, Kentucky

I. INTRODUCTION

We require the soil to perform a variety of key functions. It must provide the food, fuel, and fiber needs of the world's burgeoning population and must also regulate the quality of the air we breathe and the water we drink. We also require the soil to act as a sink for the many pollutants generated by human domestic, agricultural, and industrial activities. Because of the conflicting pressures increasingly applied to the soil resource, there is a crucial need for the capacity to assess and monitor the health or quality of soil. In 1996 the Soil Science Society of America (1) defined *soil health* as "the continued capacity of a specific kind of soil to function as a vital living system, within natural or managed ecosystem boundaries, to sustain plant and animal productivity, to maintain or enhance the quality of air and water environments, and to support human health and habitation."

The definition offered by the society provides a useful basis for considering the relevance of bioindicators and sensors for the assessment of soil health. It is clear from the definition that relevant indicators and sensors must contribute to measurement of the functional integrity of soil in order to assess whether it can sustain its key roles. As discussed in later sections in this chapter, it is unlikely that any one property or process (and therefore a single bioindicator or biosensor) is sufficient to provide a reliable measure of soil health. It is much more likely that indicators and sensors will be used in a battery of tests in which enzymes of plant, microbial, and animal origin play a part.

As well as exercising care in terms of overreliance on single bioindicators and biosensors, it has been pointed out (2) that whereas scientists select indicators for the link with functions of soil quality, others such as agriculturalists may just as validly characterize soil health by using descriptive properties such as tilth with a direct value judgment.

This chapter reviews the main biological properties/systems that can be used as indicators and sensors of soil health and the application of geostatistics for describing the spatial variability of these properties.

II. BIOINDICATORS OF SOIL HEALTH

A. Definition

A *bioindicator* is defined as "an organism, part of an organism, product of an organism (e.g., enzyme), collection of organisms or biological process which can be used to obtain information on the quality of all or part of the environment." A number of bioindicators have been suggested for monitoring soil health, and these are briefly considered.

B. Soil Microbial Biomass

Jenkinson and Rayner (3) defined the *soil microbial biomass* as the "eye of the needle" through which all organic matter in soil must eventually pass. It is therefore the key driver of ecosystem productivity and, despite the fact that the microbial biomass typically represents about 5 tons per hectare of a temperate grassland ecosystem compared to biomass of the vegetation an order of magnitude greater (4), most of the carbon/energy and nutrient flow is through the soil microbial biomass.

In the 1970s and 1980s a considerable number of methods for determining soil microbial biomass were developed. These methods have been reviewed by Sparling and Ross (5) and are dominated by two techniques: chloroform fumigation and substrate-induced respiration. The fumigation techniques are based on the susceptibility of microbial biomass to chloroform vapor. The chloroform-labile carbon mobilized is determined either by measuring the CO_2 released by mineralization when the fumigated soil is incubated or by measuring the C that can be extracted from the fumigated soil. Biomass N, P, and S can also be determined after extraction from fumigated soil. Substrate-induced respiration techniques are physiologically based; they involve providing the soil microbial biomass with a saturating concentration of a readily mineralizable substrate (usually glucose) and monitoring the respiration over a short incubation period. The substrate saturation rate of respiration represents maximal reaction velocity and is therefore proportional to the biomass. Conversion factors are available to convert the V_{max} respiration and the C, N, and P extracted from fumigated soil into a biomass value.

Because the soil microbial biomass is the main processing unit for organic matter, its size tends to be roughly proportional to the total organic matter pool. Skeletal montane soils, for example, have low organic matter and a correspondingly low microbial biomass. Deciduous woodland soils, on the other hand, have much higher organic matter status and a higher microbial biomass. Typical biomass values for a range of soils are reported in a 1997 review by Sparling (2).

Because the microbial biomass is generally related to the organic matter content of the host soil, it is not the absolute size of the biomass that indicates soil health but changes in biomass size (other than those that result from seasonal and other natural factors). The soil microbial biomass can therefore be seen as a barometer, with reductions in biomass related to either a reduction in the carbon inputs that sustain it or a toxic impact of some kind (6). A change in biomass size then heralds a later change in soil organic matter status. The predictive value of measuring soil microbial biomass as a bioindicator of soil fertility has been suggested by a number of researchers (7,8).

Although the soil microbial biomass can, as mentioned, be affected by toxic impacts, there are numerous soil contaminants that can adversely affect the biological functioning of the soil but that do not affect the size of the biomass itself. Some of these contaminants affect the respiratory quotient of the biomass (i.e., the rate of soil microbial respiration is a function of biomass size) rather than biomass alone (9), but others are more subtle and require other bioindicators in order for their impact on soil health to be evaluated.

C. Carbon and Nutrient Cycling

Mineralization reactions are vital both for the turnover of organic residue inputs to soil and for the release of bound nutrients to plants. The mineralization reactions are carried out by both soil animals and microbes. The former group may not rival the microbes in terms of total carbon/energy, nutrient flow, and breadth of their enzymatic activities, but they have a key role in comminuting organic debris and sometimes acting as vectors in inoculating the newly exposed surfaces with microbial degraders.

The measurement of rates of mineralization of organic C and associated nutrients (e.g., N, P, and S) probably targets the best overall bioindication of soil health. So many organisms are involved in these processes, however, that such measurements are unlikely to identify effects on individual species that may themselves still be of importance to soil health.

Carbon mineralization has generally been measured by loss of substrate (e.g., the traditional litter-bag techniques) or by respiration of CO_2. Measurement of carbon mineralization rates can be defined by use of C isotopes. This technique enables all mineralized C to be assessed and allows quantification of the partitioning of carbon into biomass and into cell maintenance. Various quotients can then be determined, and these can indicate stress to the microbial biomass as well as rates of C mineralization. This is because the degree to which soil microorganisms partition carbon into biomass versus maintenance of cell integrity is largely a function of environmental stress (10).

Nitrogen mineralization measurements can be made both aerobically and anaerobically. The advantage of the latter is that it precludes many of the problems of reimmobilisation of N due to microbial processing of C during cell growth and synthesis (11). As with C mineralization, the use of isotopic techniques has done a great deal to facilitate N mineralization determinations. Isotope dilution techniques involving ^{15}N enable gross mineralization to be reliably measured (12).

In contrast to the mineralization processes, nitrification is a soil N cycle flux involving very few species (Table 1). The simplified reactions illustrated indicate how potentially sensitive a bioindicator the nitrification process can be. Because the process is the domain

Table 1 Soil Microorganisms Involved in Mineralization and Nitrification N Fluxes.

N flux	Soil microorganisms involved
N mineralization Organic N $\rightarrow NH_3/NH_4^+$	Most of the heterotrophs, which dominate the soil microbial biomass.
Nitrification $NH_4 + O_2 + H^+ + 2e^- \rightarrow NO_2^- + 5H^+ + 4e^-$ $N_2O^- + H_2O \rightarrow NO_3^- + 2H$	In most soils, the first nitrification step is dominated by the genera *Nitrosolobus*, *Nitrosospira*, and to a lesser extent *Nitrosomonas*; the second by the genus *Nitrobacter*; in acid forest soils, these autotrophs are replaced by a range of heterotrophs (mainly fungi).

of a very few specialist, chemoautotrophic bacteria, any factor that adversely affects these ''keystone'' species dramatically affects the process (and hence the release of the most plant-available form of mineral N in soils, nitrate). It is for this reason that screening tests for pesticides and other agrochemicals always include assessment of impact on nitrification and why environmental risk assessments of soil pollutant also include nitrification (13). However, in many cases, good soil health does not require a high supply of available nutrients through processes such as nitrification (2).

D. Soil Enzymes

Although enzymes contribute to the part played by the other bioindicators considered in this review, it is particularly important to appreciate the invaluable integrative role of a suite of enzymes in assessing soil health. This is because of the massive array of enzyme assays that can readily be applied to soil, encompassing the hydrolases (e.g., phosphatases, sulfatases, urease, proteases, peptidases, deaminases, cellulases), the oxidoreductases (e.g., dehydrogenases, phenol oxidases, peroxidases, catalases), the lyases, and the transferases. Many soil enzymes have a functional location that is outside the cell, and the significance of these and other enzymes in soil microbial ecology has been reviewed (14). These extracellular enzymes are often relatively stable and can persist for extended periods, thereby providing a longer-term perspective than measurements involving extant soil organisms alone. The impact of pollutants on soil health has been addressed through the measurement of enzyme activity. Such an approach offers a useful soil management tool as soil enzyme activity should relate to key soil functions such as biogeochemical cycling, plant growth, and degradation of organic contaminants (15).

Enzymes that catalyze a wide range of soil biological processes offer a useful assessment of soil ''function'' (14), and common enzymes, such as dehydrogenase, urease, and phosphatase, fit into this category. Metabolic stains such as fluorescein diacetate (FDA) also provide a useful functional indicator (16). The assay works on the principle that the FDA molecule is taken by active cells and hydrolyzed by a range of enzymes, including proteases, lipases, and esterases. This releases the fluorochrome fluorescein so that enzymatically active cells can easily be distinguished with the aid of a fluorescence microscope with an ultraviolet (UV) source.

Enzymes that catalyze a narrow range of soil biological activity are useful when sensitive indicators of change, such as may result from a pollution event, are sought. Enzymes catalyzing the degradation of certain organoxenobiotics (e.g., polyaromatic hydrocarbons [PAHs], polychlorinated biphenyls [PCBs], dioxins) fall into this category. Knowledge of a reduction in a soil's capacity to act as a fully functional mineralization medium for pollutants is critical in overall soil health assessment, but particularly in waste management (17) and as an indicator of the successful bioremediation of contaminated land (16).

E. Community Structure and Biodiversity

In recent years, a great deal of research has been devoted to developing and optimizing methods to assess the structure of the soil microbial community in terms of taxonomy and in terms of function.

Developments in molecular biology have now provided soil biologists with ''off the shelf'' methods for assessing microbial diversity. This has represented nothing short of a revolution, allowing the genetic and functional diversity of the whole community, rather than just the very small percentage that can be cultured in the laboratory, to be measured for the first time. The molecular and other methods available for analysis of microbial community structure was reviewed in 1997 by White and McNaughton (18) and are briefly discussed in relation to soil health in the following section.

The genetic diversity of the soil microbial community can now be assessed by using broad screening methods as well as methods with a narrow focus. The broad screening methods, such as deoxyribonucleic acid (DNA) reanealling kinetics (i.e., the rate at which melted, single-standard DNA reaneals on cooling depends on the genetic diversity), and denaturation gradient gel electrophoresis/thermal gradient gel electrophoresis (DGGE/TGGE), methods that aim to quantify genetic diversity by exploring banding patterns of soil microbial DNA by gel electrophoresis, may have a future contributory role in soil health assessment, but probably in combination with more focused probing at the genus and the specific level. The latter gene probes use DNA and ribonucleic acid (RNA) techniques and can be linked to polymerase chain reaction (PCR) methodologies for increased sensitivity of detection. 16S-Ribosomal RNA probes are now particularly well developed for the better characterized groups of soil bacteria (19) and have contributed considerably to our understanding of genetic diversity in soil. DNA probes linked to enzymes with specific functions provide a more activity-based assessment and have, for example, been used to assess the presence of xenobiotic degraders (20) and denitrifiers (21) in soil. Messenger RNA, with its very short turnover, can be probed to provide ''real-time'' functional assessment. When such probes are linked to fluorescent tags, they can also provide spatial information on genetic/functional diversity. The RNA probes now represent a standard ecological tool that will increase in power of resolution as more and more systems are developed. This particularly applies to the soil fungi (both free-living and symbiotic), for which molecular techniques are still in their infancy; to the less well characterized bacteria; and to the microfauna.

Development of molecular probes to assess functional diversity has partly been driven by the limitations of techniques that rely on the culturability of soil microbes. Of these techniques, the most widely used is probably the BIOLOG system. This system is based on physiological profiling—the range and number of carbonaceous sole substrates utilized by the enzymatic activity of microbial communities or by individual soil microorganisms—and the data generated can be interrogated by principal component analysis to differentiate between soils or to assess changes in soil health (22).

F. Soil Animals

Because of the fundamental importance of soil animals in carbon and nutrient cycling, their abundance and diversity have been used to provide a key contribution to the overall assessment of soil health (23). There are a number of relatively simple methods for extracting the micro- and mesofauna from soil (24), although identification beyond genus level without considerable experience is difficult.

1. Microfauna and Soil Health

Numerous workers have established the potential of using protozoa and nematodes as indicators of soil health because of their tremendous abundance, their production of a wide range of enzymes for roles ranging from plant pathogenicity to mineralization of soil organic matter, and their scope for culturing the former for use in linked bioassays

(25). The diversity and abundance of soil protozoa (26) and nematodes (27) can be significantly reduced by the impact, for example, of air-borne pollutants and by heavy metal–contaminated wastes. Because of the trophic interactions that link the activity of the soil protozoa and the nematodes both to plants and to the bacteria and the fungi (4), such reductions in microfaunal abundance and diversity can have a profound effect on soil health.

2. *Mesofauna and Macrofauna and Soil Health*

That mesofaunal groups, such as the arthropods, and their associated enzymatic activities have long been used to assess ecosystem impacts of pollution suggests that they represent important bioindications of soil health. The contrasting ecophysiological characteristics of many of the soil arthropods provide the key to their value as bioindicators. For example, comparisons of the median pH preference of soil arthropods have identified the strength of the indicator value of individual arthropods with respect to this soil parameter (28). Presumably, this approach can be applied to other soil parameters such as organic matter quantity and quality, and ultimately to soil health.

The earthworms represent the most studied group of soil animals and links between earthworms and soil health have been suggested for centuries. In 1997, these links were more reliably quantified in agroecosystems with a reasonably strong correlation between the yield of a cereal crop and the biomass of earthworms in the soil supporting the crop (29). Earthworm bioindication of pollutant impacts on soil health has considerable merit and addresses pollutant bioavailability rather than total concentrations. Furthermore, it has been pointed out (30) that the different ecophysiological strategies of the earthworms provide scope for differentiating certain pollutant effects—the epigeic (surface dwelling) species tend to be directly affected by surface-deposited pollutants, whereas the endogeic (soil-dwelling) species tend to experience more chronic exposure through ingestion of soil contaminated with ''aged'' pollutants.

There are numerous advantages to the use of earthworms as bioindicators of soil health. They are relatively easy to sample and enumerate and, with some experience and care, can be readily identified. Their relatively long generation times compared to those of many other soil invertebrates also allow sampling to identify changes in soil health to be done somewhat less frequently. The use of earthworms as well as other soil animals as bioindicators of soil health must be considered carefully for soils where management has uncoupled the natural linkage between soil faunal activity and the soil's capacity to sustain crop growth as well as other soil functions. The use of pesticides and fertilizers may have this effect, for example, massively reducing the population density of the earthworms, and yet the farmer would describe the soil as fit for purpose and in good health. It has been concluded therefore that the high variability of earthworm abundance is determined by factors other than those that most influence soil health and crop yield (31).

G. Plants

The importance of plants as bioindicators of soil health has been known since ancient times (32) where the presence of a particular ''natural'' plant species or the condition of a ''crop'' species is diagnostic of soil conditions, be they physical, chemical, and/or even biological. Where a high degree of diagnostic sensitivity is required, production of particular chemicals or ''biomarkers'' by certain plant species can be used (33). These biomarkers include a range of primary and secondary metabolites, the former including the amino acid proline (34) and the latter including polyamines such as spermidine and putrescine (35). The activities of certain plant enzymes, such as peroxidases and catalases, can also be used as biomarkers, particularly for assessing pollutant impacts (36).

Plants can serve as bioindicators of toxic pollutant effects on soil health through three means: either pollutant accumulation in tissues, absence or presence of key plant species in a vegetation community, or physiological and biochemical changes to the plant. Plants that provide useful bioindicators in this regard have been proposed for different classes of pollutant. Plant response to metals is particularly well documented (plants are either metal accumulators, metal excluders, or metal indicators, depending on whether their tissue concentrations indicate accumulation or exclusion, or reflect soil concentrations, respectively) (37). This background knowledge of plant response greatly facilitates selection of plant species and the means of bioindication.

Plants have a number of major advantages as bioindicators of soil health. They are relatively cessile, they are generally easy to identify and analyze, and their root systems can integrate over space and time. This last named property is of great importance when many of the chemical and physical properties of soil are heterogeneous in distribution and can change at the microscale.

III. BIOSENSORS OF SOIL HEALTH

A. Definitions

A biosensor is ''any biological material which, when exposed to an analyte (e.g. air, soil, water), provides an information linked response via a suitable transducer'' (38).

The biological material used in a biosensor can comprise plants (whole plants, organs, or cells), vertebrates, invertebrates, microorganisms), microbial tissue, enzymes, nucleic acid probes, antibodies, as well as other kinds of biological receptor. In using biosensors to test for soil health, the analyte is the soil or soil constituents, although it may be exposed to the sensor in a number of ways. Soils may be extracted with a range of solvents and the extract used with the solid phase present, either intact as a slurry or in a procedure that more closely defines the contact with either the liquid or the solid phase of the soil (13).

The type of transducer involved in biosensing for soil quality can vary, and electrical, conductivity, acoustic, and optical transducers can be used. In Sec. III.C the emphasis is on optical transducers since the sensors being considered are light-emitting.

B. Whole Cell/Organism Sensors and Reporter Genes

Recent advances in molecular biology have allowed the introduction of reporter genes into a wide variety of soil microorganisms. These genes can provide real-time reporting on the function of the host; the nature of the function is determined by the gene promoter downstream of which the reporter gene(s) is placed in the genome. If a suitable general promoter is used, then the genes can give a signal that reports on the overall metabolic health/status of the host.

The introduction of enzyme-linked *lux*, *luc*, *gfp*, *lac*, and *xyl* reporter genes into bacteria and fungi (39) has generated a wide range of ecologically relevant whole cell reporter systems that can be used to assess soil health. Recently (C Lagido, personal communication), *luc* genes have been cloned into nematodes so that soil animals can also provide real-time reporting of soil health. The movement and greater surface area contact between a nematode and the soil environment, coupled with the key role of the soil animals in nutrient cycling, make this a particularly useful development.

C. *lux* Biosensors

The *lux* genes encode for bioluminescence in naturally luminescing marine bacteria such as *Vibrio fischeri*, *Vibrio harveyi*, and *Photobacterium phosphoreum*, and light output is expressed via the enzyme luciferase (39).

lux genes have now been cloned into a wide range of microorganisms so that bioluminescence reports on the metabolic status of each of these whole cell biosensors can be used for ecologically relevant and rapid assessment of soil health (40). Examples of these biosensors and the ecological niche they represent are provided in Fig. 1.

In addition to the "metabolic health" sensors illustrated in Fig. 1, reporter genes can be placed under the control of catabolic promoters so that catabolic activities can be monitored by the particular reporter system (luminescence, fluorescence etc.) (40). This is a particularly valuable tool in the study of the enzymological characteristics of degradation of both xenobiotics and natural soil organic constituents. Biosensors can be used in a variety of ways to assess soil health (40,41). Probably the most useful approach involves solid-phase soil health testing, although tests involving soil extracts are also used. In all cases, bioluminescence is assayed after varying periods of exposure to the soil. Acute and chronic exposures both provide important information that can contribute decision support for soil/land management (41). It has been reported (41) that *lux* bacterial biosensors may be used as a decision support tool in the management of bioremediation of a large industrially contaminated site. The sensors were used to assess whether soil health was adequate

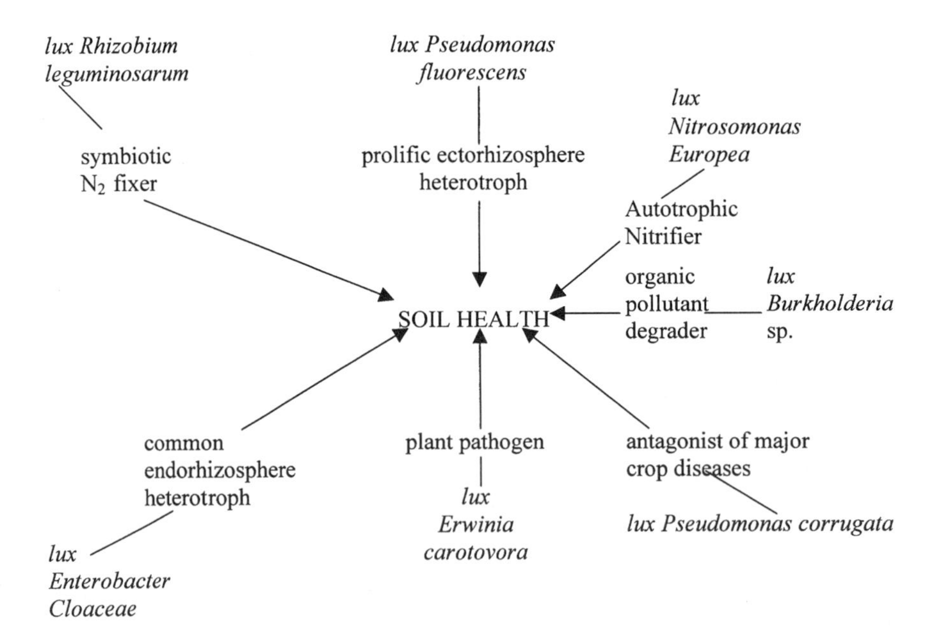

Figure 1 Examples of *lux* bacterial biosensors and the information they can provide for assessment of soil health.

for intrinsic bioremediation and, where this was not the case, what measures were required to restore soil health.

IV. GEOSTATISTICS

A. Introduction

Since the spatial variability of microbial communities and processes exists at several scales, including microsite, plot, and landscape levels (43), understanding their spatial structure is critical to understanding soil ecological processes and soil conservation efforts (44). The spatial variability of soil enzyme activities has been examined by using classical statistical approaches (45,46). However, geostatistics, which had its origins in the mining industry, is becoming increasingly popular among soil scientists for assessing spatial variability, and there are several excellent reviews of the process (47–50). Several studies have used this approach to characterize the spatial variation in soil enzyme activities (51–54). The following is a brief description of geostatistics and insights into soil enzyme ecological features it has provided. Clearly, the spatial variation of all potential bioindicators must be better understood for implementation of successful monitoring programs.

B. Definitions

Geostatistics characterizes the spatial dependence or independence of soil parameters taken at different sampling locations. It would be axiomatic to state that when soil samples are taken close together the variation (or relative lack thereof) between measured values reflects their close proximity. Such samples are said to be spatially dependent or autocorrelated since their variation reflects localized conditions. As samples are taken at increasing distances, the variation between them also increases. When the distances become large enough, the samples are independent of each other.

Geostatistics comprises two components: (1) modeling the spatial variation to create the semivariogram (Fig. 2) and (2) kriging to produce maps (Fig. 3). These studies begin by establishing sampling grids within a plot (Fig. 4). Samples are taken at each point and parameters measured. Differences in parameter values are then compared for all points. Semivariograms (Fig. 2) describe the semivariance (a measure of parameter variance)

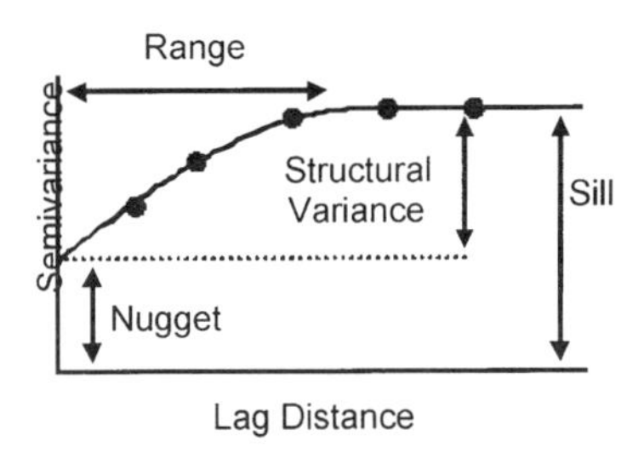

Figure 2 Example of a semivariogram. Semivariance is plotted for each log distance and a model is fitted to the points. The verge is the distance over which samples are spatially dependent. The sill represents the maximal variation in the plot. A nugget occurs when the model does not intercept at the origin and is indicative of sampling error or spatial structure between the sampling locations. Structural variance represents the proportion of the variance resulting from spatial structure.

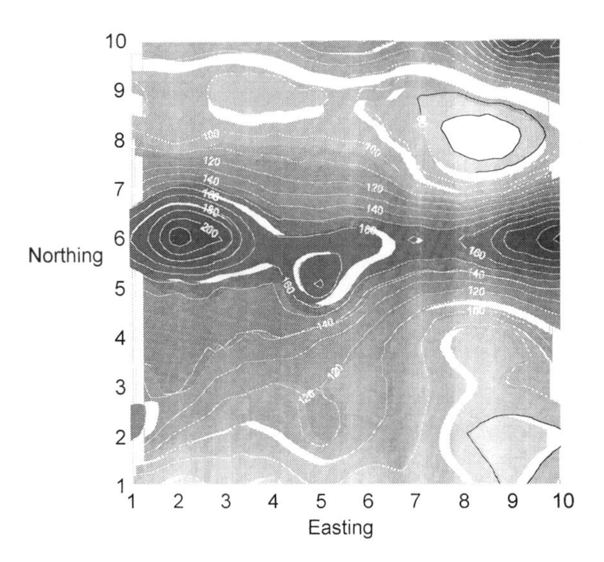

Figure 3 Map created from Kriging data. As with other interpolation techniques, the contour lines represent predicted values for a particular location. However, the values predicted by Kriging were determined by using a semivariogram, which allows errors associated with each prediction to be determined.

between sampling locations at different lag distances (Fig. 4). As one would expect, if the distance between sampling locations increases, the semivariance also increases (Fig. 2). At a certain distance, known as the range, the semivariance ceases to increase. The maximal semivariance is referred to as the sill. Soil properties that lie within the range are spatially dependent and are said to be autocorrelated. Soil samples that lie beyond the range are spatially independent. The range is important because it provides the researcher with an estimate of the area for which a sample is representative. Further, as samples taken within the range are spatially dependent, the use of classical statistics is precluded,

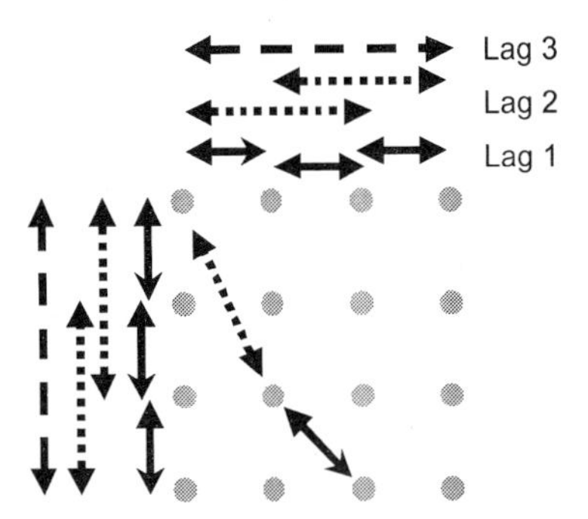

Figure 4 Grids are established in a plot and samples are taken from every point. The parameter for each sampling point is compared with those of all other sampling points. All the pairs of a given distance (known as the *lag distance*) are pooled together to give a measure of semivariance for that lag distance. Pairs that are separated by a distance that does not correspond to one of the established lag distances are assigned to the closest lag distance.

as such analyses assume sample independence. This information is valuable in the design of sampling strategies for bioindicators as an understanding of the representativeness of samples of a larger area is critical. The third important feature is the nugget. Theoretically, when the lag distance is zero (samples taken at the same point) there should be no variance. Often the semivariogram intercepts along the *Y* axis, not at the origin, and this is known as the *nugget effect*. Presence of a nugget indicates either measurement error or spatial structure over distances shorter than the intervals between sampling locations. Structural variance is the fourth property characterized by the semivariogram. This value, which is often expressed as a ratio between the variance not explained by the nugget and the total variance, quantifies the amount of variance arising from the underlying spatial structure. The greater the ratio, the more spatially dependent the soil parameter is. Information generated in the variogram is then used for kriging. Kriging allows maps (Fig. 3) that predict parameter values at unsampled locations to be drawn. What separates this approach from other interpolation techniques is that confidence in the predicted value can be assessed.

Geostatistics has gained increasing popularity in the soil sciences. Many studies have described the spatial variation of soil properties. This interest has, at least in part, been driven by the desire to develop high-precision agricultural practices. Such technologies depend on an understanding of the spatial distribution of soil properties such as nutrients and organic matter. However, soil scientists have recognized the power of this method for increasing our understanding of soil ecological characteristics at the microsite (55) and field scales (52). Comparisons of semivariograms and kirged maps allow new insights into the relationships between soil properties. It must be cautioned, however, that similarity in spatial structure does not necessarily reflect causal relationships.

C. Spatial Variability of Soil Enzyme Activity

The range over which enzyme activities are spatially dependent depends on the enzyme and localized conditions. Dehydrogenase activity was found to be moderately spatially dependent (spatial structure 37%) in a no-till field with a range that exceeded 200 m (56). In contrast, others reported that urease activity was autocorrelated over distances of <1 to 15 m, depending on the field examined (51,54). In contrasting ranges between studies, the size of the areas examined must be considered. Spatial structure can be complex, and a large area may have several sills nested within the semivariogram (50). von Steiger and associates (54) also found that organic carbon (OC) was more strongly autocorrelated than urease activity at all the sites they examined. They reasoned that OC would not fluctuate in the short term. However, urease activity reflects soil microbial biomass and nutrient status, which experience greater temporal change.

In an examination of soil enzyme activities and other soil parameters along a slope, mapping revealed similar spatial patterns for water content, OC, phosphatase, and arylsulfatase activities (52). The relationship of the two physicochemical parameters to the enzyme activities is suggestive of an underlying ecological relationship. Examination of semivariograms revealed that arylsulfatase activity was more spatially dependent (large structural variance) than either OC or phosphatase activity (low structural variance). Intriguingly, phosphatase activity showed a similar range to that of inorganic P, and the authors suggested this observation required further attention. Further, the authors found that phosphatase and arysulfatase showed similar spatial patterns. In contrast, it was demonstrated in 1999 (53) that two measures of microbial activity, fluorescein diacetate hydrolysis (FDA) and triphenyl-tetrazolium chloride (TTC) dehydrogenase activity, showed opposite trends in an agricultural plot under crop residue management. FDA activity followed a similar pattern to that of soil pH, whereas TTC activity was spatially related to organic matter and clay content.

Although several of the studies discussed have found that soil enzyme activity is spatially related to organic matter, this is not always the case. In a comparison of areas within a riparian zone that varied in drainage (51) a spatial relationship between organic matter and phosphatase was found in a moderately well drained area, but no relationship between these two parameters was noted in a poorly drained area. Again these relationships do not necessarily represent causal interactions but do enhance our understanding of soil enzyme ecological features. These insights are especially significant in the context of bioindicator development. Appropriate sampling strategies may vary for physical, chemical, and biological parameters.

D. Applications of Geostatistics

As discussed previously in this chapter, many parameters have been proposed to assess soil health. Spatial variability is a critical component in our understanding of soil quality and development of methods for its assessment. Halvorson et al. (57) described a kriging procedure that incorporated several soil parameters simultaneously, including dehydrogenase and phosphatase activities. This approach allowed maps to be drawn showing areas of potentially high and low soil health based upon several criteria. Although much attention has been paid to the spatial variability of agricultural soils, other soils would benefit from this type of analysis. For example, bioremediation is an area in which much could be learned from geostatistical approaches and bioindicators of soil health would be very important for assessing bioremediation potential and success. Spatial analysis could be useful for predicting contaminant concentrations as well as developing appropriate sampling (and then treatment) strategies. Potential studies could include examining the spatial variability of contaminant degradation, relevant enzyme activity, and survival of released or biostimulated organisms. The ability to relate such parameters to the soil properties is invaluable for the design and improvement of bioremediation strategies.

V. CONCLUSIONS

The heterogeneity of soil with respect to the chemical, physical, and biological properties and processes that contribute to soil health necessitates resolution across a range of scales. Microbial biosensors, in particular, have the power to resolve soil health from the microsite level upward. Individual *lux* biosensors can be CCD imaged and their activity monitored in situ (41), but they can also be used to assess the health of soil across large sites (42). Both scales provide invaluable information. The microsite study is essential if we are to understand contaminant bioavailability, for example, but much larger-scale resolution is required when management/decision support is required.

Although, to date, only a few papers have utilized geostatistics to examine potential bioindicators of soil health, this approach has tremendous potential. Indeed, such studies are necessary for the development of bioindicators. Knowledge of spatial variability is essential for designing appropriate sampling strategies and interpreting results of such studies.

In conclusion, this chapter reviews the exciting and rapidly developing field of bioindicators and biosensors of soil health and identifies a key role for geostatistics to help overcome the challenges of spatial heterogeneity in applying these indicators and sensors.

REFERENCES

1. JW Doran, TB Parkin. Defining and assessing soil quality. In: JW Doran, DC Coleman, DF Bezidicek, BA Stewart, eds. Defining Soil Quality for a Sustainable Environment. Soil Science Society of America Special Publication No. 35, Madison, WI:1944, 1996, pp 3–21.
2. GP Sparling. Soil microbial biomass, activity and nutrient cycling as indicators of soil health. In: CE Pankhurst, BM Doube, VVSR Gupta eds., Biological Indicators of Soil Health. Wallingford: CAB International, 1997, pp 97–120.
3. DS Jenkinson, JH Rayer. The turnover of soil organic matter in some of the Rothamsted classical experiments. Soil Sci 123:298–305, 1977.
4. K Killham. Soil Ecology. Cambridge: Cambridge University Press, 1994.
5. GP Sparling, DJ Ross. Biochemical methods to estimate soil microbial biomass: Current developments and applications. In: K Mulangoy, R Merckx, eds. Soil Organic Matter Dynamics and Sustainability of Tropical Agriculture. Chichester: Wiley, 1993, pp 21–37.
6. EA Paul. Dynamics of organic matter in soils. Plant Soil 76:275–285, 1984.
7. DS Powlson, PC Brookes, BT Christensen. Measurement of soil microbial biomass provides an early indication of changes in total soil organic matter due to straw incorporation. Soil Biol Biochem 19:159–164, 1987.
8. GP Sparling. Ratio of microbial biomass C to soil organic C as a sensitive indicator of changes in soil organic matter. Aust J Soil Res 30:195–207, 1992.
9. PC Brookes, SP McGrath. Effects of metal toxicity on the size of the soil microbial biomass. J Soil Sci 35:341–346, 1984
10. K Killham. A physiological determination of the impact of environmental stress on microbial biomass. Environ Pollut Series B 38:283–294, 1985.
11. G Gianello, JM Bremner. Comparison of chemical methods of assessing potentially available organic nitrogen in soil. Commun Soil Sci Plant Anal 17:215–236, 1986.
12. D Barraclough. The use of mean pool abundances to interpret ^{15}N tracer experiments. 1. Theory. Plant Soil 131:89–96, 1991.
13. M Richardson, ed. Environmental Toxicology Assessment. London: Taylor & Francis, 1995.
14. RG Burns, ed. Soil Enzymes. London: Academic Press, 1978.
15. RP Dick. Soil enzyme activities as integrative indicators of soil health. In: CE Pankhurst, BM Doube, VVSR. Gupta, eds. Biological Indicators of Soil Health. Wallingford: CAB International, 1997, pp 121–156.
16. HG Song, R Bartha. Effects of jet fuel spills on the microbial community of soil. Appl Environ Microbiol 56:646–651, 1990.
17. GW McCarty, R Siddaramappa, RJ Wright, EE Codling, G Gao. Evaluation of coal combustion byproducts as soil liming materials: their influence on soil pH and enzyme activities. Biol Fertil Soils 17:167–172, 1994.
18. DC White, SJ Macnaughton. Chemical and molecular approaches for rapid assessment of the biological status of soils. In: CE Pankhurst, BM Doube, VVSR. Gupta, eds. Biological Indicators of Soil Health. Wallingfored: CAB International, 1997, pp 371–396.
19. TM Embley, E Stackebrandt. The use of 16S ribosomal RNA sequences in microbial ecology. In: RW Pickup, JR Saunders, eds. Ecological Approaches to Environmental Microbiology. Oxford: Prentice Hall, 1996, pp 39–62.
20. WE Holben, BM Schroeter, VGM Calabrese, RH Olsen, JK Kukon, VO Biedirbeck, AE Smith, JM Tiedje, Gene probe analysis of soil microbial populations selected by amendment with 2,4-dichloro-phenoxyacetic acid. Appl Environ Microbiol 58:3941–3948, 1992.

21. GB Smith, JM Tiedje. Isolation and characterisation of a nitrate reductase gene and its use as a probe for denitrifying bacteria. Appl Environ Microbiol 58:376–384, 1992.
22. JL Garland, AL Mills. Classification and characterisation of heterotrophic microbial communities on the basis of patterns of community-level-sole-carbon-source utilization. Appl Environ Microb 57:2351–2359, 1991.
23. CE Pankhurst. Biodiversity of soil organisms as an indicator of soil health. In: CE Pankhurst, BM Doube, VVSR Gupta, eds. Biological Indicators of Soil Health. Wallingford: CAB International, 1997, pp 297–324.
24. F Schinner, R Ohlinger, E Kandeler, R Margesin, eds. Methods in Soil Biology. Berlin: Springer, 1996, p 426.
25. VVSR Gupta, GW Yeates. Soil microfauna as bioindicators of soil health. In: CE Pankhurst, BM Doube, VVSR Gupta, eds. Biological Indicators of Soil Health Wallingford: CAB International, 1997, pp 201–234.
26. W Foissner. Soil protozoa as bioindicators in ecosystems under human influence. In: JF Darbyshire, ed. Soil Protozoa. Wallingford: CAB International, 1994, pp 147–194.
27. Yeates, GW. Modification and qualification of the nematode maturity index. Pedobiologia 38: 97–101, 1994.
28. NM van Straalen, HA Verhoef. The development of a bioindicator system for soil acidity based on arthropod pH preferences. J Appl Ecol 34: 1997.
29. BM Doube, O Schmidt. Can the abundance or activity of soil macrofauna be used to indicate the biological health of soils? In: CE Pankhurst, BM Doube, VVSR Gupta, eds. Biological Indicators of Soil Health. Wallingford: CAB International, 1997, pp 265–296.
30. KE Lee. Earthworms: Their Ecology and Relationships with Soils and Land Use. Sydney: Academic Press, 1985.
31. CE Pankhurst, BM Doube, VVSR Gupta. Biological indicators of soil health-a synthesis. In: CE Pankhurst, BM Doube, VVSR Gupta, eds. Biological Indicators of Soil Health. Wallingford: CAB International, 1997, pp 419–436.
32. WHO Ernst. Geobotanical and biogeochemical prospecting for heavy metal deposits in Europe and Afirca. In: B Market, ed. Plants as Biomonitors: Indicators for Heavy Metals in the Terrestiral Environment. VCH Weinheim, 1993, pp 107–126.
33. MZ Hauschild. Petrescine (1,4-diaminobutane) as an indicator of pollution-induced stress in higher plants: Barley and rape stressed with Cr(III) or Cr(VI). Ecotoxicol Environ Saf 26: 228–247, 1993.
34. JM Morgan. Osmoregulation and water stress in higher plants. Ann Rev Plant Physiol 35: 299–348, 1984.
35. TA Smith. Polyamines. Ann Rev Plant Physiol 36:117–143, 1985.
36. F Van Assche, H Clijsters. Effects of metals on enzyme activity in plants. Plant Cell Environ 13:195–206, 1990.
37. Baker, AJM Accumulators and excluders strategies in the response of plants to heavy metals. J Plant Nutr 3:643–654, 1981.
38. D van der Leslie, P Cortisier, W Baeyers, S Wuertz, L Diels, M Mergeay. The use of biosensors in environmental monitoring. In: Bioremediation: Scientific and Technical Issues, 10th Forum in Microbiology, 1994.
39. JI Prosser, EAS Rattray, K Killham, LA Glover. *lux* as a marker gene to track microbes. In: ADL Akkermans, JD van Elsas, FJ de Bruijn eds. Molecular Microbial Ecology Manual. Dordrecht: Kluwer, 1996.
40. GI Paton, EAS Rattray, CD Campbell, MS Cresser, LA Glover, JCL Meeussen, K Killham. Use of genetically modified microbial biosensors for soil ecotoxicity testing. In: CE Pankhurst, BM Doube, VVSR Gupta, eds. Biological Indicators of Soil Health. Wallingford: CAB International, 1997, pp 397–418.
41. EAS Rattray, JI Prosser, K Killham, LA Glover. Luminescence-based non-extractive tech-

niques for *in situ* detection of *Escherichia coli* in soil. Appl Environ Microbiol 56:3368–3374, 1990.
42. S Sousa, C Duffy, H Weitz, LA Glover, R Henkler, K Killham. Use of a *lux*-modified bacterial biosensor to identify constraints to bioremediation of BTEX-contaminated sites. Environ Toxicol Chem 17:1039–1045, 1998.
43. TB Parkin. Spatial variability of microbial processes in soil—a review. J Environ Qual 22: 409–217, 1993.
44. JJ Halvorson, JL Smith, RI Papendick. Issues of scale for evaluating soil quality. J Soil Water Conservation 1997, 52:26–30.
45. M Bonmati, B Ceccanti, P Nanniperi. Spatial variability of phosphatase, urease, protease, organic carbon and total nitrogen in soil. Soil Biol Biochem 23:391–396, 1991.
46. TW Speir, DJ Ross, VA Orchard. Spatial variability on biochemical properties in a taxonomically-uniform soil under grazed pasture. Soil Biol Biochem 16:153–160, 1984.
47. P Goovaerts. Geostatistical tools for characterizing the spatial varaibility of microbiology and physico-chemical soil properties. Biol Fertil Soils 27:315–334, 1998.
48. RE Rossi, DJ Mulla, AG Journel, EH Franz. Geostatistical tools for modelling and interpreting ecological spatial dependence. Ecol Monogr 62:277–314, 1992.
49. BB Trangmar, RS Yost, G Uehara. Application of geostatistics to spatial studies of soil properties. Adv Agron 38:45–94, 1995.
50. R Webster. Quantitative spatial analysis of soil in the field. Adv Soil Sci 3:1–70, 1985.
51. JA Amador, AM Glucksman, JB Lyons, JH Görres. Spatial distribution of soil phosphatase activity within a riparian zone. Soil Sci 162:808–825, 1997.
52. DW Bergstrom, CM Monreal, JA Millette, DJ King. Spatial dependence of soil enzyme activities along a slope. Soil Sci Soc Am J 62:1302–1308, 1998.
53. WJ Staddon, MA Locke, RM Zablotowicz. Spatial variation of microbial and soil paremeters relevant to herbicide fate. Weed Science Society of America, 39th Annual Meeting, San Diego, 1999.
54. B von Steiger, K Nowack, R Schulin. Spatial variation of urease activity measured on soil monitoring. J Environ Quality 25:1285–1290, 1996.
55. SJ Morris. Spatial distribution of fungal and bacterial biomass in southern Ohio hardwood forest soils: Fine scale variability and microscale patterns. Soil Biol Biochem 31:1375–1386, 1999.
56. C Cambardella, TB Moorman, JM Novak, TB Parkin, DL Karlen, RF Turco, AE Konopka. Field-scale variability of soil properties in central Iowa soils. Soil Sci Soc Am J 58:1501–1511, 1994.
57. JJ Halvorson, JL Smith, RI Papendick. Integration of multiple soil parameters to evaluate soil quality: A field example. Biol Fertil Soils 21:207–214, 1996.

16

Hydrolytic Enzyme Activities to Assess Soil Degradation and Recovery

Tom W. Speir
Institute of Environmental Science and Research, Porirua, New Zealand

Des J. Ross
Landcare Research, Palmerston North, New Zealand

I. INTRODUCTION

A project of the United Nations Environmental Program on Global Assessment of Soil Degradation concluded, ''Nearly 40% of all agricultural land has been adversely affected by human-induced soil degradation, and over 6% would require major capital investment to restore its original productivity'' (1). It is, therefore, not surprising that, among regulatory authorities, there is a strong desire for the development of sensitive indicators to assess soil degradation. Properties that provide a snapshot assessment of the status of a soil can determine whether a management practice has had an adverse effect on soil ''health'' and productivity and, better still, can predict whether a practice will have an adverse effect if it is continued. This has been one of the major drivers of the worldwide research effort on *soil quality* defined as ''the capacity of a soil to function, within ecosystem and land-use boundaries, to sustain biological productivity, maintain environmental quality, and promote plant and animal health'' (1). This topic has been the subject of numerous reviews, such as those found in the Soil Science Society of America Special Publications 35 (2) and 49 (3). We do not wish to enter the debate concerning a potential role for enzyme activity measurements in the wide soil-quality context—this topic has already been reviewed (4–6)—but to focus on the application of soil enzymes to scenarios in which soil degradation is demonstrable, or at least strongly suspected to be a likely outcome of a particular land-management practice.

In this review, we do not present the many, sometimes contradictory, reports of effects of different management practices on soil enzyme activities that have already been reviewed in detail (6–14) but rather use our knowledge and perceptions of soil enzymes to try to understand what the enzyme activity measurements are telling us about the soil and how they can be used to assess soil degradation and recovery. The scenarios we cover are soil physical degradation as a result of human-induced factors, such as intensive cropping and soil compaction, and soil loss from mining. In this last example, there is no need to assess degradation at all; emphasis is on rehabilitation of the land and creation of a productive soil when the mine is closed or moved on across the landscape. We also consider soil contamination from the dumping or accidental spillage of organic and inorganic materials, e.g., hydrocarbons and heavy metals, and the application of sewage sludge and pesticides.

II. ENZYMES IN SOIL—OCCURRENCE, LOCATION, AND ASSAY

In order to use soil enzyme activity measurements to provide information that will enable us to assess the extent of soil degradation or recovery, we need to recognize the limitations of our methodology and our knowledge of the role and function of soil enzymes.

Because of the diversity of life in the soil, it is probable that most known enzymes could be found in a soil sample. However, the activities that have been measured are limited to a few oxidoreductases (EC 1), transferases (EC 2), hydrolases (EC 3), and lyases (EC 4) (11). It is impossible to extract a significant proportion of any enzyme activity from soil, unlike other living systems, and activities are therefore invariably assayed in situ. It is, consequently, not possible to assign activity to individual organisms or even to particular groups of organisms. The enzyme activity measured represents the sum of contributions from a broad spectrum of soil organisms (including plants) and also extracellular or abiontic enzymes (15,16) that retain their activities away from the living cell. For enzymes that do not require cofactors and that are not components of catabolic or anabolic sequences, a significant proportion of the total activity may be extracellular and any catalytic function performed by these particular enzymes is purely opportunistic. This does not mean that soil organisms are unable to take advantage of this catalysis, and it may be that such enzymes play an important role in the initial degradation of macromolecular substrates in soil (17,18). The most studied group of soil enzymes that are likely to have a significant active extracellular component are the hydrolases; it is generally accepted that these enzymes comprise a metabolically vital intracellular fraction and an opportunistically active extracellular fraction divided among several locations in the soil (19). The proportional size of this extracellular component is generally unknown and probably varies from enzyme to enzyme.

Most hydrolases are investigated by using artificial substrates and assay conditions that are quite foreign to those prevailing in soil. Substrates are usually small molecules, often simple esters combining the functional group of the substrate, e.g., phosphate (for phosphatase) or glucose (for β-glucosidase) with a chromophore, such as *p*-nitrophenol, for ease of extraction from soil and ease of assay. Activity normally is measured under buffered conditions at the optimal pH for the enzyme, at enzyme-saturating substrate concentrations, and usually at a temperature substantially greater than would generally prevail in soil (20). The composition and molarity of the buffer are especially important, because a buffer found suitable for some soils is not necessarily suitable for others (21). For example, a commonly used buffer (acetate-phosphate) for assays of invertase activity inhibited activity in acid grassland soils and could thereby have obscured relationships of invertase with other soil and environmental factors (22).

Under suitable assay conditions, the measured activity of an enzyme such as phosphatase, for example, represents only the potential *p*-nitrophenyl phosphate-hydrolyzing capacity of the soil. It is probable that not all of the numerous phosphatases present are assayed (all may not be active against this substrate), and it is certain that the reaction rate would be much greater than the rate of phosphate production from organic phosphorus compounds in the unamended soil. It is, therefore, difficult to see how a direct causal role in the phosphorus fertility of a soil can be ascribed to the conglomerate of phosphatase enzymes assayed in this way. Of the hydrolases, only urease and invertase are measured by using their natural substrates, viz., urea and sucrose, respectively. However, the artificial conditions used in the assay of these enzymes again preclude any direct connection between measured activity and substrate hydrolysis that occurs naturally in soil. Although starch and cellulose are used as substrates for amylase and cellulase, the chemical forms and purity of these substrates would be very different from those found in soil.

One enzyme that has been studied extensively because of its perceived close relationship with microbial activity is the oxidoreductase dehydrogenase. This enzyme, or group of enzymes, is a component of the electron transport system of oxygen metabolism and requires the organization of the living intracellular environment to express its activity. Consequently, dehydrogenase activity is not likely to be present in any of the extracellular compartments occupied by the hydrolases. The absence of an extracellular component means that dehydrogenase activity may not be well suited to assess soil degradation because it is likely to fluctuate, as does microbial activity, in response to recent management and/or seasonal (climatic) effects (5). Although the presence of dehydrogenase activity in soil should reflect the activity of physiologically active microorganisms, including bacteria and fungi (23), measured dehydrogenase activity does not correlate consistently with microbial activity (6). There are several possible reasons for this, including unsuitable assay conditions, the presence of extracellular phenol oxidases, and the presence of alternatives to the added electron acceptor (substrate) (6). These electron acceptors may be common soil constituents, such as nitrate (24) or humic acids (23). It also has been found that Cu reduces apparent dehydrogenase activity, not by inhibiting the enzyme, but by interfering with the assay procedure (25). These procedural artifacts raise questions about the accuracy of dehydrogenase activity results, especially in situations in which a management practice may be changing the amount of a soil component or adding a xenobiotic chemical that may interfere with the enzyme assay. In view of these concerns, and the likely susceptibility of dehydrogenase activity to transitory fluctuations, we focus only on the hydrolase enzymes in this review.

Obviously, at least a component of every soil enzyme has a vital metabolic role in situ, but it is most unlikely that any indication of the role(s) or even the real activity of the enzyme(s) under field conditions can be gained from the assay methods used. The assertion of Skujins that ''obtaining a fertility index by the use of abiontic soil enzyme activity values seems unlikely'' (19) applies as much today as it did over 20 years ago. It is important that these considerations be acknowledged when investigating how enzyme activities can be used to assess soil degradation and recovery.

III. SOIL HYDROLASE ENZYMES TO ASSESS PHYSICAL DEGRADATION OF SOILS

Soil degradation through loss of organic matter and structural integrity is a well known outcome of an intensive cropping regime. There have been many studies comparing the chemical, physical, and biological properties of soils subjected to conventional cultivation practices with those subjected to minimal or no tillage. When comparing conventionally ploughed and no tillage plots, Klein and Koths (26) found that urease, protease, and phosphatase activities were higher under no tillage than under ploughed treatments. Dick (27) observed the same results with acid phosphatase, arylsulfatase, invertase, amidase, and urease in the top 7.5 cm of soil and concluded that changes in activity were not attributable to long-term pesticide application. Gupta et al. (28,29) compared soils that had been under cultivation for up to 80 years and found that their arylsulfatase and phosphatase activities were considerably reduced when compared with those in native, uncultivated soils. Cultivation decreased the enzyme activities in all aggregate size fractions of a 69-year cultivated soil and decreased the Michaelis constant (K_m) and maximum reaction rate (V_m) for arylsulfatase in all cultivated soils. The authors concluded that decreased arylsulfatase activity in the cultivated soils reflected ''the reductions in organic matter content and microbial biomass and activity of the soil associated with land management'' (29). They also proposed that clearing and cultivation of native soils result in native soil organic matter being transformed into more inert forms that are less likely to form complexes with either the enzyme or its substrate; this would account for increase in substrate affinity (i.e., lower K_m) in the cultivated soils. Changes in enzyme activities in different aggregate size fractions under cultivation regimes also have been observed by Kandeler and associates (30). The effects of three different tillage systems on the total xylanase, invertase, and alkaline phosphatase activities of the 0- to 10-cm layer of soil and also on the proportions found in different particle size fractions are illustrated in Fig. 1. The authors also found that the reduced tillage and, especially, the conventional tillage treatments had decreased soil organic C content in the coarsest (>200-μm) fraction. This would have been the principal reason for the greatly reduced soil xylanase activity in the conventional tillage treatment, because a large proportion of this enzyme activity was located in this coarsely textured organic fraction (Fig. 1). The other two enzymes, invertase and alkaline phosphatase, were more closely aligned with the finer soil fractions and were less affected by tillage, although the proportions in the coarsest soil fraction also were diminished (30).

Dick et al. (31) examined skid-trail soils, i.e., soils compacted by dragging logs from forestry operations, and found that compacted soils had considerably lower phosphatase, arylsulfatase, and dehydrogenase activities than the control soil, especially in the subsoil. They also showed there was a very strong correlation between the enzyme activities and soil organic C and microbial C. They concluded that a combination of physical factors and impaired root growth was the probable reason for these compaction effects.

Sulfatase activity in arctic tundra soils also was lowered significantly after vehicle disturbance (32). The wetter, depressed portions of the vehicle tracks supported more vigorous plant growth as a result of nutrient influx caused by the channeled water flow. Sulfatase activity levels in these wet areas were considered to have become depressed because of end product inhibition or inhibition by other ions, e.g., phosphate.

Apart from the previous example, usually the main result of these, and the many other studies (33–35), is that soil enzyme activities decline in proportion to the loss of soil organic matter. This tendency does not provide any more information about potential soil degradation under a cropping regime than does the measurement of organic C alone or any information about the short-term productivity of the soil. An intensively cropped soil with lower enzyme activity and organic matter content than those of a neighboring native soil may, in fact, be far more productive because it has greater nutrient status. Many studies over the years have shown that, under intensive agriculture, in which nutrients can be added from a bag, soil enzyme activities are not good predictors of soil fertility and productivity. However, it is also generally recognized that such intensive cropping practices are not sustainable in the long term and that the soils become much more prone to erosion, waterlogging, and compaction. Residue-management trials have shown that conservation tillage and organic-residue-amendment strategies maintain soil organic matter and retain soil physical characteristics (26,33,36–38). Therefore, if a soil enzyme can tell us something about the location and perhaps the quality of soil organic matter in

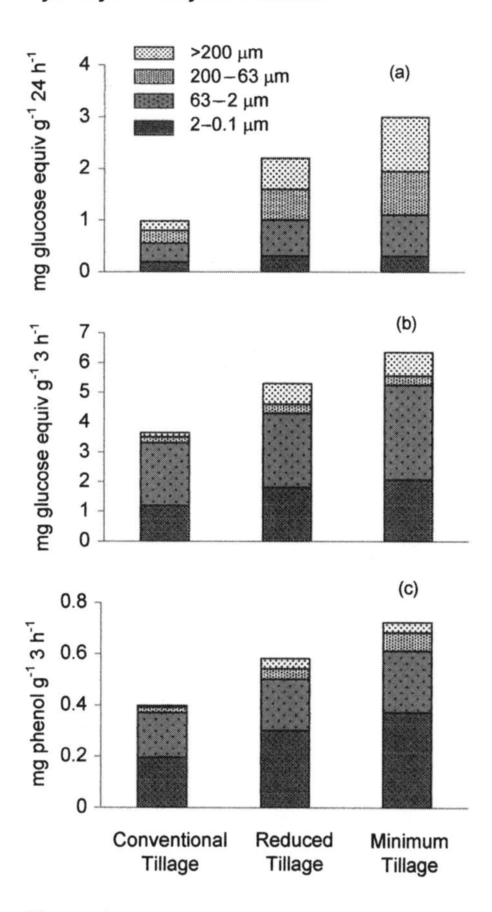

Figure 1 Xylanase (a), invertase (b), and alkaline phosphatase (c) activities in particle-size fractions of the 0- to 10-cm layer of a Haplic Chernozem soil. (Adapted, with permission, from Ref. 30.)

cropped soils, e.g., by measurement of xylanase activity in soil particle-size fractions (30), then it may be possible to use its activity as an early warning of potential structural degradation. Changes in enzyme kinetic properties, if they reflect changes in organic matter quality (29), also may provide more information about the status of a soil than can be gained from its organic matter content.

IV. SOIL HYDROLASE ENZYMES TO ASSESS SOIL RECOVERY AND DEVELOPMENT AFTER MINING

Many studies have demonstrated the decline of organic C, microbial biomass, and enzyme activities with increasing soil depth. Ross et al. (39) showed the removal of 10 cm, and especially 20 cm, of topsoil from temperate pasture plots markedly lowered activities of a number of enzymes. This finding is not at all surprising, since the top centimetres of a soil are the major loci of biological activity and organic matter. What is especially interesting, however, is that removal of 10 cm of topsoil from this pasture resulted in a new topsoil with approximately 40% less organic C, but more than 60% lower urease and phosphatase activities, 75% lower invertase and amylase activities, and more than 80% lower cellulase and xylanase activities; only sulfatase matched organic C with a 40% decline in activity (39). Speir et al. (40) showed that organic C declined relatively linearly with depth in a pasture soil, whereas most enzyme activities and soil respiratory activity and microbial biomass fell much more rapidly in the top 15 cm than in the remainder of the soil profile (Fig 2). Here again, sulfatase activity most closely matched the decline of organic C. It

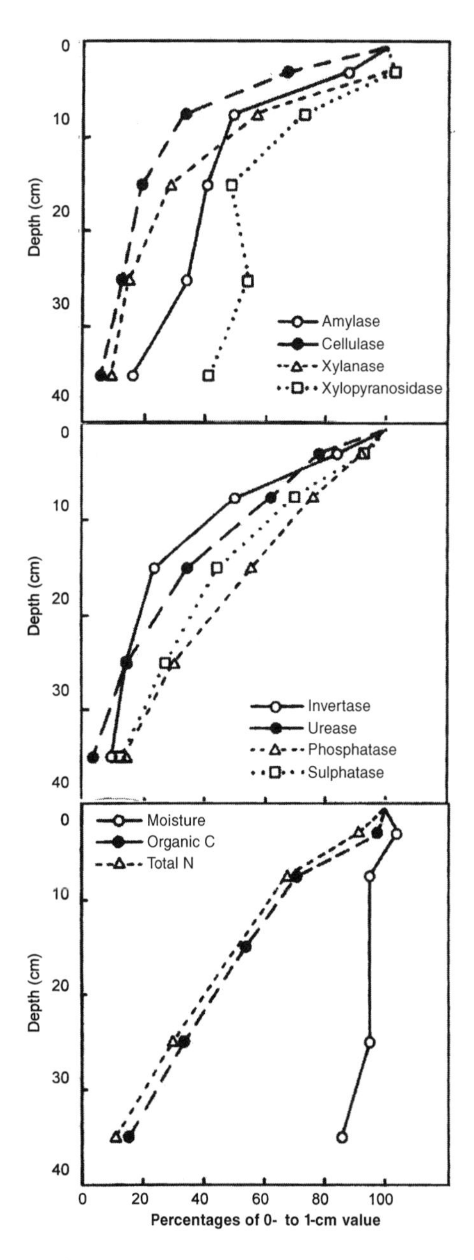

Figure 2 Influence of depth on soil chemical properties and enzyme activities. (Adapted, with permission, Ref. 40.)

was concluded that the carbohydrase enzymes (amylase, cellulase, invertase, xylanase, and, to a lesser extent, xylopyranosidase) may be closely related to current soil biological activity and be disproportionately higher than predicted from organic C content, in the topmost soil layer, because of improved aeration and substrate availability (40). On the other hand, urease, phosphatase, and especially sulfatase may be more related to total organic C because of their stabilized, extracellular, organomineral-bound component.

Technologies to recover such soils after mining and methods to assess their recovery are equally applicable to the development of landscapes reconstructed after underground, strip, or opencast mining for coal and other mineral resources. It is estimated that about 1600×10^9 m^3 of mine spoils had accumulated on the Earth's surface up to 1980 and had increased by about 40×10^9 m^3 per year by 1998 (14). Rehabilitation of these spoils and degraded landscapes is now an integral part of mining operations in many parts of the world. The enzymological characteristics of these constructed or ''technogenic'' soils have been extensively reviewed (7,11,14,41,42).

Technogenic soils may have a ''topsoil'' composed of entirely subsurface materials or the stockpiled original topsoil or some intermediate combination. Stockpiling of topsoil leads to a decline of soil biological activity (14), presumably due to the lack of replenishment of readily degradable plant residues and to factors such as compaction and reduced aeration. Speir et al. (43) found that the protease, sulfatase, and urease activities of 12 soils left fallow in a pot trial declined markedly over 5 months, whereas activities generally remained unchanged or increased if the soils were planted with perennial ryegrass. The decline in the fallow treatments was probably attributable to declining microbial activity as plant residues were degraded, leaving only more intransigent organic matter. It is probable, therefore, that the initial biological activity of the topsoil of a technogenic soil, no matter how it is constructed, is considerably lower than that of the original soil on the site. It certainly does not have the high biological and enzymatic activities found in the very surface layer of an undisturbed soil (40) (Fig. 2).

Dick and Tabatabai (9) concluded that ''in environments initially devoid of plant or microbial life, as is often found for drastically disturbed lands, a close correlation exists between plant and microbial communities and the expression of enzyme activities.'' Therefore, in the early stages of recovery of land that has had the surface soil removed (e.g., after erosion or topsoil mining), or in the early stages of development of technogenic soils from stockpiled soil and overburden materials (e.g., land reclamation after mining), a close relationship between plant productivity and soil enzyme activity might be expected.

Ross et al. (39,44) investigated the relationship between recovery of soil biochemical properties and plant productivity in a temperate pasture soil that had had 10 cm or 20 cm of topsoil removed in a trial to simulate the effects of topsoil mining. The rates of recovery of invertase, amylase, cellulase, and xylanase, but not phosphatase, sulfatase, or urease, were, after 3 years, much greater than the rate of recovery of organic C (Table 1). However, after 5 years, the recovery of all properties had slowed. During the early stages of restoration, the enzyme activities generally correlated very closely with pasture production, but in the longer term (5 years) the activities were more closely related to the recovery of organic C (Table 1). The comparatively rapid recovery of invertase activity also occurred in a temperate hill pasture (45) where the original soil had eroded in slips of up to 60-cm depth. Restoration of invertase activity in regenerating pasture was complete within 11 years, whereas phosphatase activity was then only about 36% of that of uneroded topsoil (DJ Ross, TW Speir, AW West, personal communication, 1984).

Table 1 Recovery of Organic C and Enzyme Activities, and Their Correlation with Herbage (Pasture Grasses and Clover) Production, in Soil Stripped of 20 cm of Topsoil

	Percentage of control (unstripped) soil value after			Correlation with herbage production, all data up to	
Property	0.5 year	3 years	5 years	3 years	5 years
Organic C	41	59	66	0.68*	0.41
Urease	16	46	61	0.79**	0.73**
Invertase	29	88	88	0.92***	0.51*
Amylase	34	84	101	0.70*	0.40
Cellulase	32	84	82	0.90***	0.63**
Xylanase	12	70	80	0.91***	0.71***
Phosphatase	40	61	76	0.90***	0.46
Sulfatase	19	48	62	0.91***	0.47*

*, **, *** = $P < 0.05, 0.01, 0.001$, respectively.
Source: Adapted from Refs. 39 and 44.

In an investigation of different replacement strategies in the construction of technogenic soils after simulated lignite mining, herbage yields in all replacement treatments reached the level of the temperate pasture control plots within 3 years, as long as the soil was ripped to alleviate compaction (46). Biochemical activities, including those of invertase and sulfatase, increased rapidly in all treatments in the early stages of the trial. Invertase activity reached the level of the control soil after 3 years, and sulfatase attained that level in two of the three replacement treatments after 5 years. In contrast, organic C content had increased linearly from 47% to 76% of that of the control at the start of the trial to 68%–92% after 5 years. The correlations of organic C and invertase and sulfatase activities with herbage yields, using all data over the 5 years of the trial, are shown in Table 2. The levels of soil invertase activity and, to a lesser extent, sulfatase activity provided a good indication of herbage production as restoration progressed. It was concluded that plant materials would have contributed appreciably to the rapid increase of

Table 2 Correlations of Soil Organic C and Invertase and Sulfatase Activities with Pasture Herbage Yields from Technogenic Soils Constructed Using Three Soil Replacement Strategies, Including All Data over the 5 Years of the Trial

	Soil replacement treatment		
Property	1[a]	2	3
Organic C	0.20	0.31	0.39
Invertase	0.59**	0.75***	0.55**
Sulfatase	0.37	0.59**	0.77***

[a] Treatments were 1, horizon A/B/C; 2, (A + B)/C; 3, (A + lignite overburden (O))/B + O)/C + O).
**, *** = $P < 0.01, 0.001$, respectively.
Source: Adapted from Ref. 46.

soil invertase activity. Such a rapid buildup of soil biological activity and of plant productivity is the exception rather than the rule. Most investigations have shown that the enzyme activities of technogenic soils generally were considerably lower than those of control or native soils, even after 20 or more years (11,14). It is likely that optimization of factors, such as fertilizer inputs, soil aeration, drainage, and bulk density, as well as climate, resulted in extremely favorable conditions for soil recovery in the New Zealand study (46).

It is interesting to speculate why there is a strong relationship between plant productivity and soil enzyme activity in at least the early stages of development of a fertile soil. Plants and nutrients in the soil are the drivers of the recovery, as plants provide C to enable the initially sparse microbial populations to proliferate. The microorganisms and, to a lesser extent, the plant fragments are the principal source of the enzymes. Both intracellular and extracellular enzyme concentrations increase in proportion to microbial numbers, and the extracellular enzymes are able to become bound and stabilized at the many unoccupied binding sites in the soil. As already mentioned, it is possible that an initial buildup of an extracellular enzyme component is vital during the early stages of microbial proliferation, because such enzymes may catalyze the commencement of degradation of the macromolecular plant substrates (17,18). Once these mechanisms are under way, it might be expected that the rate of recovery of soil enzyme activity would match that of plant productivity and be proportional to the input of plant residues. If nutrients and physical conditions are not limiting, plant productivity drives the process toward the levels of biological activity found in nearby undisturbed soils with the same parent materials and chemical properties. The rate of recovery of biological and enzyme activities exceeds the rate of recovery of soil organic matter content. As time passes and the sites for stabilization of extracellular enzymes become saturated, their concentrations may level off, and increases in enzyme activity with increasing microbial numbers and organic matter content may then be a function of intracellular enzymes only (microbial and plant). If the soil nutrient status and physical status are not limiting, plant productivity may still drive increased microbial numbers and organic matter content but may no longer be related directly to total soil enzyme activity.

In soil-recovery situations, such as those described, the enzyme activities do not necessarily need to be assigned a role in the recovery process. They are merely indicators that can be used to give progress reports on the rate of recovery of plant productivity and perhaps predict how long full recovery will take. Some are better indicators than others; this may be a function of the enzymes themselves or it may be specific to a site, or soil, or particular vegetation. The carbohydrase enzymes, especially those involved in the breakdown of macromolecular plant residues (e.g., xylanase), or invertase because of its relationship with plant materials (47), may be better predictors than the more often assayed phosphatase, sulfatase, and urease enzymes. As time progresses, the activities of this latter group are probably more closely related to the soil organomineral components, and their (presumably) large, stabilized, extracellular component mask more subtle changes resulting from increasing microbial and plant production. Overall, however, we do not fully understand these relationships. Therefore, predictions of productivity or recovery rates in degraded or technogenic soils from the assay of a single soil enzyme, or even of several enzymes in isolation from other soil properties, would be unwise; at this stage, a predictive role for enzymes in soil recovery is still an experimental tool.

Another approach to predicting the effects of disturbance and the success of soil rehabilitation procedures has been to use a multivariate analysis technique (48). This method uses biological properties, including the enzymes alkaline phosphatase, sulfatase, arginine deaminase, protease, invertase, and dehydrogenase, in combination with other soil properties and is able to discriminate between soils affected by oil well drilling, surface mining, hydrocarbon spills, and pipeline construction, and undisturbed soils from similar areas. Although the reason for the choice of these particular enzymes is not clear, a discriminant function combining seven properties, including alkaline phosphatase and arginine deaminase activities, correctly classified 86% of the undisturbed soils and 70% of the disturbed soils. This investigation, which comprised 68 soils covering five Canadian soil groups, appears to provide a basis for reclassifying a once-disturbed soil as having been remediated sufficiently to be equivalent to an undisturbed soil.

V. SOIL HYDROLASE ENZYMES TO ASSESS SOIL CONTAMINATION

A. Contamination by Crude Oil and Oil By-Products

Because of the huge volumes of oil and its by-products that are produced, transported, and stored, there is a very serious threat of soil contamination in the vicinity of oil fields, refineries, and storage and distribution facilities. The effects of oil pollution on the activities of soil enzymes have been extensively reviewed by Kiss et al. (14). We therefore give only a synopsis of the data presented in that review and limit our discussion to the interaction of oil products with enzymes and the capacity of enzyme activity measurements to ascertain the extent of soil degradation that has occurred.

Polar organic solvents such as ethanol and acetone destroy enzyme activity by protein denaturation. However, nonpolar organic compounds, such as hydrocarbons, are hydrophobic and do not interact significantly with proteins in solution. In soil, crude oil and some of the heavier oil fractions, if present in very high concentrations, may block the expression of enzyme activity by coating organomineral and cell surfaces and thereby prevent soluble substrates reaching the enzyme molecules. It may be concluded that the lighter petroleum products are not particularly inhibitory toward soil enzymes because of the extensive use of toluene, at concentrations up to 25% of the assay volume (19), as a microbial inhibitor in soil enzyme assays.

In the research reviewed by Kiss et al. (14), large amounts of crude oil were required to cause a significant reduction of soil enzyme activities, with concentrations as high as 100 kg m^{-2} reducing invertase, protease, and phosphatase activities by 54%, 62%, and 50%, respectively (49). Although the activity of most soil enzymes is adversely affected by crude oil, urease activity often increases (14). Different responses to crude oil were also observed in another study (50); cellulase activity declined whereas aryl-hydrocarbon hydroxylase activity increased; a shift in catabolic activity of the soil microbiota in response to the new carbon source is indicated. Important findings of Samsova et al. (51) were reduction of protease activity, increase in urease activity, and death of all plants on contamination with 8% crude oil. Other studies have shown that at moderate levels of oil contamination, some enzyme activities declined and some increased, most microbial populations increased, but plant growth was usually impaired (14). It would seem, therefore, that soil enzyme activities are less sensitive than plants to soil degradation by crude oil. In some instances, however, they may provide information about the potential for the soil microorganisms to metabolize the oil and for the contaminated soil to recover from the pollution.

Although toluene is not particularly inhibitory to soil hydrolase activities, refined oils can inhibit urease activity. In three soils, inhibition increased in the order kerosene $<$ diesel $<$ motor oil $<$ leaded gasoline, at amendment concentrations of 5%, 10%, and 25% (w/w), but only leaded gasoline at 25% resulted in more than 50% loss of urease activity (52). Amendment of soil with jet fuel at rates of 5% and 13.5% reduced the rate of fluorescein diacetate (FDA) hydrolysis (esterase activity) (53). However, if the soil was subjected to a bioremediation treatment (lime, fertilizers, and simulated tillage), FDA hydrolysis increased rapidly and markedly after a 1-week lag period. The reduced activity in the nonremediated soil was attributed to inhibition by jet-fuel degradation products. Inhibition by these fuel products may be caused by the aromatic, and not the aliphatic, components of the hydrocarbon mixtures, and possibly only by benzene (54–56).

B. Contamination by Heavy Metals and Metalloids

1. *Inhibitory Effects of Heavy Metals in Soil*

Heavy metals are toxic to living organisms primarily because of their protein-binding capacity and hence their ability to inhibit enzymes. The cationic metals are noncompetitive inhibitors, which bind irreversibly with sulfydryl and carboxylate groups and with histidine, altering protein structure and the conformation and accessibility of the enzymes' active sites. The anions of metals and metalloids, e.g., As[V], W[VI], and Mo[VI], may have analogous structures to products and/or inhibitors of certain enzymes and are, therefore, likely to be competitive inhibitors. For example, the inhibition of soil phosphatase by $HAsO_4^{2-}$, WO_4^{2-}, and MoO_4^{2-} has been attributed to the structural similarity of these anions to HPO_4^{2-} (or $H_2PO_4^-$), the product and also an inhibitor of this enzyme's activity (57). Similarly, these anions inhibit sulfatase because $HPO_4^{2-}/H_2PO_4^-$ also inhibit this enzyme (58).

In solution, cationic metal salts are effective enzyme inhibitors at very low concentrations. However, the many metal-amendment studies (e.g., 59–65) and field studies at contaminated sites (e.g., 66–70) have shown that inhibition of soil enzymes usually requires much higher heavy-metal concentrations. There are two possible explanations for this behavior:

1. The physical surroundings of the soil enzymes protect them from exposure to the metals.
2. The metals are rendered less available to the enzymes by interaction with soil constituents.

The first mechanism is possible for intracellular enzymes, via mechanisms that prevent metals from passing through cell membranes. Protective mechanisms for extracellular enzymes appear less likely since metal ions are smaller than most enzyme substrates. However, extracellular enzymes can be protected if the site of inhibition is remote from the enzyme's active site and is inaccessible to the metal ion. The second mechanism is a certainty. Heavy metals interact very strongly with soil inorganic and organic constituents through adsorption, chelation, and precipitation reactions that render them much less available. Effectively, most of the metal is ''locked-up,'' and only the small amount in soluble form at the site of enzyme activity (intracellular or extracellular) is able to interact with the enzymes. Lähdesmäki and Piispanen (71), using fractionation techniques, found a very much greater inhibitory effect of Zn and Cu salts on protease, cellulase, and amylase activities in fractions from which the clay and humus colloids had been separated out than in the original soil. One or both of the mechanisms could account for this increased inhibition.

The capacity of a soil to protect its enzymes from inhibition by heavy metals is, therefore, a function of its ability to lock up the metals; therefore, there should be a relationship of soil texture and organic matter content with enzyme inhibition. In support of this premise, it has been shown that heavy metals caused greater inhibition of enzyme activities and other biochemical properties in coarse-textured soils than in fine-textured soils (72–78). This also can be seen in the data of Tabatabai et al. (57–60) and has been attributed to the lower surface area, lower cation exchange capacity (CEC), and generally lower organic matter content of these coarse-textured soils, all of which diminish their capacity to reduce the solubility of metal ions (72). Inhibition of enzyme activity in heavy-metal-contaminated soil should, then, reflect the ''bioavailability'' of the metals, since the mechanisms that are protecting soil enzymes are likely to be the same mechanisms limiting metal uptake by plants and soil organisms. Therefore, soil enzyme activity may be considered a surrogate measurement of the impact of metals on soil biota as a whole or of their uptake by, and their toxicity to, plants. Use of an enzyme activity to assess soil degradation by heavy metals requires no knowledge of what the enzyme is doing in the soil; it is merely an indicator or biosensor of a more general effect.

2. *Dose–Response Models*

This potential ecotoxicological role for soil enzymes has been investigated in several studies to determine an ecological equivalent of LD_{50}, viz, the ecological dose 50%, ED_{50}, of heavy metal in soil. ED_{50} is defined as the concentration of a toxicant that inhibits a microbially mediated ecological process by 50% (79).

Haanstra and associates (80) developed a ''logistic response model'' to describe the observed sigmoidal relationship between biological activity (in this instance, respiration) and the natural logarithm of the toxicant (Ni) concentration (Fig 3). Although ED_{50} determined from this model was found to be a useful measure of toxicity, it provided no

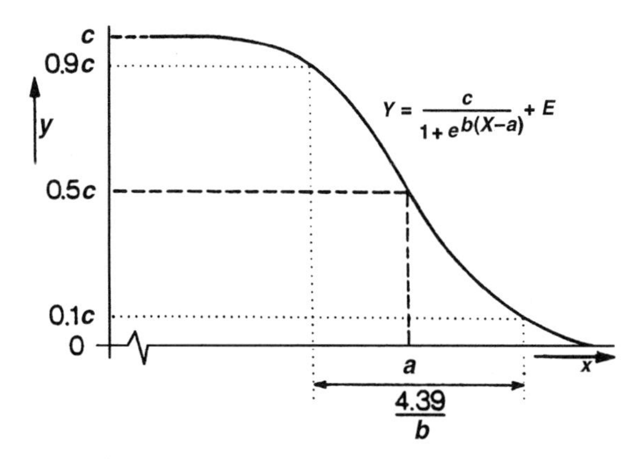

Figure 3 The logistic response curve and the relationship describing it. Parameter Y, enzyme activity; X, natural logarithm of the heavy metal concentration; c, uninhibited enzyme activity; b, slope parameter indicating the inhibition rate and equal to $4.39/(0.1c - 0.9c)$; a, logarithm of concentration at which enzyme activity is half the uninhibited level ($a = 0.5c$); E, stochastic error term. (With permission from Ref. 74.)

information about the ''suddenness'' of the decrease in activity (80). For this reason, a further measure, the ecological dose range (EDR), defined as the dose range over which activity decreases from 90% to 10% of the undisturbed activity, was proposed. Haanstra and coworkers used this approach to determine ED_{50} values and the EDR ranges for urease, phosphatase, and arylsulfatase activities in five soils 6 weeks and 18 months after amendment with six heavy metals (73–75). Generally, ED_{50} values were predictably lower in soils with low CEC and organic C content, e.g., sandy soils. ED_{50} values usually were lower after 18 months than after 6 weeks, although few of the differences were significant. These studies indicated that considerable enzyme inhibition could be expected at soil metal concentrations that were then considered acceptable under existing legislation (74,75).

More recently Speir et al. (76,77) used two Michaelis–Menten enzyme-inhibition kinetic models, in place of this sigmoidal dose–response model, to determine ED_{50} values for the inhibition of soil enzyme activities and other biological properties by heavy metals. There were two principal reasons for this different approach.

1. These Michaelis–Menten models have a physical interpretation: i.e., they can explain the behavior of an enzyme exposed to an inhibitor.
2. The logarithmic relationship represented by the sigmoidal dose–response curve is elongated to an exaggerated extent at low inhibitor concentration. Because it is not possible to fit a zero concentration to a logarithmic curve, an arbitrary value of 10^{-3} mg kg^{-1} metal (73,74) or 10^{-3} mmol kg^{-1} metal (75) was assigned to the unamended soil. In many instances, the lowest amendment concentration lay well beyond this initial part of the curve.

The first Michaelis–Menten model describes full enzyme inhibition, i.e., fully competitive, fully noncompetitive, and prescribes a linear relationship between the reciprocal of reaction rate (v) and inhibitor concentration (i) (76,77,81). This model gives a hyperbolic relationship between reaction rate and inhibitor concentration, and ED_{50} is the concentration resulting in a 50% loss of activity (Fig. 4). The second model describes partial inhibition, i.e., partially competitive, partially noncompetitive (Fig. 4). In both instances

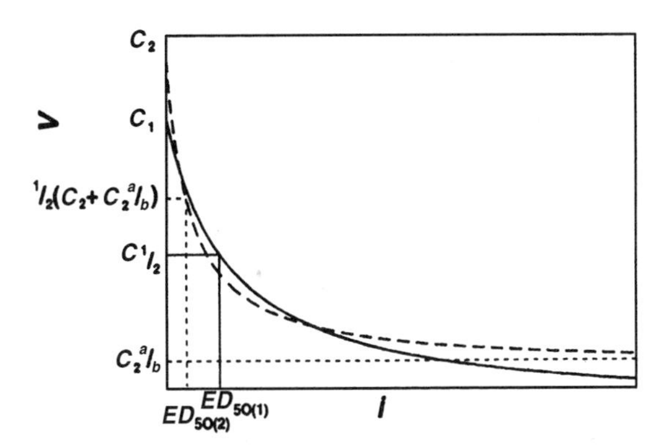

Figure 4 Relationship between reaction rate (v) and heavy metal concentration (i) as described by the full-(—) and partial-(––) inhibition models. Parameters: c_1 and c_2, $ED_{50(1)}$, and $ED_{50(2)}$ represent uninhibited rates and ED_{50} values for the full- and partial-inhibition models, respectively; c_2a/b, minimum (asymptote) for the partial-inhibition model. (With permission from Ref. 76.)

all constants, a, b, and c, are always positive and $b > c$. In the second, partial-inhibition, model, the inhibitor reduces the affinity of the enzyme for its substrate but does not prevent the enzyme-catalyzed reaction. As the inhibitor combines with the enzyme, inhibition increases to a definite limit beyond which increasing inhibitor concentration has no further effect. Therefore, the model describes a hyperbolic relationship in which activity falls to an asymptotic value as inhibitor concentration increases (Fig. 4). If this asymptote occurs at above 50% inhibition, a true ED_{50} cannot be estimated. We can, however, redefine ED_{50} in this situation, as the inhibitor concentration that results in the loss of 50% of all of the activity that can be lost, i.e., a fall to 50% of the asymptote activity value (Fig. 4).

To date, this Michaelis–Menten technique has been used for only three contrasting soils and two ''heavy metal'' species—hexavalent Cr and the metalloid As in its pentavalent oxidation state. Cr[VI] was shown to be a potent inhibitor of soil phosphatase and sulfatase activities, especially in a coarse-textured sandy soil (phosphatase ED_{50} of 0.078 mmol kg^{-1} (4 mg kg^{-1}), sulfatase ED_{50} of 0.2 mmol kg^{-1} (10 mg kg^{-1}), but a less potent inhibitor of urease activity (76). As[V], in contrast, was only a moderate inhibitor of phosphatase and sulfatase activities and was ineffective against urease (77). In almost every instance in which the inhibition data fitted both models, the second model provided the better fit, implying that the inhibition was partial.

3. *Interpretation of Dose–Response Data*

There are at least five points that need to be considered when interpreting data derived from such dose–response curves. (1) The first concerns the actual significance of ED_{50} values. If the enzyme responses are truly indicative of effects on soil organisms, then 50% loss of activity may well be unacceptable. These models do allow determination of the heavy metal concentrations causing significantly less than 50% inhibition, e.g., ED_{10}. The current trend in assessment of environmental effects of contaminants is to find the lowest observed adverse effect concentration (LOAEC) (82,83), which would obviously be much less than ED_{50}.

(2) How should ED_{50} values be interpreted when the inhibitor causes activity to fall to an asymptotic value, as predicted by the partial inhibition model (Fig. 4)? Speir and colleagues (76,77) showed that in a coarse-textured sandy soil phosphatase and sulfatase activities were reduced by only about 40% by Cr[VI] and As[V], respectively. However, in both instances, the excellent fit to model 2 indicated that most of this inhibition occurred at relatively low inhibitor concentrations. If this is partial inhibition, then all the inhibitor does is rapidly reduce the affinity of enzyme for its substrate. The reaction continues, but at a lower rate, which may not be particularly detrimental. However, in the complex soil medium, this situation also could be explained by complete inhibition of a sensitive component of the enzyme. If this happened to be the intracellular component, with the extracellular part possibly being protected from inhibition by its physical location, the consequences could be much more serious and indicate a potentially severe impact on the soil microbial population.

(3) Results to date (73–77) have revealed that soil enzymes are affected differently by different metals and respond differently in different soils. This makes it difficult to decide what enzyme(s) should be used as indicator(s). Since a component of every soil enzyme has a metabolic role in soil organisms, we probably should choose enzymes that are particularly sensitive to the metal in question. For example, phosphatase was moderately inhibited by As[V], but urease was unaffected (58,77). This is because arsenate is

a structural analog of phosphate, a known feedback inhibitor of phosphatase. On the other hand, urease was particularly sensitive to cationic forms of heavy metals (58), presumably because of the presence of sulfydryl groups in the vicinity of its active site.

(4) The experiments used to derive these dose–response relationships are very artificial and do not reflect how metals enter soil, except in a rare chemical-spill situation. Giller and associates (84) stated that this experimental approach is simplistic in that it bears little relationship to most ''real-life'' contamination of soils, in which metal concentrations are built up over many years and the metals are well equilibrated with the surrounding soil. In reality, metals generally are applied to soils in relatively small doses and often are bound strongly in organomineral complexes, e.g., in sewage sludge. In spite of this artificiality, however, the results from early amendment studies with metal salts (e.g., 57–64,73–75) have been used as supporting information to derive limits, based on LOAEC principles for heavy metals in soils, for the Danish draft soil quality standards (85). The wisdom of using these data in this way must, however, be questioned.

(5) The ED_{50} experiments conducted to date (73–77) have been relatively short term or have required storage of amended soils under artificial conditions in the absence of plants. This has allowed the assessment of acute effects, with direct inhibition of intracellular and extracellular enzyme activity by heavy metals. These experiments have not allowed a realistic assessment of long-term effects, which could be quite different because, as well as enzyme inhibition, microbial proliferation and microbial enzyme synthesis also may be adversely impacted. This limitation could be offset by a decline in the bioavailability of the metals with time, as they become adsorbed, chelated, and precipitated. As yet, there are relatively few experimental data from long-term contaminated sites. Contaminated sites, moreover, usually have the added complexity of excessive amounts of more than one contaminant, so that it is almost impossible to assign environmental effects to any one heavy metal or to determine dose parameters, such as ED_{50}.

4. Field Studies Versus Laboratory Studies

Speir and coworkers investigated the effects of contamination of a pasture soil by Cr, Cu, and As, acquired from runoff from a neighboring timber-treatment factory, on soil biological properties, including enzyme activities (68,69). Heavy metal concentrations ranged from background (<50 mg kg^{-1}) to >1200 mg kg^{-1} soil. Phosphatase, urease, and invertase activities generally were considerably lower in the contaminated soil than in the control, but differences between the lowest and highest levels of contamination (86 to 1260 mg Cr kg^{-1}) did not result in further significant changes (Cr was used here as a surrogate for all three contaminants as their concentrations correlated strongly). This implies that even low levels of metal contamination could result in reduced biological activity, an effect not always observed in short-term studies. At the locations with the highest metal concentrations, herbage yields were very low or nonexistent, possibly the cause, and/or result, of the effect of the metals on biological activity. Sulfatase, in contrast to the other enzymes, followed a hyperbolic relationship with increasing metal concentration (68), falling to about 15% of the activity found in uncontaminated soil. This suggests that this enzyme may be a good indicator of the loss of microbial activity and diminished plant yield in contaminated sites.

Kuperman and Carreiro (70) found that soil enzyme activities declined markedly with increasing metal concentration at a military site contaminated with As, Cd, Cr, Cu, Pb, Ni, and Zn. Here, in contrast to the previous study (68,69), the decline in enzyme activities generally followed a similar pattern to that of organic matter content and probably indicates a very adverse effect of the metals on soil biological activity. The authors proposed that the decrease in enzyme activity was caused primarily by direct suppression of microbial growth in the contaminated soil but considered that direct enzyme inhibition by the heavy metals may also have accounted for some of the decrease. They concluded that ''integration of microbial biomass and extracellular enzyme activity measurements into ecological risk assessment procedures would permit direct assessment of negative impacts on the structure and function of soil communities and ecosystem processes'' (70).

Precisely how the gap between short- and long-term studies, and how the complexities of multimetal-contaminated sites can be unraveled, remain challenges. Laboratory-amendment studies have a role in developing an understanding of the interactions of particular heavy metals with soil enzymes, but their results should not be overinterpreted or used in ways that might lead to erroneous or misleading conclusions. Detailed data on the toxicity of metals are, however, often difficult to obtain from field studies of long-term contaminated sites because a relevant control soil, unless planned, is usually unavailable. In addition, a gradient of metal-contaminant concentrations is rarely possible (84). Typically, in laboratory studies the response of microorganisms *adapting* to elevated metal concentrations is examined, whereas in studies of long-term metal-polluted soils the response of microorganisms already adapted to elevated metal concentrations is examined. Both the type and the sensitivity of response to these different forms of contamination may be very different, and the response obtained in the laboratory may bear little relation to that seen in the field (84).

One additional and potentially serious concern about laboratory experiments using heavy metal salts has emerged from a 1999 study that showed that the salts of cationic metals (Cd, Cr, Cu, Ni, Pb, and Zn) can cause significant acidification of soils when added at high concentrations (>10 mmol kg^{-1}), especially in soils with a coarse texture (78). Sparse attention has been paid to this process in previous publications, although reduction of pH in response to heavy metal addition has been observed in some (74,75,86,87), but not all investigations (88). It is possible that this phenomenon could, at least in part, explain a significant proportion of the apparent ''inhibition'' of soil enzymes by heavy metals. In their investigation, Speir et al. (78) included a Ca salt treatment and an acid treatment to differentiate effects attributable to osmosis and acidity from those attributable to heavy metals. They found that the loss of soil sulfatase activity, for all amendments, was almost entirely attributable to acidification of the soil during preincubation before assay (Fig. 5). In contrast, phosphatase activity was inhibited by metals but not by acid, except in a coarse-textured soil in which acidification did result in loss of activity below pH 4. They concluded that the difference in behavior of the two enzymes might be a function of their stability to changing pH during the preincubation period. Soil sulfatase has a narrow pH optimum (89,90), and the results indicated that activity was rapidly and irreversibly lost at any pH significantly below this range. In contrast, the pH optimum for phosphatase is very broad in many soils, spanning several pH units, and is probably the resultant of a mixture of acid and alkaline phosphatases (91). It appears that this broad pH optimum, within which the enzyme is obviously stable, had protected it from denaturation during the days it was exposed to acid conditions before assay. These results have considerable implications for the interpretation of previous studies. If the impact of soluble and, presumably, extremely bioavailable metals is caused by pH changes, and not the metal itself, no direct metal effect on sulfatase activity, at least in the short term, would

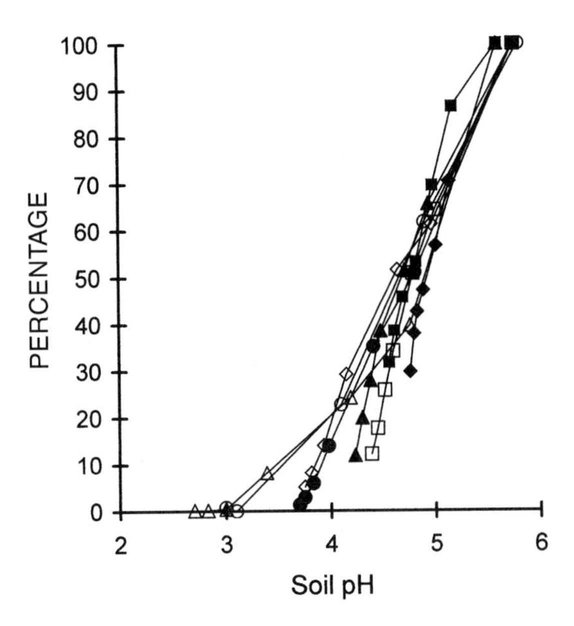

Figure 5 Effects of amendment with heavy metals or acid on sulfatase activity in a coarse-textured sandy soil, with percentage inhibition expressed as a function of soil pH. Amendments: Ca, —◆—; Cd, —▲—; Cu, —●—; Pb —◇—; Zn —□—; Ni —■—; Cr —△—; Acid—○—. (Adapted, with permission, from Ref. 78.)

be discernible at most metal-contaminated sites or after materials, such as sewage sludge had been applied to land.

C. Heavy Metals in Sewage Sludge

Land application of sewage sludge can result in the contamination of agricultural soils by heavy metals. Although this practice returns valuable nutrients and organic matter to the land, it is controlled in many countries by rules and guidelines stipulating maximal contaminant loading of the sludge, maximal annual metal application rates, and maximal soil metal-concentrations. Assessment of the impacts of sewage sludge–borne heavy metals on soil microorganisms and microbial processes is complicated by the high organic matter and nutrient contents of the sludge. Frankenberger et al. (92) found that soil urease activity was inhibited at low sludge-loading rates but markedly enhanced at higher rates. This initial reduction was attributed to the extremely high contaminant load in the sludges investigated, and the enhancement to the sludge organic matter and nutrients stimulating microbial activity and urease synthesis. Other studies, using less contaminated sludges, also have demonstrated that enzyme activities increased steadily with increasing rates of sludge application (93–95). However, in one of these studies, sewage sludge stimulated activity to a lesser extent than other organic amendments, possibly because of its metal content (94). Long-term studies are required, preferably at sites where sludge application has long ceased and the sludge nutrients and readily degradable organic matter have gone. Currently, some of the most important research on the effects of heavy metals in sewage sludge is emerging from long-term sites in the United Kingdom and Germany (reviewed in 83–85,96), but very little enzymological analysis has been conducted on their soils. Other contaminated sites, such as those investigated by Tyler (66,67), Speir et al. (68,69), and Kuperman and Carreiro (70), especially on agricultural soils, may, however, be considered as surrogates for land that had previously had sewage sludge applied.

D. Pesticides

There are large numbers of pesticides currently registered for agricultural use throughout the world and many that no longer are used but whose residues and metabolites still are found in agricultural soils. All of these have been subjected to more or less intensive investigation for the determination of potential side effects on soil organisms and biological processes. Arising from this work, soil biological criteria have been established for the routine assessment of pesticides for registration purposes (e.g., 97–100). However, these do not include assessments of impacts on soil enzyme activities, even though there have been a large number of investigations of pesticide–soil enzyme interactions (extensively reviewed in 6,10,101). One reason for this is readily apparent on reading the comprehensive review of Schäffer (10); there are very few consistent patterns of soil enzyme inhibition by pesticides, even at concentrations greatly exceeding recommended application rates. Another reason is that no attempts have been made, or could have been made, to evaluate the consequences for biodiversity, biological activity, and/or soil fertility of enzyme inhibition when it did occur. The literature on this topic is very similar to that describing the effects on soil enzymes of agricultural practices such as fertilizer application, use of other agrochemicals, and land management; pesticides inhibit/have no effect on/activate soil enzymes, the particular response depending on factors such as soil type and experimental design. For example, long-term field application of both the amine and ester formulations of the herbicide 2,4-D at normal agronomic rates, in two separate studies, led to the following conclusions: "The effects of long-term 2,4-D application (both ester and amine) were neither ecologically significant nor did they interfere with nutrient cycling" (102); "dehydrogenase activity and soil microbial respiration were reduced substantially by the ester, indicating that the ester probably interfered with nutrient cycling" (103). In the former study, urease and acid- and alkaline-phosphatase activities were temporarily reduced, whereas in the latter, urease activity was permanently depressed.

Pesticides are not designed to inhibit soil enzymes. Some may indeed be enzyme inhibitors, but their structural diversity makes it impossible to foresee direct negative effects on the hydrolase activities we measure by using artificial substrates and assay conditions. They still could have a negative effect on soil enzyme activity; obviously a fungicide or a herbicide will reduce enzyme activity, if not by direct inhibition, then by removal of a source of soil enzymes (104). However, in most instances, any depressive effect has been found to disappear after several days or weeks (10) and may be replaced by enhanced enzyme activity, possibly from enzymes released from lysed cells (105) or as a response to a flush of microbial activity as the killed organisms are metabolized. Recovery of soil biological activity some time after application is the basis for an ecological approach for the assessment of side effects of agrochemicals on soil microorganisms proposed by Domsch et al. (106). From a review of a large number of case studies, side effects were compared with the depressive effects resulting from natural stress factors, such as fluctuations of temperature, water content, pH, and physical disturbances. Such natural stresses were found to cause depressions of some properties by up to 90%, but the depressions were short-lived, and recovery phases averaged 18 days. On this basis, it was proposed that reversible side effects of agrochemicals causing delays of recovery of microbiological properties of up to 30 days are normal, those resulting in delays of 60 days are tolerable, but those with delays of more than 60 days may be critical. A minimal monitoring period of 30 days for the detection of persistent side effects was proposed. On this basis, the studies reviewed by Schäffer (10) and subsequent investigations (e.g., 105,107–111) suggest that the depressions of soil hydrolytic enzyme activities almost all fall into the ''normal'' category. This reinforces the conclusions of Ladd (112) that pesticide applications have little or no effect on enzyme activity in soils.

Is it, therefore, worthwhile to continue to test for pesticide effects on soil enzyme activities? The answer depends upon the context of the research. Existing pesticides have been studied extensively at a range of concentrations in different soil types. Further investigations may be profitable only if they concentrate on degradative enzymes from the perspective of understanding mechanisms of pesticide metabolism in soils and/or for the cleanup of contaminated soils. On balance, it probably is important that newly registered pesticides continue to be subjected to tests for effects on soil enzymes, even though they have already passed registration criteria that are arguably more stringent than these tests. It is conceivable that a new chemical or a metabolite may, by design or accident, be a particularly potent inhibitor of a soil enzyme. If the inhibition extends beyond 60 days, then according to the definition of Domsch et al. (106), a critical situation may exist and further investigation is warranted.

VI. CONCLUSIONS

There have been many studies of the impacts of soil physical and chemical degradation on soil hydrolase enzyme activities and attempts made to use enzyme activity measurements to determine or predict the extent and/or rate of soil degradation or recovery after degradation. The major barriers to progress arise from the limitations of our methodology and, therefore, our understanding of the different compartments of soil enzyme activity and their roles in soil fertility and productivity.

When soils have been subjected to physical degradation comprising loss of surface materials by erosion or mining or subjected to intensive cropping, the primary effect is loss of organic matter. Loss of enzyme activity is a consequence because there is always a strong correlation between these properties. Generally, the assay of soil enzyme activities does not provide any more information about potential soil degradation from such physical causes than does the measurement of organic C alone, nor does it give any information about the short-term productivity of the soil. However, techniques that provide insight into the various compartments of enzyme activity may provide information on soil organic matter quality that could be useful in assessing the extent of soil degradation.

In the early stages of recovery of degraded soils, or development of ''technogenic'' soils, soil enzyme activities correlate closely with plant yields and increase more rapidly than does soil organic C content. In these situations, the enzymes can be used as indicators, or biomarkers, of progress toward recovery and possibly predictors of how long recovery will take. Some enzymes, e.g., the carbohydrases, seem to be better predictors than others, but because we do not know how soil, plant, and environmental factors influence the recovery of individual enzymes, this predictive role is still only an experimental tool.

Soil enzyme activities are not particularly sensitive to contamination by oil and oil products; there is little adverse effect at concentrations of crude oil sufficient to kill all plant life. Of the fuel products tested, only the aromatic compounds, and possibly only benzene, significantly inhibited soil enzyme activity.

Soil characteristics provide considerable ''protection'' of soil enzymes from heavy metals. Because these same characteristics also determine the ''bioavailability'' of metals to plants and soil organisms, soil enzymes could have a role as ecotoxicological indicators or biosensors of environmental effects. Dose–response curves have been used to determine the impact of heavy metal salts on soil enzyme activities, but because different enzymes show different levels of inhibition, even within a single soil, it is not possible to determine what level is acceptable. Enzyme inhibition data from metal-salt amendment studies have been used to support soil quality criteria for heavy metals. It is, however, questionable whether this is appropriate, considering that metals rarely enter soils in this form or reach such high concentrations with a single application. Metal-salt amendment experiments have also generally been short-term studies that assess acute, mainly direct-inhibition, effects on soil enzyme activities. The response in long-term metal-polluted soils is likely to be considerably different, and effects on soil enzyme activities also may include those attributable to reduced enzyme synthesis and impaired microbial growth and activity.

One form of long-term contamination that has received a great deal of attention is that arising from the application of sewage sludge to land. However, the few studies that have included enzyme assays generally found that the enhancement of overall biological activity resulting from the readily available C and nutrients masked any negative impact on the enzyme activities. There are few instances of significant and prolonged adverse effects of pesticides on soil enzyme activities.

REFERENCES

1. JW Doran, TB Parkin. Quantitative indicators of soil quality: a minimum data set. In: JW Doran, AJ Jones, eds. Methods for Assessing Soil Quality. SSSA Special Publication 49. Madison, WI: Soil Science Society of America, 1996, pp 25–37.
2. JW Doran, DC Coleman, DF Bezdicek, BA Stewart, eds. Defining Soil Quality for a Sustainable Environment. SSSA Special Publication 35. Madison, WI: Soil Science Society of America, 1994.
3. JW Doran, AJ Jones, eds. Methods for Assessing Soil Quality. SSSA Special Publication 49. Madison, WI: Soil Science Society of America, 1996.
4. RP Dick. Soil enzyme activities as indicators of soil quality. In: JW Doran, DC Coleman, DF Bezdicek, BA Stewart, eds. Defining Soil Quality for a Sustainable Environment. SSSA Special Publication 35. Madison, WI: Soil Science Society of America, 1994, pp 107–124.
5. RP Dick, DP Breakwell, RF Turco. Soil enzyme activities and biodiversity measurements as integrative microbiological indicators. In: JW Doran, AJ Jones, eds. Methods for Assessing Soil Quality. SSSA Special Publication 49. Madison, WI: Soil Science Society of America, 1996, pp 247–271.
6. RP Dick. Soil enzyme activities as integrative indicators of soil health. In: C Pankhurst, BM Doube, VVSR Gupta, eds. Biological Indicators of Soil Health. Wallingford: CAB International, 1997, pp 121–156.
7. DA Klein, DL Sorensen, EF Redente. Soil enzymes: A predictor of reclamation potential and progress. In: RL Tate, DA Klein, eds. Soil Reclamation Processes: Microbiological Analyses and Applications. New York: Marcel Dekker, 1985, pp 141–171.

8. P Nannipieri, S Grego, B Ceccanti. Ecological significance of the biological activity of soil. In: J-M Bollag, G Stotzky, eds. Soil Biochemistry. Volume 6. New York: Marcel Dekker, 1990, pp 293–355.
9. WA Dick, MA Tabatabai. Significance and potential uses of soil enzymes. In: FB Metting, Jr, ed. Soil Microbial Ecology: Applications in Agricultural and Environmental Management. New York: Marcel Dekker, 1993, pp 95–127.
10. A Schäffer. Pesticide effects on enzyme activities in the soil ecosystem. In: J-M Bollag, G Stotzky, eds. Soil Biochemistry. Vol. 8. New York: Marcel Dekker, 1993, pp 273–340.
11. F Gil-Sotres, MC Leirós, MC Trasar-Cepeda, A Saá, MV González-Sangregorio. The importance of soil biochemical properties in the reclamation of lignite mining land. In: B Ceccanti, C Garcia, eds. Environment Biochemistry in Practice. Vol 1. Wastes and Soil Management. Pisa: Istituto per la Chimica del Terreno, 1994, pp 133–170.
12. P Nannipieri. The potential use of soil enzymes as indicators of productivity, sustainability and pollution. In: CE Pankhurst, BM Doube, VVSR Gupta, PR Grace, eds. Soil Biota: Management in Sustainable Farming Systems. Adelaide: CSIRO Publishing, 1994, pp 238–244.
13. L Gianfreda, J-M Bollag. Influence of natural and anthropogenic factors on enzyme activity in soil. In: G Stotzky, J-M Bollag, eds. Soil Biochemistry. Vol. 9. New York: Marcel Dekker, 1996, pp 123–193.
14. S Kiss, D Pasca, M Dragan-Bularda, eds. Developments in Soil Science 26: Enzymology of Disturbed Soils. Amsterdam: Elsevier, 1998.
15. S Kiss, M Dragan-Bularda, D Radulescu. Biological significance of enzymes accumulated in soil. Adv Agron 27:25–87, 1975.
16. J Skujins. Extracellular enzymes in soil. CRC Crit Rev Microbiol 4:383–421, 1976.
17. RG Burns. Enzyme activity in soil: Location and possible role in microbial ecology. Soil Biol Biochem 14:423–427, 1982.
18. RG Burns. Extracellular enzyme—substrate interactions in soil. In: JH Slater, R Whittenbury, JWT Wimpenny, eds. Microbes in Their Natural Environments. Symposium 34. Cambridge: Cambridge University Press, 1983, pp 249–298.
19. J Skujins. History of abiontic soil enzyme research. In: RG Burns, ed. Soil Enzymes. London: Academic Press, 1978, pp 1–49.
20. RG Burns. Enzyme activity in soil: Some theoretical and practical considerations. In: RG Burns, ed. Soil Enzymes. London: Academic Press, 1978, pp 295–340.
21. DJ Ross. Assays of invertase in acidic soils: Influence of buffers. Plant Soil 97:285–289, 1987.
22. DJ Ross. Assays of invertase and amylase activities in soil. 1. Influence of buffered and unbuffered systems with soils from grasslands. NZ J Sci 26:339–346, 1983.
23. D Rossel, J Tarradellas, G Bitton, J-L Morel. Use of enzymes in soil ecotoxicology: A case for dehydrogenase and hydrolytic enzymes. In: J Tarradellas, G Bitton, D Rossel, eds. Soil Ecotoxicology. Boca Raton, FL: CRC Press, 1997, pp 179–206.
24. JM Bremner, MA Tabatabai. Effect of some inorganic substances on TTC assay of dehydrogenase activity in soils. Soil Biol Biochem 5:385–386, 1973.
25. K Chander, PC Brookes. Is the dehydrogenase assay invalid as a method to estimate microbial activity in Cu-contaminated and non-contaminated soils? Soil Biol Biochem 23:901–915, 1991.
26. TM Klein, JS Koths. Urease, protease and acid phosphatase in soil continuously cropped to corn by conventional or no-tillage methods. Soil Biol Biochem 12:293–294, 1980.
27. WA Dick. Influence of long-term tillage and crop rotation combinations on soil enzyme activities. Soil Sci Soc Am J 48:569–574, 1984.
28. VVSR Gupta, JJ Germida. Distribution of microbial biomass and its activity in different soil aggregate size classes as affected by cultivation. Soil Biol Biochem 20:777–786, 1988.
29. RE Farrell, VVSR Gupta, JJ Germida. Effects of cultivation on the activity and kinetics of arylsulphatase in Saskatchewan soils. Soil Biol Biochem 26:1033–1040, 1994.

30. E Kandeler, S Palli, M Stemmer, MH Gerzabek. Tillage changes microbial biomass and enzyme activities in particle-size fractions of a Haplic Chernozem. Soil Biol Biochem 31: 1253–1264, 1999.
31. RP Dick, DD Myrold, EA Kerle. Microbial biomass and soil enzyme activities in compacted and rehabilitated skid trail soils. Soil Sci Soc Am J 52:512–516, 1988.
32. JL Neal, SA Herbein. Abiontic enzymes in arctic soils: Changes in sulphatase activity following vehicle disturbance. Plant Soil 70:423–427, 1983.
33. SP Deng, MA Tabatabai. Effect of tillage and residue management on enzyme activities in soils I. Amidohydrolases. Biol Fertil Soils 22:202–207, 1996.
34. SP Deng, MA Tabatabai. Effect of tillage and residue management on enzyme activities in soils II. Glycosidases. Biol Fertil Soils 22:208–213, 1996.
35. SP Deng, MA Tabatabai. Effect of tillage and residue management on enzyme activities in soils. III. Phosphatases and arylsulphatase. Biol Fertil Soils 24:141–146, 1997.
36. RP Dick, PE Rasmussen, EA Kerle. Influence of long-term residue management on soil enzyme activities in relation to soil chemical properties of a wheat-fallow system. Biol Fertil Soils 6:159–164, 1988.
37. DA Angers, N Bissonnette, A Légère, N Samson. Microbial and biochemical changes induced by rotation and tillage in a soil under barley production. Can J Soil Sci 73:39–50, 1993.
38. PL Guisquiani, G Gigliotti, D Businelli. Long-term effects of heavy metals from composted municipal waste on some enzyme activities in a cultivated soil. Biol Fertil Soils 17:257–262, 1994.
39. DJ Ross, TW Speir, KR Tate, A Cairns, KF Meyrick, EA Pansier. Restoration of pasture after topsoil removal: Effects on soil carbon and nitrogen mineralization, microbial biomass and enzyme activities. Soil Biol Biochem 14:575–581, 1982.
40. TW Speir, DJ Ross, VA Orchard. Spatial variability of biochemical properties in a taxonomically-uniform soil under grazed pasture. Soil Biol Biochem 16:153–160, 1984.
41. S Kiss, M Dragan-Bularda, D Pasca. Enzymological study of the evolution of technogenic soils. Evol Adapt (Cluj) 2:159–186, 1985.
42. S Kiss, M Dragan-Bularda, D Pasca. Enzymology of the recultivation of technogenic soils. Adv Agron 42:229–278, 1989.
43. TW Speir, R Lee, EA Pansier, A Cairns. A comparison of sulphatase, urease and protease activities in planted and in fallow soils. Soil Biol Biochem 12:281–291, 1980.
44. DJ Ross, TW Speir, KR Tate, JC Cowling, HM Watts. Restoration of pasture after topsoil removal: changes in soil biochemical properties over a 5-year period—a note. NZ J Sci 27: 419–422, 1984.
45. HG Lambert, NA Trustrum, DA Costall. Effect of slip erosion on seasonally dry Wairarapa hill pastures. NZ J Agric Res 27:57–64, 1984.
46. DJ Ross, TW Speir, JC Cowling, CW Feltham. Soil restoration under pasture after lignite mining: Management effects on soil biochemical properties and their relationships with herbage yields. Plant Soil 140:85–97, 1992.
47. T ap Rees. Pathways of carbohydrate breakdown in higher plants. In: DH Northcote, ed. Plant Biochemistry. Vol. 11. London: Butterworths, 1974, pp 89–127.
48. MJ Rowell, LZ Florence. Characteristics associated with differences between undisturbed and industrially-disturbed soils. Soil Biol Biochem 25:1499–1511, 1993.
49. SA Aliev, DA Gadzhiev. Effect of pollution by petroleum organic matter on the activity of biological processes of soils. Izv Akad Nauk AzSSR Ser Biol Nauk 2:46–49, 1977.
50. AE Linkins, RM Atlas, P Gustin. Effect of surface applied crude oil on soil and vascular plant root respiration, soil cellulase and hydrocarbon hydroxylase at Barrow, Alaska. Arctic 31:355–365, 1978.
51. SM Samsova, GP Kurbskii, GM Usacheva, TS Gubaidullina, VI Fil'chenkova, VV Abushaeva. Changes of microflora and petroleum composition in a Tatarian chernozemic soil

during the first period after pollution. In: Dobycha Poleznykh Iskopaemykh I Geokhimiya Prirodnykh Ekosistem. Moscow: Izdatel'stvo Nauka, 1982, pp 235–245.

52. WT Frankenberger Jr. Use of urea as a nitrogen fertilizer in bioreclamation of petroleum hydrocarbons in soil. Bull Environ Contam Toxicol 40:66–68, 1988.
53. H-G Song, R Bartha. Effect of jet fuel spills on the microbial community of soil. Appl Environ Microbiol 56:646–651, 1990.
54. NA Kireeva. Microbiological evaluation of the soil polluted by petroleum hydrocarbons. Bashk Khim Zh 2:65–68, 1995.
55. NA Kireeva, EI Novoselova, FK Khaziev. Phosphohydrolase activity of petroleum polluted soils. Pochvoved No 6:723–725, 1997.
56. NA Kireeva, EI Novoselova, FK Khaziev. Enzymes of nitrogen metabolism in petroleum-polluted soils. Izv Akad Nauk Ser Biol. No 6:755–759, 1997.
57. NG Juma, MA Tabatabai. Effects of trace elements on phosphatase activity in soils. Soil Sci Soc Am J 41:343–346, 1977.
58. AA Al-Khafaji, MA Tabatabai. Effects of trace elements on arylsulfatase activity in soils. Soil Sci 127:129–133, 1979.
59. MA Tabatabai. Effects of trace elements on urease activity in soils. Soil Biol Biochem 9: 9–13, 1977.
60. WT Frankenberger Jr, MA Tabatabai. Amidase activity in soils. IV. Effects of trace elements and pesticides. Soil Sci Soc Am J 45:1120–1124, 1981.
61. DS Yadav, V Kumar, M Singh. Inhibition of soil urease activity with some metallic cations. Aust J Soil Res 24:527–532, 1986.
62. F Eivazi, MA Tabatabai. Factors affecting glucosidase and galactosidase activities in soils. Soil Biol Biochem 22:891–897, 1990.
63. WT Frankenberger Jr, MA Tabatabai. Factors affecting L-glutaminase activity in soils. Soil Biol Biochem 23:875–879, 1991.
64. WT Frankenberger Jr, MA Tabatabai. Factors affecting L-asparaginase activity in soils. Biol Fertil Soils 11:1–5, 1991.
65. SK Hemida, SA Omar, AY Abdel-Mallek. Microbial populations and enzyme activity in soils treated with heavy metals. Water Air Soil Pollut 95:13–22, 1997.
66. G Tyler. Heavy metal pollution and soil enzymatic activity. Plant Soil 40:303–311, 1974.
67. G Tyler. Heavy metal pollution, phosphatase activity, and mineralization of organic phosphorus in forest soils. Soil Biol Biochem 8:327–332, 1976.
68. RD Bardgett, TW Speir, DJ Ross, GW Yeates, HA Kettles. Impact of pasture contamination by copper, chromium, and arsenic timber preservative on soil microbial properties and nematodes. Biol Fertil Soils 18:71–79, 1984.
69. GW Yeates, VA Orchard, TW Speir, JL Hunt, MCC Hermans. Impact of pasture contamination by copper, chromium, and arsenic timber preservative on soil biological activity. Biol Fertil Soils 18:200–208, 1984.
70. RG Kuperman, MM Carreiro. Soil heavy metal concentrations, microbial biomass and enzyme activities in a contaminated grassland ecosystem. Soil Biol Biochem 29:179–190, 1997.
71. P Lähdesmäki, R Piispanen. Soil enzymology: Role of protective colloid systems in the preservation of exoenzyme activities in soil. Soil Biol Biochem 24:1173–1177, 1992.
72. B Lighthart, J Baham, VV Volk. Microbial respiration and chemical speciation in metal-amended soils. J Environ Qual 12:543–548, 1983.
73. P Doelman, L Haanstra. Short- and long-term effects of heavy metals on urease activity in soils. Biol Fertil Soils 2:213–218, 1986.
74. P Doelman, L Haanstra. Short- and long-term effects of heavy metals on phosphatase activity in soils: An ecological dose-response model approach. Biol Fertil Soils 8:235–241, 1989.
75. L Haanstra, P Doelman. An ecological dose-response model approach to short- and long-term effects of heavy metals on arylsulphatase activity in soil. Biol Fertil Soils 11:18–23, 1991.

76. TW Speir, HA Kettles, A Parshotam, PL Searle, LNC Vlaar. A simple kinetic approach to derive the ecological dose value, ED_{50}, for the assessment of Cr(VI) toxicity to soil biological properties. Soil Biol Biochem 27:801–810, 1995.
77. TW Speir, HA Kettles, A Parshotam, PL Searle, LNC Vlaar. Simple kinetic approach to determine the toxicity of As[V] to soil biological properties. Soil Biol Biochem 31:705–713, 1999.
78. TW Speir, HA Kettles, HJ Percival, A Parshotam. Is soil acidification the cause of biochemical responses when soils are amended with heavy metal salts? Soil Biol Biochem 31:1953–1961, 1999.
79. H Babich, RJF Bewley, G Stotzky. Application of the "ecological dose" concept to the impact of heavy metals on some microbe-mediated ecologic processes in soil. Arch Environ Contam Toxicol 12:421–426, 1983.
80. L Haanstra, P Doelman, JH Oude Voshaar. The use of sigmoidal dose response curves in soil ecotoxicological research. Plant Soil 84:293–297, 1985.
81. M Dixon, EC Webb. Enzymes. 2nd ed. London: Longmans, 1966, pp. 315–359.
82. SP McGrath, AC Chang, AL Page, E Witter. Land application of sewage sludge: scientific perspectives of heavy metal loading limits in Europe and the United States. Environ Rev 2: 108–118, 1994.
83. SP McGrath, AM Chaudri, KE Giller. Long-term effects of metals in sewage sludge on soils, microorganisms and plants. J Ind Microbiol 14:94–104, 1995.
84. KE Giller, E Witter, SP McGrath. Toxicity of heavy metals to microorganisms and microbial processes in agricultural soils: A review. Soil Biol Biochem 30:1389–1414, 1998.
85. JJ Scott-Fordsmand, MB Pedersen. Soil Quality Criteria for Selected Inorganic Compounds. Arbejdsrapport fra Miljøstyrelsen Working Report No. 48. Copenhagen: Danish Environmental Protection Agency, Ministry of Environment and Energy, 1995.
86. M Aoyama, S Itaya, M Otowa. Effects of copper on the decomposition of plant residues, microbial biomass, and β-glucosidase activity in soils. Soil Sci Plant Nutr 39:557–566, 1993.
87. M Aoyama, S Itaya. Effects of copper on the metabolism of ^{14}C-labelled glucose in soil in relation to amendment with organic materials. Soil Sci Plant Nutr 41:245–252, 1995.
88. J-M Bollag, W Barabasz. Effect of heavy metals on the denitrification process in soil. J Environ Qual 8:196–201, 1979.
89. MA Tabatabai, JM Bremner. Arylsulphatase activity of soils. Soil Sci Soc Am Proc 34:225–229, 1970.
90. JI Thornton, AD McLaren. Enzymatic characterization of soil evidence. J Forens Sci 20: 674–692, 1975.
91. TW Speir, DJ Ross. Soil phosphatase and sulphatase. In: RG Burns, ed. Soil Enzymes. London: Academic Press, 1978, pp 197–250.
92. WT Frankenberger Jr, JB Johanson, CO Nelson. Urease activity in sewage sludge-amended soils. Soil Biol Biochem 15:543–549, 1983.
93. M Bonmati, M Pujola, J Sana, M Soliva, MT Felipo, M Garau, B Ceccanti, P Nannipieri. Chemical properties, populations of nitrite oxidizers, urease and phosphatase activities in sewage sludge-amended soil. Plant Soil 84:79–91, 1985.
94. DA Martens, JB Johanson, WT Frankenberger Jr. Production and persistence of soil enzymes with repeated addition of organic residues. Soil Sci 153:53–61, 1992.
95. MR Banerjee, DL Burton, S Depoe. Impact of sewage sludge application on soil biological characteristics. Agric Ecosys Environ 66:241–249, 1997.
96. SP McGrath. Effects of heavy metals from sewage sludge on soil microbes in agricultural ecosystems. In: SM Ross, ed. Toxic Metals in Soil—Plant Systems. Chichester: John Wiley & Sons, 1994, pp 247–274.
97. Food and Agriculture Organisation of the United Nations. Guidelines on Environmental Criteria for the Registration of Pesticides. Rome: FAO, 1985.

98. United States Environmental Protection Agency. Part 158. Data Requirements for Registration. 58 Federal Register 34203, June 23, 1993.
99. National Registration Authority. Environment. In: Guidelines for Registering Agricultural Chemicals. Canberra: Commonwealth of Australia, November 1997.
100. Ministry of Agriculture. Registration Requirements for Pesticides in New Zealand. Wellington: Pesticides Board, January 1998.
101. S Cervelli, P Nannipieri, P Sequi. Interactions between agrochemicals and soil enzymes. In: RG Burns, ed. Soil Enzymes. London: Academic Press, 1978, pp 251–293.
102. VO Biederbeck, CA Campbell, AE Smith. Effects of long-term 2,4-D field applications on soil biochemical processes. J Environ Qual 16:257–262, 1987.
103. JP Narain Rai. Effects of long-term 2,4-D application on microbial populations and biochemical processes in cultivated soil. Biol Fertil Soils 13:187–191, 1992.
104. JP Voets, P Meerschman, W Verstraete. Soil microbiological and biochemical effects of long-term atrazine applications. Soil Biol Biochem 6:149–152, 1974.
105. P Perucci, L Scarponi. Effects of the herbicide imazethapyr on soil microbial biomass and various soil enzyme activities. Biol Fertil Soils 17:237–240, 1994.
106. KH Domsch, G Jagnow, T-H Anderson. An ecological concept for the assessment of side-effects of agrochemicals on soil microorganisms. Residue Rev 86:67–105, 1983.
107. CM Tu. Effect of three newer pesticides on microbial and enzymatic activities in soil. Bull Environ Contam Toxicol 49:120–128, 1992.
108. CM Tu. Influence of ten herbicides on activities of microorganisms and enzymes in soil. Bull Environ Contam Toxicol 51:30–39, 1993.
109. CM Tu. Effect of five insecticides on microbial and enzymatic activities in sandy soil. J Environ Sci Health B30:289–306, 1995.
110. M Megharaj, HL Boul, JH Thiele. Effects of DDT and its metabolites on soil algae and enzymatic activity. Biol Fertil Soils 29:130–134, 1999.
111. M Megharaj, I Singleton, R Kookana, R Naidu. Persistence and effects of fenamiphos on native algal populations and enzymatic activities in soil. Soil Biol Biochem 31:1549–1553, 1999.
112. JN Ladd. Soil enzymes. In: D Vaughan, RE Malcolm, eds. Soil Organic Matter and Biological Activity. Boston: Martinus Nijhoff, 1985, pp 175–221.

17

Enzymatic Responses to Pollution in Sediments and Aquatic Systems

Sabine Kuhbier, Hans-Joachim Lorch, and Johannes C.G. Ottow
Justus Liebig University, Giessen, Germany

I. INTRODUCTION

Environmental pollution is defined as the introduction of any organic or inorganic substance, energy form, and/or other stresses (gases, genotoxic agents, radionuclides, etc.) to the environment at a rate faster than its accommodation by dispersion, recycling, detoxification, bioremediation, or storage in some harmless form. Pollution may change the composition, function, and trophic status of ecosystems in reversible or irreversible ways by affecting their biotic or abiotic components. Aquatic pollution comprises all allochthonous inputs and stresses that are—in contrast to natural allochthonous inputs—directly or indirectly caused by anthropogenic activities.

Pollutants can be organic wastes (1), inorganic nutrients (2), acidic ions (3,4), potentially toxic heavy metals (5,6), and/or organic contaminants such as pesticides, antibiotics, and hormones (4,5,7). Possible pathways for aquatic contamination are biotic and abiotic effluents of treated or untreated wastewater (8,9), surface runoff (3), leaf litter input contaminated with atmogenic pollutants (i.e., pollutants of atmospheric origin) (3), industrial wastes (5,6,10), and upwelling groundwater (11–13). Pollution of water, soil, or atmosphere proceeds essentially unabated, and these ecosystems serve as repositories for numerous pollutants. So far, little is known of how different pollutants may interfere with enzymatic activities in aquatic ecosystems and their sediments (14).

II. ENZYME ACTIVITIES IN NONPOLLUTED SYSTEMS

High-molecular-weight compounds from decaying plants and riparian input represent the major sources of the total organic matter in aquatic systems. Heterotrophic bacteria and aquatic hyphomycetes in the running water, on sediments, and on other surfaces are mainly responsible for their mineralization (15–17). The hydrolysis of polymeric organic compounds is the initial step in the microbial decomposition of organic matter in aquatic ecosystems (18–20). For the breakdown of high-molecular-weight organic matter, extracellular enzymes of bacteria and fungi play a key role in the initial stages of decomposition producing biomass and detritus for complex food chains (17,21–23). Whereas aquatic hyphomycetes penetrate and colonize the surface of litter, bacteria form communities in biofilms on organic and inorganic surfaces by generating extracellular polysaccharide (EPS) and secrete a great variety of depolymerizing enzymes (24–26).

Biofilms on stream-bed surfaces (stones, wood, litter, submersed macrophytes, sediment particles, etc.) are assemblages of bacterial, fungal, algal, and protozoan communities with syntrophic activities. These represent important biocoenoses of fluvial systems, because of their optimal position for photosynthetic processes as well as for scavenging of organic energy from the overlying water (27). Generally, a major part of the microbial biomass present in benthic, hyporheic (i.e., in the ecotone between groundwater and surface water) or aquifer sediments appears to be associated with biofilms (28,29). The close vicinity of different biocoenoses in biofilms facilitates the interchange and efficient recycling of nutrients and photosynthetically fixed carbon (30,31). River epilithon has been reported as an important trophic resource for invertebrate communities as well as for heterotrophic self-purification processes (30). Metabolic activities of biofilm communities apparently exceed planktonic activities (9,32), because attached cells and hyphae benefit from trapping exogenous carbon, nitrogen, and energy sources (33). Furthermore, extracellular degradative enzymes are assumed to be stabilized within the biofilm EPS matrix (27,31,34). Sinsabaugh and associates (35) failed to show a correlation between different enzyme activities and microbial biomass within epilithic biofilms. Apparently, functional effects of enzyme activities are decoupled from the producers.

Convincing examples of the enzymatic capacity of biofilms in small headstreams have been presented by Chappell and Goulder (34,36), who calculated the enzyme activity of water samples in relation to biofilm activity: 1 m^2 of epilithic biofilm had the same hydrolytic capacity as 400–10,200 L of ambient stream water or, in other dimensions, 1 m^2 biofilm was as active as a water column of 0.4–10.2 m overlying water (36). Similar results have been found on submerged macrophytes in gravel-pit ponds (37). On *Phragmites australis*, relative enzyme activities between 560–4000 L m^{-2} have been recorded, whereas the epiphyton on *Elodea canadensis* showed only a relative activity of 200–1000 L m^{-2}. In benthic biofilms of the River Horloff, a small upland stream in central Germany, relative enzyme activities of 0.6–1102 L m^{-2} have been detected (9). During low-water periods in summer the relative enzyme activity was low, whereas a significant increase was observed during winter spates, suggesting dilution of enzymes and/or substrates in the water column.

A. Natural Sources of Organic Compounds

The natural autochthonous source of organic matter in streams is photosynthesis. In anthropogenically undisturbed streams, with poor riparian inputs, bacteria and fungi utilize autochthonous organic matter derived from primary production (24,28,38). In most other streams and rivers, particularly in forested regions, organic compounds of terrestrial origin (e.g., leaf litter) dominate (23,24,28,39). Leaves that enter a stream are colonized by both fungi and bacteria (23,40). Although fungal biomass appears much greater than bacterial biomass, bacterial production may be as great as that of fungi (41).

Romani and coworkers (42) observed a significant correlation between enzymatic activities and chlorophyll *a* levels in stream-bed sediments. The detected chlorophyll was derived largely from algal detritus. Living algae were nearly absent and primary productivity was not detectable. In the biofilms of one stream significant correlations between enzyme activities and chlorophyll *a* concentrations were observed (31). In contrast, Jones and Lock (31) suggested that biofilm heterotrophs in light-grown biofilms in the same stream were supplied by actively growing photosynthetic communities and not by algal lytic products. In particular, β-glucosidase activities appeared to be linked to the hydrolysis of algal-derived organic matter (31). This finding is supported by other studies. For example, enhanced production of glucosidases and peptidases by bacteria has been postulated to occur when their growth is limited by low-molecular-weight organic matter excreted by algae (43). Overbeck and Sako (44) found an induction of β-glucosidase production caused by a high amount of polymeric carbohydrate compounds that were liberated during the breakdown of an algal bloom. In epilithic biofilms, the hydrolysis of cellobiosic polysaccharides increased in comparison to that of xylobiosic polymers as indicated by higher β-glucosidase activities (45–48). However, stream ecosystems supported with allochthonous material tend to have weaker algal–bacterial associations than do streams with instream primary production (31,38).

Natural inputs of allochthonous organic materials in streams usually derive from the riparian vegetation (i.e., leaf litter and woody debris). In the investigation of enzymes and chemical compounds of particulate organic matter, Sinsabaugh and Linkins (49; Chapter 9) found improving nutritional detritus quality (reflected by C/N and C/P ratio, lignin content, and microbial respiration) with decreasing particle size. The covarying carbohydrase-phosphatase group (α- and β-glucosidase, chitobiase, β-xylosidase, exocellulase [exo-1,4-β-glucanase] endocellulase [endo-1,4-β-glucanase], and phosphatase) was significantly correlated with microbial respiration and carbohydrate content of the particulate organic matter. The increase in microbial activity with decreasing particle size was ascribed to the enhanced surface area/volume ratio of smaller particles, as well as to an organismic succession in the microbial community. In a comparison of mineral (glass) and organic (wood) substrata, enhanced biofilm development and enzyme activity were found in the epixylon on the wood surfaces (25). The epilythic and epixylic biofilms proved to be structurally and functionally different even under similar hydrodynamic conditions. The authors attributed the more intensive epixylic development to utilization of the wood substratum as a supplemental carbon source and to better attachment to the rough surface structure.

III. ENZYME ACTIVITIES IN POLLUTED FRESHWATER SYSTEMS

Generally speaking, there are no principal differences between natural saprobic processes and those influenced by (organic) pollution (50). Different pollutants may effect enhance, inhibit, or have no effect on enzyme activities. For example, inhibition and/or stimulating effects have been reported for naturally occurring compounds, such as humic acids (28,51–54), different clay minerals (55), or heavy metals (56). Clay minerals and humus colloids may have inhibitory, stabilizing, and/or stimulating effects on enzyme activities in soils and sediments (57). The kinetic behavior of clay- or humus-sorbed enzymes usually is different from that of "free" enzymes. Generally speaking, V_{max} values tend to be lower and K_m values higher, although clay- and enzyme-specific activities may occur (58–60). Thus kinetics of enzymes immobilized to specific sediment components should be considered as site-specific and need not necessarily be affected by pollutants.

A. Effects of Wastewater Pollution

In the River Horloff, a significant increase in enzyme activity was detected in the water column and in standardized stream-bed sediment downstream of a small municipal wastewater facility with oxidation ponds (Table 1) (9,32). Similar findings are reported by Ziegelmayer and Henschel (61), who found increasing enzyme activities downstream of a wastewater discharge in the River Windach in Bavaria (Germany). During heavy rainfall events, increasing activities of enzymes processing oligomeric compounds, due to a combined-sewer overflow containing untreated domestic sewage, were detected in that stream. In the River Rhine (Germany), an increase in β-glucosidase activities also was detected after spates (62). The authors stated, in contrast to Ziegelmayer and Henschel (61), that this increase should be ascribed to the input of additional nutrients by surface runoff rather than to higher pollution with wastewater. In the River Ruhr (Germany), contaminated with domestic wastewater, decreasing esterase activities during spates and high activities in low-water periods were detected, indicating that in this case high water flows had a dilution effect on the amount of enzymes and/or organic compounds in the river (63). By normalizing enzyme activities with corresponding biomass parameters, e.g., bacterial abundance or bacterial biomass (specific activity), significant differences often become more evident.

In the Windach River, the specific esterase activity (i.e., FDA hydrolysis) was related to the abundance of saprophytic bacteria decreased downstream from the wastewater effluents, while the specific activity of the β-glucosidase increased. During rainfall events, however, all specific activities decreased downstream from the sewage treatment plant (61). In the River Horloff, the specific activity of the β-glucosidases (which was related

Table 1 Means and Standard Deviations of Extracellular Activities of Esterase, α- and β-Glucosidase, Phosphatase, and L-alanine-aminopeptidase in the Water Column and Standardized Stream-Bed Sediments of the River Horloff Downstream from a Wastewater Discharge (see footnote)

Site	Esterase	α-Glucosidase	β-Glucosidase	Phosphatase	Peptidase
			Water (μmol L^{-1} h^{-1})		
1	0.022 (0.014)	0.48 (0.52)	3.50 (2.94)	0.51 (0.31)	3.09 (2.19)
→[a]					
2	0.028 (0.015)	1.40 (2.98)	9.11 (9.86)	0.64 (0.40)	4.25 (2.92)
3	0.027 (0.013)	1.32 (2.54)	8.55 (8.82)	0.65 (0.41)	4.26 (2.82)
4	0.026 (0.014)	1.49 (3.21)	8.42 (8.25)	0.60 (0.36)	4.31 (3.20)
5	0.025 (0.014)	0.96 (1.47)	10.7 (17.8)	0.55 (0.26)	4.09 (2.44)
6	0.029 (0.014)	0.80 (0.73)	8.48 (8.25)	0.53 (0.21)	4.12 (2.64)
			Sediment (μmol kg dry matter^{-1} h^{-1})		
1	0.96 (0.34)	8.5 (4.1)	27.8 (13.7)	115.5 (45.5)	196.9 (86.0)
→					
2	0.76 (0.29)	9.4 (4.3)	27.5 (18.6)	86.7 (53.0)	244.8 (82.5)
3	0.92 (0.28)	9.2 (3.5)	23.6 (14.3)	98.7 (54.1)	263.5 (112.4)
4	0.84 (0.27)	6.5 (3.4)	19.3 (9.9)	74.7 (33.1)	190.6 (101.5)
5	0.68 (0.22)	6.0 (3.3)	17.1 (11.2)	62.6 (34.2)	147.9 (69.5)
6	0.69 (0.25)	5.9 (3.9)	19.7 (14.8)	73.2 (46.6)	137.9 (52.1)

[a] → treated wastewater discharge site
Source: Refs. 9 and 32.

to the population density of saprophytes) increased considerably downstream of the wastewater pollution in the water column, while specific FDA hydrolysis tended to decrease (Fig. 1) (32). In the sediments, both specific activities decreased after the sewage input but increased considerably in the course of self-purification. This development may be assessed as an enzyme inhibition effect downstream from the effluents and enhancement of activity during the self-purification processes (32). The specific esterase activities in the water column and in the standardized sediments were in the same range, suggesting that the single cell in the sediment samples was not more active than in the water column. The higher enzyme activity observed in the sediment per kg dry matter (Table 1) has been ascribed mainly to higher population densities of saprophytic bacteria in the sediment samples than in the water samples. In contrast to these findings, the specific enzyme activity of the β-glucosidases in the water column significantly exceeded the specific activity in the sediment samples (Fig. 1). Apparently easily degradable carbohydrates might be more important energy sources for the pelagic bacteria than for the benthic bacteria (32).

Fecal pollution may not derive only from sewage treatment plants but also from local animal breeding facilities. Studies in the River Hull (United Kingdom) showed increasing activities of alkaline phosphatases downstream of a fish farm producing rainbow trout (64). The authors assumed that the enzyme source was mainly fish feces and not a microbial origin, because the enzymes were not associated with the bacterial fraction. Therefore, the increase not only was an indicator of organic pollution, but also may have reflected a release of more biologically available inorganic phosphate that could accelerate eutrophication. In this case, enzymes themselves would act as "pollutants." Similar observations were made by Chappell and Goulder (65), who showed that enzymes discharged from wastewater treatment plants may be transported several kilometer downstream, suggesting that enzyme activity could have accumulated within biofilms below the sewage outfall. Also, Marxsen and Witzel (66) assumed there could be a considerable phosphorous flux supplying autotrophic algae due to high levels of phophatase activity.

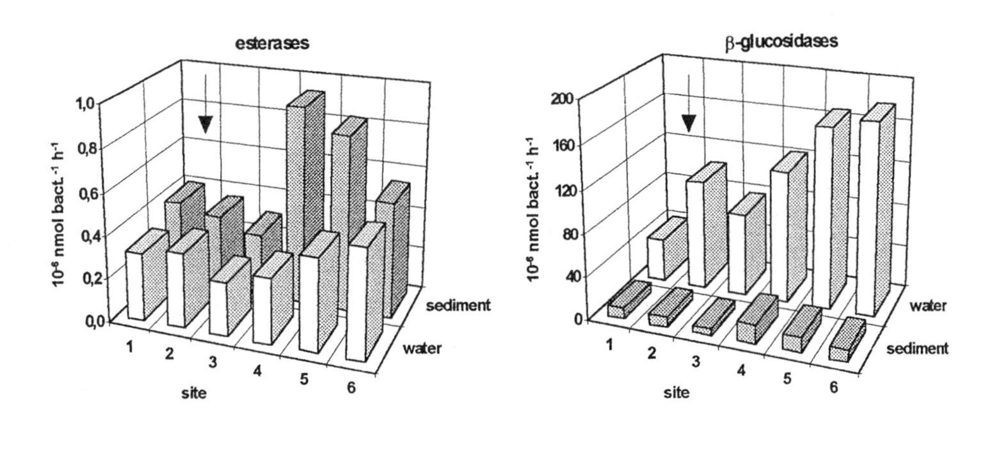

Figure 1 Specific enzyme activities of the esterase (FDA hydrolysis) and β-glucosidase in relation to densities of heterotrophic bacteria (saprophytes) in the water column and standardized streambed sediments of the River Horloff downstream from a wastewater discharge facility (arrow). (From Ref. 32.)

Moreover, the impact of wastewater may alter the ratios between enzyme activities. In an Australian lagoon, a former billabong receiving treated urban sewage, the ratio between aminopeptidase and alkaline phosphatase activities was 43, whereas the ratio in other less-polluted billabongs ranged only from 1 to 8. High protein and inorganic P concentrations in the sewage effluent enhanced peptidase activity but repressed phosphatase activity, resulting in increased aminopeptidase/phosphatase ratios (18). In the water column of the River Horloff, no changes in the ratio between aminopeptidase and phosphatase could be detected downstream of the wastewater effluents (Table 2) (32), but there was a significant decrease of the aminopeptidase/β-glucosidase ratio and an increase in the β-glucosidase/phosphatase ratio. Apparently, the β-glucosidase activity was enhanced through sewage input in comparison to that of the other enzymes. In standardized streambed sediments (sand bags), on the other hand, significant increases in the aminopeptidase/phosphatase ratio as well as in the aminopeptidase/β-glucosidase and β-glucosidase/phosphatase ratio were observed. The β-glucosidase activity seemed to be repressed in comparison to that of the water column. This finding confirms the suggestion of Kuhbier (32) that the processing of readily degradable organic compounds might be more important in the water column than in the (standardized) sediments.

In Table 3, enzymes are arranged according to their activities in different aquatic freshwater systems. In the majority of the observed aquatic systems β-glucosidase activity exceeds the activities of α-glucosidase and β-xylosidase. Vrba (22) stated that the α-glucosidase/β-glucosidase ratio often is <1.0 in freshwaters, since cellulose is a refractory compound of both autochthonous and allochthonous organic matter. In the Rimov Reservoir, an enhancement of α-glucosidase was detected during algal blooms (ratio ≈ 1.0), whereas during periods with senescent or decaying algal populations, the β-glucosidase

Table 2 Means of the Ratios Between Activities of α- and β-Glucosidase, Phosphatase, and L-alanine-aminopeptidase in the Water Column and Standardized Stream-Bed Sediments of the River Horloff Downstream from a Wastewater Discharge

Site	α-gluco/β-gluco[a]	ala-pep/phos	ala-pep/β-gluco	β-gluco/Phos
		Water		
1	0.16	6.8	1.45	7.7
→[b]				
2	0.21	6.7	0.90	13.9
3	0.20	6.5	0.89	13.1
4	0.20	7.0	0.95	14.2
5	0.15	7.3	0.89	18.4
6	0.16	7.6	0.84	16.2
		Sediment		
1	0.33	1.9	8.9	0.24
→				
2	0.46	4.3	17.5	0.31
3	0.53	3.5	23.6	0.23
4	0.41	3.1	15.4	0.25
5	0.48	3.0	15.3	0.27
6	0.34	2.3	9.7	0.26

[a] α- and β-gluco, α- and β-glucosidase; phos, phosphatase; ala-pep, L-alanine-aminopeptidase.
[b] → treated wastewater discharge site.
Source: Ref. 32.

Table 3 Ranking of Enzyme Activities in Different Polluted Aquatic Systems

Aquatic system	Status	Compartment	Ranking according to activities[a]	Reference
Columbia River Estuary	Eutrophic	Pelagic particles	leu-pep > β-gluco > excl	(81)
Gravel-pit pond	Groundwater-fed	Epiphyton	phos > β-gluco > β-galac ≈ β-xylo > sulf	(36)
Hudson River Estuary	Eutrophic	Sediments	phen > perox > β-gluco > nag > β-xylo > excl	(82)
Stockport Creek		Sediments	phen > nag > β-gluco > perox > β-xylo > excl	
Nant Waen stream	Unpolluted	Epilithon	ester > β-gluco > β-xylo > endo-pep	(31, 47)
River Clywedog	Eutrophic	Epilithon	ester > β-gluco > β-xylo > endo-pep	
Rimov Reservoir	Eutrophic	Epilimnion	β-glu > nag > α-glu	(22)
River Aber	Oligotrophic	Epilithon	ester > β-gluco > phos	(54)
River Clywedog	Eutrophic	Epilithon	ester > phos > β-gluco	
River Upper Conwy	Dystrophic, brown water	Epilithon	ester > β-gluco > phos	
River Horloff	Wastewater discharge	Water column	β-gluco > ala-pep > α-gluco > phos > ester	(9)
		Sediments	ala-pep > phos > β-gluco > α-gluco > ester	
River La Solana	Unpolluted, second-order	Stomatolytic mat	phos > β-gluco > β-xylo	(83)
River Ouse	Wastewater discharge	Epilithon	leu-pep > phos > β-gluco > β-galac ≈ β-xylo	(84)
River Derwent	Wastewater discharge	Epilithon	leu-pep > phos > β-gluco > β-galac ≈ β-xylo	
River Riera Major	Unpolluted, second-order	Surface sediments	phos > β-gluco > β-xylo	(42)
		Subsurface sediments	phos > β-gluco > β-xylo	
River Windach	Wastewater discharge	Water column	ester > ala-pep > β-gluco > α-gluco	(61)
St. Regis River	Unpolluted fourth-order	Epilithon	phos > β-gluco > β-xylo > nag	(25)
	Brown-water stream	Epixylon	phos > β-gluco > β-xylo ≈ nag	
St. Regis River		Benthic particles	phos > β-gluco > sulph ≈ perox > phen > β-xylo ≈ α-gluco ≈ nag > cbh > excl	(49)
St. Regis River		Epilithon: light grown/	phos > phen > excl > α-gluco > nag > β-gluco > β-xylo > sulf > exprot > cbh	(48)
		Slow velocity		
		Dark/slow	phen > phos > excl > α-gluco > exoprot > nag > β-gluco > β-xylo > sulf > cbh	
		Light/fast	phos > α-gluco > nag > β-gluco ≈ phen > excl > β-xyl > cbh > exoprot > sulf	
		Dark/fast	α-gluco > nag > phos > β-gluco > phen > β-xylo > excl > sulf > cbh > exoprot	
15 different streams, Northern England	From acidic to calcerous	Epilithon	phos > β-gluco > β-galac ≈ β-xylo ≈ sulf	(34)

[a] α-/β-gluco, α-/β-glucosidase; ala-pep, L-alanine aminopeptidase; leu-pep, leucine aminopeptidase; endo-pep, endopeptidase; β-galac, β-galactosidase; phos, alkaline phosphatase; β-xylo, β-xylosidase; sulf, sulfatase; ester, esterase; nag, β-*N*-acetylglucoaminidase (chitinase); phen, phenol oxidase; perox, peroxidase; excl, β-1, 4-exoglucanase (exocellulase); exprot, exoprotease, cbh, cellobiohydrolase.

activity became more important (ratio ≈ 0.2) (22). Only in the St. Regis River were distinctly higher α-glucosidase than β-glucosidase activities measured in alga-dominated epilithic biofilms (48).

B. Eutrophication

An indirect effect of sewage or agricultural fertilizer pollution on enzyme activities is algal bloom caused by eutrophication due to high supplies of inorganic nutrients (P and N in particular). Becquevort et al. (67) demonstrated that bacteria attached to *Phaeocystis* species–derived organic matter in the North Sea are severely nutrient-limited with respect to their nitrogen requirements. During years of very large algal blooms, the nutrient limitation of bacterial mineralization could lead to accumulation of alga-derived organic debris in the coastal waters of the North Sea. Thus, pollution with inorganic nutrients may overtax the hydrolytic capacity of aquatic systems, though this could be a minor problem in running waters. In lakes, however, an increase in enzyme activities of peptidases and esterases was recorded with increasing trophic status (68).

C. Acidification

Many running water systems in Central Europe show symptoms of acidification, due to natural and/or atmogenic depositions of sulfur compounds as a consequence of rain and melting snow. The proton input usually is correlated with river discharge, especially in winter (4). However, pH is known as a key variable both exerting control on stream biological characteristics as well as directly influencing the velocity of specific enzyme-catalyzed reactions (34).

In Burbage Brook (northern England) in summer, near-neutral conditions have been recorded, whereas pH decreased in winter. Compared with the River Lathkill, that showed approximately neutral pH throughout the year, there was little between-stream difference in heterotrophic activity of epilithic bacteria during summer. In winter, however, heterotrophic activity was significantly less in Burbage Brook as a result of bacterial inhibition under acid conditions (69).

Chappell and Goulder (34,36) compared diverse calcareous and acid headstreams and found a significant positive correlation between stream pH and enzyme activities of β-glucosidase, β-xylosidase, β-galactosidase, and sulfatase in stream epilithon, which was ascribed to a direct pH effect. In contrast, several reports showed enhanced phosphatase activities in epilithic biofilms (34,36,53) and stream-bed sediments (66). Freeman et al. (53) and Marxen and Witzel (66) observed lower free phosphate concentrations in the water of acid streams than in the calcareous ones and thus a greater need to release phosphate from organic matter, whereas Chappell and Goulder (34,36) suggested that the negative relationship between pH and phosphatase activities presumably indicates a shift in the relative importance of alkaline and acid phosphatases between streams of different pH.

In soils, several studies have shown that phosphatase activity is affected by soil pH even when the assay itself is carried out in a buffered medium (70–72). Phosphatase activities in soils often appear to have an acid and an alkaline pH optimum (72). Dick and Tabatabai (73) suggested that acid phosphatase dominates in acid soils and alkaline phosphatase is predominant in alkaline soils. In stream-bed sediments, however, results remain difficult to explain and final conclusions must await further study.

D. Toxic Pollution

In the Main River (Germany), a parallel development between esterase activities (FDA hydrolysis) and heterogeneous organic contaminants was detected at its junction with the Rhine River. In the River Rhine itself, contrasting conditions were observed. Generally, the activity of esterases decreased with increasing concentrations of a heterogenous group of organic substances. This may be due to the greater persistence and/or toxicity of the extremely complex organic load in the Rhine River (74).

In November 1986, large amounts of chemical pollutants (phosphate acid esters and mercury compounds) accidentally contaminated the River Rhine from a chemical industrial facility in Switzerland (Sandoz AG, Basel). This caused mass kill of fish and a broad range of aquatic macroorganisms below the city of Basel. The Rhine River water was thoroughly monitored at regular intervals at Mainz sampling station (approximately 340 km downstream of the discharges) by using enzyme and toxicity tests (74). As the wave of contaminants reached the sampling station, a significant increase in esterase activity and a decrease in urease activity were observed. Apparently, the organic P-ester pollutants were metabolized to a certain degree during the passage from Basel to Mainz, giving rise to elevated esterase activities. Urease inhibition was ascribed to heavy metals such as Hg (75). The different reactions seem to reflect the different toxic components and their concentrations in the pollution wave. However, absolute enzyme activities may not always indicate accurately inhibition or enhancements of the hydrolytic activities. Therefore, toxic effects on enzyme activities should be investigated in dilution experiments (76).

A significant decrease in hydrolytic capacities occurred with increasing portions of Rhine River water, especially during low discharge periods (Fig. 2). Additional experiments with Rhine River water and a known toxicant (formaldehyde) added to the samples showed similar results. These inhibitory influences of the river water could be traced to bank filtration wells and became weaker with increasing distance of the wells from the river. In some cases, these inhibition effects were not detectable by dilution experiments downstream of industrial wastewater treatment plants. This may have been a result of the high supplies of nutrients, which may have disguised possible inhibition of the enzyme system (62).

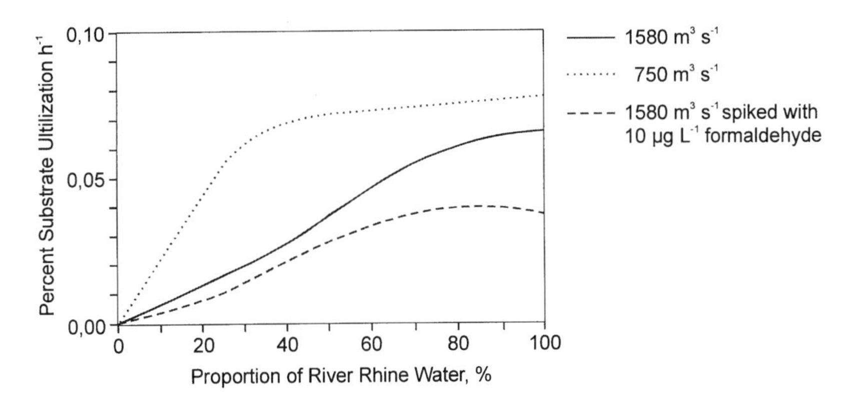

Figure 2 Enzyme inhibition in dilution experiments of River Rhine samples taken from different discharge situations and the inhibitory effect of formaldehyde. (From Ref. 76.)

In the Sudgen Beck, a small stream receiving leachate from old mine tailings contaminated by phenolic wastes, Milner and Goulder (77) observed inhibition of heterotrophic activity in the water contaminated with chlorophenols, nitrophenols, and phenoxyalkanoic acids. In contrast, the heavy metal pollution of the River Aire seemed to have no significant influence on heterotrophic activity in the pelagic compartment (78). The authors assume that bacteria in the benthic compartment probably are affected more severely as a result of continuous contact with toxics and possible accumulation of heavy metals in benthic biofilms and organic matter. Indeed, a significant accumulation of heavy metals in the river-bed sediments of the heavily polluted Indus River (Pakistan) was reported (6). A correlation between heavy metal concentrations in the sediments and in fish indirectly indicates the role of contaminated biofilms for food webs and chains.

In the stream-bed sediments of Skeleton Creek (Oklahoma), pollution with petroleum refinery and fertilizer plant effluents inhibited β-glucosidase activity, phosphatases were stimulated by refinery effluents but depressed by the fertilizer plant discharges (56). On the other hand, activities of β-galactosidases and amylases remained unaffected by petroleum pollution but were inhibited by the fertilizer inputs. Reduced phosphatase activities have been reported in areas impacted by coal coking effluents, crude oil, pesticides, and heavy metals (56). Such contradictory reports on the effects of toxics on enzymes, especially in sediments, are not surprising and can be ascribed readily to differences in sediment type and homogeneity, varying microenvironments, unknown metabolic and ecological interactions, as well as to different microbial populations, carbon sources, and other nutrients.

Enzyme inhibition may be lower in carbon-enriched sediments as a result of a population shift to more resistant microorganisms and/or because of decreasing bioavailability of the toxicants caused by sorption and immobilization to the organic matter (56). In pore waters of Elbe River sediments contaminated with heavy metals and different organic chlorine pesticides, relatively high β-glucosidase activities were found in comparison to those at an unpolluted reference site, the L-alanine aminopeptidase activity showed no significant differences. In the sediments, the variation in enzyme activities was greater than in the unpolluted sediments and was probably caused by the fluctuations in pollution through anthropogenic impacts (79). With respect to specific protease activities, the highest activities were observed at the uncontaminated sites, considerably lower specific activities could be found at the strongly contaminated Elbe River sediments, although the latter showed much higher potential substrate concentrations (80).

IV. CONCLUSIONS

It may be concluded that enzymes in the stream water as well as the sediments may be affected specifically by environmental conditions and/or by pollutants. Generally speaking, enzyme activities result from the complex interactions between various stimulating factors (type and amount of substrates, cofactors, surfaces, etc.) and environmental stresses (such as the presence and concentration of organic and/or inorganic contaminants). To demonstrate the impacts of pollution and/or pollutants with the aid of enzymes, all information about environmental parameters should be taken into consideration together with a selection of enzymes involved in N, C, S, and P turnover (9,32,75). Inhibition curves derived from dilution experiments should be used.

Future investigations should be focused on standardized surfaces (e.g., sand bags, glass beads) in comparison to other heterogeneous surfaces (stones, wood, macrophytes, sediments) in the stream bed. A number of enzymes involved in specific hydrolytic processes of C, N, S, and P cycle should be carefully selected and agreed upon, rather than an ever-changing list of tests. Standardization will facilitate interpretation, allow comparison of different aquatic systems and their zones (stream water as well as sediments), and limit analysis (saving time and costs). Microorganisms and enzymes in biofilms on surfaces may adapt to pollutants and integrate different stresses. Such parameters and conditions should be preferred to a few pelagic measurements. Last but not least, biofilms could act as ''memories'' for different contaminants in time and space as a result of their retention of enzymes and their complex microbial and process diversity (see chapter 11). This question should be addressed when comparing the suitability of different natural and introduced surfaces.

Overall, enzyme activities in aquatic systems may be useful as additional indicators for pollution and water quality following the impact of organic effluents arising from wastewater treatments as well as for potentially toxic substances such as heavy metals; enzyme activities also may be useful indicators of pH shifts and other anthropogenic stresses.

REFERENCES

1. PA Chambers, EE Prepas. Nutrient dynamics in riverbeds: The impact of sewage effluent and aquatic macrophytes. Wat Res 28:453–464, 1994.
2. H Klapper. Eutrophierung und Gewässerschutz. Jena/Stuttgart: Gustav Fischer Verlag, 1992, p 277.
3. B Wohlrab, H Ernstberger, A Meuser, V Sokollek. Landschaftswasserhaushalt. Wasserkreislauf und Gewässer im ländlichen Raum. Veränderung durch Bodennutzung, Wasserbau und Kulturtechnik. Hamburg: Paul Parey, 1992, p 351.
4. G Gunkel. Einträge von toxisch und ökotoxisch wirksamen Stoffen. In: G Gunkel, ed. Renaturierung kleiner Fließgewässer. Jena/Stuttgart: Gustav Fischer Verlag, 1996, pp 161–174.
5. P Heininger, J Pelzer, E Claus, P Tippmann. Contamination and toxicity trends for sediments–case of the Elbe river. Wat Sci Tech 37:95–102, 1998.
6. J Tariq, M Ashraf, M Jaffar, M Afzal. Pollution status of the Indus river, Pakistan, through heavy metal and macronutrient contents of fish, sediment and water. Wat Res 30:1337–1344, 1996.
7. A Hartmann, AC Alder, T Koller, RM Widmer. Identification of fluoroquinolone antibiotics as the main source of umuC genotoxicity in native hospital wastewater. Environ Toxicol Chem 17:377–382, 1998.
8. H-J Lorch. Saprobie und Selbstreinigung der Gewässer. In: G Gunkel, ed. Renaturierung kleiner Fließgewässer. Jena/Stuttgart: Gustav Fischer Verlag, 1996, pp 150–161.
9. S Kuhbier, H-J Lorch, JCG Ottow. Bakterielle Abundanz und Enzymaktivitäten im Sediment unterschiedlich belasteter Zonen eines Fließgewässers (Horloff/Vogelsberg). Verh Ges Ökol 29:473–479, 1999.
10. RA Moll, PJ Mansfield. Response of bacteria and phytoplankton to comtaminated sediments from Tenton Channel, Detroit River. Hydrobiologia 219:281–299, 1991.
11. RR Boar, DH Lister, WT Clough. Phosphorous loads in a small groundwater-fed river during the 1989–1992 East Anglian drought. Wat Res 29:2167–2173, 1995.
12. DM Fiebig, MA Lock. Immobilization of dissolved organic matter from groundwater discharging through the stream bed. Freshwater Biol 26:45–55, 1991.
13. DM Fiebig. Groundwater discharge and its contribution of dissolved organic carbon to an upland stream. Arch Hydrobiol 134:129–155, 1995.
14. L Leff. Stream bacterial ecology: A neglected field? Despite the importance of rivers, little is known about their bacterial populations. ASM News 60:135–138, 1994.

15. S Findlay, RL Sinsabaugh, DT Fischer, P Franchini. Sources of dissolved organic carbon supporting planctonic bacterial production in the tidal freshwater Hudson River. Ecosys 1: 227–239, 1998.
16. K Suberkropp, A Michelis, H-J Lorch, JCG Ottow. Effect of sewage treatment plant effluents on the distribution of aquatic hyphomycetes in the river Erms, Schwäbische Alb, F.R.G. Aquat Bot 32:141–153, 1988.
17. RJ Chróst. Ectoenzymes in aquatic environments: Microbial strategy for substrate supply. Verh Int Verein Limnol 24:2597–2600, 1991.
18. PI Boon. Organic matter degradation and nutrient regeneration in Australian freshwaters. II. Spatial and temporal variation and relation with environmental conditions. Arch Hydrobiol 117:405–436, 1990.
19. M Unanue, B Ayo, M Agis, D Slezak, GJ Herndl, J Iriberri. Ectoenzymatic activity and uptake of monomers in marine bacterioplankton described by a biphasic kinetic model. Microb Ecol 37:36–48, 1999.
20. L-A Meyer-Reil, M Köster. Microbial life in pelagic sediments: The impact of environmental parameters on enzymatic degradation of organic material. Mar Ecol Prog Ser 81:65–72, 1992.
21. H-G Hoppe, S-J Kim, K Gocke. Microbial decomposition in aquatic environments: Combined process of extracellular enzyme activity and substrate uptake. Appl Environ Microbiol 54: 784–790, 1988.
22. J Vrba Seasonal extracellular enzyme activities in decomposition of polymeric organic matter in a reservoir. Arch Hydrobiol Beih 37:33–42, 1992.
23. K Suberkropp, H Weyers. Application of fungal and bacterial production methodologies to decomposing leaves in streams. Appl Environ Microbiol 62:1610–1615, 1996.
24. AM Romaní, S Sabater. Epilithic ectoenzyme activity in a nutrient-rich Mediterranean river. Aquat Sci 61:122–132, 1999.
25. RL Sinsabaugh, SW Golladay, AE Linkins. Comparison of epilithic and epixylic biofilm development in a boreal river. Freshwater Biol 25:179–187, 1991.
26. M Ledger, AG Hildrew. Temporal and spatial variation in the epilithic biofilm of an acid stream. Freshwater Biol 40:655–670, 1998.
27. MA Lock, RR Wallace, JW Costerton, RM Ventullo, SE Charlton. River epilithon: Toward a structural-functional model. Oikos 42:10–22, 1984.
28. M Pusch, D Fiebig, I Brettar, H Eisenmann, BK Ellis, LA Kaplan, MA Lock, MW Naegeli, W Traunspurger. The role of micro-organisms in the ecological connectivity of running waters. Freshwater Biol 40:453–495, 1998.
29. WA Hamilton. Biofilms: Microbial interactions and metabolic activities. In: M Fletcher, TRG Gray, JG Jones, eds. Ecology of Microbial Environments. Cambridge: Cambridge University Press, 1987, pp 361–385.
30. A Elósegui, J Pozo. Epilithic biomass and metabolism in a north Iberian stream. Aquat Sci 60:1–16, 1998.
31. SE Jones, MA Lock. Seasonal determinations of extracellular hydrolytic activities in heterotrophic and mixed heterotrophic/autotrophic biofilms from two contrasting rivers. Hydrobiologia 257:1–16, 1993.
32. S Kuhbier. Charakterisierung der Selbstreinigungsprozesse und des Gewässerzustandes eines abwasserbelasteten Fließgewässers (Horloff/Vogelsberg) mit Hilfe von Sediment und Aufwuchs. PhD dissertation, Justus-Liebig-Universität Giessen, Giessen, 2002.
33. H-J Lorch, JCG Ottow. Scanning electron microscopy of bacteria and diatoms attached to a submerged macrophyte in an increasingly polluted stream. Aquat Bot 26:377–384, 1986.
34. KR Chappell, R Goulder. Epilitic extracellular enzyme activity in acid and calcareous headstreams. Arch Hydrobiol 125:129–148, 1992.
35. RL Sinsabaugh, D Repert, T Weiland, SW Golladay, AE Linkins. Exoenzyme accumulation in epilithic biofilms. Hydrobiologia 222:29–37, 1991.
36. KR Chappell, R Goulder. Seasonal variation of epilithic extracellular enzyme activity in three diverse headstreams. Arch Hydrobiol 130:195–214, 1994.

37. KR Chappell, R Goulder. Seasonal variation of epiphytic extracellular enzyme activity on two freshwater plants, Phragmites australis and Elodea canadensis. Arch Hydrobiol 132:237–253, 1994.
38. RM Holmes, SG Fisher, NB Grimm, BJ Harper. The impact of flash floods on microbial distribution and biogeochemistry in the parafluvial zone of a desert stream. Freshwater Biol 40:641–654, 1998.
39. CC Jenkins, K Suberkropp. The influence of water chemistry on the enzymatic degradation of leaves in streams. Freshwater Biol 33:245–253, 1995.
40. V Baldy, MO Gessner, E Chauvet. Bacteria, fungi and the decomposition of leaf litter in a large river. Oikos 74:93–102, 1995.
41. SEG Findlay, TL Arsuffi. Microbial growth and detritus transformations during decomposition of leaf litter in a stream. Freshwater Biol 21:261–269, 1989.
42. AM Romaní, B Butturini, F Sabater, S Sabater. Heterotrophic metabolism in a forest stream sediment: Surface versus subsurface zones. Aquat Microb Ecol 16:143–151, 1998.
43. RJ Chróst, H Rai. Ectoenzyme activity and bacterial secondary production in nutrient-impoverished and nutrient-enriched freshwater mesocosms. Microb Ecol 25:131–150, 1993.
44. J Overbeck, Y Sako. Ecological aspects of enzyme regulation in aquatic bacteria. Arch Hydrobiol Beih 34:81–92, 1990.
45. AM Romaní, S Sabater. Effect of primary producers on the heterotrophic metabolism of a stream biofilm. Freshwater Biol 41:729–736, 1999.
46. S Sabater, AM Romaní. Metabolic changes associated with biofilm formation in an undisturbed Mediterranean stream. Hydrobiologia 335:107–113, 1996.
47. SE Jones, MA Lock. Hydrolytic extracellular enzyme activity in heterotrophic biofilms from two contrasting streams. Freshwater Biol 22:289–296, 1989.
48. RL Sinsabaugh, AE Linkins. Exoenzyme activity associated with lotic epilithon. Freshwater Biol 20:249–261, 1988.
49. RL Sinsabaugh, AE Linkins. Enzymic and chemical analysis of particulate organic matter from a boreal river. Freshwater Biol 23:301–309, 1990.
50. W Schönborn. Fließgewässerbiologie. Jena, Stuttgart: Gustav Fischer Verlag, 1992, p 504.
51. RG Wetzel. Gradient-dominated ecosystems: Sources and regulatory functions of dissolved organic matter in freshwater ecosystems. Hydrobiologia 229:181–198, 1992.
52. RJ Dudley, PF Churchill. Effect and potential ecological significance of the interaction of humic acids with two aquatic extracellular proteases. Freshwater Biol 34:485–494, 1995.
53. C Freeman, MA Lock, J Marxsen, SE Jones. Inhibitory effects of high molecular weight dissolved organic matter upon metabolic processes in biofilms from contrasting rivers and streams. Freshwater Biol 24:159–166, 1990.
54. C Freeman, MA Lock. Recalcitrant high-molecular-weight material, an inhibitor of microbial metabolism in river biofilms. Appl Environ Microbiol 58:2030–2033, 1992.
55. RG Burns. Interaction of enzymes with soil mineral and organic colloids. In: PM Huang and M Schnitzer, eds. Interactions of Soil Minerals with Natural Organics and Microbes. Special Publication No. 17. Madison, WI: Soil Science Society of America, 1986, pp 429–451.
56. GA Burton Jr, GR Lanza. Aquatic microbial activity and macrofaunal profiles of an Oklahoma stream. Wat Res 21:1173–1182, 1987.
57. P Ruggiero, J Dec, J-M Bollag. Soil as a catalytic system. In: G Strotzky, J-M Bollag, eds. Soil Biochemistry. Vol. 9. New York: Marcel Dekker, 1996, pp 79–122.
58. HE Makboul, JCG Ottow. Clay minerals and Michaelis constant of urease. Soil Biol Biochem 11:683–686, 1979.
59. HE Makboul, JCG Ottow. Einfluß von Zwei- und Dreischichttonmineralen auf die Dehydrogenease-, saure Phosphatase-und Urease-Aktivität in Modellversuchen. Z Pflanzenernaehr Bodenkd 142:500–513, 1979.
60. HE Makboul, JCG Ottow. Michaelis Constante (K_m) of acid phosphatase as affected by montmorillonite, illite, and kaolinite clay minerals. Microb Ecol 5:207–213, 1979.
61. B Ziegelmayer, T Henschel. Änderungen der mikrobiellen Stoffwechselaktivitäten im Fließgewässer bei Stoßbelastungen. Vom Wasser 77:67–75, 1991.

62. M Wiegand-Rosinus, U Obst. Enzymatische Aktivitäten als biologische Parameter der Gewässerkontrolle. Vom Wasser 81:211–219, 1993.
63. B Kuhlmann. Enzymatische Aktivitäten in der Ruhr. Forum Städtehyg 38:206–209, 1987.
64. OJ Carr, R Goulder. Fish-farm effluents in rivers. I. Effects on bacterial populations an alkaline phosphatase activity. Wat Res 24:631–638, 1990.
65. KR Chappell, R Goulder. Enzymes as river pollutants and the response of native epilithic extracellular-enzyme activity. Environ Pollut 86:161–169, 1994.
66. J Marxsen, K-P Witzel. Significance of extracellular enzymes for organic matter degradation an nutrient regeneration in small streams. In: RJ Chróst, ed. Microbial Enzymes in Aquatic Environments. New York: Springer-Verlag, 1991, pp 270–285.
67. S Becquevort, V Rousseau, C Lancelot. Major and comparable roles for free-living and attached bacteria in the degradation of Phaeocystis-derived organic matter in Belgian coastal waters of the North Sea. Aquat Microb Ecol 14:39–48, 1998.
68. JAW Morgan, RW Pickup. Activity of microbial peptidases, oxidases and esterases in lake waters of varying trophic status. Can J Microbiol 39:795–803, 1993.
69. R Goulder. Epilithic bacteria in an acid and a calcerous headstream. Freshwater Biol 19:405–416, 1988.
70. MA Tabatabai. Soil enzymes. In: RW Weaver, JS Angle, and PS Bottomley, eds. Methods of Soil Analysis. Part 2. Microbial and Biochemical Properties. SSSA Book Series, No. 5. Madison, WI: Soil Science Society of America, 1994, pp 775–833.
71. V Acosta-Martínez, MA Tabatabai. Enzyme activities in a limed agricultural soil. Biol Fertil Soils 31:85–91, 2000.
72. L Gianfreda, J-M Bollag. Influence of natural and anthropogenic factors on enzyme activity in soil. In: G Strotzky, J-M Bollag, eds. Soil Biochemistry. Vol. 9. New York: Marcel Dekker, 1996, pp 123–193.
73. WA Dick, MA Tabatabai. Kinetic parameters of phosphatases in soils and organic waste materials. Soil Sci 137:7–15, 1984.
74. A Holzapfel-Pschorn, U Obst, B Schmitt. Beeinflussung der mikrobiellen Stoffwechselaktivität im Rhein. Vom Wasser 69:225–231, 1987.
75. U Obst, A Holzapfel-Pschorn. Enzymatische Tests für die Wasseranalytik. München: R Oldenburg Verlag, 1988, p. 86.
76. B Schmitt-Biegel, U Obst. Hemmung der mikrobiellen Reinigungsleistung im Rhein und rheinbeeinflußten Grundwasser. Vom Wasser 73:315–322, 1989.
77. CR Milner, R Goulder. The abundance, heterotrophic activity and taxonomy of bacteria in a stream subject to pollution by chlorophenols, nitrophenols and phenoxyalkanoic acids. Wat Res 20:85–90, 1986.
78. CR Milner, R Goulder. Bacterioplankton in an urban river: The effects of a metal-bearing tributary. Wat Res 18:1395–1399, 1984.
79. P Heininger, P Tippmann. Enzymaktivitäten in Poren- und Oberflächenwässern unterschiedlich belasteter Elbabschnitte. Vom Wasser 85:141–148, 1995.
80. P Heininger, P Tippmann. Determination of enzymatic activities for the characterization of sediments. Tox Environm Chem 52:25–33, 1995.
81. BC Crump, JA Baross, CA Simenstad. Dominance of particle-attached bacteria in the Columbia River estuary, USA. Aquat Microb Ecol 14:7–18, 1998.
82. RL Sinsabaugh, S Findlay. Microbial production, enzyme activity, and carbon turnover in surface sediments of the Hudson River Estuary. Microb Ecol 30:127–141, 1995.
83. AM Romani, S Sabater. A stromatolytic cyanobacterial crust in a Mediterranean stream optimizes organic matter use. Aquat Microb Ecol 16:131–141, 1998.
84. KR Chappell, R Goulder. A between-river comparison of extracellular-enzyme activity. Microb Ecol 29:1–17, 1995.

18

Microbial Dehalogenation Reactions in Microorganisms

Lee A. Beaudette, Michael B. Cassidy, Marc Habash, Hung Lee, and Jack T. Trevors
University of Guelph, Guelph, Ontario, Canada

William J. Staddon
Eastern Kentucky University, Richmond, Kentucky

I. INTRODUCTION

Haloorganics, also known as halogenated compounds, are a set of compounds that contain a carbon–halogen bond (halogens include chloride, fluoride, bromide, and iodide ions). Among enzymatic reactions, degradation of halogen-containing compounds has received considerable attention. In the environment, naturally occurring halogenated compounds are abundant (1). Natural sources of haloorganics include sea kelp, seaweed, and algae and are released during volcanic and hydrothermal vent eruptions (2,3). Quantities of methyl chloride estimated at 5–50 Mt yr^{-1} may be released from the burning of vegetation (4). Over 1500 naturally occurring halogenated organic compounds have been isolated to date. In the last 50 years synthetic (i.e., xenobiotic) haloorganics have been developed as pesticides, soil fumigants, refrigerants, solvents, warfare and disabling gases, and chemical reagents (Fig. 1). However, unlike naturally occurring halogenated compounds, many xenobiotics are toxic, are resistant to natural mechanisms of degradation, and subsequently accumulate in the environment. By their design and nature, synthetic, multihalogenated compounds created for specific uses are recalcitrant to microbial metabolism. It has been suggested that these compounds are not properly recognized by uptake proteins, regulatory proteins, or essential catabolic enzymes (5) and that a lack of a suitable set of catabolic enzymes is responsible for this recalcitrance, rather than thermodynamics (6). Recent inputs as a result of direct application, accidental spills, and inadequate disposal have resulted in levels of halogenated compounds that exceed the capacity of microorganisms to degrade them. Halogenated compounds are found in terrestrial and aquatic environments and some are thought to be potential threats to human, animal, and ecosystem health (7–10).

The susceptibility of haloorganics to degradation is linked to the type, number, and position of the halogen substituent(s), and the degree of recalcitrance (i.e., resistance to degradation) is dependent upon the halogen constituent. Compounds containing chlorine or fluorine show greater resistance to degradation than compounds containing iodine and

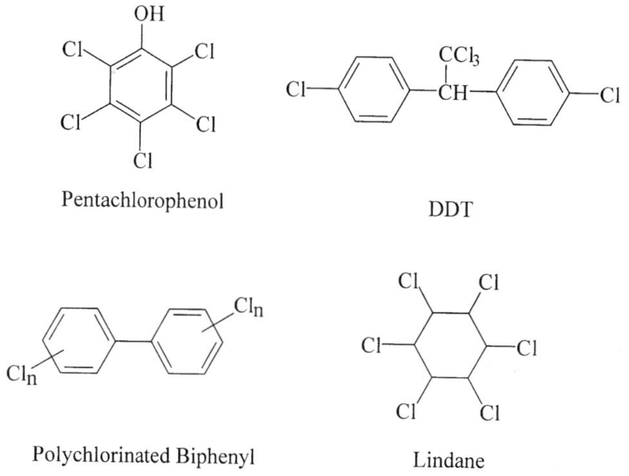

Pentachlorophenol

DDT

Polychlorinated Biphenyl

Lindane

Figure 1 Persistent halogenated compounds in the environment. DDT, dichlorodiphenyltrichloroethane, Lindane, γ-hexachlorocyclohexane.

bromine. Generally, as the number of halogens per molecule increases, so does the recalcitrance of the compound. Chlorinated aromatics, including polychlorinated biphenyls (PCBs), dioxins, pentachlorophenol (PCP), atrazine (6-chloro-*N*-ethyl-*N′*-(1-methylethyl)-1,3,5-triazine-2,4-diamine), dichlorodiphenyltrichloroethane (DDT), and 2,4-dichlorophenoxyacetate (2,4-D), can persist under aerobic conditions because chlorine atoms interfere (by steric and/or electronic effects) with the action of many microbial dioxygenase enzymes that normally initiate degradation of aromatic rings (11). This effect is problematic as many xenobiotic compounds released into the environment are multichlorinated (12).

Dehalogenation is thought to be the initial and integral step in the detoxification of these compounds. Although enzymatic, but nonorganismal dehalogenation has been demonstrated (13), the majority of dehalogenation reactions occur as a result of biotic activity. Organisms ranging from bacteria to plants have evolved dehalogenases, enzymes that exhibit carbon–halogen bond cleavage activity catalyzing the breakdown of halogenated compounds. Microorganisms have evolved several strategies of enzyme-catalyzed dehalogenation, including oxidative dehalogenation, halide elimination, substitution reactions, and reductive dehalogenation (8). The three fundamental reactions catalyzed by different dehalogenases are hydrolysis, whereby halide ions are substituted with hydroxyl groups that are derived from water; reductive dehalogenation, whereby halide ions are replaced with hydrogen atoms; and oxygen-dependent dehalogenation, whereby halide ions are replaced with hydroxyl groups derived from molecular oxygen. Hydrolysis and nearly all oxidations and reductions in the terrestrial environment result from biochemical conversions within bacteria and some plants.

Microbial enzymes may be intracellular (ectoenzymes) and associated with viable and nonviable cells or extracellular (displaced from the cell) (14). Many extracelluar enzymes in soil are stabilized by inorganic and organic soil material, components, especially clays and humics (15). As a consequence, enzymatic activity may continue after the microbial communities have been reduced by lysis or predation by soil protozoa. Indeed, it is not known how many microbial generations are necessary to create measurable increases in enzymatic activity (16). Enzymes are grouped into classes depending on their function (e.g., dehalogenases, dehydrogenases, phosphatases). Within these classes, enzymes vary in their substrate affinity. In other words, whereas all dechlorinases remove chlorine, a particular dechlorinase is only able to perform this function on one or a limited number of substrates. In the case of dehalogenases, some researchers have classified them on the basis of their activity toward certain substrate groups; others have grouped them according to their dehalogenation mechanisms (hydrolytic, oxygenolytic, or reductive) (see Table 1). Multiple pathways may be involved in the dehalogenation of a compound (10). The degradative pathway of pentachlorophenol by *Sphingomonas* spp. requires a dehalogenating oxygenase, a reductive dehalogenase, and a dehalogenating dioxygenase, which all work in concert (11) (Fig. 2), several dehalogenases from *S. paucimobilis* work together to degrade lindane (γ-hexachlorocyclohexane) (17). Also, more than one dehalogenase can degrade a particular compound (Table 1).

Systematic studies on substrate specificity, stereospecificity, and inhibition using enzymological techniques have enabled biochemists to elucidate the reaction mechanisms of dehalogenases (4). Studies on the enzymes that dehalogenate halogenated organic compounds and the regulation of these enzymes have revealed much about the molecular mechanisms and the evolution of catabolic enzymes in general. Dehalogenases are an interesting class of enzymes to study as it has been suggested they provide an environmental defense mechanism for microorganisms and are a model of enzyme evolution (4).

Trying to put the pieces of the dehalogenase's evolution puzzle together is a challenge, since the presence of halogenated xenobiotics in the terrestrial environment is a relatively recent occurrence, within the past 50 to 60 years. Thus, dehalogenases capable of degrading halogenated xenobiotics may be relatively new, evolving new sets of enzymes as a result of the recent selective pressure in the environment. Alternatively, dehalogenases may have evolved the ability to degrade halogenated xenobiotic compounds from ancient dehalogenating enzymes capable of degrading naturally occurring haloorganics.

Initial studies of organohalide degradation in the environment were undertaken as a result of concern about the environmental effects of these compounds. Recent studies have focused on the detailed mechanisms responsible for removal of halogen substituents and pathways for the subsequent metabolism of the resulting dehalogenated compounds. Current interest in dehalogenases is also aimed at elucidating structure, mechanisms of action, and active site domains, to allow subsequent protein engineering and site-directed mutagenesis to be explored. As well, adaptation of various microorganisms to halogenated compounds and the evolution of their dehalogenases are other areas of study. Results from these studies of dehalogenation reactions have important implications in terms of determining the environmental impact of xenobiotic compounds (particularly halogenated ones) and developing bioremediation technology and industrial synthesis (4).

This chapter examines microorganisms and their enzymes that have evolved to metabolize halogenated organic compounds in the environment, with a particular emphasis on soil bacteria. The major mechanisms in dehalogenation reactions are explained, and the diversity of microorganisms and enzymes is discussed. Molecular techniques that are useful for increasing our understanding of dehalogenases are presented and the information arising from this research is discussed in terms of the possible evolution of dehalogenases. Biotechnological applications and new developments are also addressed.

Table 1 Enzymes, Their Properties, and Microorganisms Involved in Dehalogenation Reactions in the Environment

Enzyme	Microorganism	Gene	Gene location (chromosome or plasmid)	Size of peptide (kD)	No. of amino acids	Substrate specificity/Substrates transformed	Reference
HYDROLYTIC							
Alkyls							
Haloacid dehalogenase DhlB	*Xanthrobacter autotrophicus* GJ10	*dhlB* inducible	Chromosome	27.4	253	Monobromoacetate	(164)
2-Haloalkanic acid dehydrogenase	*Pseudomonas* sp	*dehCI*		25.4	227-CI	Monochloroacetate	(190)
DehCI, DehCII	CBS3	*dehCII*		25.7	229-CII	(L-2MCPA) to 2-monochloropropionate	
Haloacetate dehalogenase Deh-H2	*Moraxella* sp. B	*Deh-H2* constitutive	Plasmid	26	225	Chloro-, bromo-, and iodoacetate but not fluoroacetate	(191, 192)
Hdl IVa	*Pseudomonas cepacia* MBA4	*hdl Iva* inducible		25.9	231	Broad Monobromoacetic acid (MBA) L-2MPCA	(125, 165)
Deh 109	*Pseudomonas* sp. 109	*deH109*	Plasmid	25.2	224	Broad L-2MCPA	 (193, 194)
HadL	*Pseudomonas putida* AJl	*hadL* inducible	Plasmid	25.7	227	L-2MCPA	(195)
L-DEX	*Pseudomonas* sp. YL	L-*Dex* inducible		26.1	232	Broad	(155, 196)
DL-DEX				36			
DehI/HadL	*Rhizobium* sp.				280	Dalapon (2,2-dichloropropionic acid)	(21, 197)
HadD	*Pseudomonas putida* AJl	*hadD* inducible	Plasmid	33.6	300	Narrow D-2MCPA	 (198, 199)
DehIII/HadD	*Rhizobium* sp.			29.4	266	Dalapon	(21)
DehI	*Pseudomonas putida* PP3	*dehI* inducible	Chromosome	32.7	296	2-Haloalkanoic acid	(22, 159)
DhlC	*Alcaligenes xylosoxidans* ABIV	*dhlC* inducible	Plasmid	32	296	Monochloroacetate Dalapon	 (161)
DehII	*Pseudomonas putida* PP3; *Pseudomonas* 113; *Rhizobium* species	*dehII* inducible	Chromosome			2-Haloalkanoic acid Both D- & L-2MCPA (113) Dalapon; MCA; DPA (PP3)	(197)

Monochloro acetate dehalogenase	*Pseudomonas* spp.	inducible				Broad	(82)
	Alcaligenes RC8 *Agrobacterium* 81a *Arthrobacter* 16b *Azotobacter* RC26					MCA And a wide range of others	
Alkanes							
Haloalkane dehalogenase DhlA	*Xanthobacter autotrophicus* GJ10 *Ancylobacter aquaticus* AD20/25, *Pseudomonas* sp. AD1	*dhlA* constitutive	Plasmid	35.1	310	Relatively narrow l-chloroalkanes to C4 l-bromoalkanes to C10 AD20/25 2- chloroethylvinylether	(126, 151, 173, 200)
Haloalkane dehalogenase LinB	*Pseudomonas paucimobilus* UT2	*linB*		33	295	Some Number of mono-chlorinated alkanes, 1,2-dibromoethane, lindane	(192)
Haloalkane dehalogenase Deh-Hl	*Moraxella* sp. B	*deh-Hl* constitutive	Plasmid	33	294	Narrow Monofluoroacetate	(192)
l-Chloroalkane halidohydrolase	*Rhodococcus rhodochrous* NCIMB 13064	*dhaA* inducible	Plasmid	33	292	Broad C3–C10	(129, 162, 201)
Haloalkane dehalogenase DhaA	*Pseudomonas cichorii* 170	*dhaA* constitutive	Plasmid	33	292	Broad; C2–C10 l-chloro-*n*-alkanes, Cl to Cl6 l-bromo-*n*-alkanes, C2 to C9 dihalo-*n*-alkanes	(80, 151)
Haloalkane dehalogenase	*Arthrobacter* sp. strain HAl	Inducible		37	315	Broad; over 50 halogenated compounds	(202–204)
Haloalkane dehalogenase	*Corynebacterium* sp. strain m 15-3	Inducible		36	325	Broad l-chlorobutane	(205)
Haloalkane dehalogenase	*Rhodococcus erythropolis* Y2	Inducible	Chromosome	34		Broad l-chlorobutane	(206)
Glutathione-S-transferases							
Dichloromethane dehalogenase DcmA	*Hyphomicrobium* sp. strain DM2	*dcmA* inducible	Plasmid	37.4	287	Narrow dichloromethane (DCM) (halomethanes and dihaloethanes not converted)	(23, 207)

Table 1 Continued

Enzyme	Microorganism	Gene	Gene location (chromosome or plasmid)	Size of peptide (kD)	No. of amino acids	Substrate specificity/Substrates transformed	Reference
Dichloromethane dehalogenase DcmA	*Methylobacterium* sp. strain DM4	*dcmA* inducible	Chromosome	37.4	287	Narrow; dichloromethane (DCM) (halomethanes and dihaloethanes not converted)	(208, 209)
Haloalcohol dehydrogenase DehA; DehC	*Arthrobacter erithii* H10a	*dehA dehC* constitutive				Narrow	(144)
Haloalcohol dehydrogenase	*Arthrobacter* sp. AD2	Inducible		29	264	Broad	(210)
						1,3-dichloro-2-propanol	
Aromatics							
4-Chlorobenzoyl-CoA-dehalogenase	*Acinetobacter* sp. 4-CBl	One gene in an operon of three	Plasmid	30	276	Narrow	(211)
						4-chloro-, 4-bromo-, and 4-iodobenzoate, but not 4-fluorobenzoate	
4-Chlorobenzoate dehalogenase	*Pseudomonas* sp. CBS3					Narrow; 4-chloro-, 4-bromo-, 4-iodobenzoate, but not 4-fluorobenzoate	(127)
4-Chlorobenzoyl-CoA-dehalogenase	*Arthrobacter* sp. strain SU	*fcbB* inducible	Plasmid	30	276	Narrow; 4-chloro-, 4-bromo-, 4-iodobenzoate, and 4-fluorobenzoate	(132)
4-Chlorobenzoate dehalogenase	*Acinetobacter* sp. ST1					Narrow; 4-chloro-, 4-bromo-, and 4-iodobenzoate	(212)
Heteroaromatics							
2,4-Dichlorophenol hydroxylase	*Alcaligenes eutrophus* JMPl34	*tfdB* induced	Plasmid	65			(213)
Atrazine chlorohydrolase	*Pseudomonas* sp strain ADP	*atzA*	Chromosome	60	473	Narrow	
						Dechlorinates atrazine to hydroatrazine	(128)

Dehydro-dechlorinase	*Sphingomonas paucimobilus* UT26	*linA*		66		α-, γ-, and δ-hexachlorocyclohexane (lindane) to γ-pentachlorocyclohexene to 1,3,4,6-tetrachloro-1,4-cyclohexadiene	(17)
OXYGENOLYTIC: Aromatics							
Monooxygenases							
PCP-4-monooxygenase	Several strains of *Sphingomonas* including sp. ATCC 39723 and *Mycobacterium* sp.	*pcpB* inducible	Chromosome	60	538	Broad	(67)
Dioxygenases							
2,4-Dichlorophenol monooxygenase	*Alcaligenes eutrophus* JMPl34	*tfdA*	Plasmid	32	287	Broad	(214)
2,4-Dichlorophenoxyacetate dioxygenase		*tfdA*	Plasmid				(215)
Two-component dioxygenase	*Pseudomonas* sp. CBS3					4- Chlorophenylacetate	(71)
2-Halobenzoate 1,2-dioxygenase (two-component)	*Pseudomonas cepacia* 2CBS					2-Halobenzoate	(216)
2,6-Dichloro-hydroquinone dioxygenase	*Sphingomonas* sp. ATCC 39723	*pcpA* induced	Chromosome	30	271	2,6-Dichlorohydroquinone	(76, 77, 217)
REDUCTIVE: Aromatics							
TCHQ dehalogenase	*Sphingomonas* sp. ATCC 39723	*pcpC* constitutive	Chromosome	28.2	248	Broad	
						Tetrachlorohydroquinone and trichlorohydroquinone	(40, 41)
3-Chlorobenzoate reductive dehalogenase	*Desulfomile tiedjei* DCB-1						(46)
Reductive 2,4-dichlorobenzoyl CoA dehalogenase	*Corynebacterium sepedonicum* KZ-4 & Coryneform strain NTB-1					2,4-Dichlorbenzoic acid	(44)

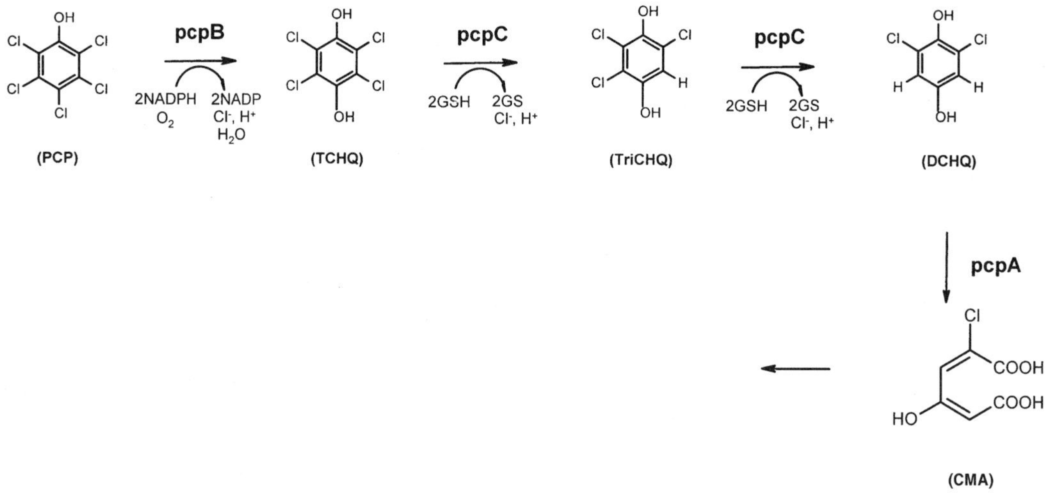

Figure 2 Degradation pathway for pentachlorophenol (PCP) by *Sphingomonas* spp. TCHQ, tetrachlorohydroquinone; TriCHQ, trichlorohydroquinone; DCHQ, 2,6-dichlorohydroquinone; CMA, 2-chloromaleylacetate; pcpA, DCHQ dioxygenase; pcpB, PCP-4-monooxygenase; pcpC, TCHQ reductive dehalogenase.

II. MECHANISMS OF DEHALOGENATION

It is recognized that compounds may be dehalogenated through a variety of pathways. For example, Ramanand et al. (18) followed the degradation in slurry microcosms of hexachlorobenzene, pentachlorobenzene, and 1,2,4-trichlorobenzene to chlorobenzene. Through an examination of intermediates that were degraded and those that remained recalcitrant, the authors argued that multiple pathways for degradation existed and that these pathways reflected the microorganisms present and the local physical and/or chemical conditions.

Much of our knowledge of soil microorganisms and their enzymes has been gained through the development of assays for specific enzymes and identification of microorganisms that produce these enzymes. This section examines dehalogenation reactions and specific mechanisms of dehalogenation in selected microorganisms.

There are six enzyme-catalyzed dehalogenation reactions in microorganisms growing on halogenated organic compounds. These include reductive, oxygenolytic, and hydrolytic (including glutathione substitution) dehalogenation, as well as hydration, intramolecular substitution, and dehydrohalogenation (19). The three best studied mechanisms, hydrolytic, reductive, and oxygenolytic dehalogenation reactions, are discussed further.

A. Hydrolytic Dehalogenation

Hydrolytic dehalogenation is a reaction catalyzed by halidohydrolases in which, a halogen substituent is replaced in a nucleophilic substitution reaction with a hydroxy group derived from water. The largest number of dehalogenases described to date are hydrolytic dehalogenases, and hydrolytic dehalogenation of aliphatic, aromatic, and heterocyclic compounds has been reported. The majority are substrate-inducible, although several are produced constitutively (Table 1). The halidohydrolases can be grouped by the substrates on which they act. This division by substrate range may be useful from a practical vantage point but has little evolutionary and mechanistic significance (19).

1. *Alkyl Halides*

Haloalkanoic acids are naturally produced at very low levels in the environment and are the most susceptible to enzymatic dehalogenation. The mechanism for dehalogenation of halogenated alkanoic acids is represented by the following formulae:

Monosubstituted:

$$R - (CHX) - COOH + H_2O \Rightarrow R - (CHOH) - COOH + H^+ + X^-$$

Disubstituted:

$$R - (CX2) - COOH + H_2O \Rightarrow R - (CO) - COOH + 2H^+ + 2X^-$$

where X represents the halogen species and R represents the side chain.

Bacterial strains that produce dehalogenases exhibiting both regio- and stereospecificity for haloalkanoic acids have been documented. These dehalogenases are called *2-haloalkanoic acid* (2HAA) *hydrolytic dehalogenases* and have been classified into two broad categories, depending upon their activity toward longer-chain halocarboxylic acids such as halopropionates (see Table 2). Category A includes haloacetate dehalogenases that convert only 2-haloacetates and exhibit no activity toward halopropionates. They are also inhibited by sulfhydryl agents. Category A can be further subdivided into two types. Type one enzymes are able to cleave carbon–fluorine bonds, whereas type two enzymes have no activity toward fluoroacetates. Category B includes enzymes with hydrolytic activity toward haloacetates and longer-chain halocarboxylic acids. These enzymes are generally inducible, with low affinities for substrates (with K_m values in the millimolar range) and broad substrates specificities. These 2-haloacid dehalogenases in category B are able to convert 2-chloropropionate to lactate and can be further classified into four groups (possibly five), which are based on their stereospecificity (12,20). Class 1L remove halides from L-2-haloalkanoic acids, invert the product configuration with respect to the substrate,

Table 2 Classification of Haloalkanoic Acid Dehalogenases

Enzyme group or class	Substrates	Substrate/product configuration	Sensitive to sulfhydryl agents
A: Haloacetate dehalogenases (E.C. 3.8.1.3)			
Group 1	Chloroacetate and fluoroacetate	NA[a]	Yes
Group 2	Chloroacetate only	NA[a]	Yes
B: 2-Haloalkanoic acid dehalogenases (E.C. 3.8.1.2)			
Class 1L	L-2-haloalkanoic acids	Inversion	Yes
Class 1D	D-2-haloalkanoic acid	Inversion	nd[b]
Class 21	D- and L-2-haloalkanoic acids	Inversion	No
Class 2R	D- and L-2-haloalkanoic acids	Retention	Yes
Class?	Elevated activity toward trichloroacetates		

[a] Not applicable.
[b] Not determined.
Source: Refs. 4, 20, and 218.

and react to sulfydryl-blocking reagents to varying degrees. There are presently 10 detailed examples of this class of enzyme, and 9 of the sequences have been determined, demonstrating nucleotide sequence similarities and defining the class as a coherent group of enzymes (12). Biochemical information strongly indicates a uniform class of proteins at the functional and mechanistic levels (12). There may be some similarities between this class and haloalkane dehalogenases, suggesting an evolutionary relationship.

Class 1D 2HAA hydrolytic dehalogenases are able selectively to dehalogenate D-isomeric substrates, such as D-2-monochloropropionic acetate (D-2MCPA), with inversion of the product configuration. Despite repeated attempts, it has been difficult to isolate microorganisms that produce this class of dehalogenase. However, *Pseudomonas putida* AJ1 was isolated from soil preenriched with racemic 2MPCA and was found to have two highly stereospecific dehalogenases: HadL, active against L-2MPCA, and HadD, active against D-2MPCA only (Table 2) (12). A *Rhizobium* species able to degrade 2,2-dichloropropionic acid (Dalapon) was found to contain two dehalogenases that were originally thought to be similar to HadL and HadD (21). But it now appears that they may only be related at the functional level (12).

Class 21 2HAA hydrolytic dehalogenases differ from class I in their ability to dehalogenate both L- and D-isomers by a mechanism that inverts substrate product configurations. Enzymes in this class are unaffected by sulfhydryl-blocking reagents. The dehalogenase synthesized by *Pseudomonas* sp. 113 dehalogenated both D- and L-2MCPA, although the L-isomer was dehalogenated more rapidly. *Pseudomonas putida* PP3 was found to synthesize two 2HAA dehalogenases, designated Dehl and Dehll (22). They were active against a similar range of HAAs, but Dehll was unaffected by sulfydryl-blocking reagents, and was classified as 21.

Class 2R 2HAA hydrolytic dehalogenases are able to dehalogenate both L- and D-isomers by a mechanisms that retains substrate-product configurations and are strongly inhibited by sulfhydryl-blocking agents. Dehl from *Pseudomonas putida* PP3 has been the best studied of the class 2R enzymes, and it has been shown that the amino acid sequence for Dehl is identical to the sequence for DhlC from *Alcaligenes xylosoxidans* (12).

The fifth group is represented by halidohydrolases that demonstrate elevated activity toward trichloroacetate (TCA) with complete dechlorination of the substrate (4,20). These dehalogenases appear to be rare, as there are few reports in the literature, and are not always classified as a separate group (8).

Although the 2HAA dehalogenases show a wide variation in the range of haloalkanoic acids that act as substrates for the different mechanistic groups (Table 1), they are not active toward the corresponding haloalkanes, amides, or aldehydes even though the amides and aldehydes are similar in size and susceptibility toward SN_1 and SN_2 reactions (4). A positively charged group at the enzyme's binding site may allow binding of a carboxyl group but not a positively charged amide group (20).

2. *Haloalkanes*

In the case of hydrolytic haloalkane dehalogenation, there are two distinct categories of haloalkane dehalogenases, glutathione-independent and glutathione-dependent (Table 1). Direct hydrolytic dehalogenation of haloalkanes (glutathione-independent) was first observed with the haloalkane dehalogenase from the nitrogen-fixing bacteria *Xanthobacter autotrophicus* GJ10 (coded for by *dhlA*). This species constitutively synthesizes two halidohydrolases (dhlA and dhlB) that allow it to utilize 1,2-dichloroethane (Fig. 3). The

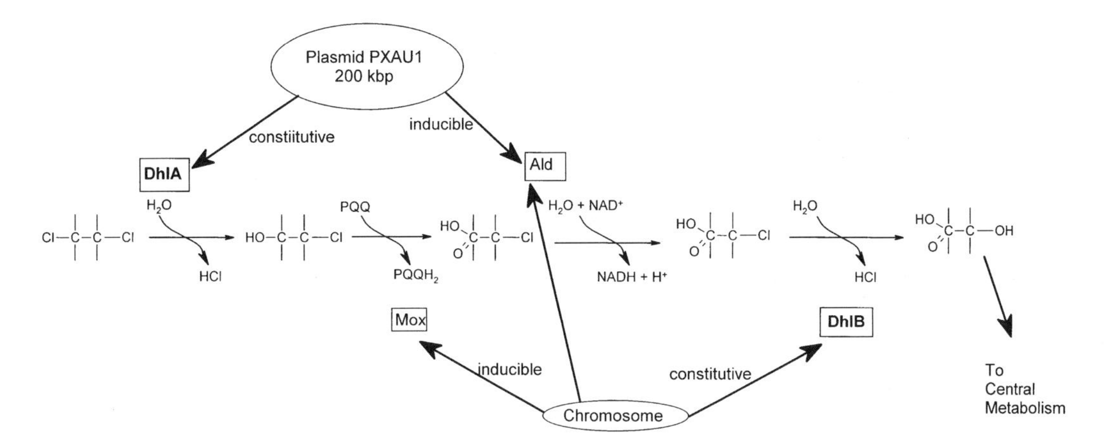

Figure 3 Catabolic pathway of 1,2-dichloroethane in *Xanthobacter autotrophicus* and *Ancylobacter aquaticus*. Both dehalogenases are constitutively produced in *Xanthobacter*; spp. DhlA, haloalkane dehalogenase; Mox, alcohol dehydrogenase; Ald, aldehyde dehydrogenase; DhlB, haloacetate dehalogenase. (Redrawn from Ref. 5.)

dhlA gene coding for haloalkane halidohydrolase was found on a 200-kb plasmid, whereas the *dhlB* gene for 2-haloalkanoic halidohydrolase was located on the chromosome. The dhlA haloalkane dehalogenase is the best studied dehalogenase to date. The enzyme is active in catalyzing more than 24 haloaliphatic compounds, including 1,2-dicholoroethane 2-chloroethylvinylether, and other haloalkanes, alcohols, ethers, amides and epoxides. Identical *dhlA* gene sequences from different strains of *X. autotrophicus* and *Ancylobacter aquaticus* have been observed (19). X-ray crystallography determined that the main domain of its structure was common to many hydrolytic proteins that are classified in the structural group of α/β hydrolase fold hydrolytic enzymes. Surrounding the main domain was a second, called the *cap domain*. The active site was located between the two domains and formed an internal hydrophobic cavity. The key residues were shown to be arranged as a catalytic triad and included Asp 124, which functions as the nucleophile. The key role of Asp 124 was demonstrated by site-directed mutagenesis, since replacement of the aspartate with alanine, glycine, or glutamic acid resulted in inactivation of the enzyme (5). The halide binding was achieved via two tryptophan residues, one located in the cap domain and one in the main domain, suggesting that both domains have specifically evolved to play a role in the catalysis of carbon–halogen bond cleavage. This demonstrates that the enzyme is a dehalogenase, and not a general broad-spectrum hydrolase that fortuitously converts chlorinated substrates (12,19). The haloalkane dehalogenase from *X. autotrophicus* GJ10 is very different from the three other known haloalkane dehalogenases, synthesized in *Arthrobacter* sp. HA1, *Rhodococcus erythropolis* Y2, and *Corynebacterium* m15-3 (4). These three species have not been as well studied, and mechanistic information is lacking. It will be interesting to see what the differences are and whether the interpretation of these differences helps us to understand the evolution of the enzymes.

The second category of hydrolytic haloalkane dehalogenases contains those that require glutathione as a cofactor. Fetzner (8) classifies the activity separately as a thiolytic dehalogenation mechanism, with an unstable glutathione intermediate that is subsequently hydrolyzed. They have different substrate specificity and distinct amino acid sequences that also differentiate them from glutathione-independent haloalkane dehalogenases. Glutathione-dependent haloalkane dehalogenases have been found in different species of bacteria from different geographical locations and have demonstrated similar substrate ranges and immunological properties and identical N-terminal amino acid sequences (23). One of the most well-studied is the dichloromethane dehalogenase (dcmA) from *Methylobacterium* sp. DM4. This particular system has a low specific activity, with good substrate binding capacity, but a slow reaction rate. Although the *dcmA* genes for DM4 were shown not to be plasmid encoded, plasmid encoding of *dcmA* genes for *Hyphomicrobium* sp. strain DM2 in this category has been observed (19).

3. Haloaromatics

In the case of haloaromatics, the most studied hydrolytic aromatic dehalogenase is 4-chlorobenzoyl coenzyme A (CoA) dehalogenase, which is found in several strains, including *Pseudomonas* sp. CBS3 and *Acinetobacter* sp. 4-CB1 (11). In this case, additional cofactors such as adenosine triphosphate (ATP), Mg^{2+}, and CoA are required. It is noteworthy that a combination of catalytic strategies is used, each of which is seen with other nondehalogenating enzymes. One strategy uses a similar mechanism to that of enoyl CoA hydratase, and sequence identity and structures between the enzymes appear to be very similar. The other strategy is similar to the mechanism used by epoxide hydrolase and haloalkane dehalogenase, but in this case there are no sequence homologies among the three enzymes, and the steric and electronic requirements are different (11). It is interesting that the same catalytic strategy is used by all three enzymes. Hydrolytic dehalogenation of other aromatics has been observed with atrazine chlorohydrolase from *Pseudomonas* sp. adenosine diphosphate (ADP) and a hydrolytic enzyme from *Rhodococcus corallinus* NRRL B-15444R that catalyzes both dechlorination and deamination of s-triazine substrates (11).

B. Reductive Dehalogenation

As the name implies, reductive dehalogenation is the removal of a halogen substituent from a molecule with concurrent reduction or addition of two electrons to the molecule (24). In most cases, the addition of electrons is in the form of a hydrogen atom. It seems logical to assume that reductive dehalogenation occurs under reducing or anaerobic conditions. Although this assumption is correct, reductive dehalogenation also occurs under aerobic conditions. For the most part, the understanding of reductive dehalogenation mechanisms is in its infancy. Over the past two decades many of the reported reductive dehalogenation studies have been on mixed cultures predominantly from soils, sediments, intestinal tracts of animals, and bioreactors. Most of the pure culture work has been on the alkyl (aliphatic) halides, mainly as a result of researchers' trying cultures from their own culture collections. Only a few pure isolates are known to dehalogenate aryl (aromatic) haloorganics reductively.

Many of the alkyl and aryl halides have been used as solvents, pesticides, heat transfer fluids, and flame retardants or generated as waste products from industrial processes. Generally the alkyl halides tend to be more water-soluble and have made their way into soil, groundwater, and river systems. Aryl halides are mostly made up of the less water-soluble benzenes, toluenes, biphenyls, and phenols and therefore remain in soil and sediments as persistent environmental hazards. Microbial reductive dehalogenation is a very important step in the degradation of many of the halogenated contaminants.

Many of the haloorganics require at least partial dehalogenation as a first step to biodegradation. If degradation does not take place, the species of haloorganic found contaminating a site is usually the dehalogenated product of the parent compound. As pointed out by Fetzner (8), alkyl and aryl reductive dehalogenation mechanisms basically fall into three categories: (a) cometabolic processes in which methanogenic, acetogenic, sulfate-, or iron-reducing bacteria dehalogenate compounds with no direct benefit to the microorganism; (b) carbon metabolism processes in which a microorganism assimilates the corresponding carboxylic acid after the halogen is removed; and (c) respiratory processes in which reduction of the haloorganic is linked to energy metabolism (i.e., used as a terminal electron acceptor).

1. *Alkyl Halides*

a. Cometabolism The methods by which various microorganisms dehalogenate many of the alkyl haloorganics vary considerably. A typical alkyl reductive dehalogenation reaction is shown in Fig. 4a. *Acetobacterium woodii*, *Desulfobacterium autotrophicum*, and *Clostridium hydrogenophilus* all use the acetyl-coenzyme A pathway for the reduction of tetrachloromethane to trichloromethane and dichloromethane (25). Along the same lines, a *Methanosarcina thermophila* dechlorinated trichloroethylene (TCE) to *cis*-dichloroethylene (*cis*-DCE), *trans*-DCE, 1,1-dichloroethylene, vinyl chloride, and ethylene by a CO-reduced CO dehydrogenase enzyme complex (a complex that cleaves the C-C and C-S bonds of acetyl-CoA) (26). The enzyme complex was divided into a two-subunit nickel/iron-sulfur (Ni/Fe-S) component and the two-unit factor III–containing corrinoid/iron sulfur (CO/Fe-S) component. The proposed pathway is shown in Fig. 5 (26).

Cobalamin and native and diepimeric forms of factor F_{430} from *Methanosarcina barkeri* catalyzed the reductive dehalogenation of 1,2-dichloroethane to ethylene or chloroethane in cell free or boiled extracts (27). In addition to factor F_{430}, vitamin B_{12} (a cobalt corrinoid) dechlorinated tetrachloroethene (PCE) to ethene (28).

Many of the cometabolic dehalogenation mechanisms are thought to involve the electron carriers of the respiratory electron transport chain. Examples include reduction of tetrachloromethane by *Escherichia coli* K12 under fumurate respiring conditions (29), by the denitrifying *Pseudomonas* sp. strain KC (30), and by *Shewanella putrifaciens* 200 (31,32).

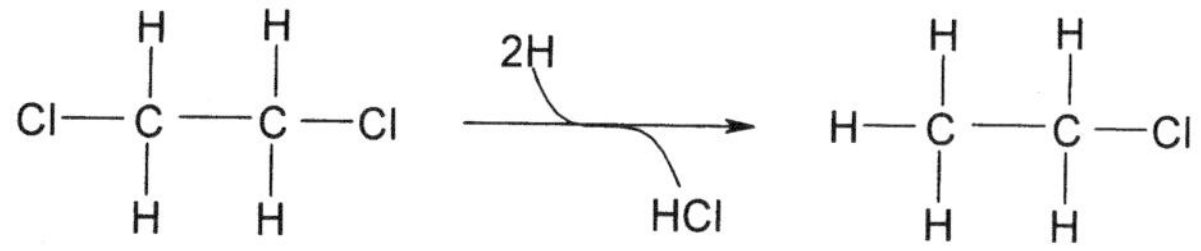

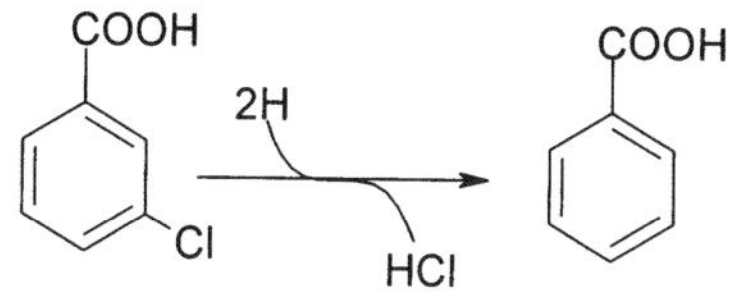

Figure 4 General reductive dehalogenation mechanisms for (a) alkyl compounds, (b) aryl compounds. (Redrawn from Ref. 24.)

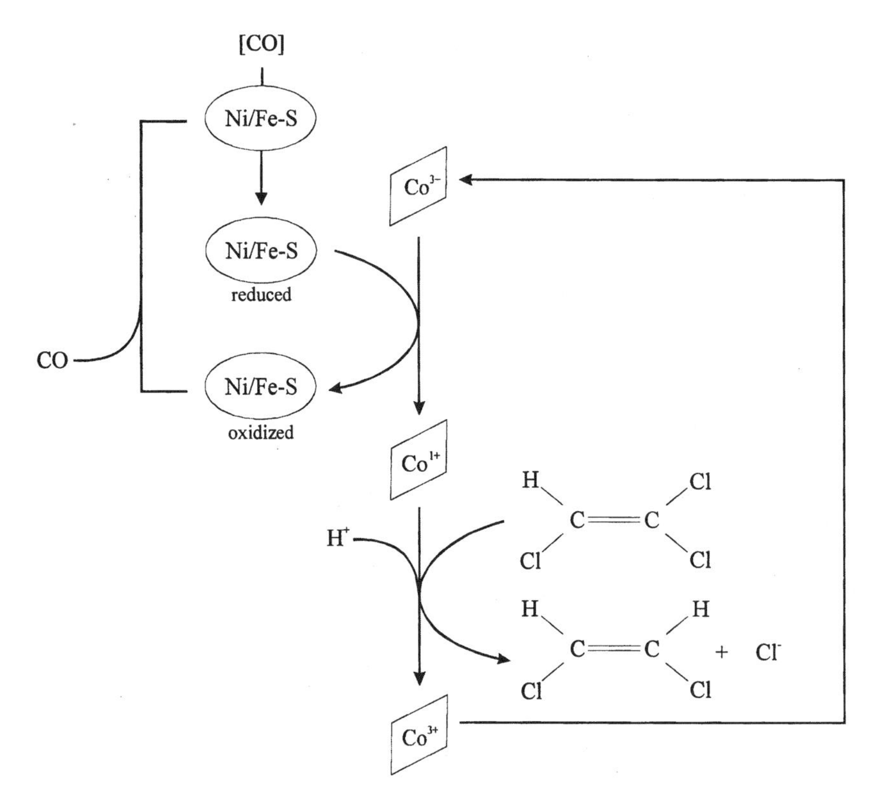

Figure 5 Enzyme complex for the reductive dechlorination of TCE to *cis*-DCE by *Methanosarcina thermophile.* (Redrawn from Ref. 26.)

b. Carbon metabolism McGrath and Harfoot (33) observed that some of the purple nonsulfur bacteria, including *Rhodospirillum rubrum*, *Rhodospirillum photometricum*, and *Rhodopseudomonas palustris*, dechlorinated and then assimilated chloroacetic, 2-bromopropionic, 2-chloropropionic, and 3-chloropropionic acids in the presence of CO_2. No mechanisms were provided.

Lindane (γ-hexachlorocyclohexane) is degraded by *Sphingomonas paucimobilis* UT26. The first enzyme (LinA) is a dehydrohalogenase (γ hexachlorocyclohexane dehydrochlorinase) that eliminates HCl from the compound and results in the formation of a double bond, giving γ-pentachlorocyclohexane followed by 1,3,4,6-tetrachloro-1,4-cyclohexadiene. LinB is a hydrolytic dehalogenase, whereas LinC is a dehydrogenase. Within the degradation pathway there is a glutathione-dependent reductive dehalogenase (LinD) that catalyzes the reductive dechlorination of 2,5-dichlorohydroquinone to chlorohydroquinone and then to hydroquinone (17). Technically the LinD enzyme should be classified as an aryl dehalogenator. However, since the parent compound is an alkyl, the LinD enzyme was allocated to the alkyl reductive dehalogenation section. Glutathione S-transferases (or GSTs) (E.C. 2.5.1.18) are a group of enzymes that attack a wide spectrum of molecules in the same manner. The glutathione characteristically attacks the electrophilic center of the molecule (34). Other glutathione-dependent reductive dehalogenases are mentioned in the discussion aryl reductive dehalogenation (sec. II. B. 2).

c. Respiratory Metabolism Most bacterial strains reported to utilize respiratory reductive dehalogenation of alkyl compounds have been tested against PCE. The strains are *Desulfomonile tiedjei* (24), *Dehalobacter restrictus* (35), *Dehalorepirillum multivorans*, and *Dehalococcoides ethanogenes* 195 (36). Strains in which there is no definitive proof of respiratory reductive dehalogenation include *Desulfitobacterium frappieri* PCP-1 (37), *Desulfuromonas chloroethenica*, *Enterobacter* MS-1, and *Enterobacter agglomerans* (35). In all cases, each of the microorganisms carried a corrinoid and an iron-sulfur cluster as cofactors. Electron donors include hydrogen, formate, acetate, and pyruvate. The only microorganism found to dechlorinate PCE to ethene was *Dehalococcoides ethanogenes* 195 (38).

2. *Aryl Halides*

a. CoMetabolism Very few aryl reductive dehalogenation reactions have been associated with cometabolism. A typical aryl reductive dehalogenation reaction is shown in Fig. 4b. The only evidence that there may be cometabolism of the aryl compounds is the aryl dehalogenation of pentachlorophenol (PCP) by vitamin B_{12S} (39). In the presence of vitamin B_{12S}, PCP was dechlorinated to a mixture of 2,3,4,6-tetrachlorophenol (TeCP) and 2,3,5,6-TeCP and subsequently dechlorinated to 2,3,6- and 2,4,6- trichlorophenol (TCP).

b. Carbon Metabolism One of the most studied of the aryl reductive dehalogenases is the glutathione-dependent tetrachlorohydroquinone dehalogenase (PcpC) (Table 1). This enzyme catalyzes reductive dechlorination of tetrachloro-*p*-hydroquinone, first to tri- and then to dichlorohydroquinone in the pentachlorophenol degradation pathway of several *Sphingomonas* spp. (Fig. 2) (40–43). PcpC is also grouped into the superfamily of GST enzymes. Interestingly, these bacteria utilizing this reductive dehalogenation mechanism are aerobic.

A reduced nicotinamide-adenine dinucleotide phosphate–(NADPH)-dependent reductive dechlorinase was found in *Corynebacterium sepedonicum* KX-4 and a coryneform strain NTB-1 (44). Both of these strains reductively *ortho*-dechlorinated 2,4-dichlorobenzoyl-CoA to 4-chlorobenzoyl-CoA.

c. Respiratory Metabolism Many of the dehalogenation reactions that take place in the environment are thought to be a result of halorespiration. Polychlorinated biphenyls (PCBs) and chlorobenzenes have found their way into the environment over the last 50 years. Samples taken from many PCB- and chlorobenzene-contaminated sites have been found to have biological dechlorinating activity under anaerobic conditions (24,45). The activity is assumed to be halorespiration. *Desulfomonile tiedjei* (45) is the most studied of only a few pure isolates that have been shown to dehalogenate by using respiratory metabolism. *D. tiedjei* is a sulfate-reducing microorganism that uses formate or hydrogen as an electron donor and 3-chlorobenzoate as the terminal electron acceptor. A cytoplasmic membrane-associated reductive dehalogenase was purified in 1995 with a heme group assumed to be part of the active site (46). In addition, several species of the genus *Desulfitobacterium* have also been found to dechlorinate aryl compounds reductively (47–51). In fact, one of the species, *Desulfitobacterium dehalogenans*, has been shown to dehalogenate hydroxylated polychlorinated biphenyls (51). This is significant as no other pure isolate has been shown to dechlorinate PCBs or hydroxylated PCBs.

C. Oxygenolytic Dehalogenation

Although substitution reactions are active in both the reductive and hydrolytic mechanisms, the presence of double bonds between carbon atoms results in halogens that are resistant to nucleophilic substitution reactions. However, the addition of oxygen atoms to the double bond is possible, and dehalogenation can ensue. The enzymes that catalyze the oxygenolytic reactions of halogenated compounds in soil can be divided into two main groups: monooxygenases and dioxygenases. A common feature of both mono- and dioxygenases is the incorporation of hydroxyl groups derived from dimolecular oxygen (O_2) into their substrates. Also, both types of oxygenases use cofactors. Transition metals, such as iron, manganese, copper, and cobalt, serve as catalysts in numerous oxygenation reactions (52). Flavins and pteridines can act as cofactors for monooxygenases (52). The oxygenases are known as ''accidental'' or ''fortuitous'' dehalogenases because their mechanism of action is to form unstable products that decompose spontaneously to release the halogen substituent (11).

1. Monooxygenases

Monooxygenases are also referred to as mixed function oxygenases (MFOs). They catalyze the incorporation of one atom of dioxygen into a substrate. The second atom of oxygen is reduced to water (H_2O) either by the substrates themselves or by a cosubstrate reductant. The general mechanisms of monooxygenases are as follows:

$$SH_2 + O_2 = SO + H_2O \tag{1}$$

$$S + O_2 + H_2X = SO + H_2O + X \tag{2}$$

where (1) represents a reaction catalyzed by an internal monooxygenase and (2) represents a reaction catalyzed by an external monooxygenase and H_2X represents a cosubstrate reductant (52). The monoxygenases act principally on aromatic compound but can also catalyze reactions with some halogenated alkyl compounds. The ability of monooxygenases to dehalogenate appears to be due to their broad substrate specificity.

a. Haloalkanes, Haloalkenes, and Haloalkynes Methane monooxygenase and ammonia monooxygenase have been shown to aid in the degradation of haloalkanes, haloalkenes, and haloalkynes (4,53–56). Typically these monooxygenases act by creating reactive species that can be degraded via chemical decomposition rather than direct removal of the halogens from the substrate.

b. Haloaromatics The predominant group of halogenated compounds degraded by monooxygenases are haloaromatics. The monooxygenase-catalyzed degradation of pentachlorophenol (PCP), a haloaromatic wood preservative, by PCP-4-monooxygenase-(encoded by the *pcpB* gene) has been shown for Gram-negative *Sphingomonas* spp. (including strains previously identified as *Arthrobacter*, *Pseudomonas*, and *Flavobacterium* spp.) (Fig. 2) and Gram-positive actinomycete *Mycobacterium* spp. (including strains previously identified as *Rhodococcus*). Degradation of PCP is initiated by *para*-hydroxylation to tetrachlorohydroquinone (TCHQ) (Fig. 2).

The monooxygenase from *Sphingomonas* ATCC 39723 (Table 1) was the first purified flavoprotein-monooxygenase shown to utilize halogenated aromatic compounds as substrates (7), catalyzing an NADPH- and oxygen-dependent dehalogenation of PCP. Verification of the origin of the oxygen necessary for the reaction was demonstrated with the incorporation of ^{18}O into PCP, providing evidence that the formation of TCHQ is catalyzed by a monooxygenase rather than a hydrolase (57). Subsequently, other microorganisms containing a PCP dehalogenating monooxygenase have been found including *Sphingomonas* sp. UG30 (58), *Sphingomonas* sp. RA2 (59), *Sphingomonas* sp. strain ATCC 33790 (60,61), and *Mycobacterium* sp. (62). The monooxygenase from *Sphingomonas* sp. ATCC 39723 and *Sphingomonas* sp. UG30 is able to catalyze the *para*-hydroxylation of a broad range of substituted phenols, including the removal of halogen, nitro, amino, and cyano groups (63–66). The monooxygenase gene (*pcpB*) found in *Sphingomonas* spp. is chromosomally located, and production of the monooxygenase is induced by the presence of PCP. A *pcpB* gene probe that hybridized with genomic DNA from five *Sphingomonas* spp. did not hybridize with the deoxyribonucleic acid (DNA) from PCP-degrading *Mycobacterium chlorophenolicus* PCP-1 (previously known as *Rhodococcus*) (67). The *Sphingomonas* spp. strains require the presence of FAD, NADPH, and molecular oxygen, the monooxygenase found in *M. chlorophenolicus* PCP-1 catalyzes the dehalogenation of PCP by a different mechanism. It has been suggested that the membrane-associated monooxygenase in *Mycobacterium* spp. may involve a cytochrome P-450–type monooxygenase (62,68), but the enzyme has yet to be purified.

Although mechanistic studies have not been carried out on the pentachlorophenol monooxygenase, it has been proposed to resemble the mechanism of tetrafluoro-*p*-hydroxybenzoate defluorination by *p*-hydroxybenzoate hydroxylase (69).

2. Dioxygenases

Dioxygenases catalyze the incorporation of both atoms of dioxygen (O_2) into their substrates. Two types of dioxygenases are aromatic ring dioxygenases and aromatic ring cleavage dioxygenases. Aromatic ring dioxygenases incorporate two hydroxyl groups on the aromatic ring utilizing dioxygen and NADPH, whereas aromatic ring cleavage dioxygenases open the aromatic ring by incorporating two atoms of dioxygen into substrates via intradiol cleavage or extradiol cleavage (Fig. 6).

The dioxygenases comprise several proteins arranged as a short electron-transfer chain necessary for the activation of O_2. A general feature of dioxygenases is the presence of a flavoprotein (either flavin mononucleotide [FMN] or flavin adenine dinucleotide [FAD]) that acts as a binding site for reduced nicotinamide adenine dinucleotide (NADH) or reduced nicotinamide adenine dinucleotide phosphate (NADPH) for the initial electron transfer. The dioxygenases also contain iron-sulfur complexes (2Fe-2S). Different types of iron complexes are found in dioxygenases, including ferredoxin and Rieske complexes. The iron-sulfur complexes act solely as electron carriers. It is believed that the iron containing centers and/or mononuclear iron allows the transfer of electrons to activate O_2, allowing its introduction into halogenated aromatic compounds. However, the precise mechanism by which this occurs has not been determined conclusively.

a. Haloaromatics Similarly to the monooxygenases, the dioxygenases predominantly act upon haloaromatic compounds, often with broad substrate specificities. The dioxygenases comprise two or three soluble proteins that can be divided into classes on the basis of the number of components and the nature of the electron transfer centers. Two-component dioxygenases contain a flavin and an iron-sulfur cluster in the same protein. Three-component dioxygenases typically contain a flavoprotein and separate iron-sulfur protein (ferredoxin); alternatively, the electron transfer chain may be made up of both a flavoprotein and a ferredoxin (70). The terminal oxygenase component of all dioxy-

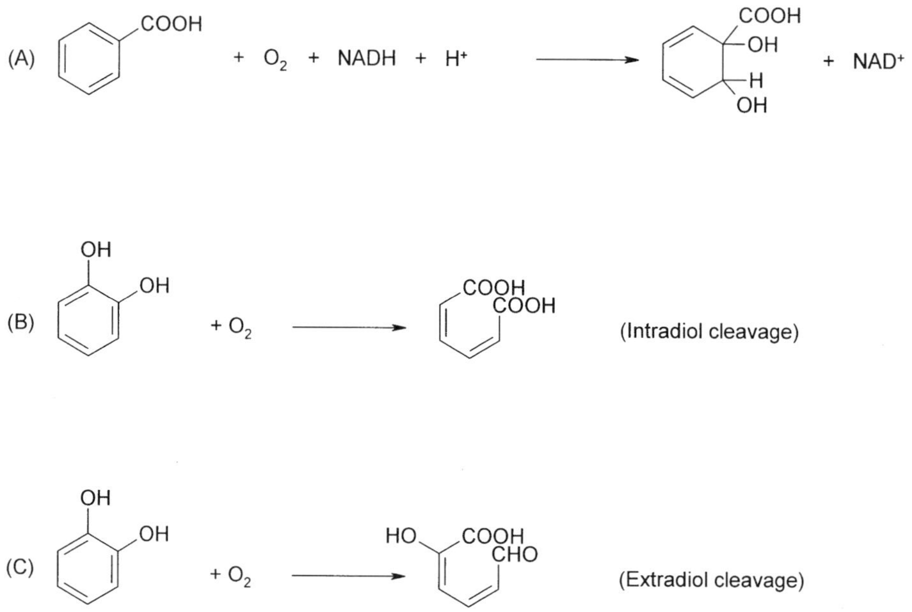

Figure 6 General mechanisms of action of various dioxygenases: (A) aromatic ring cleavage via (B) intradiol cleavage and (C) extradiol cleavage. (Redrawn from Ref. 52.)

genases contains the catalytic site that requires the presence of both O_2 and Fe^{2+} (present in the form of Rieske-type iron-sulfur complexes and as mononuclear Fe^{2+}) for the hydroxylation of halogenated aromatic compounds. Dioxygenation reactions result in the formation of a *cis*-diol product that spontaneously undergoes either HCl formation or decarboxylation coupled with chloride release, producing a catechol product (11).

Two of the best studied two-component dioxygenase systems are found in *Pseudomonas* sp. strain CBS3 and *P. cepacia* 2CBS. The dioxygenase found in strain CBS3 converts 4-chlorophenylacetate to 3,4-dihydroxyphenylacetate (71). The enzyme system contains a reductase and an oxygenase, the typical components of a two-component dioxygenase. The reductase contains a monomeric flavin mononucleotide and a (2Fe-2S) cluster. The reductase initiates the transfer of electrons from NADPH to the oxygenase component. The oxygenase component of CBS3 contains three Rieske-type iron clusters. It is also believed that the oxygenase component contains mononuclear irons for the transfer of electrons from the Rieske clusters to molecular oxygen. This would create an activated oxygenating species $[FeO_2]^+$ able to interact with the substrate (72,73). The hypothetical route of dehalogenation by the dioxygenase enzyme in strain CBS3 is shown in Fig. 7.

A similar two-component system is also found in *P. cepacia* 2CBS, containing the dioxygenase 2-halobenzoate 1,2-dioxygenase. The genes for this set of enzymes were located on a 70-kbp conjugative plasmid (74). Both the nucleotide and amino acid sequences were deduced. The terminal oxygenase component was found to contain a conserved binding motif indicating the presence of a Rieske-type (2Fe-2S) cluster (74). The reductase component of the enzyme system was found to have a putative binding site of another (2Fe-2S) cluster, and possible FAD- and NAD-binding domains (74). Interest-

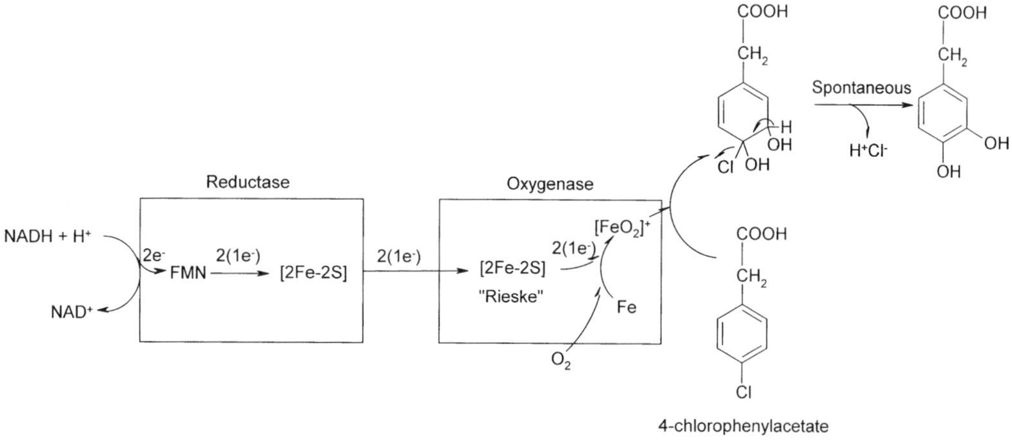

Figure 7 Proposed reaction pathway of the two-component 4-chlorophenylacetate 3,4-dioxygenase from *Pseudomonas* sp. strain CBS3. Note the presence of the [2Fe-2S] clusters in the reductase and oxygenase components. Also note the mononuclear iron (Fe) in the terminal oxygenase that reacts with O_2 to form a reactive species capable of removing chlorine from the substrate 4-chlorophenylacetate. (Redrawn from Ref. 7.)

ingly, a three-component *ortho*-halobenzoate 1,2-dioxygenase from *P. aeruginosa* 142 catalyzed the formation of catechol from 2-halobenzoates in a similar fashion to that of the two-component 2-halobenzoate 1,2-dioxygenase found in *P. cepacia* 2CBS (75). This emphasizes the typical ability of varying oxygenases to act upon a common substrate. The reaction catalyzed by the three-component dioxygenase of *P. aeruginosa* 142 and by the two-component 2-halobenzoate 1,2-dioxygenase of *P. cepacia* 2CBS is shown in Fig. 8.

The importance of continuing investigations of dehalogenating mechanisms has been realized in the case of PcpA. The enzyme encoded by *pcpA* that dehalogenates 2,6-dichloro-*p*-hydroquinone in the PCP degradation pathway (Fig. 2) in *Sphingomonas chlorophenolica* strain ATCC 39723 was originally thought to be a chlorohydrolase (76). In 1999 it was determined to be a dioxygenase (77). (Table 1); it has been shown to demonstrate novel ring-cleavage dioxygenase activity that cleaves aromatic rings with two hydroxyl groups *para* to each other (78).

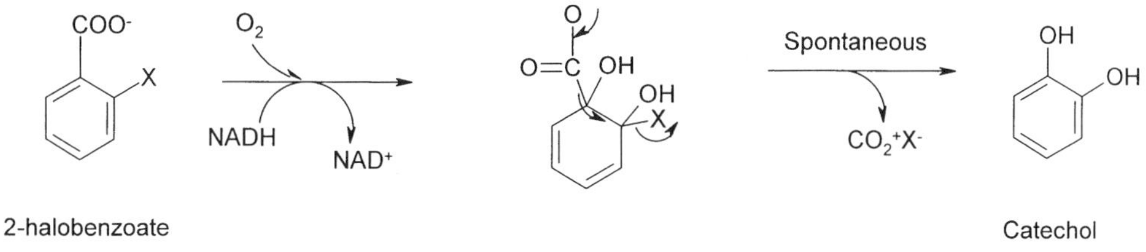

Figure 8 Reaction catalyzed by the two-component 2-halobenzoate 1,2-dioxygenase from *Pseudomonas cepacia* 2CBS and by the three-component *ortho*-halobenzoate from *Pseudomonas* sp. strain 142 for the formation of catechol from 2-halobenzoate. (Redrawn from Ref. 7.)

III. DIVERSITY AND DISTRIBUTION AMONG DEHALOGENASE-PRODUCING MICROORGANISMS

The relationship between the diversity of biological entities and the functions they perform (range, stability, resilience, and resistance) is of interest to ecologists. Diversity is thought generally to stabilize ecological functions (79). Understanding more about the diversity of dehalogenating microorganisms allows comparisons that may provide clues to the evolution and distribution of dehalogenase enzymes in the environment and may be useful in illustrating concepts in enzymology (4). The species diversity of dehalogenase-producing microorganisms is high. Different species of microorganisms can produce the same class of dehalogenase or even the same enzyme. For example, Slater et al. (12) demonstrated that the amino acid sequence for 2HAA dehalogenase coded by *dhlC* in *Alcaligenes xylosoxidans* ABIV is identical to that for the dehalogenase coded by *dehl* in *Pseudomonas putida* PP3. Poelarends et al. (80) isolated a haloalkane dehalogenase gene from a Gram-negative *Pseudomonas cichorri* strain. This gene was identical to the haloalkane dehalogenase from a strain of Gram-positive *Rhodococcus rhodochrous* NCIMB13064. Männistö et al. (81) examined bacteria from an aquifer contaminated with 2,4,6-TCP, 2,3,4,6-tetrachlorophenol (2,3,4,6-TeCP), and PCP. Of 102 isolates examined, 59 were able to degrade 2,3,4,6-TeCP and/or PCP. The authors found a diverse collection of chlorophenol-degrading bacteria belonging to the five principal lineages of the Bacteria domain. Isolates were identified as Gram-positive bacteria with high G+C content, the *Cytophaga/Flexibacter/Bacteriodes* phylum and the α-, β-, and γ-Proteobacteria. Diez et al. (82) observed the ability to dehalogenate monochloroacetate in *Pseudomonas*, *Alcaligenes*, *Agrobacterium*, *Arthrobacter*, and *Azotobacter* species.

One microorganism may produce two or three different dehalogenases, and some strains can completely degrade halogenated compounds (83). For example, *Xanthobacter autotrophicus* GJ10 constitutively synthesizes two halidohydrolases that allow it to utilize 1,2-dichloroethane. *Moraxella* sp. B produces two unrelated haloacid dehalogenases, and *Pseudomonas* sp. CBS3 produces two related haloacid dehalogenases. PCP-degrading *Sphingomonas* spp. synthesize three dehalogenases, a dehalogenating oxygenase, a reductive dehalogenase, and a dehalogenating dioxygenase (Fig. 2), which all work in concert to degrade PCP (11). A hydrolytic haloalkane dehalogenase and two 3-chloroacrylic acid dehalogenases, one specific for *cis*-3-chloroacrylic acid and the other specific for the *trans*-isomer, were synthesized by *Pseudomonas cichorii* 170 to allow the complete degradation of 1,3-dichloropropene (80).

Several strains have also been described that can dehalogenate a series of compounds. *Desulfitobacterium frappieri* PCP-1 *ortho*- and *para*-dechlorinated a variety of chlorophenols (37). *Desulfomonile tiedjei* DCB-1 *meta*-dehalogenated several halobenzoates and chlorophenols (84,85). It has been suggested that possession of more than one dehalogenase gene allows adaptability under changing environmental conditions, conferring selective advantages on the microorganism (20).

Among macroorganism ecologists there is debate as the nature of the relationship between community structure and ecological function. Ecological processes may reflect the number and/or diversity of functional groups (86). However, others argue that the ecological functions may be largely performed by a limited number of species (87). Understanding of this distinction will be important for the development of bioremediation strategies. It is possible the chances of successful bioremediation may be improved by increasing the diversity of microorganisms involved. However, it is questionable whether the introduction of multifunctional microorganisms is superior to manipulation of conditions to maximize existing microbial diversity. However, it should be cautioned that such studies will be problematic as it is difficult to manipulate diversity without affecting a number of other environmental conditions. Conversely, changes to environmental conditions also affect diversity (88).

The literature shows that there is wide diversity of bacteria that produce dehalogenases, from gram-negative to gram-positive, from aerobic to anaerobic, all found in contaminated environments widely distributed throughout the world.

A. Molecular Methods to Detect Selected Bacteria in Soil Samples

Although the isolation of microorganisms that dehalogenate and grow on xenobiotic compounds is difficult (89), many metabolically competent bacteria are present at chemically contaminated sites. The discovery of microorganisms capable of degrading halogenated compounds and the enzymes they produce is important research. Traditionally, microbial diversity has been assessed by using culture-dependent approaches (90), but these are limited as they are dependent on the ability to culture the microbes under laboratory conditions. Diversity among dehalogenating microorganisms has been demonstrated by utilizing classical microbiological techniques such as supplementing minimal media with the specific halogenated compounds (82). As our understanding of the limits of culture-dependent studies has improved, the realization of the importance of viable but nonculturable (VBNC) microorganisms has increased. Many soil microorganisms are VBNC (91), and only a fraction of the total population can be cultured in the lab (92). It is a challenge to identify microorganisms that may potentially be involved in enzymatic processes. Even if activity can be demonstrated in vitro, it is not known when or whether an enzyme will be produced in situ. Although culture-based techniques allowed a certain level of understanding, new techniques in molecular biology (Table 3; Chapter 14) have expanded our knowledge of microbial and enzyme diversity and of microbial enzyme biological characteristics (14). These molecular approaches will expand our knowledge further as increasing numbers of microorganisms and their enzymes are characterized and novel techniques continue to be developed.

Molecular biology overcomes culture-dependent limitations by detecting specific microorganisms or adding marker systems. For example, molecular approaches allow detection of microorganisms responsible for enzyme function. In 1997 El Fantroussi et al. (93) introduced *Desulfomonile tiedjei* and *Desulfitobacterium dehalogens* into nonsterile soil slurries and nested polymerase chain reaction (PCR) was used to confirm their presence and persistence. The addition of *D. tiedjei* allowed dechlorination of 3-chlorobenzoate where no activity existed before and *D. dehalogenenas* reduced the degradation time of 3-chloro-4-hydroxyphenoxyacetic acid compared to that of controls.

Alternatively, marker systems, such as the green flourescent protein (GFP), can be introduced into microorganisms to allow their detection in situ (94). The polymerase chain reaction has also been used to detect dehalogenating genes in environmental samples. Chloroaliphatics are environmentally significant compounds found in the effluent from pulp and paper mills. Fortin et al. (95) designed primers for PCR-based detection of the haloacid dehalogenase (*dhlB*) from *X. autotrophicus* and identified the gene in samples from the aerated lagoons and stabilization basins of the pulp and paper plant.

In recent years quantitative molecular approaches have been developed. Several studies have coupled PCR and most-probable-number (MPN) methods (96–98). Briefly,

Table 3 Molecular Methods for Studying Soil Microorganisms and Enzymes

Method	Measurement	Purpose
PCR of 16S rRNA genes followed by sequencing	Comparison between microorganisms to determine phlyogenetic relationships	Determination of diversity with specificity
PCR of 16S rRNA genes followed by gradient gel electrophoresis: denaturing (DGGE) or temperature (TGGE)	Allows separation of bands on gel of same size (e.g., 16s rRNA), but that may be of different sequence	Good estimate of diversity in microbial community or environmental sample
PCR of 16S rRNA genes followed by nested PCR with species specific primers	Allows determination of presence of target microorganism	Determination of specific species
PCR of specific catabolic genes	Band of same size as target sample or positive control	Determination of presence in sample
PCR of specific catabolic genes (coding for enzymes) followed by sequencing	Sequence comparison of genes	Diversity and evolutionary aspects
PCR of specific catabolic genes followed by gradient gel electrophoresis: denaturing (DGGE) or temperature (TGGE)	Allows separation of DNA on gel of same size but that may be of different sequence	Broad determination of diversity
Hybridization with labeled nucleic acid probe (of plasmid or genomic DNA, RNA, or bulk nucleic acids from environment), Restriction fragment length polymorphism (RFLP)	DNA—Southern blotting RNA—Northern blotting Direct sample—''colony,'' ''dot,'' or ''slot'' blotting Compare probed RELP patterns	Detection of target sequence; broadly used for microbial identification and study of microbial population structure and function
In situ hybridization with fluorescently labeled genetic probes (FISH), DNA, rRNA, mRNA	Visualization with epifluorescence microscopy	Detection and spatial distribution
Reverse transcriptase/PCR	Determination of mRNA for specific catabolic enzyme	Determination of enzyme activity in situ
Addition of genetic marker (GEP) linked to promoter of catabolic gene; if GEP expressed, then gene product of interest expressed	Visualization of marker denotes expression of gene of interest	Determination of gene expression in situ
Pulsed field gel electrophoresis (PFGE): genome cut with restriction endonucleases	Comparison of macro–restriction fragment fingerprints	Detailed genomic analysis of isolates, strains compared, and relationship established

Source: Refs. 219 and 220.

DNA extracts are serially diluted until the target sequence can no longer be amplified. Careful selection of primers allows this method to be species-specific (99). van Elsas et al. (100) used PCR–MPN to track the survival of *Mycobacterium chlorophenolicum* strain PCP-1 a bacterium that cannot be detected by selective plating. PCR–MPN revealed that this strain survived for longer than 14 days in soil after inoculation. This approach can also be used to characterize the effects of environmental stress. Leung et al. (97) used both PCR–MPN and traditional ^{14}C-MPN and demonstrated that *Sphingomonas* sp. UG30 cells declined after being inoculated into an agricultural soil. They reported that the PCR–MPN protocol had a detection limit of 3 colony forming units (CFU) g^{-1} dry soil.

A second quantitative approach is competitive PCR (cPCR). Briefly, a fragment of DNA, which shares homology in the primer binding sequences with the gene under study, is added to the PCR reaction and is coamplified along with the gene. This fragment competes for primers, and its amplification reflects the quantity of the target gene. Several studies have demonstrated the usefulness of this approach. In 1999 Johnsen et al. (101) reported that cPCR results correlated well to plate counts of culturable *Pseudomonas* sp. cells. In a microcosm study, Beaudet et al. (102) used competitive PCR to demonstrate that levels of *Desulfitobacterium frappieri* PCP-1, a strain capable of dechlorinating PCR to 3-chlorophenol, remained at a constant level for 21 days. Ducrocq et al. (103) were able to monitor degradative strains effectively by using competitive PCR in their toxicity study. Quantitative approaches allow temporal studies of enzyme producers and the enzymes themselves and microbial populations can be monitored as they grow and decline. By comparing enzyme activity and population dynamics, it may be possible to determine the time when enzymatic activity begins as a population develops and the duration of enzyme activity after the population declines by following the cell numbers with molecular tools. Time course studies could also provide insights into the community dynamics of specific degrading microorganisms. It is possible that the members of a microbial community responsible for a particular enzymatic reaction may change over time. Such a change has been speculated to occur in 2,4-D-degrading populations (104).

Molecular approaches can be combined with other techniques to identify dehalogenase-producing microorganisms. For example, the diversity of phospholipid composition among different species makes identification possible, and Männistö et al. (81) utilized a combination of restriction fragment length polymorphism (RFLP), 16S rRNA, and phospholipid fatty acid (PLFA) analysis to determine their diverse collection of chlorophenol-degrading bacteria.

B. Molecular Methods for Determining Microbial Diversity in Soil Communities

The determination of DNA sequences from 16S rRNA genes amplified directly from soil or sediment samples has allowed analysis of bacterial diversity in natural environments without using cultures (105). The 16S rRNA genes have sections that are conserved yet also contain sections that are variable, making them useful for identifying and comparing microorganisms. Comparisons of 16S rRNA gene sequences have revolutionized our conception of evolutionary relationships, not only between related bacteria, but through the classification of living microorganisms into the three domains Bacteria, Archaea, and Eucarya (106).

One molecular approach is the amplification of 16S rRNA genes by PCR followed by sequencing (107). Comparisons are made to known sequences to allow phylogenetic relationships to be determined. There are numerous examples in the literature making use of this approach to identify and characterize dehalogenase-producing microorganisms (108–111). The use of unique primers allows taxonomic identification of a specific genus or species. For example, LaMontagne et al. (112) examined two enrichment cultures capable of dechlorinating PCBs. Examination of the 16S rRNA genes revealed that both the *meta-* and *para*-dechlorinating enrichments were dominated by members of the genus *Clostridium.*

Community analyses of soil samples can be performed by extracting DNA and using universal 16S rRNA primers to amplify all of the 16S genes present by PCR. The resulting PCR product will contain DNA sequences of the same length that will appear as one band on an electrophoresis gel but that differ in their base composition. Further differentiation of the PCR product can be achieved by using restriction length fragment polymorphism (RFLP), denaturing gradient gel electrophoresis (DGGE), or temperature gradient gel electrophoresis (TGGE). In RFLP analysis, the amplified DNA is cut with specific restriction endonucleases, creating DNA fragments of different lengths that are separated by electrophoresis. Amplification of the 16S rRNA followed by RFLP patterns suggested at least 17 unique 2,4-D degraders from a collection of 68 isolates (113), and 4 phylotypes comprising a bromophenol-degrading consortium were determined, none of which was related to previously described aryl-dehalogenating bacteria (114). DGGE and TGGE are two approaches that allow greater resolution of PCR fragments by separating the PCR product band by using differential denaturing and temperature conditions, respectively. In 1998 Pulliam-Holoman et al. (115) described a reductive approach for analyzing the members of the microbial community that *ortho*-dechlorinate 2,3,5,6-tetrachlorobiphenyl (TCB). 16S rRNA RFLP patterns were compared for sediment microbial communities capable of degrading TCB (DEG) and nondegrading communities (NONDEG). Some bacteria RFLP patterns were unique to the DEG; others were enhanced relative to that of the NONDEG. Archaea RFLP patterns were distinct for the two communities. Microbes were removed from the sediments and grown in a defined medium to reduce the number of electron acceptors available for the TCB-degrading community. Acclimation to the new medium reduced the number of RFLP bands. Changes were noted in the RFLP patterns when different electron donors were used. The use of inhibitors also helped characterize the community. For example, sodium molybdate, which inhibits sulfate-reducing bacteria, was added, and this resulted in reduced dechlorination and diversity of RFLP bands. Kowlachuck et al. (116) used DGGE to examine differences within communities of ammonia-oxidizing bacteria from the β subdivision of *Proteobacteria.* Such approaches have tremendous potential not only to identify particular members of a dehalogenating community but to characterize their physiological properties as well.

Much can be learned about the composition of microbial communities at the microscale level with molecular tools such as fluorescently labeled genetic probes (117,118). In 1996 Mobarry et al. (119) used 16S rRNA probes to *Nitrobacter* and *Nitrosomonas* spp. to demonstrate that these members of the nitrification pathway exist in close proximity. In addition, Hodson et al. (120) have described an in situ PCR-based approach for visualizing members of a community that carry a specific gene. Probes for specific genes coding for dehalogenases can be used in Southern blotting to characterize bacterial communities, as in the detection of PCB-degrading genotypes (121). The relationship between the genes that are present and the microorganisms' dehalogenating ability can be examined. Kocabiyik and Caba (122) isolated *Pseudomonas* sp. strains capable of dechlorinating monochloroacetate. Two probes coding for 2-haloalkanoic acid dehalogenases were isolated from *Pseudomonas* sp. strain CBS3 and showed 45% DNA homology. Southern blotting with two probes for dechlorinating genes revealed that a majority of isolates did not contain sequences homologous with either probe. One probe, *dehCl*, hybridized three times as often as the other, *dehCll*. Further the *dehCl* probe was found in strains characterized as poor, moderate, or high degraders, whereas *dehCll* was only associated with poor degraders. As the researchers suggested, such studies will benefit from identification and isolation of more genes associated with a particular dehalogenating function.

Degradation of environmental chemicals is typically performed by bacterial consortia and may depend upon synergistic relationships between species. Degradation of chemicals involves complex interactions among microorganisms. For example, a Gram-positive strain, designated *DMC*, is dependent on a close association with another microbe to dehalogenate dichloromethane. The reason for this is unknown, but it is speculated that the second strain provides a growth factor needed by DMC (109). Enhanced atrazine mineralization and metabolism of a broader set of triazine compounds were observed with a two-species bacterial consortium compared to the activities of a single species capable of completely mineralizing atrazine (123). Juteau et al. (124) characterized a methanogenic consortium in a biofilm capable of degrading pentachlorophenol to phenol through a series of dechlorination reactions. Analysis with scanning electron microscopy suggested that the bacterial members of the consortia were spatially organized within the biofilm. DNA probes to specific genes could be used to investigate the spatial relationships of consortium members at the microscale.

IV. DIVERSITY AMONG DEHALOGENASE ENZYMES

The diverse nature of the dehalogenases is apparent from the information already presented. Dehalogenases clearly illustrate that enzymatic similarity and diversity exists at several levels. Certain dehalogenases have a narrow substrate specificity, including the 2HAA dehalogenases (Table 1). The presence of genetically, immunologically, and mechanistically different dehalogenases that dehalogenate the same compound illustrates the diversity of structure and function that has been observed for both the 2HAA dehalogenases and the haloalcohol dehalogenases. It is important to note, however, that there is a possibility that laboratory-induced environmental selection will result in the specialization of existing halidohydrolases (125). Many other enzymes have broad substrate ranges, such as haloalkane dehalogenase DhaA from *Rhodococcus rhodochrous* NCIMB 13064 and *Pseudomonas cichorii* 170, and the haloalkane dehalogenase from *Arthrobacter* sp. HA1, the PCP-4-monooxygenase from several strains of *Sphingomonas* sp., including sp. ATCC 39723 and *Mycobacterium* sp., and the monochloroacetate dehalogenase found in a number of different bacteria.

It has been observed there are identical or similar enzymes synthesized by different microorganisms that may be from different geographical areas. Fortin et al. (95) used primers based on the haloacetate dehalogenase gene from *Moraxella* sp. B in a PCR reaction with a sample from pulp and paper mill effluent. They amplified two dehalogenases, one identical to the haloacetate from *Moraxella* sp. and the other with 99% similarity. Enzymes identical to the DhlA enzyme from *Xanthobacter autotrophicus* have been found in *Ancylobacter aquaticus* AD20 (Fig. 3) and *Pseudomonas* sp. AD1 (126). Identical enzymes found in both Gram-positive and Gram-negative bacteria have been reported. Poelarends et al. (80) used DNA sequence analysis and determined that the haloalkane dehalogenase gene of the Gram-negative microorganism *Pseudomonas cichorii* 170 is identical to the haloalkane dehalogenase gene of the Gram-positive microorganism *Rhodococcus rhodochrous* NCIMB 13064. There is also evidence of similarities between dehalogenases from aerobic and anaerobic bacteria. The 4-chlorobenzoate-CoA-dehalogenase synthesized by *Pseudomonas* sp. CBS3 and *Arthrobacter* sp. SU dehalogenates without molecular oxygen via the production of CoA esters and may have derived from anaerobic bacteria (127). In anaerobic bacteria, degradation of aromatic acids via the production of CoA esters is a well-known process.

Some enzymes may have similar DNA sequences but be functionally different. De Souza et al. (128) described a chlorohydrolase involved in the degradation of atrazine. When the sequence was compared by using GenBank, it was found that this dehalogenase shared 41% similarity with a dechlorinating enzyme from a strain of *Rhodococcus corallinus*. However, the *R. corallinus* strain was not capable of dechlorinating atrazine.

Alternatively, functional similarity is observed between HadD and Dehlll dehalogenases in dehalogenating 2-haloalkanoic acids, without sequence similarity (12). The α/β hydrolase fold hydrolytic enzymes are known to be involved with catabolic activities. DhlA Haloalkane dehalogenase from *Xanthobacter autotrophicus* GJ10 has been compared with that of *Rhodococcus rhodochrous* NCIMB 13064 (129) and the dehCl and dehCll genes (12). Although a high degree of homology was not observed, several common important structural features were found in these α/β hydrolase fold enzymes that may be necessary for dehalogenating activity. DhlA, LinB, DhaA, and DehH1 all belong to the α/β hydrolase fold structural group of enzymes and share some sequence homology, which suggests mechanistic similarity (130). Enzyme diversity, or lack of it, may provide insights into the reaction. Highly conserved enzyme structure may reflect physiological environmental, or specificity constraints that do not permit variability.

Enzymatic function is affected by enzyme production and localized environmental conditions such as pH, moisture, and, in contaminated sites, toxic metals (131). Our understanding of this aspect of microbial enzyme ecology will be enhanced with molecular tools that allow detailed monitoring of species, genes, and mRNA synthesis.

A. Molecular Methods to Determine Enzyme Diversity

Sequencing genes allows for comparison between similar enzymes and may provide insights into their relationships and the specificity of their reactions. Schmitz et al. (132) identified an enzyme from an *Arthrobacter* sp. SU as a dehalogenase by comparing its DNA sequence to a sequence of a dehalogenase from a *Pseudomonas* species. Much can be learned about enzyme biological characteristics by examining genes encoding the enzymes. Information about enzymes at the molecular level will be invaluable in many fields, including agriculture, bioremediation, and infectious diseases caused by microorganisms. Identification of conserved regions will allow design of probes that amplify a maximal number of sequences in microorganisms. Such conserved regions may be active sites or sites necessary for cofactor utilization.

The molecular approaches utilized to determine microbial diversity can also be used to facilitate our understanding of enzymatic diversity. PCR followed by DGGE was used to show that the number of bands corresponding to [NiFe] hydrogenase genes varied between a bioreactor and a microbial mat (133). Analysis of enzymatic genes can be combined with phylogenetic information from the 16S rRNA gene. Watanabe et al. (134) examined microbial communities in a phenol-digesting activated sludge. Two phylogenetic groups were identified, and each appeared to possess a different phenol hydrolase gene. Ka et al. (135) combined the use of techniques as they grouped 2,4-D-degrading strains by fatty acid profiles with repetitive extragenic palindromic (REP) PCR analysis of *tfd* genes and revealed that 47 closely related isolates gave 30 different fingerprint patterns.

Research has illustrated the usefulness of Southern blotting and other hybridization approaches for studying enzymatic genes. With hybridization approaches a specific gene is used as a probe and relative similarities to other genes can be assessed by varying the stringency conditions during hybridization. Variation in genes among different species can also be assessed in this manner (135,136). In laboratory studies, Holben et al. (137) exposed soils to 2,4-D and found that 2,4-D degraders increased in number by using an MPN test and that hybridization of specific probes for genes in the 2,4-D degrading pathway to total genomic DNA confirmed this increase. Greater hybridization was observed with probes for the 2,4-D monooxygenase (*tfdA*) and 2,4-dichlorophenol hydroxylase (*tfdB*). However, changes in other genes known to be involved in 2,4-D degradation were not detected, suggesting these genes products were not involved. The hypothesis that degradation can be performed by different populations in different soils was supported by a further study (138). After application of 2,4-D to Saskatchewan soils, hybridization of the *tfdA* gene probe was enhanced. Soils from Michigan did not show an increase in *tfdA* hybridization; however, *Spa*, a probe from another 2,4-D-degrading species, *Sphingomonas paucimobilis*, did show an increase in hybridization.

A field study revealed that application of 2,4-D resulted in an increase of 2,4-D degraders (139), and Southern blots, using 16S rRNA probes, revealed that soil microbial communities appeared to be dominated by a limited number of strains, probably 2,4-D degraders. In a separate study, Ka et al. (135) grouped 47 bacteria able to degrade 2,4-D into separate groups on the basis of their hybridization patterns to a series of probes involved in the degradation of this herbicide. Two groups were more frequently detected in fields with greater previous 2,4-D exposure, suggesting that members of this group were superior competitors for this herbicide. These researchers further demonstrated that certain 2,4-D-degrading isolates were better competitors when added to new environments than indigenous strains (140).

Vallaeys et al. (113) characterized microbial and enzymatic diversity simultaneously. They examined a community of 2,4-D degraders by analyzing 16S rRNA genes with PCR–RFLP and *tfdA* and *tfdB* genes. A variety of 2,4-D-degrading isolates were identified by amplification of the 16S rRNA. However, the ability to utilize *tfdA* and *tfdB* primers for amplification of the respective genes was not consistent among these isolates, and two *tfdB* PCR–RFLP patterns were observed. These results suggest a diverse 2,4-D-degrading community.

There are a number of halogenated compounds for which researchers have not been able to isolate microorganisms that can grow on and degrade them, such as tri- and tetrachloroethylene, chloroform, and 1,1,1-trichloroethane. The diversity of existing dehalogenases serves as a good starting point for the evolution of enzymes able to degrade these compounds (5).

Molecular tools may assist us in developing a better understanding of dehalogenase diversity in microorganisms. While these tools will increase our knowledge of enzyme biological characteristics they should also assist in our attempts to manipulate enzymes to perform new functions (141). It is important to realize that although these new molecular methods are proving to be very useful in determining microbial and enzymatic diversity in environmental samples, there are some pitfalls and potential biases of nucleic acid extraction, PCR, and DGGE that need to be considered in the interpretation of the data generated (142) and appropriate protocols and controls must be followed.

V. MICROBIAL GENE EXPRESSION

The use of molecular probes to detect certain catabolic genes may be a valuable tool, but it is important to note that the presence of a gene does not indicate expression of that gene. Saboo and Gealt (143) used primers specific for a 507-bp region of the *pcpB* gene from PCP-degrading *Sphingomonas chlorophenolica* to assess the presence of the *pcpB* gene in isolates that could grow on 100 $\mu g\ ml^{-1}$ PCP. They observed some isolates that did not contain the predicted gene, but also found some that did. Those isolates that contained the gene had sequences with 99% similarity to the sequence from *S. chlorophenolica*. However, the isolates could not metabolize PCP. In this case, detection of this segment of the *pcpB* gene would not serve as an indicator of PCP degradative capability.

When dehalogenases are expressed, microorganisms often express large quantities of the enzymes to allow growth on a xenobiotic compound as a sole carbon source, such as dichloromethane (DCM) dehalogenase (141). Increased haloalkanoic dehalogenase expression has been associated with gene duplication, decryptification of previously cryptic genes, and gene acquisition via plasmid vectors (4). Overexpression of haloacid dehalogenase (DhlB) from mutants of *Xanthobacter autotrophicus* was the result of a small DNA fragment insertion upstream of the *dhlB* gene (19). Many of the dehalogenases reported to date are inducible, and although haloalkanoic dehalogenases are produced both constitutively and inducibly, induction appears to be the main method of expression. However, constitutive dehalogenases have been reported in dehalogenating strains of *Moraxella* spp. (191), *Xanthobacter autotrophicus*, and *Ancylobacter aquaticus* (DhlA and DhlB) (19) (see Fig. 3), and *Arthrobacter* (144). If a microorganism produces more than one dehalogenase, individual changes in the amounts produced have occurred in response to changes in chemical concentration of the halogenated compound in the outer environment (4).

The development of mRNA probes may be a useful technique to determine specific microbial enzyme activity. Relating mRNA synthesis to specific enzyme activity by using mRNA/DNA hybridization followed by PCR (145) would be interesting to explore. Although utilization of this technique can be limited in situ, much could be learned from laboratory systems. This could provide insights into the environmental conditions and other factors, such as chemical concentrations, that influence gene expression. More information about mRNA probing methods can be found in a review by Gottschal et al. (146). Studies of environments that had previous exposure to halogenated compounds and those without prior exposure (147,148) illustrate the need for quantitative approaches. In pristine soils or those with no previous exposure to 2,4-D, the isolation of stable 2,4-D-degrading microorganisms is difficult (113,149). It would be intriguing to quantify the levels of dehalogenase genes and gene products in these environments. One would expect that in cases like this lower levels of dehalogenase genes and gene products would be observed. The use and development of molecular diagnostics with respect to bioremediation are discussed in an excellent review by Stapleton et al. (150), which includes three case studies.

VI. EVOLUTION OF DEHALOGENATING ENZYMES

How did the genes that code for dehalogenases evolve and become distributed in the environment? It is possible that the dehalogenases are at an early stage of evolution driven by the use of large quantities of both natural and synthetic halogenated organic compounds in the environment over the last 50 years. Are there different evolutionary explanations for various dehalogenases? Given the information that we have determined to date, there are no absolute answers to these questions. Four possible explanations for dehalogenase evolution have been suggested (Table 4).

Since there is a limited fossil record for microorganisms, molecular methods may be the only effective means for resolving evolutionary relationships among bacteria, and the use of 16S rRNA gene sequences for determining phylogenies has been of great value (105). However, there is more to the study of bacterial evolution than the compilation of these phylogenies, and the role of recombination and gene transfer needs to be considered, particularly in bacteria, in which these events permit rapid adaptation to environmental pressures (105).

In discussing long-term evolutionary events, consideration of the original environment that the bacteria had to survive in is important (105). Although there is considerable evidence to demonstrate that many organohalogens are produced naturally in the environment, there is little evidence to suggest that levels produced would exert the selection pressure necessary for microorganisms to evolve mechanisms to degrade them. In addition, the constitutive production of dehalogenases suggests that their evolution may have been recent, since it would take time to develop the type of regulatory mechanisms that are seen with more evolved enzyme systems that can metabolize sugars and amino acids in common metabolic pathways such as the tricarboxylic acid (TCA) cycle. It would not be metabolically or bioenergetically beneficial for microorganisms to produce a dehalogenase continuously (even at low levels) without the presence of high levels of halogenated compounds.

A. Mechanisms of Genetic Adaptation

Studies have demonstrated there are some broad-substrate enzymes able to metabolize halogenated compounds that are active toward other nonhalogenated forms (e.g., monooxygenases or glutathione transferases), or toward other halogenated compounds (e.g., halo-

Table 4 Possible Routes of Dehalogenase Evolution

1. Organisms adapt to a halogenated compound due to the natural presence of the compound in the environment.
2. The current functional degradative pathway was formed by a set of enzymes which evolved for a different function, possibly utilization of a naturally-produced organohalogen.
3. An industrially emitted halogenated compound may have been present at high concentrations in some areas, which stimulated rapid selection of genetically-modified strains that degrade the particular compound.
4. Another synthetic compound other than the one of interest may have played a role in creating a selection pressure for dehalogenating activity towards the compound of interest.

Source: Ref. 4.

alkane dehalogenase DhlA). The basic compounds, resulting from halogenated compounds after the halogen moieties have been removed, are metabolized by pathways that have evolved to channel metabolites into a central metabolic cycle (e.g., TCA). Many of the dehalogenases provide the crucial first step of halogen removal to allow further catabolism of the compound and likely initially evolved for natural, particularly monosubstituted halogenated compounds. It is possible that further adaptation to metabolize xenobiotic oganohalogens may have occurred with a few additional mutations (151).

A distinction should be made between vertical and horizontal evolutionary processes that may be involved in the evolution of dehalogenases. Vertical processes lead to a divergence in DNA sequence in daughter cells, which is due to an accumulation of mutations. The mutations could be single-base pair changes or larger changes due to deletions, duplications, or slipped strand mispairing (152). Horizontal processes are those that cause an exchange of DNA sequences between the genome of two different microorganisms, or between different DNA molecules (e.g., between chromosome and extrachromosomal elements) within one microorganism. Horizontal transfer of DNA may occur by recombination or by other mechanisms of gene exchange such as conjugation, transduction, or transformation. Conjugation has been observed to occur in both in vitro and in vivo settings between very distantly related bacterial species (153). Plasmid encoding of genes that are under recent evolutionary pressure has been observed in other systems, such as resistance to antibiotics and heavy metals, and adaptation to contaminants is evidenced by selective enrichment of antibiotic- and metal-resistance microorganisms. It has been observed over time that horizontal gene exchange plays an important role in adaptation of microorganisms to xenobiotic compounds (152,154).

B. Horizontal Gene Transfer

Many genes coding for catabolic enzymes are found on plasmids, and the dehalogenases are no exception (Table 1). Greater numbers of plasmid-containing bacteria have often been isolated from polluted areas than from pristine control sites (155). Many of the genes responsible for dehalogenation are plasmid-borne, including dehalogenase genes for dichloromethane, 4-chlorobenzoic acid, and 2,4-dichlorophenoxyacetate (156), and many of the haloalkane dehalogenases (4). In 1994 Top et al. (157) used genetic conjugation to demonstrate that soils exposed to 2,4-D contained a diverse variety of plasmids with degradative genes. However, it is important to remember that the same degradative genes may be located on plasmids or in the chromosome in different strains (135). Even when located on the chromosome, chlorobenzoate dehalogenation genes, for example, have been observed on transposons (158), or in chromosomal gene clusters as for chlorobiphenyl degradation (156). Other microbial degradative pathway genes are typically clustered in operons and often on mobile elements (154). The mobile genetic element DEH has been characterized; it contains one of the 2HAA dehalogenases (*dehl*) and a regulatory component (22,159,160). It has also been established that bacterial insertion sequences are involved in horizontal gene transfer (154). This hypothesis that dehalogenation ability could be conferred to other microorganisms by gene transfer is supported by examples of all of these mechanisms.

Transfer of plasmid-encoded degrading genes has been documented in microcosm studies (158). The *dhlC* gene transferred from *Alcaligenes xylosoxidans* ABIV to *Pseudomonas fluorescens* and other soil bacteria in sterile soil microcosms (161), and transfer was observed for the *dhaA* gene (located on a 100-kbp plasmid) between *Rhodococcus* sp. strains (162). However, determination of gene flow under natural environmental conditions can be challenging, and there is little information on the frequency and importance of in situ gene transfer (155). Ka and Tiedje (112,163) isolated two strains, *Alcaligenes paradoxus* 2811P and *Pseudomonas pickettii* 712, from the same location but at different times. Each contained plasmids for 2,4-D degradation. The finding that the plasmids were similar suggested horizontal gene transfer. Further, an interesting study by De Souza et al. (123) examined five different species of atrazine-degrading bacteria from several locations across the United States. They sequenced three genes, atrazine chlorohydrolase, hydroxyatrazine ethylaminohydrolase, and *N*-isopropylammelide isopropylaminohydrolase, and found high degrees of similarity (>99%) among these bacteria. They suggested the genes may have a common origin and may also be transmissible by plasmids. Verhagen et al. (148), examined the *dhlA* gene, which codes for a haloalkane dehalogenase enzyme that is involved in the degradation of the nematocide *cis*-1,3-dichloropropene (DCPe). They demonstrated that some of the strains carried *dhlA* on transferable plasmids. Transconjugants that not only contained the *dhlA* gene but were capable of degrading DCPe were created. Researchers have reported much lower G+C ratios for some dehalogenase structural genes, such as *dhlA* (164), *hld IVa* (162,165), 2,4,5-trichlorophenoxyacetate-degrading genes (166), and the regulatory component of *dehl*; the lower ratios suggest the genes originated in other species and were transferred to *Pseudomonas* species later in their evolution (12). The diverse range of the five mechanistic groups of hydrolytic haloacid hydrolases with activity toward halopropionates (Group B, Table 2) illustrates convergent evolution of enzymes. Their geographical pattern of distribution suggests the particular enzymes have evolved independently and horizontal gene transfer has then distributed the genes among different species. Kohler-Staub et al. (23) compared dichloromethane dehalogenases from four different bacterial isolates that showed broadly similar substrate ranges, immunological properties, and identical N-terminal amino acid sequences. Similarity suggests either vertical evolution from a common ancestor or horizontal transfer of the coding sequence. Although the first to be cloned was found to be chromosomally encoded, the others were plasmid-encoded.

Moraxella sp. B produces two unrelated haloacid dehalogenases, whereas *Pseudomonas* sp. CBS3 produces two related haloacid dehalogenases. Comparisons of the *dehCl* gene (227 amino acids) and the *dehCll* gene (229 amino acids) from *Pseudomonas* sp. CBS3 resulted in a 45% nucleotide sequence homology, which corresponded to a 37.5% amino acid sequence identity and greater than 70% amino acid similarity. The high degree of similarity is interesting with respect to evolution. Slater et al. (12) suggested two possibilities: first, a common origin from an ancestral gene, in which the observed pattern may have resulted from gene duplication in the ancestral species followed by separate, but parallel evolutionary events, or, second, possible parallel gene evolution in another host microorganism, followed by transfer to CBS3.

Hübner et al. (167) studied *Burkholderia cepacia* AC 1100, a strain that can utilize 2,4,5-trichlorophenoxyacetic acid (2,4,5-T) as a sole source of carbon and energy. They demonstrated that several genes are involved in the degradation of 2,4,5-T. Five replicons were identified through pulse-field gel electrophoresis of the genomic DNA. Degradative genes were found on different replicons, indicating that the capacity to degrade 2,4,5-T arose through several independent gene transfers.

The study of the catabolic genes of 2,4-D-degrading bacteria provides some interesting information about DNA rearrangements. A comparison of the genes of 32 degraders demonstrated that these microorganisms appear to have pieces of the *tfd* or *tfd*-like genes in various combinations and positions on the genome like pieces of a mosaic (168). The microorganisms had a high diversity of these genes, some of which were similar in sequence to the well-studied *tfd* genes on the catabolic plasmid pJP4 and others that were dissimilar. It was shown that each gene cassette in the 2,4-D pathway could be recruited separately and independently of each other (169), supporting the hypothesis that they may have evolved as part of separate pathways and were modified and recombined at a later time in response to selective pressure from 2,4-D (113). Plasmid pJP4, which is distributed globally, appears to be just one of the possible mosaic constructions that allow 2,4-D degradation to occur. This movement of genetic elements in a modular fashion has been seen in other pathways, including the orders of dehalogenation genes for 4-chlorobenzoate degradation (132), and seems to be a plasmid biological and evolutionary characteristic (153). Another interesting point is that gene duplication has been observed for different genes (*tfdA*, *tfdD*, and *tfdC*) of the 2,4-D degradation pathway on pJP4. Studies involving the DEH transposon carrying the *dehl* gene and its regulatory unit $dehR_1$ demonstrated that when transposed to target DNA, elements of sizes varying from 6 to 13 kb were observed. Major rearrangement of DNA was also observed with mutants containing remnants of DEH that had significantly variable restriction enzyme digest patterns (22). Results from a study using Gram-positive *Rhodococcus rhodochrous* NCIMB 13064 indicated genomic rearrangements that may have occurred in the region of the haloalkane dehalogenase locus (129).

Other genetic rearrangements, such as the presence of a bacterial insertion sequence (IS) in the plasmid pJP4 from *Ralstonia eutropha* JMP134, may indicate a possible role of the IS in the formation of the *tfd* pathway genes (170). Another study with *Pseudomonas* sp. PP1 (which could not metabolize chlorinated compounds) in continuous culture with 2,2-dichloropropionic (2DCPA) acid resulted in a mutant with two differently mechanistic dehalogenases that allowed the bacteria to degrade the 2DCPA. Cryptic dehalogenase genes were possibly activated by selective pressure (12).

In contrast, genes for PCP mineralization that have been identified, cloned, and sequenced for a number of bacteria capable of PCP degradation are interesting in that they are not found on the plasmid (if the bacteria have one, e.g., *Sphingomonas chlorophenolica* sp. ATCC 39723) nor together in an operon on the chromosome. The *pcpB* gene that encodes the PCP-4-monooxygenase is not associated with any other dechlorination function, and Lange et al. (171) speculated that the isolated nature of the gene in relation to other dechlorinating functions could be attributed to a general detoxification mechanism in the cell.

C. Vertical Evolution Through Mutations

The ability of dehalogenases to degrade recently introduced compounds in the environment could be due to the mutation of preexisting enzymes. Examples of single-site mutations that can alter substrate specificities of enzymes have been recorded (154). The stress of chemical pollutants, such as organohalides, may stimulate error-prone DNA replication and accelerate DNA evolution. As with many monooxygenases, PCP-4-monooxygenase has a very broad substrate range, but PCP has been shown to be a poor substrate for this enzyme (65). Thus, it has been suggested that PCP-4-monooxygenase could have evolved from an adaptation of a previously existing flavin monooxygenase enzyme in order to be able to degrade PCP, a recently introduced environmental contaminant (11). Another example is the repositioning of the catalytic triad of haloalkane dehalogenase. Movement of the catalytic aspartate residue, which was confirmed by creation of a double mutant enzyme, may have been an important event in the adaptation of DhlA to dehalogenate 1,2-dichloromethane (130). This enzyme, a member of the α/β hydrolases, has an anomolous positioning of the third member of the catalytic triad (Asp260). This repositioning was critical for creation of an enzyme able to dehalogenate chlorinated substrates (141). It is also thought that point mutations of duplicated genes are responsible for the evolution of the ability to broaden the range of substrates that can be degraded (172).

D. Are Some Dehalogenases Ancient Enzymes?

There are two hypotheses regarding the time frame of evolution of dehalogenases. Dehalogenases may be new enzymes, evolved as a result of selective pressure resulting from the input of halogenated compounds into the environment in the last five decades. One piece of evidence is the observation that the activity of many dehalogenases are slow (173) and microorganisms often express large quantities of these enzymes to allow growth on a xenobiotic compound as a sole carbon source (141). So, in terms of the functions dehalogenases undertake, they may not be ideally suited for their roles (i.e., more adaptation may be necessary). For example, the low observed specific activity of dichloromethane dehalogenases, which resulted from a slow reaction rate (although good binding capacity was exhibited), suggested that they may be at an early stage of evolution (4). More evidence suggesting recent evolution of dehalogenases is that a number of environmentally important chlorinated compounds, including chloroform, carbon tetrachloride, 1,1-dichloroethane, 1,1,1-trichloroethane, the three dichloroethylenes, trichloroethylene, tetrachloroethylene, 1,2-dichloropropane, and 1,2,3-trichloropropane, have not demonstrated the ability to support growth under aerobic conditions (5). It is possible that although dehalogenation mechanisms have evolved for a number of halogenated compounds to date, perhaps it is too early in evolutionary terms for enzyme systems to have evolved for every one.

The development of a regulatory system is suggestive of a more evolved enzyme system. The dehalogenases that are produced constitutively may indicate enzymes that have evolved more recently. The suggestion that the evolution of haloalkane dehalogenase DhlA from *X. autotrophicus* (19) is possibly recent is supported by the fact that the *dhlA* gene product is constitutively expressed (Fig. 3). The evolution of a regulatory system requires a regulatory protein that recognizes and binds the substrate, but no evidence of such a protein has been observed for the bacteria known to degrade 1,2-dichloroethane (5). Studies on adaptation of DhlA to novel substrates indicated that the generation of short repeats in the N-terminal part of the cap domain plays an important role in adaptation that suggests recent evolution, possibly from enzymes involved in degradation of naturally occurring organohalides (19), since many of the naturally produced halogenated compounds are low-molecular-weight aliphatics (e.g., dibromomethane and methyl chloride).

However, evidence to support the second hypothesis that dehalogenases may be evolutionarily old enzymes has also been forthcoming. For example, the fact that haloacids do occur naturally and the diversity of related haloacid dehalogenases that is observed suggest that significant evolutionary divergence has occurred (19). The dichloromethane dehalogenase from *Methylobacterium* sp. strain DM4 has a regulatory protein, and expression is efficiently induced by dihalomethanes and may have developed from an evolutionarily older glutathione transferase (5).

Vallaeys et al. (169) used a phytogenetic analysis approach to demonstrate diverse and divergent *tfd*-like genes from 2,4-D-degrading bacteria, which suggest an ancient origin of the genes. As well, these 2,4-D catabolic enzymes were shown to be related to other catabolic enzymes with different substrate specificities, particularly ones that degrade natural, nonhalogenated substrate derivatives.

There is a possible link between group 1 and group 2 haloacid dehalogenases in the critical regions of the active sites that would suggest a deep and conserved evolutionary link within this region of the proteins (12), but this has been questioned by Janssen et al. (19) and will require further study to confirm this hypothesis. Sadowsky et al. (174) found that three genes involved in the degradation of atrazine in the species *Pseudomonas* sp. strain ADP, *atzA* (hydrolytic dechlorination), *atzB* (hydroyatrazine deamination), and *atzC* (aminohydrolase), shared a conserved region of metal-coordinating histidines. They suggested these genes may have diverged from a common ancestor.

Dehalogenase evolution is an area that still has more questions than answers at the present time. However, with the information forthcoming from current studies, and the possibilities afforded by new molecular techniques, answers to the evolutionary puzzle may be determined in the near future.

VII. BIOTECHNOLOGICAL APPLICATIONS

The understanding of dehalogenation reactions catalyzed by microorganisms through fundamental studies has provided a basis for developing industrial and environmental applications (20). The technological applications of microbial dehalogenation can be classified into two broad categories: the commercial generation of halogenated synthetic intermediates or synthesis of novel halocompounds and the detoxification and degradation of halogenated compounds in the environment.

Derivatives of chlorinated propionic acid are important building blocks in the synthesis of chiral agrochemicals, such as Dichloroprop and Fluazifop (175). The potential exploitation of the properties of the various dehalogenases in biotransformations will become economically attractive if the microorganisms themselves or crude preparations of them can be used as cheap catalysts. Certain dehalogenases are highly stereospecific, for example, the 2HAA hydrolytic dehalogenase, which selectively dehalogenates D-isomeric substrates such as D-2MPCA. One commercial application utilizing this property is the development of a novel herbicide using L-2MPCA as the starting material. The inexpensive racemic mixture of D,L-2MCPA only had half of the biologically active material, so an initial treatment to remove the unwanted D-2MCPA was developed (12). Another example of exploiting the enantioselectivity of dehalogenases is solvent engineering, employing the solvent-mediated inversion of enzyme enantioselectivity. Vyazmensky and Geresh (175) investigated the use of intact cells, disrupted cells, and crude extracts of *Pseudomonas putida* 109 in polar organic solvents such as DMSO with 2-bromopropionic acid as a substrate. Inversion of the product configuration is observed when the reaction occurs in an aqueous system, but the authors observed no inversion of the configuration in DMSO (in the presence of stoichiometric amounts of water). This type of solvent engineering to produce preferential isomers may be an area of applied enzymology to be developed further. A 1999 review by Swanson (176) on the application of dehalogenases for industrial-scale biocatalysis is recommended for additional information.

The addition of microorganisms to enhance a specific biological activity (bioaugmentation) has been practiced for years in agriculture, forestry, and wastewater treatment, and this technology has been explored for the bioremediation of contaminated soil

(177,178). Bioremediation has been developed and used to transform or mineralize halogenated contaminants into less harmful metabolites that can be integrated into biogeochemical cycles. Numerous examples of bioremediation of contaminated soil and contaminated waters exist in the literature (20; Chapter 19). Many are described for bioreactors and others for in situ applications. There have been a number of successes using bioaugmentation in the field with chlorinated compounds where microbial populations capable of completely degrading contaminants were not present (10), but that is not always the case. Alternatively, the process of biostimulation, creating an optimal environment to enhance the intrinsic bioremediation potential of indigenous microorganisms, has been demonstrated to be effective in some chlorinated pesticide-contaminated soils (179). Other applications of bacterial inoculants include the detoxification of compounds (a) in the rhizosphere mediated by plant-growth-promoting rhizobacteria (180) and (b) by bacteria that could rapidly detoxify pesticides as protectants for susceptible plant species (181).

The immobilization of whole cells or enzymes (the entrapment of biocatalysts within a porous polymeric matrix such as alginate, carrageenan, polyacrylamide, or polyurethane) has been a topic of research since the early 1960s (182). Studies on the degradation of chlorinated phenols using immobilized cells demonstrated that immobilized cells can degrade these compounds at higher concentrations, more effectively, and with a shorter lag phase when compared with the same density of freely suspended cells (183,184). Immobilized *Sphingomonas* sp. UG30 cells have also proved to be effective for use in soil microcosms to degrade PCP (185). Immobilization has been used as a tool for retaining enzymes in bioreactors, allowing continuous operation of enzymic processes that could be useful for degradation of halogenated wastewater. The immobilization process may also modify the enzyme activity in an advantageous manner, including increasing the pH range of activity or increasing the metabolic activity of enzymes as compared to when they are freely suspended (182).

Enzymes have been used for a number of biotechnological applications; one important requirement for many is that they are stable under a variety of conditions. Some testing of stability has been done on certain dehalogenases. Diez et al. (82) demonstrated high stability of monochloroacetate (MCA) dehalogenase activities against thermal denaturation, pH variation from 6 to 10, and solvents at 1- or 10-m*M* concentrations. If required for biotechnological applications, the stability of these enzymes would be a useful characteristic.

VIII. CONCLUSIONS

As our database of information about dehalogenases increases, more light will be shed on enzyme pathways and evolution. New methods, from rapid ways to screen dehalogenation reactions (186) to developments in molecular procedures, will allow confirmation or nullification of existing hypotheses and development of new hypotheses. To increase the usefulness of dehalogenase enzymes, potential enhancement of enzymatic activity or increased substrate range through genetic manipulation is being investigated. The following are some of the approaches (a) elimination of the necessity for cofactors in those dehalogenation reactions that require them; (b) achievement of a significant increase in enzymatic activity that translates to the field; (c) expansion of the substrate range of existing pathways through the recruitment of isofunctional enzymes (horizontal expansion) and the grafting of additional enzyme activities (vertical expansion) (58); (d) restructuring of existing pathways to prevent the generation of deleterious metabolites; (e) construction of new metabolic activities by the patchwork assembly of genes from different microorganisms that encode desirable enzymatic conversions; and (f) construction of new sequences by enhancing natural genetic exchanges among dissimilar microorganisms with imposed selection pressure (molecular breeding) (155). It may be possible to realize these beneficial alterations and optimize enzymes with recent molecular biology techniques such as site-directed mutagenesis (187) or by DNA shuffling of existing dehalogenases. Enhanced degradation of polychlorinated biphenyls by using directed evolution of the biphenyl dioxygenase has been successful (58). More information on directed evolution of enzymes can be found in two 1997 reviews (188,189). The discovery of new enzymes with these desirable properties or with enzymes that will become precursors of new genetic modifications is also still important.

The dehalogenases are a group of relatively well-studied enzymes because of the toxic effects of halogenated compounds in the environment. The knowledge garnered to date has provided a glimpse of the possible recent evolution of enzymes and has provided insight on microbial defense systems against toxic xenobiotics. Recent innovations in molecular biology have provided, and will continue to provide, important information on expression, diversity, and evolution. Larger sequence databases will allow more useful comparisons and recognition of important domains.

A better understanding of dehalogenases, and enzyme biological characteristics in general, will be useful in biotechnological applications, from industrial processing to bioremediation of chemically contaminated environmental sites.

ACKNOWLEDGMENTS

Research by H.L. and J.T.T. is supported by NSERC operating and strategic grants. They are also funded under a Canada Foundation for Innovation (CFI) Institutional Award on Biological Systems for Terrestrial and Space Applications.

REFERENCES

1. CE Castro. Environmental dehalogenation: Chemistry and mechanism. Rev Environ Contam Toxicol 155:1–67, 1993.
2. GW Gribble. The natural production of chlorinated compounds. Environ Sci Technol 28: 310A–319A, 1994.
3. GW Gribble. Naturally occurring organohalogen compounds—a comprehensive survey. Prog Organization Nat Products 68:1–498, 1996.
4. JD Allpress, PC Gowland. Dehalogenases: Environmental defence mechanism and model of enzyme evolution. Biochem Educ 26:267–276, 1998.
5. F Pries, JR van der Ploeg, J Dolfing, DB Janssen. Degradation of halogenated aliphatic compounds: The role of adaptation. FEMS Microbiol Rev 15:279–295, 1994.
6. J Dolfing, DB Janssen. Estimates of the Gibbs free energies of formation of chlorinated aliphatic compounds. Biodegradation 6:237–246, 1994.
7. S Fetzner, F Lingens. Bacterial dehalogenases: Biochemistry, genetics and biotechnological applications. Microbiol Rev 58:641–685, 1994.
8. S Fetzner. Bacterial dehalogenation. Appl Microbiol Biotechnol 50:633–657, 1998.

9. MM Häggblom. Microbial breakdown of halogenated aromatic pesticides and related compounds. FEMS Microbiol Rev 103:29–72, 1992.
10. MD Lee, JM Odom, RJ Buchanan Jr. New perspectives on microbial dehalogenation of chlorinated solvents: Insights from the field. Annu Rev Microbiol 52:423–452, 1998.
11. SD Copley. Diverse mechanistic approaches to difficult chemical transformations: Microbial dehalogenation of chlorinated aromatic compounds. Chem Biol 4:169–174, 1997.
12. JH Slater, AT Bull, DJ Hardman. Microbial dehalogenation of halogenated alkanoic acids, alcohols and alkanes. Adv Microb Physiol 38:133–176, 1997.
13. J Dec, J-M Bollag. Dehalogenation of chlorinated phenols during binding to humus. In: TA Anderson, JR Coats, eds. Bioremediation Through Rhizosphere Technology. ACS Symposium Series 563. Washington, DC: American Chemical Society, 1994, pp 102–111.
14. RL Sinsabaugh. Enzymic analysis of microbial pattern and processes. Biol Fertil Soils 17: 69–74, 1994.
15. WA Dick, MA Tabatabai. Significance and potential uses of soil enzymes. In: FB Metting ed. Soil Microbial Ecology. New York: Marcel Dekker, 1993, pp 95–126.
16. TW Speir. Soil phosphatase and sulfatase. In: DJ Ross, RG Burns, eds. Soil Enzymes. Academic Press, New York, 1978, pp 198–235.
17. K Miyauchi, S-K Suh, Y Nagata, M Takagi. Cloning and sequencing of a 2,5-dichlorohydroquinone reductive dehalogenase gene whose product is involved in degradation of y-hexachlorocyclohexane by *Sphingomonas paucimobilis*. J Bacteriol 180:1354–1359, 1998.
18. K Ramanand, MT Balba, J Duffy. Reductive dehalogenation of chlorinated benzenes and toluenes under methanogenic conditions. Appl Environ Microbiol 59:3266–3272, 1993.
19. DB Janssen, F Pries, JR van der Ploeg. Genetics and biochemistry of dehalogenating enzymes. Annu Rev Microbiol 48:163–191, 1994.
20. DJ Hardman. Biotransformation of halogenated compounds. Crit Rev Biotechnol 11:1–40, 1991.
21. SS Cairns. The cloning and analysis of *Rhizobium* dehalogenase genes. PhD thesis, University of Leicester, Leicester, 1994.
22. AW Thomas, JH Slater, AJ Weightman. The dehalogenase gene *dehl* from *Pseudomonas putida* PP3 is carried on an unusual mobile genetic element designated *DEH*. J Bacteriol 174:1932–1940, 1992.
23. D Kohler-Staub, S Hartmans, R Gälli, F Suter, T Leisinger. Evidence for identical dichloromethane dehalogenases in different methylotrophic bacteria. J Gen Microbiol 132:2837–2843, 1986.
24. WW Mohn, JM Tiedje. Microbial reductive dehalogenation. Microbiol Rev 56:482–507, 1992.
25. C Egli, T Tschan, R Scholtz, AM Cook, T Leisinger. Transformation of tetrachloromethane to dichloromethane and carbon dioxide *by Acetobacterium woodii*. Appl Environ Microbiol 54:2819–2824, 1988.
26. PE Jablonski, JG Ferry. Reductive dechlorination of trichloroethylene by the CO-reduced CO dehydrogenase enzyme complex from *Methanosarcina thermophila*. FEMS Microbiol Lett 96:55–60, 1992.
27. C Holliger, G Schraa, E Stupperich, AJM Stems, AJB Zehnder. Evidence for the involvement of corrinoids and factor F_{430} in the reductive dechlorination of 1,2-dichloroethane by *Methanosarcina barkeri*. J Bacteriol 174:4427–4434, 1992.
28. CJ Gantzer, LP Wackett. Reductive dechlorination catalyzed by bacterial transition-metal coenzymes. Environ Sci Technol 25:715–722, 1991.
29. CS Criddle, JT DeWitt, PL McCarty. 1990. Reductive dehalogenation of carbon tetrachloride by *Escherichia coli* K-12. Appl Environ Microbiol 56:3247–3254, 1990.
30. MJ Dybas, GM Tatara, CS Criddle. Localization and characterization of the carbon tetrachloride transformation activity of *Pseudomonas* sp. strain KC. Appl Environ Microbiol 61:758–762, 1995.

31. FW Picardel, RG Arnold, H Couch, AM Little, ME Smith. Involvement of cytochrome in the anaerobic biotransformation of tetrachloroethane by *Shewanella putrefaciens* 200. Appl Environ Microbiol 59:3763–3770, 1993.
32. FW Picardel, RG Arnold, BB Huey. Effects of electron donor and acceptor conditions on reductive dehalogenation of tetrachloroethane by *Shewanella putrefaciens* 200. Appl Environ Microbiol 61:8–12, 1995.
33. JE McGrath, CG Harfoot. Reductive dehalogenation of halocarboxylic acids by the phototrophic genera *Rhodospirillum* and *Rhodopseudomonas*. Appl Environ Microbiol 63:3333–3335, 1997.
34. WB Jakoby, WH Habig. Glutathione transferases, p. 63–94. In: WB Jacoby, (ed.) Enzymatic Basis of Detoxification. Vol. 2. New York: Academic Press, 1980.
35. C Holliger, G Wohlfarth, G Diekert. Reductive dechlorination in the energy metabolism of anaerobic bacteria. FEMS Microbiol Rev 22:383–398, 1999.
36. X Maymo-Gatell, T Anguish, SH Zinder. Reductive dechlorination of chlorinated ethenes and 1,2-dichloroethene by ''*Dehalococcoides ethenogenes*'' 195. Appl Environ Microbiol 65:3108–3113, 1999.
37. D Dennie, I Gladu, F Lépine, R Villemur, J-G Bisaillon, R Beaudet. Spectrum of the reductive dehalogenation activity of *Desulfitobacterium frappieri* PCP-1. Appl Environ Microbiol 64: 4603–4606, 1998.
38. JK Magnuson, RV Stern, JM Gosset, SH Zinder, DR Burris. Reductive dechlorination of tetrachloroethene to ethene by a two-component enzyme pathway. Appl Environ Microbiol 64:1270–1275, 1998.
39. MH Smith, SL Woods. Regiospecificity of chlorophenol reductive dechlorination by vitamin B_{12s}. Appl Environ Microbiol 60:4111–4115, 1994.
40. DL McCarthy, S Navarrete, WS Willett, PC Babbitt, SD Copley. Exploration of the relationship between tetrachlorohydroquinone dehalogenase and the glutathione S-transferase superfamily. Biochemistry 35:14634–14642, 1996.
41. CS Orser, J Dutton, C Lange, P Jablonski, L Xun, M Hargis. Characterization of a *Flavobacterium* glutathione-S-transferase gene involved in reductive dechlorination. J Bacteriol 175: 2640–2644, 1993.
42. L Xun, E Topp, CS Orser. Glutathione is the reducing agent for the reductive dehalogenation of tetrachloro-*p*-hydroquinone by extracts from a *Flavobacterium* sp. Biochem Biophys Res Commun 182:361–366, 1992.
43. L Xun, E Topp, CS Orser. Purification and characterization of a tetrachloro-*p*-hydroquinone reductive dehalogenase from a *Flavobacterium* sp. J Bacteriol 174:8003–8007, 1992.
44. V Romanov, RP Hausinger. NADPH-dependent reductive *ortho* dehalogenation of 2,4-dichlorobenzoic acid in *Corynebacterium sepedonicum* KZ-4 and coryneform bacterium strain NTB-1 via 2,4-dichlorobenzoyl coenzyme A. J Bacteriol 178:2656–2661, 1996.
45. DA Abramowicz. Aerobic and anaerobic biodegradation of PCBs: A review. Crit Rev Biotechnol 10:241–251, 1990.
46. S Ni, JK Fredrickson, L Xun. Purification and characterization of a novel 3-chlorobenzoate-reductive dehalogenase from the cytoplasmic membrane of *Desulfomonile tiedjei* DCB-1. J Bacteriol 177:5135–5139, 1995.
47. B Bouchard, R Beaudet, R Villemur, G McSween, F Lépine, and JG Bisaillon. Isolation and characterization of *Desulfitobacterium frappieri* sp. nov., an anaerobic bacterium which reductively dechlorinates pentachlorophenol to 3-chlorophenol. Int J System Bacteriol 46: 1010–1015, 1996.
48. J Gerritse, V Renard, TM Pedro Gomes, PA Lawson, MD Collins, JC Gottschal. *Desulfitobacterium* sp. strain PCE1, an anaerobic bacterium that can grow by reductive dechlorination of tetrachloroethene or *ortho*-chlorinated phenols. Arch Microbiol 165:132–140, 1996.
49. RA Sanford, JR Cole, FE Loffler, JM Tiedje. Characterization of *Desulfitobacterium chloro-*

respirans sp. nov., which grows by coupling the oxidation of lactate to the reductive dechlorination of 3-chloro-4-hydroxybenzoate. Appl Environ Microbiol 62:3800–3808, 1996.
50. I Utkin, C Woese, J Wiegel. Isolation and characterization of *Desulfitobacterium dehalogenans* gen. nov., sp. nov., an anaerobic bacterium which reductively dehalogenates chlorophenolic compounds. Int J System Bacteriol 44:612–619, 1994.
51. J Wiegel, X Zhang, Q Wu. Anaerobic dehalogenation of hydroxylated polychlorinated biphenyls by *Desulfobacterium dehalogenans*. Appl Environ Microbiol 65:2217–2221, 1999.
52. S Harayama, M Kok, EL Niedle. Functional and evolutionary relationships among diverse oxygenases. Annu Rev Microbiol 46:565–601, 1992.
53. D Jahng, TK Wood. Trichloroethylene and chloroform degradation by a recombinant pseudomonad expressing soluble methane monooxygenase from *Methylosinus trichosporium* OB3b. Appl Environ Microbiol 60:2473–2482, 1994.
54. ME Rasche, RE Hicks, MR Hyman, DJ Arp. Oxidation of monohalogenated ethanes and n-chlorinated alkanes by whole cells of *Nitrosomonas europaea*. J Bacteriol 172:5368–5373, 1993.
55. ME Rasche, MR Hyman, DJ Arp. Biodegradation of halogenated hydrocarbon fumigants by nitrifying bacteria. Appl Environ Microbiol 56:2568–2571, 1990.
56. JET van Hylckama-Vlieg, W De Koning, DB Janssen. Transformation kinetics of chlorinated ethenes by *Methylosinus trichosporium* OB3b and detection of unstable epoxides by on-line gas chromatography. Appl Environ Microbiol 62:3304–3312, 1996.
57. L Xun, E Topp, CS Orser. Confirmation of oxidative dehalogenation of pentachlorophenol by a *Flavobacterium* pentachlorophenol hydroxylase. J Bacteriol 174:5745–5747, 1992.
58. T Kumamaru, H Suenaga, M Mitsuoka, T Watanabe, K Furukawa. Enhanced degradation of polychlorinated biphenyls by directed evolution of biphenyl dioxygenase. Nat Biotechnol 16:663–666, 1998.
59. PM Radehaus, SK Schmidt. Characterization of a novel *Pseudomonas* sp. that mineralizes high concentrations of pentachlorophenol. Appl Environ Microbiol 58:2879–2885, 1992.
60. RU Edgehill, RK Finn. Isolation, characterization and growth kinetics of bacteria metabolizing pentachlorophenol. Eur J Appl Microbiol Biotechnol 16:179–184, 1982.
61. T Schenk, R Muller, F Lingens. Mechanism of enzymatic dehalogenation of pentachlorophenol by *Arthrobacter* sp. strain ATCC 33790. J Bacteriol 171:5487–5491, 1990.
62. JS Uotila, VH Kitunen, T Saastamoinen, T Coote, MM Haggblom, MS Salkinoja-Salonen. Characterization of aromatic dehalogenases of *Mycobacterium fortuitum* CG-2. J Bacteriol 174:5669–5675, 1992.
63. KT Leung, O Tresse, D Errampalli, H Lee, JT Trevors. Mineralization of *p*-nitrophenol by pentachlorophenol degrading *Sphingomonas* spp. FEMS Microbiol Lett 173:247–253, 1997.
64. KT Leung, S Campbell, Y Gan, DC White, H Lee, JT Trevors. The role of the *Sphingomonas* species UG30 pentachlorophenol-4-monooxygenase in *p*-nitrophenol degradation. FEMS Microbiol Lett 173:247–253, 1999.
65. L Xun, E Topp, CS Orser. Diverse substrate range of a *Flavobacterium* pentachlorophenol hydrolase and reaction stoichiometries. J Bacteriol 174:2898–2902, 1992.
66. RM Zablotowicz, KT Leung, T Alber, MB Cassidy, JT Trevors, H Lee, L Veldhuis, JC Hall. Degradation of 2,4-dinitrophenol and selected nitroaromatic compounds by *Sphingomonas* sp. UG30. Can J Microbiol 45:840–848, 1999.
67. CS Orser, CC Lange, L Xun, TC Zahrt, BJ Schneider. Cloning, sequence analysis, and expression of the *Flavobacterium* pentachlorophenol-4-monooxygenase gene in *Escherichia coli*. J Bacteriol 175:411–416, 1993.
68. JS Uotila, VH Kitunen, JHA Apajalahti, MS Salkinoja-Salonen. Environment-dependent

mechanism of dehalogenation by *Rhodococcus chlorophenolicus* PCP-1. Appl Microbiol Biotechnol 38:408–412, 1993.
69. M Husain, B Entsch, DP Ballou, V Massey, PJ Chapman. 1980. Fluoride elimination from substrates in hydroxylation reactions catalyzed by *p*-hydroxybenzoate hydroxylase. J Biol Chem 255:4189–4197, 1980.
70. JR Mason, R Cammack. The electron-transport proteins of hydroxylating bacterial dioxygenases. Annu Rev Microbiol 46:277–305, 1992.
71. A Markus, U Klages, S Krauss, F Lingens. Oxidation and dehalogenation of 4-chlorophenylacetate by a two-component enzyme system from *Pseudomonas* sp. Strain CBS3. J Bacteriol 160:618–621, 1984.
72. A Markus, D Krekel, F Lingens. Purification and some properties of component A of the 4-chlorophenylacetate 3,4-dioxygenase from *Pseudomonas* sp. Strain CBS3. J Biol Chem 261:12883–12888, 1986.
73. D Schweizer, A Markus, M Seez, HH Ruf, F Lingens. Purification and some properties of component B of the 4-chlorophenylacetate 3,4-dioxygenase from *Pseudomonas* sp. strain CBS3. J Biol Chem 262:9340–9346, 1987.
74. B Haak, S Fetzner, F Lingens. Cloning, Nucleotide sequence and expression of the plasmid-encoded genes for the two component 2-halobenzoate 1,2-dioxygenase from *Pseudomonas cepacia* 2CBS. J Bacteriol 177:667–675, 1995.
75. V Romanov, RP Hausinger. *Pseudomonas aeruginosa* 142 uses a three component ortho-halobenzoate 1,2dioxygenase for metabolism of 2,4-dichloro- and 2-chlorobenzoate. J Bacteriol 176:3368–3374, 1994.
76. L Xun, CS Orser. Purification and properties of pentachlorophenol hydroxylase, a flavoprotein from *Flavobacterium* sp. strain ATCC 39723. J Bacteriol 173:4447–4453, 1991.
77. L Xu, K Resing, SL Lawson, PC Babbitt, SD Copley. Evidence that *pcpA* encodes 2,6-dichlorohydroquinone dioxygenase, the ring cleavage enzyme required for pentachlorophenol degradation in *Sphingomonas chlorophenolica* strain ATCC 39723. Biochemistry 38:7659–7669, 1999.
78. Y Ohtsubo, K Miyauchi, K Kanda, T Hatta, H Kiyohara, T Sends, Y Nagata, Y Mitsui, M Takagi. PcpA, which is involved in the degradation of pentachlorophenol in Sphingomonas chlorophenolica ATCC39723, is a novel type of ring-cleavage dioxygenase. FEBS Lett 459: 395–398, 1999.
79. AC Kennedy, VL Gewin. Soil microbial diversity: Present and future considerations. Soil Sci 162:607–617, 1997.
80. GJ Poelarends, Wilkens, MJ Larkin, JD van Elsas, DB Janssen. Degradation of 1,3-dichloropropene by *Pseudomonas cichorii* 170. Appl Environ Microbiol 64:2931–2936, 1998.
81. MK Männistö, MA, Tiirola, MS Salkinoja-Salonen, MS Kulomaa and JA Puhakka. Diversity of chlorophenol-degrading bacteria isolated from contaminated boreal groundwater. Arch Microbiol 171:189–197, 1999.
82. A Diez, MJ Alvarez, MI Prieto, JM Bautista, A Garrido-Pertierra. Monochloroacetate dehalogenase activities of bacterial strains isolated from soil. Can J Microbiol 41:730–739, 1995.
83. V Andreoni, G Baggi, M Colombo, L Cavalca, M Zangrossi, S Bernasconi. Degradation of 2,4,6-trichlorophenol by a specialized organism and by indigenous soil microflora: Bioaugmentation and self-remediability for soil restoration. Lett Appl Microbiol 27:86–92, 1998.
84. TG Linkfield, JM Tiedje. Characterization of the requirements and substrates for reductive dehalogenation by strain DCB-1. J Ind Microbiol 5:9–16, 1990.
85. WW Mohn, KJ Kennedy. Reductive dehalogenation of chlorophenols by *Desulfomonile tiedjei* DCB-1. Appl Environ Microbiol 58:1367–1370, 1992.
86. D Tilman, J Knops, D Wedin, P Reich, M Ritchie, E Siemann. The influence of functional diversity and composition on ecosystem processes. Science 277:1300–1302, 1997.
87. DU Hooper, PM Vitousek. The effects of plant composition and diversity of ecosystem processes. Science 277:1302–1305, 1997.

88. BS Griffiths, K Ritz, RE Wheatley. Relationship between functional diversity and genetic diversity in complex microbial communities. In: H Insam, A Rangger, eds. Microbial Communities: Functional Versus Structural Approaches. New York: Springer-Verlag, 1997, pp 1–9.
89. LCM Commandeur, JR Parsons. Biodegradation of halogenated aromatic compounds. In: C Rutledge, ed. Netherlands Kluwer Academic: Biochemistry of Microbial Dehalogenation. 1994, pp 423–458.
90. RM Atlas. Diversity of microbial communities. In: KC Marshall, ed. Advances in Microbial Ecology. Vol. 7. New York: Plenum Press, 1998, pp 1–47.
91. D McDougald, SA Rice, D Weichart, S Kjelleberg. Nonculturability: Adaptation or debilitation? FEMS Microbiol Ecol 25:1–9, 1998.
92. RM Atlas, R Bartha. Microbial Ecology: Fundamentals and Applications. 3rd ed. New York: Benjamin/Cummings 1993.
93. SE El Fantroussi, J Mahillon, H Naveau, SN Agathos. Introduction of anaerobic dechlorinating bacteria in soil slurry microcosms and nested-PCR monitoring. Appl Environ Microbiol 63:806–811, 1997.
94. D Errampalli, K Leung, MB Cassidy, M Kostrzynska, M Blears, H Lee, JT Trevors. Applications of the green flourescent protein as a molecular marker in environmental microorganisms. J Microbiol Methods 35:187–199, 1999.
95. N Fortin, RR Fulthorpe, DG Allen, CW Greer. Molecular analysis of bacterial isolates and total community DNA from kraft pulp mill effluent treatment systems. Can J Microbiol 44: 537–546, 1998.
96. V Degrange, R Bardin. Detection and counting of *Nitrobacter* populations in soil by PCR. Appl Environ Microbiol 61:2093–2098, 1995.
97. KT Leung, A Watt, H Lee, JT Trevors. Quantitative detection of pentachlorophenol-degrading *Sphingomonas* sp. UG30 in soil by most-probable-number/polymerase chain reaction protocol. J Microbiol Methods 31:59–66, 1997.
98. C Picard, C Ponsonnet, E Paget, X Nesme, P Simonet. Detection and enumeration of bacteria in soil by direct DNA extraction and polymerase chain reaction. Appl Environ Microbiol 58:2717–2722, 1992.
99. AS Rosado, L Seldin, AC Wolters, JD van Elsas. Quantitative 16S rDNA-targeted polymerase chain reaction and oligonucleotide hybridization for the detection of *Paenibacillus azotofixans* in soil and the wheat rhizosphere. FEMS Microbiol Ecol 19:153–164, 1996.
100. JD van Elsas, V. Mäntynen, AC Wolters. Soil DNA extraction and assessment of the fate of *Mycobacterium chlorophenolicum* strain PCP-1 in different soils by 16S ribosomal RNA gene sequence based most-probable-number PCR and immunofluorescence. Biol Fertil Soils 24:188–195, 1997.
101. K Johnsen, ø Enger, CS Jacobsen, L Thirup, V Torsvik. Quantitative selective PCR of 16s ribosomal DNA correlates well with selective agar plating in describing population dynamics of indigenous *Pseudomonas* spp. in soil hot spots. Appl Environ Microbiol 65:1786–1789, 1999.
102. R Beaudet, M-J Lévesque, R Villemur, M Lanthier, M Chénier, F Lépine, J-G Bisaillon. Anaerobic biodegradation of pentachlorophenol in a contaminated soil inoculated with a methanogenic consortium or with *Desulfitobacterium frappieri* strain PCP-1. Appl Microbiol Biotechnol 50:135–141, 1998.
103. V Ducrocq, P Pandard, S Hallier-Soulier, E Thybaud, N Truffaut. The use of quantitative PCR, plant and earthworm bioassays, plating and chemical analysis to monitor 4-chlorobiphenyl biodegradation in soil microcosms. Appl Soil Ecol 12:15–27, 1999.
104. G Soulas, Evidence for the existence of different physiological groups in the microbial community responsible for 2,4-D mineralization in soil. Soil Biol Biochem 25:443–449, 1993.
105. AM Osborn, KD Bruce, P Strike, DA Ritchie. Distribution, diversity and evolution of the bacterial mercury resistance (*mer*) operon. FEMS Microbiol Rev 19:239–262, 1997.

106. CR Woese. Bacterial evolution. Microbiol Rev 51:221–271, 1987.
107. G Muyzer, EC De Waal, G Uitterlinden. Profiling of complex microbial communities by denaturing gradient gel electrophoresis analysis of polymerase chain reaction-amplified genes coding for 16S rRNA. Appl Environ Microbiol 59:695–700, 1993.
108. AW Boyle, CD Phelps, LY Young. Isolation from estuarine sediments of a *Desulfovibrio* strain which can grow on lactate coupled to the reductive dehalogenation of 2,4,6-tribromophenol. Appl Environ Microbiol 65:1133–1140, 1999.
109. A Mägli, FA Rainey, T Leisinger. Acetogenesis from dichloromethane by a two-component mixed culture compromising a novel bacterium. Appl Environ Microbiol 61:2943–2949, 1995.
110. PK Sharma, PL McCarty. Isolation and characterization of a facultatively aerobic bacterium that reductively dehalogenates tetrachloroethene to *cis*-1,2- dichloroethene. Appl Environ Microbiol 62:761–765, 1996.
111. C Wischnak, FE Löffler, J Li, JW Urbance, R Müller. *Pseudomonas* sp. Strain 273, an aerobic α,ω-dichloroalkane-degrading bacterium. Appl Environ Microbiol 64:3507–3511, 1998.
112. MG LaMontagne, GJ Davenport, L-H Hou, SK Dutta. Identification and analysis of PCB dechlorinating anaerobic enrichments by amplification: Accuracy of community structure based on restriction analysis and partial sequencing of 16S rRNA genes. J Applied Microbiol 84:1156–1162, 1998.
113. T Vallaeys, F Persello-Cartieaux, N Rouard, C Lors, G Laguerre, G Soulas. PCR-RFLP analysis of 16s RNA, *tfdA* and *tfdB* genes reveals a diversity of 2,4-D degraders in soil aggregates. FEMS Microbiol Ecol 24:269–278, 1997.
114. VK Knight, LJ Kerkhof, MM Häggblom. 1999. Community analyses of sulfidogenic 2-bromophenol-dehalogenating and phenol-degrading microbial consortia. FEMS Microbiol Ecol 29:137–147, 1998.
115. TR Pulliam-Holoman, MA Elberson, LA Cutter, HD May, KR Sowers. Characterization of a defined 2,3,5,6-tetrachlorobiphenyl-*ortho*-dechlorinating microbial community by comparative sequence analysis of genes coding for 16S rRNA. Appl Environ Microbiol 64:3359–3367, 1998.
116. GA Kowlachuk, JR Stephen, De Boer, W, JI Prosser, TM Embley, JW Woldendorp. Analysis of ammonia-oxidizing bacteria of the β subdivision of the class *Proteobacteria* in coastal sand dunes by denaturing gradient gel electrophoresis and sequencing of PCR-amplified 16S ribosomal DNA fragments. Appl Environ Microbiol 63:1489–1497, 1997.
117. RI Ammann, L Krumholz, DA Stahl. Fluorescent-oligonucleotide probing of whole cells for determinative, phylogenetic, and environmental studies in microbiology. J Bacteriol 172: 762–770, 1990.
118. H Christensen, M Hansen, Jørensen. Counting and size classification of active soil bacteria by fluorescence in situ hybridization with an rRNA oligonucleotide probe. Appl Environ Microbiol 65:1753–1761, 1999.
119. BK Mobarry, M Wagner, V Urbain, BE Rittmann, DA Stahl. Phylogenetic probes for analyzing abundance and spatial organization of nitrifying bacteria. Appl Environ Microbiol 62: 2156–2162, 1996.
120. RE Hodson, WA Dustman, RP Garg, MA Moran. In situ PCR for visualization of microscale distribution of specific genes and gene products in prokaryotic communities. Appl Environ Microbiol 61:4074–4082, 1995.
121. S Walia, A Khan, N Rosenthal. Construction and applications of DNA probes for detection of polychlorinated biphenyl-degrading genotypes in toxic organic-contaminated soil environments. Appl Environ Microbiol 56:254–259, 1990.
122. S Kocabiyik, E Caba. Prevalence of monochloroacetate degrading genotypes among soil isolates of *Pseudomonas* sp. Biodegradation 8:371–377, 1998.
123. ML De Souza, J Steffernick, B Martinez, MJ Sadowsky, LP Wackett. The atrazine catabol-

ism genes *atzABC* are widespread and highly conserved. J Bacteriol 180:1951–1954, 1998.
124. P Juteau, R Beaudet, G McSween, F Lépine. Anaerobic biodegradation of pentachlorophenol by a methanogenic consortium. Appl Microbiol Biotechnol 44:218–224, 1995.
125. JSH Tsang, PJ Sallis, AT Bull, DJ Hardman. A monobromoacetate dehalogenase from *Pseudomonas cepacia* MBA4. Arch Microbiol 150:441–446, 1988.
126. AJ van den Wijngaard, J Prins, AJAC Smal, DB Janssen. Degradation of 2-chloroethylvinylether by *Ancylobacter aquaticus* AD25 and AD27. Appl Environ Microbiol 59:2777–2783, 1993.
127. F Löffler, R Müller, F Lingens. Dehalogenation of 4-chlorobenzoate by 4-chlorobenzoate dehalogenase from *Pseudomonas* sp. CBS3: An ATP/coenzyme A dependent reaction. Biochem Biophys Res Commun 176:1106–1111, 1991.
128. ML De Souza, MJ Sadowsky, LP Wackett. Atrazine chlorohydrolase from *Pseudomonas* sp strain ADP: Gene sequence, enzyme purification and protein characterization. J Bacteriol 178:4894–4900, 1996.
129. AN Kulakova, MJ Larking, LA Kulakov. The plasmid-located haloalkane dehalogenase gene from *Rhodococcus rhodochrous* NCIMB 13064. Microbiol 143:109–115, 1997.
130. GH Krooshof, EM Kwant, J Damborský, J. Koča, DB Janssen. Repositioning the catalytic triad aspartic acid of haloalkane dehalogenase: effects on stability, kinetics, and structure. Biochemistry 36:9571–9580, 1997.
131. C-W Kuo, BR Sharak Genthner. Effect of added heavy metal ions on biotransformation and biodegradation of 2-chlorophenol and 3-chlorobenzoate in anaerobic bacterial consortia. Appl Environ Microbiol 62:2317–2323, 1996.
132. A Schmitz, K-H Gartemann, J Fiedler, E Grund, R Eichenlaub. Cloning and sequence analysis of genes for dehalogenation of 4-chlorobenzoate from *Arthrobacter* sp. strain SU. Appl Environ Microbiol 58:4068–4071, 1992.
133. C Wawer, G Muyzer. Genetic diversity of *Desulfovibrio* spp. in environmental samples analyzed by denaturing gradient gel electrophoresis of [NiFe] hydrogenase gene fragments. Appl Environ Microbiol 61:2203–2210, 1995.
134. K Watanabe, M Teramoto, H Futamata, S Harayama. Molecular detection, isolation, and physiological characterization of functionally dominant phenol-degrading bacteria in activated sludge. Appl Environ Microbiol 64:4396–4402, 1998.
135. JO Ka, WE Holben, JM Tiedje. Genetic and phenotypic diversity of 2,4-dichlorophenoxyacetic acid (2,4-D)-degrading bacteria isolated from 2,4-D-treated field soils. Appl Environ Microbiol 60:1106–1115, 1994.
136. MA Bruns, MR Fries, JM Tiedje, EA Paul. Functional gene hybridization patterns of terrestrial ammonia-oxidizing bacteria. Microbial Ecol 36:293–302, 1998.
137. WE Holben, BM Schroeter, VGM Calabrese, RH Olsen, JK Kukor, VO Biederbeck, AE Smith, JM Tiedje. Gene probe analysis of soil microbial populations selected by amendment with 2,4-dichlorophenoxyacetic acid. Appl Environ Microbiol 58:3941–3948, 1992.
138. JO Ka, WE Holben, JM Tiedje. Use of gene probes to aid in recovery and identification of functionally dominant 2,4-dichlorophenoxyacetic acid-degrading populations in soil. Appl Environ Microbiol 60:1116–1120, 1994.
139. JO Ka, P Burauel, JA Bronson, WE Holben, JM Tiedje. DNA probe analysis of microbial communities selected in field by long-term 2,4-D application. Soil Sci Soc Am J 59:1581–1587, 1995.
140. JO Ka, WE Holben, JM Tiedje. Analysis of competition in soil among 2,4-dichlorophenoxyacetic acid-degrading bacteria. Appl Environ Microbiol 60:1121–1128, 1994.
141. SD Copley. Microbial dehalogenases: Enzymes recruited to convert xenobiotic substrates. Curr Opin Chem Biol 2:613–617, 1998.
142. A Gelsomino, Keijzer-Wolters, AC, G Cacco, JD van Elsas. Assessment of bacterial commu-

nity structure in soil by polymerase chain reaction and denaturing gradient gel electrophoresis. J Microbiol Methods 38:1–15, 1999.

143. VM Saboo, MA Gealt. Gene sequences fo the *pcpB* gene of pentachlorophenol-degrading *Sphingomonas chlorophenolica* found in nondegrading bacteria. Can J Microbiol 44:667–675, 1998.
144. HMS Assis, PJ Sallis, AT Bull, DJ Hardman. Biochemical characterization of a haloalcohol dehalogenase from *Arthrobacter erithii* H10a. Enzyme Microbial Technol 22:568–574, 1998.
145. RT Lamar, B Schoenike, AV Wymelenberg, P Stewart, DM Dietrich, D Cullen. Quantitation of fungal mRNAs in complex substrates by reverse transcription PCR and its application to *Phanerochaete chrysoprium*-colonized soil. Appl Environ Microbiol 61:2122–2126, 1995.
146. JC Gottschal, WG Meijer, Y Oda. Use of molecular probing to assess microbial activities in natural ecosystems. In: H Insam, A Rangger, eds. Microbial Communities: Functional Versus Structural Approaches. New York: Springer-Verlag, 1997, pp 10–18.
147. GM Klečka, CL Carpenter, SJ Gonsior. Biological transformation of 1,2-dichloroethane in subsurface soils and groundwater. J Contaminant Hydrol 34:139–154, 1998.
148. C Verhagen, E Smit, B Janssen, JD van Elsas. Bacterial dichloropropene degradation in soil: Screening of soils and involvement of plasmids carrying the *dhlA* gene. Soil Biol Biochem 27:1547–1557, 1995.
149. RR Fulthorpe, Rhodes, AN, JM Tiedje. Pristine soils mineralize 3-chlorobenzoate and 2,4-dichlorophenoxyacetate via different microbial populations. Appl Environ Microbiol 62: 1159–1166, 1996.
150. RD Stapleton, S Ripp, L Jimenez, S Cheol-Koh, JT Fleming, IR Gregory, GS Sayler. Nucleic acid analytical approaches in bioremediation: Site assessment and characterization. J Microbiol Methods 32:165–178, 1998.
151. DB Janssen, T Bosma, GJ Poelarends. Diversity and mechanisms of bacterial dehalogenation reactions. In: DB Janssen, K Sodq, P Wever, eds. Mechanisms of biohalogenation and dehalogenation. Proceedings of KNAW colloquium. Amsterdam, The Netherlands, 1997, pp. 119–129.
152. JR van der Meer. Genetic adaptation of bacteria to chlorinated aromatic compounds. FEMS Microbiol Rev 15:239–249, 1994.
153. CF Amábile-Cuevas, ME Chicurel. Bacterial plasmids and gene flux. Cell 70:189–199, 1992.
154. JR van der Meer, WM De Vos, S Harayama, AB Zehnder. Molecular mechanisms of genetic adaptation to xenobiotic compounds. Microbiol Rev 56:677–694, 1992.
155. SL Liu, JM Suflita. Ecology and evolution of microbial populations for bioremediation. Trends Biotechnol 11:344–352, 1993.
156. I Singleton. Microbial metabolism of xenobiotics: Fundamental and applied research. J Chem Technol Biotechnol 59:9–23, 1994.
157. EM Top, WE Holben, LJ Forney. Characterization of diverse 2,4-dichlorophenoxyacetic acid-degradative plasmids isolated from soil by complementation. Appl Environ Microbiol 61:1691–1698, 1995.
158. RR Fulthorpe, RC Wyndham. Involvement of chlorobenzoate-catabolic transposon, Tn*5271*, in community adaptation to chlorobiphenyl, chloroaniline, and 2,4-dichlorophenoxyacetic acid in a freshwater ecosystem. Appl Environ Microbiol 58:314–325, 1992.
159. AW Thomas, AW Topping, JH Slater, AJ Weightman. Localization and functional analysis of structural and regulatory dehalogenase genes carried on DEH from *Pseudomonas putida* PP3. J Bacteriol 174:1941–1947, 1992.
160. AW Thomas, J Lewington, S Hope, AW Topping, AJ Weightman, JH Slater. Environmentally directed mutations in the dehalogenase system of *Pseudomonas putida* strain PP3. Arch Microbiol 158:176–182, 1992.

161. A Brokamp, FRJ Schmidt. Survival of *Alcaligenes xylosoxidans* degrading 2,2-dichloropropionate and horizontal transfer of its halidohydrolase gene in a soil microcosm. Curr Microbiol 22:299–306, 1991.
162. AN Kulakova, TM Stafford, MJ Larking, LA Kulakov. Plasmid pRTL1 controlling 1-chloroalkane degradation by Rhodococcus rhodochrous NCIMB13064. Plasmid 33:208–217, 1995.
163. JO Ka, JM Tiedje. Integration and excision of a 2,4-dichlorophenoxyacetic acid-degradative plasmid in *Alcaligenes paradoxus* and evidence of its natural intergeneric transfer. J Bacteriol 176:5284–5289, 1994.
164. J van der Ploeg, GV Hall, DB Janssen. Characterization of the haloacid dehalogenase from *Xanthobacter autotrophicus* GJ10 and sequencing of the *dhlB* gene. J Bacteriol 173:7925–7933, 1991.
165. U Murdiyatmo, W Asmara, JSH Tsang, AJ Baines, AT Bull, DJ Hardman. Molecular biology of the 2-haloacid halidohydrolase IVa from *Pseudomonas cepacia* MBA4. Biochem J 284: 87–93, 1992.
166. PH Tomasek, B Franz, UMX Sangodkar, RA Haugland, AM Chakrabarty. Characterization and nucleotide sequence determination of a repeat element isolated from a 2,4,5-T degrading strain of *Pseudomonas cepacia*. Gene 76:227–238, 1989.
167. A Hübner, CE Danganan, L Xun, AM Chakrabarty, W Hendrickson. Genes for 2,4,5-trichlorophenoxyacetic acid metabolism in *Burkholderia cepacia* AC1100: Characterization of the *tft*C and *tft*D genes and locations of the *tft* operons on multiple replicons. Appl Environ Microbiol 64:2086–2093, 1998.
168. RR Fulthorpe, C McGowan, OV Maltseva, WE Holben, JM Tiedje. 2,4-dichlorophenoxyacetic acid-degrading bacteria contain mosaics of catabolic genes. Appl Environ Microbiol 61: 3274–3281, 1995.
169. T Vallaeys, L Courde, C McGowan, AD Wright, RR Fulthorpe. Phylogenetic analyses indicate independent recruitment of diverse gene cassettes during assemblage of the 2,4-D catabolic pathway. FEMS Microbiol Ecol 28:373–382, 1999.
170. JHJ Leveau, JR van der Meer. Genetic characterization of insertion sequence ISJP4 on plasmid pJP4 from *Ralstonia eutropha* JMP134. Gene 202:103–114, 1997.
171. CC Lange, BJ Schneider, CS Orser. Verification of the role of PCP-4-monooxygenase in chlorine elimination from pentachlorophenol by *Flavobacterium* sp. strain ATCC 39723. Biochem Biophys Res Commun 219:146–149, 1996.
172. L Cavalca, A Hartmann, N Rouard, G Soulas. Diversity of *tfd*C genes: Distribution and polymorphism among 2,4-dichlorophenoxyacetic acid degrading soil bacteria. FEMS Microbiol Ecol 29:45–48, 1999.
173. AJ van den Wijngaard, KWHJ van der Kamp, J van der Ploeg, F Pries, B Kazemier, DB Janssen. Degradation of 1,2-dichloroethane by *Ancylobacter aquaticus* and other facultative methylotrophs. Appl Environ Microbiol 58:976–983, 1992.
174. MJ Sadowsky, Z Tong, M De Souza, LP Wackett. AtzC is a new member of the amidohydrolase protein superfamily and is homologous to other atrazine-metabolizing enzymes. J Bacteriol 180:152–158, 1998.
175. M Vyazmensky, S Geresh. Substrate specificity and product stereochemistry in the dehalogenation of 2-haloacids with the crude enzyme preparation from *Pseudomonas putida*. Enzyme Microb Technol 22:323–328, 1998.
176. PE Swanson, Dehalogenases applied to industrial-scale biocatalysis. Curr Opin Biotechnol 10:365–369, 1999.
177. TM Vogel. Bioaugmentation as a soil bioremediation approach. Curr Opin Biotechnol 7: 311–316, 1996.
178. EM Top, MP Malia, M Clerinx, J Goris, P De Vos, W Verstraete. Methane oxidation as a method to evaluate the removal of 2,4-dichlorophenoxyacetic acid (2,4-D) from soil by plasmid-mediated bioaugmentation. FEMS Microbiol Ecol 28:203–213, 1999.

179. RM Zablotowicz, RE Hoagland, MA Locke. Biostimulation: enhancement of cometabolic processes to remediate pesticide-contaminated soils. In: P Kearner and T Roberts, eds. Pesticide Remediation in Soils and Water. New York: John Wiley, 1998, pp 217–250.
180. DC Yee, JA Maynard, TK Wood. Rhizoremediation of trichloroethylene by a recombinant, root-colonizing *Pseudomonas fluorescens* strain expressing toluene *othro*-monooxygenase constitutively. Appl Environ Microbiol 64:112–118, 1998.
181. RM Zablotowicz, RE Hoagland, MA Locke, WJ Hickey. Glutathione-S-transferase activity and metabolism of glutathione conjugates by rhizosphere bacteria. Appl Environ Microbiol 61:1054–1060, 1995.
182. DS Clark. 1994. Can immobilization be exploited to modify enzyme activity? Trends Biotechnol 12:439–443, 1994.
183. MB Cassidy, KW Shaw, H Lee, JT Trevors. Enhanced mineralization of pentachlorophenol by k-carrageenan-encapsulated *Pseudomonas* sp. UG30. Appl Microbiol Biotechnol 47:108–113, 1997.
184. CM Lee, CJ Lu MS Chuang. Effects of immobilized cells on the biodegradation of chlorinated phenols. Water Sci Technol 30:87–90, 1995.
185. MB Cassidy, H Mullineers, H Lee, JT Trevors. Mineralization of pentachlorophenol in a contaminated soil by *Pseudomonas* sp. UG30 cells encapsulated in k-carrageenan. J Ind Microbiol Biotechnol 19:43–48, 1997.
186. P Holloway, JT Trevors, H Lee. A colorimetric assay for detecting haloalkane dehalogenase activity. J Microbiol Methods 32:31–36, 1998.
187. P Holloway, KL Knoke, JT Trevors, H Lee. Alteration of substrate range of haloalkane dehalogenase by site-directed mutagenesis. Biotechnol Bioengineer 59:520–523, 1998.
188. FH Arnold, JC Moore. Optimizing industrial enzymes by directed evolution. Adv Biochem Engin Biotechnol 58:1–14, 1997.
189. O Kuchner, FH Arnold. Directed evolution of enzyme catalysts. Trends Biotechnol 15:523–530, 1997.
190. B Schneider, R Müller, R Frank, F Lingens. Complete nucleotide sequences and comparison of the structural genes of two 2-haloalkanoic acid dehalogenases from *Pseudomonas* sp. strain CBS3. J Bacteriol 173:1530–1535, 1991.
191. H Kawasaki, N Tone, K Tonomura. Purification and properties of haloacetate halidohydrolase specified by plasmid from *Moraxella* sp. strain B. Agric Biol Chem 45:35–42, 1981.
192. H Kawasaki, K Tsuda, I Matsushita, K Tonomura. Lack of homology between two haloacetate dehalogenase genes encoded on a plasmid from *Moraxella* sp. strain B. J Gen Microbiol 138:1317–1323, 1992.
193. H Kawasaki, T Toyama, T Maeda, H Nishino, K Tonomura. 1994. Cloning and sequence analysis of a plasmid-encoded 2-haloacid dehalogenase gene from *Pseudomonas putida* no. 109. Biosci Biotechnol Biochem 58:160–163, 1994.
194. K Motosugi, N Esaki, K Soda. Purification and properties of 2-halo acid dehalogenase from *Pseudomonas putida*. Agric Biol Chem 46:837–838, 1982.
195. DHA Jones, PT Barth, D Byrom, CM Thomas. Nucleotide sequence of the structural gene encoding a 2-haloalkanoic acid dehalogenase of *Pseudomonas putida* strain AJ1 and purification of the encoded protein. J Gen Microbiol 138:675–683, 1992.
196. V Nardi-Dei, T Kurihara, T Okamura, J-Q Liu, H Koshikawa, H Ozaki, Y Terashima, N Esaki, K Soda. Comparative studies of genes encoding thermostable L-2-halo acid dehalogenase from *Pseudomonas* sp. strain YL, other dehalogenases, and two related hypothetical proteins from *Escherichia coli*. Appl Environ Microbiol 60:3375–3380, 1994.
197. JA Leigh, AJ Skinner, RA Cooper. Partial purification, stereospecificity and stoichiometry of three dehalogenases from a *Rhizobium* species. FEMS Microbiol Lett 49:353–356, 1988.
198. PT Barth, L Bolton, JC Thomson. Cloning and partial sequencing of an operon encoding two *Pseudomonas putida* haloalkanoate dehalogenases of opposite stereospecificity. J Bacteriol 174:2612–2619, 1992.

199. JM Smith, K Harrison, J Colby. Purification and characterization of D-2-haloacid dehalogenase from *Pseudomonas putida* strain AJ1/23. J Gen Microbiol 136:881–886, 1990.
200. DB Janssen, F Pries, JR van der Ploeg, B Kazemier, P Terpstra, B Witholt. Cloning of 1,2-dichloromethane degradation genes of *Xanthobacter autotrophicus* GJ10 and expression and sequencing of the *dhlA* gene. J Bacteriol 171:6791–6799, 1989.
201. H Curragh, O Flynn, MJ Larkin, TM Stafford, JTG Hamilton, DB Harper. Haloalkane degradation and assimilation by *Rhodococcus rhodochrous* NCIMB 13064. Microbiology 140: 1433–1442, 1994.
202. R Scholtz, A Schmuckle, AM Cook, T Leisinger. Degradation of eighteen 1-monohaloalkanes by *Arthrobacter* sp. strain HA1. J Gen Microbiol 133:267–274, 1987.
203. R Scholtz, T Leisinger, F Suter, AM Cook. Characterization of 1-chlorohexane halidohydrolase, a dehalogenase of wide substrate range from an *Arthrobacter* sp. J Bacteriol 169:5016–5021, 1987.
204. R Scholtz, F Messi, T Leisinger, AM Cook. Three dehalogenases and physiological restraints in the biodegradation of haloalkanes by *Arthrobacter* species strain HA1. Appl Environ Microbiol 54:3034–3038, 1988.
205. T Yokota, T Omori, T Kodama. Purification and properties of haloalkane dehalogenase from *Corynebacterium* sp. strain m15-3. J Bacteriol 169:4049–4054, 1987.
206. PJ Sallis, SJ Armfield, AT Bull, DJ Hardman. Isolation and characterization of a haloalkane halidohydrolase from *Rhodococcus erythropolis* Y2. J Gen Microbiol 136:115–120, 1990.
207. D Kohler-Staub, T Leisinger. Dichloromethane dehalogenase of *Hyphomicrobium* sp. strain DM2. J Bacteriol 162:676–681, 1985.
208. R Gälli, T Leisinger. Plasmid analysis and cloning of the dichloromethane-utilization genes of *Methylobacterium* sp. DM4. J Gen Microbiol 134:943–952, 1988.
209. SD LaRoche, T Leisinger. Sequence analysis and expression of the bacterial dichloromethane dehalogenase structural gene, a member of the glutathione-s-transferase supergene family. J Bacteriol 172:164–171, 1990.
210. AJ van den Wijngaard, PTW Reuvekamp, DB Janssen. Purification and characterization of haloalcohol dehalogenase from *Arthrobacter* sp. strain AD2. J Bacteriol 173:124–129, 1991.
211. SD Copley, GP Crooks. Enzymatic dehalogenation of 4-chlorobenzoyl coenzyme A in *Acinetobacter* sp. strain 4-CB1. Appl Environ Microbiol 58:1385–1387, 1992.
212. K Kobayashi, Katayama-Hirayama, K, S Tobita. Hydrolytic dehalogenation of 4-chlorobenzoic acid by an *Acinetobacter* sp. J Gen Appl Microbiol 43:105–108, 1995.
213. EJ Perkins, MP Gordon, O Caceres, PF Lurquin. Organization and sequence analysis of the 2,4-dichlorophenol hydroxylase and dichlorocatechol oxidative operons of plasmid pJP4. J Bacteriol 172:2351–2359, 1990.
214. WR Streber, KN Timmis, MH Zenk. Analysis, cloning, and high-level expression of 2,4-dichlorophenoxyacetate monooxygenase gene *tfdA* of *Alcaligenes eutrophus* JMP134. J Bacteriol 169:2950–2955, 1987.
215. F Fukumori, RP Hausinger. *Alcaligenes eutrophus* JMP134 '2,4-dichlorophenoxyacetate monooxygenase' is an α-ketoglutarate-dependent dioxygenase. J Bacteriol 175:2083–2086, 1993.
216. S Fetzner, S Müller, F Lingens. Purification and some properties of 2-halobenzoate 1,2-dioxygenase, a two-component enzyme system from *Pseudomonas cepacia* 2CBS. J Bacteriol 174:279–290, 1992.
217. S Chanama, RL Crawford. Mutational analysis of *pcpA* and its role in pentachlorophenol degradation by *Sphingomonas (Flavobacterium) chlorophenolica* ATTC 39723. Appl Environ Microbiol 63:4833–4838, 1997.
218. T Leisinger, R Bader. Microbial dehalogenation of synthetic organohalogen compounds: Hydrolytic dehalogenases. Chimia 47:116–121, 1993.
219. ADL Akkermans, MS Mirza, HJM Harmsen, HJ Blok, PR Herron, A Sessitsch, WM Akker-

mans. Molecular ecology of microbes: A review of promises, pitfalls and true progress. FEMS Microbiol Rev 15:185–194, 1994.

220. J Sanseverino, JT Fleming, A Heitzer, BM Applegate, GS Sayler. Applications of environmental biotechnology to bioremediation. In: DL Wise, DJ Trantolo, ed. Remediation of Hazardous Waste Contaminated Soils. New York: Marcel Dekker, USA. 1994. pp 97–123.

19

Isolated Enzymes for the Transformation and Detoxification of Organic Pollutants

Liliana Gianfreda
University of Naples Federico II, Portici, Naples, Italy

Jean-Marc Bollag
Center for Bioremediation and Detoxification, The Pennsylvania State University, University Park, Pennsylvania

I. INTRODUCTION

A. Pollution of the Environment

Pollution of the environment has been one of the largest concerns for science and the general public in the last 50 years. The rapid industrialization of agriculture, expansions in the chemical industry, and the need to generate cheap forms of energy have all caused the continuous release of anthropogenic organic chemicals into the biosphere. Consequently, the atmosphere, the hydrosphere, and many soil environments have become polluted to a lesser or greater extent by a large variety of xenobiotic compounds (1). Some toxic compounds are resistant to physical, chemical, or biological degradation and thus constitute an environmental burden of considerable magnitude (2). At high concentration, or after prolonged exposure, some xenobiotics have the potential to produce adverse effects in humans and other organisms; these include acute toxicity, carcinogenesis, mutagenesis, and teratogenesis. Moreover, substances usually found at very low concentrations, and not considered as pollutants, may become contaminants because of their bioaccumulation in food chains to high concentrations (3).

Some of the principal sources of environmental contamination with organic chemicals are current or decommissioned industrial sites where there has been spillage of wastes of various origins. Thus both petroleum- and coal-derived fossil fuel–related materials, effluents from vehicle and equipment cleaning and maintenance, wood-preserving chemicals, paper mill effluents, and a host of pesticides are organic chemicals that find their way into the environment (3). In addition, feedlot operations and landfill sites generate potential pollutants of soil, water, and atmosphere. Finally, industrial sites that produce, store, and distribute organic chemicals may be considered one of the largest sources of environmental pollutants.

Soils and groundwater are natural and preferred sinks for contamination, and their pollution represents an important concern for human and environmental health. Table 1

Table 1 Common Pollutants in Soil and Water

Chemicals	Potential source	Location
Pesticides	Pesticide manufacture; pesticide application	Soil, water, sediments
Volatile organic compounds (e.g., benzene, toluene, xylenes, dichloromethane, trichloroethane, trichloroethylene)	Industrial and commercial wastes	Soil
Polychlorinated biphenyls (PCBs) (100 different isomers in commercial use)	Insulators in electrical equipment	Soil, water
Chloroderivatives (e.g., chlorophenols, chlorobenzene)	Paper mill effluents; bleached kraft mill effluents; wood industry; solvents, intermediates in pesticide manufacture	Soil, water
Polycyclic aromatic hydrocarbons (PAHs) (e.g., naphthalene, anthracene, fluorene, phenanthrene)	Creosote waste sites; fossil fuel wastes; by-products of old gas manufacturing	Soil, water

summarizes some of the more common pollutants of soil and water and lists their potential origins and possible locations in the environment. Among these chemicals, some, such as the polycyclic aromatic hydrocarbons (PAHs), have long been recognized as a worldwide environmental contamination problem because of their intrinsic chemical stability, high resistance to all types of degradation, and carcinogenic and genotoxic effects (4). PAHs, which occur ubiquitously in the environment as complex mixtures, are formed mainly during industrial combustion such as coke production, catalytic cracking, or power generation by fossil fuels. *N*-Heterocyclic compounds, which constitute the second largest group of chemicals present in coal-derived material; chlorophenols, largely used as wide-spectrum biocides for the control of bacteria, fungi, algae, molluscs, and insects (5); and nitrophenols, which are widely used in the chemical industry, all accumulate in soils, sediments, natural waters, and animals because of their long-term usage and recalcitrant nature.

Trichloroethylene (TCE), as well as other halogenated compounds, may be persistent in the environment, partly because of its physical properties (i.e., high density and water solubility and low chemical reactivity) and partly because of its biological recalcitrance (i.e., ability to resist microbial degradation).

B. Remediation Technologies

The U.S. Environmental Protection Agency (EPA) has classified those pollutants whose removal from soil and water is considered an indispensable priority for environmental cleanup and human health. As a result, research groups, not only in the United States but also in other countries, are putting great effort into the exploration of new strategies directed at remediating contaminated systems. Remediation is a general term that indicates the use of techniques suitable for partial or total recovery of a polluted system. In other words, physical, chemical, and biological treatments are applied to remove as much as possible of the contaminant(s) from the site or to transform it into an innocuous or less-toxic compound.

A complete remediation program usually requires more than one step, including (1) knowledge of the past history of the polluted area and activities leading to the contamination of the site, (2) examination and quantification of the severity of the contamination problem, (3) development of the remediation action program to target the specific contaminant or group of contaminants, and (4) development of a treatment sequence suited for the wastes and the site.

With regard to soil, remediation techniques can be distinguished as hard or soft techniques depending on the intensity of chemical and/or physical manipulations required for the contaminated area and the expected cost of the operation. Bioremediation, typically, is thought of as a soft technique requiring less equipment and cost than many other methods. Bioremediation usually refers to the use of biological processes that transform pollutants into innocuous products, and it is the activities of biological agents (microorganisms, plants and their enzymes) that account for most of the transformation. However, physical and chemical processes contribute directly or indirectly to the removal of some compounds under certain environmental conditions (6).

In natural systems, bacteria, fungi, and yeasts and their enzymatic components biodegrade a large variety of hazardous compounds. However, such natural processes can be very slow, and consequently certain chemicals may persist for years. Bioremediation technologies usually help natural biodegradation processes work faster, or they may provide additional, exogenous biological agents to polluted systems and improve the transformation processes.

Depending on the type of remediation strategy used, treatments for the recovery and restoration of both aquatic and terrestrial polluted systems may be carried out in situ or ex situ. Where appropriate, in situ strategies usually are preferred because they generally do not require expensive and sometimes dangerous manipulations of the environment. Ex situ treatments require the excavation of the polluted material; its transfer to another location, where remediation takes place; and its return to the original site.

C. Potential for Using Isolated Enzymes

As previously stated, microorganisms are among the main agents for the transformation and degradation of organic chemicals. A large number of genera and species of aerobic and anaerobic bacteria (including actinomycetes) and fungi have all been shown to transform, cometabolize, and metabolize many anthropogenic organic compounds. The complete mineralization of organics to simple nontoxic compounds (e.g., CO_2, NH_4, H_2O) may occur, and this is usually the desired outcome in bioremediation.

One of the most utilized techniques for in situ bioremediation of contaminated sites is the enhancement (or biostimulation) of the indigenous microbial activity by removal of existing constraints (e.g., supplying necessary nutrients, electron acceptors, moisture, or aeration). An alternative methodology is the inoculation of laboratory cultures of known degraders into the contaminated environment. However, this technique, referred to as bioaugmentation, has often been unsuccessful (7) because (1) the concentration of pollutant in the site is too low to sustain the growth of the microbial inoculant; (2) inoculated microorganisms are inactivated by toxins or natural predators in the environment; (3) inoculated microorganisms prefer centrally metabolized other (natural) organic substrates rather than the pollutants as their substrates; (4) the movement of microorganisms to sites containing pollutants (which are usually discontinuously distributed) is hindered in solid environments such as soil (8); (5) inoculants are outcompeted by indigenous microorganisms that are highly evolved and well adapted to the nutrient status of the environment; and (6) the expression of the pollutant-degrading property (demonstrated in vitro) may not be induced in the natural environment. The regulating and control mechanisms of organisms with novel degradative abilities are poorly understood, and unpredictable processes may occur under natural conditions. Furthermore, bioaugmentation methodology requires continuous analysis and monitoring of microbial population dynamics to define the persistence and activity of the inoculant(s) from both efficacy and risk assessment perspectives. Other factors involving the intrinsic toxicity and solubility of the compounds or the type or nature of the microbial strain may influence the efficacy of a microbial inoculum.

One strategy for overcoming the limitations to the use of microorganisms in the detoxification of organic-polluted sites (and particularly the processes underlined in point [6]) is the use of cell-free enzyme preparations; a review of research aimed at developing the use of isolated enzymes in solid, liquid, and hazardous waste treatment has been published (9). These authors examined the potential application of several enzymes according to categories of specific waste types (e.g., phenols and related compounds, pulp and paper wastes, pesticides, food processing wastes, solid waste and sludge treatment, and heavy metals).

D. Origin of Enzymes

As outlined in the previous paragraphs, microorganisms degrading various aliphatic, alicyclic, aromatic, and heterocyclic compounds have been identified and isolated. In many cases, the detailed biochemical pathways and enzymes responsible for the main reactions of the degradation pattern have been characterized. Microorganisms produce enzymes able to react with chemicals different from those being utilized as primary carbon and energy sources (9), thus becoming a potential source of a large array of enzymes useful for the transformation of various xenobiotic compounds.

Although most research has centered on bacterial processes, fungal enzymes have been shown to be involved in the transformation of many toxic organic compounds (10). For example, many white rot fungi, such as basidiomycetes, secrete enzymes (e.g., ligninase, manganese peroxidase, laccase) involved in lignin degradation. These enzymes seem to be nonspecifically reactive toward many organic pollutants (11,12).

In a review dedicated to fundamental and applied research in the microbial metabolism of xenobiotics, Singleton (13) reported that relatively few enzymes capable of degrading xenobiotics have been studied and purified. A pentachlorophenol monooxygenase, a dichloromethane dehalogenase, and a 2-hydroxy-6-*oxo*-6-phenylhexa-2,4-dienoic acid–(HOPDA)-reducing enzyme are cited as examples of the few enzymes purified from bacteria. Ligninases and lignin peroxidases are used as examples of fungal enzymes. Nonetheless, the feasibility of using fungal enzymes in the decontamination of soil containing phenolic and anilinic compounds has been suggested (14).

II. APPLICABLE ENZYMES AND THEIR PROPERTIES

A. Hydrolases

Some organic pollutants, especially pesticides, may lose their toxic properties after hydrolytic reactions catalyzed by nonspecific hydrolases (Table 2). For example, the breakdown of esteric, amidic, and peptidic bonds by esterases, amidases, and proteases may lead to products with little or no toxicity. However, in some cases toxic by-products may be formed.

The hydrolysis of halogen-carbon bonds also may occur and is catalyzed by a group

Table 2 Examples of Hydrolases Purified from Microorganisms Active Toward Organic Chemicals

Enzyme	Microrganism	Substrate	Reference
Acylamidases	*Fusarium oxysporum*	Acylanilides	44
2-Ketocyclohexanecarboxyl Coenzyme-A hydrolase	*Rhodopseudomonas palustris*	2-Ketocyclohexanocarboxyl coenzyme A	40
Chlorohydrolase	*Pseudomonas* spp.	Atrazine	38, 39
Malathion hydrolase	Soil microorganisms	Malathion	32
N-methylcarbamate hydrolase	*Achromobacter* spp., *Pseudomonas* spp. strain *CRL-OK*	Carbaryl, carbofuran, aldicarb	41,42
Parathion hydrolase	*Pseudomonas, Brevibacterium, Azotomonas, Xanthomonas* spp.	Parathion, triazophos, paraoxon, diazinon	17–19
	Flavobacterium spp.	Parathion; methyl parathion	20,29
	Pseudomonas spp.	Organophosphates	22,24
	Nocardia spp.	Coumaphos, Parathion	31
Permethrinase	*Bacillus cereus*	Pyrethroids	37

of halidohydrolases. These enzymes are part of a larger group of dehalogenases that are discussed later (see B3. Hydrolytic dehalogenation; b. Halidohydrolases, p. 509).

1. Parathion Hydrolases

A large number of aquatic and terrestrial species (microorganisms and animals) are known to produce enzymes capable of hydrolyzing organophosphorus compounds. The enzymes are generally called organophosphorus acid anhydrases, although other names such as phosphotriesterases, parathion hydrolases, somanases, and paraoxonase have been used.

Parathion hydrolases are among the most studied classes of enzymes that possess pesticide-hydrolyzing abilities. They have received great attention from researchers because of their presence in organophosphorus insecticide-resistant organisms (15,16). The results obtained in several studies seem to indicate that this enzyme and similar hydrolases are responsible for parathion and methyl parathion resistance in some strains of insects. It is possible to obtain parathion hydrolase in large quantities from various bacterial species that help to make the enzyme commercially viable for the detoxification of pesticide-polluted sites.

Munnecke (17–19) reported that a crude, cell-free parathion hydrolase preparation isolated from a bacterial mixture that included *Pseudomonas* sp., *Brevibacterium* sp., *Azotomonas* sp., and *Xanthomonas* sp. was able to hydrolyze several organophosphates, such as parathion, triazophos, paraoxon, diazinon, methyl parathion, chlorpyrifos, fenitrothion, and cyanophos. The enzymatic hydrolysis of almost all the pesticides occurred at rates 40 to 1000 times more rapid than those of chemical methods. Furthermore, the enzyme was stable in the presence of the detergents and solvents used to solubilize and prepare pesticide mixtures (19).

Three unique parathion hydrolases were purified from Gram-negative bacterial strains (20). Two membrane-bound hydrolases and one cytosolic hydrolase were purified from a *Flavobacterium* sp. strain ATCC 27551, a strain SC, and a strain B1, respectively. The purified proteins demonstrated similarities in their affinity for ethyl parathion and their broad temperature optimum around 40°C. Conversely, they differed in their composition (e.g., number and molecular mass of proteic subunits), their response to sulfhydryl reagents (dithiothreitol) and to metal salts ($CuCl_2$), and their substrate range. The B1 hydrolase showed equal affinity for parathion and the related organophosphate insecticide *O*-ethyl-*O*-4-nitrophenyl phenylphosphonothionate (EPN), whereas the *Flavobacterium* sp. enzyme displayed twofold lower affinity for EPN than for parathion and the SC-produced enzyme was not active toward EPN as a substrate.

The capability of 18 Gram-negative bacterial isolates to hydrolyze the organophosphorus compound diisopropyl fluorophosphate (a structural analog of the nerve gas agents soman and sarin) was investigated (21). The authors compared the detoxifying ability of crude enzyme preparations (frozen cell sonicates and acetone powders) by measuring the hydrolytic release of fluoride and the disappearance of acetyl cholinesterase inhibition in vivo. The highest activities were present in acetone powder preparations from a strain with known parathion hydrolase activity. The results demonstrated that parathion hydrolase was not specific with regard to its phosphotriesterase activity and showed a significant detoxifying activity at low concentrations (<1 enzymatic unit g^{-1} protein, e.g., 1 μmol of substrate transformed min^{-1} g^{-1} protein) and near-neutral pH.

A phosphotriesterase (a parathion hydrolase) from *Pseudomonas diminuta* was characterized for its kinetic behavior toward different organophosphorus compounds (22). To obtain the enzyme in a purified form, the organophosphate degradation gene, *opd*, was cloned in an *Escherichia coli* strain. At pH 7.0 and 25°C, the enzyme showed a higher affinity for paraoxon (K_m value = 0.012 m*M*) than for diisopropyl fluorophosphate (K_m = 0.12 m*M*) and behaved at pH 9.0 as a competitive inhibitor of paraoxon with an inhibition constant K_i = 0.32 m*M*. In addition, an enzyme capable of degrading parathion and methyl parathion at pH 8.0 and fenitrothion at pH 8.5 was obtained from an isolated strain, YF11. For the three insecticides, the optimal reaction temperature was 32.5°C and the K_m values ranged from 18.7 to 212.4 μ*M* for parathion and fenitrothion, respectively (23).

Crude enzyme extracts from a *Pseudomonas* sp. and a *Xanthomonas* sp., both isolated from a pesticide disposal site in northern Israel, also displayed parathion hydrolase activity (24). Crude enzyme preparations degraded parathion but showed different sensitivity to cations such as Cu^{2+}, Fe^{2+}, Ca^{2+}, Mn^{2+}, Al^{3+}, Zn^{2+}, and sodium ethylenediaminetetraacetic acid (EDTA). Cu^{2+} strongly inhibited the *Pseudomonas* sp. enzyme, but it had a stimulatory effect on parathion degradation by the *Xanthomonas* sp. enzyme. A significant inhibition of the *Xanthomonas* sp. hydrolase but not of the *Pseudomonas* sp. hydrolase was also reported for NaEDTA.

Great efforts have been made to apply recombinant deoxyribonucleic acid (DNA) technology to the production of parathion hydrolase. Typical genetic studies, involving two bacteria, *Pseudomonas diminuta* and *Flavobacterium* sp., were carried out by Mulbry et al. (25,26). Cloning experiments, using DNA–DNA hybridization and restriction mapping techniques, indicated that two discrete plasmids from the encoding parathion-hydrolyzing soil bacteria possess a common but limited region of sequence homology within potentially nonhomologous plasmid structures.

Genetic engineering was also applied by Coppella et al. (27) to obtain large amounts of parathion hydrolase. When the gene encoding the enzyme was cloned into *Streptomyces lividans*, the transformed bacterium was able to express and secrete the enzyme. Fermentation conditions were investigated to improve and enhance enzyme production (27,28). The fermentation was first carried out in the presence of large quantities of nutrients supplied throughout the fermentation period and by sparging with oxygen-enriched gas. Enzyme activity and production did not further increase when cultivation was prolonged more than 90 hours. The authors concluded that some enzyme deactivation occurred and that the rates of enzyme synthesis and deactivation were balanced after 90-h periods.

In another study, parathion hydrolase cloned from a *Flavobacterium* sp. into *S. lividans* was produced in large amounts (milligrams of protein) and purified to homogeneity (29). The enzyme, characterized for its structural and catalytic features, was a single-polypeptide chain with an apparent molecular mass of 35 kD. The enzyme had an affinity in the order of *O*-ethyl-*O*-*p*-nitrophenyl phenylphosphothionate > parathion > and *p*-nitrophenyl ethyl(phenyl) phosphinate > methyl parathion, as assessed by K_m values. An optimal pH of 9.0 and an optimal temperature of 45°C were determined. As observed for other parathion hydrolases, the enzyme was inhibited by dithiothreitol and $CuSO_4$. These results demonstrated that the purified recombinant enzyme presented the same characteristics as those of the protein produced by the donor *Flavobacterium* sp. strain. Further studies suggested that the use of a native *Streptomyces* sp. signal sequence may have resulted in more efficient secretion of the heterologous protein (30).

In 1998 another bacterium, *Nocardia* sp. B1, was reported to hydrolyze organophosphate insecticides, such as coumaphos and parathion (31). The enzyme organophosphorus hydrolase (OPH) was isolated and shown to be active toward organophosphate insecticides. However, it was demonstrated, and explained by genetic studies, that OPH activity in *Nocardia* sp. often was spontaneously lost during growth in the laboratory (31).

2. Other Hydrolases

There is relatively little evidence of other isolated esterases suitable for the detoxification of organic pollutants. In addition, where it is reported, there is little detailed information on the metabolites and their toxicity. For example, Getzin and Rosefield (32) described the properties of a partially purified enzyme extracted from soil that degraded the insecticide malathion to its monoacid derivative. The enzyme had a high resistance to both thermal and microbial deactivation, probably because of a carbohydrate moiety attached to the protein (33).

A HOPDA-hydrolyzing enzyme involved in the degradation of biphenyl was purified to homogeneity from *Pseudomonas cruciviae* S-93-B1 that was grown on biphenyl as the sole carbon source (34–36). The hydrolytic reaction occurred between the C5-C6 bond of HOPDA, to produce benzoic acid and 2-oxopent-4-enoic acid. Further studies showed the production of three HOPDA-reducing enzymes (I,II,III) in the bacterium, having different catalytic and structural properties. Experiments performed with methylated-HOPDA derivatives and ring-fission products as substrates allowed new metabolic divergence of biphenyl and related compounds to be proposed (35,36).

An enzyme responsible for hydrolyzing second- and third-generation synthetic pyrethroids and producing noninsecticidal metabolites was isolated from a pyrethroid-transforming strain of *Bacillus cereus* (37). The enzyme, named permethrinase, was purified as a single protein chain of 61-kD molecular mass. It had a pH optimum of 7.5 and a temperature optimum of 37°C. Several characteristics (i.e., no requirement for cofactors or coenzymes; sensitivity to tetraethylpyrophosphate; protection by dithiothreitol against the inhibition effects of sulfhydryl agents, *p*-chloromercuribenzoate, and *N*-ethylmaleimide) suggested that the microbial esterase was a carboxylesterase.

The gene, sequence, enzyme purification, and characterization of a novel enzyme involved in the metabolic transformation of atrazine to carbon dioxide and ammonia via the intermediate hydroxyatrazine by a *Pseudomonas* sp. strain ADP were studied and described by de Souza et al. (38). Genetic studies previously performed on the bacterium allowed the production of hydroxyatrazine to be ascribed to a specific DNA fragment (39). Furthermore, sequence analysis of the fragment indicated that a single open-reading frame, named *atzA*, encoded an activity transforming atrazine to hydroxyatrazine.

The protein was purified and characterized and had an oligomeric structure with a molecular mass of 245 kD. Chlorohydrolase rather than oxygenase activity was attributed to the purified enzyme by studies performed with $H_2{}^{18}O$ that was converted to ^{18}O-hydroxyatrazine. The enzyme that dechlorinated atrazine, simazine, and desethylatrazine (but not melamine, terbutylazine, or desethyldesisopropylatrazine) was apparently a novel enzyme that also participated in the hydrolysis of atrazine in soils and groundwaters (38). As discussed later, this enzyme could also be considered a dehalogenase acting on haloaromatic compounds.

A hydrolase-catalyzing ring cleavage reaction (which is very uncommon) during anaerobic degradation of benzoate was isolated from the anaerobic bacterium *Rhodopseudomonas palustris* and purified (40). The enzyme 2-ketocyclohexanecarboxyl coenzyme A (2-ketochc-CoA) hydrolase catalyzed the hydration of 2-ketochc-CoA to pimelyl-CoA. The native protein was a homotetramer of 34-kD subunits. The enzyme had no sensitivity to oxygen, and its production was induced by growing the bacterium on benzoate and other benzoate intermediates.

Enzymes capable of hydrolyzing the carbamate linkage of the pesticides carbofuran and carbaryl were purified from an *Achromobacter* sp. (41) and *Pseudomonas* sp. strain

CRL-OK (42), respectively. The *Pseudomonas* sp.–extracted enzyme was demonstrated to be a unique cytosolic enzyme, able to hydrolyze carbofuran and aldicarb but not the phenylcarbamate isopropyl *m*-chlorocarbinilate, the thiocarbamate *S*-ethyl-*N,N*-dipropylthiocarbamate, or the dimethylcarbamate *O*-nitrophenyldimethylcarbamate (42).

Early papers reported the purification and properties of acylamidases responsible for the hydrolysis of acylamides and/or phenyl ureas (43,44). However, to our knowledge, no further studies have been carried out with these enzymes aimed specifically at their application in the detoxification of organic pollutants in the environment.

A lot of literature is available on other hydrolases such as proteases, lipases, and cellulases. All of these enzymes have been isolated and purified from several sources, their properties have been well characterized, and most are available commercially. However, there is no direct evidence that these enzymes are involved in the transformation of organic pollutants. Nonetheless, proteases could be considered a group of hydrolases particularly useful for the treatment of wastes derived from food processing (e.g., fish and meat wastes). Indeed, they can solubilize proteins present in waste streams and generate products with added nutritional value (9). Similarly, cellulases may hydrolyze lignocellulose and cellulose present in municipal solid wastes or paper industry wastes. The possibility of obtaining energy sources such as fermentable sugars, biogas, and end products such as ethanol has attracted the attention of several researchers, and numerous studies have been conducted in this field (9).

The interest in lipases stems mainly from the capability of these enzymes to be agents of nonconventional enzymatic transformations such as synthetic rather than hydrolytic catalytic reactions. Esterification and transesterification can be achieved if the process is performed in organic or water/organic solvents. Similar synthetic reactions can be catalyzed by proteases as well. Several investigations have been conducted on the properties of proteases and lipases and their involvement in the formation of synthetic products. For example, a protease from *Bacillus licheniformis* was shown in anhydrous organic solvents to catalyze the polytransesterification of a diester of glutaric acid with aromatic diols such as benzene dimethanol (45).

Studies also were aimed at defining prochiral selectivity of both lipases and proteases when involved in the organic solvent–mediated transformation of 2-substituted 1,3-propanediol or its diester (46). A mechanistic model was proposed that predicted an inverse correlation between lipase's prochiral selectivity and solvent hydrophobicity as well as particular effects of substrate structure variation.

In the context of this chapter, the importance of these enzymes and their varying activity in soil have to be mentioned with regard to different management and/or fertilization treatments. For example, protease activities strongly increased when solid urban wastes were applied to three different semiarid area soils, thus showing the response of the soil microbial population to the applied organic matter (47).

B. Dehalogenases

Halogenated compounds are present in the environment as either naturally occurring or synthetic introduced compounds. A detailed overview of microorganisms capable of metabolizing halogenated compounds is provided in Chapter 18.

The removal of halogen atoms from aliphatic and aromatic halogen-carbon-substituted compounds is an essential step in the biochemical transformation of pollutants; the reaction reduces or eliminates toxicity. The cleavage of carbon-halogen bonds may occur by (1) enzymatic dehalogenation catalyzed by specific enzymes (dehalogenases), (2) a fortuitous reaction catalyzed by enzymes with a broad substrate specificity and acting on halogenated analogs of their natural substrate (discussed later), or (3) spontaneous dehalogenation of unstable intermediate products of unrelated enzymatic reactions.

Dehalogenases have received continuous interest because they not only may play an important role in the remediation of the environment, but also may be applied in biotechnological transformations for producing biologically active compounds (discussed late). Slater and coworkers (48) classified dehalogenases according to dehalogenation mechanisms in three groups, namely: hydrolytic dehalogenases, haloalcohol dehalogenases (hydrogen halide lyases), and cofactor-dependent dehalogenases. A further classification may be made in relation to substrate specificity. In an exhaustive review of the mechanisms by which halogenated compounds may lose their halogen substituents, the enzymes involved in the reactions, the products, and the biotechnological applications of these enzymes, Fetzner and Lingens (49) identified seven dehalogenation mechanisms, taking into account both the mechanism of the enzymatic reaction and the substrate involved in it.

Both types of classification are summarized in Fig. 1 and discussed for the following six dehalogenation mechanisms:

1. Reductive Dehalogenation

In the reductive dehalogenation mechanism, halogen-carbon bonds are replaced by hydrogen-carbon bonds, with a concomitant release of halogen ions. The process can be per-

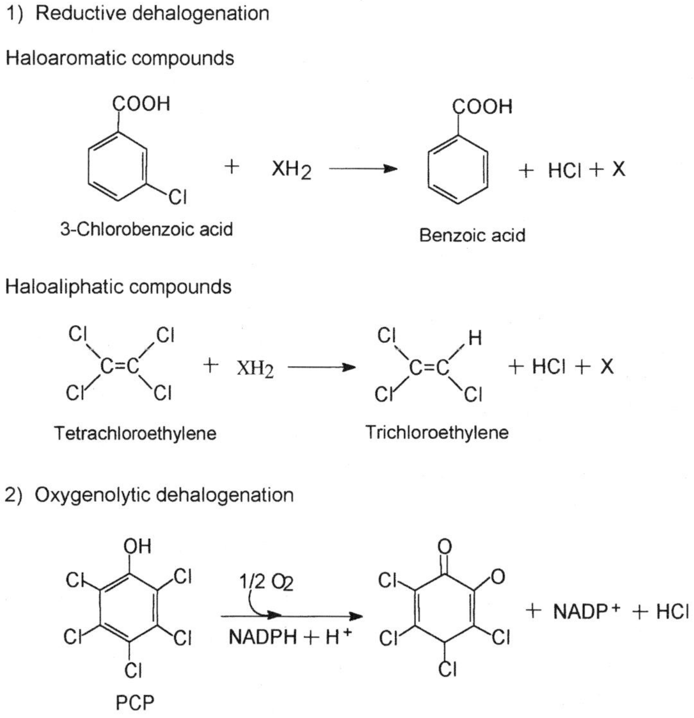

1) Reductive dehalogenation

Haloaromatic compounds

3-Chlorobenzoic acid + XH_2 → Benzoic acid + HCl + X

Haloaliphatic compounds

Tetrachloroethylene + XH_2 → Trichloroethylene + HCl + X

2) Oxygenolytic dehalogenation

PCP + 1/2 O_2 (NADPH + H^+) → + $NADP^+$ + HCl

Figure 1 Reactions catalyzed by dehalogenases.

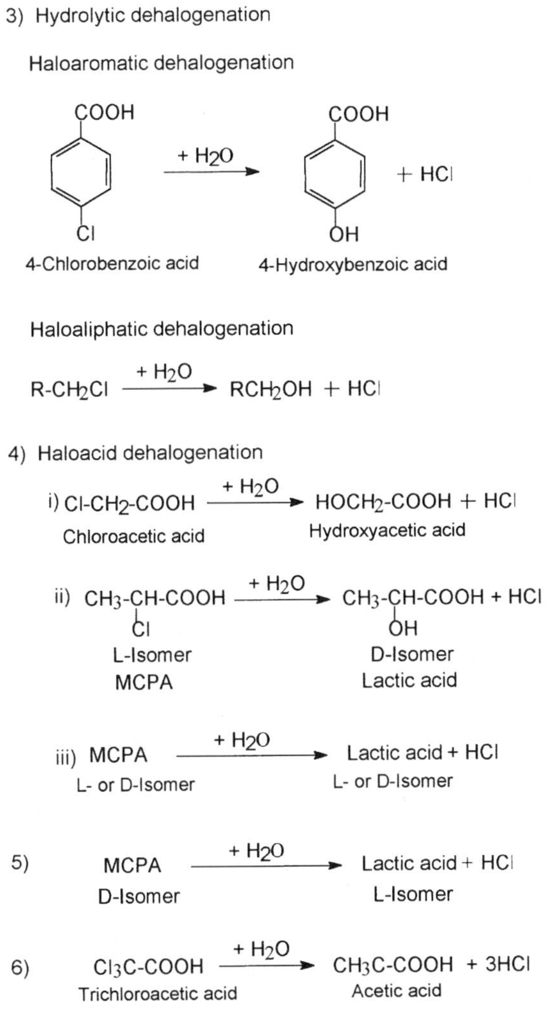

3) Hydrolytic dehalogenation

Haloaromatic dehalogenation

$$\text{4-Chlorobenzoic acid} \xrightarrow{+H_2O} \text{4-Hydroxybenzoic acid} + HCl$$

4-Chlorobenzoic acid 4-Hydroxybenzoic acid

Haloaliphatic dehalogenation

$$R\text{-}CH_2Cl \xrightarrow{+H_2O} RCH_2OH + HCl$$

4) Haloacid dehalogenation

i) $Cl\text{-}CH_2\text{-}COOH \xrightarrow{+H_2O} HOCH_2\text{-}COOH + HCl$

Chloroacetic acid Hydroxyacetic acid

ii) $CH_3\text{-}CH(Cl)\text{-}COOH \xrightarrow{+H_2O} CH_3\text{-}CH(OH)\text{-}COOH + HCl$

L-Isomer MCPA D-Isomer Lactic acid

iii) $\text{MCPA} \xrightarrow{+H_2O} \text{Lactic acid} + HCl$

L- or D-Isomer L- or D-Isomer

5) $\text{MCPA} \xrightarrow{+H_2O} \text{Lactic acid} + HCl$

D-Isomer L-Isomer

6) $Cl_3C\text{-}COOH \xrightarrow{+H_2O} CH_3C\text{-}COOH + 3HCl$

Trichloroacetic acid Acetic acid

Figure 1 Continued

formed by several anaerobic bacteria. It is coupled to energy conservation and therefore usually is termed dehalorespiration (50,51). A reduced organic substrate or H_2 is required as the source of the two electrons and the protons. In vitro studies of reductive dehalogenation usually have used methyl viologen as the artificial electron donor. Methyl viologen is transformed from a reduced blue to an oxidized colorless form, thus allowing the progress of the reaction to be recorded by spectrophotometric measurements.

A more convenient, rapid, and quantitative pH indicator dye-based colorimetric assay for detecting reductive dehalogenase activity has been developed. The assay is based on the decrease in the pH of a weakly buffered medium that occurs when protons and chloride ions are released by enzymatic activity (52).

Reductive dehalogenation may occur on both haloaromatic and haloaliphatic compounds, including several halogenated pesticides, which are transformed by microorganisms through an initial reductive dehalogenation step. A 3-chlorobenzoate-reductive dehalogenase was isolated and purified from the cytoplasmic membrane of the bacterium *Desulfomonile tiedjei* DCB-J (53). The reducing agent was methyl viologen, and the reaction produced benzoate. Structural studies demonstrated that the enzyme was probably an oxygen-stable, heme protein with two subunits of 66- and 37-kD molecular mass (53). The aryl reductive dehalogenation reaction was inhibited by sulfur oxyanions, as demonstrated by studies performed on whole cells and extracts of *D. tiedjei* cells. Sulfate, sulfite, and thiosulfate showed separate mechanisms of inhibition, depending on the growing conditions and the presence or absence of substrate and/or inducer for the dehalogenation activity (54).

Enzymatic activity capable of sequentially dehalogenating the haloaliphatic compound tetrachloroethene (or perchloroethene [PCE]) to trichloro- and dichloroethylene had been observed previously in cell extracts of the same bacterium (55). The reductive dehalogenation of haloaromatic and haloaliphatic compounds showed several similarities such as sensitivity to heat, stability toward oxygen, a similar inhibition pattern, and induction by 3-chlorobenzoate. These findings suggested that 3-chlorobenzoate and PCE in *D. tiedjei* are transformed by the same reductive enzyme (51).

A glutathione-dependent reductive dehalogenase was isolated from *Flavobacterium* sp. ATCC39723 (56,57). The enzyme transformed tetrachloro-*p*-hydroquinone, formed from pentachlorophenol (PCP) by the action of a PCP-hydrolase (discussed late), to trichloro-*p*-hydroquinone and dichloro-*p*-hydroquinone but was not active in the presence of reduced nicotinamide-adenine dinucleotide phosphate (NADPH), reduced nicotinamide-adenine dinucleotide (NADH), dithiothreitol, or ascorbic acid as a reducing agent. A similar dehalogenating activity was isolated and purified from *Sphingosomonas chloro-fenolica*, a soil bacterium that degrades pentachlorophenol (58,59). The reductive dehalogenation reaction requires two molecules of glutathione. Detectable amounts of 2,3,5-trichloro-6-*S*-glutathionyl hydroquinone (GS-TriCHQ) and an unidentified isomer of dichloro-*S*-glutathionyl hydroquinone (GS-DCHQ) are produced as aberrant byproducts. However, almost no GS-TriCHQ or GS-DCHQ is produced either by the enzyme in freshly prepared crude extracts or after treatment of the purified enzyme with dithiothreitol. The enzyme probably suffers oxidative damage during purification, which is reversibly repaired by treatment with dithiothreitol. These results are consistent with the hypothesis that a cysteine or methionine residue is required to act as a nucleophile during the conversion of tetrachloro-*p*-hydroquinone to trichloro-*p*-hydroquinone (58). Using electrospray liquid chromatography/mass spectrometry and treating the enzyme with hydrogen peroxide, Willett et al. (59) supported this hypothesis by demonstrating that oxidation of the enzyme actually occurs and a sulfenic acid forms at Cys13 position. Further oxidation to sulfinic acid was observed.

A cytoplasmatic membrane preparation from cells of *Desulfitobacterium chlororespirans* was shown to dechlorinate 3-Cl-4-hydroxybenzoate reductively (60,61). The enzyme was active with methyl viologen as the artificial electron donor but not with reduced benzyl viologen, NADH, NADPH, $FMNH_2$, or $FADH_2$. The membrane-bound enzyme system was stable to oxygen and to a temperature of 57°C. Inhibition by sulfite and the capability of dehalogenating several chlorophenols in the *ortho* position also were demonstrated (60,61).

An NADPH-dependent reductive *ortho*-dehalogenating enzyme was found in the soluble fraction of *Corynebacterium sepedonicum* KZ-4 and Coryneform bacterium strain NTB1 cell extracts (62). The enzymatic fraction was active on 2,4-dichlorobenzoate, catalyzing the NADPH-dependent *ortho*-dehalogenation to 4-Cl-benzoyl-CoA of the intermediate 2,4-dichlorobenzoyl-CoA, which had formed in the first step of the metabolic degradation by the action of 2,4-dichlorobenzoyl ligase and required magnesium adenosine triphosphate (Mg ATP) and CoA. An enzymatic fraction, capable of catalyzing the direct hydrolytic removal of chlorine from the *para* position of 4-chlorobenzoyl and producing 4-hydroxybenzoyl CoA, was also found (discussed later).

Several examples of reductive dehalogenation of haloaliphatic compounds are reported in the literature (49). However, the exact mechanism of the reaction is obscure. Magnuson et al. (63) purified a reductive dehalogenase from a *Dehalococcoides ethanogen* 195 grown under anaerobic conditions. The dehalogenase was capable of transforming tetrachloroethene (or perchloroethene [PCE]) to trichloroethene (TCE) in the presence of titanium (III), citrate, or methyl viologen as the reductant. Reductive dechlorination of PCE also was catalyzed by a dehalogenase isolated from a *Desulfitobacterium* sp. strain PCE-S, growing on PCE as the carbon source and using methyl viologen as the artificial electron donor (64). Another anaerobic microorganism, *Dehalospirillum multivarians*, was previously demonstrated to produce cytoplasmic PCE- and TCE-dehalogenases (65,66). The enzymes were produced under strictly anaerobic conditions and were rapidly inactivated by propyl iodide.

PCE- and TCE-dechlorinating enzymes are usually characterized by light-reversible inhibition by iodo derivatives (propane and ethane), thus suggesting the involvement of a Co(1) corrinoid cofactor in the dechlorination mechanism. In further studies, Neumann et al. (67) purified the enzyme to apparent homogeneity. The enzyme was active with PCE and TCE and reduced methyl viologen at a specific activity of 2.6 microkatal mg^{-1} protein. Gel filtration and SDS-gel electrophoresis revealed a single protein band with a molecular mass of 57 kD. The enzyme showed an optimal pH of 8.0 and temperature of 42°C. It was stimulated by ammonium ions, was stable to oxygen and temperature up to 50°C, and contained 1.0 mol corrinoid, 9.8 mol iron, and 8.0 mol acid-labile sulfur/mol protein (67).

A PCE-reductive dehalogenase was purified to homogeneity from a TCE-reducing anaerobic strain PCES (68). The enzyme was an omo-oligomeric protein with a whole molecular mass of 200 kD and three subunits of 65 kD each. The protein contained corrinoid (0.7 mol), cobalt (1 mol), iron (7.8 mol), and acid-labile sulfur (10.3 mol) in each subunit, and it demonstrated a pH optimum of 7.2 and a temperature optimum of about 50°C. A high oxygen sensitivity (half-life = 50 min) was demonstrated. No significant similarity to the amino acid sequence of the PCE-dehalogenase from *Dehalospirillum multivarians* was found. The presence of corrinoid cofactors (0.68 mol corrinoid, 12 mol iron, and 13 mol acid-labile sulfur/mol subunit) was also demonstrated in a 3-chloro-4-hydroxyphenyl acetate reductive dehalogenase purified from a *Desulfitobacterium hafniense*. The enzyme was a single protein with a molecular mass of 46.5 kD, as revealed by sodium dodecyl sulfate-polyacrylamide gel electrophoresis (SDS-PAGE) (69).

2. *Oxygenolytic Dehalogenation*

The oxygenolytic dehalogenation reaction is catalyzed by monooxygenase or dioxygenases NAD(P)H and is oxygen-dependent. One or two OH groups are inserted on the aromatic ring of the haloaromatic compound with release of the corresponding halogenated acid and production of CO_2.

Haloalkanes may also be dehalogenated by an oxygenase-mediated reaction. However, no experimental evidence is provided for the presence of oxygenolytic dehalogenases specific for halogenated alkanes, although long-chain haloalkanes were shown to be dehalogenated oxidatively (49). Oxidation of some chloroalkanes may occur by the action of nonspecific bacterial oxygenases. Toluene 2-monooxygenase, purified from *Burkholderia cepacia* G4, acted directly on the alternative substrate trichloroethylene (70). The latter, however, caused inactivation of toluene 2-monooxygenase activity that was partly protected by the addition of cysteine to the reaction mixtures. Diffusible intermediates may have contributed to the inactivation of the oxidoreductive enzyme.

A well-described monooxygenase is pentachlorophenol (PCP) 4-hydrolase from *Flavobacterium* sp. strain ATCC 39723, a monomeric FAD-protein, requiring NADPH and oxygen and active toward a very large range of substrates, including halogen-, nitro-, amino-, and cyano-substituted phenols (71,72).

A complex dioxygenase system capable of converting 4-chlorophenylacetate to 3,4-dihydroxyphenylacetate was isolated and purified from *Pseudomonas* sp. strain CBS3 (73). The system consisted of two reductase components, one containing a monomeric flavin mononucleotide and the other containing a homotrimeric iron-sulfur protein, iron-sulfur clusters (2Fe-2S), and mononuclear iron centers. A similar two-component dioxygenase system active against 2-halobenzoates was isolated and purified from *P. cepacia* 2CBS (74,75).

A new kind of extradiol dioxygenase, 3-chlorocatechol 2,3-dioxygenase, capable of transforming 3-chlorocatechol to 2-hydroxymuconate, was purified and characterized from *Pseudomonas putida* GJ31 (76,77). The enzyme was a homotetrameric protein (4 × 33.4 kD), showed an isoelectric point of 7.1, and, as do other catechol 2,3-dioxygenases, required Fe(II) as a cofactor. Conversely, it was heat-sensitive, being unstable at temperatures above 40°C.

3. *Hydrolytic Dehalogenation*

The hydrolytic dehalogenation reaction is catalyzed by halidohydrolases, which replace the halogen substituent with a hydroxy group derived from a water molecule. This group of dehalogenases is probably the most abundant, and several studies have been carried out to isolate, purify, identify, and characterize the enzymatic proteins. Hydrolytic dehalogenation may occur on haloaromatic compounds, haloalkanes, and haloacid compounds. The enzymes catalyzing the hydrolysis of the three groups of compounds usually present different characteristics and properties, thus forming distinct classes of halidohydrolases.

a. Halidohydrolases Acting on Haloaromatic Compounds Knackmuss (78) considered bacteria to be unable to perform the direct hydrolytic cleavage of aromatic carbon-halogen bonds because of the high stability of the aromatic ring system. However, several findings show that a direct hydrolytic dehalogenation may occur in the degradation of various haloaromatic compounds (49).

Enzymatic dehalogenation of 4-chlorobenzoate was shown to occur by a hydrolytic reaction in extracts of an *Arthrobacter* sp. SU-DSM 20407 (79). An enzyme capable of transforming 4-chlorobenzoate to 4-hydroxybenzoate in the absence of oxygen was detected in cell extracts. The enzyme was very labile: its activity was completely lost after 5 min of boiling. It was active toward 4-benzoate and 4-I-benzoate but not toward 4-F-benzoate, 4-chlorophenylacetate, or 4-chlorocinnamic acid. The maximal activity occurred at pH 7 to 7.5 and at 16°C. The enzyme was a single-protein subunit of 45-kD molecular mass; was inhibited by Zn^{2+}, Cu^{2+}, and EDTA; and was activated by Mn^{2+}.

Hydrolytic dehalogenation of 4-chlorobenzoate was also discovered in *Pseudomonas* sp. strain CBS3 (80–82). Dechlorination of 4-chlorobenzoate to 4-hydroxybenzoate in *Pseudomonas* sp. required Mg^{2+}, adenosine triphosphate ATP, and CoA, showing that a more complex ATP/coenzyme A–dependent reaction was involved. Löffler and Müller (83) determined that a 4-chlorobenzoyl-CoA was formed as an intermediate in the dehalogenation reaction. A 4-chlorobenzoate-CoA ligase was found to be responsible for the adenylation of the carboxyl group of 4-chlorobenzoate, with a concomitant pyrophosphoric cleavage of ATP to adenosine monophosphate (AMP) and pyrophosphate.

The intermediate thioester 4-chlorobenzoyl-CoA is a more active substrate because the substituent in the *para* position is activated by the formation of the CoA ester and then is more susceptible to attack. This intermediate 4-hydroxybenzoyl-CoA is transformed by 4-chlorobenzoyl-CoA dehalogenase, and successively by a third enzyme, 4-hydroxybenzoyl-CoA thioesterase, to the final compound, 4-hydroxybenzoate. This reaction mechanism, involving three distinct enzymes (4-chlorobenzoate-CoA ligase, 4-chlorobenzoyl-CoA dehalogenase, and 4-hydroxybenzoyl-CoA thioesterase), was confirmed in other bacteria such as *Arthrobacter* sp. strain SV (79), *Arthrobacter* sp. strain 4-CB1 (previously named *Acinetobacter* sp. strain 4-CB1) (84), *Corynebacterium sepedonicum* KZ-4, and coryneform bacterium strain NTB-1 (62).

The dehalogenase from a *Pseudomonas* sp. strain CBS3 was purified and found to be a tetramer of 120-kD molecular mass, made of four identical subunits. Further structural, crystallographic investigations performed by Benning et al. (82) indicated that the enzyme is a trimer rather a tetramer and provided significant insight into the reaction mechanism.

The 4-chlorobenzoyl dehalogenase purified to homogeneity from *Arthrobacter* sp. strain 4-CB1 was a homotetramer with subunits of 33-kD molecular mass and was stable between pH 6.5 and 10, with an isoelectric point of 6.1 and maximal activity at pH 8.0. The enzyme may dechlorinate *p*-F-, Cl-, Br-, and I-benzoyl-CoA but is not active for *ortho*- or *meta*-substituted halogen derivatives.

b. Halidohydrolases Acting on Haloalkanes Direct hydrolytic dehalogenation has been demonstrated for haloalkanes. Much experimental evidence seems to indicate that a single enzyme able to react with a broad range of haloalkanes (e.g., C-1 to C-4 1-Cl or 1-Br *n*-alkanes, C-2 to C-5 1-halo-*n*-alkanes, C-3 to C-5 Br-*n*-alkanes, C-2 to C-6 halogenated alcohols, halomethanes) is produced by haloalkane-degrading microorganisms (49).

In the case of dichloromethane, the hydrolysis to formaldehyde and inorganic chloride may also occur through a thiolytic cleavage catalyzed by a dehalogenating glutathione transferase that requires glutathione. The enzyme is produced by various *Pseudomonas* sp. strains, *Hyphomicrobium* sp. strains, and *Methylobacterium* sp. strains (49) and is involved in the detoxification of electrophilic compounds. Experiments performed with dideuterodichloromethane contributed to the elucidation of the reaction mechanism. It seems that a nucleophilic substitution by glutathione occurs, forming an *S*-chloromethyl glutathione conjugate that rapidly hydrolyzes in aqueous solvents.

A single halidohydrolase type of haloalkane dehalogenase was characterized from *Rhodococcus erythropolis* Y2, which was isolated from soil by enrichment culture using 1-chlorobutane (85). The enzyme was a monomeric protein (34 kD) and showed a broad range of substrate specificity, including haloalkanes, haloalcohols, and haloethers, with the highest affinity to α-, ω-di-Cl, and di-Br C2 to C6 alkanes.

The gene of a halidohydrolase able to transform 1,3,4,6-tetrachloro-1,4-cyclohexadiene, the unstable intermediate produced by dehydrohalogenation of the insecticide Lindane (see Sect. V.) in *Pseudomonas paucimobilis* UT26, was cloned in *Escherichia coli*. The enzyme was overproduced by *E. coli* and was purified to homogeneity; it was a monomeric protein of 30- to 32-kD molecular mass that was active with monochloroalkanes (C3 to C10), dichloroalkanes, dibromoalkanes, and chlorinated aliphatic alcohols. The enzyme showed several similarities to various haloalkane dehalogenases (86).

c. Halidohydrolases Acting on Haloacid Compounds As suggested by Hardman (87) and summarized by Fetzer and Lingens (49), dehalogenases catalyzing the hydrolysis of halide ions from haloalkanoic acids may be classified into six groups, depending on their biochemical features (Table 3). The first group, comprising halidohydrolases active on haloacetate but not haloproprionate, can be further subdivided, depending on the ability, or lack thereof, to catalyze the hydrolytic cleavage of the carbon-fluoride bond of fluoroacetate. Groups 2 to 5 are clearly distinguished from each other by substrate specificity and sensitivity to sulfhydryl-blocking agents: (1) enzymes active only toward the L-isomer of 2-monochloropropionate (2-MCPA), producing D-lactate as product and being uninhibited by SH-blocking agents (group 2); (2) enzymes using both D- and L-isomers of 2-MCPA as substrate, producing compounds with an inverse optical configuration and being unaffected by SH-blocking agents (group 3); (3) enzymes still active toward both D- and L-2-MCPA isomers, but with retention of configuration and inhibition by SH-blocking agents (group 4); and (4) enzymes capable of transforming only D-2-MCPA isomer, yielding inversion of configuration and being unaffected by SH-agents (group 5). Finally, group 6 includes enzymes particularly active toward trichloroacetate and catalyzing the complete hydrolytic dechlorination of this substrate.

Several studies have been aimed at elucidating the mechanism by which the inversion of configuration occurs and three possible mechanisms are suggested: (1) a simple displacement of the halide by a hydroxy ion, (2) a nucleophilic attack by activated water, or (3) an attack by a carboxylate group of the enzyme followed by ester hydrolysis.

Further investigations supported the hypothesis that an esteric intermediate is formed by a nucleophilic attack of a carboxylic group of the enzyme on the α-carbon of L-2-haloalkanoic acids and its subsequent hydrolysis by nucleophilic attack of a water molecule (88,89). For a dehalogenase purified from *Pseudomonas* sp. YL, aspartate at position 10 in the protein sequence was proposed to be responsible for the ester intermediate formation (89,90).

Haloalkanoic acid dehalogenases were produced as either constitutive or inducible enzymes. Inducible dehalogenase activities showing a fairly broad substrate specificity were separated and partially purified from a bacterium capable of utilizing the selective herbicide 2,2-dichloropropionate (dalapon) as the sole source of carbon and energy (91,92). Other haloalkanoic acids behaved as possible inducers and were C-substituted

Table 3 Biochemical Features of Halidohydrolases Acting on Haloacid Compounds

Group	Substrate Product Specificity and/or Stereospecificity	Sensitivity to SH-agents
1	Active on haloacetate but not on haloproprionate	—
2	Active on L-isomers, producing D-isomers	No
3	Active on D- and L-isomers, producing L- and D- isomers (inversion of configuration)	No
4	Active on D- and L-isomers, producing L- and D-isomers (retention of configuration)	Yes
5	Active on D-isomers, producing L-isomers	No
6	Active on trichloroacetate	—

more efficiently than Br-substituted compounds (91). Two types of D- and L-specific 2-haloalkanoic acid dehalogenases producing inversion of configuration were purified and characterized (92). They differed from each other in their molecular mass and sensitivity to thiol reagents (92). By contrast, a single halidohydrolase-type dehalogenase was induced by *Pseudomonas cepacia* MBA4 and was able to grow on monobromoacetic acid as a sole C and energy source (93). The enzyme was purified and was found to be a dimeric protein of the 23-kD molecular mass subunit. The enzyme was L-isomer-specific, able to transform only L-2-MCPA.

Dehalogenase activities, induced under different growing conditions and showing different catalytic behavior, were detected by using monochloroacetate as the sole carbon source in crude extracts from seven soil bacterial strains (94). The bacteria were three *Pseudomonas* spp., an *Alcaligenes* sp., an *Agrobacterium* sp., an *Arthrobacter* sp., and an *Azotobacter* sp. The crude enzymatic preparations showed a relatively wide range of substrate specificity, had different affinities for monochloroacetate (as estimated by K_m values), and had different thermal stability. Conversely, a similar optimal pH range (8 to 10) and pH stability profile were determined for all active fractions.

A thermostable L-2-haloacid dehalogenase and a nonthermostable D-, L-2-haloacid dehalogenase were synthesized in *Pseudomonas* sp. strain YL by 2-chloropropionate and 2-chloroacrylate, respectively (95). The two enzymes were purified to homogeneity and showed different biochemical and structural properties, but both catalyzed halide hydrolysis with inversion of product configuration. The L-2-haloacid dehalogenase was a dimer with two 27-kD subunits active toward short- and long-carbon-chain haloacids. It showed pH and temperature optimal values of 9.5 and 65°C, respectively, and was particularly thermostable: it was fully active after 30-min heating at 60°C. By contrast, D- and L-2-haloacid halogenase was a 36-kD monomer that was active at pH 10.5 and 45°C toward D- and L-isomers of 2-chloropropionate and had a high sensitivity to thermal treatment (95). A relatively thermostable monomeric 2-haloacid dehalogenase specific for the L-isomer of optically active haloacids and producing the inversion of the product configuration was also purified from *Azotobacter* sp. strain RC26. The enzyme was active up to 60°C and showed a high substrate affinity and resistance to enzyme inhibitors (96,97).

Two dehalogenase enzymatic fractions, both active on D- and L-isomers of 2-MCPA, had been isolated previously and purified from *Pseudomonas putida* PP3 (98). One yielded products with the same optical configuration (fraction I), and the other yielded products with the opposite optical configuration (fraction II). The two dehalogenases showed other differences in their responses to SH-blocking agents and their efficiency in dechlorinating L- and D-2-MCPA isomers. Conversely, a unique DL-2-haloacid dehalogenase catalyzing the hydrolytic dehalogenation of both D- and L-2-haloalkanoic acids to yield the corresponding L- and D-2-hydroxy derivatives was purified from *Pseudomonas* sp. strain 113 (99).

4. *Haloalcohol Dehalogenation*

Vicinal haloalcohols are converted to the corresponding epoxides by a class of dehalogenases called haloalcohol dehalogenases. The reaction is an intramolecular nucleophilic substitution of the halogen with an oxygen atom and subsequent release of HCl.

Enzymes able to catalyze this kind of reaction have been identified in *Flavobacterium* sp. (100), *Pseudomonas* sp. strain AD1, *Arthrobacter* sp. strain AD2, and a coryneform strain AD3 (101), *Arthrobacter erithii* H10a (102,103), and *Corynebacterium* sp. strain N-1074 (104–106). The enzymes were purified and characterized from *Flavobacte-*

rium sp. (100), *Arthrobacter* sp. strain AD2 (101), *Corynebacterium* sp. strain N-1074 (104–106), and *Arthrobacter erithii* H10a (103). The enzyme purified from *Arthrobacter* sp. strain AD2 was a dimer with a subunit molecular mass of 29 kD. It was active toward C-2 and C-3 bromo- and chloroalcohols, producing epoxides as products. The intramolecular substitution involved in the reaction mechanisms was confirmed by the lack of a requirement for cofactors or oxygen for the dehalogenation of substrates and by the lack of immunological cross-reactions with 2-haloalkanoic acid dehalogenases from other bacterial strains (101).

By contrast, two enzymes catalyzing the dehalogenation of vicinal halohydrins were purified from *Corynebacterium* sp. strain N-1074 and were named *la* and *lb* (104–106); two enzymes from *Arthrobacter erithii* H10a were named *DehA* and *DehC* (103). When the gene from *Corynebacterium* sp. strain N-1074 was cloned in *E. coli*, a single dehalogenating enzyme that consisted of four identical subunits (28 kD), had no metals, and was not inhibited by thiols or carbonyl reagents (104) was expressed.

The native *DehA* enzyme from *Arthrobacter erithii* H10a is a hexamer protein made by two subunits with different molecular masses (31.5 and 34 kD) that were combined in the ratio 1:1. Five dehalogenase-active bands were obtained with SDS-PAGE; two-dimensional PAGE results indicated that different combinations of each or both subunits gave rise to the protein bands: 1,3-dichloro-2-propanol (1,3-DCP), 3-chloro- 1,2-propanediol (3-CPD), and brominated alcohols were substrates of the enzyme that showed a greater affinity for 1,3-DCP, as assessed by the lowest value of the Michaelis–Menten constant. Enzyme activity reached its maximum at 50°C and at pH between 8.5 and 10.5. The enzyme was inactivated at temperatures above 50°C and was inhibited with a mixed-type inhibition mechanism by 2-chloroacetic acid and 2,2-dichloroacetic acid. Punctual amino acid modification studies indicated the importance of one or more cysteine and arginine residues in the catalysis and the stability of the protein structure (103).

The lb enzyme isolated from *Corynebacterium* strain N-1070 also showed five protein bands. It was a 115-kD tetramer in which two subunits of 32 and 35 kD were present in different combinations (105). With respect to *A. erithii* DehA dehalogenase, lb enzyme showed lower pH and temperature optima, less affinity for 1,3-DCP than for 3-CPD and brominated alcohols, and slight inhibition by SH agents (104,106).

When substrate profiles of *A. erithii* DehC and *Corynebacterium* sp. la dehalogenases were compared, they were very similar to each other and to the dehalogenase of the enzyme purified from *Arthrobacter* sp. AD2 (101). Furthermore, other similarities (e.g., comparable molecular mass subunits, N-terminal amino acid sequence, sensitivity to SH-reagents, optimal pH and temperature around 8.0 to 9.0 and 50°C to 55°C, respectively) were displayed by dehalogenase la and that from *Arthrobacter* sp. AD2.

These results suggest the existence of two types of haloalcohol dehalogenases. The first includes *A. erithii* DehA and *Corynebacterium* lb and is represented by more complex structural proteins with a restricted substrate specificity and higher affinity for 1,3-DCP. The second, including *Corynebacterium* la and *Arthrobacter* AD2 dehalogenases, consists of multimeric protein having less structural complexity and showing a broader substrate specificity (103).

5. Dehydrohalogenation

In the dehydrohalogenation mechanism, the removal of an HCl molecule gives rise to the formation of a double bond. This kind of dehydrohalogenating reaction has been recognized to be involved in the two first steps of the metabolic pathway of the insecticide γ-hexachlorocyclohexane (γ-HCH) (Lindane) in *Pseudomonas paucimobilis* UT26 (107–110). The enzyme, named *LinA*, is responsible for the dehalogenation of γ-HCH to γ-pentachlorocyclohexene, and then to an unstable intermediate, 1,3,4,6-tetrachloro 1,4-cyclohexadiene, which can be converted by further spontaneous and/or enzymatic reactions to different dead-end products (see previous sections).

The gene expressing *LinA* in *P. paucimobilis* UT26 was identified and sequenced (108). A comparison was made with the well-known and well-characterized dehydrohalogenase responsible for the monodehydrodechlorination of DDT in *Musca domestica* (111,112). Neither homology of amino acid sequences nor similar catalytic properties were found for the two enzymes (108).

The enzyme from the gene cloned and expressed in *E. coli* (109) showed a very narrow substrate specificity; was active only toward α-, δ-, and γ-HCH and toward α- and γ-pentachlorocyclohexene; was a tetramer composed of four subunits (molecular mass 16.5 kD); and did not require cofactors. A release of three chloride ions per molecule of γ-HCH was found. Genes encoding for an extracelluar LinA enzyme were characterized by using polymerase chain reaction (PCR) strategy from a novel γ-HCH-degrading bacterium isolated from a contaminated soil (113).

6. *Common Features of Dehalogenases*

Some dehalogenases have been purified to homogeneity and characterized for their molecular, catalytic, and genetic properties. In some cases, crystals have been obtained and analyzed, thus allowing the three-dimensional structure of the enzyme to be determined (82,114,115).

Dehalogenases present some common molecular features. They are usually monomeric small proteins with a molecular mass around 30 to 40 kD. Oligomeric structures (two to four subunits) have also been found with higher molecular mass (64 to 200 kD). Moreover, they are often active at alkaline pH and between 40°C and 50°C. Numerous protein-engineering studies have been performed to determine substrate specificity and reaction mechanisms. The genes encoding purified enzymes have been identified, sequenced, and cloned in hosting organisms (76,86,88,99,116–121).

To study the properties and structure of the thermostable L-2-haloacid dehalogenase from *Pseudomonas* sp. strain YL, the enzyme was purified from *E. coli*, in which the gene encoding the enzyme was easily overexpressed (88). The purified enzyme was crystallized and the crystal structure determined (114,122). Two structurally distinct domains, the core domain and the subdomain, with an active site between them, were identified (122). A similar three-dimensional structure, but with an extra dimerization domain, was found by using sophisticated methods to determine the crystal structure of an L-2-haloacid dehalogenase isolated from the 1,2-dichloroethane-degrading bacterium *Xanthobacter autotrophicus* GJ10 (115).

Site-directed mutagenesis studies were applied to haloalkane dehalogenass to identify the catalytic mechanism of the enzyme. Punctual mutations in the protein sequence provided more insights into the role of some specific amino acids in the reaction mechanism. Krooshof et al. (123) showed that the catalytic triad (Aspl24, His 289, and Asp260) is important for the catalytic performance of the enzyme. Mutation of Asp260 to asparagine produced a catalytically inactive enzyme. Furthermore, the authors recognized that the presence of an aromatic residue at position 175 was essential for the formation of the enzyme–substrate complex (124). A detailed description of the reaction mechanism was provided by the authors (125). Release of the halide ions from the hydrolysis of short-chain haloalkanes can proceed via a two-step or a three-step route, both containing a slow enzyme isomerization step. Thermodynamic analysis of halide binding and release suggested that the three-step route involves larger conformational changes than the two-step route. A more open configuration of the active site from which the halide ion can readily escape may result.

Detailed analysis of the rate-determining step in the hydrolysis of various substrates by haloalkane dehalogenases was performed by Schanstra et al. (126–129). The values of the kinetic constant k_{cat} for the natural substrates 1,2-dichloroethane and the brominated analog and nematocide 1,2-dibromoethane confirmed that the rate-limiting step in the conversion of the compounds is the release of the charged ion out of the active site (127,128). Using stopped-flow fluorescence techniques and different concentrations of halide, Schanstra and Janssen (126) also proposed the existence of two parallel routes for halide binding. The routes involved, in this case, slow conformational changes.

The roles of phenylalanine 172 (128) and valine 226 (129) in dehalogenase function were established by experiments performed by mutational analysis of position 172 and by study of the kinetics and X-ray structure of the Phe172/Trp (128) and Val226/Ala, Val226/Gly, and Val226/Leu enzymes (129). The key role of the amino acid in position 172 was confirmed by studying quantitative structure–function and structure–stability relationships of 15 mutants in position 172 of the haloalkane dehalogenase (130,131). The computational site-directed mutagenesis of the set of single-point mutants of the enzyme allowed the researchers to distinguish between protein variants with high activity and mutant proteins with low activity (131).

Protein engineering studies, directed at elucidating the role of amino acids located at specific positions of the protein sequence, were performed on other dehalogenases, including fluoroacetate dehalogenase from *Moraxella* sp. B (90), 4-chlorobenzoyl-coenzyme A dehalogenase from *Pseudomonas* sp. CBS-3 (82,132), dichloromethane dehalogenase/glutathione transferase from *Methylophilus* sp. DM11 (133), *Hyphomicrobium* sp. DM2 and *Methylobacterium* sp. DM4 (134), L-2-haloacid dehalogenase from *Pseudomonas* sp. YL (89,122,135,136), and D- and L-2-haloacid dehalogenase from *Pseudomonas* sp. 113 (99).

The role of aspartate 145 was identified as being essential to dehalogenase catalysis by 4-chlorobenzoyl-coenzyme A dehalogenase from *Pseudomonas* sp. strain CBS-3; this was confirmed by structural studies performed on a substrate (4-hydroxybenzoyl-coenzyme A complex) (82,132). Sophisticated techniques (i.e., multiple isomorphous replacement, solvent flattering, and molecular averaging) and site-directed mutagenesis studies demonstrated that not only aspartate, but also histidine and tryptophan residues, are essential to 4-chlorobenzoyl-coenzyme A dechlorination.

Site-directed mutagenesis investigations were also carried out on a haloalkane dehalogenase from *Xanthobacter autotrophicus* GJIO to enlarge the range of chlorinated solvents degraded by the bacterium. Replacement of some amino acids with alanine produced enzymes with lower activity toward 1,2-dichloroethane than shown by the naturally occurring enzymes. Valuable information on the reaction mechanism and the limited substrate specificity of the *Xantobacter* sp. dehalogenase was derived from the three-dimensional structure of the enzyme complexed with the substrate-analog formate (115). Furthermore, modeling studies performed on mutant haloalkane dehalogenases suggested that coupling of the dehalogenation reaction with hydrogenation of the halide ion formed during the reaction in the active site could improve the catalytic activity of the enzyme (131).

As outlined by Fetzner and Lingens (49), the possibility of modifying the substrate specificity of an enzyme by site-directed mutagenesis may be regarded as a useful strategy for the preparation of bacterial strains tailored to the bioremediation of recalcitrant compounds. This strategy could improve the catabolic activity of a simple organism or develop catabolic enzymes with new or higher activities than those of the wild ones.

C. Oxidoreductive Enzymes

Some oxidoreductases have been proved to be active toward a large variety of potentially polluting compounds or their direct derivatives. Oxygenases, phenoloxidases, and peroxidases are able to oxidize a broad range of aromatic compounds such as phenols and substituted phenols, anilines, PAHs, as well as nonaromatic compounds, including alkanes and substituted alkanes. The cleavage of the aromatic ring as well as the formation of unstable substrate cation radicals with subsequent nonenzymatic transformation (e.g., C-C or ether cleavage or oxidative coupling) and polymerization are reactions that may occur. The involvement of oxygen or hydrogen peroxide as an oxidative agent is required. A common feature of these enzymes is their capability to promote, in some cases, a spontaneous or fortuitous dehalogenation of halogenated compounds as a result of chemical decomposition of unstable primary products.

The biochemical properties, the catalytic features, and the potential use of these enzymes for practical applications have been exhaustively studied, and numerous reviews have been published (11,137–154).

In the following subsections, the principal characteristics of these enzymes are summarized and recent findings are presented.

1. *Oxygenases*

Mono- and dioxygenases catalyze essential steps of the metabolic degradative pathway of several toxic compounds. Alkanes, aromatics, chlorophenols and nitrophenols, and PAHs are oxidatively degraded by bacteria and mediated by either specific or nonspecific oxygenases (49,139). Compared to the phenoloxidases and peroxidases, relatively few oxygenases have been isolated from their producing organisms and characterized for their molecular and kinetic properties.

Usually, oxygenases are multimeric proteins, comprising two, three, up to six enzymatic components. No component alone is able to oxidize the substrate; only when the components are combined can oxidation be detected. Iron, a coenzyme (NAD(P)H or $FADH_2$), and oxygen typically are required as cofactors. The enzymes present a wide substrate specificity and a pH optimum near neutral, and they often are inactivated by their substrate or substrate-like compounds. The products of the oxygenase-assisted reaction usually undergo further transformation to simpler compounds in vivo or are completely mineralized to carbon dioxide and water.

One of the first oxygenases extensively studied in a purified form was a pyrazon (4-amino-5-chloro-1-phenyl-6-pyridazinone) dioxygenase (155). The protein produced by pyrazon-degrading bacteria was purified to homogeneity and consisted of three different enzyme components. By electron paramagnetic resonance spectroscopy (EPR) and ultraviolet (UV) measurements, it was demonstrated that the first component is a Fe-protein containing 2 mol Fe and 2 mol inorganic sulfur/mol protein; the second has flavin adenine dinucleotide (FAD) as a prosthetic group; the third is a ferredoxin-type protein. All three components are required for the oxidation of pyrazon (155).

A catechol 2,3-dioxygenase was isolated and purified to homogeneity from a pyrazon-degrading bacterium (156). The enzyme, which was active against catechol, was composed of six identical subunits, had an optimal pH in the range 7 to 8, and required F^{2+} as a cofactor. A chlorocatechol 2,3-dioxygenase (part of the degradative pathway used for growth of *Pseudomonas* sp. with chlorobenzene) was isolated and purified from *Pseudomonas putida* GJ31 and showed similar biochemical and kinetic features (76,77) (see Sect. II.B.2).

As previously reported, TCE may be fortuitously oxidized by enzymes involved in the degradation of aromatics (toluene-2-monooxygenase) or alkanes (methane monooxygenase) in *Pseudomonas* (*Burkholderia*) species. The enzyme toluene 2-monooxygenase is usually very sensitive to the presence of other compounds that can cause a partial or total inactivation of the protein. Yeager et al. (157) have demonstrated that the enzyme from *Burkholderia cepacia* G4 is efficiently inactivated by low concentrations of longer-chain alkynes (C5 to C10). The toluene- and *o*-cresol-dependent O_2 activities were irreversibly lost when the bacterium was exposed to alkynes. Experiments performed in the presence of increasing concentrations of toluene or oxygen (supplied as H_2O_2) suggest that alkynes are specific, mechanism-based inactivators of the enzyme. Ethylene and propylene were not inactivators; they behaved as substrates and were oxidized to their respective epoxides.

A toluene/*o*-xylene monooxygenase (ToMO) from *Pseudomonas stutzeri* OX1 was shown to be able to degrade several environmental pollutants, including TCE, 1,1-dichloroethylene (1,1-DCE), *cis*-1,2-DCE, *trans*-1,2-DCE, chloroform, dichloromethane, phenol, 2,4-dichlorophenol, 2,4,5-trichlorophenol, 2,4,6-trichlorophenol, 2,3,5,6-tetrachlorophenol, and 2,3,4,5,6-pentachlorophenol (158). The ToMO genes were cloned in *E. coli* and the expressed enzyme degraded TCE, 1,1-DCE, and chloroform very efficiently (i.e., very low K_m values and high turnover numbers: number of substrate molecules transformed per second per enzyme (159–161).

Another oxygenase, purified from *Burkholderia* sp. strains, catalyzes the removal of the nitro group from 4-methyl-5-nitrocatechol, an intermediate product of 2,4-dinitrotoluene degradation by the bacterium; 2-hydroxy-5-methylquinone is formed as the product. The enzyme is a monomeric protein of 65 kD, contains 1 mol FAD/mol protein, and requires NADPH and oxygen (162). A dimeric nitroalkane-oxidizing enzyme, containing flavin mononucleotide rather than flavin dinucleotide as the prosthetic group, was purified to homogeneity from *Neurospora crassa* (163). The enzyme catalyzes the oxidation of nitroalkanes, producing the corresponding carbonyl compounds, and is not active on aromatic compounds.

The degradation of PAHs, naphthalene, and phenanthrene may occur by oxygenase action. A biphenyl 2,3-dioxygenase isolated and purified from *Pseudomonas* sp. strain LB400 (164) oxidized naphthalene to *cis*-dihydrodiol but with a lower efficiency than in the oxidation of biphenyl. Haddock et al. (165) demonstrated that the purified enzyme also is able to oxidize chlorinated biphenyls, producing dihydrodiols and dechlorination of the substrate. A purified dioxygenase, namely, a 1,2-dihydroxy-naphthalene dehydrogenase, involved in the naphthalene degradation pathway, was able to catalyze the *meta* cleavage of dihydroxy and/or polychlorinated biphenyls (166).

2. Phenoloxidases and Peroxidases

Phenoloxidases and peroxidases are two groups of oxidoreductases produced by a large number of living cells (microorganisms, plants, and animals). They are classified also as oxygenases, but they are described separately for their peculiar characteristics. The main producers of both groups are white rot fungi, suggesting a primary role of these enzymes in lignin transformation (11,12,138,140,144,145,147,148,152–154).

Phenoloxidases, including tyrosinases and laccases, require molecular oxygen for activity, whereas peroxidases that comprise horseradish peroxidase, ligninases (i.e., lignin- and manganese-peroxidases), and chloroperoxidases utilize hydrogen peroxide. In some, e.g., manganese-peroxidases, the reaction depends on the presence of other components such as divalent manganese and particular types of buffers. Both groups catalyze, by different mechanisms, the oxidation of phenolic and nonphenolic aromatic compounds through an oxidative coupling reaction that results in the formation of polymeric products of increasing complexity. Cross-coupling may occur in reactions between substrates of different nature. Oxidation of relatively inert substrates by the copresence of more reactive molecules also may occur. As a consequence of the oxidative coupling reaction, a ''spontaneous'' or ''fortuitous'' dehalogenation of halogenated compounds may be promoted. Polyphenoloxidases may be subdivided into two subclasses: tyrosinases and laccases.

Tyrosinases are copper-containing monooxygenases, usually named *polyphenoloxidases*, *phenolases*, or *catecholases*. Tyrosinases catalyze two types of reaction that occur sequentially. The first reaction is an *o*-hydroxylation of phenols with molecular oxygen to produce catechols (cresolase activity); the second reaction is the subsequent oxidation of catechols with oxygen to form *o*-quinones (catecholase activity); *o*-quinones are unstable compounds and spontaneously polymerize in a nonenzymatic reaction to produce melanin like, insoluble products (137–139,144,145,152). In 1999 the substrate range of tyrosinases was extended to other compounds such as *p*-hydroxy- and 3,4-dihydroxyphenylpropionic acid (167) and caffeic acid (168).

Experiments performed with guaiacol and catechol have provided insights into the nature and structural characteristics of products formed by tyrosinase action (169,170). Seven guaiacol-derived oligomers formed by tyrosinase were found to be similar to those obtained by laccase and/or peroxidase action (169). Further investigations carried out with catechol as the substrate demonstrated that reaction products formed after catalysis by tyrosinase were catechol-melanin, brown-colored polymers with a high degree of aromatic ring condensation (170). Tyrosinases have been proved to be stable in organic solvents, provided that a small amount of water is present in the reaction system to maintain conformational flexibility in the protein molecules (171–173). This property makes the enzyme suitable for producing some compounds such as quinones that are difficult to obtain, rather than completely removing phenols (173).

Laccases are also multicopper proteins, produced primarily by fungi but also by bacteria and higher plants. An extensive literature is available on this group of enzymes (11,137–139,142–144,146–148,152–154). Current knowledge on the origin and distribution of laccases, the requirements and characteristics for their production at high yields, and their properties at molecular and kinetic levels has been reviewed (152). The authors also examined applications of laccases for both environmental and nonenvironmental purposes. Other progress in the study of laccases has also been published (154).

One of the most interesting aspects of these enzymes is their capability to transform relatively recalcitrant compounds if additional, cosubstrates or proper redox, mediators are present (174–178). Chemical mediators are believed to be oxidized by the enzyme and then undergo oxidative coupling with laccase substrates. This process seems to be essential to the laccase-mediated delignification of chemical pulp. For example, 2,2′-azino-*bis*(3-ethyl-benzthiazoline-6s-sulfonic acid) (ABTS) was one of the first reported mediators for laccase. The presence of ABTS allowed the enzyme to oxidize various benzyl alcohols to the corresponding aldehydes, thus demonstrating a novel catalytic action of the enzyme (175). In 1998 (177) it was shown that *N*-hydroxybenzotriazole is a more effective mediator for the laccase-mediated transformation of chemical pulp.

Mediators also allow the oxidation of PAHs by laccases. Collins and colleagues (176) provided evidence that laccase from *Trametes versicolor* was able to oxidize anthracene and benzo(*a*)pyrene, and ABTS showed a significant stimulatory effect when added to the reaction mixture. In further studies conducted with 14 different PAHs of environmental relevance, Majcherczyk et al. (178) demonstrated that both ABTS and 1-hydroxybenzotriazole (HBT) increased oxidation of almost all the PAHs to their complete removal from the reaction mixture.

The herbicides triazine and prometryn, which are not substrates for laccase, showed an inhibiting effect on laccase activity when assayed using 2,4-dichlorophenol and catechol as substrates (179). This effect might be a drawback to the possible use of laccase for the removal of phenols from polluted systems. Indeed, agricultural sites are usually treated with several organochemicals, so that triazine compounds may be present at phenol-contaminated sites.

Peroxidases are another group of oxygenases that differ from phenoloxidases and common oxygenases because they do not require coenzymes and are H_2O_2- (but not O_2-) dependent. Alkyl hydroperoxides, such as methyl peroxide or ethyl peroxide, may act as hydrogen acceptors but with a lower specificity. Peroxidases are found ubiquitously in nature (i.e., in microorganisms, plants, and animals), and they catalyze the oxidation of a large array of natural and synthetic substrates forming polymeric products. The most studied peroxidase is that produced by horseradish (HRP), although there are other fungal peroxidases such as lignin and manganese peroxidases (139–141,149–151).

Phenols, biphenols, anilines, benzidines, and related heteroaromatic compounds, as well as arylamine carcinogens (benzidine and naphthylamines), behaved as peroxidase substrates (180–183). Compounds such as PCBs, naphthalene, and anthracene, which themselves are not substrates for the enzyme, were efficiently removed by coprecipitation with phenols and anilines used as the substrate (184–186). The formation of azo dye compounds, occurring by oxidative coupling between phenols and hydrazone derivatives, was demonstrated and proposed as a method for determining peroxidase activity at very low (picomolar) levels (187). Moreover, enzymes demonstrated catalytic activity over a wide range of pH values, temperatures, and substrate concentrations (188).

The enzymes are classical heme proteins in which ferric protoporphyrin IX is the prostethic group. Ca^{2+} often is present. Another characteristic of this group of enzymes is their occurrence as a large family of isoenzymes. For example, HRP enzyme may exist in >10 isoenzymic forms (some sources estimate approximately 40 isoenzymes) (140).

Peroxidase-catalyzed reactions of aromatic compound AH_2 proceed through a one-electron oxidation usually described by the following mechanism:

$$E + H_2O_2 \rightarrow E_1 + H_2O$$

$$E_1 + AH_2 \rightarrow E_2 + \cdot AH$$

$$E_2 + AH_2 \rightarrow E + \cdot AH + H_2O$$

where E indicates the native enzyme; E_1 is the first intermediate, i.e., an oxidized state of the enzyme, able to accept the substrate and to oxidize it; and E_2 is the second intermediate enzymic form that oxidizes a second molecule of the substrate and regenerates the native enzyme E. The free radicals produced may diffuse into solution and generally undergo nonenzymatic transformation to polyaromatic, nonradical products by pathways characteristic of each substrate (coupling, dismutation, etc.).

Some side reactions that can influence the efficiency of the enzyme may occur. A free radical may bind to or near the active site of the enzyme and permanently inactivate the enzyme's catalytic ability (189). In addition, the intermediate E_2 may react with H_2O_2 (if present in excess) and transform into a third, catalytically inactive form E_3 of the enzyme. This latter may spontaneously regenerate the native enzyme E and produce O_2^- (190); the process, however, is so slow as to hamper the catalytic oxidation of aromatic substrates. Furthermore, losses in catalytic efficiency may result from adsorption of the oxidative enzyme onto or entrapment within polymeric aggregates as they form (191). As postulated by Nakamoto and Machida (192), this process represents the main mechanism of inactivation because it hinders contact between the substrate and the enzyme.

Efforts have been devoted to finding useful strategies to reduce the inactivation of the enzyme during the reaction. One of the most promising breakthroughs is the use of additives with high hydrophilicity. Such compounds, including polyethylene glycol (PEG) and gelatin, may have a greater affinity than the enzyme for the hydroxyl groups that are present on the growing polymers. Consequently, they preferentially bind to the polymers and allow the enzyme to stay free in solution and catalyze further reactions (192–194). Experiments performed with HRP and devoted to elucidating the inactivation kinetics indicate that additives couple with most of the polymer products so that less enzyme is subtracted during the reaction. The enzyme still may combine with polymers and become inactivated, but at a much slower rate when additives are present, thus showing that the additives have a protective effect (194).

III. IMMOBILIZATION OF ENZYMES

A. Advantages of Immobilization

Enzymes, as globular protein macromolecules, are characterized by high solubility in aqueous and/or diluted salt solutions. This solubility implies that enzymes are homogeneous catalysts: i.e., they exert their catalytic action in a single isotropic soluble phase in which all components (enzymes, hydrogen ions, substrates, products, inhibitors, activators, cofactors, etc.) are simultaneously present. All experimental procedures used to isolate and purify enzymes from living organisms are based on the solubility of proteins in water solutions.

Compared to inorganic catalysts, enzymes have unusual properties (e.g., high substrate catalytic power, high substrate specificity, regulated by small ions or other molecules) that make them versatile agents for biotechnological and environmental applications. In fact, their high degree of substrate specificity largely eliminates the production of undesirable by-products and thus decreases not only material costs but downstream environmental burdens. Their high reaction velocities reduce manufacturing costs, and the mild conditions usually utilized in enzymatic processes substantially decrease the possibility of damage to heat-sensitive substrates and reduce the energy requirements and corrosion effects of the process.

However, the use of a catalyst for applied purposes requires that the catalyst must (1) be reused; (2) be recovered at the end of the process; (3) be used in continuous processes; and (4) have long-term and operational stability. These properties are not found in an enzyme used in soluble form, which: (1) is not reusable; (2) is difficult or impossible to recover; (3) usually has low stability toward several denaturing agents; (4) cannot be used for prolonged periods; and (5) if used as a commercial preparation, may have adverse sensory or toxicological properties. In the early 1970s, scientists circumvented some of these effects by attaching enzymatic proteins onto solid supports. In other words, they converted an enzyme from a homogeneous to a heterogeneous catalyst, developing a new technology based on enzyme immobilization.

Many definitions of immobilization have been proposed. One involves the conversion of enzymes from a water-soluble mobile state to a water-insoluble immobile state. Thus, immobilization implies that the mobility of protein molecules or their fragments is limited in space or in relation to each other.

During the last few decades, much research effort has been devoted to producing satisfactory immobilized enzymes for large-scale application. Several techniques have been elaborated, and extensive literature is available on methods of immobilization, types of immobilizing supports, mechanisms of immobilization, properties of immobilized enzymes, and use of immobilized enzymes in several areas.

B. Techniques of Immobilization

To attach enzymes to solid supports, different techniques may be applied. Enzymes may be bound either to carriers or to each other, or physically confined in a restricted space by entrapment or encapsulation. The most widely used immobilization techniques can be classified according to whether the protein becomes immobilized by chemical binding or by physical retention. Immobilization methods include (1) binding of enzyme molecules to carriers by covalent bonds; (2) binding of enzyme molecules to carriers by cooperative adsorptive interactions; (3) entrapment into gels, beads, or fibers; (4) encapsulation in stable aggregates; and (5) cross-linking or co-cross-linking with bifunctional agents. These methods may be used individually or in combination (Figs. 2 and 3). Most of the immobilization mechanisms are present in nature, as enzymes often work as immobilized rather than free catalysts. Soil enzymes can be considered a type of naturally immobilized enzymes.

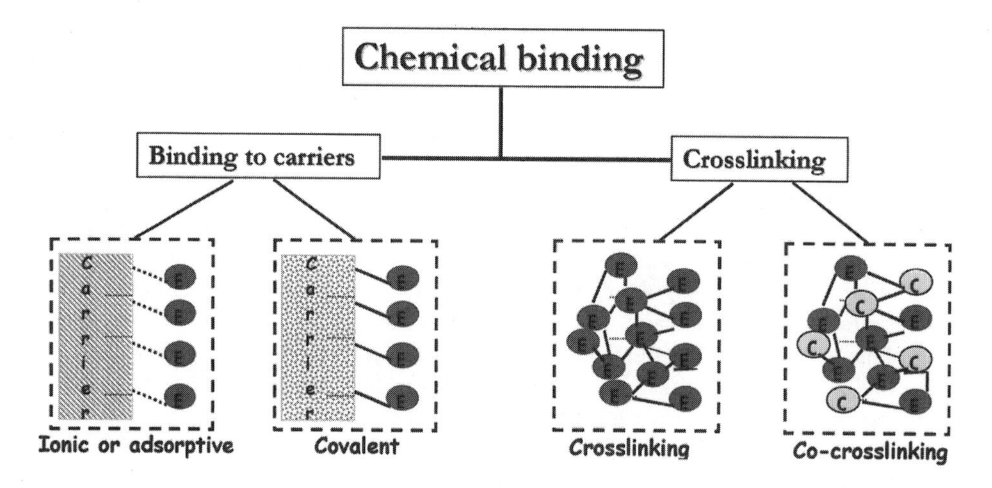

Figure 2 Immobilization of enzymes by chemical techniques.

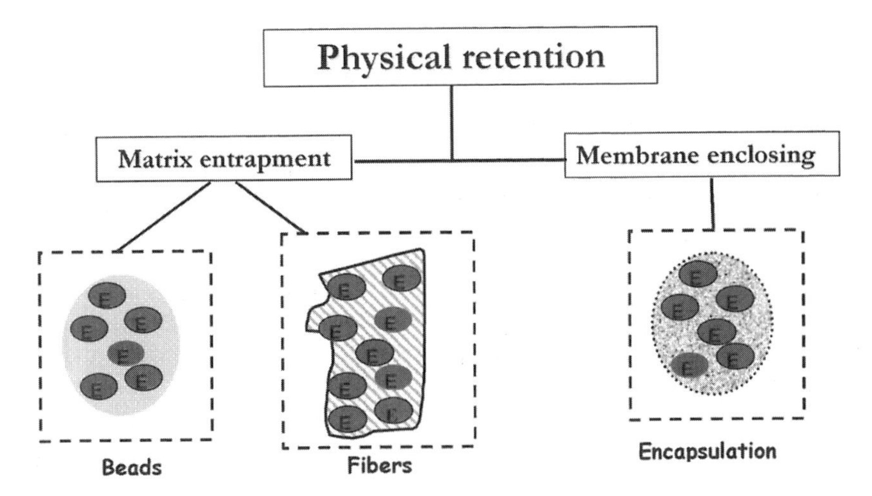

Figure 3 Immobilization of enzymes by physical techniques.

The art of immobilization is to balance the loss of initial activity (that usually occurs) with long-term retention of activity. In general, continuing activity is more valuable than high initial activity. Adsorption is probably the simplest of all the techniques, but it is less convenient because the binding forces between the enzyme and the carrier may be so weak that the enzyme can be easily desorbed under operational conditions. By contrast, physical entrapment in polymeric gels or membranes prevents the escape of the enzyme while allowing passage of substrate and products. However, continuous decrease of activity due to loss of enzyme molecules through large pores, as well as diffusional barriers to high-molecular mass substrates and/or products, may constitute major drawbacks to this immobilization technique.

Covalent binding, accomplished through functional groups on both enzyme and carrier, probably is a more advantageous method, although some disadvantages may be present. These include difficulty of preparation, no possibility for regenerating the carrier, high costs, and the possibility of direct involvement of the active site in the binding.

The following criteria are used to select the most suitable immobilization technique to use an enzyme for practical applications: simplicity of the method, low cost, general applicability to a large number of enzymes, and capability of easy scale-up. The choice of carrier should be made according to some general principles that should take into account its binding capacity, influence on enzyme performance, operational stability, chemical inertness, and cost.

C. Properties of Immobilized Enzymes

Catalysis by immobilized enzymes is generally different from that of free enzymes. The catalysis acquires a "heterogeneous nature" brought about by a "static enzyme" that represents an individual phase separated from the outer solution.

The main two properties of immobilized enzymes are (1) increased stability toward physical, chemical, and biological denaturing agents due to minimization or prevention of conformational changes, and (2) altered catalytic behavior caused by the heterogeneous nature of the catalysis associated with a static enzyme. These properties are strictly dependent on the major components of the immobilized enzyme system (the enzyme and the carrier) and their mode of association. Additional variables, such as pH, temperature, ionic strength, pressure, agitation, need for cofactors, substrate delivery, and product removal, may all contribute to the environment and thus to the performance of the enzyme product removal. The surface of the carrier, whether inorganic or organic in composition, may affect the activity of the enzyme by means of microenvironmental variables such as pH effects and/or buffering, hydrophobicity and hydrophilicity, and redox surface effects. The reduced activity of enzymes attached to or entrapped within an insoluble carrier may be ascribed to possible conformational changes occurring within the tertiary structure of the protein after immobilization. Denaturing effects can be due to the great number of linkages between the enzyme and the support and to the interference of hydrophilic residue of the carrier with the hydrophobic groups of the enzymatic protein. The mode of attachment may involve groups that are part of the active site. Furthermore, immobilization of the protein may alter the binding between the various subunits of the enzymes (195).

Steric limitations on the penetration of the substrate into the enzyme active site may be the result of restricted movement of the substrate to the otherwise available active site of the enzyme or of the orientation of the enzyme in relation to the carrier surface. In other words, the active site may be partially or completely inaccessible to the substrate. These effects are more evident for high-molecular-mass substrates and may account for the increased stability of immobilized enzymes to the attack of proteases, which are high-molecular-weight proteins. Their attack can be prevented by the shield formed by the support around the enzyme. The shield may hinder the ability of proteases to move closer to the enzyme without hindering the diffusion of the substrate to, and of the product from, the active site. Several hydrolases immobilized on solid supports may exhibit markedly lower enzymatic activity toward polymeric substrates (e.g., proteins, polysaccharides, and nucleic acids) than expected on the basis of their reactivity to small substrates.

The rate of transformation by immobilized enzymes is influenced by the availability of the substrate and effectors. With regard to the diffusion of reactants, five distinct steps in the overall enzymatic process may be identified: (1) diffusion of the substrate from the bulk solution to the immobilized biocatalyst across a boundary layer of water; (2) diffusion of the substrate inside the immobilized enzyme particle, i.e., transport of the substrate from the carrier surface to the active site; (3) enzymatic conversion of the substrate into a product; (4) transport of the product to the external surface of the immobilized particle; and (5) diffusion of the product into the bulk solution (196).

Immobilized enzymes often show different pH-activity profiles that are usually attributed to microenvironmental effects. In the case of charged supports, the pH activity may result in the partitioning of charged substrates and products between the domain of the immobilized enzyme and the outer solution (197).

Immobilized enzymes in the proximity of the support exist in an environment that is different from the environment of the solution. The concentrations of substrate, effectors, and/or hydrogen or hydroxyl ions (if the support is charged) within the immobilizing support may be different from those in the bulk solution. Electrostatic or hydrophobic interactions between the matrix and species may occur, and a consequent modification of the microenvironment (e.g., pH values, composition, dimensions) surrounding the enzyme may result.

The temperature–activity profile of an immobilized enzyme usually shows higher activity at high temperatures than that of the corresponding free enzyme. The increased activity indicates increased stability achieved by the enzyme upon immobilization. The increased thermal stability is usually attributed to minimization or prevention of conformational changes in the protein's tertiary structure induced by increased temperatures. Enzyme adsorption to soil surfaces and the consequences for activity are discussed in detail in Chaper 11.

IV. EXAMPLES OF USING ENZYMES FOR BIOREMEDIATION

As evident from the previous subsections, many enzymes that are able to transform hazardous substances are now available in a purified form. Catalytic, molecular, and biochemical features are known for several of these enzymes. Studies have been dedicated to the use of these enzymes for practical applications. Both free enzymes and enzymes immobilized on diverse matrices by different means have been applied for the transformation of polluting and nonpolluting compounds. Batch as well as reactor systems have been devised and studied to optimize the use of these enzymes as practical catalysts. However most investigations have been conducted at laboratory scale. Relatively few examples of enzyme applications for pollutant removal at pilot and/or field scale are available. Moreover, the most abundant studies have dealt with the treatment of polluted aquatic systems. Few examples are available for terrestrial systems.

First Munnecke (18,19) then Caldwell and Raushel (198) investigated the feasibility of using an immobilized enzymatic system to detoxify organophosphate pesticides. A parathion hydrolase was immobilized successfully on porous glass or silica beads and showed good operational performance (i.e., high residual activity and long-term stability) (18,19). When applied in a fluidized-bed reactor, the hydrolase was able to remove more than 90% of the parathion from parathion-polluted water. A similar reactor was utilized by Caldwell and Raushel (198) with a phosphotriesterase from *Pseudomonas putida* adsorbed on trityl agarose. The chemical and kinetic parameters of the immobilized enzyme were comparable to those of the free enzyme. Furthermore, the immobilized system was able to hydrolyze, to an elevated extent ($> 90\%$), several organophosphate pesticides, including methyl parathion, paraoxon, ethyl parathion, diazinon, and coumaphos (198).

An enzymatic strategy for waste treatment and minimization based on the use of parathion hydrolase was examined by Smith et al. (199) and Grice et al. (200). In a large-scale experiment the enzyme was used selectively to hydrolyze potasan, a dechlorination product of coumaphos that usually accumulates in cattle-dipping liquid to concentrations hazardous to the cattle (199). The studies showed that the toxic product potosan was hydrolyzed selectively while preserving coumaphos when very small amounts of parathion hydrolase were used. The process allowed the useful lifetime of coumaphos to be extended by reducing the amount of waste generated (199,200).

Typical examples of using free and immobilized enzymes for waste treatment are those dealing with polyphenoloxidases and peroxidases for the removal of phenols and related compounds from industrial effluents and wastewaters of different types (189,193,194,201–206). Tyrosinase was used as a free enzyme by Atlow and associates (201) to dephenolize two different actual industrial wastewater samples, one from a plant producing triarylphosphates and the other from a coke plant. The enzymatic treatment completely dephenolized water (201). Tyrosinase immobilized on different supports was effective in removing several phenols from the wastewater samples. The enzyme showed different degrees of efficiency, and its activity was reduced very little even after repeated treatments (203,204). In these studies, the immobilized enzyme was used to treat small volumes of polluted effluents operating as batch systems.

A new technological approach was reported by Edwards and coworkers (206), who immobilized polyphenoloxidase on single capillary membranes in a small-scale bioreactor and used it in a continuous process to treat two phenol-containing effluents from an industrial coal-gas conversion plant. Almost complete removal of phenolics was observed, thus demonstrating that membrane-immobilized polyphenoloxidase in a capillary bioreactor can potentially be used in an effective bioremediation process (206).

Wu et al. (194) and Nicell et al. (202) performed investigations to optimize the reaction conditions and to develop the most suitable reactor configuration for peroxidase-mediated removal of phenols from wastewaters. Optimal conditions for pH, molar ratio of hydrogen peroxide to phenol, HRP and additive concentrations, as well as reaction times were determined to achieve at least 95% removal of phenols from synthetic wastewater (194). The enzyme-catalyzed process was implemented in a continuously stirred tank reactor (CSTR) configuration (202). In such a configuration the inactivation of the enzyme through free radicals (see previous section) was reduced by lowering the reactant and enzyme concentrations immediately upon entering the reactor. Single- and multiple-CSTR performances were proved to be superior to batch reactor performance (202). In further studies, the feasibility of the enzyme process to treat a foundry wastewater was evaluated (193). The treated sample was a composite wastewater containing phenolic pollutants typical of high-concentration streams (scrubber waters) found in steel foundries. Phenolic contaminants were removed up to 99%, despite the presence of other contaminants such as organic compounds and iron in the waste matrix (193).

Cooper and Nicell (193) compared enzyme treatment to chemical oxidation using Fenton's reagent (ferrous iron and hydrogen peroxide) and concluded that the enzymatic process was more expensive. The authors, however, referred to current-time (1996) prices of the purified enzyme. Studies performed with minced roots of horseradish (207,208) and with olive husk (a by-product of olive oil production showing phenoloxidase activity). Greco et al. (209) have demonstrated the effectiveness (and comparative inexpensiveness) of crude enzymes in the removal of several phenols from wastewaters. The direct use of such plant material as enzymatic sources may represent an interesting, low-cost alternative to the use of purified enzymes.

Various chlorophenols, including *p*-chlorophenols, 2,4-dichlorophenol, 2,4,5-trichlorophenol, 2,4,6-trichlorophenol, 2,3,4,6-tetrachlorophenol, and pentachlorophenol, as well as total organic carbon and adsorbable organic halogens were almost 100% removed from an aqueous solution by HRP physically adsorbed onto magnetic particles (205). The particles, as a support, seem to offer some advantages in that, after the process, they were easily separable from the system through the use of magnetic devices.

There is limited number of papers dealing with the use of isolated enzymes for the detoxification of polluted soils. Ruggiero et al. (210) and Gianfreda and Bollag (211) evaluated the capability of free and immobilized laccase and peroxidase to detoxify soils polluted by 2,4-dichlorophenols. Investigations were performed under laboratory conditions with soils differing in their chemical and physical properties. Besides some interesting properties shown by the immobilized enzymes (e.g., high residual activity and elevated stability to thermal denaturation, storage, and proteolytic attack), both free and immobilized enzymes displayed catalytic activity in soil environments. Laccase and peroxidase were able to remove up to 60% of the initial substrate concentration in the presence of soil having a low content of organic matter. Increasing the organic matter content had an inhibitory effect on the activity of free and immobilized enzymes.

To simulate the soil environment, Masaphy et al. (212) studied the degradation of parathion by a crude enzyme extract from *Xanthomonas* sp. in clay suspensions. The experiments were carried out with a montmorillonite and albumin to determine the influence of both inorganic and organic macrospecies on the availability of parathion enzymatic degradation. The substrate was adsorbed strongly on both matrices, thus inhibiting enzymatic transformation. A two-enzyme biocatalyst, consisting of β-glucosidase adsorbed on a sintered clay copolymerized with acid phosphatase, was added to both a partially and a totally sterilized soil (213). In both soil samples, more than 80% of enzymatic activity of the two enzymes was retained eight days after the addition. These experiments suggest that enzymes could be used to decontaminate soil.

V. CONCLUSIONS

A. Potential for Applying Enzymes for Environmental Cleanup

Extensive research efforts have been directed toward the search for novel methods to remediate polluted environments. Biological agents such as microorganisms and plants have received intense attention because of their potential to remove pollutants from the environment without harsh side effects (1). However, ultimately it is the enzymes that transform pollutants during bioremediation procedures based on microbial activity or plant treatment. Therefore, it should be of major interest to investigate the possibility of directly applying enzymes to contaminated environmental sites.

The application of enzymes to transform or degrade pollutants is an innovative treatment technique for the removal of these chemicals from polluted environments. However, a better understanding of the relationship between the activity, kinetic properties, and stability of enzymes in soil and the characteristics of immobilizing soil colloids is required.

As previously outlined, extension of the catalytic life of an enzyme (that is, its ability to retain its activity for prolonged periods) may be obtained by immobilizing the enzyme on a solid support. Indeed, in most cases, immobilized enzymes have been found to be more efficient than free enzymes. The advantages of using immobilized enzymes include improved stability and the possibility of continuous application as well as enzyme reuse. Further studies are necessary to identify additional enzymes that may be able to transform the increasing number of chemicals polluting the environment.

In spite of the promising potential applications, only a few enzymes have been tested as tools for the detoxification of pollutants. Numerous investigations have documented the degradation of pollutants by microorganisms, but few studies have examined which enzymes in those microorganisms were responsible. This lack of study of practical applications for pollutant-detoxifying enzymes appears to be related to the limited knowledge of biochemical characteristics of the degradation processes rather than to an inability to overcome factors that restrict the use of the enzymes themselves. Thus, future research concerning the biochemical processes of pollutant degradation is greatly needed and should be aimed not only at defining the types of enzymes involved in the various metabolic steps, but also at characterizing their kinetic properties and resistance to adverse environmental conditions.

B. Comparison Between the Use of Enzymes and Microorganisms

Applying isolated enzymes for the removal of pollutants from contaminated environmental sites appears to offer several advantages over using microorganisms, as described by Nannipieri and Bollag (214). The following is a summary of their conclusions and some additional factors:

- Enzymes possess unique and specialized substrate-transforming capabilities.
- Enzymes may be active under a wide range of environmental conditions that could be detrimental to active microbial cells.
- Enzymes are usually not affected by the presence of predators or certain inhibitors of microbial metabolism.
- The mobility of enzymes in free form in soil is greater than that of the larger microbial cells.
- Enzymes are not susceptible to microbial competition.
- Microbial cultures capable of degrading xenobiotics generally are less effective than the specific enzyme because, at low xenobiotic concentrations, other soluble carbon sources are utilized.

For these reasons, the use of cell-free enzyme preparations for the detoxification of pollutants should continue to be explored.

Conversely, several factors, such as difficulty in extracting and purifying the enzyme, as well as the potential instability of the purified protein, could restrict the use of enzymes. The isolation and production of enzymes are relatively expensive, and decontamination of a full-scale polluted system (aquatic or terrestrial) could require massive amounts of enzyme. Thus, even the use of crude enzyme preparations might prove too costly. However, this fact should not hamper efforts to carry out research to identify the most promising enzymes and determine the optimal conditions for their use. In fact, the results of such research might provide the incentive for commercial development to achieve large-scale production of enzymes at a much lower cost.

Successful application of deoxyribonucleic acid (DNA) recombination technology will result in both increased enzyme production and improved catalytic performance. Significant progress has been made in producing enzymes at industrial levels. High secretion capacity combined with the presence of low toxin levels of selected microbial hosts could be employed in industrial fermentation on a large scale by economical and environment-friendly means. In addition to enzyme production, DNA recombination technology can affect the structure–function relationship of an enzyme and its substrates by generating enzymes whose performance can be adapted to specific applications (153).

Results from 1998 demonstrated that it is possible to alter enzyme properties (such as specific activity or optimal pH) by site-directed mutagenesis, for instance, those of a laccase (215); 1998 success in obtaining the crystallographic structure of laccase (216) will further assist researchers in designing new laccases with better substrate specificity/affinity, catalytic efficiency, or stability (under adverse pH and temperature conditions and counteracting inhibition or other inactivators) for improved performance in bioremediation.

C. Using a Combination of Methods for Bioremediation

Enzyme technology has been refined to permit the use of cell-free systems under circumstances in which active microbial cells have been employed previously (217). Treatments that require multistep transformations, involving a number of different enzymes acting sequentially, and/or requiring the regeneration of cofactors or coenzymes (218–220) have necessitated the use of intact cells. However, in 1988 Gu and Chu (217) succeeded in encapsulating several different enzymes and a cofactor bound to a soluble dextran and employing the whole multienzyme system to convert urea to L-glutamic acid. The cofactor–dextran complex was free to move and react with enzymes inside the microcapsules but because of its size could not pass through the membrane surrounding the system. Similar multienzyme systems may be prepared and proven useful in degrading pollutants in the environment.

The most desired end result of enzymatic activity is complete biodegradation or mineralization of pollutants. In general, mineralization cannot be achieved after application of a single enzyme; however, it might be possible if several enzymes are applied simultaneously. Selection of the most suitable enzymes depends on requirements for their specificity, their need for cofactors, and their capacity to retain a substantial amount of their activity for an extended period. The specificity of the enzyme is important; an enzyme with too broad a specificity might have reduced efficiency. Enzymes that require inexpensive cofactors (such as O_2) or that do not require any cofactors obviously are preferred.

Caution is required in predicting the fate of pollutants in situ from enzymatic studies under laboratory conditions, and pilot-scale studies on wastewater treatments and field experiments on polluted soils are of great importance. An economic analysis of these applications should be made and compared with currently available methods for the detoxification of pollutants. Such experiments and analyses should allow a determination of the feasibility of using enzyme amendments to detoxify pollutants in soil and water and may prove that enzymatic decontamination methods may be of great importance in future attempts to control environmental pollution.

REFERENCES

1. M Alexander. Biodegradation and Bioremediation. San Diego: Academic Press, 1999.
2. M Alexander. Biodegradation: Problems of molecular recalcitrant and microbial fallibility. Adv Appl Microbiol 7:35–80, 1965.
3. GM Pierzynski, JT Sims, GF Vance. Soils and Environmental Quality. Boca Raton, FL: Lewis, 1994.
4. S Harayama. Polycyclic aromatic hydrocarbon bioremediation design. Curr Opin Biotechnol 8:268–273, 1997.
5. P-Y Lu, RL Matcalf, LK Cole. Pentachlorophenol: Chemistry, Pharmacology and Environmental Toxicology. New York: Plenum Press, 1978.
6. EL Bouwer, AJB Zehnder. Bioremediation of organic compounds—putting microbial metabolism to work. Tibtech 11:360–365, 1993.
7. JT Trevors, P Kuikman, JD van Elsas. Release of bacteria into soil: Cell numbers and distribution. J Microbial Methods 19:247–259, 1994.
8. RM Goldstein, LM Mallory, M Alexander. Reasons for possible failure of inoculation to enhance biodegradation. Appl Environ Microbiol 50:967–983, 1985.
9. J Karam, JA Nicell. Potential applications of enzymes in waste treatment. J Chem Tech Biotechnol 69:141–153, 1997.
10. A Paszczynski, RL Crawford. Recent advances in the use of fungi in environmental remediation and biotechnology. In: J-M Bollag, G Stotzky, ed. Soil Biochemistry. Vol. 10. New York: Marcel Dekker, 2000, pp 379–422.
11. HE Hammel. Organopollutant degradation by ligninolytic fungi. Enzyme Microb Technol 11:776–787, 1989.
12. A Hatakka. Lignin-modifying enzymes from selected white-rot fungi: Production and role in lignin degradation. FEMS Microbiol Rev 13:125–135, 1994.

13. I Singleton. Microbial metabolism of xenobiotics: Fundamental and applied research. J Chem Tech Biotechnol 59:9–23, 1994.
14. J-M Bollag. Decontaminating soils with enzymes. Environ Sci Technol 26:1876–1881, 1992.
15. T Konno, Y Kasai, RL Rose, E Hodgson, WC Dauterman. Purification and characterization of a phosphotriester hydrolase rom methyl parathion-resistant *Heliotis virescens*. Pest Biochem Physiol 36:1–13, 1990.
16. BD Siegfrid, M Ono, JJ Swanson. Purification and characterization of a carboxylesterase associated with organophosphate resistance in the greenburg *Schizaphis graminum* (Homoptera: Aphidiae). Arch Ins Biochem Physiol 36:229–240, 1997.
17. DM Munnecke. Enzymatic hydrolysis of organophosphate insecticides, a possible pesticide disposal method. Appl Environ Microbiol 32:7–13, 1976.
18. DM Munnecke. Chemical, physical and biological methods for the disposal and detoxification of pesticides. Residue Rev 70:1–26, 1979.
19. DM Munnecke. Enzymatic detoxification of waste organophosphate pesticides. J Agric Food Chem 28:105–111, 1980.
20. WW Mulbry, JS Karns. Purification and characterization of three parathion hydrolases from Gram-negative bacterial strains. Appl Environ Microbiol 55:289–293, 1989.
21. H Attaway, JO Nelson, AM Baya, MJ Voll, WE White, DJ Grimes, RR Colwell. Bacterial detoxification of DFP. Appl Environ Microbiol 53:1685–1689, 1987.
22. DP Dumas, JR Wild, FM Raushel. Diisopropylfluorophosphate hydrolysis by a phosphotriesterase from *Pseudomonas diminuta*. Biotechnol Appl Biochem 11:235–243, 1989.
23. Y Yu, C Hexin, F Defang, L Bin. Enzymatic degradation of O-phosphothionate insecticides. Huanjing Kexue 19:58–61, 1998.
24. R Tehelet, D Levanon, U Mingelgrin, Y Henis. Parathion degradation by a *Pseudomonas* sp. and a *Xanthomonas* sp. and by their crude enzyme extracts as affected by some cations. Soil Biol Biochem 25:1665–1671, 1993.
25. WW Mulbry, JS Karns, PC Kearney, JO Nelson, CS McDaniel, JR Wild. Identification of a plasmid-borne hydrolase gene from *Flavobacterium*-sp by southern hybridization with opd from *Pseudomonas putida*. Appl Environ Microbiol 51:926–930, 1986.
26. WW Mulbry, PC Kearney, JO Nelson, JS Karns. Physical comparison of parathion hydrolase plasmids from *Pseudomonas diminuta* and *Flavobacterium*-sp. Plasmid 18:173–177, 1987.
27. SJ Coppella, ND Cruz, GF Payne, BM Pogell, MK Speedie, JS Karns, EM Sybert, MA Connor. Genetic engineering approach to toxic waste management: A case study for organophosphate waste treatment. Biotechnol Prog 6:76–81, 1990.
28. N Delacruz, GF Payne, JM Smith, SJ Coppella. Bioprocess development to improve foreign protein production from recombinant *Streptomyces*. Biotechnol Prog 8:307–315, 1992.
29. SS Rowland, MK Speedie, BM Pogell. Purification and characterization of a secreted recombinant phosphotriesterase (parathion hydrolase) from *Streptomyces lividans*. Appl Environ Microbiol 57:440–444, 1991.
30. SS Rowland, JJ Zulty, M Sathyamoorthy, BM Pogell, MK Speedie. The effect of a signal sequence on the efficiency of secretion of a hetrologous phosphotriesterase by *Streptomyces lividans*. Appl Environ Microbiol 38:94–100, 1992.
31. W Mulbry. Selective deletions involving the organophosphorus hydrolase gene adpB from *Nocardia* strain B-1. Microbiol Res 153:213–217, 1998.
32. LW Getzin, I Rosefield. Partial purification and properties of a soil enzyme that degrades the insecticide malathion. Biochim Biophys Acta 235:442–453, 1971.
33. T Satayanarayana, LW Getzin. Properties of a stable cell-free esterase from soil. Biochemistry 12:1566–1572, 1973.
34. T Omori, K Sugimura, H Ishigooka, Y Minoda. Purification and some properties of a 2 hydroxy-6-oxo-6-phenylhexa-24-dienoic-acid hydrolysing enzyme from *Pseudomonas*-

cruciviae S-93 B-1 involved in the degradation of biphenyl. Agric Biol Chem 50:931–938, 1986.

35. T Omori, H Ishigooka, Y Minoda. Purification and some properties of 2 hydroxy-6-oxo-6-phenylhena-24-dienoic-acid reducing enzyme from *Pseudomonas-cruciviae* S-93B-1 involved in the degradation of biphenyl. Agric Biol Chem 50:1513–1518, 1986.
36. T Omori, H Ishigooka, Y Minoda. A new metabolic pathway for meta ring-fission compounds of biphenyl. Agric Biol Chem 52:503–510, 1988.
37. SE Maloney, A Maule, ARW Smith. Purification and preliminary characterization of permethrinase from a pyrethroid-transforming strain of *Bacillus cereus*. Appl Environ Microbiol 59:2007–2013, 1993.
38. ML de Souza, MJ Sadowsky, LP Wackett. Atrazine chlorohydrolase from *Pseudomonas* sp. strain ADP: Gene sequence, enzyme purification, and protein characterization. J Bacteriol 178:4894–4900, 1996.
39. ML de Souza, LP Wackett, KL Boundy-Mills, RT Mandelbaum, MJ Sadowsky. Cloning, characterization, and expression of a gene region from *Pseudomonas* sp. strain ADP involved in the dechlorination of atrazine. Appl Environ Microbiol 61:3373–3378, 1995.
40. DA Pelletier, CS Harwood. 2-Ketocyclohexanecarboxyl coenzyme A hydrolase, the ring cleavage enzyme required for anaerobic benzoate degradation by *Rhodopseudomonas palustris*. J Bacteriol 180:2330–2336, 1998.
41. MK Derbyshire, JS Karns, PC Kearney, JO Nelson. Purification and characterization of an N-methylcarbamate pesticide hydrolyzing enzyme. J Agric Food Chem 35:871–877, 1987.
42. WW Mulbry, RW Eaton. Purification and characterization of the N-methylcarbamate hydrolase from *Pseudomonas* strain CRL-OK. Appl Environ Microbiol 57:3679–3682, 1991.
43. G Engelhardt, PR Wallnöfer, R Plapp. Purification and properties of an aryl acylamidase of *Bacillus sphaericus*, catalyzing the hydrolysis of various phenylamide herbicides and fungicides. Appl Microbiol 26:709–718, 1973.
44. J Blake, DD Kaufman. Characterization of acylanilide-hydrolyzing enzyme(s) from *Fusarium oxysporum* Schlecht. Pest Biochem Physiol 5:305–313, 1975.
45. HG Park, HN Chang, JS Dordick. Enzymatic polytransesterification of aromatic diols in organic solvents. Biotechnol Lett 17:1085–1090, 1995.
46. F Terradas, M Teston-Henry, PA Fitzpatrick, AM Klibanov. Marked dependence of enzyme prochiral selectivity on the solvent. J Am Chem Soc 11:390–396, 1993.
47. C Fortun, A Fortun, B Ceccanti. Changes in water-soluble and EDTA-extractable cations and enzymatic activities of soils treated with a composted solid urban waste. Arid Soil Res Rehab 11:65–276, 1997.
48. JH Slater, AT Bull, DJ Hardman. Microbial dehalogenation. Biodegradation 6:181–189, 1995.
49. S Fetzner, F Lingens. Bacterial dehalogenases: Biochemistry, genetics, and biotechnological applications. Microbiol Rev 58:641–685, 1994.
50. SEI Fantroussi, H Naveau, SN Agathos. Anaerobic dechlorinating bacteria. Biotechnol Prog 14:167–188, 1998.
51. C Holliger, G Wohlfarth, G Diekert. Reductive dechlorination in the energy metabolism of anaerobic bacteria. FEMS Microbiol Rev 22:383–398, 1999.
52. P Holloway, J Trevors, T L Hung. A colorimetric assay for detecting haloalkane dehalogenase activity. J Microbiol Methods 32:31–36, 1998.
53. S Ni, JK Fredrickson, L Xun. Purification and characterization of a novel 3-chlorobenzoate-reductive dehalogenase from the cytoplasmic membrane of *Desulfomonile tiedjei* DCB-1. J Bacteriol 177:5135–5139, 1995.
54. GT Townsend, JM Suflita. Influence of sulfur oxyanions on reductive dehalogenation activities in *Desulfomonile tiedjei*. Appl Environ Microbiol 63:3594–3599, 1997.
55. GT Townsend, JM Suflita. Characterization of chloroethylene dehalogenation by cell extracts

of *Desulfomonile tiedjei* and its relationship to chlorobenzoate dehalogenation. Appl Environ Microbiol 62:2850–2853, 1996.
56. L Xun, E Topp, CS Orser. Confirmation of oxidative dehalogenation of pentachlorophenol by a *Flavobacterium* pentachlorophenol hydrolase. J Bacteriol 174:5745–5747, 1992a.
57. L Xun, E Topp, CS Orser. Glutathione is the reducing agent for the reductive dehalogenation of tetrachloro-*p*-hydroquinone by extracts from *Flavobacterium* sp. Biochem Biophys Res Commun 182:361–366, 1992b.
58. DL McCarthy, S Navarrete, WS Willett, PC Babbitt, SD Copley. Exploration of the relationship between tetrachlorohydroquinone dehalogenase and the glutathione S-transferase superfamily. Biochem 35:14634–14642, 1996.
59. W Willett, S Copley, D Shelley. Identification and localization of a stable sulfenic acid in peroxide-treated tetrachlorohydroquinone dehalogenase using electrospray mass spectrometry. Chem Biol (London) 3:851–857, 1996.
60. FE Löffler, RA Sanford, JM Tiedje. Characterization of a reductive dehalogenase from *Desulfitobacterium* strain Co23. Gen Meeting Am Soc Microbiol 96:1–412, 1996.
61. FE Löffler, RA Sanford, JM Tiedje. Initial characterization of a reductive dehalogenase from *Desulfitobacterium chlororespirans* Co23. Appl Environ Microbiol 62:3809–3813, 1996.
62. V Romanov, RP Hausinger. NADPH-dependent reductive ortho dehalogenation of 2,4-dichlorobenzoic acid in *Corynebacterium sepedonicum* KZ-4 and *Coryneform* bacterium strain NTB-1 via 2,4-dichlorobenzoyl coenzyme. J Bacteriol 178:2656–2661, 1996.
63. JK Magnuson, RV Stem, JM Gossett, SH Zinder, DR Burris. Reductive dechlorination of tetrachloroethene to ethene by a two-component enzyme pathway. Appl Environ Microbiol 64:1270–1275, 1998.
64. E Miller, G Wohlfarth, G Diekert. Comparative studies on tetrachloroethene reductive dechlorination mediated by *Desulfitobacterium* sp. strain PCE-S. Arch Microbiol 168:513–519, 1997.
65. A Neumann, G Wohlfarth, G Diekert. Properties of tetrachloroethene and trichloroethene dehalogenase of *Dehalospirillum multivorans*. Arch Microbiol 163:276–281, 1995.
66. E Miller, G Wohlfarth, G Diekert. Studies on tetrachloroethene respiration in *Dehalosprillum multivorans*. Arch Microbiol 166:379–387, 1996.
67. A Neumann, G Wohlfarth, G Diekert. Purification and characterization of tetrachloroethene reductive dehalogenase from *Dehalospirillum multivorans*. J Biol Chem 271:16515–16519, 1996.
68. E Miller, G Wohlfarth, G Diekert. Purification and characterization of the tetrachloroethene reductive dehalogenase of strain PCE-S. Arch Microbiol 169:497–502, 1998.
69. N Christiansen, BK Ahring, G Wohlfarth, G Diekert. Purification and characterization of the 3-chloro-4-hydroxy-phenyacetate reductive dehalogenase of *Desulfitobacterium hafniense*. FEBS Lett 436:159–162, 1998.
70. LM Newman, LP Wackett. Trichloroethylene oxidation by purified toluene 2-monooxygenase: Products, kinetics, and turnover-dependent inactivation. J Bacteriol 179:90–96, 1997.
71. L Xun, CS Orser. Purification and properties of a pentachlorophenol hydrolase a flavoprotein from a *Flavobacterium* sp. strain ATCC 39723. J Bacteriol 173:4447–4453, 1991.
72. L Xun, E Topp, CS Orser. Purification and characterization of a tetrachloro-*p*-hydroquinone reductive dehalogenase from a *Flavobacterium* sp. J Bacteriol 174:8003–8007, 1992.
73. A Markus, U Klages, S Krauss, F Lingens. Oxidation and dehalogenation of 4-chlorophenylacetate by a two-component enzyme system from *Pseudomonas* sp. strain CBS3. J Bacteriol 160:618–621, 1984.
74. S Fetzner, R Müller, F Lingens. Degradation of 2-chlorobenzoate by *Pseudomonas cepacia* 2CBS. Biol Chem Hoppe Seyler 370:1173–1182, 1989.
75. S Fetzner, R Müller, F Lingens. Purification and some properties of 2-halobenzoate 1,2-dioxygenase, a two-component enzyme system from *Pseudomonas cepacia* 2CBS. J Bacteriol 174:279–290, 1992.

76. AE Mars, J Kingma, SR Kaschabek, DB Janssen. Conversion of 3-chlorocatechol by various catechol 2,3-dioxygenases and sequence analysis of the chlorocatechol dioxygenase region of *Pseudomonas putida* GJ31. J Bacteriol 181:1309–1318, 1999.
77. SR Kaschabek, T Kasberg, D Muller, AE Mars, DB Janssen, W Reineke. Degradation of chloroaromatics: purification and characterization of a novel type of chlorocatechol 2,3-dioxygenase of *Pseudomonas putida* GJ31. J Bacteriol 180:296–302, 1998.
78. HJ Knackmuss. Degradation of halogenated and sulfonated hydrocarbons. FEMS Symp 12: 189–212, 1981.
79. R Müller, RH Oltmanns, F Lingens. Enzymatic dehalogenation of 4-chlorobenzoate by extracts from *Arthrobacter*-SP SU-DSM 20407. Biol Chem Hoppe Seyler 369:567–571, 1988.
80. A Elsner, F Löffler, K Miyashita, R Müller, F Lingens. Resolution of 4-chlorobenzoate dehalogenase from *Pseudomonas* sp. strain CBS3 into three components. Appl Environ Microbiol 57:324–326, 1991.
81. F Löffler, R Müller, F Lingens. Dehalogenation of 4-chlorobenzoate by 4-chlorobenzoate dehalogenase from *Pseudomonas* sp. CBS3: an ATP/coenzyme A dependent reaction. Biochem Biophys Res Commun 176:1106–1111, 1991.
82. MM Benning, KL Taylor, RQ Liu, G Yang, H Xiang, G Wesenberg, D Dunaway-Mariano, HM Holden. Structure of 4-chlorbenzoyl coenzyme A dehalogenase determined to 1.8 A resolution: An enzyme catalyst generated via adaptive mutation. Biochemistry 35:8103–8109, 1996.
83. F Löffler, R Müller. Identification of 4-chlorobenzoyl-coenzyme A as intermediate in the dehalogenation catalyzed by 4-chlorobenzoate dehalogenase from *Pseudomonas* sp. CBS3. FEBS Lett 290:224–226, 1991.
84. GP Crooks, SD Copley. Purification and characterization of 4-chlorobenzoyl CoA dehalogenase from *Arthrobacter* sp. strain 4-CB1. Biochemistry 33:11645–11649, 1994.
85. PJ Sallis, SJ Armfield, AT Bull, DJ Hardman. Isolation and characterization of a haloalkane halidohydrolase from *Rhodococcus erythropolis* Y2. J Gen Microbiol 136:115–120, 1990.
86. Y Nagata, K Miyauchi, J Damborsky, K Manova, A Ansorgova, M Takagi. Purification and characterization of a haloalkane dehalogenase of a new substrate class from a gamma-hexachlorocyclohexane-degrading bacterium, *Sphingomonas paucimobilis* UT26. Appl Environ Microbiol 63:3707–3710, 1997.
87. DJ Hardman. Biotransformation of halogenated compounds. Crit Rev Biotechnol 11:1–40, 1991.
88. J-Q Liu, T Kurihara, V Nardi-Dei, T Okamura, N Esaski, K Soda. Overexpression and feasible purification of thermostable L-2-halo acid dehalogenase of *Pseudomonas* sp. YL. Biodegradation 6:223–227, 1995.
89. J-Q Liu, T Kurihara, M Miyagi, S Tsunasawa, M Nishihara, N Esaki, K Soda. Paracatalytic inactivation of L-2-haloacid dehalogenase from *Pseudomonas* sp. YL by hydroxylamine: Evidence for the formation of an ester intermediate. J Biol Chem 272:3363–3368, 1997.
90. J-Q Liu, T Kurihara, MS Ichiyama, T Masaru. Reaction mechanism of fluoroacetate dehalogenase from *Moraxella* sp. J Biol Chem 273:30897–30902, 1998.
91. N Allison, AJ Skinner, RA Cooper. The dehalogenases of a 2,2-dichloropropionate degrading bacterium. J Gen Microbiol 129:1283–1294, 1983.
92. R Schwarze, A Brokamp, FRJ Schmidt. Isolation and characterization of dehalogenases from 2,2-dichloropropionate-degrading soil bacteria. Curr Microbiol 34:103–109, 1997.
93. JSH Tsang, PJ Sallis, AT Bull, DJ Hardman. A monobromoacetate dehalogenase from *Pseudomonas cepacia* MBA4. Arch Microbiol 150:441–446, 1988.
94. A Diez, MJ Alvarez, JM Prieto, JM Bautista, A Garrido-Pertierra. Monochloroacetate dehalogenase activities of bacterial strains isolated from soil. Can J Microbiol 41:730–739, 1995.
95. J-Q Liu, T Kurihara, AKMQ Hasan, V Nardi-Dei, H Koshikawa, N Esaki, K Soda. Purification and characterization of thermostable and nonthermostable 2-haloacid dehalogenases with

different stereospecificities from *Pseudomonas* sp. strain YL. Appl Environ Microbiol 60: 2389–2393, 1994.
96. A Diez, MI Prieto, MJ Alvarez, JM Bautista, A Garrido, A Puyet. Improved catalytic performance of a 2-haloacid dehalogenase from *Azotobacter* sp. by ion-exchange immobilisation. Biochem Biophys Res Commun 220:828–833, 1996.
97. A Diez, MI Prieto, MJ Alvarez, JM Bautista, A Puyet, A Garrido-Pertierra. Purification and properties of a high-affinity L-2-haloacid dehalogenase from *Azotobacter* sp; strain RC26. Lett Appl Microbiol 23:279–282, 1996.
98. AJ Weightman, AL Weightman, JH Slater. Stereospecificity of 2 mono chloro propionate dehalogenation by the 2 dehalogenases of *Pseudomonas putida* PP-3: Evidence for 2 different dehalogenation mechanisms. J Gen Microbiol 128:1755–1762, 1982.
99. V Nardi-Dei, K Tatsuo, PC Esaki, NS Kenji. Bacterial DL-2-haloacid dehalogenase from *Pseudomonas* sp. strain 113:Gene cloning and structural comparison with D- and L-2-haloacid dehalogenase. J Bacteriol 179:4232–4238, 1997.
100. CE Castro, EW Bartnicki. Biodehalogenation: Epoxidation of halohydrins, epoxide opening, and transhalogenation by a *Flavobacterium* sp. Biochemistry 7:3213–3218, 1968.
101. AJ Van den Wijngaard, PTW Reuvekamp, DB Janssen. Purification of dehalogenase from *Arthrobacter* sp. strain AD2. J Bacteriol 173:124–129, 1991.
102. HMS Assis, PJ Sallis, AT Bull, DJ Hardman. Synthesis of chiral epihalohydrins using haloalcohol dehalogenase A from *Arthrobacter erithii* H10a. Enzyme Microb Technol 22:545–551, 1998.
103. HMS Assis, PJ Sallis, AT Bull, DJ Hardman. Biochemical characterization of a haloalcohol dehalogenase from *Arthrobacter erithii* H10a. Enzyme Microb Technol 22:568–574, 1998.
104. T Nagasawa, T Nakamura F Yu, I Watanabe, H Yamada. Purification and characterization of halohydrin hydrogenhalide lyase from a recombinant *Escherichia coli* containing the gene from a *Corynebacterium* sp. Appl Microbiol Biotechnol 36:478–482, 1992.
105. T Nakamura, T Nagasawa, F Yu, I Watanabe, Y Yamada. Resolution and some properties of enzymes involved in enantioselective transformation of 1,3-dichloro-2-propanol to 3-3-chloro-1,2-propanediol by *Corynebacterium* sp. Strain N-1074. J Bacteriol 174:7613–7619, 1992.
106. T Nakamura, T Nagasawa, F Yu, I Watanabe, H Yamada. Characterization of a novel enantioselective halohydrin hydrogen-halide-lyase. Appl Environ Microbiol 60:1297–1301, 1994.
107. R Imai, Y Nagata, K Semnoo, H Wada, M Fukuda, M Takagi, K Yano. Dehydrochlorination of γ-hexachlorocyclohexane (γ-BHC) by γ-BHC *Pseudomonas paucimobilis*. Agric Biol Chem 53:2015–2017, 1989.
108. R Imai, Y Nagata, M Fukuda, M Takagi, K Yano. Molecular cloning of a *Pseudomonas paucimobilis* gene encoding a 17-kilodalton polypeptide that eliminates HCI molecules from γ-hexachlorocyclohexane. J Bacteriol 173:6811–6819, 1991.
109. Y Nagata, T Hatta, R Imai, K Kimbara, M Fukuda, K Yano, M Takagi. Purification and characterization of γ-hexachlorocyclohexane (γ-HCH) dehydrochlorinase (LinA) from *Pseudomonas paucimobilis*. Biosci Biotechnol Biochem 57:1582–1583, 1993.
110. Y Nagata, T Nariya, R Ohtomo, M Fukuda, K Yano, M Takagi. Cloning and sequencing of a dehalogenase gene encoding an enzyme with hydrolase activity involved in the degradation of γ-hexachlorocyclohexane in *Pseudomonas paucimobilis*. J Bacteriol 175:6403–6410, 1993b.
111. H Lipke, CW Kearns. DDT dehydrochlorinase. I. Isolation, chemical properties, and spectrophotometric assay. J Biol Chem 234:2123–2128, 1959.
112. H Lipke, CW Kearns. DDT dehydrochlorinase. II. Substrate and cofactor specificity. J Biol Chem 234:2129–2132, 1959b.

113. JC Thomas, F Berger, M Jacquier, D Bernillon, F Baud-Grasset, N Truffaut, P Normand, TM Vogel, P Simonet. Isolation and characterization of a novel gamma-hexachlorocyclohexane-degrading bacterium. J Bacteriol 178:6049–6055, 1996.
114. TH Hisano, FT Yasuo, JQ Liu, T Kurihara, N Esaki, K Soda. Crystallization and preliminary X-ray crystallographic studies of L-2-haloacid dehalogenase from *Pseudomonas* sp. YL. Protein Struct Function Genet 24:520–522, 1996.
115. IS Ridder, HJ Rozeboom, KH Kalk, DB Janssen, BW Dijkstra. Three-dimensional structure of L-2-haloacid dehalogenase from *Xanthobacter autotrophicus* GJ10 complexed with the substrate-analogue formate. J Biol Chem 272:33015–33022, 1997.
116. CS Orser, J Dutton, C Lange, P Jablonski, L Xun, M Hargis. Characterization of a *Flavobacterium* glutathione S-transferase gene involved in reductive dechlorination. J Bacteriol 17: 2640–2644, 1993.
117. DB Janssen, JR Van Der Ploeg, F Pries. Genetics and biochemistry of 1,2-dichloroethane degradation. Biodegradation 5:249–257, 1994.
118. M Widersten. Heterologous expression in *Escherichia coli* of soluble active-site random mutants of haloalkane dehalogenase from *Xanthobacter autorophicus* GJ10 by coexpression of molecular chaperonins GroEL/ES. Protein Exp Purif 13:389–395, 1998.
119. S Beil, JR Mason, KN Timmis, DH Pieperl. Identification of chlorobenzene dioxygenase sequence elements involved in dechlorination of 1,2,4,5-tetrachlorobenzene. J Bacteriol 180: 5520–5528, 1998.
120. A Neumann, G Wohlfarth, G Diekert. Tetrachloroethene dehalogenase from *Dehalospirillum multivorans*: Cloning, sequencing of the encoding genes, and expression of the pceA gene in *Escherichia coli*. J Bacteriol 180:4140–4145, 1998.
121. Y Rozen, A Nejidat, KH Gartemann, S Belkin. Specific detection of a *p*-chlorobenzoic acid by *Escherichia coli* bearing a plasmid-borne fcbA': lux fusion. Chemosphere 38:633–641, 1999.
122. T Hisano, H Yasuo, F Tomomi, JQ Liu, K Tatsuo, E Nobuyoshi, S Kenji. Crystal structure of L-2 haloacid dehalogenase from *Pseudomonas* sp. YL: AN alpha/beta hydrolase structure that is different from the alpha/beta hydrolase fold. J Biol Chem 271:20322–20330, 1996.
123. GH Krooshof, EM Kwant, J Damborsky, J Koca, DB Janssen. Repositioning the catalytic triad aspartic acid of haloalkane dehalogenase: Effects on stability, kinetics, and structure. Biochemistry 36:9571–9580, 1997.
124. GH Krooshof, IS Ridder, AWJW Tepper, J Vos Gerda, HJ Rozeboom, KH Kalk, BW Dijkstra, DB Janssen. Kinetic analysis and X-ray structure of haloalkane dehalogenase with a modified halide-binding site. Biochemistry 37:15013–15023, 1998.
125. GH Krooshof, R Floris, AWJW Tepper, DB Jannssen. Thermodynamic analysis of halide binding to haloalkane dehalogenase suggest the occurrence of large conformational changes. Protein Sci 8:355–360, 1999.
126. JP Schanstra, DB Janssen. Kinetics of halide release of haloalkane dehalogenase: Evidence for a slow conformational change. Biochem 35:5624–5632, 1996.
127. JP Schanstra, J Kingma, DB Janssen. Specificity and kinetics of haloalkane dehalogenase. J Biol Chem 271:14747–14753, 1996.
128. JP Schanstra, A Ridder, G Heimeriks, JR Rick, JG Poelaarends, HK Kalk, WB Dijstra, DB Janssen. Kinetic characterization and X-ray structure of a mutant of haloalkane dehalogenase with higher catalytic activity and modified substrate range. Biochemistry 35:13186–13195, 1996.
129. JP Schanstra, A Ridder, J Kingma, DB Janssen. Influence of mutation of Val226 on the cataklytic rate of haloalkane dehalogenase. Protein Eng 10:53–61, 1997.
130. J Damborsky. Quantitative structure-function and structure-stability relationships of purposely modified proteins. Protein Eng 11:21–30, 1998.

131. J Damborsky, M Bohac, M Prokop, M Kuty, J Koca. Computational site-directed mutagenesis of haloalkane dehalogenase in position 172. Protein Eng 11:901–907, 1998.
132. Y Liu, R-Q Guang, KL Taylor, H Xiang, J Price, D Dunaway-Mariano. Identification of active site residues essential to 4-chlorobenzoyl-coenzyme A dehalogenase catalysis by chemical modification and site directed mutagenesis. Biochem 35:10879–10885, 1996.
133. S Vuilleumier, T Leisinger. Protein engineering studies of dichloromethane dehalogenases/glutathione S-transferases from *Methylophilus* sp. strain DM11: Ser12 but not Tyr6 is required for enzyme activity. J Biochem 239:410–417, 1996.
134. S Vuilleumier, H Sorribas, T Leisinger. Identification of a novel determinant of glutathione affinity in dichloromethane dehalogenases/glutathione S-transferases. Biochem Biophys Res Commun 238:452–456, 1997.
135. Y-FH Li, Y Fujii, T Hisano, T Nishihara, M Kurihara, TE Nobuyoshi. X-ray structure of a reaction intermediate of L-2-haloacid dehalogenase with L-2-chloropropionamide. J Biochem 124:20–22, 1998.
136. Y-FH Li, Y Fujii, T Hisano, T Nishihara, M Kurihara, TE Nobuyoshi. Crystal structures of reaction intermediates of L-2-haloacid dehalogenase and implications for the reaction mechanism. J Biol Chem 273:15035–15044, 1998b.
137. B Reinhammar, BG Malmström. Blue copper-containing oxidases. In: TG Spiro ed. Copper Proteins. New York: Wiley-Interscience, 1981, pp 109–149.
138. AM Mayer. Polyphenol oxidases in plants-recent progress. Phytochem 26:11–20, 1987.
139. FS Sariaslani. Microbial enzymes for oxidation of organic molecules. Crit Rev Biotechnol 9:171–257, 1987.
140. J Everse, KF Everse, MR Grisham. Peroxidases in Chemistry and Biology. New York: CRC Press, 1991.
141. L Casella, S Poli, M Gullotti, C Selvaggini, T Beringhelli, A Marchesini. The chloroperoxidase-catalyzed oxidation of phenols: Mechanisms, selectivity and characterization of enzyme-substrate complexes. Biochemistry 33:6377–6386, 1994.
142. AI Yaropolov, OV Skorobogat'ko, SS Vartanov, SV Varfolomeyev. Laccase properties, catalytic mechanism, and applicability. Appl Biochem Biotechnol 49:257–280, 1994.
143. CF Thurston. The structure and function of fungal laccases. Microbiology 140:19–26, 1994.
144. A Messerschmidt. Blue copper oxidases. Adv Inorg Chem 40:121–185, 1994.
145. A Sanchez-Ferrer, JN Rodriguez, F Garcia-Cànovas, F Garcia-Carmona. Tyrosinase: A comprehensive review of its mechanism. Biochim Biophys Acta 1247:1–11, 1995.
146. DR Mcmillin, MK Eggleston. Bioinorganic chemistry of laccase. In: A Messerschmidt, ed. Multicopper oxidases. Singapore: World Scientific, 1997, pp 129–166.
147. M Smith, CF Thurston DA Wood. Fungal laccases: Role in delignification and possible industrial applications. In: Multi-copper oxidases. A Messerschmidt, ed. Singapore: World Scientific, 1997, pp 201–224.
148. LJ Jonsson, E Palmqvist, NO Nilvebrant, B Hahn-Hagerdal. Detoxification of wood hydrolysates with laccase and peroxidase from the white-rot fungus *Trametes versicolor*. Appl Microbiol Biotechnol 49:691–697, 1998.
149. M Sundaramoorthy, J Terner, TL Poulos. Stereochemistry of the chloroperoxidase active site: Crystallographic and molecular-modeling studies. Chem Biol 5:461–473, 1998.
150. ZR Akhmedova. Ligninolytic enzymes of basidiomycetes: Lignin peroxidases from the fungus *Pleurotus ostreatus* UzBI-Zax 108. II. Isolation, purification, and characterization of isozymes. Biochem (Moscow) 61:981–987, 1998.
151. S Colonna, N Gaggero, C Richelmini, P Pasta. Recent biotechnological developments in the use of peroxidases. Trends Biotechnol 17:163–168, 1999.
152. MA McGuirl, DM Dooley. Copper-containing oxidases. Curr Opin Chem Biol 3:138–144, 1999.
153. L Gianfreda, XF Xu, J-M Bollag. Laccases: A useful group of oxidoreductive enzymes. Bioremediation J 3:1–25, 1999.

154. F Xu. Recent progress in laccase study: Properties, enzymology, production, and applications. In MC Flickinger, SW Drew, eds. The Encyclopedia of Bioprocessing Technology: Fermentation, Biocatalysis, and Bioseparation. New York: John Wiley & Sons, 1999, pp 1545–1554.
155. K Sauber, C Froehner, G Rosenberg, J Eberspaecher, F Lingens. Purification and properties of pyrazon di oxygenase EC-1.14.12. from pyrazon degrading bacteria. Eur J Biochem 74: 89–98, 1977.
156. R Müller, S Huang, J Eberspaecher, F Lingens. Catechol 2,3-dioxygenase EC-1.13.11.2 from pyrazon degrading bacteria. Hoppe Seyler's Zeits Physiol Chemie 358:797–806, 1977.
157. CM Yeager, PJ Bottomley, DJ Arp, MR Hyman. Inactivation of toluene 2-monooxygenase in *Burkholderia cepacia* G4 by alkynes. Appl Environ Microbiol 65:632–639, 1999.
158. S Chauhan, P Barbieri, TK Wood. Oxidation of trichloroethylene, 1,1-dichloroethylene, and chloroform by toluene/o-xylene monooxygenase from *Pseudomonas stutzeri* OX1. Appl Environ Microbiol 64:3023–3024, 1998.
159. S Grosse, L Laramee, KD Wendlandt, IR McDonald, CB Miguez, HP Kleber. Purification and characterization of the soluble methane monooxygenase of the type II methanotrophic bacterium *Methylocystis* sp. strain WI 14. Appl Environ Microbiol 65:3929–3935, 1999.
160. JH Maeng, Y Sakai, Y Tani, N Kato. Isolation and characterization of a novel oxygenase that catalyzes the first step on n-alkane oxidation in *Acinetobacter* sp. Strain M-1. J Bacteriol 178:3695–3700, 1996.
161. V Kadiyala, JC Spain. A two-component monoxgenase catalyzes both the hydroxylation of *p*-nitrophenol and the oxidative release of nitrite from 4-nitrocatechol in *Bacillus sphaericus* JS905. Appl Environ Microbiol 64:2479–2484, 1998.
162. BE Haigler, WC Suen, JC Spain. Purification and sequence analysis of 4-methyl-5-nitrocatechol oxygenase from *Burkholderia* sp. strain DNT. J Bacteriol 178:6019–6024, 1996.
163. N Gorlatova, M Tchorzewski, T Kurihara K Soda N Esaki. Purification, characterization, and mechanism of a flavin mononucleotide-dependent 2-nitropropane dioxygenase from *Neurospora crassa*. Appl Environ Microbiol 64:1029–1033, 1998.
164. JD Haddock, DT Gibson. Purification and characterization of the oxygenase component of biphenyl 2,3-dioxygenase from *Pseudomonas* sp. strain LB400. J Bacteriol 177:5834–5839, 1995.
165. JD Haddock, JR Horton, DT Gibson. Dihydroxylation and dechlorination of chlorinated biphenyls by purified biphenyl 2,3-dioxygenase from *Pseudomonas* sp. strain LB400. J Bacteriol 177:20–26, 1995.
166. D Barriault, J Durand, H Maaroufi, LD Eltis, M Sylvestre. Degradation of polychlorinated biphenyl metabolites by naphthalene-catabolizing enzymes. Appl Environ Microbiol 64: 4637–4642, 1998.
167. V Kahn, N Bem-Shalom, V Zakin. p-Hydroxyphenylpropionic acid (PHPPA) and 3,4-dihydroxyphenylpropionic acid (2,4-DPPA) as substrate from mushroom tyrosinase. J Food Biochem 23:75–94, 1999.
168. A Rompel, H Fischer, D Meiwes, K Buldt-Karentzopoulos, A Margrini, C Eichen, C Gerdemann, B Krebs. Substrate specificity of catechol oxidase from *Lycopus europaeus* and characterization of the by-products of enzymic caffeic oxidation. FEBS Lett 445:103–110, 1999.
169. J-M Bollag, C Myers, S Pal, PM Huang. The role of abiotic and biotic catalysts in the transformation of phenolic compounds. Environ Impact Soil Component Interact 1:299–309, 1995.
170. A Naidja, PM Huang, J-M Bollag. Comparison of reaction products from the transformation of catechol catalyzed by birnessite or tyrosinase. Soil Sci Soc Am J 62:188–195, 1998.
171. A Zaks, AM Klibanov. The effect of water on enzyme action in organic media. J Biol Chem 263:8017–8021, 1988.
172. JS Dordick. Enzymic catalysis in monophasic organic solvents. Enzyme Microbiol Technol 1:194–211, 1989.

173. SG Burton, JR Duncan, PT Kaye, PD Rose. Activity of mushroom polyphenol oxidase in organic medium. Biotechnol Bioeng 42:938–944, 1993.
174. JC Roper, JM Sarkar, J Dec, J-M Bollag. Enhanced enzymatic removal of chlorophenols in the presence of co-substrates. Water Res 29:2720–2724, 1995.
175. A Potthast, T Rosenau, CL Chen, JS Gratz. A novel method for the conversion of benzyl alcohols to benzaldeydes by laccase-mediated oxidation. J Mol Catal 108:5–9, 1996.
176. PJ Collins, MJJ Kottermann, JA Field, ADW Dobson. Oxidation of anthracene and benzo[a]-pyrene by laccases from *Trametes versicolor*. Appl Environ Microbiol 62:4563–4567, 1996.
177. J Sealey, AJ Ragauskas. Residual lignin studies of laccase-delignified kraft pulp. Enzyme Microbiol Technol 23:422–426, 1998.
178. A Majcherczyk, C Johannes, A Hüttermann. Oxidation of polycyclic aromatic hydrocarbons (PAH) by laccase of *Trametes versicolor*. Enzyme Microbiol Technol 22:335–341, 1998.
179. MT Filazzola, F Sannino, MA Rao, L Gianfreda. Effect of various pollutants and soil-like constituents on laccase from *Cerrena unicolor*. J Environ Qual 28:1929–1938, 1999.
180. AM Klibanov, BN Alberti, ED Morris, LM Felshin. Enzymatic removal of toxic phenols and anilines from waste waters. J Appl Biochem 2:414–421, 1980.
181. AM Klibanov, ED Morris. Horseradish peroxidase for the removal of carcinogenic aromatic amines from water. Enzyme Microbiol Technol 3:119–122, 1981.
182. T Sawahata, RA Neal. Horseradish peroxidase-mediated oxidation of phenol. Biochim Biophys Res Commun 109:988–994, 1982.
183. PD Josephy, T Eling, RP Mason. The horseradish peroxidase-catalyzed reaction of 3,5,3′,5′-tetramethylbenzidine. J Biol Chem 257:3669–3675, 1982.
184. KE Hammel, B Kalyanaraman, TK Kirk. Oxidation of polycyclic aromatic hydrocarbons and dibenzo(*p*)dioxins by *Phanerochaete chrysosporium* lignininase. J Biol Chem 261: 16948–16952, 1986.
185. KB Male, RS Brown, JHT Luong. Enzymatic oxidation of water-soluble cyclodextrin-polynuclear aromatic hydrocarbon inclusion complexes usign lignin peroxidase. Enzyme Microbiol Technol 17:607–614, 1995.
186. JA Field, RH Vledder, JG van Zelst, WH Rulkens The tolerance of lignin peroxidase and manganese-dependent peroxidase to miscible solvents and the in vitro oxidation of anthracene in solvent: water mixtures. Enzyme Microbiol Technol 18:300–308, 1996.
187. L Setti, S Scali, I Degli Angeli, PG Pifferi. Horseradish peroxidase-catalyzed oxidative coupling of 3-methyl-2-benzothiazolinone hydrazone and methoxyphenols. Enzyme Microbiol Technol 22:656–661, 1998.
188. JA Nicell, JK Bewtra, KE Taylor, N Biswas, C St Pierre. Enzyme catalyzed polymerization and precipitation of aromatic compounds from wastewater. Water Sci Technol 25:157–164, 1992.
189. AM Klibanov. Enzymatic removal of hazardous pollutants from industrial aqueous effluents. Enzyme Eng 6:319–323, 1982.
190. MB Arnao, M Acosta, JA del Rio, R Varon, F Garcia-Canovas. A kinetic study on the suicide inactivation of peroxidase by hydrogen peroxide. Biochim Biophys Acta 1041:43–47, 1990.
191. L Gianfreda, F Sannino, MT Filazzola, A Leonowicz. Catalytic behavior and detoxifying ability of a laccase from the fungal strain *Cerrena unicolor*. J Mol Cat B Enz 14:13–23, 1998.
192. S Nakamoto, N Machida. Phenol removal from aqueous solutions by peroxidase-catalyzed reaction using additives. Water Res 26:49–54, 1992.
193. VA Cooper, JA Nicell. Removal of phenols from a foundry wastewater using horseradish peroxidase. Water Res 30:954–964, 1996.
194. Y Wu, KE Taylor, N Biswas, JK Bewtra. A model for the protective effect of additives on the activity of horseradish peroxidase in the removal of phenol. Enzyme Microbiol Technol 22:315–322, 1998.

195. K Martinek, VV Mozhaev. Immobilization of enzymes: an approach to fundamental studies in biochemistry. Adv Enzymol 57:179–249, 1985.
196. KJ Laidler, PS Bunting. The kinetics of immobilized enzyme systems. In: DL Putich, ed. Methods in Enzymology. Vol. 64. New York: Academic Press, 1980, pp 227–248.
197. E Katchalski, I Silman, R Goldman. Effect of the microenvironment on the mode of action of immobilized enzymes. Adv Enzymol 28:506–511, 1971.
198. SR Caldwell, FM Raushel. Detoxification of organophosphate pesticides using an immobilized phosphotriesterase from *Pseudomonas diminuta*. Biotechnol Bioeng 37:103–109, 1991.
199. JM Smith, GF Payne, JA Lumpkin, JS Kams. Enzyme-based strategy for toxic waste treatment and waste minimization. Biotechnol Bioeng 39:741–752, 1992.
200. K Grice, J Payne, F Gregory, JS Karns. Enzymatic approach to waste minimization in a cattle dipping operation: Economic analysis. J Agric Food Chem 44:351–357, 1996.
201. SC Atlow, L Bonadonna-Aparo, AM Klibanov. Dephenolization of industrial wastewaters catalyzed by polyphenol oxidase. Biotechnol Bioeng 26:599–603, 1984.
202. JA Nicell, JK Bewtra, N Biswas, KE Taylor. Reactor development for peroxidase catalyzed polymerization and precipitation of phenols from wastewater. Water Res 27:1629–1639, 1993.
203. S Wada, H Ichikawa, K Tatsumi. Removal of phenols from wastewater by soluble and immobilized tyrosinase. Biotechnol Bioeng 42:854–858, 1993.
204. C Crecchio, P Ruggiero, MDR Pizzigallo. Polyphenoloxidases immobilized in organic gels: Properties and applications in the detoxification of aromatic compounds. Biotechnol Bioeng 48:585–591, 1995.
205. K Tatsumi, S Wada, H Ichikawa. Removal of chlorophenols from wastewater by immobilized horseradish peroxidase. Biotechnol Bioeng 51:126–130, 1996.
206. W Edwards, R Bownes, WD Leukes, EP Jacobs, R Sanderson, PD Rose, SG Burton. A capillary membrane bioreactor using immobilized polyphenol oxidase for the removal of phenols from industrial effluents. Enzyme Microbiol Technol 24:209–217, 1999.
207. J Dec, J-M Bollag. Use of plant material for the decontamination of waters polluted with phenols. Biotechnol Bioeng 44:1132–1139, 1994.
208. JC Roper, J Dec, J-M Bollag. Using minced horseradish roots for the treatment of polluted waters. J Environ Qual 25:1242–1247, 1996.
209. G Greco Jr, G Toscano, M Cioffi, L Gianfreda, F Sannino. Dephenolisation of olive mill waste-waters by olive husk. Water Res 33:3046–3050, 1999.
210. P Ruggiero, JM Sarkar, J-M Bollag. Detoxification of 2,4-dichlorophenol by a laccase immobilized on soil or clay. Soil Sci 147:361–370, 1989.
211. L Gianfreda, J-M Bollag. Effect of soils on the behavior of immobilized enzymes. Soil Sci Soc Am J 58:1672–1681, 1994.
212. SF Masaphy, T Fahima, D Levanov, Y Henis, U Mingelgrin. Parathion degradation by *Xanthomonas* sp. and its crude enzyme extract in clay suspensions. J Environ Qual 25:1248–1255, 1996.
213. S Cervelli, D Perret. The immobilized enzymes as biotechnological suggestion for soil decontamination. Anal Chim 88:509–515, 1998.
214. P Nannipieri, J-M Bollag. Use of enzymes to detoxify pesticide-contaminated soils and waters. J Environ Qual 20:510–517, 1991.
215. F Xu, RM Berka, JA Wahleithner, BA Nelson, JR Shuster, SH Brown, A Palmer, EI Solomon. Site-directed mutations in fungal laccase: Effect on redox potential, activity, and pH profile. Biochem J 334:63–70, 1998.
216. V Ducros, AM Brzozowski, KS Wilson, SH Brown, P Ostergaard, AH Pedersen, P Schneider, DS Yaver, AH Pedersen, GI Davies. Crystal structure of the type-2-Cu depleted laccase from *Coprinus cinereus* at 2.2 resolution. Nature Struct Biol 5:310–316, 1998.
217. KF Gu, TMS Chu. Immobilization of a multienzymic system and dextran-NAD+ in semiper-

meable microcapsules for use in a bioreactor to convert urea into L-glutamic acid. In: M Mog ed. Bioreactor immobilized enzymes and cells: Fundamentals and applications. London: Elsevier Applied Science, 1988, pp 59–62.

218. DM Munnecke, LM Johnson, HW Talbot, S Barik. Microbial metabolism and enzymology of selected pesticides. In AM Chakrabarty, ed. Biodegradation and Detoxification of Environmental Pollutants. Boca Raton, FL: CRC Press, pp 1–33, 1982.

219. LM Johnson, HW Talbot. Detoxification of pesticides by microbial enzymes. Experientia 39:1236–1246, 1983.

220. R Honeycutt, L Ballatine, H LeBaron, D Paulson, V Seim, C Ganz, G Milad. Degradation of higher concentrations of a phosphoric ester by hydrolase. In RF Krueger, N Seiber, ed. Treatment and Disposal of Pesticide Wastes. ACS Symp Ser 259:343–352, 1984.

20

Enzyme-Mediated Transformations of Heavy Metals/Metalloids

Applications in Bioremediation

Robert S. Dungan
George E. Brown, Jr. Salinity Laboratory, USDA–ARS, Riverside, California

William T. Frankenberger, Jr.
University of California–Riverside, Riverside, California

I. INTRODUCTION

A major emphasis has been placed on the bioremediation of organic compounds (1) and their fate and transport throughout the environment (2,3). However, another important class of chemicals polluting our environment are inorganic, particularly heavy metals and metalloids. Heavy metals are elements of the periodic table with a density of more than 5 g cm^{-3}. Although this encompasses a large percentage of the metals, only several heavy metals/metalloids are regarded as of environmental concern, including selenium (Se), arsenic (As), chromium (Cr), and mercury (Hg). In the United States, more than 50% of the National Priority (Superfund) sites ranked on the National Priorities List (NPL) contain heavy metals that are designated as a threat or problem to the environment (4). Since heavy metals/metalloids cannot be degraded (i.e., biologically or chemically) they are among the most intractable pollutants to remediate.

The contamination of soils and waters with heavy metals/metalloids usually occurs by direct application from sources, including mine waste, atmospheric deposition (a result of metal emissions to the atmosphere from metal smelting, fossil fuel combustion, and other industrial processes), animal manure, and sewage sludge (5). Surprisingly, some inorganic fertilizers contain significant quantities of heavy metal impurities. Sewage sludge, which is often used as a soil conditioner, contains useful quantities of organic matter, N, and P; however, it often contains heavy metals. The metals are chelated by the organic matter and are released upon its decomposition. Heavy metal/metalloid cations in soil may be present as several different forms: (1) ions in soil solution; (2) easily exchangeable ions; (3) organically bound; (4) coprecipitated with metal oxides, carbonates, phosphates, or secondary minerals; or (5) ions in primary minerals (6). As a result, the heavy metal form is highly influenced by soil properties such as pH, oxidation–reduction (redox) state, clay content, iron oxide content, and organic matter content.

Although the ultimate goal is the complete removal of heavy metals and metalloids from water, this is not necessarily the case with contaminated soil. The most commonly used remedial techniques to deal with heavy metal/metalloid-contaminated soil are landfilling and solidification (7). Solidification involves a process in which the contaminated soil is stabilized, fixed, solidified, or encapsulated into a solid material by the addition of a resin or some other chemical compound that acts as a cement. However, although the contaminants are immobilized in the matrix, they are not destroyed, and, as a result, there is major concern over the stability of the contaminants in the solidified matrix. Additional remedial technologies include soil washing, soil flushing, acid extraction, and vitrification.

In an effort to find economically viable remedial technologies, much attention is focused on bioremedial approaches. Investigations have shown that microbiological metal transformations may be applicable in remediating heavy metals and metalloids in soil as well as in water. Novel applications in bioremediation have been designed for aquatic systems; unfortunately, relatively few applications are available for contaminated soils. Nonetheless, it is well known that the fate and transport of inorganic solutes in soils and waters can be controlled by biochemical processes such as oxidation, reduction, methylation, and demethylation (8). As a consequence of these reactions mediated by microorganisms, heavy metals and metalloids can exist in chemical states (i.e., soluble phase, insoluble nonaqueous phase such as mineral precipitants, or gaseous phase) that are biologically less toxic or more easily removed from the environment or both.

In natural soil and aquatic systems, heavy metal/metalloid transformations are generally carried out as a direct result of microbial activities (e.g., respiration and detoxification mechanisms). However, extracellular enzymes and enzymes not directly associated with the soil and aquatic microbiota may also contribute to these transformations. In soil, a number of extracellular enzymes are produced by microorganisms; other enzyme sources include plant seeds, fungal and bacterial endospores, protozoan cysts, and plant roots, all of which contribute to free enzymes found in soil and in some cases water. Free enzymes can be inactivated by adsorption to organic and inorganic particles, can be denatured by physical and chemical factors, or can serve as growth substrates for other microorganisms. Although many background enzymes can be found in natural soil and aquatic systems, very little research has been conducted on their involvement in heavy metal/metalloid transformations. Further research attention should be applied to this area, especially with regard to bioremediation of heavy metals/metalloids. The purpose of this chapter is to review microbially mediated transformations of Se, As, Cr, and Hg and discuss, where applicable, how they are currently being applied in bioremediation approaches to detoxify soils and waters.

II. SELENIUM

Selenium belongs to group VIA of the periodic table and has been classified as a metalloid. In the environment Se exits in four oxidation states, +2, 0, +4, and +6, forming a variety of compounds. Selenate (SeO_4^{2-}, Se^{6+}) and selenite (SeO_3^{2-}, Se^{4+}) are the most common ions found in soil solution and natural waters. Organic Se-containing compounds include Se-substituted amino acids, such as selenomethionine, selenocysteine, and selenocystine, and volatile methyl species such as dimethylselenide (DMSe, $[CH_3]_2Se$), dimethyldiselenide (DMDSe, $[CH_3]_2Se_2$), methaneselenol (CH_3SeH), and dimethylselenenylsulfide

(DMSeS, $[CH_3]_2SeS$). Inorganic reduced forms include mineral selenides and hydrogen selenide (H_2Se). The environmental threat of elevated levels of Se in soils and waters has been recognized in many locations throughout the western United States (9). In California's San Joaquin Valley, elevated levels of Se in agricultural drainage water have been linked to the death and deformity of aquatic birds (10).

Selenium is predominantly cycled via biological pathways similar to that of sulfur. Like sulfur, Se undergoes various oxidation and reduction reactions that directly affect its oxidation state and, hence, its chemical properties and behavior in the environment. To date, most work has focused on reduction and methylation/volatilization reactions of Se because of their potential application in remediating seleniferous environments. Currently, bioremediation strategies for Se are much further along in terms of implementation than methods for As, Cr, and Hg, which are largely still in the experimental stage.

A. Reduction of Selenium(VI)

The bioreduction of Se to insoluble Se^0 has been extensively investigated as a technique for removing Se from contaminated water. Selenium undergoes dissimilatory microbial reduction, whereby SeO_4^{2-} is reduced to Se^0 as the terminal electron acceptor in respiratory metabolism. Macy (11) isolated *Thauera selenatis*, a SeO_4^{2-}, NO_3^-, and NO_2^- respiring bacterium, from seleniferous sediments. The reduction of SeO_4^{2-} to SeO_3^{2-} and NO_3^- to NO_2^- by *T. selenatis* occurs through the use of separate terminal reductases, a SeO_4^{2-} and NO_3^- reductase, respectively. The complete reduction of SeO_4^{2-} to Se^0 only occurs when the organism is grown in the presence of both SeO_4^{2-} and NO_3^-. During the tricarboxylic acid (TCA) cycle, reduced nicotinamide-adenine dinucleotide (NADH) and succinate are used as electron donors to reduce SeO_4^{2-} and SeO_3^{2-}. The electrons are then transferred via an electron transport system that is part of a dehydrogenase, which is loosely bound to the cytoplasmic membrane. Selenate reduction to SeO_3^{2-} involves a periplasmic SeO_4^{2-} reductase, whereas SeO_3^{2-} produced during the respiration of SeO_4^{2-} and NO_3^- is believed to be reduced via a periplasmic NO_2^- reductase (Fig. 1).

Enterobacter cloacae strain SLD1a-1, a facultative anaerobe isolated by Losi and Frankenberger (12), operates under mechanisms very similar to that of *T. selenatis. E. cloacae* strain SLD1a-1 uses SeO_4^{2-} and NO_3^- as terminal electron acceptors during anaerobic growth and can reduce SeO_4^{2-} to Se^0 under growth conditions and in washed-cell suspensions under microaerophilic conditions. Although strain SLD1a-1 respires SeO_4^{2-} anaerobically, the complete reduction of SeO_3^{2-} to Se^0 does not occur unless NO_3^- is present, suggesting that NO_3^- is necessary for the reduction of SeO_3^{2-} to Se^0 (13). Oremland and associates (14) isolated a strictly anaerobic motile vibrio (*Sulfurospirillum barnesii* strain SES-3) that grows in the presence of either SeO_4^{2-} or NO_3^- while using lactate as an electron donor. It was determined that the reduction of SeO_4^{2-} and NO_3^- ions is achieved by separate inducible enzyme systems. Although growth was not observed on SeO_3^{2-}, washed-cell suspensions of SES-3 could reduce SeO_3^{2-} to Se^0. A *Pseudomonas stutzeri* isolate could only reduce SeO_4^{2-} and SeO_3^{2-} to Se^0 under aerobic conditions, a limitation that was speculated to be a detoxification mechanism (15).

The biological reduction of SeO_3^{2-} to Se^0 also occurs, but only the reduction of SeO_4^{2-} supports anaerobic growth. Although a number of SeO_3^{2-} reducing bacteria have been isolated and described metabolically, it is still unclear which reductive processes are involved. In the literature it has been reported that SeO_3^{2-} can be reduced anaerobically by a periplasmic NO_2^- reductase (16) or reduced aerobically as a detoxification mechanism,

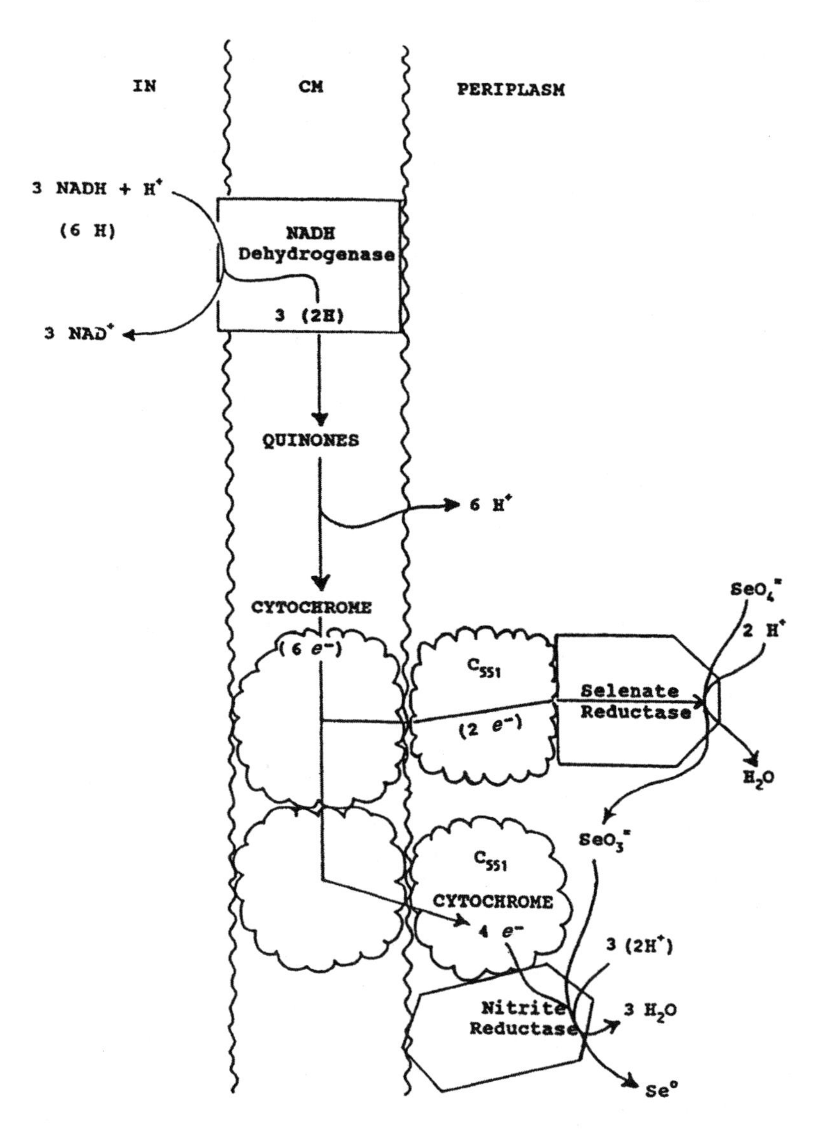

Figure 1 Hypothetical model of selenate reduction to elemental selenium by *Thauera selenatis* involving a periplasmic selenate reductase, cytochrome C_{551}, and nitrite reductase. (From Ref. 11.)

independently of dissimilatory reduction (15). However, in a 1998 study, the reduction of SeO_3^{2-} by *Bacillus selenitireducens* was linked to its respiration (17). Selenite was reduced to Se^0 by aerobically grown *Salmonella heidelberg* (18) and by resting cells of *Streptococcus faecalis* and *Streptococcus faecium* (19). Two common soil bacterial strains, *Pseudomonas fluorescens* and *Bacillus subtilis*, apparently reduced SeO_3^{2-} to Se^0 via a detoxification mechanism independent of NO_2^- and SeO_3^{2-} (20,21). Yanke and coworkers (22) found that *Clostridium pasteurianum* utilized the constitutive enzyme hydrogenase (I) as a SeO_3^{2-} reductase. In addition, the enzyme was found to reduce not only SeO_3^{2-}

but also tellurite (TeO_3^{2-}). Selenite reduction ceased when the enzyme was exposed to O_2 and $CuSO_4$, potent inhibitors of hydrogenase (I) activity.

B. Methylation of Selenium

The methylation of Se is a biological process and is thought to be a protective mechanism used by microorganisms to detoxify their surrounding environment. The methylation and subsequent volatilization of Se may constitute important steps in the transport of Se from contaminated terrestrial and aquatic environments. Bacteria and fungi are the predominant Se-methylating organisms isolated from soils, sediments, and waters (23). The predominant Se gas produced by most microorganisms is DMSe (24), although other volatile Se compounds, such as DMDSe, DMSeS, and methaneselenol, may also be produced in lesser amounts. Although the biological significance of Se methylation is not clearly understood, once volatile Se compounds are released to the atmosphere and diluted, Se has lost its hazardous potential.

The first report of microbially derived gaseous Se was discovered by Challenger and North (25) during their studies of pure cultures of *Penicillium brevicaule* (previously named *Scopulariopsis brevicaulis*). They found that the fungus was able to convert both SeO_4^{2-} and SeO_3^{2-} to DMSe while growing on bread crumbs. Several reports that followed over the years identified many other fungi capable of methylating Se, including *Penicillium* sp., *Fusarium* sp., *Schizopyllum commune, Aspergillus niger, Alternaria alternata*, and *Acremonium falciforme* (26). Abu-Erreish et al. (27) noticed the production of volatile Se in seleniferous soils appeared to be related to fungal growth. The addition of a fungal inoculum, *Candida humicola*, to soil caused the rate of Se volatilization to double (28). However, the addition of chloramphenicol to soil reduced the amount of Se volatilized from a soil by 50%, suggesting that bacteria also play an important role in Se methylation.

To date, only a few bacterial genera capable of methylating Se have been identified. Chau et al. (29) isolated three bacteria (*Aeromonas* sp., *Flavobacterium* sp., and *Pseudomonas* sp.) from lake sediment that were capable of methylating SeO_3^{2-} to DMSe and DMDSe. A strain of *Corynebacterium* sp., isolated from soil, formed DMSe from SeO_4^{2-}, SeO_3^{2-}, Se^0, selenomethionine, selenocystine, and methaneseleninate (methaneseleninic acid) (30). *Aeromonas veronii*, isolated from seleniferous agricultural drainage water, was active in volatilizing DMSe and lesser amounts of methaneselenol, DMSeS, and DMDSe (31). McCarty et al. (32) identified two phototrophic bacterial species, *Rhodospirillum rubrum* S1 and *Rhodocyclus tenuis*, that produced DMSe and DMDSe in the presence of SeO_4^{2-}. *Enterobacter cloacae* SLD1a-1, the SeO_4^{2-} and SeO_3^{2-} reducing bacterium, produces DMSe from SeO_4^{2-}, SeO_3^{2-}, Se^0, dimethylselenone [$(CH_3)_2SeO_2$], selenomethionine, 6-selenopurine, and 6-selenoinosine (33). The methylation of Se by algae has also been confirmed by Fan et al. (34), who isolated a euryhaline green microalga species of *Chlorella* from a saline evaporation pond that was able to transform SeO_3^{2-} aerobically into DMSe, DMDSe, and DMSeS.

In general, the formation of alkylselenides from Se oxyanions involves a reduction and methylation step; however, the pathway by which these reactions occur is still highly debated. Challenger (35) postulated that the formation of DMSe occurs through successive methylation and reduction steps, in which dimethylselenone was suspected to be the last intermediate prior to the formation of DMSe (Fig. 2). Reamer and Zoller (36) identified DMDSe and dimethylselenone in addition to DMSe as products from soil and sewage sludge amended with either SeO_3^{2-} or Se^0. It was then suggested that Challenger's pathway

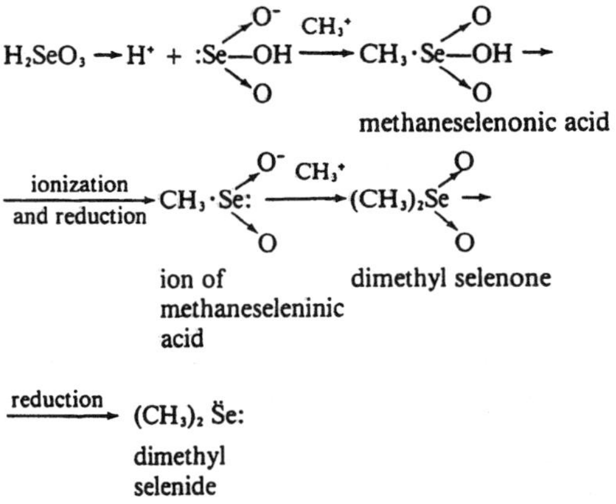

$$H_2SeO_3 \rightarrow H^+ + :Se(O^-)(O)\text{—}OH \xrightarrow{CH_3^+} CH_3\cdot Se(O)(O)\text{—}OH \rightarrow$$

methaneselenonic acid

$$\xrightarrow[\text{and reduction}]{\text{ionization}} CH_3\cdot Se(O^-)(O): \xrightarrow{CH_3^+} (CH_3)_2Se(O)(O) \rightarrow$$

ion of methaneseleninic acid — dimethyl selenone

$$\xrightarrow{\text{reduction}} (CH_3)_2\ddot{S}e:$$

dimethyl selenide

Figure 2 Proposed mechanism for the methylation of selenium by fungi. (From Ref. 35.)

could be modified to include the production of DMDSe through an alternate pathway whereby methaneseleninic acid is reduced to methaneselenol or methaneselenenic acid or both, to produce DMDSe. Doran (24) proposed that SeO_3^{2-} is reduced via Se^0 to a selenide from before it is methylated to form methaneselenol and finally DMSe. Although methaneselenol and methaneselenide were not tested for as intermediates, evidence in support of Doran's pathway comes from Bird and Challenger (37), who detected small amounts of methaneselenol emitted from actively methylating fungal cultures. Doran's pathway is also markedly similar to findings of studies conducted with mammals, which demonstrated that methaneselenol is an intermediate in the methylation of Se to DMSe (38,39). Cooke and Bruland (40) proposed a pathway for the formation of DMSe from SeO_4^{2-} and SeO_3^{2-} in natural waters. Apparently both Se oxyanions are reduced and assimilated into the intermediate selenomethionine [$CH_3Se(CH_2)_2CHNH_2COOH$], which is then methylated to produce methylselenomethionine [$(CH_3)_2Se^+(CH_2)_2CHNH_2COOH$]. Finally, methylselenomethionine is hydrolyzed to DMSe and homoserine.

The biosynthesis of methionine from homocysteine is an important transformation in the methylation of Se. During the activated methyl cycle homocysteine is methylated via the coenzyme methylcobalamin (CH_3B_{12}; derivative of vitamin B_{12}), yielding methionine. Methylcobalamin has been isolated from bacteria (41) and is believed to donate methyl groups to Se, resulting in the formation of volatile alkylselenides. Thompson-Eagle et al. (42) found that the addition of methylcobalamin promoted the methylation of SeO_4^{2-}. McBride and Wolfe (43) found that cell-free extracts of a *Methanobacterium* sp. methylated SeO_4^{2-} when methylcobalamin was present. Cell-free extracts of *E. cloacae* SLD1a-1 catalyzed the formation of DMSe from SeO_3^{2-} or Se^0 when methylcobalamin was the methyl donor (33). In addition, *S*-adenosylmethionine has been identified as a cofactor in the microbial methylation of inorganic Se (30). Doran (24) found that cell-free extracts of the soil *Corynebacterium* sp. were able to methylate SeO_3^{2-} or Se^0 when *S*-adenosylmethionine was present. Drotar and associates (44) identified an *S*-adenosylmethionine-dependent selenide methyltransferase in cell-free extracts of *Tetrahymena thermophila*, which reportedly produced methaneselenol from Na_2Se. Although there is some understanding of the pathway by which Se oxyanions are transformed to DMSe, neither of the pathways elucidates the mechanism of the reaction. Clearly more work is needed to understand the biochemical characteristics of Se methylation.

C. Bioremediation of Seleniferous Water and Sediment

Since the 1990s attention has been given to the development of an effective remediation technology for the permanent removal of Se oxyanions from seleniferous soil and water. A majority of the focus has been applied to contaminated agricultural drainage water, which has been responsible for a number of well-documented ecotoxicological problems. Since Se undergoes microbial transformations, their application may be potentially useful as bioremediation strategies. Several different bioremedial approaches have been or are being developed; they include a variety of bioreactors utilizing bacteria with the ability to reduce the toxic, soluble Se oxyanions to insoluble Se^0. These systems are designed to remove Se from contaminated wastewater (industrial or agricultural) before release into the environment. Because of the high SeO_4^{2-} to SeO_3^{2-} ratio of most agricultural drainage waters of the western United States, removal of mainly SeO_4^{2-} must be considered in these systems. Another means to remove Se from contaminated soil and water involves stimulation of the indigenous microorganisms that volatilize Se. This process has proved effective as an in situ treatment for seleniferous soils and sediments in the San Joaquin Valley, California (45,46).

1. Bioreduction of Selenium Oxyanions to Elemental Selenium

The use of *Thauera selenatis*, a SeO_4^{2-} respiring bacterium, in a biological reactor system to remediate both SeO_4^{2-} and SeO_3^{2-} ions from contaminated water has been described by Macy and associates (47), Lawson and Macy (48), and Cantafio and coworkers (49). The latest pilot scale system, which consisted of a series of four medium-packed tanks, was used to treat seleniferous agricultural drainage water (49). Using acetate as the electron donor, Se oxyanion and NO_3^- concentrations were reduced by 98%. An earlier system included the use of two bioreactors in series; the first was an aerobic sludge blanket reactor and the second a fluidized bed reactor (47). Once again acetate was used as the electron donor and the growth of the organism was found to be dependent on the presence of NH_4Cl. The SeO_4^{2-}, SeO_3^{2-}, and NO_3^- levels were all reduced by 98% in the influent. A similar system, later used to remediate SeO_3^{2-} from oil refinery wastewater, reduced the Se oxyanion concentration by 95%. Although Macy (11) has shown that this organism can reduce both SeO_4^{2-} and NO_3^- simultaneously, NO_3^- must be present in the system for SeO_4^{2-} to be completely reduced to Se^0, since the NO_2^- reductase only catalyzes the reduction of SeO_3^{2-} when denitrification is occurring.

The algal–bacterial selenium removal system (ABSRS) is another process used to remove soluble Se and NO_3^- from drainage water (50). The influent is first directed toward high-rate ponds where microalgae are grown; removal of some NO_3^- results. About 10% of the N is removed in the high-rate ponds, a proportion that supports that algae are made up of 9.2% N by dry weight (51). After this step, the biomass suspension is discharged into an anoxic unit where bacteria use the algae as a C and energy source and subsequently reduce the SeO_4^{2-} and SeO_3^{2-} to Se^0, and NO_3^- to N_2 gas. Although near-complete removal of SeO_4^{2-} and NO_3^- occurred at times in field experiments, it was speculated that since the project was run for an insufficient amount of time, steady-state reducing conditions could not be established. Since substantial reduction of SeO_4^{2-} to SeO_3^{2-} was occurring, use of $FeCl_3$ was applied to precipitate out inorganic SeO_3^{2-}, thereby reducing the soluble Se levels.

Oremland (52) has also described a process similar to the ABSRS. This process involves using a two-stage reaction, which uses algae in the first aerobic stage to deplete the NO_3^- concentrations below 62 mg L^{-1}. The water is then transferred to an anoxic reactor containing SeO_4^{2-} reducing bacteria where SeO_4^{2-} is reduced to insoluble Se^0. In 7 days, the influent SeO_4^{2-} concentration of 56 mg Se L^{-1} was reduced by more than 99%.

EPOC AG (Binnie California) conducted studies on the removal of Se from agricultural drainage water using a pilot-scale two-stage biological process (53). The system consisted of an upflow anaerobic sludge blanket reactor followed by a fluidized-bed reactor. A crossflow microfilter was used after the biological reactors for the removal of particulate Se. The effluent concentration from the system averaged less than 30 μg L^{-1} of soluble selenium. When the effluent was further processed through a soil column the soluble Se concentration was less than 10 μg L^{-1}.

Owens (53) describes a pilot-scale biological system that utilized an upflow anaerobic sludge blanket reactor. The C source used in the system was methanol, which was added at a dosage of 250 mg L^{-1}. Most of the C added to the system was used during denitrification; thus, enough methanol must be added to support both denitrification and Se reduction. Denitrification is important to the process since Se reduction does not occur until the NO_3^- is removed. It was reported that the reactor was able to remove 94% of the soluble Se, with a final effluent concentration of 29 μg L^{-1} obtained.

Adams et al. (54) conducted a pilot study in which *Escherichia coli* was used to treat a weak acid effluent from a base metal smelter containing 30 mg Se L^{-1}. The bioreactor system consisted of a rotating biological contactor (RBC) and was able to remove 97% of the Se within 4 hours. A bench-scale RBC system was also tested on mining process waters, and using *Pseudomonas stutzeri*, with molasses (1 g L^{-1}) as the C source, 97% of the Se was removed in 6-hour retention time.

2. Selenium Volatilization in the Field

Field studies were performed on the Sumner Peck Ranch (Fresno County, California) evaporation pond water in an effort to determine whether the addition of casein would stimulate Se volatilization (55). Water columns in the evaporation ponds were treated with a single casein application of 0.2 g L^{-1} pond water. The evaporation pond water Se concentration was reported as high as 2.9 mg L^{-1}. Unamended pond water evolved volatile Se at low rates of 0.1 μg Se L^{-1} d^{-1}, whereas casein amended pond water produced emission rates of 2.2 μg Se L^{-1} d^{-1}. After 142 days, the casein amended pond water lost 38% of the initial Se inventory.

In dewatered evaporation pond sediments at the Sumner Peck Ranch, 32% of the Se in the top 15 cm was removed with the application of water plus tillage alone; the addition of cattle manure resulted in the removal of 58% after 22 months (56). The initial mean plot soil Se concentration in the top 15 cm was 11.4 mg kg^{-1}. The background emission rate of volatile Se averaged 3.0 μg Se m^{-2} h^{-1}, whereas the cattle manure treated plot promoted an average emission rate of 54 μg Se m^{-2} h^{-1}. As reported in other Se volatilization studies, the parameters that enhanced Se volatilization were moisture, high temperatures, aeration, and an available C source. The highest gaseous Se flux was recorded in the summer months and the lowest flux occurred in the winter.

Over a 100-month period at Kesterson Reservoir, 68%–88% of the total Se was dissipated from the top 15 cm of seleniferous soil (46). The soil Se concentration varied in each of the plots from approximately 40 to 60 mg Se kg^{-1}. Since no pattern of Se depletion was correlated with rainfall events or temperature, it was speculated that leaching dominated during the winter months, because most rainfall occurred during the winter, whereas volatilization was dominant during the summer months. The addition of C amendments had no significant effect greater than that of the moisture-only treatment, a finding that suggests that tillage and irrigation prevailed over the effects of the amendments. However, cattail roots providing C were disked into all plots at the onset of this investigation (57).

3. Cell-Free Systems

Adams et al. (54,58) treated mine water containing 0.62 mg L^{-1} SeO_4^{2-} by using an immobilized cell-free preparation of *Pseudomonas stutzeri*. Tests were conducted by using a single-pass bioreactor with a retention time of 18 hours. The cell-free extracts were prepared by disrupting the cells then immobilizing the lysate in calcium alginate beads. The immobilized enzyme preparation performed for approximately 4 months, achieving effluent levels below 10 μg L^{-1}. Another cell-free system was used to treat mining process solution containing cyanide and Se. The system contained cell-free extracts of *P. pseudoalcaligenes, P. stutzeri*, CN-oxidizing, and Se-reducing microbes combined and immobilized in calcium alginate beads. Tests were conducted in single-pass 1-in-diameter columns with a retention time of 9 to 18 hours. The system was capable of simultaneously removing cyanide and Se (initial concentrations of 102 and 31.1 mg L^{-1}, respectively) to concentrations of 1.0 and 1.6 mg L^{-1}, respectively.

III. ARSENIC

Arsenic (As) is a metalloid of group VA of the periodic table and exists in four oxidation states, +5, +3, 0, and −3. It occurs naturally in the environment as well through anthropogenic discharge in a variety of chemical states. Arsenic forms alloys with various metals and covalently bonds with carbon, hydrogen, oxygen, and sulfur (59). Arsenate (AsO_4^{3-}), a biochemical analog of phosphate, is transported by highly specific energy-dependent membrane pumps into the cell during assimilation of phosphate, whereas arsenite (AsO_2^-) has a high affinity for thiol groups of proteins, resulting in the inactivation of many enzymes. Its similarity to phosphorus and its ability to form covalent bonds with sulfur are two reasons for As toxicity. The poisonous character of As make it an effective herbicide and insecticide. The ubiquity of As in the environment, its biological toxicity, and its redistribution are factors evoking public concern.

Both oxidation and methylation are microbial transformations involved in the redistribution and global cycling of As. Oxidation involves the conversion of toxic AsO_2^- to less toxic AsO_4^{3-}. Arsenite is much more toxic to aquatic microbiota of agricultural drainage water and evaporation pond sediments than any other As species (60). Bacterial methylation of inorganic As is coupled to the formation of methane in methanogenic bacteria and may serve as a detoxification mechanism. The mechanism involves the reduction of AsO_4^{3-} to AsO_2^-, followed by methylation to dimethylarsine. Fungi are also able to transform inorganic and organic As compounds into volatile methylarsines. The pathway proceeds aerobically by AsO_4^{3-} reduction to AsO_2^- followed by several methylation steps producing trimethylarsine. Currently, a number of microbially mediated oxidation and methylation reactions are being studied in the interest of developing bioremediation techniques for detoxifying As-contaminated soil and water.

A. Reduction of Arsenic(V)

It is known that a certain number of bacteria reduce As^{5+} to As^{3+} as a detoxification mechanism; however, the significance in the biogeochemical cycling of As is not clear. In addition to reductive detoxification, which may occur under both aerobic and anaerobic conditions, dissimilatory reduction of AsO_4^{3-} may contribute to the reduction of As^{5+} to As^{3+} in anaerobic sediments (61,62). Dowdle et al. (61) found that As^{5+} was reduced to As^{3+} in anoxic salt marsh sediment slurries when the electron donor was lactate, H_2, or glucose. The addition of the respiratory inhibitor/uncoupler dinitrophenol, rotenone, or 2-heptyl-4-hydroxyquinoline *N*-oxide blocked the reduction of As^{5+}, suggesting that the reduction of As^{5+} in sediments proceeds through a dissimilatory process.

To date, several $SAsO_4^{3-}$ respiring organisms have been isolated and characterized: *Sulfurospirillum arsenophilus* strain MIT-13 (63), *S. barnesii* strain SES-3 (14,64), *Desulfotomaculum auripigmentum* strain OREX-4 (65), and *Chrysiogenes arsenatis* strain BAL-1T (66). The only common electron acceptor of these organisms is fumurate, and studies have shown that strain MIT-3, strain SES-3, and strain BAL-1T respire NO_3^- and AsO_4^{3-} but not SO_4^{2-}, whereas strain OREX-4 can grow on SO_4^{2-} but not NO_3^-. The mechanisms by which electrons are passed to AsO_4^{3-} during dissimilatory reduction and reductive detoxification differ.

The reductive detoxification of AsO_4^{3-} occurs when reduced dithiols transfer electrons for the ArsC enzymes (67), whereas the respiratory AsO_4^{3-} reductase in strain SES-3 appears to utilize prosthetic groups such as Fe:S clusters (68). Additionally, a *b*-type cytochrome is present in the membrane when it is grown on AsO_4^{3-}, and it may be involved in the transfer of electrons. Fig. 3 is the proposed biochemical pathway by which strain

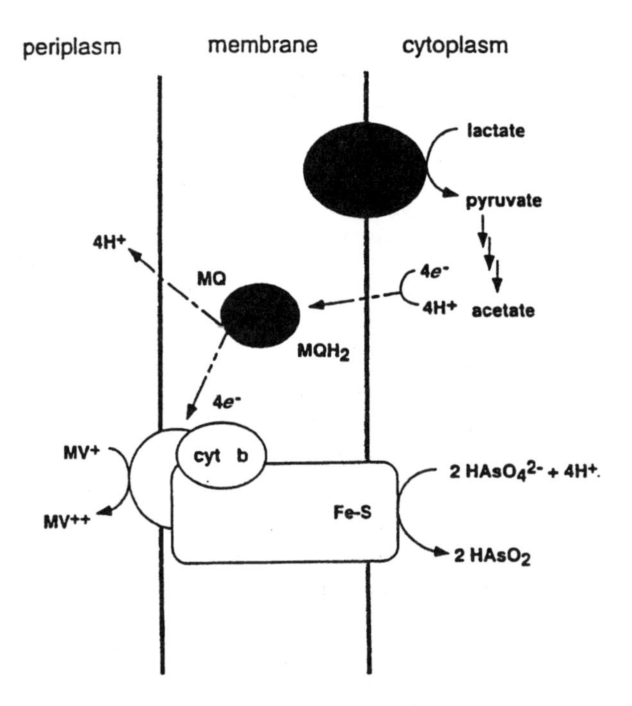

Figure 3 Biochemical model of arsenate respiration in *Sulfurospirillum barnesii* strain SES-3. (From Ref. 68.)

SES-3 reduces AsO_4^{3-} to AsO_2^- when grown on lactate. It was postulated that strain SES-3 contains an AsO_2^--efflux system similar to that of other bacteria, which would enable strain SES-3 to cope with the AsO_2^- produced in the cytoplasm. Therefore, the flow of electrons could be generated from a cytoplasmically oriented lactate dehydrogenase to the AsO_4^{3-} reductase and could occur through the use of a proton-pumping intermediate (e.g., menaquionone) or through diffusion of H_2 (formed by a cytoplasmic hydrogenase) through the membrane to the outside. Hydrogen in the periplasm would be oxidized by a hydrogenase, allowing electrons to flow back to the AsO_4^{3-} reductase through membrane-bound electron carriers. White (69) proposed a similar model for the generation of the proton motive force during dissimilatory reduction of SO_4^{2-}. Although the reduction of As^{5+} is of environmental interest because As^{3+} is more mobile and toxic than As^{5+}, additional work is clearly needed to understand the environmental significance of dissimilatory AsO_4^{3-} reduction.

B. Oxidation of Arsenic(III)

Currently, a number of microbially mediated oxidation reactions are being studied in the interest of developing bioremediation techniques for detoxifying As-contaminated soil and water. *Bacillus, Thiobacillus*, and *Pseudomonas* spp. that have been isolated can oxidize AsO_2^- to the less toxic AsO_4^{3-}. In addition, a strain of *Alcaligenes faecalis* obtained from raw sewage was capable of oxidizing AsO_2^- (70). Osborne and Ehrlich (71) isolated a similar AsO_2^--oxidizing soil strain of *A. faecalis* whose oxidation process was induced by AsO_2^- and AsO_4^{3-}. The use of respiratory inhibitors prevented further oxidation of AsO_2^-, indicating that oxygen served as the terminal electron acceptor. Studies suggested that the oxidation of AsO_2^- involved an oxidoreductase with a bound flavin that passed electrons from AsO_2^- to O_2 by way of cytochrome *c* and cytochrome oxidase (71). Indirect evidence suggested that the organism may be able to derive maintenance energy from the oxidation of AsO_2^- (72). Ilyaletdinov and Abdrashitova (73) isolated *Pseudomonas arsenitoxidans* from a gold and arsenic ore deposit that was capable of growing autotrophically with AsO_2^- as the soil energy source.

Anderson et al. (74) found that *A. faecalis* strain NCIB 8687 contained an inducible AsO_2^--oxidizing enzyme that was located on the outer surface of the plasma membrane, a finding that suggested that AsO_2^- oxidation occurred in its periplasmic space. The 85-kD enzyme was a molybdenum-containing hydroxylase with a pterin cofactor and inorganic sulfide, one atom of molybdenum, and five or six atoms of iron. The enzyme catalyzed the oxidation of AsO_2^- when both azurin and cytochrome *c* were used as electron acceptors. Oxidation of AsO_2^- by heterotrophic bacteria plays an important role in detoxifying the environment, catalyzing as much as 78% to 96% of the AsO_2^- to AsO_4^{3-} (75).

C. Methylation of Arsenic

1. Bacterial Methylation

Bacterial methylation of inorganic As has been studied extensively using methanogenic bacteria. Methanogenic bacteria are a morphologically diverse group consisting of coccal, bacillary, and spiral forms but are unified by the production of methane as their principal metabolic end product. They are present in large numbers in anaerobic ecosystems, such as sewage sludge, freshwater sediments, and composts where organic matter is decomposing (76). Under anaerobic conditions, the biomethylation of As only proceeds to dimethylarsine, which is stable in the absence of O_2 but is rapidly oxidized under aerobic conditions. It has been shown that at least one *Methanobacterium* sp. is capable of methylating inorganic As to produce volatile dimethylarsine. Arsenate, AsO_2^-, and methylarsonic acid (methanearsonic acid) can serve as substrates in dimethylarsine formation. Inorganic arsenic methylation is coupled to the CH_4 biosynthetic pathway and may be a widely occurring mechanism for As detoxification.

Cell-free extracts of *Methanobacterium* sp. strain MOH, when incubated under anaerobic conditions with AsO_4^{3-}, methylcobalamin, H_2, and adenosine triphosphate (ATP), produced volatile dimethylarsine (43). Fig. 4 shows the pathway by which dimethylarsine is produced by *Methanobacterium* sp., which involves the reduction of AsO_4^{3-} to AsO_2^- with subsequent methylation by a low-molecular-weight cofactor coenzyme M (CoM). CoM has been found in all methane bacteria examined and chemically is 2,2′-dithiodiethane sulfonic acid (76). Methylarsonic acid added to cell-free extracts is not reduced to methylarsine but requires an additional methylation step before reduction. However, dimethylarsinic acid (cacodylic acid) is reduced to dimethylarsine even in the absence of a methyl donor (43). Under anaerobic conditions, whole cells of methanogenic bacteria also produce dimethylarsine as a biomethylation end product of As, but not heat-treated cells, indicating that this is a biotic reaction. Cell-free extracts of *Desulfovibrio vulgaris* strain 8303 also produced a volatile As derivative, presumably an arsine, when incubated with AsO_4^{3-} (43). The reaction occurred in the absence of exogenous methyl donors; however, the addition of methylcobalamin greatly stimulated the reaction.

Interestingly, another study indicated that resting cell suspensions of *Pseudomonas* and *Alcaligenes* spp. incubated with either AsO_4^{3-} and AsO_2^- under anaerobic conditions produced arsine, but no other As intermediates were formed (77). An *Aeronomonas* sp. and a *Flavobacterium* sp. isolated from lake water were capable of methylating As to dimethylarsinic acid, and *Flavobacterium* sp. additionally methylated dimethylarsinic acid to trimethylarsine oxide (78).

2. *Fungal Methylation*

In addition to bacteria, several fungi have demonstrated the ability to transform As. It is well established the fungi are able to volatilize As as methylarsine compounds, which are

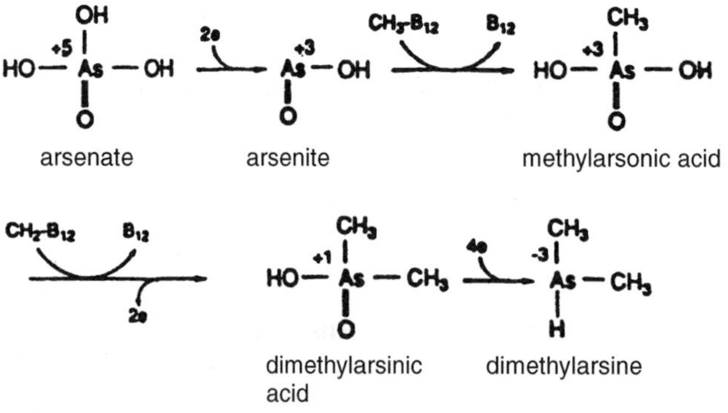

Figure 4 Anaerobic biomethylation pathway for dimethylarsine production by *Methanobacterium* sp. (From Ref. 43.)

derived from inorganic and organic As species. The volatilized As dissipates from the cells, effectively reducing the As concentration to which the fungus is exposed. The importance of fungal metabolism of As dates back to the early 1800s, when a number of poisoning incidents in Germany and England were caused by trimethylarsine gas (35). Since then, several species of fungi that are able to volatilize As have been identified. The fungus *Penicillium brevicaule* produces trimethylarsine when grown on bread crumbs containing either methylarsonic acid or dimethylarsinic acid. A biochemical pathway for trimethylarsine production was proposed by Challenger in 1945 (Fig. 5).

In 1973 studies, three different fungal species, *Candida humicola, Gliocladium roseum*, and *Penicillium* sp., were reported as capable of converting methylarsonic acid and dimethylarsinic acid to trimethylarsine (79). In addition, *C. humicola* used AsO_4^{3-} and AsO_2^- as substrates to produce trimethylarsine. Cell-free extracts of *C. humicola* transformed AsO_4^{3-} into AsO_2^-, methylarsonic acid to dimethylarsinic acid and trimethylarsine oxide, and dimethylarsinic acid to methylarsonic acid and trimethylarsine oxide (80). Although trimethylarsine formation from inorganic As and methylarsonic acid is inhibited by the presence of phosphate, its synthesis from dimethylarsinic acid is increased in the presence of phosphate (81). More recently, Huysmans and Frankenberger (82) isolated a *Penicillium* sp. from agricultural evaporation pond water capable of producing trimethylarsine from methylarsonic acid and dimethylarsinic acid.

Methylation of As is thought to occur via the transfer of the carbonium ion from *S*-adenosylmethionine (SAM) to As. Incubation of cells with an antagonist of methionine inhibited the production of arsines, thus supporting the role of methionine as a methyl donor (83). The addition of either methanearsonic acid or dimethylarsinic acid to cell-free extracts yields trimethylarsine oxide (80). Further reduction of trimethylarsine oxide to trimethylarsine requires the presence of intact cells (84). Various arsenic thiols (cysteine, glutathione, and lipoic acid) are thought to be involved in the reduction step of trimethylarsine oxide to trimethylarsine (85,86). The final reduction step is inhibited by several electron transport inhibitors and uncouplers of oxidative phosphorylation (84,87). Preincu-

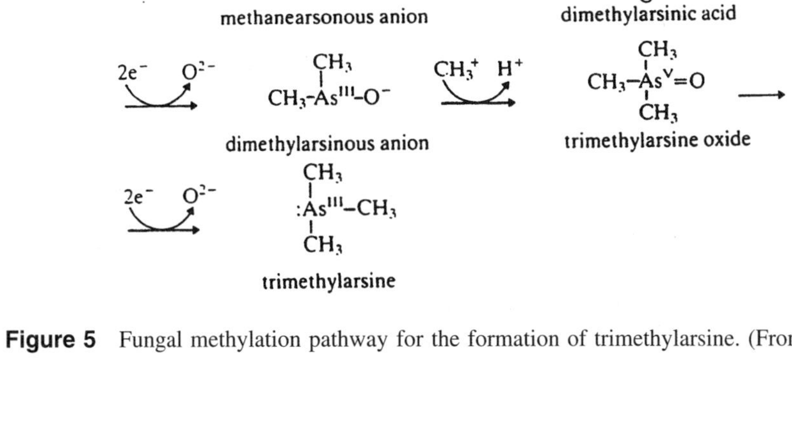

$H_2As^VO_4^-$ (arsenate) $\xrightarrow[\text{reduction}]{2e^-,\ O^{2-}}$ $As^{III}O(OH)_2^-$ (arsenite) $\xrightarrow[\text{methylation}]{CH_3^+,\ H^+}$ $CH_3\text{-}As^V(OH)(O^-){=}O$ (methanearsonic acid) $\rightarrow$

$\xrightarrow{2e^-,\ O^{2-}}$ $CH_3\text{-}As^{III}(O^-){=}O$ (methanearsonous anion) $\xrightarrow{CH_3^+,\ H^+}$ $CH_3\text{-}As^V(CH_3)(O^-){=}O$ (dimethylarsinic acid) $\rightarrow$

$\xrightarrow{2e^-,\ O^{2-}}$ $CH_3\text{-}As^{III}(CH_3)\text{-}O^-$ (dimethylarsinous anion) $\xrightarrow{CH_3^+,\ H^+}$ $CH_3\text{-}As^V(CH_3)_2{=}O$ (trimethylarsine oxide) $\rightarrow$

$\xrightarrow{2e^-,\ O^{2-}}$ $:As^{III}(CH_3)_3$ (trimethylarsine)

Figure 5 Fungal methylation pathway for the formation of trimethylarsine. (From Ref. 35.)

bation of cells with trimethylarsine oxide increases the rate of conversion to trimethylarsine, suggesting an inducible system (84). In addition, the rate of transformation of AsO_4^{3-} to trimethylarsine is increased by preconditioning the cells with dimethylarsinic acid (87). The compounds isolated during the reduction of AsO_4^{3-} by *C. humicola* are consistent with the intermediates reported in the pathway for methylation of As as proposed by Challenger (35).

3. Algal Methylation

Arsenic is also metabolized into various methylated forms by freshwater algae; however, there is some question about whether biomethylation of As in freshwater is a widespread, common process. Arsenite is methylated by at least four freshwater species of green algae: *Ankistrodesmus* sp., *Chlorella* sp., *Selenastrum* sp., and *Scenedesmus* sp. (88). All four species methylated AsO_2^- when present in media at 5000 mg L^{-1}, approximately the same level of AsO_2^- used to control aquatic plants in lakes (89). The levels of recovered methylated As species were quite high on a per gram dry weight basis. Each of these organisms transformed AsO_2^- to methylarsonic acid and dimethylarsinic acid, and all, except *Scenedesmus* sp., produced detectable levels of trimethylarsine oxide. Unlike for fungi, volatile methylarsines were not produced (89); instead, limnetic (freshwater) algae like marine algae synthesize lipid-soluble As compounds. Freshwater algae grown in media amended with 1.0 to 3.0 μg L^{-1} AsO_4^{3-} synthesized lipid-soluble As compounds to levels approximately equal to those of marine algae (90,91).

IV. CHROMIUM

Chromium (Cr) has many industrial uses, and, as a result, large volumes of Cr waste in various chemical forms are discharged into the environment. Chromium can exist in oxidation states ranging from −2 to +6. However, only +3 and +6 are normally found within the range of pH and redox potentials common in environmental systems. Hexavalent Cr (Cr^{6+}) forms chromate (CrO_4^{2-}) and dichromate ($Cr_2O_7^{2-}$), which are toxic and mutagenic, soluble over a wide pH range, and mobile in the environment. Hexavalent Cr can easily cross the membranes of eukaryotic and prokaryotic cells, and enters the cells by SO_4^{2-} transport pathways (92). Once in the cytosol, Cr^{6+} may be reduced to Cr^{3+}, which in turn reacts with deoxyribonucleic acid (DNA) (93). The trivalent form (Cr^{3+}) is virtually nonmobile, largely because it precipitates as oxides and hydroxides at $pH > 5$ (94,95), and considerably less toxic, since biological membranes do not allow its passing (96). Thus, the reduction of Cr^{6+} to Cr^{3+} represents a viable remediation and detoxification strategy.

Traditional techniques for remediating chromate (CrO_4^{2-}) from contaminated water involved reducing Cr^{6+} to Cr^{3+} by chemical or electrochemical means at $pH > 5$, followed by precipitation and finally filtration or sedimentation (97). The discovery of microorganisms that preferentially reduce Cr^{6+} has led to applications in the bioremediation field that are potentially more cost-effective than traditional techniques. Russian researchers were the first to propose using Cr^{6+}-reducing bacterial isolates in the removal of CrO_4^{2-} from various industrial effluents (98,99). Since then a wide variety microorganisms that reduce Cr^{6+} have been isolated from CrO_4^{2-}-contaminated waters and sediments. Through the use of these organisms, research is presently under way to develop commercially viable bioremediation techniques.

A. Reduction of Chromium(VI)

Although the microbial reduction of Cr^{6+} to Cr^{3+} is not known to be a plasmid-determined process, it may act as an additional mechanism of resistance to CrO_4^{2-}. Bopp and coworkers (100) isolated a CrO_4^{2-}-resistant strain of *Pseudomonas fluorescens* and determined that the resistance was plasmid-conferred but later reported (101) that CrO_4^{2-} reduction was associated with a constitutive, membrane-associated enzyme. It was subsequently shown that CrO_4^{2-} reduction and plasmid-mediated CrO_4^{2-} resistance were independent processes, since CrO_4^{2-}-sensitive and -resistant *P. fluorescens* were equally able to reduce CrO_4^{2-} (93,101).

The direct biological reduction of CrO_4^{2-} is an enzyme-mediated reaction that takes place under both aerobic (102–104) and anaerobic (99,105,106) conditions. Some bacteria are capable of reducing CrO_4^{2-} under both aerobic and anaerobic conditions (107,108). Although it was reported that some facultative anaerobic bacteria are capable of using Cr^{6+} as the sole terminal electron acceptor during respiratory metabolism (99,106), Lovley (109) concluded that Cr^{6+} reduction had not been shown to yield energy to support growth. To date, enzymes involved in the microbial reduction of CrO_4^{2-} have not been identified, but a few have been characterized in species of *Bacillus* (104,108), *Enterobacter* (110), *Streptomyces* (111), and *Pseudomonas* (101,103,112).

Chromate reductase activities have been associated with the cell membrane (101,110) and with the soluble protein fraction (103,104,108,111). Crude cell-free extracts of *Pseudomonas fluorescens* LB300 readily reduced Cr^{6+} with glucose as the electron donor (101). The addition of NADH to the S_{32} supernatant fraction (32,000 × g for 20 min) prepared from the crude cell-free extract exhibited increased CrO_4^{2-}-reducing activity. However, no CrO_4^{2-} reductase activity occurred in the S_{150} fraction (150,000 × g for 40 min), suggesting that some or all of the enzymes necessary for the transfer of electrons from NADH to CrO_4^{2-} are membrane-bound. Wang et al. (110) found the addition of NADH did not increase the CrO_4^{2-} reductase activity of *Enterobacter cloacae* HO1, probably because the membrane vesicles were not accessible to NADH. It was observed that membrane vesicles reduced by NADH and later exposed to Cr^{6+} oxidized the *c*- (c_{548}, c_{549}, and c_{550}) and *b*- (b_{555}, b_{556}, and b_{558}) type cytochromes (113). The cytochrome c_{548} was found to be involved specifically in the electron transfer to CrO_4^{2-}. In *Desulfovibrio vulgaris*, the c_3 cytochrome reportedly functioned as a Cr^{6+} reductase when H_2 was used as the electron donor (114). Chromate reducing activity occurred in the membrane-bound and soluble protein fraction; the fastest Cr^{6+} reduction occurred in the soluble protein fraction. The Cr^{6+} reductase activity was lost when the soluble protein fraction was passed over a cation-exchange column that specifically removed the c_3 cytochrome. The ability to reduce Cr^{6+} was restored when cytochrome c_3 was added back to the soluble protein fraction.

Chromate-reducing activity in *Pseudomonas putida* PRS2000 was associated with the soluble protein fraction, but the addition of either NADH or reduced nicotinamide-adenine dinucleotide phosphate (NADPH) to cell-free extracts was required for Cr^{6+} reduction. The Cr^{6+}-reducing enzyme in cell-free extracts of *Pseudomonas ambigua* G-1 required NADH but not NADPH as the hydrogen donor (102). Similar results were obtained by Garbisu et al. (104), who found that CrO_4^{2-} reducing activity of *Bacillus subtilis* was associated with the soluble protein fraction, and the addition of NADH or NADPH greatly increased the reduction of CrO_4^{2-}. The reduction of CrO_4^{2-} by *Bacillus* sp. strain QC1-2 was NADH-dependent and highest in the soluble protein fraction, whereas lower activity was detected in the membrane fraction (108). In addition, the reduction of CrO_4^{2-} by cell suspensions was dependent upon glucose, and SO_4^{2-}, a competitive inhibitor of CrO_4^{2-} transport, had no effect on the rate of reduction. In *Streptomyces* sp. 3M, reduction of CrO_4^{2-} was observed in the soluble protein fraction and was NADH- and NADPH-dependent; however, higher reduction rates were obtained with NADH (111).

B. Bioremediation of Chromium-Contaminated Water

The bioremediation of CrO_4^{2-}-contaminated water is currently proceeding along two fronts: indirect microbial reduction, by stimulating sulfur-reducing bacteria that produce H_2S, which subsequently serves as the reductant; and direct, enzyme-mediated reduction carried out by various Cr^{6+}-reducing bacteria. Whereas indirect microbial reduction of Cr^{6+} is largely an anaerobic process, direct reduction can occur under both aerobic and anaerobic conditions.

1. *Bioreduction of Chromium*

The major focus of Cr^{6+} bioreduction is aimed at developing bioreactors, which contain Cr^{6+}-reducing bacteria immobilized on inert matrices within the reactor. After the bioreduction phase, Cr^{3+} precipitates are removed by a settling or filtration phase. With this type of system, CrO_4^{2-}-contaminated water is pumped through the reactor along with various C sources and nutrient supplements. Advantages include its potential cost-effectiveness and absence of chemical reductants; the major disadvantage is that the lowest achievable Cr concentration obtained in batch experiments so for is around 1 mg L^{-1} (115,116). Unfortunately, an effluent concentration of 1 mg L^{-1} Cr would be 20 times higher than the Environmental Protection Agency (EPA) drinking water standard. Also reaction rates may also be slow, since CrO_4^{2-} must diffuse into direct contact with the cells, which maximizes its toxicity effects as well. Since optimal bioreduction has been correlated with the exponential growth phase of the bacteria (115), conditions within the bioreactor must be adjusted to maintain high growth rates. As a result, bioreactors of this design are probably better suited to treatment of wastewater prior to discharge rather than in situ applications because flow rates are expected to be relatively slow, and pumping of groundwater can significantly raise costs.

Losì et al. (117) in 1994 investigated a land application method for the treatment of Cr^{6+}-contaminated groundwater. The method consisted of using contaminated water high in Cr^{6+} to irrigate an alfalfa field (supplemented with an organic amendment), where reduction, precipitation, and immobilization would take place. In a greenhouse experiment, Cr^{6+}-contaminated water was passed through soil columns that were either amended with cattle manure or grown with alfalfa. The influent concentration to the columns was 1 mg L^{-1} Cr^{6+}, and effluent levels were consistently below 0.02 mg L^{-1}. It was found that the dominant removal mechanism was reduction, followed by precipitation. Lower O_2 levels favored the reduction reaction, and subsequent experiments determined that indigenous microorganisms, in conjunction with a readily available C source (i.e., cattle manure), were largely responsible for the Cr^{6+} reduction reactions.

Another application of bioreduction was demonstrated by Komori et al. (118). In this study, anaerobic Cr-reducing bacteria were contained within dialysis tubing and subsequently submerged in contaminated water. Chromate that diffused through the tubes was reduced and precipitated within the dialysis tubing. A laboratory study employing this technique demonstrated 90% removal efficiency when the initial Cr^{6+} concentration was 208 mg L^{-1}. A major disadvantage of this system is that the process is gradient-driven, and therefore, diffusion of Cr into the dialysis tubing decreases as the Cr^{6+} concentrations are lowered. As a result, relatively long residence times are needed to attain acceptable removal rates.

2. Gaseous Bioreduction of Chromium

DeFilippi and Lupton (119) designed an anaerobic bioreactor utilizing marine SO_4^{2-}-reducing bacteria, which are immobilized as a biofilm on gravel. Their experiments have shown that Cr effluent levels as low as 0.01 mg L^{-1} may be attainable (115). One advantage of this type of system over direct bioreduction is that the bacterial cells do not come in direct contact with CrO_4^{2-}; instead, the H_2S diffuses out into the medium, and, as a result, the cells are protected from the toxic effects of the Cr^{6+}. This may explain why effluent Cr^{6+} concentrations are lower than concentrations reported by Apel and Turik (116) utilizing direct bacterial reduction. A potential disadvantage of this type of system is that anaerobic conditions must be maintained, requiring additional energy costs.

A version of this technique is also potentially useful for the in situ immobilization of Cr in chromate-contaminated soils and groundwater. Such an application involves the production of H_2S in groundwater or deep within the soil profile by in situ stimulation of SO_4^{2-}-reducing bacteria through additions of SO_4^{2-} and other nutrients. This type of process was effective in the removal of Cr from Cr^{6+}-containing tannery wastes entering Otago Harbor in New Zealand (120).

V. MERCURY

Mercury (Hg) is noted for being one of the few metals that exist as a liquid at ambient temperatures and for being a potent human neurotoxin. Natural (e.g., volcanic eruptions) and anthropogenic (e.g., fossil fuel combustion) activities release large amounts of metallic Hg (Hg^0) into the biosphere; however, it readily undergoes biotic and abiotic conversion to organic forms such as methylmercury (MeHg, CH_3Hg^+). Methylmercury is water-soluble and fat-soluble; therefore it poses a threat to aquatic organisms such as fish and especially to fish consumers. Mercury pollution first received a great deal of publicity from the infamous Minamata Bay incident in Japan, where direct discharge of Hg-contaminated industrial waste led to extremely high levels of MeHg in fish, which, when eaten by humans, caused physical impairments and death. Aquatic organisms found in surface waters in the United States also have elevated levels of Hg, particularly in the Florida Everglades, which are believed to be the result of bioaccumulation within the food chain. Biomagnification can result when Hg concentrations in lakes and streams may be undetectable. Concentrations in fish tissues can be far greater than levels in surrounding water, particularly when MeHg is present.

In the environment, microorganisms are involved in the transformation of inorganic and organic mercury compounds, mostly as a detoxification mechanism. The microbial formation of volatile Hg^0 or dimethylmercury [$(CH_3)_2Hg$], neither of which is soluble, ensures the removal of Hg from the environment through atmospheric dissipation. Current studies are now focusing on biological reduction and methylation reactions as a remedial approach to immobilize Hg.

A. Reduction of Mercury(II)

Numerous microorganisms avoid Hg toxicity by reducing ionic Hg (Hg^{2+}) to volatile Hg^0, a potentially useful application to remove Hg from Hg-contaminated water. The reduction of Hg^{2+} to Hg^0 can be mediated by a number of microorganisms, including enteric bacteria, *Pseudomonas* sp., *Staphylococcus aureus, Thiobacillus ferrooxidans, Streptomyces* sp., and *Cryptococcus* sp. (121). The ability of bacteria to reduce Hg^{2+} is linked to Hg resistance (*mer*) operons (122). The hypothesized plasmid-mediated detoxification mechanism is shown in Fig. 6. The plasmid codes for a protein (merP) that initially binds to Hg^{2+} in the periplasm. The Hg^{2+} is then transported through the inner membrane to the cytoplasm by the membrane-bound protein merT. In the cytoplasm, Hg^{2+} is reduced to Hg^0 by a soluble, NADH-dependent, flavin adenine dinucleotide (FAD)-containing mercuric reductase. Mercuric reductase is active in the presence of excess thiols (R-SH) such as mercaptotethanol, dithiothreitol, glutathione, or cysteine. Intracellular Hg^0 is then eliminated from the cell by enhanced diffusion.

B. Methylation of Mercury

It was initially believed that methanogens were important methylators of Hg (123); however, this may not be the case after all. The methylation of Hg was not detected in whole cells of methanogens or in methanogenic sewage sludge (124), and the addition of a specific methanogen inhibitor to lake sediments did not alter the production of MeHg (125), suggesting that methanogens are not active in Hg methylation.

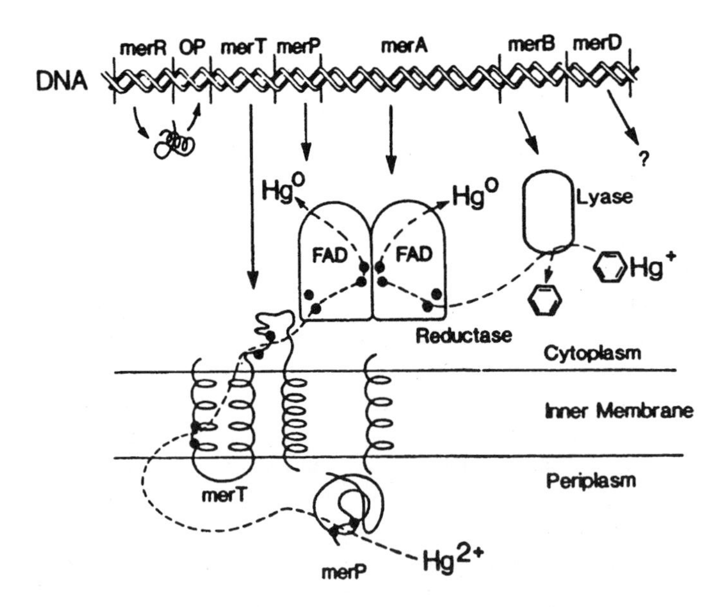

Figure 6 Model of the mercury detoxification system. (From Ref. 122.)

Although a number of microorganisms are capable of methylating Hg under both aerobic and anaerobic conditions (126), field studies suggest that Hg methylation occurs most rapidly under anoxic conditions (127,128). Evidence from 1985 and 1992 studies suggests that SO_4^{2-}-reducing bacteria are the dominant Hg methylators in estuarine (129) and lacustrine (130) anoxic sediments. Aerobic bacteria that are active in methylating Hg^{2+} include *Pseudomonas* sp., *Bacillus megaterium, Escherichia coli*, and *Enterobacter aerogenes*; fungi include *Aspergillus niger, Neurospora crassa, Scopulariopsis brevicaulis*, and *Saccharomyces cerevisiae*. The anaerobic bacterium *Clostridium cochlearium* was found to methylate a variety of Hg compounds, including HgO, $HgCl_2$, $Hg(NO_3)_2$, $Hg(CN)_2$, $Hg(SCN)_2$, and $Hg(OOCCH_3)_2$ (131).

Methylcobalamin (derivative of vitamin B_{12}) has been identified in bacteria (41) and has long been suspected as a cofactor in the microbial methylation of Hg, since it is known to donate methyl groups to Hg^{2+} (132). It is also an important coenzyme in the biosynthesis of methionine (133). In *E. coli*, methylcobalamin was found to catalyze the transfer of a methyl group to homocysteine, resulting in the formation of methionine (134). In *Neurospora crassa*, the formation of MeHg was stimulated by the addition of homocysteine but was inhibited by the addition of methionine (135). It was proposed that the fungus first complexes Hg^{2+} with homocysteine or cysteine, which is then methylated and finally cleaved by a transmethylase to produce MeHg. Since vitamin B_{12} was not known to be involved in the metabolism of *N. crassa*, methylcobalamin was not suspected to be the likely methyl donor. On the basis of this evidence, Hg methylation by *N. crassa* could be regarded as an incorrect synthesis of methionine. Choi et al. (136) confirmed in 1994 that the SO_4^{2-} -reducing bacterium *Desulfovibrio desulfuricans* methylates Hg^{2+} via the coenzyme methylcobalamin. It was additionally proposed that Hg^{2+} might be methylated during the acetyl coenzyme A synthase reaction. Clearly, very little is known about the biochemical mechanisms involved in microbial methylation of Hg, and further research is certainly required to elucidate these mechanisms.

C. Bioremediation of Mercury

1. *Bioaccumulation*

A number of naturally occurring Hg biotransformations may have applications in the bioremediation of Hg-contaminated soil and water. Mercury-resistant bacteria have been isolated by Glombitza at the Akademie der Wissenschaften (137,138). These Hg-resistant bacteria can accumulate up to 2%–4% by weight of nonvolatile Hg from aerated solutions containing 50–100 mg L^{-1} Hg^{2+} (139). Biomass can be suspended in a feed solution within a bioreactor and provided a C source (e.g., methanol or acetate) plus additional nutrients. The Hg-laden biomass is separated from solution and thermally decomposed at 400°–500°C to recover the Hg distillate.

2. *Biosorption*

Another technique being investigated to remediate Hg is biosorption. One such technology, which employs immobilized algae, is marketed by Bio-Recovery Systems under the name *Alga-SORB* (139). The product is prepared by heating the algae (*Chlorella vulgaris*) to 300°–500°C and immobilizing them on a silicate-based matrix. Alga-SORB shows a strong adsorption for Hg^{2+} independent of pH over a range of 2–6. Mercury adsorption occurs as a result of interactions with ''soft'' ligands (forms covalent complexes with ''soft'' functional groups that contain N or S) present in the cell wall. In an E.P.A.-

sponsored SITE demonstration, Alga-SORB columns treated with 0.14–1.6 mg L^{-1} Hg resulted in an effluent consistently below 10 μg L^{-1}. Unfortunately, biosorption is not effective in removing organically bound Hg compounds such as phenylmercuric acetate or MeHg (139). The biosorption systems are only effective on Hg in its cationic form (i.e., Hg^{2+}). Therefore, Hg covalently bound to organic carbon such as MeHg must be hydrolyzed to the free Hg ion prior to its removal by biosorption.

3. Bioreduction

The microbiological reduction of Hg^{2+} to Hg^0, followed by volatilization, is another potentially useful process to remediate Hg-contaminated waste streams. To be used as a method to remove Hg, volatile Hg^0 must be captured, requiring a containment system (see proposed bioreactor, Fig. 7). Unfortunately, this design would be limited in the volume of water that could be treated.

Some of the more well-characterized Hg treatment systems have been developed with *Chlorella emersonii* (140) and Hg-reducing bacteria, including *Xanthomonas maltophilia, Aeromonas hydrophila, Alcaligenes eutrophus, Pseudomonas paucimobilis*, and *Microbacterium* sp. (Gesellschaft für Biotechnologische Forschung [GBF], Institute of Biotechnological Research, Braunschweig, Germany; 139). When *Chlorella emersonii* was immobilized on calcium alginate beads it was capable of removing 99% of the Hg from a 1-mg L^{-1} solution in 12 days. The system devised by GBF worked by coating a porous support with bakers yeast and then exposing it to nonsterile Hg feed. Mercury-resistant bacteria then colonized the particles. Loading rates of 2–48 mg Hg^{2+} h^{-1} resulted

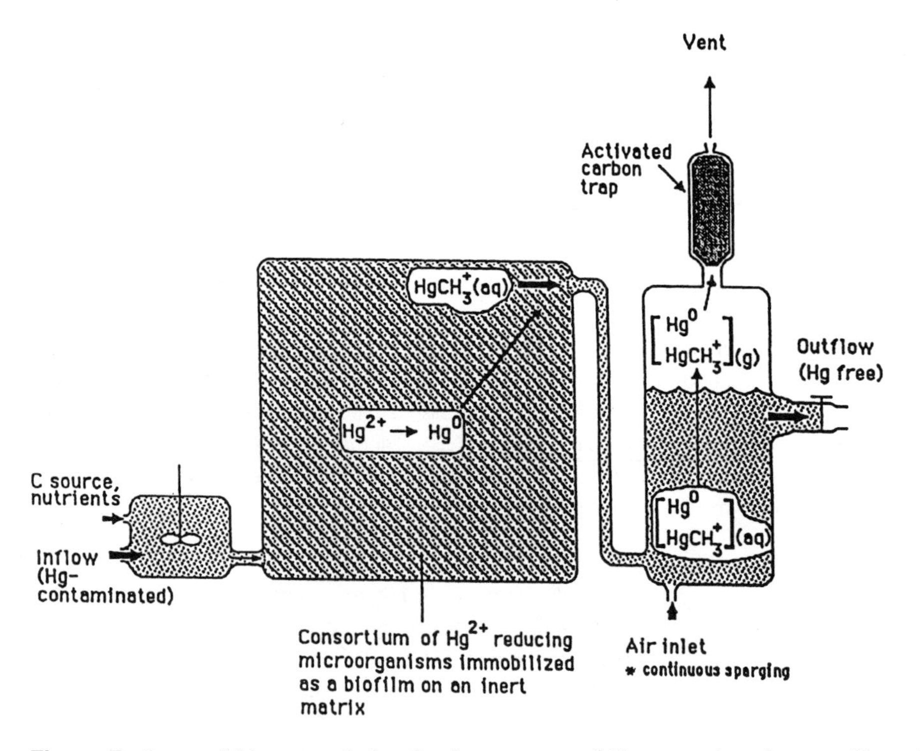

Figure 7 Proposed bioreactor design for the treatment of Hg-contaminated water. (From Ref. 142.)

in effluent concentrations of 50–100 μg L^{-1}. Another system, created by C. Hansen (Department of Nutrition and Food Science, Utah State University), utilized a bacterial consortium isolated from municipal activated sludge (139). In this system, a feed solution containing Hg^{2+} is added to a fluidized bed, where the Hg^{2+} is reduced to Hg^0 as it contacts biofilm-coated sand. Effluent concentrations of 10 μg L^{-1} Hg^{2+} were achieved when the influent concentration was 2 mg L^{-1}.

VI. CONCLUSION

Bioremediation is a proven technology in the removal and detoxification of many environmental contaminants, especially organic compounds. Contamination by heavy metals/metalloids is a widespread environmental problem throughout the world. The microbial transformation of heavy metals and metalloids into insoluble or volatile forms, which often represents a detoxification mechanism, may have applications in remediating contaminated soil and water systems. There are a number of advantages of bioremediation, including cost-effectiveness and potential for in situ applications. Currently application of bioremediation principles to heavy metal/metalloid contamination is largely in the experimental stage, but a growing body of evidence favors employing these biotechnologies. However, it is clear that information gaps exist with respect to the microbial transformation of Se, As, Cr, and Hg. Further investigations will certainly be required, especially if heavy metal/metalloid bioremediation processes are to be implemented successfully in the near future.

REFERENCES

1. M Alexander. Biodegradation and Bioremediation. San Diego: Academic Press, 1994.
2. Z Gerstle, Y Chen, U Mingelgrin, B Yaron. Toxic Organic Chemicals in Porous Media. New York: Springer-Verlag, 1989.
3. O Richter, B Diekkrüger, P Nörtersheuser. Environmental fate modeling of pesticides: From the laboratory to the field scale. New York: VCH, 1996.
4. U.S. Environmental Protection Agency. Internet site: www.epa.gov/superfund/
5. JE Rechcigl. Soil Amendments and Environmental Quality. Boca Raton, FL: CRC Press, 1995.
6. FG Viets. Chemistry and availability of micronutrients in soils. J Agric Food Chem 10:174–178, 1962.
7. U.S. Environmental Protection Agency, Office of Solid Waste and Emergency Response. Cleaning up the nation's waste sites: Markets and technology trends. EPA 542-R-96-005, April 1997.
8. HL Ehrlich. Geomicrobiology. 3rd ed. New York: Marcel Dekker, 1996.
9. RA Engberg, DW Westcot, M Delamore, DD Holz. Federal and state perspectives on regulation and remediation of irrigation-induced selenium problems. In: WT Frankenberger, Jr, RA Engberg, eds. Environmental Chemistry of Selenium. New York:Marcel Dekker, 1998, pp 1–25.
10. HM Ohlendorf, DJ Hoffman, MK Saiki, TW Aldrich. Embryonic mortality and abnormalities of aquatic bird: Apparent impact of selenium from irrigation drain water. Sci Total Environ 52:49–63, 1986.
11. JM Macy. Biochemistry of selenium metabolism by *Thauera selenatis* gen. nov. sp. nov. and use of the organism for bioremediation of selenium oxyanions in San Joaquin Valley drainage water. In: WT Frankenberger, Jr, S Benson, eds. Selenium in the Environment. New York: Marcel Dekker, 1994, pp 421–444.

12. ME Losi, WT Frankenberger, Jr. Reduction of selenium oxyanions by *Enterobacter cloacae* strain SLD1a-1: Isolation and growth of the bacterium and its expulsion of selenium particles. Appl Environ Microbiol 63:3079–3084, 1997.
13. RS Dungan, WT Frankenberger Jr. Reduction of selenite to elemental selenium by *Enterobacter cloacae* SLD1a-1. J Environ Qual 27:1301–1306, 1998.
14. RS Oremland, J Switzer Blum, CW Culbertson, PT Visscher, LG Miller, P Dowdle, FE Strohmaier. Isolation, growth, and metabolism of an obligately anaerobic selenate-respiring bacterium, strain SES-3. Appl Environ Microbiol 60:3011–3019, 1994.
15. L Lortie, WD Gould, S Rajan, RGL McCready, K-J Cheng. Reduction of selenate and selenite to elemental selenium by a *Pseudomonas stutzeri* isolate. Appl Environ Microbiol 58:4042–4044, 1992.
16. H DeMoll-Decker, JM Macy. The periplasmic nitrite reductase of *Thauera selenatis* may catalyze the reduction of selenite to elemental selenium. Arch Microbiol 160:241–247, 1993.
17. JA Switzer Blum, JA Burns Bindi, J Buzzelli, JF Stolz, RS Oremland. *Bacillus arsenicoselenatis* sp. nov., and *Bacillus selenitireducens* sp. nov.: Two alkaliphiles from Mono Lake, California that respire oxyanions of selenium and arsenic. Arch Microbiol 171:1505–1510, 1998.
18. RG McCready, JN Campbell, JI Payne. Selenite reduction by *Salmonella heidelberg*. Can J Microbiol 12:703–714, 1966.
19. RC Tilton, HB Gunner, W Litsky. Physiology of selenite reduction by Enterococci. I. Influence of environmental variables. Can J Microbiol 13:1175–1182, 1967.
20. C Garbisu, S Gonzalez, W-H Yang, BC Yee, DL Carlson, A Yee, NR Smith, R Otero, BB Buchanan, T Leighton. Physiological mechanisms regulating the conversion of selenite to elemental selenium by *Bacillus subtilis*. Biofactors 5:29–37, 1995.
21. C Garbisu, T Ishii, T Leighton, BB Buchanan. Bacterial reduction of selenite to elemental selenium. Chem Geol 132:199–204, 1996.
22. LJ Yanke, RD Bryant, EJ Laishley. Hydrogenase (I) of *Clostridium pasteurianum* functions as a novel selenite reductase. Anaerobe 1:61–67, 1995.
23. ET Thompson-Eagle, WT Frankenberger Jr. Bioremediation of soils contaminated with selenium. In: R Lal, BA Stewart, eds. Advances in Soil Science. New York: Springer-Verlag, 1992, pp 261–310.
24. JW Doran. Microorganisms and the biological cycling of selenium. In: KC Marshall, ed. Advances in Microbial Ecology. New York: Plenum Press, 1982, pp 1–32.
25. F Challenger, HE North. The production of organometalloidal compounds by microorganisms. Part II. Dimethylselenide. J Chem Soc 1934:68–71, 1934.
26. U Karlson, WT Frankenberger Jr. Effects of carbon and trace element addition on alkylselenide production by soil. Soil Sci Soc Am J 52:1640–1644, 1988.
27. GM Abu-Erreish, EI Whitehead, OE Olson. Evolution of volatile selenium from soils. Soil Sci 106:415–420, 1968.
28. R Zieve, PJ Peterson. Factors influencing the volatilization of selenium from soil. Sci Total Environ 19:277–284, 1981.
29. YK Chau, PTS Wong, BA Silverberg, PL Luxon, GA Bengert. Methylation of selenium in the aquatic environment. Science 192:1130–1131, 1976.
30. JW Doran, M Alexander. Microbial transformations of selenium. Appl Environ Microbiol 33:31–37, 1977.
31. RM Rael, WT Frankenberger Jr. Influence of pH, salinity, and selenium on the growth of *Aeromonas veronii* in evaporation agricultural drainage water. Water Res 30:422–430, 1996.
32. S McCarty, T Chasteen, M Marshall, R Fall, R Bachofen. Phototrophic bacteria produce volatile, methylated sulfur and selenium compounds. FEMS Microbiol Lett 112:93–98, 1993.
33. RS Dungan. Microbial transformations of selenium by *Enterobacter cloacae* SLD1a-1 and measurement of selenium volatilization under field-like conditions. PhD dissertation, University of California, Riverside, 1999.

34. TW-M Fan, AN Lane, RM Higashi. Selenium biotransformations by a euryhaline microalga isolated from a saline evaporation pond. Environ Sci Technol 31:569–576, 1997.
35. F Challenger. Biological methylation. Chem Rev 36:315–361, 1945.
36. DC Reamer, WH Zoller. Selenium biomethylation products from soil and sewage sludge. Science 208:500–502, 1980.
37. ML Bird, F Challenger. Studies in biological methylation. Part IX. The action of *Scopulariopsis brevicaulis* and certain penicillia on salts of aliphatic seleninic and selenonic acids. J Chem Soc 1942:574–577, 1942.
38. J Bremer, Y Natori. Behavior of some selenium compounds in transmethylation. Biochim Biophys Acta 44:367–370, 1960.
39. HS Hsieh, HE Ganther. Acid-volatile selenium formation catalyzed by glutathione reductase. Biochemistry 14:1632–1636, 1975.
40. TD Cook, KW Bruland. Aquatic chemistry of selenium: Evidence of biomethylation. Environ Sci Technol 21:1214–1219, 1987.
41. K Linstrand. Isolation of methylcobalamin from natural source material. Nature 204:188–189, 1964.
42. ET Thompson-Eagle, WT Frankenberger Jr, U Karlson. Volatilization of selenium by *Alternaria alternata*. Appl Environ Microbiol 55:1406–1413, 1989.
43. BC McBride, RS Wolfe. Biosynthesis of dimethylarsine by a methanobacterium. Biochemistry 10:4312–4317, 1971.
44. A Drotar, LR Fall, EA Mishalanie, JE Tavernier, R Fall. Enzymatic methylation of sulfide, selenide, and organic thiols by *Tetrahymena thermophila*. Appl Environ Microbiol 53:2111–2118, 1987.
45. WT Frankenberger Jr, U Karlson. Volatilization of selenium from dewatered seleniferous sediment: A field study. J Ind Microbiol 14:226–232, 1995.
46. M Flury, WT Frankenberger, Jr, WA Jury. Long-term depletion of selenium from Kesterson dewatered sediments. Sci Total Environ 198:259–270, 1997.
47. JM Macy, S Lawson, H DeMoll-Decker. Bioremediation of selenium oxyanions in San Joaquin drainage water using *Thauera selenatis* in a biological reactor system. Appl Environ Microbiol 40:588–594, 1993.
48. S Lawson, JM Macy. Bioremediation of selenite in oil refinery wastewater. Appl Environ Microbiol 43:762–765, 1995.
49. AW Cantafio, DK Hagen, GE Lewis, TL Bledsoe, KM Nunan, JM Macy. Pilot-scale selenium bioremediation of San Joaquin drainage water using with *Thauera selenatis*. Appl Environ Microbiol 62:3298–3303, 1996.
50. TJ Lundquist, F Baily Green, R Blake Tresan, RD Newman, WJ Oswald, MB Gerhardt. The algal–bacterial selenium removal system: Mechanisms and field study. In: WT Frankenberger Jr, S Benson, eds. Selenium in the Environment. New York: Marcel Dekker, 1994, pp 251–278.
51. WJ Oswald. Large-scale algal culture systems (engineering aspects). In: MA Borowitzka, LJ Borowitzka, eds. Micro-Algal Biotechnology, Cambridge: Cambridge University Press, 1988, p 357.
52. RS Oremland. Selenate removal from wastewater. 1991. U.S. patent 5,009,786.
53. LP Owens. Bioreactors in removing selenium from agricultural drainage water. In: WT Frankenberger Jr, RA Engberg, eds. Environmental Chemistry of Selenium. New York: Marcel Dekker, 1998, pp 501–514.
54. DJ Adams, PB Altringer, WD Gould. Bioremediation of selenate and selenite. In: AE Torma, ML Apel, CE Brierley, eds. Biohydrometallurgy. Warrendale, PA: Minerals, Metals, and Materials Society, 1993, pp 755–771.
55. ET Thompson-Eagle, WT Frankenberger Jr. Protein-mediated selenium biomethylation in evaporation pond water. Environ Tox Chem 9:1453–1462, 1990.
56. WT Frankenberger Jr, U Karlson. Volatilization of selenium from dewatered seleniferous sediments: A field study. J Ind Microbiol 14:226–232, 1995.

57. WT Frankenberger, Jr, U Karlson. Dissipation of soil selenium by microbial volatilization at Kesterson Reservoir. US Dept. of the Interior, Bureau of Reclamation, Contract no. 7-NC-20-05240, 1988.
58. DJ Adams, TM Pickett, JR Montgomery. Biotechnologies for metal and toxic inorganic removal from mining process and waste solutions. Olympic Valley, CA: Randol Gold Forum Proceedings, 1996.
59. JF Ferguson, J Gavis. A review of the arsenic cycle in natural waters. Water Res 6:1259–1274, 1972.
60. KD Huysmans, WT Frankenberger Jr. Arsenic resistant microorganisms isolated from agricultural drainage water and evaporation pond sediments. Water Air Soil Pollut 53:159–168, 1990.
61. PR Dowdle, AM Laverman, RS Oremland. Bacterial dissimilatory reduction of arsenic(V) to arsenic(III) in anoxic sediments. Appl Environ Microbiol 62:1664–1669, 1996.
62. JM Harrington, SE Fendørf, RF Rosenzweig. Biotic generation of arsenic(III) in metal(loid)-contaminated freshwater lake sediments. Environ Sci Technol 32:2425–2430, 1998.
63. D Ahmann, AL Roberts, LR Krumholz, FMM Morel. Microbe grows by reducing arsenic. Nature 372:750, 1994.
64. AM Laverman, J Switzer Blum, JK Schaefer, EJP Philips, DR Lovley, RS Oremland. Growth of strain SES-3 with arsenate and diverse electron acceptors. Appl Environ Microbiol 61: 3556–3561, 1995.
65. DK Newman, EK Kennedy, JD Coates, D Ahmann, DJ Ellis, DR Lovley, FMM Morel. Dissimilatory arsenate and sulfate reduction in *Desulfotomaculum auripigmentum* sp. nov. Arch Microbiol 168:380–388, 1997.
66. JM Macy, K Nunan, KD Hagen, DR Dixon, PF Harbour, M Cahill, LI Sly. *Chrysiogenes arsenatis* gen. nov., sp. nov., a new arsenate-respiring bacterium isolated from gold mine wastewater. Int J Sys Bacteriol 46:1153–1157, 1996.
67. BP Rosen, S Silver, TB Gladysheva, G Ji, KL Oden, S Jagannathan, W Shi, Y Chen, J Wu. The arsenite oxyanion-translocating ATPase: Bioenergetics, functions, and regulation. In: A Torriani-Gorini, E Yagil, S Silver, eds. Phosphate in Microorganisms. Washington DC:ASM Press, 1994, pp 97–107.
68. DK Newman, D Ahmann, FMM Morel. A brief review of microbial arsenate respiration. Geomicrobiology 15:255–268, 1998.
69. D White. The physiology and biochemistry of prokaryotes. New York: Oxford University Press, 1995, pp 226–228.
70. SE Phillips, ML Taylor. Oxidation of arsenite to arsenate by *Alcaligenes faecalis*. Appl Environ Microbiol 32:392–399, 1976.
71. FH Osborne, HL Ehrlich. Oxidation of arsenite by a soil isolate of *Alcaligenes*. J Appl Bacteriol 41:295–305, 1976.
72. HL Ehrlich. Inorganic energy sources for chemolithotrophic and mixotrophic bacteria. Geomicrobiol J 1:65–83, 1978.
73. AN Ilyaletdinov, SA Abdrashitova. Autotrophic oxidation of arsenic by a culture of *Pseudomonas* arsenitoxidans. Mikrobiologiya 50:197–204, 1981.
74. GL Anderson, J Williams, R Hiller. The purification and chracterisation of arsenite oxidase from *Alcaligenes faecalis*, a molybdenium-containing hydroxylase. J Biol Chem 267:23674–23682, 1992.
75. N Wakao, H Koyatsu, Y Komai, H Shimokawara, Y Sakurai, H Shiota. Microbial oxidation of arsenate and occurrence of arsenite-oxidizing bacteria in acid mine water from a sulfur-pyrite mine. Geomicrobiology J 6:11–24, 1988.
76. BC McBride, H Merilees, WR Cullen, W Pickett. Anaerobic and aerobic alkylation of arsenic. In: FE Brickman, JM Bellama, eds. Organometals and organometalloids: Occurrence and fate in the environment. Am Chem Soc Symp Ser 82:94–115, 1978.

77. CN Cheng, DD Focht. Production of arsine and methylarsines in soil and in culture. Appl Environ Microbiol 38:494–498, 1979.
78. PTS Wong, YK Chau, L Luxon, GA Bengert. Methylation of arsenic in the aquatic environment. In: DD Hemphill, ed. Trace Substances in Environmental Health. Part X. Columbia: University of Missouri, 1977, pp 100–105.
79. DP Cox, M Alexander. Production of trimethylarsine gas from various arsenic compounds by three sewage fungi. Bull Environ Contam Toxicol 9:84–88, 1973.
80. WR Cullen, BC McBride, AW Pickett. The transformation of arsenicals by *Candida humicola*. Can J Microbiol 25:1201–1205, 1979.
81. DP Cox, M Alexander. Effect of phosphate and other anions on trimethylarsine formation by *Candida humicola*. 25:408–413, 1973.
82. KD Huysmans, WT Frankenberger Jr. Evolution of trimethylarsine by a *Penicillium* sp. isolated from agricultural evaporation pond water. Sci Total Environ 105:13–28, 1991.
83. WR Cullen, CL Froese, A Lui, BC McBride, DJ Patmore, M Reimer. The aerobic methylation of arsenic by microorganisms in the presence of L-methionine-methyl-d3. J Organometal Chem 139:61–69, 1977.
84. AW Pickett, BC McBride, WR Cullen, H Manji. The reduction of trimethylarsine oxide by *Candida humicola*. Can J Microbiol 27:773–778, 1981.
85. WR Cullen, BC McBride, J Reglinski. The reaction of methylarsenicals with thiols: some biological implications. J Inorg Biochem 21:148–154, 1984.
86. WR Cullen, BC McBride, J Reglinski. The reduction of trimethylarsine oxide to trimethylarsine by thiols: A mechanistic model for the biological reduction of arsenicals. J Inorg Biochem 21:45–60, 1984.
87. RA Zingaro, NR Bottino. Biochemistry: Recent developments. In: WH Lederer, RJ Fensterheim, eds. Arsenic: Industrial, Biomedical, Environmental Perspectives. New York: Van Nostrand Reinhold, 1983, pp 327–347.
88. MD Baker, PTS Wong, YK Chau, CI Mayfield, WE Inniss. Methylation of arsenic by freshwater green algae. Can J Fish Aquat Sci 40:1245–1257, 1983.
89. RD Hood and Associates. Cacodylic acid: Agricultural uses, biological effects, and environmental fate. Superintendent of Documents, US Government Printing Office, Washington DC, 1985, p 164.
90. G Lunde. The analysis of arsenic in the lipid phase from marine and limnetic algae. Acta Chem Scand 26:2642–2644, 1972.
91. G Lunde. The synthesis of fate and water-soluble arseno-organic compounds in marine and limnetic algae. Acta Chem Scand 27:1586–1594, 1973.
92. H Ohtake, C Cervantes, S Silver. Decreased chromate uptake in *Pseudomonas fluorescens* carrying a chromate resistance plasmid. J Bacteriol 169:3853–3856, 1987.
93. CA Miller, III, M Costa. Analysis of proteins cross-linked to DNA after treatment of cells with formaldehyde, chromate, and *cis*-diamminechloroplatinum(II). Mol Toxicol 2:11–26, 1989.
94. EE Cary. Chromium in air, soils, and natural waters. In: S Langard, ed. Biological and Environmental Aspects of Chromium. Amsterdam: Elsevier Biomedical Press, 1982, pp 49–64.
95. SP McGrath, S Smith. Chromium and nickel. In: BJ Alloway, ed. Heavy Metals in Soils. New York: John Wiley & Sons, 1990, p 137.
96. AG Levis, V Bianchi. Mutagenic and cytogenic effects of chromium compounds. In: S Langard, ed. Biological and Environmental Aspects of Chromium. Amsterdam: Elsevier Biomedical Press, 1982, pp 171–208.
97. LE Eary, D Rai. Chromate removal from aqueous waste by reduction with ferrous iron. Environ Sci Technol 22:972–977, 1988.
98. VI Romanenko, SI Dusnetsov, VN Koren'kov. Koren'kov method for biological purification of wastewater. USSR Patent SU 521,234, 1976.

99. VI Romanenko, VN Koren'kov. A pure culture of bacteria utilizing chromates and bichromates as hydrogen acceptors in growth under anaerobic conditions. Institute of Biology of Inland Water, Academy of Sciences of the USSR. Mikrobiologiyia 46:414–417, 1977.
100. LH Bopp, AM Chakrabarty, HL Ehrlich. Chromate resistance plasmid in *Pseudomonas fluorescens*. J Bacteriol 155:1105–1109, 1983.
101. LH Bopp, HL Ehrlich. Chromate resistance and reduction in *Pseudomonas fluorescens* strain LB300. Arch Microbiol 150:426–431, 1988.
102. H Horitsu, S Futo, Y Miyazawa, S Ogai, K Kawai. Enzymatic reduction of hexavalent chromium tolerant *Pseudomonas ambigua* G-1. Agric Biol Chem 51:2417–2420, 1987.
103. Y Ishibashi, C Cervantes, S Silver. Chromium reduction in *Pseudomonas putida*. Appl Environ Microbiol 56:2268–2270, 1990.
104. C Garbisu, I Alkorta, MJ Llama, JL Serra. Aerobic chromate reduction by *Bacillus subtilis*. Biodegradation 9:133–141, 1998.
105. K Komori, P-C Wang, K Toda, H Ohtake. Factors affecting chromate reduction in *Enterobacter cloacae* strain HO1. Appl Microbiol Biotechnol 31:567–570, 1989.
106. P-C Wang, T Mori, K Komori, M Sasatsu, K Toda, H Ohtake. Isolation and characterization of an *Enterobacter cloacae* strain that reduced hexavalent chromium under anaerobic conditions. Appl Environ Microbiol 55:1665–1669, 1989.
107. S Llovera, R Bonet, MD Simon-Pujol, F Congregado. Chromate reduction by resting cells of *Agrobacterium radiobacter* EPS-916. Appl Environ Microbiol 59:3516–3518, 1993.
108. J Campos, M Martinez-Pacheco, C Cervantes. Hexavalent-chromium reduction by a chromate-resistant *Bacillus* sp. strain. Antonie van Leeuwenhoek 68:203–208, 1995.
109. DR Lovley. Dissimilatory metal reduction. Annu Rev Microbiol 47:263–290, 1993.
110. P-C Wang, T Mori, K Toda, H Ohtake. Membrane-associated chromate reductase activity from *Enterobacter cloacae*. J Bacteriol 172:1670–1672, 1990.
111. S Das, AL Chandra. Chromate reduction in *Streptomyces*. Experientia 46:731–733, 1990.
112. T Suzuki, N Miyata, H Horitsu, K Kawai, K Takamizawa, Y Tai, M Okazaki. NAD(P)H-dependent chromium (VI) reductase of *Pseudomonas ambigua* G-1:A Cr(V) intermediate is formed during the reduction of Cr (VI) to Cr (III). J Bacteriol 174:5340–5345, 1992.
113. P-C Wang, K Toda, H Ohtake. Membrane-bound respiratory system of *Enterobacter cloacae* HO1 grown anaerobically with chromate. FEMS Microbiol Lett 78:11–16, 1991.
114. DR Lovley, EJP Phillips. Reduction of chromate by *Desulfovibrio vulgaris* and its c_3 cytochrome. Appl Environ Microbiol 60:726–728, 1994.
115. PL Mattison. Chromium. In: Bioremediation of Metals—Putting It to Work. Santa Rosa, CA: Cognis, 1992.
116. WA Apel, CE Turick. Bioremediation of hexavalent chromium by bacterial reduction. In: RW Smith, ed., Mineral Bioprocessing. Warrendale, PA: The Minerals, Metals, and Materials Society, 1991, pp 377–387.
117. ME Losi, C Amrhein, WT Frankenberger Jr. Bioremediation of chromate contaminated groundwater by reduction and precipitation in surface soils. J Environ Qual 23:1141–1150, 1994.
118. K Komori, A Rivas, K Toda, H Ohtake. A method for removal of toxic chromium using dialysis-sac cultures of a chromate-reducing strain of *Enterobacter cloacae*. Appl Microbiol Biotechnol 33:117–119, 1990.
119. LJ DeFilippi, FS Lupton. Bioremediation of soluble Cr(VI) using anaerobic sulfate reducing bacteria. San Francisco, CA. Feb 4–6. National R&D Conference on the Control of Hazardous Materials, 1992, pp 138–141.
120. RH Smillie, K Hunter, M Loutit. Reduction of Cr(VI) by bacterially produced hydrogen sulfides. Water Res 15:1351–1354.
121. HL Ehrlich. Geomicrobiology. 3rd ed. New York: Marcel Dekker, 1996, pp 295–311.
122. S Silver, TK Mirsa. Plasmid-mediated heavy metal resistances. Annu Rev Microbiol 42: 717–743, 1988.

123. JM Wood, F Scott Kennedy, CG Rosen. Synthesis of methyl-mercury compounds by extracts of a methanogenic bacterium. Nature 220:173–174, 1968.
124. BC McBride, TL Edwards. Role of methanogenic bacteria in the alkylation of arsenic and mercury. In: H Drucker, RE Wildung, eds. Biological Implications of Metals in the Environment. ERDA Symposium Series 42. Washington, DC: Department of Energy, 1977, pp 1–19.
125. A Kerry, PM Welbourn, B Prucha, G Mierle. Mercury methylation by sulphate-reducing bacteria from sediments of an acid stressed lake. Water Air Soil Pollut 56:565–575, 1991.
126. F Baldi, M Filippelli, GJ Olson. Biotransformations of mercury by bacteria isolated from a river collecting cinnabar mine waters. Microbiol Ecol 17:262–274, 1989.
127. G Compeau, R Bartha. Methylation and demethylation of mercury under controlled redox, pH, and salinity conditions. Appl Environ Microbiol 48:1203–1207, 1984.
128. O Regnell, A Tunlid, G Ewald, O Sangfors. Methyl mercury production in freshwater microcosms affected by dissolved oxygen levels: Role of cobalamin and microbial community composition. Can J Fish Aquat Sci 53:1535–1545, 1996.
129. G Compeau, R Bartha. Sulphate-reducing bacteria: Principal methylators of mercury in anoxic estuarine sediments. Appl Environ Microbiol 50:498–502, 1985.
130. CG Gilmour, EA Henry, R Mitchell. Sulfate stimulation of mercury methylation in freshwater sediments. Environ Sci Technol 26:2281–2287, 1992.
131. M Yamada, K Tonomura. Formation of methylmercury compounds from inorganic mercury compounds by *Clostridium cochlearium*. J Ferment Technol 50:159–166, 1972.
132. L Bertilsson, HY Neujahr. Methylation of mercury compounds by methylcobalamin. Biochemistry 10:2805–2808, 1971.
133. L Stryer. Biochemistry. 4th ed. New York: WH Freeman, 1995, p 722.
134. MA Foster, MJ Dilworth, DD Woods. Cobalamin and the synthesis of methionine by *Escherichia coli*. Nature (London) 201:39–42, 1964.
135. L Lander. Biochemical model for the biological methylation of mercury suggested from methylation studied in vivo with *Neurospora crassa*. Nature 230:452–454, 1971.
136. S-C Choi, T Chase, R Bartha. Metabolic pathways leading to mercury methylation in *Desulfovibrio desulfuricans* LS. Appl Environ Microbiol 60:4072–4077, 1994.
137. F Glombitza, U Iske, C Gwenner. Biosorption of mercury by microorganisms. Acta Biotechnol 4:281–284, 1984.
138. F Glombitza, U Iske, C Gwenner. Cultivation of mercury-resistant microorganisms. Acta Biotechnol 4:285–288, 1984.
139. PL Mattison. Mercury. In: Bioremediation of Metals—Putting It to Work. Santa Rosa, CA: Cognis, 1992.
140. SC Wilkinson, KH Goulding, PK Robinson. Mercury accumulation and volatilization in immobilized algal cell systems. Biotechnol Lett 11:861–864, 1989.
141. PJ Craig. Organometallic Compounds in the Environment: Principles and Reactions. Essex, England: Longman Group, 1986.
142. WT Frankenberger Jr, ME Losi. Applications of bioremediation in the cleanup of heavy metals and metalloids. In: HD Skipper, RF Turco, eds. Bioremediation: Science and Applications. Special Publication 43. Madison, WI: Soil Science Society of America, 1995, pp 173–210.

21

Enzymes in Soil

Research and Developments in Measuring Activities

M. Ali Tabatabai
Iowa State University, Ames, Iowa

Warren A. Dick
The Ohio State University, Wooster, Ohio

I. INTRODUCTION

Reactions in the environment involve chemical, biochemical, and physical processes. It is well known that most biochemical reactions are catalyzed by enzymes, which are proteins with catalytic properties. Catalysts are substances that, without undergoing permanent alteration, cause chemical reaction to proceed at faster rates. In addition, enzymes are specific for the type of chemical reactions in which they participate. All living systems, ranging from bacteria to the animal kingdom, from algae and molds to the higher plants, contain a vast number of enzymes catalyzing both simple and complex networks of chemical reactions. Enzymes also are found in ponds, lakes, rivers, water treatment plants, animal manures, and soils and exist either as extracellular forms separated from their origins or as intracellular forms as part of the living biomass. These enzymes are involved in the synthesis of proteins, carbohydrates, nucleic acids, and other components of living systems and also in the degradation and essential cycling of carbon, nitrogen, phosphorus, sulfur, and other nutrients.

The study of enzymes, in general, is a subject of interest to many disciplines ranging from biology to the physical sciences. This is not surprising as enzymes have a central place in biology, and life depends on a complex network of chemical reactions facilitated by specific enzymes. Any alterations in the enzyme protein structures might have far-reaching consequences for the living organism. It is safe to say that soils would remain lifeless and basically unaltered without enzymatic reactions. Within the past five decades enzymology, the science of studying enzymes, has developed rapidly. This field of science has connections with many other sciences and has contributed to our understanding of microbiology, biochemistry, molecular biology, botany, soil science, toxicology, animal science, pharmacology, pathology, medicine, and chemical engineering.

II. ENZYMES IN SOILS

The first known report on enzymes in soils was presented by Woods at the 1899 Annual Meeting of the American Association for the Advancement of Science in Columbus, Ohio (1). However, little significant progress occurred in the area of soil enzymology until the 1950s. This was mainly due to a lack of appropriate methodologies and understanding of the true nature of enzymes. Although Sumner first isolated urease in crystalline form from jack bean (*Canavalia ensiformis*) meal in 1926, for which he received a Nobel Prize, this field of biochemistry took several decades to mature. Questions asked during the early years of soil enzyme research dealt with the origin, stabilization, importance in plant nutrition, and role of soil enzymes in organic matter turnover. Many of these questions remain to be answered.

A wealth of information about various enzymatic reactions in soils has been collected since 1950, and theoretical approaches and methods have been developed. A history of abiotic soil enzyme research has been prepared by Skujins (1), and reviews of recent advances and the state of knowledge in this field are presented in a book edited by Burns (2) and in a number of book chapters (3–10) by other researchers. No specific review article has been prepared on the progress that has been made in the different chemical and instrumental methods used in measuring enzyme activities in soils. This chapter focuses on the various techniques used in assay of enzyme activities of soils.

Soil can be thought of as a biological entity (i.e., a living tissue with complex biochemical reactions) (11). Soil contains free enzymes, immobilized extracellular enzymes stabilized by a three-dimensional network of macromolecules, and enzymes within microbial cells. Each of the organic and mineral fractions in both bulk soil and the rhizosphere has a special influence on the total enzymatic activity of that soil (12,13).

Enzymes are protein catalysts, and physicochemical measurements indicate that enzyme-catalyzed reactions in soils have lower activation energies than non-enzyme-catalyzed reactions and, therefore, have faster reaction rates (14–17). Enzymes in soil are similar to enzymes in other systems, in that their reaction rates are markedly dependent on pH, ionic strength, temperature, and the presence or absence of inhibitors (8,18).

A. Sources

Both microorganisms and plants release enzymes into the soil environment (Fig. 1). It has long been known that ribonucleases and alkaline phosphatase, for example, are excreted by *Bacillus subtilis* under certain conditions (19,20), and pyrophosphatase and acid phosphatase may exist extracellularly on the surface of cell walls of *Saccharomyces mellis* (21,22). A number of bacteria release phosphatases (23) and other microbial extracellular enzymes with important commercial applications, including proteases, amylases, glucose isomerases, pectinases, and lipases (24). Therefore, it is not surprising that microorganisms are the logical choice to account for most of the soil enzyme activity (25). This is because of their large biomass, high metabolic activity, and short lifetimes, which allow them to produce and release relatively large amounts of extracellular enzymes in comparison to plants or animals.

The effect of microorganisms in supplying phosphatase activity to soils, however, seems temporary and short-lived. Ladd and Paul (26) incubated a soil with glucose and sodium nitrate at 22°C and found that bacterial numbers increased almost 2-fold in 36 hours, accompanied by a 3.2-fold increase in phosphatase activity. However, the new

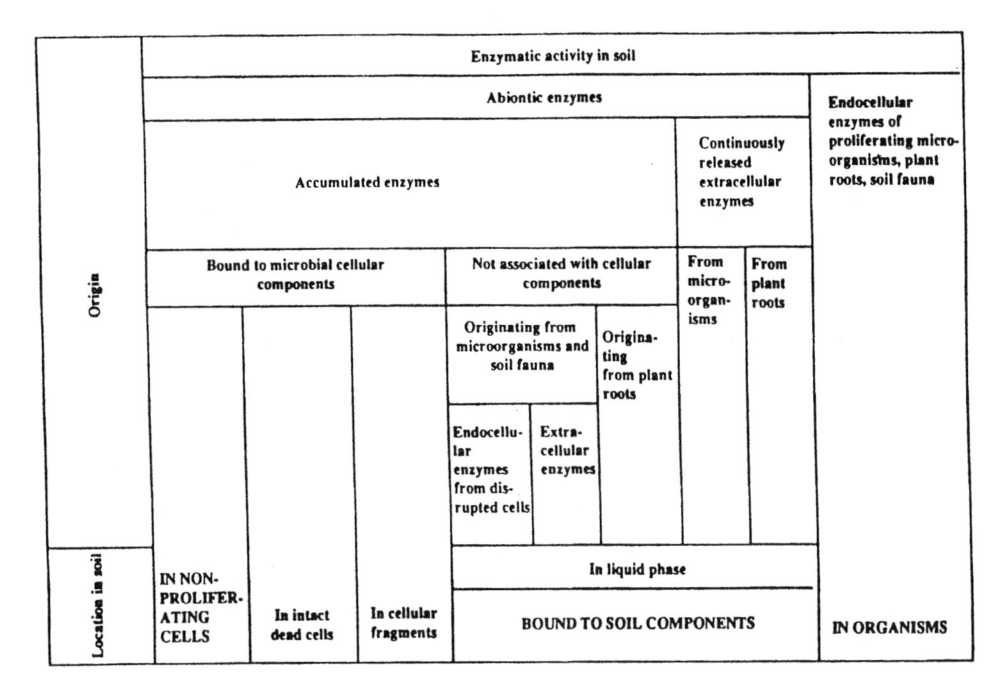

Figure 1 Conceptual scheme of the composition of soil enzyme activities. (From Ref. 1.)

activity was rapidly lost and after 21 days, activity had returned to its original level. It has been theorized (25) that the increased activity of phosphatase produced during incubation of soil with glucose and nitrate was largely lost during the microbial proliferation–dying–lysing cycle and that many cycles of microbial activity may be required to obtain a permanent increase in the extracellular level of phosphatase activity.

Plants have also been considered a source of extracellular enzymes in soils. Estermann and McLaren (27) found that barley (*Hordeum vulgare*) root caps possess phosphatase activity. Other studies showed that a variety of plants have amidase and urease activities (28) and that sterile corn (*Zea mays*) and soybean (*Glycine max*) roots contain acid phosphatase, but no alkaline phosphatase activity (29). In other work (30) it was demonstrated that sterile corn and soybean roots could exude acid phosphatase into a solution that surrounded them. Roots, placed into sterile buffer or water for 4–48 hours, released acid phosphatase into the solution. Greater amounts of acid phosphatase were released into water than into the buffered solution.

Major amounts of enzymes introduced into the soil environment by microorganisms or plant roots are inhibited by soil constituents, rapidly degraded by soil protease, or both. Work by Dick et al. (31) showed that when 10 mg corn root homogenate was mixed with 1 g of soil, the inhibition of corn root acid phosphatase and pyrophosphatase by 12 soils was 43–63% (average = 52%) and 11–62% (average = 44%), respectively. The inhibition of a similar amount of partially purified acid phosphatase from wheat (*Triticum aestivum*) germ was 88–95% (average = 92%). Inhibition by steam-sterilized soils was less than that by nonsterilized soils, suggesting that the observed inhibition was due, at least partly, to heat-labile organic constituents. Also, the magnitude of inhibition by nonsterilized soil increased as the quantity of soil added was increased from 0.1 to 1.0 g. No such increase in inhibition was observed with sterilized soils. Soil extracts also inhibited acid phosphatase from corn roots and wheat germ, but this inhibition was most likely due to inorganic or heat-resistant organic compounds, because sterilized and nonsterilized soil extracts yielded similar results.

Although most enzymes introduced into soil are rendered inactive, a small percentage of active enzyme protein may become stabilized in the soil. Kinetic studies indicate that the activity of clay-enzyme complexes formed in soil is greatly reduced, but not totally eliminated (32,33).

Plants are able to synthesize many enzymes. These enzymes, added to soil as plant residues, may remain active. Phosphatase activity in soil has been observed to be associated with intact cell walls of plant tissue, with cell wall fragments, and with amorphous organic material (34,35). Thus, it is not surprising that the type of vegetation added to soil can greatly affect soil enzyme activity. Plants also influence soil enzyme activity by indirect means. Enzyme activity is considerably greater in the rhizosphere of plants than in ''bulk'' soil, and this increased activity is due to either a specific flora or the plant root, or most likely, to the relationship between both (4). Another indirect influence of plants on enzyme activity is the increased number of microorganisms present upon addition of plant litter to soils. Examples of plants affecting soil enzyme activities can be found in the review chapters in the book Soil Enzymes edited by Burns (2) and in Chapters 4 and 6 of this volume.

B. States or Locations of Enzymes in Soils

The term state of enzymes in soils has been used by Skujins (4,5) to describe the phenomenon whereby enzymes exist in soils. Characterization of the state of an enzyme in soil entails the attempt to describe the location and microenvironment in which it functions and the way the enzyme is bound or stabilized within that microhabitat (36). As indicated by Burns (37), the activity of any particular enzyme in soils is a composite of activities associated with various biotic and abiotic components. Burns envisaged 10 distinct categories of enzymes in soils, ranging from enzymes associated with proliferating microbial, animal, and plant cells (located in the cytoplasm, periplasmic space, on outer surface or secreted) to extracellular enzymes associated with humic colloids and clay minerals.

Extracellular enzyme accumulation in soils was clearly demonstrated by Ramirez-Martinez and McLaren (38), who reported that the amount of phosphatase activity in 1 g soil was equivalent to 10^{10} bacteria or 1 g of fungal mycelia. If we assume that nonprotein soil components do not catalyze hydrolysis of organic P compounds in soils, then it can be concluded that a certain portion of phosphatase activity in soils is no longer associated with living tissue.

Several theories have been proposed to explain the protective influence of soil on extracellular enzyme activity. The dominant mechanisms of enzyme immobilization and stabilization (Fig. 2) have been summarized by Weetall (39). These include microencapsulation, cross-linking, copolymer formation, adsorption, entrapment, ion exchange, adsorption and cross-linking, and covalent attachment.

Early work by Ensminger and Gieseking (40) provided evidence that protein adsorbed to montmorillonite was stabilized against microbial attack. Haig (41) also found that acetylesterase activity in a fine sandy loam soil—fractionated into sand, silt, and clay sizes—was associated primarily with the clay fraction. McLaren (42) observed that kaolinite adsorbed trypsin and chymotrypsin, and Mg-bentonite was shown to adsorb pep-

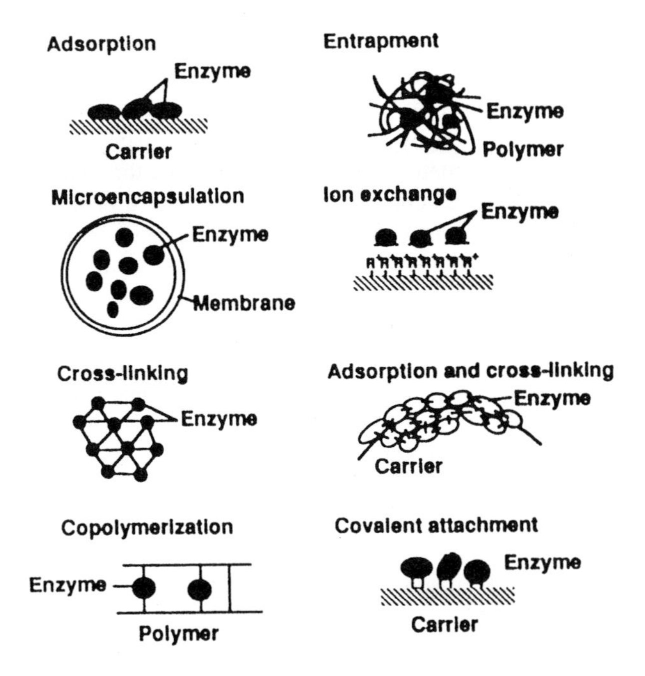

Figure 2 Schematic representation of methods of immobilizing enzymes. (From Ref. 39.)

sin and lysozyme (43). This adsorption occurred rapidly and was 90% complete after 2–3 min. The mechanisms by which clay minerals bind proteins are not always clearly understood. Albert and Harter (44) reported that adsorption of lysozyme and ovalbumin by Na–clay minerals caused an increase in sodium ion concentration of the clay-protein suspension. They interpreted this result as evidence that a cation-exchange adsorption mechanism was occurring.

More recently, Ruggiero et al. (10) have summarized a large amount of work that has been done to enhance understanding of the clay–enzyme complex. In soils, clay surfaces are constantly being renewed or altered by environmental factors, and this condition makes it difficult to extrapolate results obtained by using relatively pure clay minerals saturated with a specific cation to soil conditions.

Stabilization of enzymes in the soil environment by soil organic matter, rather than by inorganic components, has also been suggested. Much of the information dealing with this hypothesis has been obtained by studies involving synthetic polymer–enzyme complexes (45). Early studies by Conrad (46) showed that native soil urease was more stable than urease added to soils. He concluded that organic soil constituents protect urease against microbial degradation and other processes, leading to decomposition or inactivation. Since then, numerous studies have supported this conclusion by showing that enzyme activities in soils are significantly correlated with organic matter content (47–50). A study by Burns and coworkers (51) found that an organic fraction extracted from soil, which was free of clays (confirmed by x-ray analysis), contained urease activity. Supporting results indicating that enzyme activity is associated with humus–enzyme complexes have been reported (52–55) for protease, phosphatase, tyrosinase, peroxidase, and catalase.

Ladd and Butler (45) suggested that enzymes bind to soil humus by hydrogen, ionic, or covalent bonding. The extent that enzymes are bound by each of these mechanisms is difficult to determine. Work by Simonart and associates (56) suggests that hydrogen bonding may be only a minor factor in enzyme stabilization in soils. By using phenol to break hydrogen bonds, they were able to dissolve only a small amount of proteinaceous material.

Enzymes may be bound to organic matter by ionic-bonding mechanisms. Butler and Ladd (57) proposed that enzyme–organic matter complexes are formed through the formation of amino–carboxyl salt linkages. Such complexes, however, should be easily broken by many of the extraction reagents (i.e., urea and pyrophosphate) used to remove active enzyme materials from soils. The small yields of active enzyme materials that have been extracted from soils indicate that ionic-bonding mechanisms may be only partially responsible for enzyme stabilization (58–60). However, Burns et al. (61) extracted approximately 20% of the original soil urease activity by using urea (urea hydrolyzed subsequently by the extracted urease). The clay-free precipitate has urease activity that was not destroyed by the addition of the proteolytic enzyme pronase. The native soil urease was thought to be located in organic colloidal particles that contained pores large enough to allow water, urea, ammonia, and carbon dioxide to pass freely, but small enough to exclude pronase.

A clear hypothesis explaining enzyme stabilization by means of covalent attachment has yet to be proposed. Ladd and Butler (45) suggested that the linkage of soil quinones by nucleophilic substitution to sulfhydryl and to terminal and ε-amino groups of enzyme proteins may lead to active organoenzyme derivatives, provided these groups do not form a part of the active site of the enzyme.

One hypothesis that has as yet received little attention is that enzymes exist in soils as glycoproteins. Malathion esterase, extracted from soils by Satyanarayana and Getzin (62), was thought to be a glycoprotein because of the following evidence: (1) amino acids constituted only 65% of the purified enzyme; and (2) a carbohydrate-splitting enzyme, hyaluronidase, enhanced the catalytic effect of the esterase, presumably by loosening the carbohydrate shield and allowing the protein core to gain easier access to the substrate. The evidence gained by incubating the esterase with hyaluronidase suggested that the carbohydrate–protein linkage occurs through *N*-acetylhexoseamine–tyrosine bonds. Mayaudon et al. (63) drew similar conclusions when they observed that diphenol oxidase activity was not affected by pronase alone, but was destroyed when incubated in the presence of both lysozyme and pronase.

In soils, a strong association exists between clay and humus. Each does not separately influence enzyme stabilization; rather, Paul and McLaren (64) postulated, a three-dimensional network of clay and humus complexes exists in which active enzyme becomes incorporated (Fig. 3). A study by Burns et al. (51) supported this hypothesis when they observed that a bentonite–lignin complex protected urease from degradation much more effectively than did bentonite alone.

C. Stability

Most of the information available on the stability of enzymes in soils is derived from work on urease, acid phosphatase, and arylsulfatase. The first evidence that soil enzymes are more stable than those added to soils was obtained by Conrad (46) in his work on urease; Conrad concluded that organic matter in soils protect-enzymes (urease) against microbial degradation. Support for this conclusion has been provided by numerous studies showing that enzyme activities are significantly correlated with organic C in surface soils and soil profiles (8,50). Further evidence supporting this conclusion is provided by work

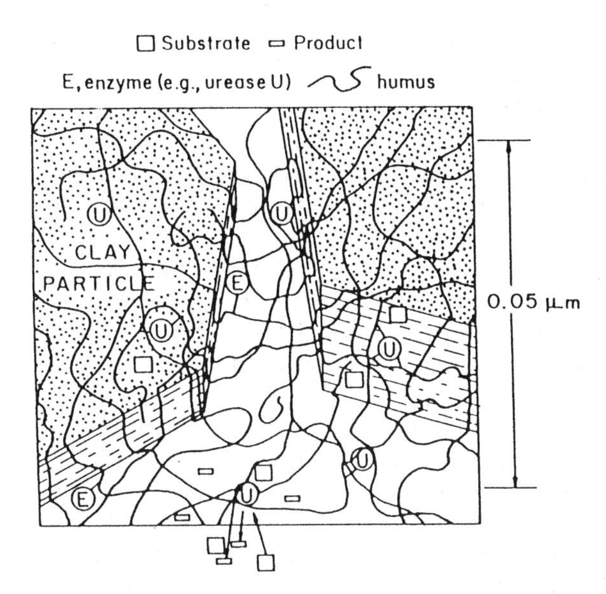

Figure 3 Model for soil enzyme location and activity consisting of enzyme embedded in, and perhaps chemically attached to, a humus polymer network in contact with clay particles. Substrates, such as urea, can reach the enzyme by diffusion through pores too small for enzymes to penetrate. (From Ref. 64.)

showing that soil enzymes are stable for many months and years in air-dried soils (4). Now it is generally accepted that enzymes in soils are immobilized within a network of organomineral complexes (13,65).

III. ROLE OF CHEMISTRY IN ENZYME ACTIVITY MEASUREMENT

One of the fundamental requirements of enzyme measurements is a thorough understanding of the reactions involved, quantitative extraction of the product(s) released, and a suitable analytical method for measuring quantitatively the extracted compound. Therefore, knowledge of analytical chemistry and chemical kinetics are essential in soil enzyme research. In addition, because soils contain both organic constituents and mineral components, a thorough understanding of the potential reactions between the substrate, and more importantly the product released, and the soil constituents is a prerequisite for any methods development.

The detailed study of an enzyme reaction in soils involves characterization and measurement, if possible, of several properties, some of which cannot be obtained for enzyme in soils. One, therefore, has to rely on the biochemical literature for the information required.

1. *Protein properties*: Even though it is difficult to extract and purify enzyme proteins from soils, information about the enzyme molecular weight, isoelectric point, electrophoretic mobility, and stability to pH, heat, and oxidation can be obtained from the biochemical literature. Some of these properties can be ob-

tained from direct experiments by using soil samples, as has been demonstrated for a number of soil enzymes (8).

2. *Structure*: Many of the structural features of purified enzymes are useful in classical biochemistry, but for soil enzyme research it is primarily important to know whether the presence of a prosthetic group or special group of metal atoms is required for activity and to know the effect of chemical reagents on such activity.
3. *Enzyme properties*: It is important to know the nature of the reaction being catalyzed; whether a coenzyme is involved, and its nature and mode of action; specificity for substrate; nature of the chemical structure; and specificity to inhibitors.
4. *Active center*: Knowledge of the nature and composition of the enzyme active center is required.
5. *Thermodynamics*: Because the exact molecular weights of soil enzymes are not known, several properties such as free energies and entropies of enzyme–substrate combination cannot be determined, other properties, however, can be. They include the activation energy of an enzyme-catalyzed reaction, affinity of the enzyme for its substrate, Michaelis–Menten constant, effect of pH on affinity of the enzyme for its substrate, affinities for inhibitors, inhibitor constants, and competition of inhibitors with the substrate.

   The temperature dependence of the rate constant, at a temperature below that which results in activation of the enzyme activity, can be described by the Arrhenius equation ($k = A \cdot \exp(-Ea/RT)$, where k is the rate constant, A is the preexponential factor, Ea is the activation energy, R is the gas constant, and T is the Kelvin (K) temperature. The Arrhenius equation, when expressed in its log form ($\log k = (-Ea/2.303\ RT) + \log A$), allows calculation of Ea by plotting log of the initial rate of the reaction *vs* $1/T$. The slope of the resulting line is equal to $-Ea/2.303R$. The enthalpy of activation (ΔHa) can be calculated from $\Delta Ha = Ea - RT$.
6. *Kinetics*: Characterization of the kinetic parameters of the Enzyme–substrate reaction is important because anyone who is concerned with catalysis in soil is most certainly concerned with the velocities of chemical reactions (chemical kinetics). The usual way to follow an enzyme-catalyzed reaction is by measuring the amount of reactant remaining or the product formed. By contrast, most kinetic models are formulated in terms of rates of reaction. Traditionally enzyme kinetic studies have focused on the initial rates of reactions by measuring tangents to the reaction curves (i.e., by measuring the linear portion of the reaction curve at the time the reaction is initiated). In soil, kinetic data have only progressed to the point where we can study simple one-substrate systems, which react with a single enzyme. The Michaelis–Menten equation does an excellent job of describing this type of kinetics and there are several assumptions that are made when applying this equation to soil systems. The enzyme reaction is expressed by the following equation:

$$\mathrm{E} + \mathrm{S} \underset{k_2}{\overset{k_1}{\rightleftharpoons}} \mathrm{ES} \xrightarrow{k_3} \mathrm{E} + \mathrm{P}$$

   The assumptions made in deriving the Michaelis–Menten equation are as follows: (1) The rate of reaction of the enzyme-catalyzed system changes from

Table 1 Parameters of Linear Equations Describing Inhibition of Enzyme–Substrate Interactions

Inhibitor type	Slope	*Y* intercept	*X* intercept
No competition	K_m/V_{max}	$1/V_{max}$	$-1/K_m$
Linear competitive	$K_m/V_{max}(1 + [I]/K_i)$	$1/V_{max}$	$-1/K_m(1 + [I]/K_i)$
Linear noncompetitive	$K_m/V_{max}(1 + [I]/K_i)$	$(1/V_{max})(1 + [I]/K_i)$	
Linear uncompetitive	K_m/V_{max}	$(1/V_{max})(1+ [I]/K_i)$	

first-order to zero-order kinetics; (2) enzyme (E) reversibly binds with substrate (S) to form an intermediate enzyme–substrate (ES) complex, which then breaks down to form product (P). Each reaction is described by a specific rate constant: k_1, k_2, k_3; (3) a steady-state equilibrium between the rate of formation of ES and the rate of degradation of ES is rapidly achieved; (4) total enzyme concentration is defined as that in the free state and in the enzyme–substrate complex; (5) the initial rate-limiting parameter is the decomposition of the enzyme–substrate (ES) complex to form the product (or k_3); and (6) V_{max} is achieved when ES complex concentration reaches a maximum equal to the total enzyme concentration: i.e., there is no free enzyme.

Much of what we know about biological systems is based on more complex enzyme systems that have an inhibitor present. For example, it is well known that the presence of inorganic phosphate in solution strongly inhibits phosphatase activity in soils (8). The simplest systems are those in which there is a single substrate, single enzyme, and a single inhibitor (I). In general, the type of inhibition could be one of the following: (1) linear competitive inhibition, (2) linear noncompetitive inhibition, or (3) linear uncompetitive inhibition. The parameters of the linear equations are shown Table 1.

7. *Biological properties or role of soil enzymes in metabolic reactions*: Information on the occurrence and distribution of the enzyme among different species and associated with different plant and microbial tissues that are deposited in soils is important. Much of the information can be obtained from the biochemistry literature.

IV. SUBSTRATE STRUCTURE, ENZYME SPECIFICITY, AND ACTIVITY MEASUREMENT

Substrate structure has a significant effect on the reaction rate, and the structure of the product released markedly affects its extractability from soils and the potential for its quantitative determination by any procedures or techniques. Detailed discussion of enzyme specificity is beyond the scope of this chapter, but it should be made clear that specificity is one of the most striking properties of the enzyme molecule. It depends on the particular atomic structure and configuration of both the substrate and the enzyme. There are three types of enzyme specificity. The first is absolute specificity, which is rare and describes a reaction in which a single member of a substrate class is attacked by an enzyme. An example is urease. Relative specificity describes a situation in which an enzyme acts preferentially on one class of compounds but will attack a member of another class to a certain extent. This term may also be used to illustrate the different rates of reactions within a given class. The third type is optical specificity, which is a common property of some yeast enzymes, which act on optically active substrate. Stereochemical specificity is strikingly illustrated by the action of glycosidases. Maltase hydrolyzes maltose and several other α-glucosides to glucose but not β-glucosides. Emulsin contains a β-glucosidase, which acts only on β-glucosides but not on α-glucosides. Both α- and β-glucosidases are present in soils (66). Other similar examples are the D- and L-specific amino acid oxidases (67).

An essential step in enzyme activity measurement requires the availability of chemical methods and instrumental techniques for determination of the reaction product formed. Almost all the methods developed by biochemists for enzyme assay are useful as guides for assay on enzyme activities in soils, but caution should be exercised to be sure that the product formed is determined quantitatively. This is because many of the methods are not compatible with the complex chemical characteristics of the soil system.

V. ENZYME PROTEIN CONCENTRATION IN SOILS

Numerous attempts have been made to extract pure enzymes from soils, but in reality the best that has been achieved is the extraction of enzyme-containing substances and complexes (68). The reagents used in the extraction procedures range from water to salt solutions or buffers to strong organic matter-solubilization reagents, such as NaOH or sodium pyrophosphate. The extracted activities are usually associated with carbohydrate–enzyme protein complexes and are often difficult to purify. Modern biochemical techniques have been used in the purification of the extracted enzymes, but little progress has been made in obtaining pure enzyme proteins from soils. Several of the enzymes extracted from soils could be present in soils as glycoproteins. Although many investigators have demonstrated that clay-free extracts could be obtained from soils, the major problem appears to be the strong affinity of the carbohydrate–enzyme complexes for chromatographic columns, which makes the separation difficult. It appears that various carbohydrates in soils adsorb the enzyme proteins and are responsible for their stabilization against denaturation or proteolysis.

A. Estimation of Concentrations of Enzyme Proteins in Soils

Enzyme activities are associated with active microorganisms because the microbial biomass is considered the primary source of enzymes in soils. Nevertheless, there is no direct correlation between the size of the microbial biomass and its metabolic state (69). One approach to estimate the metabolic state of microbial populations in soils is to differentiate between intra- and extracellular enzyme activities. Among the many attempts that have been made to determine the state of enzymes in soils are techniques that employ elevated and decreased temperatures; antiseptic agents such as toluene, ethanol, Triton X-100, dimethyl sulfoxide; irradiation with gamma rays or electron beams; and fumigation with compounds such as chloropicrin, methylbromide, and chloromycetin (2,3,8,70,71). None of these methods can distinguish between intracellular (activity associated with the microbial biomass) and extracellular activity (that portion stabilized in the three-dimentional network of clay–organic matter complexes) (Fig. 3), because all these techniques also denature the enzyme proteins. Another suggested approach is plotting enzyme activity against the number of ureolytic microorganisms (in the case of urease) or adenosine triphosphate (ATP) concentration (in the case of phosphomonoesterases). The extrapolation to zero population or ATP concentration produces a positive intercept, which is assumed to be the extracellular component of the enzyme activity (72,73).

B. Estimation of Active Enzyme Protein Equivalent

Studies in 1998 and 1999 by Klose and Tabatabai (74–76) have estimated the concentrations of 12 enzyme proteins in soils. The averages of the concentrations in 10 Iowa surface soils ranged from 0.014 mg protein kg^{-1} soil for β-glucosidase to 22.5 mg protein kg^{-1} soil for acid phosphatase (Table 2). These estimates were done by analysis of reference protein materials by sodium dodecyl sulfate-polyacrylamide gel electrophoresis (SDS-PAGE) and by calculation of the specific activity of the reference proteins. From the specific activities of the reference proteins and activities of the enzymes in soils in the presence of toluene, the enzyme protein equivalents in soils were calculated. These calculations were not intended to give an accurate concentration of enzyme proteins in soils but instead to provide some quantitation of enzyme protein equivalent. Actual concentrations of enzyme proteins in soils are undoubtedly much greater than those calculated because many soil components can inhibit activity, and structural stabilization also leads to decreases in the activity. However, the calculations illustrate one reason for the difficulties encountered in the extraction and purification of enzyme from soils (68). From the results reported in Table 2, it is clear that the small concentrations of enzyme proteins in soils either are denatured during extraction or bond tightly with the soluble carbohydrates, making their separation very difficult.

VI. TYPES OF ENZYMES AND SUBSTRATES

To date, soil enzyme studies have been primarily restricted to hydrolases but with additional effort also put into measuring specific types of oxidoreductases and lyases. This emphasis on hydrolases is understandable because of the need for microorganisms in soil to degrade a complex variety of substrates in soil. Many of these substrates are polymeric and can only be degraded by enzymes secreted into soil. The fate of the secreted or extracellular enzymes is still not well understood, but it is probably safe to assume that most of them are rapidly degraded by proteases and/or inactivated in soil. However, some may become immobilized and stabilized in soil through a variety of mechanisms (7) so that their activity continues long after they are first introduced into soil. Hydrolases are also relatively simple enzyme systems, which generally do not require cofactors; have multiple subunits; and are small in size. Thus they are much more resistant to denaturation by temperature, desiccation, sorption, or other physical factors of the soil environment.

The specificity of an enzyme reaction is difficult to assess in soils. For example, total cellulase activity in soils may be due to wide variety of extracellular enzymes from fungi and bacteria including those stabilized by association with the organic and mineral components. An accurate description of all enzymes in all locations, for example, that contribute to measured cellulase activity is not possible at this time. In addition, cellulases are endocellular or ectocellular (i.e., cleave internal or exposed β1-4 linkages) and interact in a synergistic way that would make it very difficult to distinguish between the individual components of ''total'' cellulase activity.

In Table 3, we have described some of the major polymeric substrates that are naturally introduced to soil as plant, animal, or microbial products. The enzymes involved and

Table 2 Estimated Enzyme Protein Equivalents in Soils

	Enzyme protein equivalent (mg protein kg^{-1} soil)[a]												
	Glycosidases[b]				Amidohydrolases[c]					Phosphatases[d]		Arylsulfatase[e]	
Soil	α-Gal	β-Gal	α-Glu	β-Glu	L-Asg	L-Glu	Amid	Urea	L-Asp	Acid-P	Alk-P	A	B
Harps	0.031	1.6	4.8	0.021	2.4	1.2	4.2	3.6	8.5	12.2	5.2	9.0	37.6
Okoboji	0.038	2.1	4.8	0.018	0.82	0.69	4.3	2.6	3.3	21.5	3.2	7.5	31.4
Muscatine	0.018	1.2	3.6	0.014	0.84	0.58	3.6	0.95	2.8	16.3	3.3	7.3	30.7
Grundy	0.022	1.1	3.6	0.014	0.56	0.45	2.8	0.87	1.8	25.7	1.8	4.2	17.7
Gosport	0.035	2.5	3.9	0.014	0.59	0.57	3.5	1.4	2.8	29.2	1.8	4.6	19.5
Clinton	0.033	2.8	4.2	0.019	0.80	0.71	7.3	2.6	2.9	33.5	2.1	5.7	24.1
Pershing	0.028	1.4	3.9	0.013	0.35	0.22	2.1	0.95	1.1	34.5	0.97	3.20	13.3
Luther	0.010	0.29	1.6	0.005	0.41	0.10	0.65	0.87	0.40	8.8	0.36	0.10	0.42
Grundy	0.033	1.5	3.3	0.012	0.37	0.38	3.3	1.4	1.6	21.0	1.1	3.3	13.9
Weller	0.020	1.2	3.1	0.011	0.19	0.14	2.0	0.57	0.78	22.7	0.97	1.70	7.2
average	0.027	1.56	3.7	0.014	0.73	0.50	3.4	1.6	2.6	22.5	2.1	4.7	19.6

[a] Calculated for the nonfumigated soils (except for urease and arylsulfatase, which were based on fumigated samples) on the basis of their activity values and specific activities of the purified reference enzyme proteins.
[b] α-Gal, α-galactosidase; β-Gal, β-galactosidase, α-Glu, α-glucosidase; β-Glu, β-glucosidase.
[c] L-Asg: L-asparaginase; L-Glu, L-glutaminase; Amid, amidase; urea, urease; L-Asp, L-aspartase.
[d] Acid-P, acid phosphatase; Alk-P, alkaline phosphatase.
[e] A, Helix *pomatia* as a reference protein; B, Patella *vulgata* as a reference protein.
Source: Ref. 74.

Table 3 Summary of Some of the Major Polymeric Substrates, the Enzymes Involved, and Techniques Used in Their Assay

Polymeric material	Enzymes Involved	Assay Methods	Comments
Cellulose—a crystalline polymer associated with lignin and hemicellulose	Many cellulases of three main types: 1) exocellulohydrolase (EC 3.2.1.91) (2) endo-1,4-β-D-glucanase (EC 3.2.1.4) 3) β-glucosidase (EC 3.2.1.21). These exist extracellularly in soil.	Measure release of glucose using a glucose oxidase reaction. Measure release of *p*-nitrophenol (many substrates available with this chromophore).	Various pretreatments may be helpful to expose the cellulose to enzyme. The major barriers are association of the cellulose with lignin and the crystalline structure of the cellulose.
	not possible	Fluorometric substrates (e.g., 4-methylumbelliferyl).	
Proteins—a polymer that comprises amino acids bound together by peptide bonds	A large variety of proteinases and peptidases most of which exist extracellularly in soil.	The three most commonly used substrates are (1) peptide 4-nitroanilines (2) peptide thioesters (3) peptide derivatives of 7-amino-4-methylcoumarin. Prepare proteins with a fluorogenic label attached to individual amino acids. After incubation, precipitate proteins and measure fluorescence in the solution phase.	Not all of the methods for using these substrates in soil have been worked out. The peptide thioesters provide a sensitive assay because they have high k_{cat}/k_m values and low background hydrolysis and the thiol leaving group can be detected at low concentrations.
Lipids—derived from a large number of cell membranes.	Lipases of many types including the phospholipases. Lipases exist extracellularly in soil	Assays can measure the organic leaving group or, in the case of phospholipids, the phosphate released. Methods used include titrimetric, radiometric, colorimetric, and fluorometric procedures.	Many natural lipid substrates are not soluble in water and this makes the design of an assay difficult. Short-chain phospholipids that are water soluble can be used as synthetic substrates and choice of substrate can often distinguish the different types of lipases and phospholipases.

Table 3 Summary of Some of the Major Polymeric Substrates, the Enzymes Involved, and Techniques Used in Their Assay

Polymeric material	Enzymes Involved	Assay Methods	Comments
Lignin—a nonrepeating polymer of sinapyl, coniferyl, and caumaryl alchols that is part of cell walls and is a major part of soil humus and more resistant to degradation than most substrates.	Ligninases,(e.g., phenol oxidases, peroxidases) of fungi are the best studied	Polymeric dyes are substrates that have been developed for assays of ligninases. Assays generally require days, not hours, to complete. Different dyes that vary in absorbance intensity and wavelength can be used. ^{14}C-Labeled materials can also be used.	Only lignin degrading fungi can decolorize many of the dyes used to measure ligninases. The polymeric dyes are inexpensive and stable, can be obtained commercially, have high purity, are water-soluble, and have high extinction coefficients.
Chitin, pectin, and other polymers in soils. Chitin is a mucopolysaccharide often intimately associated with calcareous shell material. Pectin polymers are chains of predominantly 1,4-linked-α-D-galacturonic acid and methylated derivatives.	Degradation of most polymers is due to extracellular enzymes as these molecules are too large to be taken into microbial cells. Pectinases (especially) and chitinases have various forms.	Chitinase is most commonly assayed by measuring N-acetylglucosamine release by using a spectrometric procedure when chitin is incubated with soil. Tritiated chitin can also be prepared and used as a substrate. Pectinases are traditionally assayed by using a viscosity reduction and by measuring reaction products by a variety of chemical and biochemical methods.	Chitin is insoluble. A pure form of chitin can be purchased. Tritiated chitin must be prepared in the laboratory, and the amount of tritium that remains in solution after centrifugation is a measure of chitinase activity.

the techniques used to measure these enzymes are summarized. Detailed accounts of the procedures for individual enzymes assay can be found in a book chapter by Tabatabai (8) and a manual edited by Alef and Nannipieri (77).

VII. MEASUREMENT METHODS

Techniques to measure enzyme activity in soils are primarily derived from the biochemical literature. However, because the soil system is generally much more complex than many systems studied by biochemists, most of these methods and techniques require modifications. Advances in our scientific understanding of many subjects is directly linked to our ability to develop methods to measure what we are attempting to study. For many years, progress in soil enzymology was hampered by a lack of standard methods and development of new methods was difficult because of the great complexity of soils. There have been many advances in analytical capabilities, but application of these new procedures to soil enzymology has not kept pace. The reader is referred to the series Methods in Enzymology for up-to-date accounts. Major published works related specifically to describing methods of soil enzymes include those by Alef and Nannipieri (77) and Tababatai (8). The factors that limit advances in soil enzyme research are related to our inability (1) to separate extracellular from intracellular enzyme activity, (2) to extract and purify enzymes from soil, and (3) to extract the many products of enzyme reactions from soil quantitatively. Depending on whether a decrease in the substrate concentration or an increase in the concentration of the product released is to be measured, the method selected for quantitatively following any enzyme reaction may be one of many analytical techniques:

A. Spectrophotometric Methods

Many substrates and the products of enzymatic reactions absorb light, either in the visible or in the ultraviolet region of the spectrum. Most often the change in the concentration of the substrate or the product is followed colorimetrically after extraction from a soil sample incubated with the substrate at specific temperature, pH, and time. Here we summarize some of the most commonly used methods.

Numerous colorimetric procedures for analysis of urea with diacetylmonoxime have been developed (78); most of these methods are actually variations of that developed by Fearon (79). One of the procedures has been evaluated for the determination of urea (extracted from soils with 2 *M* KCl-phenylmercuric acetate) with a reagent containing diacetylmonoxime and thiosemicarbazide in a boiling water bath for 30 min (80). The absorbance of the chromogen complex is measured at 550 nm. The main disadvantages of most of the procedures available for colorimetric determination of urea are (1) lack of sensitivity, (2) lack of linearity at low urea concentration, (3) lengthiness of procedure, (4) low precision, and (5) instability of some of the reagents or the chromogen compound. Because of these problems, attempts have been made to automate the development of color (81,82). Caution should be exercised in using this method, however, because no buffer is used to control the pH of the incubation mixture.

Following are examples of methods involving colorimetric determination of the products released in assays of arylsulfatase and arylamidase activities in soils.

Arylsulfatase (EC 3.1.6.1) is the enzyme that catalyzes the hydrolysis of organic sulfate ester ($R \cdot O \cdot SO_3 \cdot + H_2O \rightarrow R \cdot OH + H^+ + SO_4^{2-}$). This enzyme has been detected in plants, animals, and microorganisms, and it was first detected in soils by Tabatabai and Bremner (83). This enzyme hydrolyzes a number of organic sulfate esters. Among those *p*-nitrophenyl sulfate, 4-nitrocatechol sulfate, and phenolphthalein sulfate have been tested as substrates for this enzyme in soils.

OSO_3K … NO_2 $\xrightarrow[+ H_2O]{\text{ARYLSULFATASE}}$ OH … NO_2 $+ KHSO_4$

POTASSIUM P-NITROPHENYL SULFATE P-NITROPHENOL

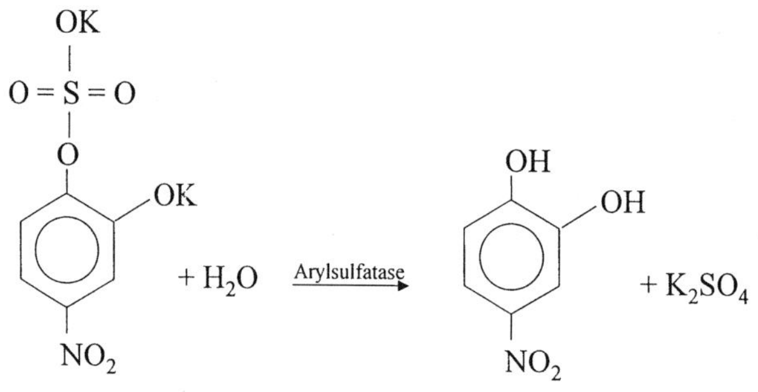

2-Hydroxy-5-nitrophenyl sulfate dipotassium salt (4-Nitrocatechol sulfate)

4-Nitrocatechol (Red under alkaline conditions; 510 nm)

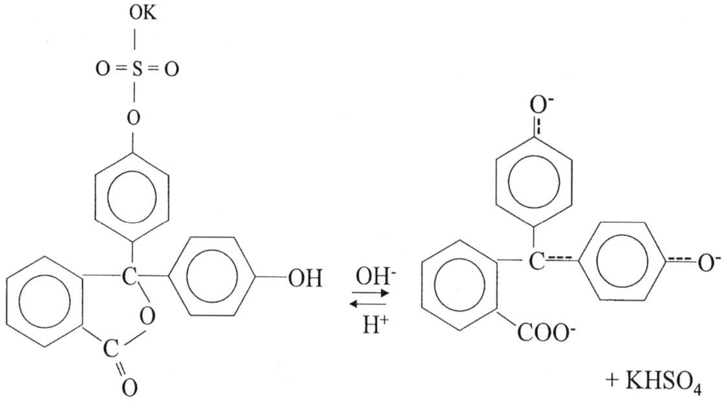

Colorless

Red phenolphthalein (Max. color at pH 10—10.4; 550 nm)

Although results showed that all these compounds are hydrolyzed in soils, only the product of *p*-nitrophenyl sulfate (*p*-nitrophenol) is quantitatively extractable from soils (83). The organic moieties of the other two substrates are not quantitatively extracted, because they are highly reactive with phenolic compounds in soils. This is unfortunate, because the organic moiety of 4-nitrocatechol sulfate (4-nitrocatechol) is red under alkaline conditions and gives a very sensitive color reaction. The same is true with phenolphthalein sulfate, which is hydrolyzed to phenolphthalein and sulfate. Even though phenolphthalein is pink under alkaline conditions and, therefore, should be useful for assay of arylsufatase activity in soils, it is not extractable from soils.

Since the introduction of *p*-nitrophenolates as substrates for assay of acid phosphatase in soils, esters of *p*-nitrophenol have been used to assay numerous enzymes in soils (8). Other substrates with chromophore moieties have also been investigated as potential substrates in assay of soil enzymes. For example, arylamidase (EC 3.4.11.2) readily hydrolyzes neutral amino acid arylamides: i.e., it hydrolyzes amino acids attached to β-naphthylamine and *p*-nitroaniline according to the following reaction (using the amino acid L-leucine as an example):

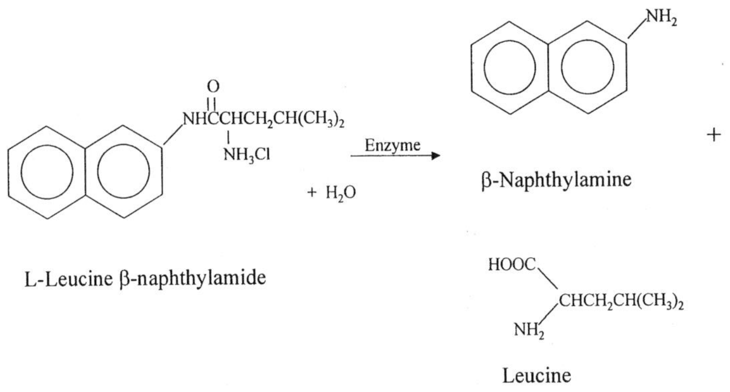

NH_2

O

$NHCCHCH_2CH(CH_3)_2$

NH_3Cl

Enzyme

$+ H_2O$

β-Naphthylamine

+

L-Leucine β-naphthylamide

HOOC

$CHCH_2CH(CH_3)_2$

NH_2

Leucine

The β-naphthylamine released can be extracted from soils and determined colorimetrically after its reaction with *p*-dimethylaminocinnamaldehyde as follows (84).

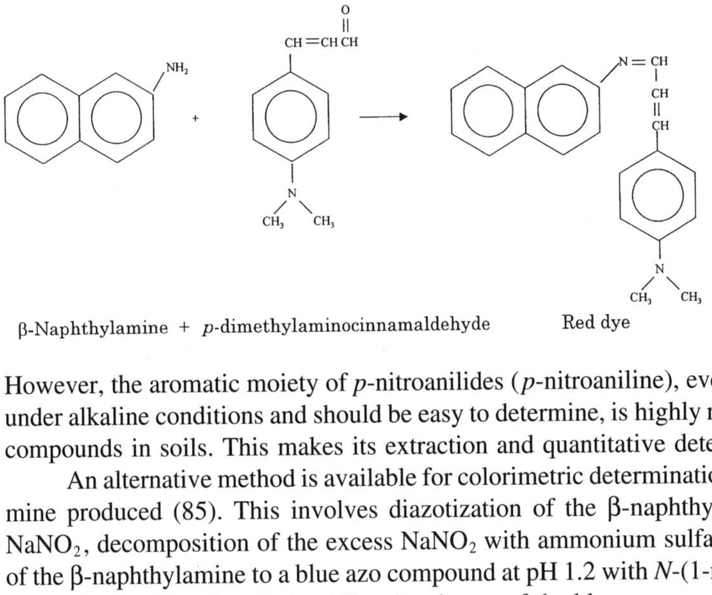

O

CH = CH CH

NH_2

$N = CH$

CH

CH

+

N

CH_3 CH_3

N

CH_3 CH_3

β-Naphthylamine + *p*-dimethylaminocinnamaldehyde

Red dye

However, the aromatic moiety of *p*-nitroanilides (*p*-nitroaniline), even though it is yellow under alkaline conditions and should be easy to determine, is highly reactive with phenolic compounds in soils. This makes its extraction and quantitative determination difficult.

An alternative method is available for colorimetric determination of the β-naphthylamine produced (85). This involves diazotization of the β-naphthylamine released with $NaNO_2$, decomposition of the excess $NaNO_2$ with ammonium sulfamate, and conversion of the β-naphthylamine to a blue azo compound at pH 1.2 with *N*-(1-naphthyl)ethylenediamine dihydrochloride solution. The absorbance of the blue azo compound is measured at 700 nm. This method, however, is complicated and tedious.

B. Titrimetric Methods

Several amidohydrolases are present in soils. All are involved in hydrolysis of specific native and added organic N compounds in soils. Among these, L-asparaginase, L-glutaminase, L-aspartase, amidase, and urease are the most important in the biogeochemical context. In assaying the activity of these enzymes, the soil is incubated with the substrate in an appropriate buffer and the NH_4^+ produced is determined. The reaction is stopped by adding 2*M* KCl containing Ag_2SO_4. Because a simple distillation apparatus is available, normally an aliquot of the incubated soil–solution mixture is distilled with MgO, and the NH_3 released is collected in boric acid containing bromcresol green and methyl red indicators and titrated with standard H_2SO_4. Details of the methods available are reported elsewhere (8,17).

C. Fluorescence Methods

Several fluorimetric techniques have been reported for assay of enzyme activities in soils. One of the early techniques is that described by Ramirez-Martinez and McLaren (38) which uses Na-β-naphthyl-phosphate as a fluorogenic substrate for assay of acid phosphatase activity in soils based on the measurement of β-naphthol released.

$$\text{Na-}\beta\text{-naphthyl-phosphate} + H_2O \xrightarrow[\text{phosphatase}]{\text{Acid}} \beta\text{-naphthol} + NaH_2PO_4$$

In using such substrates, however, retention by soil of the hydrolysis product must be measured and accounted for when the phosphatase activity of soils is expressed quantitatively.

A similar approach was used by Pancholy and Lynd (86) for assaying soil lipase activity by using the nonfluorescent butyryl ester of 7-hydroxy-4-methylcoumarin to the highly fluorescent 7-hydroxy-4-umbelliferone.

$$\text{Triglyceride} + H_2O \xrightarrow{\text{Lipase}} \text{Diglyceride} + \text{Fatty Acid (RCOOH)}$$

Triglyceride   Diglyceride   Fatty Acid

7-Hydroxy- 4- methylcoumarin (R-C(=O)-O- ester; Nonfluorescent) → Lipase → 7-Hydroxy- 4- umbelliferone (HO-; Highly fluorescent)

7-Hydroxy- 4- methylcoumarin
Nonfluorescent

7-Hydroxy- 4- umbelliferone
Highly fluorescent

Other fluorogenic model substrates have been used for the assay of β-glucosidase, phosphatase, and arylsulfatase activities in peat (87) and for assay of β-cellobiase, β-galactosaminemidase, β-glucosidase, and β-xylosidase, arylsulfatase, and alkaline phosphatase activities in soils (88). In all these methods, either a very small amount of the soil sample (in milligrams) must used or the capacity of the soil to sorb the fluorogenic compound released must be determined and the assay results corrected for.

A microplate assay to screen soils for hydrolytic enzymes based on methylumbelliferyl (MUB) substrates was developed by Marx et al. (89). Fluorescence production was measured by a computerized microplate fluorometer. This method offers increased sensitivity and the possibility of estimating enzyme kinetics. If it is successful, the advantages of this technology are (1) speed of operation (less than 1 hour), (2) simultaneous analysis of a large number of samples, (3) simultaneous use of a range of MUB conjugates, (4) measurement under standard conditions, and (5) automatic calculation of reaction rates. This method, which requires only milligram quantities of homogenous soil samples, has been used to measure the activities of β-D-glucosidase, β-D-galactosidase, *N*-acetyl-β-D-glucosaminidase, β-cellobiase, β-xylosidase, acid phosphatase, and arylsulfatase in a sandy loam and a silty clay loan soil. Marx and coworkers (89) reported that the results confirmed the potential usefulness of this technique and demonstrated the precision of the MUB microplate assay.

D. Radioisotope Methods

Among the many enzymes assayed in soils, only a limited number allow for use of substrates labeled with radioisotopes. This is due to the problems of isolating the labeled substrates or products from the soil and from each other. One of the most widely used assays involves the use of ^{14}C urea and determination of the $^{14}CO_2$ released (90–93). Such methods, however, have the disadvantage of incomplete release of the $^{14}CO_2$ unless the incubation mixture contains acidic (pH 5.5) buffer (92,93). This approach has the disadvantage of assaying urease activity under acid conditions, which are known to reveal only a small fraction of the total activity. The optimal urease activity in soils occurs using THAM buffer pH 9.0, which would not allow for CO_2 evolution (94). Another problem associated with the use of ^{14}C-labeled urea is the possibility of isotope effects, including isotope exchange. This isotope effect has not been studied in the assay of soil urease activity, but the information available suggests that ^{12}C urea hydrolysis by jack bean urease is about 10% faster than ^{14}C urea hydrolysis at 37°C (95).

E. Manometric Methods

The manometric methods are simple, accurate, convenient, and inexpensive, provided that one of the reaction components is a gas. These methods, therefore, are well adapted for assay of oxidases (O_2 uptake) or decarboxylation (CO_2 release), as well as assays of hydrogenase and urease. Their use in studies of enzymatic reactions in soils, however, is very limited because soils contain a variety of organic and inorganic compounds that could involve the release or consumption of gases.

F. Electrode Methods

Several electrodes are available that have been considered for the assay of enzyme activities in soils. The glass electrode can be used to follow reactions that involve the production of acids, but there are two problems with such approaches. The first is that the change in pH during incubation alters the reaction rate of the enzyme. The second problem is that the rate of change of pH depends not only on the reaction but also on the buffering poise of the soil and soil solution. Therefore, such methods are not useful for assaying enzyme in soils directly, as the buffer capacity of the soil sample and the buffer used complicate the titration procedure. However, such methods may be useful for comparing reaction rates in a single unbuffered soil extract containing enzyme activities.

One ion-specific electrode that might be useful for enzyme assay is the ammonia electrode. This electrode has been shown to be useful for the determination of ammonium N in soils and water samples (94,96), and it is possible to apply this technique for quantitative determination of the ammonium N released from urea by the urease activity. Any electrode used for detection of the product formed in the enzyme assay must be compatible with the chemical properties of soils, i.e, must give quantitative results for compounds being determined when added to soils. Another ion-specific electrode that has potential in enzyme assay is the nitrogen oxide electrode. This electrode has been shown to be useful for determining NO_2^- in soils (97), which is the product of nitrate reductase, provided the nitrite reductase is specifically inhibited to allow accumulation of NO_2^- in the incubation mixture (98).

G. Chromatographic Methods

Several chromatographic techniques that are available have potential for application in enzyme research. These range from thin layer chromatography (TLC), to high-performance liquid chromatography (HPLC), to ion chromatography (IC), to gas chromatography (GC). Chromatographic methods are separation procedures that are useful for removing interfering substances so that the compounds of interest, generally the product of an enzymatic reaction, can be measured. The enzymatic degradation of xenobiotics is measured frequently with HPLC and GC. A wide diversity in enzymatic assay procedures can be achieved by combining various types of chromatographic and detection systems. For example, IC combines anion exchange chromatography with electrical conductivity detection for analyses of simple anions such as phosphate, nitrate, sulfate, and chloride, and even simple organic acids are available (99).

H. Capillary Electrophoresis

Capillary electrophoresis (CE) is a technique that uses a narrow-bore (a typical length is 60 cm with 75 μm inner diameter) fused-silica capillary with optical viewing to perform high-efficiency separations of both large and small molecules, or ions. The separation is facilitated by using high voltages, which may generate electroosmotic and electrophoretic flow of buffer solution and ionic species, respectively, within the capillary. The properties of the separation, and the electropherogram produced, have characteristics resembling a cross between traditional polyacrylamide gel electrophoresis and high-performance liquid chromatography. The instrument consists of a fused-silica capillary with an optical viewing window (the outer protective coating scraped off to allow a window for detection), a controllable high-voltage power supply, two electrode assemblies, two buffer reservoirs, and an ultraviolet-visible light detector. The ends of the capillary are placed in the buffer reservoirs and the optical viewing window is aligned with the detector. The capillary is filled with a suitable electrolyte, usually in aqueous solution. Electrodes are placed at each end of the capillary and a large voltage is applied, typically in the range of 15–30 kV. A "positive" power supply is used for the separation and determination of cations so that the cathode (−) is near the detector. A "negative" power supply can be used in which the polarity of the electrodes is reversed. After filling the capillary with buffer and visualization reagent (e.g., UV-Cat1, developed by Waters), the sample is introduced by dipping the end of the capillary into the sample solution for a fixed time. Automated instruments are available that feature computer control of all operations, pressure and electrokinetic injection, an autosampler and fraction collector, automated methods development, precise temperature control, and an advanced heat dissipation system. Automation is critical to CE because repeatable operation is required for precise quantitative analysis (100).

The history of the development of CE is summarized by Jandik et al. (101). The details of the theory of CE operation are not within the scope of this review, but information is available (100–103). A recent review on CE shows that it has been applied to the analysis of a variety of samples in several fields, including organic acids, carbohydrates, pharmaceutical compounds, clinical analyses, catecholamines, biological samples, pesticides and proteins, oligosaccharides, and biotechnology-derived samples (102–103).

In an attempt to address the potential diversity of extracellular enzymes in the environment, Arrieta and Hernd (104) developed a method to characterize the different β-glucosidases from seawater samples by means of CE. The different isozymes were detected and their activities quantified by combining CE separation and on-column hydrolysis of fluorogenic substrate analogues. Trace amounts of fluorescent products can be detected with laser-induced fluoresence, resulting in a fingerprint of the isozymes present in the sample. Efficient separation of β-glucosidase isozymes was achieved in mixtures of environmental bacteria isolates and in natural seawater communities, and the response of the different isozymes to different concentrations of substrate analogues was determined. This technique allows the determination and characterization of isozymes present in aquatic systems. Results indicated that in surface layers of the North Sea, at least two different isozymes with β-glucosidase activity are present. The CE instrument has not been used for the assay of soil enzymes, but seems to be a promising analytical tool for such a purpose.

XIII. ESTABLISHMENT OF ASSAY CONDITIONS AND EFFECTS OF PLASMOLYTIC AGENTS AND SOIL STERILIZATION

Several questions must be addressed by the soil biochemist prior to making enzyme measurements in soils. These provide the guiding principles needed to make an appropriate and valid interpretation of results. These are outlined in Table 4.

A number of methods that have been proposed for soil sterilization or inhibition of microbial proliferation allow assay of enzyme activities (71). An ideal sterilization agent for extracellular enzyme detection in soil would be one that would completely inhibit all microbial activities, but not lyse the cells and not affect the extracellular enzymes. Unfortunately, no such universal agent is available.

Toluene has been the most widely used as a microbial inhibitor, but its usefulness is limited to assay procedures that require only a few hours of incubation. Nonetheless, assay procedures involving long incubation times with or without toluene should be avoided, because the risk of microbial proliferation increases as the time of incubation increases. It is believed that in assay procedures involving short incubation times, toluene inhibits the synthesis of enzymes by living cells and prevents assimilation of the reaction products. Toluene has also been shown to be a plasmolytic agent in certain groups of microorganisms in which it apparently induces the release of intracellular enzymes (4). A critical examination of the effect of toluene on soil microorganisms has been made by Beck and Poschenrieder (105). Who showed that the inhibitory effect and concentration of toluene needed to suppress microbial activity are strikingly dependent on the pretreatment and moisture content of a particular soil. To suppress microbial proliferation in an

Table 4 Important Questions That Must Be Addressed When Designing Soil Enzyme Assays and Measuring Activities

Question asked	Relevance of question
What is the source of activity?	Reaction may be only associated with living cells or due to accumulated enzymes in soil that retain activity but no longer are part of a viable cell.
Should I use a buffer?	pH is a major factor controlling enzyme activity. Using a buffer to maintain optimal pH allows measurement of total potential enzyme activity. This is preferred when comparing results among soils. Using an unbuffered solution provides a measure of the real activity of the soil but only for the time the sample was taken.
Has a method already been developed for soils?	Using a well-established method saves time. Thus, it is important to determine how much work has been done to standardize the assay using a range of soils.
What temperature should I use?	Enzyme assays have traditionally been measured at the temperature of the human body (approx. 37°C.) This is higher than temperatures normally found in soils but does provide for greater rates of activity and thus sensitivity.
What concentration of substrate should I use?	Substrate concentration should be great enough to achieve zero-order kinetics with regard to the substrate concentration (i.e., substrate concentration should not limit the reaction rate). The only limitation to reaction rate should be enzyme concentration, which should be directly proportional to the reaction rate.
What is the level of sensitivity required?	The greater the sensitivity required, the more care is needed in conducting the assay and choosing the assay to be used. Radiolabeled substrates or substrates yielding fluorescence are generally very sensitive. Sensitivity can also be increased by using a larger soil sample size or increasing the time of incubation.
What equipment is available?	Must I use wet chemistry for my measurement? Are automated procedures available? Do I have the proper equipment to do the analytical part of the assay?
Can I quantitatively extract either the substrate or the product after the reaction is ended?	There are many good substrates that generate products that have excellent properties for analyses. However, if they cannot be quantitatively extracted from soil, there is no way of knowing the actual in situ enzyme reaction rate. In most cases, sensitivity is much greater if the reaction product is measured instead of the disappearance of substrate.
How should I prepare and store soil samples prior to enzyme assay?	Air drying, grinding, or other sample preparation steps almost always change the enzyme activity measured. For comparative purposes and for experiments in which storage of samples is required, air-dry soils work well, although field-moist soils stored at 4°C are preferable. Samples should not be oven-dried as this process inactivates soil enzymes.

Table 4 Continued

Question asked	Relevance of question
When should I sample the soil for enzyme measurements?	Enzymes are very sensitive to changes in climate, residue inputs, pH, plant rhizospheres, soil water content, etc. A well-thought-out sampling scheme is important in assessing soil enzyme activity.
What enzyme activity should I measure?	There are more than 100 enzymes reported in soils, and the choice of an enzyme to use as an indicator of some soil function or response can be difficult. Often enzymes that can be measured inexpensively and rapidly are chosen, but these criteria may not always yield the most meaningful results.
Should I use toluene or some other plasmolytic agent?	This has been an important topic in soil enzyme assay development because of the requirement to restrict microbial growth and new enzyme synthesis during the assay. (see Sec. IX). In general, it is best to test whether a plasmolytic agent increases activity. If it does not or inhibits the reaction, it may be inappropriate. If the plasmolytic agent increases activity, it can be used, but its use must be noted.
Is a cofactor required?	Some oxidation–reduction enzymes need coenzymes for activity. Other enzymes require addition of metal ion cofactors to achieve maximal activity.

air-dried, naturally moist, or dried and remoistened soil, at least 20% volume/weight (V/W) toluene is necessary. In a soil suspension, however, 5% to 10% (V/W) toluene is sufficient. Gram-positive bacteria and *Streptomyces* spp. are considerably more resistant to toluene treatment than are gram-negative bacteria. In addition to the effect of toluene, the effect of dimethyl sulfoxide, ethanol, and Triton X-100 on soil enzyme activities has been evaluated by Frankenberger and Johanson (71). They reported that toluene, a plasmolytic agent as well as an antiseptic, had little effect or only slightly inhibited purified preparations of acid and alkaline phosphatases, α-glucosidase, and invertase but severely inhibited catalase and dehydrogenase activities. The soil enzyme activities of arylsulfatase and urease were enhanced (1.30- to 1.34-fold) in the presence of toluene, suggesting that its plasmolytic character was affecting the intercellular enzyme contribution to the measured activities. However, recent work by Acosta-Martinez and Tabatabai (106) showed that toluene should not be used for assay of arylamidase activity, because it is a noncompetitive (mixed-type) inhibitor of the activity of this enzyme in soils.

Irradiation of soil with high-energy radiation is another method used for soil sterilization. Dunn et al. (107) introduced the electron beam for heatless sterilization of soil. McLaren et al. (108) showed that soils can be sterilized by an electron beam of sufficient energy and intensity. They found that a 2×10^6 roentgen-equivalent-physical (rep) dose was necessary to sterilize a 1-g sample. Urease activity in this sterilized sample was retained. The effects of ionizing radiation on soil constituents have been reviewed (4,109,110). Generally, the relationship between microorganisms killed and enzymes inactivated is an exponential function of dose of radiation. However, the dose required depends on the soil type, soil moisture, and genus of the organism.

IX. POTENTIAL USEFUL TECHNIQUES IN LOCALIZING ENZYMES PROTEINS IN ENVIRONMENTAL MATRICES

Several methods and techniques that are available have potential for detecting and localizing enzyme protein in environmental matrices, including soils.

A. Thin Section Techniques

The combination of thin section techniques and histochemical and imaging techniques has a long history related to the study of enzymes in soils. The techniques have been successfully used in localizing specific compounds in animal and plant tissues (111,112). Early work in visualizing the location of enzymes in soil was reported by Foster and Martin (113) and Foster et al. (35). They combined the use of electron microscopy and ultrathin sections of undisturbed soil and root samples. Samples were prepared for study by using either transmission electron microscopy (TEM) or scanning electron microscopy (SEM) (114). The location of acid phosphatase, peroxidase, and succinic dehydrogenase in root rhizosphere, bacterial cell walls, or nonidentifiable organic matter in soil was clearly shown. Novel imaging techniques such as confocal microscopy (115) and isolation approaches by using laser-capture or other types of microdissection (116,117) that are being developed should continue to aid our understanding about the identification and localization of enzyme proteins in soils.

B. Atomic Force Microscopy

Atomic force microscopy is a method of measuring surface topographical features on the nanometer and micrometer scale. A probe with a radius of 20 nm is held several nanometers above the surface using feedback mechanisms, which measure surface-tip interactions on the scale of nanonewtons. The tip is scanned across the sample while the height of the tip is recorded, resulting in a topographical image of the surface. Although this technique should be useful for studies that involve enzyme protein–clay interaction or complexation, it has not so far been used in soil biochemistry research.

C. Confocal Microscopy

The confocal microscopy technique allows for production of three-dimensional fluorescent images. Resolution of up to 0.6 μm in the axial direction and 0.25 μm in the lateral direction can be achieved. By using specialized image processing software, the detailed three-dimensional distribution of the observed protein(s) can be reconstructed. The opacity of soil minerals can interfere with or greatly obstruct visualization of sample details. However, when combined with thin section and fluorescence techniques, it should be potentially useful for localization and identification of enzyme proteins in soils. As is the case for the other techniques mentioned, confocal microscopy is a relatively new technique, but its use in soil biochemistry and microbiology research is being developed at a fast rate.

X. CONCLUSIONS

Soil enzyme research is progressing rapidly and, as the number of enzymes in soils and natural waters for which assay procedures have been carefully developed increases and more scientists participate in soil enzyme research, the significance of their contribution to the total biological activity in soil–plant–water environment becomes more evident. This review was written to provide the reader with a better understanding of the advantages and limitations of the various methods, procedures, and instrumental techniques used for assaying the activities of enzymes in a complex system such as soil. Because of the trace concentrations of enzyme proteins in soils, it would be extremely difficult, if not impossible, to extract and purify soil enzymes. Nevertheless, it is possible to extract enzyme-active organic complexes from soils with a variety of reagents (68).

Advancement of a scientific discipline is often closely tied to the development of new and improved methods and equipment. The explosion of electronic devices and large-scale processing of samples needs to be applied to the study of enzymes in soils. This chapter has emphasized existing methods and summarized some of the new and exciting development in analytical chemistry and imaging technologies that offers a whole new window of opportunity to provide increased understanding of the distribution, reactions, activation, inhibition, and kinetic and thermodynamic properties of enzymes in environmental matrices. The result will be better management of ecosystem health and quality.

ACKNOWLEDGMENTS

This work was supported partly by the Biotechnology By-Products Consortium of Iowa.

REFERENCES

1. JJ Skujins. History of abiontic soil enzyme research. In RG Burns, ed. Soil Enzymes. New York: Academic Press, 1978, pp 1–49.
2. RG Burns. Soil Enzymes. New York: Academic Press, 1978.
3. S Kiss, M Dragan-Bularda, D Radulescu, Biological significance of enzymes accumulated in soil. Adv Agron 27:25–87, 1975.
4. JJ Skujins. Enzymes in soil. In AD McLaren and GH Peterson, eds. Soil Biochemistry. Vol. 1. New York: Marcel Dekker, 1967, pp 371–414.
5. JJ Skujins. Extracellular enzymes in soil. CRC Crit Rev Microbiol 4:383–421, 1976.
6. JN Ladd. Soil enzymes. In D. Vaughan, RE Malcolm, eds. Soil Organic Matter and Biological Activity. Boston: Martinus Nijhoff, 1985, pp 175–221.
7. WA Dick, MA Tabatabai. Significance and potential uses of soil enzymes. In: F Metting, Jr ed. Soil Microbial Ecology. New York: Marcel Dekker, 1992, pp 95–127.
8. MA Tabatabai, 1994. Soil enzymes. In: RW Weaver, JS Angel, PS Bottomley, eds. Methods of Soil Analysis. Part 2. Microbiological and Biochemical Properties. Soil Science Society of America Book Series no. 5, SSSA, Madison, WI: 1994, pp 775–833.
9. L Gianfreda, JM Bollag, Influence of natural and anthropogenic factors on enzyme activity in soils. In G Stotzky, JM Bollag, eds. Soil Biochemistry. Vol. 9. New York: Marcel Dekker, 1996, pp 123–193.
10. P Ruggiero, J Dec, JM Bollag, Soil as a catalyst system. In: G Stotzky, JM Bollag, eds. Soil Biochemistry. Vol. 9. New York, Marcel Dekker, 1996, pp 79–122.
11. JH Quastel. Soil Metabolism. London: The Royal Institute of Chemistry of Great Britain and Ireland, 1946.
12. RG Burns. Microbial and enzymic activities in soil biofilms. In: WG Characklis, PA Wilderer, eds. Structure and Function of Biofilms. London: John Wiley & Sons, 1989, pp 333–349.

13. AD McLaren. Soil as a system of humus and clay immobilized enzymes. Chem Screpta 8: 97–99, 1975.
14. MG Browman, MA Tabatabai. Phosphodiesterase activity of soils. Soil Sci Soc Am J 42: 284–290, 1978.
15. WA Dick, MA Tabatabai. Inorganic pyrophosphatase activity of soils. Soil Biol Biochem 10:59–65, 1978.
16. MA Tabatabai, BB Singh. Rhodanese activity of soils. Soil Sci Soc Am J 40:381–385, 1976.
17. ZN Senwo, MA Tabatabai. Aspartase activity of soils. Soil Sci Soc Am J 60:1416–1422, 1986.
18. RG Burns. Enzyme activity in soil: Some theoretical and practical considerations. In RG Burns, ed. Soil Enzymes. New York: Academic Press, 1978, pp 295–340.
19. M Cashel, E Freese. Excretion of alkaline phosphatase by *Bacillus subtilis*. Biochem Biophys Res Commun 16:541–544, 1964.
20. S Nishimura, M Nomura. Ribonuclease of *Bacillus subtilis*. J Biochem 46:161–167, 1959.
21. R Weimberg, WL Orton. Repressible acid phosphomonoesterase and constitutive pyrophosphatase of *Saccharomyces mellis*. J Bacteriol 86:805–813, 1963.
22. R Weimberg, WL Orton. Evidence for an exocellular site for the acid phosphatase of *Saccharomyces mellis*. J Bacteriol 88:1743–1754, 1964.
23. J Jacquet, O Villette, R Richou. Les phosphatases des filtrats microbiens. I. *Schizomycetes*. Rev Immunol 20:189–206, 1956.
24. RM Atlas. Microbiology: Fundamentals and Applications. 2nd ed. New York: Macmillan, 1988.
25. TW Speir, DJ Ross. Soil phosphatase and sulphatase. In: RG Burns, ed. Soil Enzymes. New York: Academic Press, 1978. pp 197–250.
26. JN Ladd, EA Paul. Changes in enzymic activity and distribution of acid-soluble amino acid-nitrogen in soil during nitrogen immobilization and mineralization. Soil Biol Biochem 5: 825–840, 1973.
27. EF Estermann, AD McLaren. Contribution of rhizoplane organisms to the total capacity of plants to utilize organic nutrients. Plant Soil 15:243–260, 1961.
28. WT Frankenberger Jr, MA Tabatabai. Amidase and urease activities in plants. Plant Soil 64: 153–166, 1982.
29. NG Juma, MA Tabatabai. Phosphatase activity in corn and soybean roots: Conditions for assay and effects of metals. Plant Soil 107:39–47, 1988.
30. NG Juma. Phosphatase activity of soils and of corn and soybean roots. MS Thesis, Iowa State University, Ames, 1976.
31. WA Dick, NG Juma, MA Tabatabai. Effects of soils on acid phosphatase and inorganic pyrophosphatase of corn roots. Soil Sci 136:19–25, 1983.
32. WA Dick, MA Tabatabai. Kinetics and activities of phosphatase-clay complexes. Soil Sci 143:5–15, 1987.
33. CM Lai, MA Tabatabai. Kinetic parameters of immobilized urease. Soil Biol Biochem 24: 225–228, 1992.
34. JN Ladd. Origin and range of enzymes in soil. In: RG Burns, ed. Soil Enzymes. New York: Marcel Dekker, 1978. pp 51–96.
35. RC Foster, AD Rovira, TW Cock. Ultrastructure of the Root-Soil Interface. St. Paul, MN: American Phytopathological Society, 1983.
36. RG Burns. Interaction of enzymes with soil mineral and organic colloids. In: PM Huang, M Schnitzer, eds. Interactions of Soil Minerals with Natural Organics and Microbes. SSSA Spec Publ Serv 17:429–451, 1986.
37. RG Burns. Enzyme activity in soil. Location and a possible role in microbial ecology. Soil Biol Biochem 14:423–428, 1982.
38. JR Ramirez-Martinez, AD McLaren. Some factors influencing the determination of phospha-

tase activity in native soils and in soils sterilized by irradiation. Enzymologia 31:23–28, 1966.
39. HH Weetall. Immobilized enzymes and their application in the food and beverage industry. Process Biochem 10:3–24, 1975.
40. LE Ensminger, SE Gieseking. Resistance of clay-adsorbed proteins to proteolytic hydrolysis. Soil Sci 53:205–209, 1942.
41. AD Haig. Acetyl-esterase activity in soil. PhD Dissertation. University of California, Davis, 1955.
42. AD McLaren. The adsorption and reactions of enzymes and protein on kaolinite. II. The action of chymotrypsin on lysozyme. Soil Sci Soc Am Proc 18:170–174, 1954.
43. DE Armstrong, G Chesters. Properties of protein-benetonite complexes as influenced by equilibrium conditions. Soil Sci 98:39–52, 1964.
44. JT Albert, RD Harter. Adsorption of lysozyme and ovalbumin by clay: Effect of clay suspensions pH and clay mineral type. Soil Sci 115:130–136, 1973.
45. JN Ladd, JAH Butler. Humus-enzyme systems and synthetic, organic polymer-enzyme analogs. Soil Biochem 4:143–194, 1975.
46. JP Conrad. The nature of the catalyst causing the hydrolysis of urea in soils. Soil Sci 54: 367–380, 1940.
47. WA Dick. Influence of long-term tillage and crop rotation combinations on soil enzyme activities. Soil Sci Soc Am J 48:569–574, 1984.
48. WT Frankenberger Jr, MA Tabatabai. Amidase activity in soils. III. Stability and distribution. Soil Sci Soc Am J 45:333–338, 1981.
49. NG Juma, MA Tabatabai. Distribution of phosphomono-esterases in soils. Soil Sci 126:101–108, 1978.
50. MA Tabatabai. Effects of trace elements on urease activity in soils. Soil Biol Biochem 9: 9–13, 1977.
51. RG Burns, AH Pukite, AD McLaren. Concerning the location and persistence of soil urease. Soil Sci Soc Am Proc 36:308–311, 1972.
52. B Ceccanti, M Bonmati-Pont. Nannipieri. Microdetermination of protease activity in humic bands of different sizes after analytical isoelectric focusing. Biol Fertil Soils 7:202–206, 1989.
53. P Nannipieri, B Ceccanti, D Bianchi. Characterization of humus-phosphatase complexes extracted from soil. Soil Biol Biochem 20:683–691, 1988.
54. P Ruggiero, VM Radogno. Tyrosinase activity in a humus-enzyme complex. Sci Total Environ 62:365–366, 1988.
55. A Serhan, A Nissenbaum. Humic acid association with peroxidase and catalase. Soil Biol Biochem 18:41–44, 1986.
56. P Simonart, L Batistic, J Mayaudon. Isolation of protein from humic acid extracted from soil. Plant Soil 27:153–161, 1967.
57. JHA Butler, JN Ladd. The effect of methylation of humic acids and their influence on proteolytic enzyme activity. Aust J Soil Res 7:263–268, 1969.
58. B Ceccanti, P Nannipieri, S Cervelli, P Sequi. Fractionation of humus-urease complexes. Soil Biol Biochem 10:39–45, 1978.
59. K Hayano. Extraction and properties of phosphodiesterase from a forest soil. Soil Biol Biochem 9:221–223, 1977.
60. K Hayano, A Katami. Extraction of β-glucosidase activity from pea field soil. Soil Biol Biochem 9:349–351, 1977.
61. RG Burns, MH El-Sayed, AD McLaren. Extraction of an urease-active organo-complex from soil. Soil Biol Biochem 4:107–108, 1972.
62. T Satyanarayana, LW Getzin. Properties of a stable cell-free esterase from soil. Biochemistry 12:1566–1572, 1973.
63. J Mayaudon, L Batistic, and JM Serbor. Propriétés des activités protéolytiques extraites des sols frais. Soil Biol Biochem 7:281–286, 1975.

64. EA Paul, AD McLaren. Biochemistry of the soil system. In: EA Paul, AD MaLaren, eds. Soil Biochemistry. Vol. 3. New York: Marcel Dekker, 1975, pp 1–36.
65. NM Pettit, ARJ Smith, RB Freedman, RG Burns. Soil urease: Activity, stability and kinetic properties. Soil Biol Biochem 8:479–484, 1976.
66. F Eivazi, MA Tabatabai. Glucosidases and galactosidases in soils. Soil Biol Biochem 20: 601–606, 1988.
67. JB Neiland, RK Stumpf. Outlines of Enzyme Chemistry. 2nd ed. New York: Wiley 1958.
68. MA Tabatabai, MH Fu. Extraction of enzymes from soils. In: G Stotzky, J-M Bollag, eds. Soil Biochemistry. Vol. 7. New York: Marcel Dekker, 1992, pp 197–227.
69. P Nannipieri, S Grego, B Ceccanti. Ecological significance of biological activity in soil. In: J-M Bollag, G Stotzky, eds. Soil Biochemistry. Vol. 6. New York: Marcel Dekker, 1990, pp 293–355.
70. JN Ladd, PG Brisbane, JHA Butler, A Amato. Studies on soil fumigation. III. Effects on enzyme activities, bacterial numbers and extractable ninhydrin reactive compounds. Soil Biol Biochem 8:255–260, 1976.
71. WT Frankenberger Jr, JB Johanson. Use of plasmolytic agents and antiseptics in soil enzyme assays. Soil Biol Biochem 18:209–213, 1986.
72. KN Paulson, LT Kurtz. Locus of urease activity in soil. Soil Sci Soc Am Proc 33:879–901, 1969.
73. P Nannipieri, I Sastre, L Landi, MC Lobo, G Pietramellara. Determination of extracellular neutral phosphomonoesterase activity in soil. Soil Biol Biochem 28:107–112, 1966.
74. S Klose, MA Tabatabai. Effect of chloroform fumigation on enzyme activities in sois. Agron Abstr 1998, p 211.
75. S Klose, MA Tabatabai. Urease activity of microbial biomass in soils. Soil Biol Biochem 31:205–211, 1999.
76. S Klose, MA Tabatabai. Arylsulfatase activity of microbial biomass in soils. Soil Sci Soc Am J 63:569–574, 1999.
77. F Alef, P Nannipieri. Methods in Applied Soil Microbiology and Biochemistry. New York: Academic Press, 1995.
78. J Yashphe. Estimation of micro amounts of urea and carbamyl derivatives. Anal Biochem 52:143–153, 1973.
79. WR Fearon. The carbamido diacetyl reaction: A test for citruline. Biochem J 33:902–907, 1939.
80. LA Douglas, JM Bremner. Extraction and colorimetric determination of urea in soils. Soil Sci Soc Am Proc 34:859–862, 1970.
81. PL Searle, WT Speir. An automated colorimetric method for the determination of urease activity in soil and plant material. Commun Soil Sci Plant Anal 7:365–374, 1976.
82. LA Douglas, H Sochtig, W Flaig. Colorimetric determination of urea in soil extracts using an automated system. Soil Sci Soc Am J 42:291–292, 1978.
83. MA Tabatabai, JM Bremner. Arylsulfatase activity of soils. Soil Sci Soc Am J 34:225–229, 1970.
84. V Acosta-Martinez. Arylamidase activity of soils. PhD dissertation, Iowa State University, Ames, 2000.
85. JA Goldbarg, AM Rutenburg. The colorimetric determination of leucine aminopeptidase in urine and serum of normal subjects and patients with cancer and other diseases. Cancer 11: 283–291, 1958.
86. SK Pancholy, JQ Lynd. Quantitation of soil lipase with 4-MUB fluorescence. Agron Abstr 1971, p. 83.
87. C Freeman, G Liska, NJ Ostle, SE Jones, MA Lock. The use of fluorogenic substrates for measuring enzyme activity in peatlands. Plant Soil 175:147–152, 1995.
88. PR Darrah, PJ Harris. A fluorimetric method for measuring the activity of soil enzymes. Plant Soil 92:81–88, 1986.

89. C-M Marx, M Wood, SC Jarvis. A microplate assay for soil enzyme studies based on methylumbelliferyl substrates. Abstracts, Enzymes in the Environment: Activity, Ecology & Applications. Granada, Spain. p. 84.
90. DMH Simpson, SW Melsted. Urea hydrolysis and transformation in some Illinois soils. Soil Sci Soc Am Proc 27:48–50, 1963.
91. JJ Skujins. $^{14}CO_2$ detection chamber for studies in soil metabolism. Biol du Sol 4:15–17, 1965.
92. JJ Skujins, AD McLaren. Persistence of enzymatic activities in stored and geologically preserved soils. Enzymologia 34:213–225, 1968.
93. JJ Skujins, AD McLaren. Assay of urease activity using ^{14}C-urea in stored, geologically preserved, and in irradiated soils. Soil Biol Biochem 1:89–99, 1969.
94. MA Tabatabai, JM Bremner. Assay of urease activity in soils. Soil Biol Biochem 4:479–487, 1972.
95. JL Rabinowitz, T Sall, JN Bierly, O Oleksyshyn. Carbon isotope effects in enzyme systems. I. Biochemical studies with urease. Arch Biochem Biophys 53:437–445, 1956.
96. WL Banwart, MA Tabatabai, JM Bremner. Determination of ammonium and soil extracts and warer samples by an ammonia electrode. Commun Soil Sci Plant Anal 3:449–458, 1972.
97. MA Tabatabai. Determination of nitrite in soil extracts and water samples by a nitrogen oxide electrode. Commun Soil Sci Plant Anal 5:569–578, 1974.
98. HM Abdelmagid, MA Tabatabai. Nitrate reductase activity of soils. Soil Biol Biochem 19: 421–427, 1987.
99. MA Tabatabai, WT Frankenbeger Jr. Liquid chromatography. In DL Sparks, ed. Methods of Soil Analysis: Chemicall Methods, SSSA Book Series No. 5. Madison, WI: Soil Science Society of America, 1996, pp 225–245.
100. Beckman Instruments. Introduction to Capillary Electrophoresis. Fullerton, CA: Bioanalytical Systems Group, Beckman Instruments, 1991.
101. P Jandik, WR Jones, A Weston, PR Brown. Electrophoretic capillary ion analysis: origins, principles, and applications. LC-GC Mag 9:634–645, 1991.
102. JD Olechno, JMY Tso, J Thayer, A Wainright. Capillary electrophoresis: A multiaced technique for analytical chemistry. Am Lab 22:51–59, 1990.
103. WR Jones, P Jandik, R Pfeifer. Capillary ion analysis: An innovative technology. Am Lab 23:40–46, 1991.
104. JM Arrieta, GJ Herndl. Variability of β-glucosidase in seawater. Abstracts, Enzymes in the Environment: Activity, Ecology & Applications. Granada, Spain. p. 83.
105. T Beck, H Poschenrieder. Experiments on the effect of toluene on the soil microflora. Plant Soil 18:346–357, 1963.
106. V Acosta-Martinez, MA Tabatabai. Arylamidase activity of soils. Soil Sci Soc Am J 64: 215–221, 2000.
107. CG Dunn, WL Campbell, H Fram, A Hutchins. Biological and photochemical effect of high energy, electrostatically produced roentgen rays and cathode rays. J Appl Phys 19:605–616, 1948.
108. AD McLaren, L Reshetko, W Huber. Sterilization of soil by irradiation with an electron beam and some observations on soil enzyme activity. Soil Sci 83:497–502, 1957.
109. AD McLaren. Radiation as a technique in soil biology and biochemistry. Soil Biol Biochem 1:63–73, 1969.
110. PA Cawse. Microbiology and biochemistry of irradiated soils. In: EA Paul, AD McLaren, eds. Soil Biochemistry. Vol. 3. New York: Marcel Dekker, 1975, pp 213–267.
111. J McLean, PB Cahan. The distribution of acid phosphatases and esterases in differentiating roots of *Vicia faba*. Histochemie 24:41–49, 1970.
112. JL Hall, CA Davie. Localization of acid hydrolase activity in Zita mays L. root tips. Ann Bot 35:849–855, 1971.

113. RC Foster, JK Martin. In situ analysis of soil components of biological origin. In Soil Biochemistry. Vol. 5. New York: Marcel Dekker, 1981, pp 75–111.
114. TR McKee, JL Brown. Preparation of specimens for electron microscopic examination. In: JB Dixon, SB Weed, eds. Minerals in Soil Environments. Madison, WI: Soil Science Society of America, 1977, pp 809–846.
115. PC DeLeo, P Baveye, WC Ghiorse. Use of confocal laser scanning microscopy on soil thin-sections for improving characterization of microbial growth in unconsolidated soil and aquifer materials. J Microbiol Methods 30:193–203, 1997.
116. K Schutze, G Lahr. Use of laser technology for microdissection and isolation. Am Biotech Lab March: 24–26, 1999.
117. B Sinclair. The cell in my test tube. The Scientist 13:17–19, 1999.

Index